NUCLEAR STRUCTURE

NUCLEAR STRUCTURE

WILLIAM F. HORNYAK

Department of Physics
University of Maryland
College Park, Maryland

ACADEMIC PRESS New York San Francisco London 1975
A Subsidiary of Harcourt Brace Jovanovich, Publishers

ACADEMIC PRESS, INC.
111 Fifth Avenue, New York, New York 10003

United Kingdom Edition published by
ACADEMIC PRESS, INC. (LONDON) LTD.
24/28 Oval Road, London NW1

Library of Congress Cataloging in Publication Data

Hornyak, W F (date)
Nuclear structure.

Includes bibliographies and index.
1. Nuclear physics. I. Title.
QC776.H6 539.7'4 74-10201
ISBN 0-12-356050-0

PRINTED IN THE UNITED STATES OF AMERICA

CONTENTS

II—Nuclear Shape and Nuclear Moments

III—Nuclear Matter Characteristics

IV—Single-Particle Shell Model

V—Individual-Particle Model

VI—Collective Nuclear Effects

VII—Electromagnetic Interactions with Nuclei

VIII—Beta-Decay

Appendix A—Coupling of Two Angular Momenta, Clebsch–Gordan Coefficients

Appendix B—The Wigner–Eckart Theorem

PREFACE

This text is an outgrowth of a course conducted by the author for several years at the University of Maryland. It covers material normally discussed in courses relating to nuclear structure. The text is basically designed for a second-year graduate student, preferably but not necessarily having had some introductory nuclear physics at an elementary or undergraduate level. The presentation while relying only slightly on such a background does, however, require a good knowledge of the elements of quantum mechanics including an introduction to Dirac theory, knowledge commonly gained from the usual one-year graduate level course in the subject.

One motivation for writing this text has been to present the subject in a manner offering the realistic possibility that an average student with proper preparation could, in fact, absorb this material. Perhaps the author should immediately interject the comment that he is aware of the limited technical facility with quantum mechanics of the average student. This, of course, is due to lack of time for the student to absorb the full implications of the theory, which is unavoidably presented to him in rather concentrated doses. This "working text" allows for this limitation and is intended for the student. However, the research scientist will also benefit from numerous current reviews and up-to-date references.

A limited number of selected topics are treated with some completeness using techniques that are only simple extensions of those reasonably expected to have been introduced in the prerequisite quantum mechanics course. Perforce this implies the use of somewhat "old fashioned" methods resulting in some loss of the elegance inherent in current treatments but which have the advantage of throwing some light on the historical development of the subject. The development of the newer techniques and their application to nuclear theory is left to more specialized works. Emphasis is placed on establishing the basic concepts with specialized and detailed applications only sparingly considered in contradistinction to the more usual treatment. Whenever possible, such concepts are reinforced by examples employing direct calculations with simple model wave functions, operators, etc., partly to convert in the student's mind his knowledge of elementary quantum mechanics into a workaday tool. In large part, this is made possible by the adopted philosophy of relying on (and *using*) only elementary theoretical methods in the text.

While the above philosophy may indeed reduce the *reflection coefficient* at the student interface, the author is well aware that the price paid may be to blur the available accuracy of the description of nature by denying the advantage that the precision of modern techniques is, in fact, designed to offer. Although a strong effort has been made to minimize this problem, some ambiguities undoubtedly remain.

The first chapter, Nucleon–Nucleon Forces, is in some ways different in character from the remaining chapters. In view of the recent encouraging progress made in relating both the free nucleon–nucleon interaction and the complex nuclear many-body problems to meson theory, this chapter attempts to develop a background in the relevant portions of elementary particle physics adequate to provide the student with a basis for understanding these relationships. Indeed, it may be argued that nuclear physics has arrived at the point where the inclusion of elementary particle interactions is in fact essential. This chapter inevitably tends to be in the nature of a survey with a rather large number of reference citations. Introducing the student to periodical literature may also be a service in itself. A subsidiary consideration motivating the breadth of this chapter is the recognition that classes devoted to the topics of this text are also attended by students of elementary particle physics as well as of nuclear physics. The relevant connections between these basic fields are generally appreciated by all the students.

The remaining chapters are more self-contained and generally give a step-by-step derivation of the important equations (sometimes provided in an Exercise). The "it can be shown" presentation is kept to a minimum and over 125 exercises are provided along with references to additional profitable reading. A special effort has also been made to provide numerous reference citations in those areas in which there is current active research. To provide coherence and continuity a number of topics are illustrated by selecting examples from the 1P-shell nuclei. The treatment of these somewhat simpler cases permits a sharper focus on the theoretical point in question without any particular loss in essential generality.

Somewhat more material is presented than can be thoroughly covered in a typical semester to allow the instructor the freedom to select his own desired emphasis. The topics covered—nucleon–nucleon forces, nuclear shape and nuclear moments, nuclear matter characteristics, single-particle shell model, individual-particle model, collective nuclear effects, electromagnetic interactions with nuclei, and beta-decay—emphasize only topics directly related to the properties of the nucleus as such. Topics such as ionization, stopping of charged particles, and molecular beam techniques are not treated. Virtually no description is given of the experiments or the apparatus used to obtain the empirical evidence cited, in the belief that devoting valuable space to such an exposition tends only to emphasize the historical fact that the development of nuclear physics consists of a collection of topics rather than a unified discipline leaning on an all-embracing theoretical foundation. It should also be noted that even a cursory presentation of experimental techniques would add considerably to the length of the text and would be perhaps of only limited value since these techniques become rapidly dated. It is hoped that a prior course at an elementary level would have given the student some appreciation for the many interesting experimental techniques that have been employed, even if only in a historical context. Instead, an effort is made to connect the *analyzed* empirical data with the relevant theory.

ACKNOWLEDGMENTS

The author is particularly indebted to Professor Edward F. Redish for his many useful and critical comments. Professors Manoj K. Banerjee and Gerard Stephenson, as well as other members of the University of Maryland nuclear physics group, have been very helpful in numerous discussions. Secretaries Marie Daston and Mary Ann DeMent have tirelessly contributed their labors. Finally, I am most grateful for the long hours my wife Eva devoted to proofreading and translating my efforts from "archaic Hungarian" to English.

Chapter I

NUCLEON–NUCLEON FORCES

A. INTRODUCTION

A central problem of nuclear physics is to understand the nature of the isolated nucleon–nucleon interaction and to explain the properties of complex nuclei in terms of these nuclear forces. The description of nuclear systems can be attempted in a fundamental or *microscopic* sense by explicitly accounting for the motion of each nucleon. This approach is generally quite complex, occasionally even to the point of obscuring the "physics" of the problem by the presence of mathematical or calculational difficulties. The advent of fast computer technology has in many instances materially aided in carrying out this program. However, the penetrating insight generally offered by closed-form analytic expressions is seldom available. This situation is hardly surprising when the large number of degrees of freedom involved is considered.

Alternately, one might develop relevant *macroscopic* or many-body concepts, models, and parameters in terms of which a satisfactory treatment of complex nuclei could be sought. The reduced number of collective variables, the principal economy of this approach, usually results in a more manageable, if not always transparent, grasp of the problem. When desirable, such model calculations can be augmented by specific corrections for aspects of the fundamental nucleon interactions that have been either omitted or inadequately included in the model. These corrections, if small enough, can be successfully considered in the framework of perturbation theory.

It not infrequently happens that the added understanding offered by a macroscopic view merits considering the development of suitable models as an additional valid goal in itself. In a number of the following chapters, our main purpose will be to dwell on the progress that has been made in this direction. It should also be pointed out from the outset that, while many remarkably accurate models for the behavior of complex nuclei exist, the derivative connection to the nucleon–nucleon interaction in a fundamental sense is only in its infancy in many cases.

The objective of this chapter is to discuss the basic nature of the nucleon–nucleon force from an elementary particle point of view. This task is materially aided by describing the nature of the nucleons in terms of any possible substructure they may possess and by discussing the relationship of the nucleons to other closely associated elementary particles. These topics are discussed in this chapter largely in a survey format and only to the extent necessary to provide useful nuclear physics background material. Indeed, a detailed treatment would require the full theoretical apparatus of elementary particle physics, which would take us far beyond the intended scope of this text. Such treatment is left to the numerous excellent texts on the subject and the conference reports in this rapidly developing field.

Notwithstanding the now apparent composite structure of both the free neutron and free proton as they are encountered individually in the laboratory, complex nuclei can be conveniently considered to consist of Z protons strongly interacting with $N = A - Z$ neutrons. For example, the isotope of beryllium ^{10}Be with atomic number $Z = 4$ and mass number $A = 10$ is considered to consist of 4 protons strongly interacting with 6 neutrons. When ^{10}Be undergoes β^- radioactive decay to its neighboring isobar of boron ^{10}B (with 5 protons and 5 neutrons)[1] by the nuclear emission of an electron (and antineutrino), one of the 6 neutrons is imagined to decay into a proton (viz., $n \to p + e^- + \bar{\nu}$). Thus the observation of electrons being emitted from the nuclear interior does not require the prior or steady-state existence of electrons within the nucleus.

The basic properties of the free neutron and proton are (Particle Data Group, 1973)

mass:[2]

$$m_p c^2 = (938.2592 \pm 0.0052)\ \text{MeV}$$

$$m_n c^2 = (939.5527 \pm 0.0052)\ \text{MeV}$$

$$m_n - m_p = 2.531 m_e;$$

[1] Elements with the same Z but different A are called *isotopes*. Elements with the same A but different Z are called *isobars*.

[2] The electron mass is $m_e c^2 = (0.5110041 \pm 0.0000016)$ MeV.

magnetic moment: $\mu_p = (2.792782 \pm 0.000017)\mu_0$

$$\mu_n = -(1.913148 \pm 0.000066)\mu_0$$

where $\mu_0 = e\hbar/2m_p c$ is the nuclear magneton. In addition, both are fermions, i.e., have intrinsic spin angular momentum $\frac{1}{2}\hbar$ and obey Fermi–Dirac statistics. By convention, both are taken to have the same intrinsic parity (defined as even). The free neutron is unstable and β-decays to the proton with a mean life of (15.6 ± 0.2) min providing $(2.531 - 1)m_e c^2 = 0.782$ MeV of kinetic energy for the decay products.

B. FUNDAMENTALS

In this section we introduce a number of concepts that we shall find useful later. The first of these, the *isospin* variable, allows for a more compact formulation of nucleon wave function symmetries and leads to a generalized Pauli principle in the extended nucleon space–spin–isospin degrees of freedom. In addition, the fact that isospin in many instances is almost a good quantum number allows a useful first-order classification of nuclear system characteristics. The symmetry property of nuclear wave functions associated with *parity* also plays an important rôle, since *strong interactions* between elementary particles, such as in the dominant nucleon–nucleon interaction, conserve parity. On the other hand, the fact that *weak interactions* (responsible, for example, for nuclear β-decay) violate parity conservation has far-reaching consequences.

A brief description of the systematics of the elementary particles closely associated with the nucleons (such as the π-mesons or pions) is given in the following subsections. Since the present emphasis is on nuclear physics, a large amount of important and interesting material relating to elementary particle physics per se is omitted. Thus many of the fundamental experiments involving elementary particles, such as the "missing mass" experiments, are eschewed in favor of those more closely associated with the nucleons themselves. Because high-energy electron scattering offers particularly striking evidence for the existence of nucleon substructure, it is discussed at length. In the next chapter we shall find that such experiments also relate importantly to our knowledge of the charge density distribution in complex nuclei.

The stage is set for the presentation of modern meson-exchange models of the nucleon–nucleon force in the final section of this chapter by first discussing in detail a simplified one-boson exchange model (OBEM) involving only single pions (ignoring such vital but unfortunately complicated factors as pion–pion interactions and resonances, etc.). A brief discussion of some of the more complicated two-pion exchange effects follows.

Finally, the *quark model* of elementary particles is introduced, not only to give a coherent and comprehensive view of the possible nature of a more fundamental underlying substructure of elementary particles, but also to shed light on the nature of the strong interactions.

Some aspects of the preceding topics, when amenable to simple theoretical analysis, are explicitly treated; however, much of the presentation is relegated perforce to a descriptive survey.

1. Isospin

Soon after the discovery of the neutron, Heisenberg (1932) speculated, on the basis of the similar properties of the proton and the neutron, that they represent two different charge states of the same particle, referred to generically as the nucleon. The new internal variable distinguishing this nearly degenerate mass doublet, called isospin (earlier designations were isotopic spin, isobaric spin, and i-spin) is associated with a vector operator $\tilde{\mathbf{t}}$.[1] We wish to describe the two-nucleon charge states as discrete projection states of the eigenvectors of this operator. Thus a formal analogy with the two discrete m-states or space projection states of the (mechanical) spin-$\frac{1}{2}$ operator suggests itself. In analogy to the Pauli matrices for $\tilde{\boldsymbol{\sigma}}$ and its relationship to the spin operators $\tilde{\mathbf{s}} = \frac{1}{2}\hbar\tilde{\boldsymbol{\sigma}}$, we introduce $\tilde{\mathbf{t}} = \frac{1}{2}\tilde{\boldsymbol{\tau}}$. In terms of Cartesian coordinates in isospin space (unit vectors $\hat{1}$, $\hat{2}$, and $\hat{3}$) we write[2]

$$\tilde{\boldsymbol{\tau}} = \tau_1\hat{1} + \tau_2\hat{2} + \tau_3\hat{3}$$

$$\tau_1 = \begin{pmatrix} 0 & 1 \\ 1 & 0 \end{pmatrix}, \qquad \tau_2 = \begin{pmatrix} 0 & -i \\ i & 0 \end{pmatrix}, \qquad \tau_3 = \begin{pmatrix} 1 & 0 \\ 0 & -1 \end{pmatrix}. \tag{1-1}$$

It then follows, in analogy to $\tilde{\mathbf{s}}^2 = \tilde{\mathbf{s}}\cdot\tilde{\mathbf{s}} = \frac{3}{4}\hbar^2 I$, that

$$\tilde{\mathbf{t}}^2 = \tilde{\mathbf{t}}\cdot\tilde{\mathbf{t}} = \tfrac{1}{4}(\tau_1{}^2+\tau_2{}^2+\tau_3{}^2) = \tfrac{3}{4}I,$$

where I is the identity matrix $\left(\begin{smallmatrix}1&0\\0&1\end{smallmatrix}\right)$. This can be used to define the isospin quantum number t, viz., $t(t+1) = \frac{3}{4}$, giving $t = \frac{1}{2}$.[3] The two-component isospinors $\pi = \left(\begin{smallmatrix}1\\0\end{smallmatrix}\right)$ and $\nu = \left(\begin{smallmatrix}0\\1\end{smallmatrix}\right)$ are immediately seen to be eigenspinors of the diagonal matrix τ_3 with eigenvalues $+1$ and -1, respectively, corresponding to $\pm\frac{1}{2}$ for the eigenvalues of $\tilde{t}_3$. The identification can then be made that nucleons are particles with isospin $\frac{1}{2}$, the isospin polarization "up" or π-state representing the proton and the polarization "down" or ν-state representing the neutron.[4] One can then write proton and neutron wave functions in terms

[1] We shall employ a tilde over a symbol to call attention explicitly to an operator quantity when required for clarity.

[2] Following general custom, a caret over a symbol will be used to designate an appropriate unit vector.

[3] The solution $t = -\frac{3}{2}$ is unacceptable, since $t \geqslant 0$ is required by convention.

[4] This convention conforms to modern usage, while earlier literature in nuclear physics generally used reversed designations.

of isospinors in isospin space as

$$|\mathrm{p}\rangle = \psi_{\mathrm{p}}(\mathbf{r},\boldsymbol{\sigma}) = \psi(\mathbf{r},\boldsymbol{\sigma})\pi = \psi(\mathbf{r},\boldsymbol{\sigma})\begin{pmatrix}1\\0\end{pmatrix}$$

$$|\mathrm{n}\rangle = \psi_{\mathrm{n}}(\mathbf{r},\boldsymbol{\sigma}) = \psi(\mathbf{r},\boldsymbol{\sigma})\nu = \psi(\mathbf{r},\boldsymbol{\sigma})\begin{pmatrix}0\\1\end{pmatrix}. \tag{1-2}$$

The systematics of elementary particle properties has led to the introduction of new quantum numbers and selection rules. One of these relationships relates the particle charge Q to a generalized t_3, the hypercharge Y, the baryon number B, and the strangeness S.[1,2] The relevant relationships are[3]

$$Q = e[t_3 + \tfrac{1}{2}Y], \qquad Y = B + S. \tag{1-3}$$

For nucleons $B = 1$ and $S = 0$, thus giving a hypercharge $Y = 1$, and generating a charge operator

$$\tilde{Q} = e[\tilde{t}_3 + \tfrac{1}{2}I] = e\begin{pmatrix}1 & 0\\0 & 0\end{pmatrix}. \tag{1-4}$$

We immediately see that the eigenvalue problem $\tilde{Q}|\zeta\rangle = \lambda|\zeta\rangle$ is satisfied by the isospinor solution $|\zeta\rangle = |\mathrm{p}\rangle$ with $\lambda = +e$ and $|\zeta\rangle = |\mathrm{n}\rangle$ with $\lambda = 0$.

While spin angular momentum states having a polarization orientation in any arbitrary direction can be constructed in the form of a spinor in spin space $\binom{a}{b} = a\binom{1}{0} + b\binom{0}{1} = a\alpha + b\beta$, *mixtures* of isospinors in isospin space such as $a\pi + b\nu$ are taken as unphysical. The net positive charge of the proton is known to have a magnitude equal to that carried by the electron to a precision better than $5 \times 10^{-19}e$ and the neutron net charge is known to be zero to better than $2 \times 10^{-15}e$, attesting to the lack of mixing of the states (1-2) in nature.

Another particle that will be of interest to us is the π-meson (or pion), which has quantum numbers $t = 1$, $S = 0$, $B = 0$, and $Y = 0$. It is thus seen

[1] It is an unfortunate circumstance that the number of parameters needed in nuclear physics and particle physics far exceeds the availability of convenient symbols. In order to facilitate the comparison of text material to the literature, we shall use those symbols that custom has accepted. It will be left to the context to determine the proper identification. Thus at this point we use S to represent the strangeness quantum number; however, it is also used later as the spin quantum number as well as other quantitites. The reader must beware.

[2] Numerous texts on elementary particle physics at the elementary to intermediate level are available, e.g., Gasiorowicz (1966), Frazer (1966), Feld (1969), and Perkins (1972).

[3] In (1-3) and in subsequent equations, the symbol e stands for the positive quantity equal to the magnitude of the electronic charge. A subscripted quantity such as e_i or q_i requires specifying both the magnitude and sign of the charge.

to be an isospin triplet (or *isovector*) for which the charge operator is

$$\tilde{Q} = e\tilde{\tau}_3 = e\begin{pmatrix} 1 & 0 & 0 \\ 0 & 0 & 0 \\ 0 & 0 & -1 \end{pmatrix}. \tag{1-5}$$

The eigenvalues of (1-5) are just the diagonal elements $e, 0, -e$.[1] Experimentally, pions do in fact form a nearly degenerate mass triplet distinguished by these three charge states, namely,

$$\begin{aligned} m_\pi{}^{\pm}c^2 &= (139.5688 \pm 0.0064)\ \text{MeV} \\ m_\pi{}^{0}c^2 &= (134.9645 \pm 0.0074)\ \text{MeV}. \end{aligned} \tag{1-6}$$

Thus, while at first the simple need to distinguish the two charge states of the nucleon led to the introduction of an apparently more than necessary complexity of a spinor operator in an abstract three-dimensional isospin (or charge) space, this concept is seen to have the advantage of offering a natural extension to a generalized vector operator useful in discussing the pion fields. We now turn to other advantages of this point of view. Returning to the isospin formalism for the nucleons, it is convenient to introduce raising and lowering operators and projection operators. As in the case of spin angular momentum, we write for raising and lowering operators[2]

$$t_{\pm} = \tfrac{1}{2}(\tau_1 \pm i\tau_2) \qquad \text{or} \qquad t_+ = \begin{pmatrix} 0 & 1 \\ 0 & 0 \end{pmatrix}, \qquad t_- = \begin{pmatrix} 0 & 0 \\ 1 & 0 \end{pmatrix}. \tag{1-7}$$

We note that $t_+\pi = 0$ and $t_+\nu = \pi$; thus t_+ converts a neutron into a proton. Also, since $t_-\nu = 0$ and $t_-\pi = \nu$, the operator t_- converts a proton into a neutron. These operators will prove useful in discussing β-decay. Projection operators

$$\Lambda^\pi = \tfrac{1}{2}(I + \tau_3) = \begin{pmatrix} 1 & 0 \\ 0 & 0 \end{pmatrix}, \qquad \Lambda^\nu = \tfrac{1}{2}(I - \tau_3) = \begin{pmatrix} 0 & 0 \\ 0 & 1 \end{pmatrix} \tag{1-8}$$

allow for a simple phenomenological way to write operators for such quantities as, for example, mass, magnetic moment, Coulomb interaction, etc., viz.,

$$\tilde{m} = \Lambda^\pi m_\text{p} + \Lambda^\nu m_\text{n}, \qquad \tilde{\mu} = \Lambda^\pi \mu_\text{p} + \Lambda^\nu \mu_\text{n}, \qquad \tilde{V}_{\text{c},ij} = (e^2/|\mathbf{r}_i - \mathbf{r}_j|)\Lambda_i{}^\pi \Lambda_j{}^\pi. \tag{1-9}$$

[1] The representation used here is one in which t_3 is diagonal. The usual rotational (or spherical) states used for an intrinsic spin of unity are related through a simple transformation [see relationships (1-33)–(1-36)].

[2] For spin angular momentum $\boldsymbol{\sigma}$, we distinguish between the raising and lowering operators $\sigma_{\pm} = \sigma_x \pm i\sigma_y$, and the spherical tensor components $\sigma_{\pm 1} = \mp(1/\sqrt{2})(\sigma_x \pm i\sigma_y)$ and $\sigma_0 = \sigma_z$. A similar distinction is implied for the isospin $\boldsymbol{\tau}$ and the related quantity $\mathbf{t} = \frac{1}{2}\boldsymbol{\tau}$. Care must be exercised to distinguish among τ_+, τ_{+1}, and τ_1.

▶**Exercise 1-1** If the interaction Hamiltonian of a nucleon and the electromagnetic field ($\nabla \cdot \mathbf{A} = 0$ and taking $m_p = m_n = m$) can be written as

$$H = -(e_\alpha/mc)\,\mathbf{A}\cdot\mathbf{p}_\alpha - (e\hbar/2mc)\,\boldsymbol{\mu}_\alpha \cdot (\nabla \times \mathbf{A}),$$

show that one can write $H = H_s + H_v$, where

$$H_s = -(e/2mc)[\mathbf{A}\cdot\mathbf{p}_\alpha + \tfrac{1}{2}\hbar(\boldsymbol{\mu}_p + \boldsymbol{\mu}_n)\cdot(\nabla \times \mathbf{A})]$$

$$H_v = -[e\tau_3(\alpha)/2mc][\mathbf{A}\cdot\mathbf{p}_\alpha + \tfrac{1}{2}\hbar(\boldsymbol{\mu}_p - \boldsymbol{\mu}_n)\cdot(\nabla \times \mathbf{A})].$$

The index α is used to identify the particle under discussion. H_s is referred to as the isoscalar interaction, while H_v is referred to as the isovector interaction since it contains τ_3 (the third component of a vector).

Identify the commuting and noncommuting parts of H with $\mathbf{t}$ (i.e., with t_1, t_2, and t_3 separately). ◀

The intrinsic isospin vectors for the individual particles comprising a complex system can be added in isospin space to generate the isospin for the system in a manner precisely analogous to spin angular momentum.[1] The coupling of isospin vectors has the same coupling algebra as angular momenta; thus a system of two nucleons (i.e., the deuteron, or one of the scattering configurations of two nucleons) could generate either a $T = 0$ or $T = 1$ state from the coupling of the two $t = \frac{1}{2}$ vectors.[2] Using Appendix A, we write in obvious notation the possible two-nucleon states as

(isospin singlet state, $T = 0$)

$$\begin{aligned} {}^1\mathrm{T}_0 &= \zeta_0{}^0(1,2) \\ &= (1/\sqrt{2})[\pi(1)\,\nu(2) - \nu(1)\,\pi(2)] \\ &= (1/\sqrt{2})[\pi_1\nu_2 - \nu_1\pi_2) \end{aligned}$$

(isospin triplet state, $T = 1$)

$$\left\{\begin{aligned} {}^3\mathrm{T}_1 &= \zeta_1{}^1(1,2) = \pi_1\pi_2 \\ {}^3\mathrm{T}_0 &= \zeta_1{}^0(1,2) = (1/\sqrt{2})[\pi_1\nu_2 + \nu_1\pi_2] \\ {}^3\mathrm{T}_{-1} &= \zeta_1^{-1}(1,2) = \nu_1\nu_2. \end{aligned}\right. \tag{1-10}$$

The *direct product* space implied in (1-10) requires defining operators in the usual way, so that, for example, the charge operator is

$$\tilde{Q} = \tilde{Q}_1 I_2 + I_1 \tilde{Q}_2. \tag{1-11}$$

[1] The isopin operators, while satisfying angular momentum commutation relations (and therefore coupling in like manner to angular momenta), do not operate in coordinate space but only in a formal vector space associated with the symmetry of the particle states. Thus, while spin angular momentum could be coupled to orbital angular momentum and physical reality could be ascribed to such quantities as $\boldsymbol{\sigma}\cdot\mathbf{L}$, $\boldsymbol{\sigma}\cdot\mathbf{r}$, etc., similar coupling of coordinate space and isopin space is devoid of physical meaning, at least at this stage.

[2] Generally, capital letters are used to designate system quantities and lowercase letters for individual particle quantities.

It is understood that the first operator in this *direct product operator* operates only on $|\zeta_1\rangle$ and the second only on $|\zeta_2\rangle$ in the product $\zeta(1,2) = |\zeta_1\rangle|\zeta_2\rangle$. The states (1-10) form a particularly convenient basis since they possess well-defined symmetry in the exchange of the isospin coordinates of nucleons 1 and 2.

▶**Exercise 1-2** (a) Using (1-11), verify formally the obvious charge state of the composite nucleon states (1-10).

(b) What is the symmetry of each of the states (1-10) to exchange of the isospin coordinates of nucleons 1 and 2?

(c) Recalling that $\mathbf{A}\cdot\mathbf{B} = A_x B_x + A_y B_y + A_z B_z$, construct the operator $\tilde{\mathbf{t}}_1\cdot\tilde{\mathbf{t}}_2$ in isospin space. (Exercise caution in distinguishing the subscripts 1, 2 used as labels for particles 1 and 2 and the subscripts 1, 2, 3 referring to the Cartesian coordinates in isospin space.) Determine by direct calculation the result of operating on the states (1-10) with $\tilde{\mathbf{t}}_1\cdot\tilde{\mathbf{t}}_2$ and verify your results by using the vector identity

$$\mathbf{t}_1\cdot\mathbf{t}_2 = \tfrac{1}{2}(\mathbf{T}^2-\mathbf{t}_1{}^2-\mathbf{t}_2{}^2), \qquad \text{with} \quad \mathbf{T} = \mathbf{t}_1+\mathbf{t}_2. \qquad ◀$$

The isospin formalism (ignoring the mass difference between the neutron and proton, i.e., assuming mass degeneracy) permits a view of nucleons as being identical Fermi–Dirac particles appropriate to their intrinsic spin angular momentum of $\hbar/2$. A system of nucleons, therefore, is expected to obey Fermi–Dirac statistics with only totally antisymmetrized wave functions being physically admissible. Total antisymmetry must, however, be extended to mean that the wave function of the system changes sign when *all* coordinates ($\mathbf{r}_i$, s_{zi}, and t_{3i}) of any two nucleons are simultaneously interchanged.

Recall that for the two-nucleon system, (n, n), (p, p), or (n, p), the intrinsic spins can be coupled to give $S = 0$ or 1, designated in a notation similar to (1-10) as

$$\begin{aligned}
&(\text{spin singlet state, } S=0) \qquad {}^1\Sigma_0 = \chi_0{}^0(1,2)\\
&\qquad\qquad = (1/\sqrt{2})[\alpha(1)\beta(2)-\beta(1)\alpha(2)]\\
&\qquad\qquad = (1/\sqrt{2})(\alpha_1\beta_2-\beta_1\alpha_2)\\
&(\text{spin triplet state, } S=1) \quad \begin{cases} {}^3\Sigma_1 = \chi_1{}^1(1,2) = \alpha_1\alpha_2\\ {}^3\Sigma_0 = \chi_1{}^0(1,2) = (1/\sqrt{2})(\alpha_1\beta_2+\beta_1\alpha_2)\\ {}^3\Sigma_{-1} = \chi_1^{-1}(1,2) = \beta_1\beta_2\end{cases}
\end{aligned} \tag{1-12}$$

where $\alpha = \binom{1}{0}$ gives $s_z = \frac{1}{2}\hbar$ or "spin up," and $\beta = \binom{0}{1}$ gives $s_z = -\frac{1}{2}\hbar$ or "spin down." Suitable antisymmetrized basis states can be formed, viz.:

$$
\begin{array}{lll}
L \text{ even} & \begin{cases} R_{n,L}(r)\, Y_L{}^M(\theta,\varphi)\; {}^1\mathrm{T}_0\; {}^3\Sigma_{1,0,-1}, & T=0,\quad S=1 \\ R_{n,L}(r)\, Y_L{}^M(\theta,\varphi)\; {}^3\mathrm{T}_{1,0,-1}\; {}^1\Sigma_0, & T=1,\quad S=0 \end{cases} \\
L \text{ odd} & \begin{cases} R_{n,L}(r)\, Y_L{}^M(\theta,\varphi)\; {}^1\mathrm{T}_0\; {}^1\Sigma_0, & T=0,\quad S=0 \\ R_{n,L}(r)\, Y_L{}^M(\theta,\varphi)\; {}^3\mathrm{T}_{1,0,-1}\; {}^3\Sigma_{1,0,-1}, & T=1,\quad S=1. \end{cases}
\end{array}
\tag{1-13}
$$

In the above we have set $\mathbf{r} = \mathbf{r}_{12} = \mathbf{r}_2 - \mathbf{r}_1$ and $\mathbf{r}_1 = -\mathbf{r}/2$, $\mathbf{r}_2 = \mathbf{r}/2$, with $\mathbf{r}_1$ and $\mathbf{r}_2$ measured from the center of mass of the two-nucleon system. Such states would, of course, obey the Pauli exclusion principle requiring $(-1)^{L+S+T} = -1$. Linear combinations of the states (1-13) may form allowable physical states; however, charge conservation requires mixing only states with the same T_3.

If we further require the spatial portion of the wave function to have definite symmetry under space inversion, e.g., under the transformation $\mathbf{r}_1 \to -\mathbf{r}_1$ and $\mathbf{r}_2 \to -\mathbf{r}_2$, only all $L =$ odd or all $L =$ even states can be combined. This follows quite simply by noting that in a center-of-mass (c.m.) description of *two equal-mass* particles, $\mathbf{r}_1 \to -\mathbf{r}_1 = \mathbf{r}_2$ and $\mathbf{r}_2 \to -\mathbf{r}_2 = \mathbf{r}_1$ just exchange the space coordinates of the particles 1 and 2. Now $\mathbf{r}$ $(= \mathbf{r}_{12}) \to -\mathbf{r}$ in spherical coordinates corresponds to $\theta \to \pi - \theta$ and $\varphi \to \pi + \varphi$ and, since $Y_L{}^M(\pi-\theta, \pi+\varphi) = (-1)^L Y_L{}^M(\theta,\varphi)$, the space inversion symmetry requirement that either $\Psi(\mathbf{r}) = +\Psi(-\mathbf{r})$ or $\Psi(\mathbf{r}) = -\Psi(-\mathbf{r})$ then yields the stated result.

We note that for the (p, p) or (n, n) system $T = 1$, since $T_3 = \pm 1$, and hence for these pairs the spectroscopic states[1] such as ${}^3\mathrm{S}_1$, ${}^1\mathrm{P}_1$, ${}^3\mathrm{D}_{1,2,3}$, etc., are ruled out. These two systems in any event must obey the Fermi–Dirac statistics of identical particles. However, the point of view requiring the particles of the (n, p) system to be after all distinguishable, if the concept of isospin formalism is suppressed, must be compatible with any conclusion drawn from (1-13).

Consider taking the attitude that the neutron and proton are distinguishable, but *only* by virtue of the difference in charge. Suppose $\Psi'(1,2) = \Phi(1,2)\,\pi_1\,\nu_2$ is an eigenfunction of the Hamiltonian $H(1,2)$, with $\Phi(1,2)$ a combined spin and space function possessing no definite symmetry on simultaneous spin and space coordinate exchange (i.e., nonidentical-particle point of view) and $\pi_1\,\nu_2$ is simply used to identify the label 1 with the proton and label 2 with the neutron. Since neutron and proton differ only in charge, $H(1,2) = H(2,1)$; hence exchange degeneracy exists with the result that $\Psi''(1,2) = \Phi(2,1)\,\nu_1\,\pi_2$

[1] The notation for spectroscopic states is ${}^{2S+1}L_J$. The angular momenta obey the triangle condition $\mathbf{J} = \mathbf{L} + \mathbf{S}$. When $L \geqslant S$ there are $2S+1$ different values of J possible, and when $S \geqslant L$ there are $2L+1$ possibilities. Also $L = 0$ is called the S state, $L = 1$ the P state, $L = 2$ the D state, $L = 3$ the F state, and thereafter $\mathrm{G} \to L = 4$, $\mathrm{H} \to L = 5$, and so on alphabetically.

must also be an eigensolution, as indeed must any linear combination of Ψ' and Ψ''. Instituting isospin formalism coupled with Fermi–Dirac statistics requires the selection

$$\Psi(1,2) = (1/\sqrt{2})[\Phi(1,2)\pi_1 \nu_2 - \Phi(2,1)\nu_1 \pi_2], \tag{1-14}$$

which is clearly antisymmetric in the interchange of the two nucleons. It is now no longer possible for us to say which particle is the neutron or proton, only that one is a proton and the other surely a neutron. We can write

$$\Phi(1,2) = \frac{1}{\sqrt{2}}\left[\frac{\Phi(1,2)+\Phi(2,1)}{\sqrt{2}} + \frac{\Phi(1,2)-\Phi(2,1)}{\sqrt{2}}\right]$$

or (1-15)

$$\Phi(1,2) = \frac{1}{\sqrt{2}}[\Phi_S(1,2)+\Phi_A(1,2)],$$

where $\Phi_S(1,2)$ is a symmetric function in spin and space and $\Phi_A(1,2)$ is antisymmetric. Clearly, we also have

$$\Phi(2,1) = (1/\sqrt{2})[\Phi_S(1,2)-\Phi_A(1,2)]. \tag{1-16}$$

Substituting (1-15) and (1-16) into (1-14) and rearranging readily gives

$$\Psi(1,2) = \frac{1}{\sqrt{2}}\left\{\Phi_S(1,2)\left[\frac{\pi_1 \nu_2 - \nu_1 \pi_2}{\sqrt{2}}\right] + \Phi_A(1,2)\left[\frac{\pi_1 \nu_2 + \nu_1 \pi_2}{\sqrt{2}}\right]\right\}. \tag{1-17}$$

We see that (1-14), rewritten in the form (1-17), is simply various largely arbitrary linear combinations of the terms appearing in (1-13), with, however, the isospin portions of (1-17) always for $T_3 = 0$. In contradistinction to the (n, n) and (p, p) systems, *all* spectroscopic states such as $^1S_0$, $^3S_1$, $^1P_1$, $^3P_{0,1,2}$, $^1D_2$, $^3D_{1,2,3}$, etc., are possible for the (n, p) system under discussion. The $T = 1$ portion of (1-17) is coupled to the symmetric space–spin spectroscopic states (i.e., $^3S_1$, $^1P_1$, $^3D_{1,2,3}$, etc.), while the $T = 0$ portion is coupled to the antisymmetric space–spin spectroscopic states (i.e., $^1S_0$, $^3P_{0,1,2}$, $^1D_2$, etc.). Thus, insofar as allowable states are concerned, we have complete equivalence of the isospin treatment and the charge-distinguishable point of view.

The treatment of the coupling of the isospin in the two-nucleon system can be readily extended to the system of A nucleons by defining a total isospin operator and its 3-component as

$$\mathbf{T} = \sum_{\alpha=1}^{A} \mathbf{t}_\alpha, \quad \text{and} \quad T_3 = \sum_{\alpha=1}^{A} t_{\alpha 3}. \tag{1-18}$$

The eigenvalues of $\mathbf{T}^2 = \mathbf{T}\cdot\mathbf{T}$ are $T(T+1)$, where T can assume the values between $(0, \frac{1}{2})$ and $A/2$. For each value of T, there are $2T+1$ eigenvalues for T_3, either all integers or half-integers, ranging as usual from $-T$ to T. Since

$t_{\alpha 3}$ is $+\frac{1}{2}$ for the proton and $-\frac{1}{2}$ for the neutron, clearly

$$T_3 = \tfrac{1}{2}(Z-N) = \tfrac{1}{2}(2Z-A) \tag{1-19}$$

and

$$Q = e\sum_{\alpha=1}^{A} [T_{\alpha 3}+\tfrac{1}{2}] = e[T_3+\tfrac{1}{2}A] = eZ,$$

as expected, giving the proper nuclear charge.

If only strong interactions are included in the Hamiltonian for a system of nucleons, the isospin **T** of the system would be conserved; however, as Exercise 1-1 shows, the inclusion of electromagnetic interactions breaks isospin conservation. In this latter event, charge conservation still holds, and hence T_3 remains a constant of the motion.

For a noteworthy and comprehensive review of isospin impurities in nuclear states, refer to Bertsch and Mekjian (1972).

2. Space Inversion Symmetry

Invariance of the description of physical systems to coordinate transformations and inversions requires a relativistic treatment for full generality. However, in the nonrelativistic domain and for "strong" interactions,[1] a simpler discussion paralleling classical physics will suffice. Rotational invariance of physical wave functions leading to the conservation of angular momentum is one example of a universal symmetry requirement. Another important requirement involves the parity operator $\tilde{P}$, which performs the inversion of space coordinates. In classical physics we write

$$\tilde{P}_c f(x,y,z) = f(-x,-y,-z), \tag{1-20}$$

essentially converting a right-handed coordinate system into a left-handed one. In quantum mechanics we formally associate $\tilde{P}$ with a unitary transformation that takes **r** into $-\mathbf{r}$, viz.,

$$\tilde{P}\mathbf{r}\tilde{P}^{-1} = -\mathbf{r}, \tag{1-20a}$$

and in coordinate representation[2]

$$\tilde{P}\psi(\mathbf{r}) = \psi(-\mathbf{r}), \tag{1-20b}$$

where this latter is clearly the analog of (1-20). If the Hamiltonian for a system expressed in terms of the canonical momentum and coordinate **p** and **q** is such that $H(\mathbf{p},\mathbf{q}) = H(-\mathbf{p},-\mathbf{q})$, then clearly P and H commute and parity becomes a constant of the motion, and its eigenvalues are "good quantum

[1] The many-nucleon system for low individual momenta sensibly obeys the Schrödinger equation and classical symmetry requirements. The exception of β-decay will be examined separately later.

[2] If the wave function or coordinate representation of the Dirac state vector $|A\rangle$ is denoted by $\psi_A(\mathbf{r}) = \langle\mathbf{r}|A\rangle$, then (1-20b) can be written $\langle\mathbf{r}|\tilde{P}|A\rangle = \langle -\mathbf{r}|A\rangle$.

numbers." Since $P^2 = PP$ returns the coordinates to their original values, the eigenvalue problem $P\psi(\mathbf{r}) = \pi\psi(\mathbf{r})$ has a ready solution with $\pi = \pm 1$ [viz., $P^2\psi(\mathbf{r}) = \pi P\psi(\mathbf{r}) = \pi^2\psi(\mathbf{r})$, which is also just $\psi(\mathbf{r})$]. In representations that are simultaneous eigenfunctions of P and H, states with $\pi = +1$ are referred to as having even parity and those with $\pi = -1$ are referred to as having odd parity.

Operators $O(\mathbf{p}, \mathbf{q})$ are also described as being even or odd depending on whether $PO(\mathbf{p}, \mathbf{q})P^{-1} = O(-\mathbf{p}, -\mathbf{q}) = \pm O(\mathbf{p}, \mathbf{q})$. In analogy with classical physics all additive terms in O must have the same behavior under coordinate inversion as well as under rotation (i.e., be tensors of the same order and parity).[1]

Operators or dynamical variables are thus conveniently categorized as follows:

(a) Vectors such as $\mathbf{r}, \mathbf{v} = \dot{\mathbf{r}}, \mathbf{p}$ (and also ∇_r) and $\mathscr{E}$ (the electric field intensity) that are odd under space inversion (e.g., $P\mathbf{r}P^{-1} = -\mathbf{r}$, $P\nabla_r P^{-1} = -\nabla_r$, etc.) are called "*polar vector*" operators.

(b) Vectors such as $\mathbf{L} = \mathbf{r} \times \mathbf{p}$, $\mathbf{S}$, $\mathbf{J} = \mathbf{L} + \mathbf{S}$, torque, $\mathscr{H}$ [magnetic field intensity, since $\mathscr{H} = \nabla \times \mathbf{A}$ or $d\mathscr{H} = i(d\mathbf{l} \times \mathbf{r})/r^3$] that are even under space inversion [e.g., $P\mathbf{L}P^{-1} = P\mathbf{r} \times \mathbf{p}P^{-1} = -\mathbf{r} \times (-\mathbf{p}) = \mathbf{r} \times \mathbf{p}$] are called "*axial vector*" or "*pseudovector*" operators.

(c) Scalars formed as scalar products of either two polar vectors or two axial vectors, and therefore even under space inversion, are called "*true*" *scalar* (or just *scalar*) operators. Examples are $\mathbf{L} \cdot \mathbf{L}$, $\mathbf{L} \cdot \mathbf{S}$, $\mathbf{r} \cdot \mathbf{p}$, $\mathscr{E} \cdot \mathscr{E}$, or $\mathscr{H} \cdot \mathscr{H}$ (electromagnetic field energy densities), etc.

(d) Scalars formed as scalar products of one polar vector and one axial vector, and therefore odd under space inversion, are called "*pseudoscalar*" operators. Examples are $\mathbf{r} \cdot \{\mathbf{r} \times \mathbf{p}\}$, $\mathbf{v} \cdot \boldsymbol{\sigma}$, $\mathbf{p} \cdot \mathbf{L}$, etc.

A very important consequence of these considerations is that parity-conserving Hamiltonians must then ultimately involve only true scalars. Another important consequence is that, for representations that are simultaneous eigenfunctions of H and P, expectation values for polar vector and pseudoscalar operators must vanish, as must all their matrix elements between states of the same parity.[2]

[1] The same obviously also applies to state functions. These properties are characteristic of systems conserving total angular momentum and parity.

[2] Parity-conserving Hamiltonians coupled with only convergence requirements on wave functions may generate eigenstates for a specific energy, degenerate or not, that have well-defined parity. It may also happen, as for the states of the nonrelativistic (i.e., no Lamb shift) hydrogen atom that resultant degeneracies involve states of both parities. The non-vanishing linear Stark effect for the first excited state of the hydrogen atom is the consequence of this latter behavior.

For states with well-defined parity, the generalized parity selection rule on matrix elements follows from $\langle B|O|A\rangle = \langle B|P^{-1}POP^{-1}P|A\rangle = \pi_B \pi_O \pi_A \langle B|O|A\rangle$. Thus unitarity of P (i.e., $P^{-1} = P^\dagger$ or $PP^\dagger = P^\dagger P = I$) yields $\pi_B \pi_O \pi_A = +1$.

One early experimental verification of parity violation in weak interactions was the discovery that the expectation value of the pseudoscalar $\boldsymbol{\sigma}\cdot\mathbf{v}$ in β-decay was *nonzero*.

▶**Exercise 1-3** (a) Determine the expectation value of the momentum $\mathbf{p}$ (i.e., p_x, p_y, p_z) for the three-dimensional, linear, isotropic harmonic oscillator and for the plane wave e^{ikz}.

(b) Discuss your findings in terms of the concept of parity. ◀

▶**Exercise 1-4** The scattering of spin-$\frac{1}{2}$ nucleons by spin-0 target nuclei often discloses the presence of a spin–orbit potential term. Show that if the only suitable vectors in the problem are $\boldsymbol{\sigma}$, $\mathbf{p}$, and $\nabla\rho(r)$ [where $\rho(r)$ is the density distribution of the struck nucleus, taken as isotropic], and the nucleon is treated as a pointlike particle, the requirement that only true scalars appear in the Hamiltonian leads to a term

$$V_{LS} = g\frac{1}{r}\frac{dV_c(r)}{dr}\mathbf{L}\cdot\mathbf{S},$$

where we have further assumed that the nuclear potential $V_c(r)$ is proportional to $\rho(r)$. ◀

Parity nonconservation due to the weak nucleon–nucleon interaction in nuclei has also been experimentally established. These experiments involve either the observation of a forward–backward asymmetry in the γ-radiation from polarized nuclei or the observation of the circular polarization of γ-radiation from unpolarized nuclei. The work of Lobashov and co-workers (Lobashov *et al.*, 1971, 1972) is typical. They detect a small circular polarization, $P_\gamma = -(6.1 \pm 0.7)\times 10^{-6}$, for the γ-radiation emitted by unpolarized ^{181}Ta nuclei and $P_\gamma = -(1.30 \pm 0.45)\times 10^{-6}$ for radiative thermal neutron capture by protons. A precise theoretical interpretation of such experimental values is rather difficult [e.g., see Eman and Tadic (1971), Fischbach and Tadic (1973), Gari (1973), Henley (1973); for additional references see Krane *et al.* (1971), Alberi *et al.* (1972), Krane *et al.* (1972)].

As an example of parity violation involving only nucleons, the parity-forbidden α-particle decay of the 8.87-MeV 2^- state in ^{16}O has been extensively investigated. A decay width of $\Gamma_\alpha = (1.03 \pm 0.28)\times 10^{-10}$ eV is observed [refer to Jones *et al.* (1970), Hättig *et al.* (1970), Fox and Robson (1969, pp. 198, 199), and Neubreck *et al.* (1974)]. Recently (Adelberger *et al.*, 1974), parity violation has been observed in the parity mixing of the $J = \frac{1}{2}^-$ first excited state and the $J = \frac{1}{2}^+$ ground state of ^{19}F.

The parity of the nonrelativistic state representing orbital motion of an elementary particle

$$\psi(\mathbf{r}, s_z, t_3) = u_{nl}(r)\, Y_l^m(\theta, \varphi)\, \chi_s^{s_z} \zeta_t^{t_3} \tag{1-21}$$

can be considered to be the product of the particle intrinsic parity and the spatial parity associated with $Y_l^m(\theta, \varphi)$. This concept can be generalized to a system of elementary particles possessing well-defined orbital angular momenta with respect to a common reference center (usually the center of mass of the system). Thus the parity of simple product states of the type (1-21) is

$$\pi = (-1)^{\sum_{i=1}^{A} l_i} \prod_{i=1}^{A} \pi_i, \tag{1-22}$$

where π_i represents the intrinsic parity of the ith particle and we have noted that $Y_l^m(\pi-\theta, \varphi+\pi) = (-1)^l Y_l^m(\theta, \varphi)$, since $\mathbf{r} \to -\mathbf{r}$ in spherical polar coordinates requires $r \to r$, $\theta \to \pi-\theta$, and $\varphi \to \varphi+\pi$.[1] Since nucleons are defined to have even intrinsic parity, the product of intrinsic parities in (1-22) is always $+1$, and this factor, therefore, can be omitted for convenience for nuclear systems. It should also be evident that a completely antisymmetrized (or symmetrized) nuclear wave function formed from products (1-21) also has the parity $(-1)^{\sum_{i=1} l_i}$.

▶**Exercise 1-5** Verify the last statement. ◀

3. Hadron Fields

A convenient classification of elementary particles distinguishes three categories based on the characteristic nature of the interactions involving the particle in question. In increasing order of rest mass the first category is occupied by the *photon*, with zero rest mass and involved as the basic agent of the electromagnetic interaction. The next category is that of the *lepton* and includes the neutrino, electron, and muon (and their antiparticles). These have baryon number $B = 0$. A muon number can also be introduced to distinguish the two types of neutrino (those involved in β-decay and those involved in muon decay). Charged members can interact through the electromagnetic interaction, but their characteristic "hallmark" is their role in processes involving weak interactions. Finally, we have the *hadrons*; these include the nucleons, pions, kaons, etc. They are heavier than the leptons and have baryon numbers $B = 0, 1, 2, \ldots$ (their antiparticles have appropriate negative baryon

[1] While the simple treatment given here is adequate for our purposes, many complications arise in relativistic elementary particle physics when decay modes, production processes, and composite structure problems are considered. Additional relevant quantum numbers are needed, such as G-parity and charge conjugation parity C. Also, we have conceptually taken the elementary particle entities described in (1-22) to remain intact, i.e., they undergo only coherent interactions. When models explicitly exhibiting virtual substructure are employed, conservation rules operate to leave the total parity of the system the same as inferred from (1-22). Conservation of parity for a system initially characterized by (1-22) leads to a *multiplicative* conservation law, requiring only that π(initial) equal π(final).

numbers). The characteristic interaction for hadrons is through the strong interaction. Weak and electromagnetic interactions involving hadrons can also be observed under suitable conditions. The relative strengths of these interactions are approximately: hadronic/electromagnetic/weak $\sim 1/10^{-2}/10^{-13}$.

All the above particles can be further designated as fermions if their intrinsic mechanical spin is an odd half-integer, or as bosons if their intrinsic spin is an integer (zero included).[1] Thus photons with spin $1\hbar$ are bosons, all the leptons are fermions since all have spin $\frac{1}{2}\hbar$, and hadrons exist both as bosons (the pions with spin 0, for example) and as fermions (the nucleons with spin $\frac{1}{2}\hbar$, for example).

Other intrinsic quantum numbers, such as charge, isopin, parity, G-parity, and strangeness, further classify the elementary particles. The parity designation of the hadrons requires the arbitrary fixing of three of them. The three so selected are the two nucleons and the lambda particle, and the intrinsic parity assigned to them is positive. All systems of these, nuclei and hypernuclei, etc., then have well-defined parities in view of (1-22).

The theory of strongly interacting particles, the hadrons, has a fascinating and rapidly developing history. Without wishing to detail this history, understanding current developments insofar as they affect nucleon behavior does require some historical perspective.

The first attempt to give a field-theoretic treatment of nuclear forces was due to Yukawa (1935). This view, and its subsequent development, is perhaps best understood, at least at an elementary level, by considering analogies with the more familiar electromagnetic interactions.

The electromagnetic field in Maxwell's theory is given in terms of the vector potential $\mathbf{A}$ and the scalar potential ϕ that satisfy the equations[2]

$$\nabla^2 \mathbf{A} - \frac{1}{c^2}\frac{\partial^2 \mathbf{A}}{\partial t^2} = -\frac{4\pi}{c}\mathbf{j}, \qquad \nabla^2 \phi - \frac{1}{c^2}\frac{\partial^2 \phi}{\partial t^2} = -4\pi\rho. \tag{1-23}$$

It is well known that the electromagnetic interaction between two charges can be viewed as the interaction of one charge with the field produced by the other particle. The quantized, source-free ($\mathbf{j} = 0$, $\rho = 0$) solution of (1-23) is recognized to give the field equations of the free photon. In a somewhat pictorial, although essentially correct, sense the electromagnetic interaction between two charges can be viewed as the exchange of a virtual photon, radiated by one particle and absorbed by the other.[3] The relevant Feynman

[1] These assertions follow from the spin-statistics theorem, relating basic wave function symmetry and spin.

[2] In the gauge $\nabla \cdot \mathbf{A} + (1/c)\,\partial\phi/\partial t = 0$ (i.e., the Lorentz gauge).

[3] We limit ourselves in this elementary discussion to only the first-order interactions and also omit self-energy terms.

diagram[1] exhibiting this interpretation for the interaction (scattering) of two electrons is shown in Fig. 1.1a.

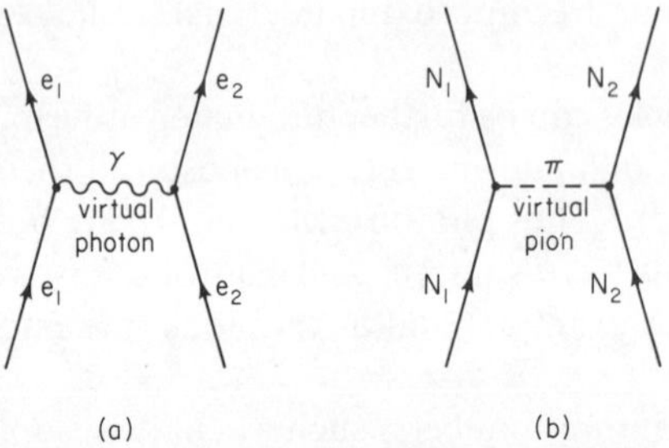

FIG. 1.1. Single quantum exchange Feynman diagrams: (a) virtual photon exchange in electron–electron scattering, (b) virtual pion exchange in nucleon–nucleon scattering. First-order diagrams involving nucleon or electron coordinate exchange are also possible, leading in the electron–electron interaction to Møller scattering.

The quadratic relativistic wave equation (Klein–Gordon equation) for a free particle with rest mass m

$$(\Box^2 - \mu^2)\,\phi = 0 \tag{1-24}$$

results from

$$E^2 = p^2c^2 + m^2c^4$$

by taking[2]

$$E \to -\frac{\hbar}{i}\frac{\partial}{\partial t}, \qquad \mathbf{p} \to \frac{\hbar}{i}\nabla, \qquad \Box^2 \to \nabla^2 - \frac{1}{c^2}\frac{\partial^2}{\partial t^2}, \qquad \text{and} \qquad \mu = \frac{mc}{\hbar}.$$

Yukawa assumed that the nucleon–nucleon interaction process can be viewed as in Fig. 1.1b to be analogous to the electromagnetic interaction of Fig. 1.1a with, however, the exchanged field quantum having finite rest mass m_π. This field quantum or pion would in the source-free case obey (1-24). It now remains to modify (1-24) to include the nucleon-derived source term for the pion field in analogy to the source terms (right-hand sides) of (1-23) or the generation of the *vertex* interactions of Fig. 1.1. The simplest model results

[1] A useful introduction to Feynman diagrams with particular emphasis on nuclear physics is given by Baranger (Italian Phys. Soc., 1969, p. 511 ff.). Another good reference at a comparable level is Mattuck (1967). We shall use such diagrams only in their most elementary context.

[2] Some authors define $\Box^2 = (1/c^2)(\partial^2/\partial t^2) - \nabla^2$, which introduces an occasional sign inconsistency in the literature for the source term when it appears on the right-hand side of (1-24).

when the vertex term is a true scalar in the static approximation. It is instructive to examine the solution of (1-23) for the equivalent case, namely $\mathbf{A} = 0$ and $\rho(\mathbf{r})$ independent of time.[1] The most general well-behaved solution to the resulting Poisson's equation is

$$\phi(\mathbf{r}) = \int \frac{\rho(\mathbf{r}')\, d^3 r'}{|\mathbf{r}-\mathbf{r}'|}. \tag{1-25}$$

We recall that a powerful technique for solving equations of the form

$$(\nabla_r^2 - \mu^2)\, \phi(\mathbf{r}) = U(\mathbf{r}) \tag{1-25a}$$

results by using the Green's function $G(\mathbf{r}, \mathbf{r}')$ defined by

$$(\nabla_r^2 - \mu^2)\, G(\mathbf{r}, \mathbf{r}') = -4\pi\, \delta(\mathbf{r}-\mathbf{r}'). \tag{1-25b}$$

The formal solution to (1-25a) then follows,

$$\phi_G(\mathbf{r}) = -(1/4\pi) \int G(\mathbf{r}, \mathbf{r}')\, U(\mathbf{r}')\, d^3 r', \tag{1-25c}$$

as can be readily verified by substituting into (1-25a) and using (1-25b). The most general solution is to add to $\phi_G(\mathbf{r})$ any homogeneous solution $\phi_H(\mathbf{r})$ obtained from

$$(\nabla_r^2 - \mu^2)\, \phi_H(\mathbf{r}) = 0. \tag{1-25d}$$

We should note that (1-25b) does not completely specify $G(\mathbf{r}, \mathbf{r}')$. A complete specification requires a knowledge of the physical boundary conditions $\phi(\mathbf{r})$ must satisfy. A convenient form for $G(\mathbf{r}, \mathbf{r}')$, for a function $\phi(\mathbf{r})$ that vanishes at infinity and corresponds to the right-hand side of (1-25a) constituting a source, is

$$G_+(\mathbf{r}, \mathbf{r}') = [\exp(-\mu|\mathbf{r}-\mathbf{r}'|)]/|\mathbf{r}-\mathbf{r}'|. \tag{1-25e}$$

We considered here the general case $\mu \neq 0$ since this is required to solve equations of the form (1-24) that we shall encounter, and since (1-25e) is valid for $\mu = 0$ as well.

If $\rho(\mathbf{r}')$ in (1-25) is the charge density to be attributed to a single particle (say an electron), it can be assumed to be concentrated in a very small region of space in the neighborhood of its central coordinate $\mathbf{r}_1$, essentially $\rho_1(\mathbf{r}') = e_1\, \delta(\mathbf{r}' - \mathbf{r}_1)$. Thus the point-charge solution of (1-25) is simply the classical static potential of the charge e_1 located at $\mathbf{r}_1$,

$$\phi_1(r) = e_1/|\mathbf{r}-\mathbf{r}_1|. \tag{1-26}$$

The interaction energy with another pointlike particle located at the coordinate

[1] This is actually an unphysical set of requirements, ignoring as it does recoil effects. The situation described would only be appropriate for infinitely massive charge centers.

$\mathbf{r}_2$ is[1]

$$V_{12} = \int e_2\,\delta(\mathbf{r}-\mathbf{r}_2)\,\phi_1(\mathbf{r})\,d^3r = e_2\phi_1(\mathbf{r}_2) = e_1 e_2/|\mathbf{r}_2-\mathbf{r}_1|. \tag{1-27}$$

The analogous construction for the pion field equation would be (again for the static case, when $\Box^2 \to \nabla^2$)

$$(\nabla^2-\mu^2)\phi = -4\pi\eta(\mathbf{r}). \tag{1-28}$$

The most general physical Green's function solution to (1-28) is [see (1-25a)–(1-25e)]

$$\phi(\mathbf{r}) = \int \frac{\exp(-\mu|\mathbf{r}-\mathbf{r}'|)}{|\mathbf{r}-\mathbf{r}'|}\,\eta(\mathbf{r}')\,d^3r'. \tag{1-29}$$

If again the source density $\eta(\mathbf{r}')$ attributed to the nucleon is confined to a compact region of space such that the delta-function approximation $\eta_1(\mathbf{r}') = g\,\delta(\mathbf{r}'-\mathbf{r}_1)$ can be used, we obtain

$$\phi_1(\mathbf{r}) = g[\exp(-\mu|\mathbf{r}-\mathbf{r}_1|)]/|\mathbf{r}-\mathbf{r}_1|. \tag{1-30}$$

The interaction potential energy, using $\eta_2(\mathbf{r}) = g\,\delta(\mathbf{r}-\mathbf{r}_2)$, becomes[2,3]

$$V_{12} = \int \phi_1(\mathbf{r})\,\eta_2(\mathbf{r})\,d^3r = -g^2[\exp(-\mu|\mathbf{r}_2-\mathbf{r}_1|)]/|\mathbf{r}_2-\mathbf{r}_1|. \tag{1-31}$$

[1] An equivalent but more symmetric classical calculation, which in a sense simulates the two vertices in Fig. 1.1a, would use the basic equation for total electrostatic energy $V = (1/8\pi)\int \mathscr{E}\cdot\mathscr{E}\,d^3r = \frac{1}{2}\int \rho(\mathbf{r})\phi(\mathbf{r})\,d^3r$. For linear field quantities and two "particles" $\rho(\mathbf{r}) = \rho_1(\mathbf{r})+\rho_2(\mathbf{r})$ and $\phi(\mathbf{r}) = \phi_1(\mathbf{r})+\phi_2(\mathbf{r})$, we have

$$\begin{aligned} V &= \tfrac{1}{2}\int \rho_1(\mathbf{r})\phi_1(\mathbf{r})\,d^3r + \tfrac{1}{2}\int [\rho_2(\mathbf{r})\phi_1(\mathbf{r})+\rho_1(\mathbf{r})\phi_2(\mathbf{r})]\,d^3r + \tfrac{1}{2}\int \rho_2(\mathbf{r})\phi_2(\mathbf{r})\,d^3r \\ &= V_{11} + V_{12} + V_{22}. \end{aligned}$$

Ignoring the troublesome self-energy terms V_{11} and V_{22}, we have for the interaction potential, if $\rho_i(r) = e_i\delta(\mathbf{r}-\mathbf{r}_i)$ and $\phi_i = e_i/|\mathbf{r}_i-\mathbf{r}|$, precisely V_{12} of (1-27).

[2] There is an additional complication in comparing the pion field, here taken to be a true scalar field, and the electrostatic field, which is properly a component of the generalized four-vector $A_\mu = (A_x, A_y, A_z, i\phi)$.

The current density can also be expressed as a four-vector $j_\mu = (j_x, j_y, j_z, ic\rho)$, which for a moving charge (with charge $+e$) is $j_\mu = (e/m)(p_x, p_y, p_x, imc)$. The interaction Hamiltonian of the particle and field is $H = -\sum_\mu j_\mu(A_\mu/c) = -(e/mc)\mathbf{p}\cdot\mathbf{A}+e\phi$. This introduces the factor $i^2 = -1$ in the electrostatic case, which is not present for the scalar case.

[3] There is a vexing difference in the literature concerning the introduction of g into the theory. Three conventions can be discerned, "rationalized coupling," "nonrationalized coupling," and, for want of a better term, "analog coupling" (referring to the analogy with the electromagnetic field). Analog coupling is the scheme we use. In rationalized units the 4π in (1-28) is omitted, while in nonrationalized units it is replaced by $(4\pi)^{1/2}$. In the analog convention the field Hamiltonian has a factor $1/8\pi$ (in analogy to the electromagnetic field), while in both the rationalized and nonrationalized conventions this factor is $\frac{1}{2}$. This, coupled with the different convention on the source terms, leads to variants on the interaction energy V_{12} (1-31), in which g^2 appears as in (1-31) for *both* the nonrationalized and analog conventions, while an additional 4π appears in the denominator for the rationalized case; hence $g_A{}^2 = g_N{}^2 = g_R{}^2/4\pi$. Unfortunately, there are also differences in the normalization of ϕ.

This, of course, is the Yukawa potential and is seen to differ from the electromagnetic interaction energy (1-27) in an essential way by having a well-defined range $\mu^{-1} = \hbar/m_\pi c$ (the pion Compton wavelength). The short-range character of the interaction potential (1-31) and the virtual nature of the pion are, of course, related to the uncertainty condition $\Delta E\,\Delta t \approx \hbar$. The creation of a virtual pion involves an energy uncertainty $\Delta E = m_\pi c^2$; hence the virtual state can exist for a time no longer than $\Delta t \approx \hbar/m_\pi c^2$. The range of the pion cloud (equivalent to the range of the pion field) is approximately $c\,\Delta t$ or $\hbar/m_\pi c$. It should also be observed that allowing $\mu \to 0$ in (1-31) appropriately reduces V_{12} to a simple Coulomb-type potential.

It was this relationship between the pion mass and the range of nuclear forces that led Yukawa to predict the existence of the meson. Since it was known that $\mu^{-1} \approx 1$ fm,[1] he predicted $m_\pi \approx 350 m_e$. We need not chronicle here the confusion that resulted from the fact that the first meson to be discovered was the muon, which, while having a similar mass to the pion, interacts only weakly with nucleons.

It is evident from (1-31) that the dimensions of g are the same as that for charge, and we can construct the dimensionless parameter $g^2/\hbar c$, called the coupling constant,[2] in analogy with $\alpha = e^2/\hbar c = 1/137$. While the singlet spin state nucleon–nucleon scattering problem at low energy can in fact be fitted by an interaction potential energy of the form (1-31) leading to a numerical value for $g^2/\hbar c \approx \frac{1}{4}$, the agreement is entirely fortuitous. As we shall see presently, a realistic meson-theoretic explanation of the nucleon–nucleon force is far more complicated. The dominant coupling constant that appears is of the order $g^2/\hbar c \approx \frac{1}{10}$. In any event the meson coupling constant is seen to be at least an order of magnitude greater than α.

Before proceeding further, it might be instructive to see how (1-31) can be generated by a simple second-order perturbation calculation. We follow the treatment given by Fermi (1951, Chapter 4),[3] which gives an equivalent precursor calculation based on Fig. 1.1 without the full apparatus of Feynman diagram physics. The nuclear interaction between the proton and neutron is imagined to be given in terms of a transition probability from an initial state $\mathbf{k}_p = -\mathbf{k}_n$ to a final state $\mathbf{k}_p' = -\mathbf{k}_n'$ proceeding through an intermediate state with the virtual pion present ($\hbar\mathbf{k}_\pi = \mathbf{p}_\pi$ and $|\mathbf{k}_p| = |\mathbf{k}_p'|$ for elastic scattering). We write the *two-step* process schematically as $p + n \to n' + \pi^+ + n \to n' + p'$ for the case with charge exchange or $p + n \to$

[1] fm stands for fermi or femtometer and is 10^{-13} cm $= 10^{-15}$ m (femten is fifteen in Danish).

[2] The literature in elementary particle physics generally employs units with $\hbar = c = 1$; thus both g^2 in those units and $g^2/\hbar c$ as here are referred to as the coupling constant.

[3] Although this reference is by now quite dated, it does contain an excellent introduction to the general field of elementary particles in a particularly digestible if old-fashioned manner.

$p' + \pi^0 + n \to p' + n'$ for the case without charge exchange. The appropriate "Golden Rule" for the transition probability per unit time for two-step processes involving intermediate states is[1]

$$W_{ks} = \frac{2\pi}{\hbar} |H_{ks}|^2 \rho_s(E_s^{(0)}) = \frac{2\pi}{\hbar} \left| \sum_m \frac{H_{km} H_{ms}}{E_s^{(0)} - E_m^{(0)}} \right|^2 \rho_s(E_s^{(0)}).$$

The subscripts k and s refer to the initial and final states, while m refers to an intermediate state. Note that the denominator involves $E_s^{(0)} - E_m^{(0)}$; here $E_k^{(0)} = E_s^{(0)}$. The matrix elements for the creation and destruction of a pion can be estimated to be $-g\hbar c/(2\Omega E)^{1/2}$,[2] where E is the total pion energy ($E^2 = p_\pi{}^2 c^2 + m_\pi{}^2 c^4$) and Ω is the normalization volume confining the pion field. Thus the effective transition matrix element becomes

$$H_{ks} = -\frac{2}{E}\left(\frac{g\hbar c}{(2\Omega E)^{1/2}}\right)^2 = -\frac{g^2 \hbar^2 c^2}{\Omega(p_\pi{}^2 c^2 + m_\pi{}^2 c^4)},$$

where the factor 2 takes into account that for each two-step process the pion field could originate as a creation process with either nucleon (i.e., corresponding to the first or charge-exchange process given above, we could also have $p + n \to p + \pi^- + p' \to n' + p'$) and we have further assumed charge independence of g. Now the transition matrix element can also be calculated from the Born approximation if we know the potential or mutual energy of the interaction $V_{12}(\mathbf{r})$. In terms of the momentum transfer $\mathbf{p}_\pi$ we can then write

$$-\frac{g^2 \hbar^2 c^2}{\Omega(p_\pi{}^2 c^2 + m_\pi{}^2 c^4)} = \frac{1}{\Omega} \int V_{12}(\mathbf{r}) \exp\left(-i \frac{\mathbf{p}_\pi \cdot \mathbf{r}}{\hbar}\right) d^3 r.$$

Or, taking the Fourier transform of both sides,

$$V_{12}(\mathbf{r}) = -\frac{g^2}{8\pi^3 \hbar} \int \frac{\exp(+i\mathbf{p}_\pi \cdot \mathbf{r}/\hbar)}{p_\pi{}^2 + m_\pi{}^2 c^2} d^3 p_\pi = -\frac{g^2}{4\pi r} \exp\left(-\frac{m_\pi c}{\hbar}\right) r.$$

Except for the 4π factor (see previous footnote on factors of 4π) this is just the result (1-31).

Unfortunately, the simple *scalar* Yukawa theory presented above fails to account for even the low-energy nuclear interaction. A proper meson theory for the nucleon–nucleon interaction requires the participation of the full range of meson fields, including the π, η, η', ρ, ω, ..., etc., mesons. The exchange forces generated by such mesons must also include the effects of the interactions among the fields themselves, i.e., the mesons themselves undergo strong interactions appropriate to their hadronic character.

[1] This is also (18.109) in Merzbacher (1970).

[2] See also Messiah (1966, p. 983 ff.).

A mass spectrum of the presently known mesons is given in Fig. 1.8 and will be the subject of discussion later. The conclusion drawn from the simplest application of the finite-mass field equations (1-24) and (1-28) that the resulting potential has a range $\mu^{-1} = \hbar/mc$ is, however, quite generally true. As a reference to Fig. 1.8 will reveal, the pions are the lightest mass mesons. Thus, for a treatment that first examines the behavior of the nuclear potential at the largest distances ($r \gtrsim 1$–2 fm), the pion is in fact the appropriate meson to consider. For example, proton–proton elastic scattering below 10 MeV, while largely independent of potential shape, is known with sufficient precision to reveal the existence of the pion field through the slight energy dependence it produces (see later). The pions of (1-6), however, are *not* scalar particles since they are known to have opposite intrinsic parity from the nucleons (i.e., are of odd intrinsic parity). This, combined with their intrinsic spin angular momentum of zero and the three possible charge states, means that they are *pseudoscalar isovector* particles. Other relevant mesons considered later will include vector and scalar mesons in combination with the pseudoscalar pions.

It is instructive at this point to proceed with a plausible heuristic treatment of a pseudoscalar isovector meson theory, here introduced ad hoc but discussed more carefully later.

The nuclear two-body force is spin dependent [the bound state of the (n, p) system, the deuteron, is a triplet spin state, while the singlet spin state is unbound]. The nonzero quadrupole moment of the deuteron requires a tensor component to the force. The nuclear force has charge exchange components made possible by the pion possessing charge. The nuclear force is, however, largely charge independent [the singlet spin states of the (n, n), (n, p), and (p, p) systems have approximately equal forces apart from Coulomb effects]. Setting aside for the moment the property of charge independence, we take as the simplest linear pseudoscalar field equations in the static approximation the expression

$$\nabla^2 \phi_k - {\mu_k}^2 \phi_k = -4\pi (f_k/\mu_k)\, \tau_{k,i}^{\dagger}\, \boldsymbol{\sigma}_i \cdot \boldsymbol{\nabla}_i\, \rho(\mathbf{r}-\mathbf{r}_i). \qquad (1\text{-}32)$$

A number of comments are in order concerning (1-32).

The fact that (1-32) is a linear equation implies a non-self-interacting meson field. This means that (correlated) multiple-pion exchange is not included. Thus the important "resonances" for the interacting pions are not considered. The model derived from (1-32) is referred to as the OPE model (one-pion-exchange model).

The source density $\eta(\mathbf{r})$ in (1-28) was treated up to now as a scalar function. Noting that V_{12}, (1-31), is a potential energy operator in the Schrödinger equation of the two-nucleon system, we can extend the concept of the source density to include various operators intended in the resulting expression of

$\tilde{V}_{12}$ to operate on the two-nucleon wave function. Thus we can write $\eta_1(\mathbf{r}) = \tilde{\theta}_1 f(\mathbf{r})$ and $\eta_2(\mathbf{r}) = \tilde{\theta}_2 f(\mathbf{r})$, giving $V_{12} = \tilde{\theta}_1 \tilde{\theta}_2 J(\mathbf{r})$, with $\tilde{\theta}_1$ and $\tilde{\theta}_2$ appropriate Pauli operators (or in the relativistic case, Dirac operators). The operator $\tau_{k,i}^\dagger \boldsymbol{\sigma}_i \cdot \boldsymbol{\nabla}_i$ is just such an operator $\tilde{\theta}_i$. We also note that the static approximation $\nabla^2 \phi(\mathbf{r}) - \mu^2 \phi(\mathbf{r}) = -4\pi\eta(\mathbf{r})$ transforms under space inversion into $-\nabla^2\phi(\mathbf{r}) + \mu^2\phi(\mathbf{r}) = -4\pi\eta(-\mathbf{r})$, since $\phi(-\mathbf{r}) = -\phi(\mathbf{r})$ for a pseudoscalar field. Thus $\eta(-\mathbf{r}) = -\eta(\mathbf{r})$, showing that the proper source density of a pseudoscalar field is also a pseudoscalar. The inner product $\boldsymbol{\sigma}_i \cdot \boldsymbol{\nabla}_i \rho(\mathbf{r} - \mathbf{r}_i)$ is clearly a pseudoscalar as required. If we specify $\rho(\mathbf{r})$ for a particlelike source with respect to its center of mass $\mathbf{r}_1$, we would write $\boldsymbol{\sigma}_1 \cdot \boldsymbol{\nabla}_1 \rho_1(\mathbf{r} - \mathbf{r}_1)$, with the subscript on the gradient operator denoting differentiation with respect to the source center $\mathbf{r}_1$. Since nucleon 2 will also give rise to a similar operator, V_{12} will again be a true scalar operator as required.

The selection of the suitable required polar vector to couple with $\boldsymbol{\sigma}$ to yield the overall pseudoscalar behavior is difficult in the static approximation. If the source particle were in motion, we could have used its momentum $\mathbf{p}_1$. Using the gradient of the source center, as we have, simulates this behavior to some extent. An additional neglected consideration involves the momentum recoil effects associated with the pion emission and absorption; recall that $m_\pi/m_\mathrm{N} \approx 1/7$. We defer until later the discussion of the relativistic origin of the terms in (1-32). We also defer until later showing that this form leads to a two-body force involving $\boldsymbol{\sigma}_1 \cdot \boldsymbol{\sigma}_2$ and the tensor operator S_{12}.

The index k can take the four values 0, +1, −1, or 4 corresponding to four possible isospin operators $\tau_{k,i}^\dagger$ that operate on the *nucleon* source state identified by the subscript i. The operators necessary to generate the physical charge states ϕ_k of the meson field are[1]

$$\tau_{+1}^\dagger = -(1/\sqrt{2})(\tau_1 + i\tau_2)^\dagger = -\sqrt{2}\begin{pmatrix} 0 & 1 \\ 0 & 0 \end{pmatrix}^\dagger = -\sqrt{2}\begin{pmatrix} 0 & 0 \\ 1 & 0 \end{pmatrix},$$

$$\tau_{-1}^\dagger = (1/\sqrt{2})(\tau_1 - i\tau_2)^\dagger = \sqrt{2}\begin{pmatrix} 0 & 0 \\ 1 & 0 \end{pmatrix}^\dagger = \sqrt{2}\begin{pmatrix} 0 & 1 \\ 0 & 0 \end{pmatrix},$$

$$\tau_0^\dagger = \tau_3^\dagger = \tau_3 = \begin{pmatrix} 1 & 0 \\ 0 & -1 \end{pmatrix}, \qquad \tau_4^\dagger = I = \begin{pmatrix} 1 & 0 \\ 0 & 1 \end{pmatrix}.$$

With these definitions of $\tau_k^\dagger$ it is easily seen that for four corresponding meson fields ϕ_k have suitable charges. The equation with $k = +1$ involves the operator $\left(\begin{smallmatrix} 0 & 0 \\ 1 & 0 \end{smallmatrix}\right)$, which converts a proton nucleon state to a neutron and when operating on a neutron state gives zero. The conversion of a proton to a neutron, in view

[1] As noted in connection with (1-7), the operators $\tau_{\pm 1}$ are related to the raising and lowering operators, viz., $\tau_{\pm 1} = \mp(1/\sqrt{2})\tau_\pm$.

of charge conservation, must generate the positively charged meson field ϕ_{+1}. A similar consideration holds for $k=-1$ generating the negatively charged meson field ϕ_{-1}. The operators τ_0 and τ_4 ,while leaving the nucleon charge state unaltered, and hence corresponding to neutral (or zero charge) meson fields, are not equivalent since τ_0 changes the *sign* of the neutron state. The three meson states ϕ_{+1}, ϕ_{-1}, and ϕ_0 form an isospin triplet, which we associate with the pions of (1-6).

The meson state ϕ_4 is an isospin singlet (zero charge) and has as a possible corresponding candidate the η-meson of mass $m_\eta c^2 = (548.8 \pm 0.6)$ MeV (Particle Data Group, 1973), with $t=0$, zero intrinsic spin, and odd intrinsic parity. The η-meson decays with a total width of only 2.6 keV into $\pi^+\pi^0\pi^- + \pi^+\pi^-\gamma$ and is the lowest mass pseudoscalar isoscalar meson.

Returning to the pion triplet states $\phi_{\pm 1}, \phi_0$, these are taken as the eigenfunctions of the *pion* isospin operator t_3 appropriate for an isovector with $t=1$. For the standard spherical coordinates, we use

$$t_1 = \begin{pmatrix} 0 & 0 & 0 \\ 0 & 0 & -i \\ 0 & i & 0 \end{pmatrix}, \qquad t_2 = \begin{pmatrix} 0 & 0 & i \\ 0 & 0 & 0 \\ -i & 0 & 0 \end{pmatrix}, \qquad t_3 = \begin{pmatrix} 0 & -i & 0 \\ i & 0 & 0 \\ 0 & 0 & 0 \end{pmatrix}. \tag{1-33}$$

Note that in this representation t_3 is not diagonal. The eigenvalue problem diagonalizing $\tilde{t}_3$, namely

$$\tilde{t}_3 \mathbf{u} = \begin{pmatrix} 0 & -i & 0 \\ i & 0 & 0 \\ 0 & 0 & 0 \end{pmatrix} \begin{pmatrix} u_1 \\ u_2 \\ u_3 \end{pmatrix} = \lambda \begin{pmatrix} u_1 \\ u_2 \\ u_3 \end{pmatrix}, \tag{1-34}$$

yields the normalized unit vectors

$$\begin{aligned}
\lambda = +1: &\quad \hat{u}_{+1} = -\frac{1}{\sqrt{2}} \begin{pmatrix} 1 \\ i \\ 0 \end{pmatrix} = -\frac{1}{\sqrt{2}}(\hat{1} + i\hat{2}) \\
\lambda = -1: &\quad \hat{u}_{-1} = \frac{1}{\sqrt{2}} \begin{pmatrix} 1 \\ -i \\ 0 \end{pmatrix} = \frac{1}{\sqrt{2}}(\hat{1} - i\hat{2}) \qquad (1\text{-}35) \\
\lambda = 0: &\quad \hat{u}_0 = \begin{pmatrix} 0 \\ 0 \\ 1 \end{pmatrix} = \hat{3},
\end{aligned}$$

with

$$\hat{1} = \begin{pmatrix} 1 \\ 0 \\ 0 \end{pmatrix}, \quad \hat{2} = \begin{pmatrix} 0 \\ 1 \\ 0 \end{pmatrix}, \quad \hat{3} = \begin{pmatrix} 0 \\ 0 \\ 1 \end{pmatrix}.$$

The pion field components are thus related in the two representations (unit vectors $\hat{u}_{+1}, \hat{u}_{-1}, \hat{u}_0$ and $\hat{1}, \hat{2}, \hat{3}$) by

$$\begin{aligned} \phi_{+1} &= -\frac{1}{\sqrt{2}}(\phi_1 + i\phi_2), & \phi_1 &= -\frac{1}{\sqrt{2}}(\phi_{+1} - \phi_{-1}) \\ \phi_{-1} &= \frac{1}{\sqrt{2}}(\phi_1 - i\phi_2), & \phi_2 &= \frac{i}{\sqrt{2}}(\phi_{+1} + \phi_{-1}) \\ \phi_0 &= \phi_3, & \phi_3 &= \phi_0. \end{aligned} \tag{1-36}$$

The Cartesian components ϕ_1, ϕ_2, and ϕ_3 are seen to be linear combinations of the physical charged states ϕ_{+1}, ϕ_{-1}, and ϕ_0. It is also quickly seen that the operators τ_{+1}, τ_{-1}, and τ_0 appearing in (1-32) are just the spherical components of $\boldsymbol{\tau}$ (the Cartesian components of which are τ_1, τ_2, and τ_3). This also makes it evident why ϕ_0 (coupled to operator τ_0) and not ϕ_4 is associated with $\phi_{\pm 1}$ to be the pion triplet. The quantity $\mu_k^{-1} = \hbar/m_k c$ (Compton wavelength of the appropriate meson of mass m_k) has been inserted on the right-hand side of (1-32) to preserve the chargelike dimensions of f_k (playing the role of g in the scalar theory).

As stated earlier, the presence of the rest mass term $\mu_k{}^2\phi_k$ on the left-hand side of (1-32) results in the range of the various meson fields varying inversely with the mass m_k. The behavior of the long-range nuclear potential ,therefore, should be dominated by the pion fields since $m_\eta/m_\pi \approx 4$. Thus, even though in the bewildering array of meson types (Particle Data Group, 1973),[1] the pion and η-meson are the lightest nonstrange mesons, we restrict our present attention to the pion fields and rewrite (1-32) with mass degeneracy (i.e., $\mu_k = \mu_\pi = \mu$) as

$$\nabla^2\phi_k - \mu^2\phi_k = -4\pi(f_k/\mu)\,\tau_{k,i}^\dagger\,\boldsymbol{\sigma}_i \cdot \nabla_i\,\rho(\mathbf{r}-\mathbf{r}_i), \tag{1-37}$$

with $k = 0, +1, -1$ and ϕ_k of the form (1-36).

The general Green's function solution of (1-37) is

$$\phi_k(\mathbf{r}) = \frac{f_k}{\mu}\tau_{k,i}^\dagger \int \frac{\exp(-\mu|\mathbf{r}-\mathbf{r}'|)}{|\mathbf{r}-\mathbf{r}'|}\,\boldsymbol{\sigma}_i \cdot \nabla_i\,\rho(\mathbf{r}'-\mathbf{r}_i)\,d^3r'. \tag{1-38}$$

[1] In passing, we should note that the scalar isoscalar (or neutral scalar) meson theory of Yukawa, (1-28)–(1-31), now has a candidate meson, albeit with a different coupling constant, namely, the ε-meson, $m_\varepsilon c^2 \lesssim 700$ MeV, with $t = 0$, zero charge, even parity, and zero intrinsic spin. We relegate a discussion of this and other mesons to later.

If charge is to be conserved at the second vertex as well, the operator τ_k must be involved in that case rather than $\tau_k^\dagger$, so that [using labels 1 and 2 for the two vertex interactions, and for convenience writing $\tau_{k,1} \equiv \tau_k(1)$ and $\tau_{k,2} \equiv \tau_k(2)$][1]

$$V_{12,k}(\mathbf{r}_1, \boldsymbol{\tau}_1, \boldsymbol{\sigma}_1; \mathbf{r}_2, \boldsymbol{\tau}_2, \boldsymbol{\sigma}_2; k)$$

$$= \int \phi_{k,1}(\mathbf{r})\, \eta_{k,2}(\mathbf{r})\, d^3r$$

$$= -\frac{f_k^{\,2}}{\mu^2} \tau_k(2)\, \tau_k^\dagger(1)\, (\boldsymbol{\sigma}_1 \cdot \boldsymbol{\nabla}_1)(\boldsymbol{\sigma}_2 \cdot \boldsymbol{\nabla}_2)$$

$$\cdot \iint \rho_2(\mathbf{r}-\mathbf{r}_2) \frac{\exp(-\mu|\mathbf{r}-\mathbf{r}'|)}{|\mathbf{r}-\mathbf{r}'|} \rho_1(\mathbf{r}'-\mathbf{r}_1)\, d^3r\, d^3r', \qquad (1\text{-}39)$$

and

$$V_{12} = \sum_{k=+1,0,-1} V_{12,k}.$$

This last sum involves the factor $\sum_{k=+1,0,-1} f_k^{\,2} \tau_k(2)\, \tau_k^\dagger(1)$. Taking $f_{+1}^2 = f_{-1}^2$ results in a charge symmetric theory and incidentally ensures that V_{12} is Hermitian. We further note that

$$\tau_{+1}(2)\,\tau_{+1}^\dagger(1) + \tau_{-1}(2)\,\tau_{-1}^\dagger(1) = -2[t_{+1}(2)\,t_{-1}(1) + t_{-1}(2)\,t_{+1}(1)] \qquad (1\text{-}40)$$

and using (1-7),

$$-2[t_{+1}(2)\,t_{-1}(1) + t_{-1}(2)\,t_{+1}(1)] = \tau_1(1)\,\tau_1(2) + \tau_2(1)\,\tau_2(2). \qquad (1\text{-}41)$$

Hence, if in addition $f_0^{\,2} = f_{\pm 1}^2$, we would have a multiplicative factor in (1-39)

$$\tau_{+1}(2)\,\tau_{+1}^\dagger(1) + \tau_{-1}(2)\,\tau_{-1}^\dagger(1) + \tau_0(2)\,\tau_0^\dagger(1) = \boldsymbol{\tau}_1 \cdot \boldsymbol{\tau}_2, \qquad (1\text{-}42)$$

and could therefore write, with $f^2 = f_0^{\,2} = f_{+1}^2 = f_{-1}^2$,

$$V_{12} = -\frac{f^2}{\mu^2} \boldsymbol{\tau}_1 \cdot \boldsymbol{\tau}_2 (\boldsymbol{\sigma}_1 \cdot \boldsymbol{\nabla}_1)(\boldsymbol{\sigma}_2 \cdot \boldsymbol{\nabla}_2)$$

$$\cdot \iint \rho_2(\mathbf{r}-\mathbf{r}_2) \frac{\exp(-\mu|\mathbf{r}-\mathbf{r}'|)}{|\mathbf{r}-\mathbf{r}'|} \rho_1(\mathbf{r}'-\mathbf{r}_1)\, d^3r\, d^3r'. \qquad (1\text{-}43)$$

Such a form is charge independent. One speaks of the theories in which the *only nonzero* values of f_k are f_4 as neutral, $f_{+1} = f_{-1}$ as charged, and $f_0 = f_{+1} = f_{-1}$ as symmetric.

[1] The tacit assumption throughout is that the "k-charge" distribution of the source is proportional to the matter density and simply scales with the appropriate coupling constant f_k. This is analogous to using for the regular charge density $\rho_\alpha(\mathbf{r}) = \psi^*(\mathbf{r})\, e_\alpha \psi(\mathbf{r})$.

▶**Exercise 1-6** Show that $\boldsymbol{\tau}_1 \cdot \boldsymbol{\tau}_2$ gives charge independence, while still permitting a spin dependence for the two-nucleon system. Use the suitable states from (1-13) and apply your results from Exercise 1-2. ◀

While (1-43) possesses divergence difficulties when the δ-function approximation of pointlike nucleons is used, a formal solution can nonetheless be effected. Taking $\rho_i(\mathbf{r}-\mathbf{r}_i) = \delta(\mathbf{r}-\mathbf{r}_i)$ in (1-43) gives

$$V_{12} = -\frac{f^2}{\mu^2}\boldsymbol{\tau}_1 \cdot \boldsymbol{\tau}_2 (\boldsymbol{\sigma}_1 \cdot \boldsymbol{\nabla}_1)(\boldsymbol{\sigma}_2 \cdot \boldsymbol{\nabla}_2)\frac{\exp(-\mu|\mathbf{r}_2-\mathbf{r}_1|)}{|\mathbf{r}_2-\mathbf{r}_1|}. \tag{1-44}$$

Performing the indicated gradient operations gives

$$V_{12} = \frac{f^2}{3}\boldsymbol{\tau}_1 \cdot \boldsymbol{\tau}_2 \left\{\left[\left(1+\frac{3}{\mu r_{12}}+\frac{3}{\mu^2 r_{12}^2}\right)S_{12} + \boldsymbol{\sigma}_1 \cdot \boldsymbol{\sigma}_2\right]\frac{\exp(-\mu r_{12})}{r_{12}} - \frac{4\pi}{\mu^2}\boldsymbol{\sigma}_1 \cdot \boldsymbol{\sigma}_2\, \delta(r_{12})\right\} \tag{1-45}$$

with $r_{12} = |\mathbf{r}_2-\mathbf{r}_1|$ and $S_{12} = 3(\boldsymbol{\sigma}_1 \cdot \hat{\mathbf{r}}_{12})(\boldsymbol{\sigma}_2 \cdot \hat{\mathbf{r}}_{12}) - \boldsymbol{\sigma}_1 \cdot \boldsymbol{\sigma}_2$.

▶**Exercise 1-7** Show that for any function of r, say $F(r)$, we have

$$\boldsymbol{\nabla}(\boldsymbol{\sigma}_2 \cdot \boldsymbol{\nabla} F) = \hat{\mathbf{r}}(\boldsymbol{\sigma}_2 \cdot \hat{\mathbf{r}})\left[\frac{\partial^2 F}{\partial r^2} - \frac{1}{r}\frac{\partial F}{\partial r}\right] + \frac{1}{r}\frac{\partial F}{\partial r}\boldsymbol{\sigma}_2$$

and hence that

$$(\boldsymbol{\sigma}_1 \cdot \boldsymbol{\nabla})(\boldsymbol{\sigma}_2 \cdot \boldsymbol{\nabla}) F = \frac{1}{3}[3(\boldsymbol{\sigma}_1 \cdot \hat{\mathbf{r}})(\boldsymbol{\sigma}_2 \cdot \hat{\mathbf{r}}) - \boldsymbol{\sigma}_1 \cdot \boldsymbol{\sigma}_2]\left[\frac{\partial^2 F}{\partial r^2} - \frac{1}{r}\frac{\partial F}{\partial r}\right] + \frac{1}{3}\boldsymbol{\sigma}_1 \cdot \boldsymbol{\sigma}_2 \nabla^2 F.$$

Then noting that $\boldsymbol{\nabla}_2 = -\boldsymbol{\nabla}_1 = \boldsymbol{\nabla}$, writing $\mathbf{r} = |\mathbf{r}_2 - \mathbf{r}_1|$, and taking $F(r) = e^{-\mu r}/r$, show that the result (1-45) follows from (1-44). The last term in (1-45) requires recalling that $\nabla^2(1/|\mathbf{r}|) = -4\pi\,\delta(\mathbf{r})$. ◀

▶**Exercise 1-8** Derive the expression $[3(\boldsymbol{\mu}_1 \cdot \hat{\mathbf{r}})(\boldsymbol{\mu}_2 \cdot \hat{\mathbf{r}}) - \boldsymbol{\mu}_1 \cdot \boldsymbol{\mu}_2](1/r^3)$ for the interaction energy of two particles with intrinsic magnetic moments $\boldsymbol{\mu}_1$ and $\boldsymbol{\mu}_2$ at a separation $\mathbf{r}$. (*Hint:* Use $\mathbf{A} = \boldsymbol{\mu} \times \mathbf{r}/r^3$, $\mathscr{H} = \boldsymbol{\nabla} \times \mathbf{A}$, and finally $V_{12} = \boldsymbol{\mu}_2 \cdot \mathscr{H}_1$).

Estimate the strength of the nucleon–nucleon magnetic moment interaction at a separation equal to the range of nuclear forces ($R \approx \hbar/m_\pi c$) in units of e^2/R. Comment on the results. ◀

The form of (1-45) is satisfying in that it contains the necessary tensor term S_{12}, spin dependence through $\boldsymbol{\sigma}_1 \cdot \boldsymbol{\sigma}_2$, charge symmetry through $\boldsymbol{\tau}_1 \cdot \boldsymbol{\tau}_2$, and, of course, the short-range Yukawa dependence on r_{12}. This potential also generates attractive forces for both the triplet spin as well as singlet spin states of the (n, p) system in space symmetric (L-even) configurations (the force in the triplet spin state being stronger). This is in agreement with empirical evidence, at least at low energy. The spin-dependent core term involving $\delta(r_{12})$ and the r_{12}^{-3} dependence in the ground-state deuteron problem are, however, serious difficulties. These can be avoided if, for such a static model, we simulate the omitted recoil effects by converting the theory into a nonlocal theory spreading the nucleon source density over a small region (approximately the size of the Compton wavelength of the nucleon $\hbar/m_{\mathrm{N}}c$) and using (1-45) for $r_{12} \geqslant r_{\min}$ with $r_{\min}$ describing the source "size." Fortunately, the detailed nature of $\rho_i(\mathbf{r}-\mathbf{r}_i)$ is found not to be important and the nonlocal nature can be parameterized through a single quantity, the "cutoff momentum" $p_{\max} = \hbar k_{\max}$ corresponding to $r_{\min}$.[1]

The nonrelativistic discussion of the pseudoscalar meson theory presented above can be considered the static limiting case of a more general relativistically invariant theory. In such relativistic theories the nucleon–pion interaction Lagrangian and Hamiltonian densities are taken as

$$\mathscr{L}_{\mathrm{N}\pi} = -H_{\mathrm{N}\pi} \approx -iG\psi^{\dagger}\gamma_4\gamma_5\boldsymbol{\tau}\cdot\boldsymbol{\phi}\psi. \tag{1-46}$$

Here $\boldsymbol{\tau}$ is the nucleon isospin and ψ the four-component Dirac spinor referring to the nucleon. The pion field is described by three independent real fields ϕ_1, ϕ_2, and ϕ_3 taken as the $\hat{1}$, $\hat{2}$, and $\hat{3}$ components of the vector $\boldsymbol{\phi}$ in *isospin space*. Each component in turn is a function of the space–time coordinate x_λ. The interaction (1-46) is one of two Lorentz-invariant bilinear forms that are pseudoscalar in nature. A second form, $-i(G'/\mu)\psi^{\dagger}\gamma_4\gamma_5\sum_\lambda\gamma_\lambda[(\partial/\partial x_\lambda)\boldsymbol{\tau}\cdot\boldsymbol{\phi}]\psi$, also leading to a pseudoscalar theory, is possible. The form (1-46) is referred to as a pseudoscalar theory with pseudoscalar coupling, while this second form is a pseudoscalar theory with axial vector coupling, also sometimes referred to as "derivative" coupling for obvious reasons [$\sum_\lambda(\partial/\partial x_\lambda)\phi$ behaves like the gradient of a pseudoscalar and is therefore axial vector in character].

In a formal procedure, a Lagrangian density involving (1-46) is constructed and the resulting equations of motion for the system are deduced, viz.,[2]

$$(\Box^2-\mu^2)\boldsymbol{\phi} = +4\pi iG\boldsymbol{\tau}\psi^{\dagger}\gamma_4\gamma_5\psi. \tag{1-47}$$

A proper discussion of the Lagrangian field-theoretic formalism involved in (1-46) and (1-47) would take us too far afield for our purposes. A condensed treatment can be found in DeBenedetti (1964, Chapter 7) and the Handbuch

[1] One can write the Fourier transform of $\rho(\mathbf{r})$ as $v(k^2) = (2\pi\hbar)^{-3/2}\int[\exp(-i\mathbf{k}\cdot\mathbf{r})]\rho(\mathbf{r})\,d^3r$, noting that v is only a function of k^2 if $\rho(\mathbf{r})$ is spherically symmetric. Conversely we have $\rho(\mathbf{r}) = (2\pi)^{-3/2}\int[\exp(i\mathbf{k}\cdot\mathbf{r})]\,v(k^2)\,d^3k$. It is common to use $v(k^2) = \text{const}$ for $k < k_{\max}$ and $v(k^2) = 0$ for $k > k_{\max}$.

[2] The additional equations of motion for the nucleon mechanical spin and isospin, $i\dot{\boldsymbol{\sigma}} = [\boldsymbol{\sigma}, H]$ and $i\dot{\boldsymbol{\tau}} = [\boldsymbol{\tau}, H]$, are also required.

article by Hulthén and Sugawara (1957, p. 14 ff.). Comprehensive treatments can be found in Schweber *et al.* (1955) and in a somewhat more recent reference by Nishijima (1964).

At nonrelativistic energies and in the static approximation (all nucleon coordinates including $\boldsymbol{\tau}$ and $\boldsymbol{\sigma}$ constant in time) we can approximate the nucleon state by that appropriate for a freely propagating Dirac (spin-$\frac{1}{2}$) nucleon spinor. Using Appendix C, we write the coupled equations

$$-i\hbar c\boldsymbol{\sigma}^{\mathrm{P}} \cdot \nabla\psi_{\mathrm{s}} + m_{\mathrm{N}}c^2\psi_l = E\psi_l, \qquad -i\hbar c\boldsymbol{\sigma}^{\mathrm{P}} \cdot \nabla\psi_l - m_{\mathrm{N}}c^2\psi_{\mathrm{s}} = E\psi_{\mathrm{s}}, \tag{1-48}$$

where the four-component Dirac spinor has been partitioned into two Pauli spinors. The Pauli spinors ψ_l and ψ_{s} stand for the "large" and "small" components, respectively. In the nonrelativistic approximation $E = m_{\mathrm{N}}c^2$; hence we have

$$\psi_{\mathrm{s}} \approx -(i\hbar/2m_{\mathrm{N}}c)\boldsymbol{\sigma}^{\mathrm{P}} \cdot \nabla\psi_l. \tag{1-49}$$

In this representation

$$\gamma_4 = \begin{pmatrix} 1 & 0 \\ 0 & -1 \end{pmatrix} \quad \text{and} \quad \gamma_5 = -\begin{pmatrix} 0 & 1 \\ 1 & 0 \end{pmatrix},$$

and therefore we have for the source term $\eta(\mathbf{r})$ on the right-hand side of (1-47)

$$\begin{aligned} -iG\boldsymbol{\tau}\psi^\dagger\gamma_4\gamma_5\psi &= iG\boldsymbol{\tau}(\psi_l^\dagger, \psi_{\mathrm{s}}^\dagger)\begin{pmatrix} 0 & 1 \\ -1 & 0 \end{pmatrix}\begin{pmatrix} \psi_l \\ \psi_{\mathrm{s}} \end{pmatrix} = iG\boldsymbol{\tau}(\psi_l^\dagger, \psi_{\mathrm{s}}^\dagger)\begin{pmatrix} \psi_{\mathrm{s}} \\ -\psi_l \end{pmatrix} \\ &= iG\boldsymbol{\tau}(\psi_l^\dagger\psi_{\mathrm{s}} - \psi_{\mathrm{s}}^\dagger\psi_l) \end{aligned} \tag{1-50}$$

or

$$\eta(\mathbf{r}) = (G\hbar/2m_{\mathrm{N}}c)\boldsymbol{\tau}\,\nabla \cdot (\psi_l^\dagger\boldsymbol{\sigma}^{\mathrm{P}}\psi_l). \tag{1-51}$$

In the last step we use (1-49) and the fact that $\boldsymbol{\sigma}$ is a constant matrix commuting with ∇ and $\boldsymbol{\sigma} = \boldsymbol{\sigma}^\dagger$.

Our overall intention is to arrive at the *one-meson*-exchange potential V_{12} operating between two nucleons in the static approximation. To this end we have introduced the *auxiliary* equations (1-48) to determine the approximate nature of the source term (right-hand side) of (1-47). The *single-particle* nucleon field ψ of (1-48) is simply the lowest order (or noninteracting) solution to the problem in the coordinate of one of the two nucleons, carrying the label i. We seek by the use of V_{12} [e.g., (1-39)] to arrive at the next order of approximation for the behavior of the two-nucleon system. In the static approximation, ψ_i of (1-48) in effect determines the nonlocality of the potential V_{12} by ascribing a possible spatial extension to $\psi_i^\dagger\psi_i$ in terms of $\mathbf{r}-\mathbf{r}_i$, with $\mathbf{r}_i$ the coordinate of the "nucleon center." In next order, the Schrödinger equation of the two-nucleon system (if the nonrelativistic approximation is used) will be in terms of the relative coordinate $(\mathbf{r}_1 - \mathbf{r}_2)$, with $\mathbf{r}_1$ and $\mathbf{r}_2$ referring to the "nucleon centers." The spin dependence of ψ_i in (1-48) can be taken to be the same as that of nucleon i as it will occur in this next higher approximation.[1] Thus we write

$$\eta_i(\mathbf{r}) = (G\hbar/2m_\mathrm{N} c)\,\boldsymbol{\tau}_i(\boldsymbol{\sigma}_i^\mathrm{P}\cdot\nabla_i)\,\rho_i(\mathbf{r}-\mathbf{r}_i). \tag{1-51a}$$

Here $\mathbf{r}$ refers to the coordinate of the meson field, $\rho_i(\mathbf{r}-\mathbf{r}_i)$ measures the nucleon i density in the vicinity of $\mathbf{r}$, and $\boldsymbol{\sigma}_i^\mathrm{P}$ and $\boldsymbol{\tau}_i$ are left to form part of the operator $\tilde{\theta}_i$ that will be incorporated into $\tilde{V}_{12}$. The nonlocality of $\tilde{V}_{12}$ is clearly evident in (1-39) and (1-43) and is controlled by ρ_1 and ρ_2. This nonlocality vanishes in the limit of δ-function behavior for ρ_1 and ρ_2 as V_{12} of (1-45) shows. We shall discuss local and nonlocal potentials again later.[2]

Comparing (1-51) and (1-32) for the charge symmetric case,[3] we see that

$$G = (2m_\mathrm{N}/m_\pi)\,f. \tag{1-52}$$

Since experimental quantities that largely depend on the pion field suggest

$$f^2/\hbar c \approx 0.1,$$

we have

$$G^2/\hbar c \approx 16. \tag{1-53}$$

The static approximation, when applied to the pseudoscalar theory with the axial vector interaction, leads to the same form as (1-51)! The one difference is that the factor $G\hbar/2m_\mathrm{N}c$ is replaced by G'/μ; thus $G' = f$. This situation is referred to as the pseudoscalar–axial vector coupling equivalence.[4] At low velocities the one form involves the nucleon velocity, and the other involves the pion velocity. Thus equivalence is related to the conservation of momentum. In a previous footnote in referring to the "cutoff" problem, we loosely expressed matters in terms of $k_{\max}$. For the one-nucleon source problem with recoil, $\mathbf{k}_{\max}(\text{pion}) = -\mathbf{k}_{\max}(\text{nucleon})$. In applying the static model, correct kinematic factors are still required and the model cannot be taken too literally.

The generally valid equation (1-50) shows that G represents the coupling strength between ψ_l and ψ_s (more precisely, the positive- and negative-energy nucleon components, which at low energy become the "large" and "small" components). Thus the meson field interaction (1-46) couples antinucleon

[1] The reduction to the static limit performed here is a simplified treatment of the general case briefly discussed later in connection with (1-145).

[2] For present purposes the potential $V_{12}(\mathbf{r}_1, \mathbf{r}_2)$ is referred to as nonlocal if its form cannot be reduced to *only* a function of $\mathbf{r}_{12} = \mathbf{r}_2 - \mathbf{r}_1$.

[3] With $\sum_{+1,0,-1} f_k \tau_k^\dagger = f\boldsymbol{\tau}$, when $f = f_0 = f_{\pm 1}$.

[4] The full relativistic source terms without going to the static limit are different however [e.g., Moravcsik (1963, p. 82)].

and nucleon with a coupling constant $G^2/\hbar c \approx 16$, which may be called "superstrong," while in the nonrelativistic limit (eliminating as it does the antinucleon state) the nucleon–nucleon coupling is measured by $f^2/\hbar c \approx 0.1$, which is still "strong" when contrasted to $e^2/\hbar c = 1/137$ for the electromagnetic interaction.

Using the term $\boldsymbol{\tau}\cdot\boldsymbol{\phi}$ in (1-46) guarantees charge symmetry in (1-51), in which $\boldsymbol{\tau}$ appears. We note that in Cartesian coordinates

$$\boldsymbol{\tau}\cdot\boldsymbol{\phi} = \tau_1\phi_1 + \tau_2\phi_2 + \tau_3\phi_3, \tag{1-54}$$

thus associating the nucleon operator τ_1 with ϕ_1, etc., when (1-32) is interpreted in Cartesian coordinates.[1] In spherical coordinates the association with $k = 0, +1, -1$ is that stated in the discussion of (1-32).

▶**Exercise 1-9** If in Cartesian coordinates $\mathbf{A}\cdot\mathbf{B} = A_x B_x + A_y B_y + A_z B_z$ and with $A_{\pm 1} = \mp(1/\sqrt{2})(A_x \pm iA_y)$ and $A_0 = A_z$ as the corresponding spherical components (and similarly for B), show that

$$\mathbf{A}\cdot\mathbf{B} = -A_{+1}B_{-1} + A_0 B_0 - A_{-1}B_{+1} = \sum_{\mu=+1,0,-1} (-1)^\mu A_\mu B_{-\mu}.$$

Applying the result to $\boldsymbol{\tau}\cdot\boldsymbol{\phi}$, show that the appropriate nucleon isospin operator to associate with ϕ_{+1} is $-\tau_{-1} = \tau_{+1}^\dagger$, etc., as stated in connection with (1-32).

These results explain why using $f_0 = f_{\pm 1}$ in (1-32) leads to the charge symmetric form involving $\boldsymbol{\tau}_1\cdot\boldsymbol{\tau}_2$ as in V_{12} of (1-43). ◀

We have restricted our discussion to the lowest mass meson (the pion) and therefore have dealt only with the pseudoscalar ($J^\pi = 0^-$) meson theory. While in our introductory remarks we also dealt briefly with the scalar meson field ($J^\pi = 0^+$), two other important cases remain: the vector meson field ($J^\pi = 1^-$) and the axial vector meson field ($J^\pi = 1^+$). We shall not treat these cases in detail, but refer the reader to DeBenedetti (1964, Chapter 7) and Brink (1965, Chapter 6), where these topics are discussed at a level comparable to our discussion.

4. Anomalous Nucleon Magnetic Moments

If the free proton and neutron were Fermi–Dirac particles obeying the linearized Dirac equations (see Appendix C), they would be predicted to have intrinsic angular momentum $\frac{1}{2}\hbar$ and to have magnetic moments $\mu_0 = +e\hbar/2m_p c$ for the proton and zero for the neutron. The observed anomalous values of $\mu_p = 2.793\mu_0$ and $\mu_n = -1.913\mu_0$ are in marked contrast to this prediction. Clearly, composite structure for the nucleons is indicated. The possibility of a virtual decay of the nucleon into a state with a pion and nucleon present suggests a model for generating the anomalous behavior.

[1] Of course $\tau_k^\dagger = \tau_k$ $(k = 1, 2, 3)$ as can be readily verified using the definitions (1-1).

Considering only the lowest one-pion-field contributions, we view the experimentally observed nucleon (or "laboratory" particle, designated by subscript E) as a superposition of various Dirac nucleon (or "bare" nucleon, designated by the subscript D) and pion states. We should note that this view is consistent with use of the linearized Dirac equation (1-48) for the nucleon state. Accordingly, we write the lowest angular momentum states in obvious spectroscopic notation and consistent with charge symmetry as

$$\begin{aligned} |\mathrm{n_E}, \tfrac{1}{2}\rangle &= C_0 |\mathrm{n_D}, \tfrac{1}{2}\rangle + C_1 |(\mathrm{p_D}, \pi^-)\ {}^2\mathrm{P}_{1/2}\rangle + C_2 |(\mathrm{n_D}, \pi^0)\ {}^2\mathrm{P}_{1/2}\rangle \\ |\mathrm{p_E}, \tfrac{1}{2}\rangle &= C_0 |\mathrm{p_D}, \tfrac{1}{2}\rangle + C_1 |(\mathrm{n_D}, \pi^+)\ {}^2\mathrm{P}_{1/2}\rangle + C_2 |(\mathrm{p_D}, \pi^0)\ {}^2\mathrm{P}_{1/2}\rangle. \end{aligned} \qquad (1\text{-}55)$$

In the spirit of the static model (no nucleon recoil), the expectation value of the magnetic moment (in units of the nuclear magneton μ_0) for normalized states (1-55) would be

$$\begin{aligned} \mu(\mathrm{n_E}) &= (1 - |C_1|^2 - |C_2|^2)(0) + |C_1|^2(-1 + \mu_\pi^{\ -}) + |C_2|^2(0) \\ \mu(\mathrm{p_E}) &= (1 - |C_1|^2 - |C_2|^2)(1) + |C_1|^2(\mu_\pi^{\ +}) + |C_2|^2(-1). \end{aligned} \qquad (1\text{-}56)$$

In (1-56) the contribution of the charged pion currents can be thought of as giving rise to the quantity $|C_1|^2 \mu_\pi^{\ \pm}$.

We must review briefly the ensuing consequences of making the plausible assumption of isospin conservation. Taking the Dirac nucleon isospin $t(\mathrm{n, p}) = \frac{1}{2}$ and the pion isospin $t(\pi) = 1$, states with $T = \frac{1}{2}$ or $\frac{3}{2}$ can be formed. Using Appendix A, we readily see that the various isospin states resulting from the coupling of $\mathbf{t}(\mathrm{n, p})$ and $\mathbf{t}(\pi)$ are

	$T = \frac{3}{2}$	$T = \frac{1}{2}$
$T_3 = -\frac{3}{2}$:	$\lvert\frac{3}{2}, \frac{3}{2}\rangle = \lvert\mathrm{p}\pi^+\rangle$	—
$T_3 = \frac{1}{2}$:	$\lvert\frac{3}{2}, \frac{1}{2}\rangle = \sqrt{\frac{2}{3}}\lvert\mathrm{p}\pi^0\rangle + \sqrt{\frac{1}{3}}\lvert\mathrm{p}\pi^+\rangle$	$\lvert\frac{1}{2}, \frac{1}{2}\rangle = -\sqrt{\frac{1}{3}}\lvert\mathrm{p}\pi^0\rangle + \sqrt{\frac{2}{3}}\lvert\mathrm{n}\pi^+\rangle$
$T_3 = -\frac{1}{2}$:	$\lvert\frac{3}{2}, -\frac{1}{2}\rangle = \sqrt{\frac{2}{3}}\lvert\mathrm{n}\pi^0\rangle + \sqrt{\frac{1}{3}}\lvert\mathrm{p}\pi^-\rangle$	$\lvert\frac{1}{2}, -\frac{1}{2}\rangle = \sqrt{\frac{1}{3}}\lvert\mathrm{n}\pi^0\rangle - \sqrt{\frac{2}{3}}\lvert\mathrm{p}\pi^-\rangle$
$T_3 = -\frac{3}{2}$:	$\lvert\frac{3}{2}, -\frac{3}{2}\rangle = \lvert\mathrm{n}\pi^-\rangle$	—

(1-57)

Since the experimental nucleons are components of the $T = \frac{1}{2}$ state, we see that we should use $|C_1|^2 = 2|C_2|^2$ in (1-56).

It might be expected that the contribution of the charged pion currents $|C_1|^2 \mu_\pi^{\ \pm}$ could be calculated from the pion field (1-38). Indeed, the lowest-order orbital angular momentum implied in (1-38) is $l = 1$ and hence represents a possible charged pion current. A current density (up to constant factors) can be defined as

$$\mathbf{j} \sim ie\tau_3(\pi)(\phi_{+1}\nabla\phi_{-1} - \phi_{-1}\nabla\phi_{+1}) \qquad (1\text{-}58)$$

and a resulting magnetic moment as

$$\boldsymbol{\mu} = \tfrac{1}{2}\int (\mathbf{r}\times\mathbf{j})\, d^3r. \tag{1-59}$$

Unfortunately, the straightforward calculation based on (1-59) diverges for a pointlike source term $\rho(\mathbf{r}) = \delta(\mathbf{r}-\mathbf{r}_i)$, again suggesting the need for a cutoff approximation for the static model.

A simple numerical solution to (1-56) that gives the observed magnetic moments yields

$$|C_1|^2 = 0.040, \qquad |C_1|^2\mu_\pi{}^{\pm} = \pm 1.873. \tag{1-60}$$

▶**Exercise 1-10** Verify that (1-55) couples terms implying conservation of baryon number, charge, total angular momentum, and parity in the virtual processes involved. Verify (1-56).

Contrast the present pseudoscalar isovector model and a possible scalar isovector meson model patterned after (1-30) in lowest order. ◀

▶**Exercise 1-11** Show that a simple semiclassical model of two charged particles (m, q) and (m', q') having total orbital angular momentum $[L(L+1)]^{1/2}\hbar$ would produce an orbital contribution to the system magnetic moment $|\boldsymbol{\mu}| = [L(L+1)]^{1/2}\hbar\bar{q}/2c\bar{m}$ with $\bar{m} = mm'/(m+m')$ and $\bar{q} = (m'^2q+m^2q')/(m+m')^2$.

Estimate the orbital motion contributions that would appear in (1-56) in a nonstatic treatment by taking $m_N/m_\pi \approx 7$ in the above and noting that, in conformity with the definition of "magnetic moment," we would write $\mu \equiv \mu_{z,\max} = (\hbar\bar{q}/2c\bar{m})L$. Obtain a best value for $|C_1|^2$ and contrast this model with the static model results (1-60). Even without knowing quite how to take recoil effects into account, is the symmetric "effective" static model expression (1-60) reasonable? Although, as we shall see, the attempt to view the physical or laboratory nucleon as a bare nucleon with its complement of meson clouds is no longer the modern approach, reference to articles using this point of view might be instructive [e.g., Sachs (1952, 1954), Feynman and Speisman (1954), and Holladay and Sachs (1954)]. ◀

Perhaps the two most influential reasons for a modified and more phenomenological approach to current hadron physics (and indeed most of elementary particle physics) are the convergence difficulties encountered in perturbation calculations using the field-theoretic approach involving "bare" hadrons (nucleons, pions, kaons) and the growing evidence that more fundamental particle "building blocks" may be involved in the basic interactions.

Perturbation calculations in atomic problems involving the electromagnetic interaction are manageable, as is well known, due to the small numerical value of $\alpha = e^2/\hbar c = 1/137$, which determines the ratio of successive contributions in the attendent expansions. In hadron interactions "bare" hadrons would involve coupling constants $\gtrsim 1$, leading to divergent perturbation calculations when all orders are included. A "renormalization" procedure in which a cut-off momentum is used with an effective coupling constant $f^2/\hbar c$ has been shown to give consistent results in cases involving long-range interactions. In this procedure, the vertex involving the physical or laboratory nucleon, sometimes referred to as the "clothed" nucleon, is described only to spatial dimensions greater than $r_{\min} = (k_{\max})^{-1} \approx 0.2$ fm. The corresponding effective coupling constant is empirically determined to be $f^2/\hbar c = 0.081 \pm 0.004$. This should be contrasted to the value, for example, required to have (1-45) fit the spin singlet nucleon–nucleon interaction for $L = 0$ [the term in S_{12} contributes zero and assuming the δ-function term can also be dropped yields the same expression as (1-31)], giving $f^2/\hbar c \approx \frac{1}{4}$ with no cutoff.

A rough nonquantized field calculation in this spirit, attributing the anomalous portion of the nucleon magnetic moment to the outer pion field (i.e., the dominant meson field beyond $r_{\min} \approx 0.2$ fm) has been performed by Jackson (1958, p. 44 ff.). Using a value of 0.080 for the coupling constant and a square momentum cut-off calculation, he predicts $\mu_{\mathrm{ap}} = \mu_{\mathrm{p}} - 1 \approx 1.76$ and $\mu_{\mathrm{an}} \approx -1.76$. The fact that these are close to $\mu_{\mathrm{p}} - 1 = 1.793$ and $\mu_{\mathrm{n}} = -1.913$ is encouraging. This estimate, relying as it does on (1-58), attributes the static magnetic moment entirely to the isovector meson contribution.

Recently, observation of the atomic fine-structure splitting in atoms containing antiprotons has yielded a value for the magnetic moment of the antiproton equal to $(-2.83 \pm 0.10)\mu_0$ [see Fox *et al.* (1972)]. According to the TCP conservation theorem the proton and antiproton magnetic moments should be equal in magnitude but opposite in sign, since both the Dirac and anomalous portions of the magnetic moment change sign for $\bar{\mathrm{p}}$ (the antiproton). The quoted experimental value agrees with this prediction.

5. High-Energy Electron–Nucleon Scattering

The detailed behavior of electron–nucleon scattering gives striking evidence for the nature of the electromagnetic structure of the nucleons. In the low-energy limit these experiments relate to the macroscopic static nucleon charge and magnetic moment. In the GeV (i.e., 10^9 eV) range, however, many interesting phenomena appear germane to the microscopic structure of nucleons.

A very thorough and excellent collection of papers on the subject arranged in historical context (up to year 1962) is given by Hofstadter (1963). More recent results are discussed in Chan *et al.* (1966), Wilson and Levinger (1964), and Gasiorowicz (1966, Chapter 26). A rather complete and instructive monograph on the electromagnetic structure of nucleons and their form factors is given by Drell and Zachariasen (1961).

The concept of the form factor can be readily appreciated from a simple nonrelativistic derivation. Consider the elastic scattering of a pointlike charge from a charge distribution $\rho(\mathbf{r})$. We divide the total Hamiltonian into a term H_0, containing the kinetic energy of *all* components of the problem (including the scattered pointlike charge) and all interaction potentials acting within the charge distribution, and the term ΔH driving the scattering interaction. Thus we have

$$H_0 = KE_\rho + KE_e + V_{\rho\rho}, \qquad \Delta H = V_{\rho e},$$

with the subscript e designating the pointlike charge and ρ designating the distributed target charge (here taken to be an infinitely massive scattering center). The initial and final eigenstates of H_0 are

$$|\mathrm{i}\rangle = (2\pi)^{-3/2}[\exp(i\mathbf{k}\cdot\mathbf{r}_e)]\psi_n(\mathbf{r}_\rho), \qquad |\mathrm{f}\rangle = (2\pi)^{-3/2}[\exp(i\mathbf{k}'\cdot\mathbf{r}_e)]\psi_n(\mathbf{r}_\rho).$$

In first Born approximation, we have $d\sigma/d\Omega = |f(\theta)|^2$ with

$$f(\theta) = -(m_e/2\pi\hbar^2)\int [\exp(-i\mathbf{k}'\cdot\mathbf{r}_e)]\, V_{\rho e}(\mathbf{r}_e)\exp(i\mathbf{k}\cdot\mathbf{r}_e)\, d^3r_e$$

if ψ_n is properly normalized and m_e is the scattered particle's mass. If V is not velocity dependent, and if we define $\mathbf{q} = \mathbf{k} - \mathbf{k}'$, or since $k = k'$, $q = 2k\sin(\theta/2)$, then

$$f(0) = -(m/2\pi\hbar^2)\int [\exp(i\mathbf{q}\cdot\mathbf{r})]\, V(\mathbf{r})\, d^3r,$$

where we have dropped the subscripts e and ρ.

The potential energy $V(\mathbf{r})$ for electron scattering can be taken to be $V(\mathbf{r}) = -e\phi(\mathbf{r})$, with $\phi(\mathbf{r})$ obeying Poisson's equation $\nabla^2\phi(\mathbf{r}) = -4\pi\rho(\mathbf{r})$ in the static approximation. Define

$$W(\mathbf{q}) = (2\pi)^{-3/2}\int [\exp(i\mathbf{q}\cdot\mathbf{r})]\, V(\mathbf{r})\, d^3r$$

so that

$$f(\theta) = -[m(2\pi)^{1/2}/\hbar^2]\, W(\mathbf{q}).$$

By taking the Fourier transform of $W(\mathbf{q})$ and using Poisson's equation,

$$4\pi e\rho(\mathbf{r}) = \nabla^2 V(\mathbf{r}) = (2\pi)^{-3/2}\nabla^2\int [\exp(-i\mathbf{q}\cdot\mathbf{r})]\, W(\mathbf{q})\, d^3q$$

$$= -(2\pi)^{-3/2}\int [\exp(-i\mathbf{q}\cdot\mathbf{r})]\{q^2 W(\mathbf{q})\}\, d^3q.$$

If a Fourier transform is taken of this result,

$$q^2 W(\mathbf{q}) = -4\pi e(2\pi)^{-3/2} \int [\exp(i\mathbf{q}\cdot\mathbf{r})]\,\rho(\mathbf{r})\,d^3r,$$

whence

$$f(\theta) = (2me^2/\hbar^2)(1/q^2)\,F(\mathbf{q}),$$

with

$$F(\mathbf{q}) = (1/e)\int [\exp(i\mathbf{q}\cdot\mathbf{r})]\,\rho(\mathbf{r})\,d^3r.$$

Also,

$$d\sigma/d\Omega = (4m^2e^4/\hbar^4q^4)\,|F(\mathbf{q})|^2.$$

The quantity $F(\mathbf{q})$, which up to constants is the Fourier transform of $\rho(\mathbf{r})$, is called the *form factor*.

In this derivation the electron was considered a true pointlike particle. To anticipate a bit, the diagram representation of elastic electron scattering from nucleons is pictured in Figs. 1.3d and 1.3e. Although the discussion of this figure will center on the photon–nucleon vertex, the photon–electron vertex, however, has quantum electrodynamic corrections as well. Such corrections can in effect be considered as giving rise to an effective "size" to the interacting particles. These effects at the photon–electron vertex involve powers of $\alpha = 1/137$ and can generally be neglected in contrast to the photon (hadron)–nucleon vertex behavior, which involves strong interaction corrections.[1]

The nonrelativistic expression derived is, of course, the well-known Rutherford scattering formula. A relativistic derivation has been given by Mott. In the extreme relativistic version, the Mott scattering formula becomes

$$\frac{d\sigma}{d\Omega} = \left(\frac{e^2}{2E}\right)^2 \frac{\cos^2(\theta/2)}{\sin^4(\theta/2)}\,|F(\mathbf{q})|^2,$$

where $\hbar kc \approx E$ and the $\cos^2(\theta/2)$ term arises from an approximation to $1-\beta^2\sin^2(\theta/2)$ for $\beta = v/c \to 1$.

We take as our starting point for discussing electron–nucleon scattering the appropriate expression for elastic scattering with form factors for both the extended charge and magnetic moment densities included, namely,

$$\left(\frac{d\sigma}{d\Omega}\right)_\alpha = \left(\frac{d\sigma}{d\Omega}\right)_{\text{Mott}} \left\{F_{\alpha 1}^2 + \left(\frac{\hbar q}{2m_\alpha c}\right)^2 \left[2(F_{\alpha 1} + \mu_{a\alpha}F_{\alpha 2})^2 \tan^2\frac{\theta}{2} + \mu_{a\alpha}^2 F_{\alpha 2}^2\right]\right\}. \tag{1-61}$$

[1] Current theoretical speculation would ascribe an extended structure to the leptons, perhaps of order 10^{-16} cm [e.g., see Greenberg and Yodh (1974)].

Here $(d\sigma/d\Omega)_{\text{Mott}}$ is the Mott elastic differential cross section for scattering an electron from a spinless point charge having a charge $+e$. The quantity $\mu_{a\alpha}$ (α standing for either the neutron or proton) is the *anomalous portion* of the nucleon magnetic moment in units of the nuclear magneton (i.e., as above $\mu_{ap} = +1.793$, $\mu_{an} = -1.913$). The four form factors $F_{p1}, F_{n1}, F_{p2}, F_{n2}$ are approximately

$$F_{\alpha 1}(q^2) = \frac{1}{e}\int \rho_{\alpha c}(\mathbf{r}) \exp(i\mathbf{q}\cdot\mathbf{r})\, d^3r, \qquad F_{\alpha 2}(q^2) = \frac{1}{\mu_{a\alpha}}\int \rho_{\alpha\mu}(\mathbf{r}) \exp(i\mathbf{q}\cdot\mathbf{r})\, d^3r, \tag{1-62}$$

with $q = 2|k_i k_f|^{1/2}\sin(\theta/2)$.[1] At low energy $F_{p1} \to 1$, $F_{n1} \to 0$, and $F_{\alpha 2} \to 1$; thus the form factors $F_{\alpha 1}(0)$ and $F_{\alpha 2}(0)$ relate to the static electromagnetic properties of the physical nucleons (i.e., charge and moment measured in a macroscopic experiment involving the exchange of essentially zero-energy photons). Of course, for delta-function densities involving $\delta(\mathbf{r})$, we also get these limiting values. When these values are substituted into (1-61), the so-called Rosenbluth cross section results.

▶**Exercise 1-12** Expanding $\exp(i\mathbf{q}\cdot\mathbf{r})$ in the power series $1 + i\mathbf{q}\cdot\mathbf{r} - \frac{1}{2}(\mathbf{q}\cdot\mathbf{r})^2 + \cdots$, determine the first two terms in the form factors for spherically symmetric density functions. Determine the derivatives $dF_{\alpha 1,2}/dq^2$ for $q^2 \to 0$ in terms of the various rms values for the density distributions. ◀

Primarily to make evident the operating universality, it is more convenient to express (1-61) in terms of new form factors

$$\begin{aligned} G_{\alpha E}(q^2) &= F_{\alpha 1}(q^2) - (\hbar q/2m_\alpha c)^2 \mu_{a\alpha} F_{\alpha 2}(q^2) \\ G_{\alpha M}(q^2) &= F_{\alpha 1}(q^2) + \mu_{a\alpha} F_{\alpha 2}(q^2) \end{aligned} \tag{1-63}$$

with $G_{pE}(0) = 1$, $G_{nE}(0) = 0$, $G_{pM}(0) = 2.793$, $G_{nM}(0) = -1.913$. Recalling that the projection operators for neutron and proton states Λ^ν and Λ^π involve $1 \pm \tau_3$ [see (1-8)], it is clear that the densities appearing in (1-62) can be written as

$$\rho_{\alpha c}(\mathbf{r}) = \rho_{sc}(\mathbf{r}) + \tau_3(\alpha)\,\rho_{vc}(\mathbf{r}), \qquad \rho_{\alpha\mu}(\mathbf{r}) = \rho_{s\mu}(\mathbf{r}) + \tau_3(\alpha)\,\rho_{v\mu}(\mathbf{r}). \tag{1-64}$$

In this form (as in Exercise 1-1) we can identify the first terms in (1-64) with isoscalar contributions and the second with isovector contributions. The definitions (1-64) also permit $F(q^2)$ and $G(q^2)$ to be reduced to isoscalar and

[1] Actually $\hbar^2 q^2 = (\mathbf{q}\cdot\mathbf{q} - q_0{}^2)\hbar^2$ is the four-momentum transfer. For the approximation of zero nucleon recoil $\mathbf{q} \approx \mathbf{k}_i - \mathbf{k}_f$, details are given in Hofstader (1963, Paper 42). The quantity q^2 can be expressed either in fm^{-2} or $(1/\hbar^2)(\text{GeV}/c)^2$. The conversion from one set of units to the other is $(1/\hbar^2)(\text{GeV}/c)^2 = 25\ \text{fm}^{-2}$.

The quantities $\rho_{\alpha c}$ and $\rho_{\alpha\mu}$ are the charge and magnetic moment density distributions in appropriate units.

isovector components, which give rise to form factors for the interaction with the corresponding components of the electromagnetic field (or photon) (see Exercise 1-1). The G form factors can be written

$$G_{\alpha E} = G_{sE} + \tau_3(\alpha)\, G_{vE}, \qquad G_{\alpha M} = G_{sM} + \tau_3(\alpha)\, G_{vM}. \tag{1-65}$$

Following the semiempirical guidance based on a dispersion theory estimate, Bergia *et al.* (Hofstader, 1963, Paper 68, or Bergia *et al.*, 1961) suggest expanding the isoscalar and isovector components of G in mass resonances, viz.,

$$\begin{aligned}
G_{sE}(q^2) &= G_{sE}(0)\left(1 - \alpha_1 - \alpha_2 - \cdots + \frac{\alpha_1}{1+q^2/q_{s1}^2} + \frac{\alpha_2}{1+q^2/q_{s2}^2} + \cdots\right)\\
G_{sM}(q^2) &= G_{sM}(0)\left(1 - \beta_1 - \beta_2 - \cdots + \frac{\beta_1}{1+q^2/q_{s1}^2} + \frac{\beta_2}{1+q^2/q_{s2}^2} + \cdots\right)\\
G_{vE}(q^2) &= G_{vE}(0)\left(1 - \gamma_1 - \gamma_2 - \cdots + \frac{\gamma_1}{1+q^2/q_{v1}^2} + \frac{\gamma_2}{1+q^2/q_{v2}^2} + \cdots\right)\\
G_{vM}(q^2) &= G_{vM}(0)\left(1 - \delta_1 - \delta_2 - \cdots + \frac{\delta_1}{1+q^2/q_{v1}^2} + \frac{\delta_2}{1+q^2/q_{v2}^2} + \cdots\right).
\end{aligned} \tag{1-66}$$

The corresponding masses are given by $(m_i c/\hbar)^2 = q_i{}^2$. Early efforts (Hofstader, 1963, Paper 81) obtained a reasonable fit to the data below ~ 1 GeV/c with one isovector meson $q_v{}^2 = (9.0 \pm 1)\ \mathrm{fm}^{-2}$ (or mass $m_v/m_\pi = 1.42(9.0)^{1/2} = 4.3$, hence $m_v c^2 \approx 580$ MeV) and one isoscalar meson $q_s{}^2 = (11.5 \pm 2)\ \mathrm{fm}^{-2}$ [or mass $m_s/m_\pi = 1.42(11.5)^{1/2} = 4.8$, hence $m_s c^2 \approx 660$ MeV]. The values of $\alpha_1, \beta_1, \gamma_1$, and δ_1 were close to unity. A more accurate fit good up to $q^2 \approx 50\ \mathrm{fm}^{-2} \approx (2/\hbar^2)(\mathrm{GeV}/c)^2$ is given by Hand *et al.* (1963). They give the following results:

$$\begin{aligned}
G_{sE}(q^2) &= \frac{1}{2}\left[\frac{-0.80}{1+q^2/30} + \frac{1.80}{1+q^2/15.8}\right]\\
G_{vE}(q^2) &= \frac{1}{2}\left[\frac{-1.68}{1+q^2/30} + \frac{2.68}{1+q^2/14.5}\right]\\
G_{sM}(q^2) &= 0.44\left[\frac{-1.7}{1+q^2/30} + \frac{2.7}{1+q^2/15.8}\right]\\
G_{vM}(q^2) &= 2.35\left[\frac{-1.0}{1+q^2/30} + \frac{2.0}{1+q^2/14.5}\right].
\end{aligned} \tag{1-67}$$

It is clear that these values, when substituted into (1-65), reduce to the proper conditions cited under (1-63) for $q = 0$.

By selecting an ad hoc "soft-core" term $q_c^2 = 30\ \text{fm}^{-2}$ corresponding to a mass of $m_c c^2 \approx 1070$ MeV, the remaining resonances correspond to $m_v c^2 = m_\rho c^2 = 740$ MeV for the isovector meson and $m_s c^2 = m_\omega c^2 = 770$ MeV for the isoscalar meson. Other fits (Gasiorwicz, 1966, p. 444), including a "hard core" $m_{hc} c^2 \to \infty$, have identified isoscalar masses $m_\omega c^2 = 790$ MeV and $m_\phi c^2 \approx 1030$ MeV and isovector masses $m_\rho c^2 = 750$ MeV and $m_{\rho'} c^2 \approx 1220$ MeV.[1] While the low-energy data (below 2 GeV) pose some difficulties in correctly accounting for the core term, the need to include the ρ and ω mesons and possibly the ϕ meson seems established. Recent theory would suggest very little ϕ-meson contribution (see later). The best characterization of these mesons is deduced from inelastic pion scattering on nucleons producing multiple pions in which the ρ appears as a strong two-pion resonance and the ω as a three-pion resonance.

Best values (Particle Data Group, 1973) give $m_\rho c^2 = (770 \pm 5)$ MeV. It is established to be a charge triplet $t = 1$ (charge equal to $\pm e, 0$) and therefore has the required isovector character. However, only the ρ^0 contributes in the electron–nucleon scattering. It has intrinsic mechanical spin $1\hbar$ and odd parity with a rapid decay rate into two pions giving a width of ~ 150 MeV.

The ω-meson is now established to be a charge singlet $t = 0$ (zero charge) as required above with a mass $m_\omega c^2 = (783.8 \pm 0.3)$ MeV. It also has intrinsic mechanical spin $1\hbar$ and odd parity. It largely ($\sim 90\%$) decays into three pions π^+, π^0, π^-, and has a total width ~ 10 keV. The ϕ-meson has mass $m_\phi c^2 = (1019.6 \pm 0.3)$ MeV and has a total width ~ 4 MeV. It decays ($\sim 85\%$) to two kaons and ($\sim 15\%$) to three pions π^+, π^-, π^0. It is an isoscalar with $t = 0$ (zero charge) and intrinsic mechanical spin $1\hbar$. It has intrinsic odd parity. The above values are seen to be in good agreement with the elastic electron scattering requirements.

In an earlier fit using a soft core with a range equal to the Compton wavelength of the nucleon (i.e., $m_c c^2 \approx 940$ MeV) and simple isoscalar and isovector meson clouds, Hofstadter (1963, Paper 74) determined the density decompositions of (1-64) shown in Fig. 1.2. The rms radii found for the distributions shown in Fig. 1.2 are $\langle r_{pc}^2 \rangle^{1/2} = 0.88$, $\langle r_{p\mu}^2 \rangle^{1/2} = 0.95$, $\langle r_{nc}^2 \rangle^{1/2} = 0.00$, and $\langle r_{n\mu}^2 \rangle^{1/2} = 0.87$ (in units of fm). These radii are considerably smaller than the pion range $\hbar / m_\pi c = 1.42$ fm and, of course, are consistent with the view requiring heavier meson contributions. The zero value found for the rms charge radius for the neutron is perhaps not inconsistent with a value of

[1] When $\sum_{i=1}^{\infty} \alpha_i \neq 1$ (and similarly for β, γ, and δ), a constant term equivalent to $m_i \to \infty$ or ~ 0 Compton wavelength is left in (1-66). A fit of this form is said to involve a *hard-core* term. When m_i is taken large but not infinite, i.e., Compton wavelength $\lesssim \frac{1}{20}$ fm, one speaks of a *soft-core* term.

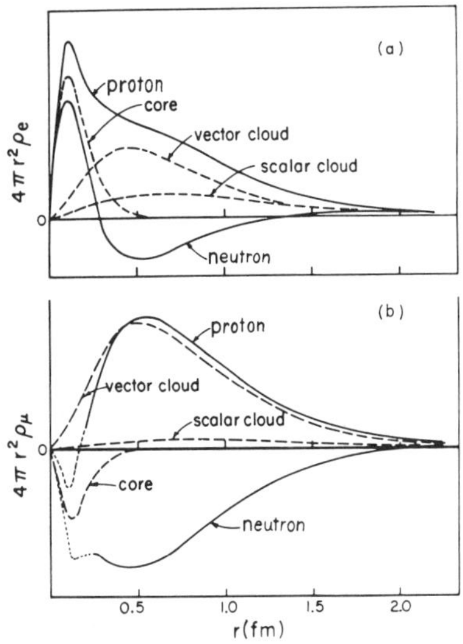

FIG. 1.2. Spatial distribution of charge (a) and anomalous magnetic moment (b) for proton and neutron, according to the core model [taken from (Littauer *et al.*, 1961)].

(0.33 ± 0.01) fm found in low-energy neutron interactions with atoms.[1] Better values will be quoted later.

We note from Fig. 1.2 the strong role the isovector "cloud" plays ("ρ-dominance"). It is instructive to compare the simplified diagrams for the dominant contributions to the static magnetic moments discussed earlier with the magnetic electron–nucleon scattering considering only the core and ρ-meson contributions. Figure 1.3 illustrates these interactions for the case of the proton. The essentially classical treatment of the static magnetic moment interaction has major contributions from Fig. 1.3a–c, where the substructure of the physical or experimental proton p_E is specifically exhibited. The major high-energy electron–proton elastic scattering contributions come from Fig. 1.3d, e. The interaction at the nucleon vertex is imagined to be "obscured" by a box of spacelike dimension equal to the cutoff radius of ~ 0.2 fm and an effective interaction is employed with the actual physical proton, endowed with its anomalous magnetic moment and observed mass, etc. No attempt is made in (1-62)–(1-67) to detail the behavior for $q = 0$ (the static case). The various factors are simply "normalized" to give the known experimental values for the physical particle. If one were to probe the interior interaction corresponding to Fig. 1.3e′, it would require all the diagrams such as Fig. 1.3e″, e‴, which in totality give serious divergence problems. Figure 1.3e″, e‴

[1] See relevant discussion in Chan *et al.* (1966).

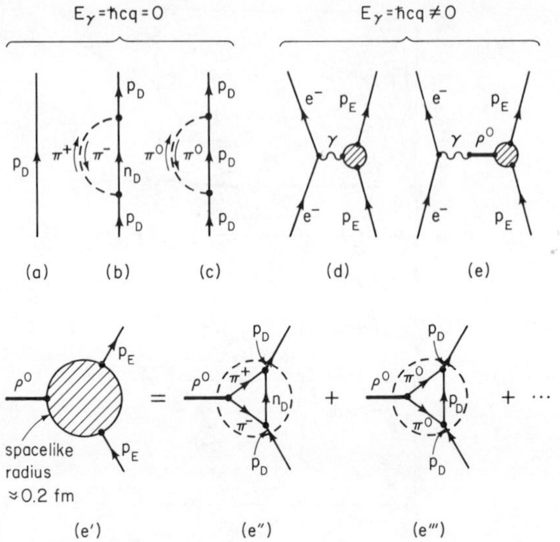

FIG. 1.3. Highly simplified diagrams contrasting the major contributions to the static magnetic moment of the proton [the sum (a)+(b)+(c)] and the core and isovector contributions to the magnetic portion of the elastic electron scattering [the sum (d)+(e)]. A detailing of some of the interior contribution to (e′) is given in (e″) and (e‴). Diagrams (d)+(e″)+(e‴) for $q = 0$ are linear combinations of diagrams (a)+(b)+(c).

corresponds to coupling to the pion current of the physical proton. In some sense Fig. 1.3e″, e‴ combined with Fig. 1.3d are dominant diagrams in that they directly correspond to linear combinations of Fig. 1.3a–c at $q = 0$. In this comparison, we must note that downward directed trajectories represent particles traveling backward in time and are equivalent to the corresponding antiparticles traveling forward in time and vice versa. Thus, a downward directed π^- is the equivalent to a $\bar{\pi}^- = \pi^+$ (the antiparticle to the π^- is the π^+)[1] trending in the reverse time sense. Since $\bar{\pi}^0 = \pi^0$, an appropriately similar remark pertains for π^0 trajectories. However, the correspondence should not be taken too literally in that effects of the superstrong coupling $G^2/\hbar c$ are not displayed. In spite of this, the view expressed by Fig. 1.3d, e would still be valid while the others, of course, would not.

Figure 1.3e shows the virtual photon coupling to the ρ-meson. While virtual transitions of this sort need not conserve energy (and hence mass), they must conserve charge, angular momentum, and parity. We note that the photon behaves as if it were a massless, chargeless particle with intrinsic mechanical spin $1\hbar$, odd intrinsic parity, and mixed isospin $t = 0, 1$ (recall Exercise 1-1;

[1] A bar over the particle symbol signifies the antiparticle, thus $\bar{\pi}^-$ is the antiparticle of the π^-.

the spin-$1\hbar$ behavior will be demonstrated at a later point; see Chapter VII). Thus virtual coupling to the vector mesons (the name implying mechanical spin $1\hbar$) ρ, ω, and ϕ is possible, the first through the isovector component of the photon, the second and third through the isoscalar component of the photon. The coupling constants $em_i/2\gamma_i$ have been found experimentally to be $(75\pm2)\%$ for the ρ, $(10\pm2)\%$ for the ω, and $(15\pm2)\%$ for the ϕ, leading to what has been called ρ-dominance. These relatively strong couplings are said to exhibit the hadronic character of high-energy virtual photon interactions, since the ρ-, ω-, and ϕ-hadrons strongly interact with other hadrons (except for special cases such as the weak coupling of the ϕ-meson to the nucleons, which of course is quite relevant in the present context) and lead to a two-step process interaction of greater strength than a single, direct electromagnetic interaction.[1] In this regard note that fit (1-67) contains no contributions corresponding to Fig. 1.3d although other parameterizations do have small contributions.

Examining the two-pole fit (1-67), we note that the first terms are opposite in sign from the second and have strengths $(\alpha, \beta, \gamma, \delta)$ approximately inversely weighted to the square of the resonant masses. Thus within the accuracy of the fit, combining the two terms would sensibly lead to a single term with a q dependence of dipole character with an effective mass factor between $q_c^{\,2} = 30\ \mathrm{fm}^{-2}$ and $q_\rho^{\,2} \approx q_\omega^{\,2} \approx 15\ \mathrm{fm}^{-2}$. In fact, a surprisingly simple relationship of this dipole character

$$
\begin{aligned}
G_{\mathrm{pE}}(q^2) &= \frac{G_{\mathrm{pM}}(q^2)}{\mu_{\mathrm{p}}} = \frac{G_{\mathrm{nM}}(q^2)}{\mu_{\mathrm{n}}} = \left(\frac{2m_{\mathrm{n}}c}{\hbar q}\right)^2 \frac{G_{\mathrm{nE}}(q^2)}{\mu_{\mathrm{n}}} \\
&= \left(\frac{1}{1+q^2/18.1}\right)^2, \qquad q^2 \ \text{ in fm}^{-2} \qquad\qquad (1\text{-}68) \\
&= \left(\frac{1}{1+q^2/0.71}\right)^2, \qquad q^2 \ \text{ in } \frac{1}{\hbar^2}\left(\frac{\mathrm{GeV}}{c}\right)^2
\end{aligned}
$$

fits the resonance region (up to $\sim 2\ \mathrm{GeV}/c$) quite well [see Chan *et al.* (1966), Bartel *et al.* (1972), and Bloom *et al.* (1973)]. In this region $G_{\mathrm{nE}} \approx 0$ is also consistent, except that $(dG_{\mathrm{nE}}/dq^2)|_{q=0} = \frac{1}{6}\langle r_{\mathrm{nc}}^2\rangle$ should give $+(0.0178\pm0.0009)$ fm^2, a point, however, which is open to discussion [see Chan *et al.* (1966) and Bartel *et al.* (1972)]. In view of (1-68), $\langle r_{\mathrm{pc}}^2\rangle^{1/2} = \langle r_{\mathrm{p}\mu}^2\rangle^{1/2} = \langle r_{\mathrm{n}\mu}^2\rangle^{1/2}$. The

[1] We should note that the present discussion involves a modification of statements made at the beginning of this section comparing weak, electromagnetic, and strong interactions. Because of the hadronic character of short-range (high-energy) electromagnetic interactions, no electromagnetic theory can be complete without being a unified *strong-EM* theory. For typical papers discussing unified approaches to strong, weak, and electromagnetic interactions see Bars *et al.* (1972), Weinberg (1972), and Pati and Salam (1974).

best value for these is (0.813 ± 0.025) fm (Frerejacque *et al.*, 1966). These authors also give $G_{pM}/\mu_p G_{pE} = 1.001 \pm 0.013$. If the interpretation of the atomic neutron experiments is correct, $\langle r_{nc}^2 \rangle^{1/2} = (0.33 \pm 0.01)$ fm. Figure 1.4a shows the validity of the equivalences claimed in (1-68) for momentum transfers up to $(3/\hbar^2)\,(\mathrm{GeV}/c)^2$.

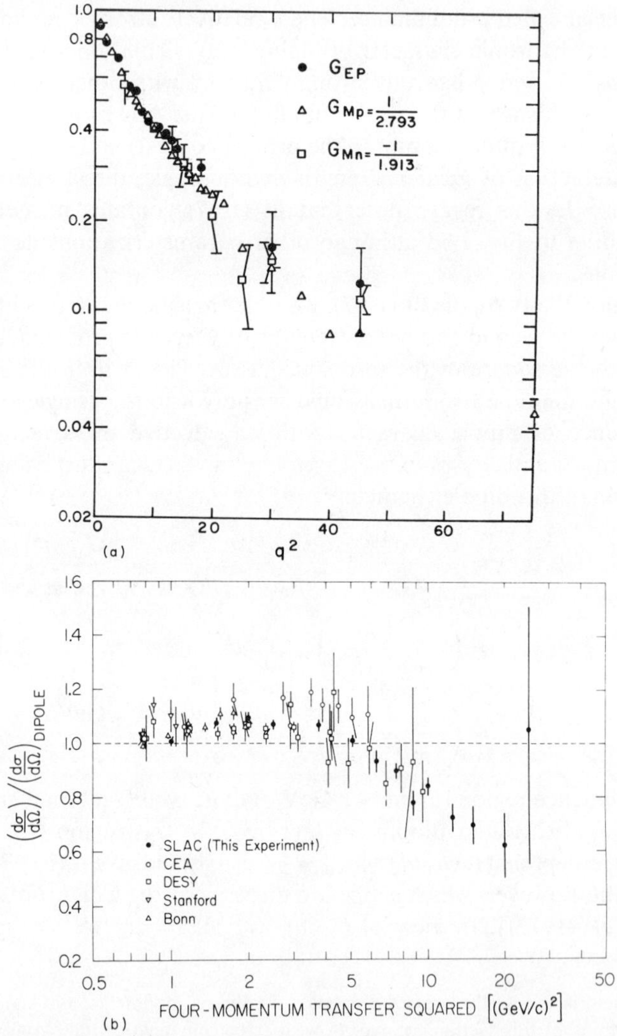

FIG. 1.4. (a) Semilogarithmic plot of form factors vs. momentum transfer in $(\mathrm{fm})^{-2}$ [taken from Wilson and Levinger (1964)]. (b) World data for the elastic electron–proton scattering differential cross section divided by the dipole prediction based on (1–68). Figure taken from Kirk *et al.* (1973).

Perhaps the most interesting result of recent measurements at higher momentum transfers, up to $(25/\hbar^2)$ $(\text{GeV}/c)^2$, indicates an approximate q^{-4} dependence, and $G_{nE} \approx 0$ consistent with (1-68). Without a precise ad hoc adjustment of the constants, (1-67) would eventually give a high-energy behavior corresponding to q^{-2}. In actual fact, the SLAC–MIT high-energy experiments suggest a magnetic form factor dropoff as a function of q, which is somewhat *steeper* than (1-68) [see Kirk *et al.* (1973)]. These results are conveniently displayed by taking the ratio of the observed differential cross section to that predicted by using the "dipole" form factor (1-68). [The differential cross section is proportional to $G_{pE}^2(q^2)$, see (1-61).] The "world data" comparing various experimental results are shown in Fig. 1.4b. Various theoretical interpretations have been offered [e.g., see DeRujula (1974) and Gross and Treiman (1974)].

There has also been an extensive study of inelastic processes at these higher energies. The measurements of the various relevant cross sections are found to be characterizable in a particularly simple manner known as *scaling* or *scale invariance*. High-energy inelastic electron scattering probes different aspects of nucleon structure than elastic scattering; "deep" inelastic scattering, i.e., highly inelastic scattering, can be associated with interaction centers (called "partons") much smaller in size than nucleons. The reason for the precision of the dipole fit rather than the dispersion-relationship-derived resonance fit, as well as the other features of "deep" inelastic scattering, is still under theoretical investigation. An excellent set of review papers dealing with the high-energy electron–nucleon interaction is contained in the proceedings of the 1967 Stanford symposium (Proc. Int. Conf., 1967). There is a rather complete although sophisticated review paper discussing the parton model in a CERN lecture by Smith (1970). For a thorough review of deep electron scattering emphasizing the experimental results refer to Friedman and Kendall (1972). A recent reference on an interesting parton model and deep electron scattering is by Wilson (1971). For typical recent communications (and references) see, for example, Shei and Tow (1971) and West (1973).

▶**Exercise 1-13** Evaluate and plot (on a single semilog graph) the quantities $G_{pE}(q^2)$, $G_{pM}(q^2)/2.793$, and $-G_{nM}(q^2)/1.913$, both from (1-66) and (1-67) and from (1-68) for values of q^2 up to $(25/\hbar^2)(\text{GeV}/c)^2$ or 625 fm^{-2}. Attempt to adjust the constants in (1-66) and (1-67) to provide a better match. ◀

6. Pion–Nucleon Interactions

Earlier, in connection with (1-57), we coupled the pion isospin $t(\pi) = 1$ with that of the nucleon $t(\text{n}, \text{p}) = \frac{1}{2}$ to generate the $T = \frac{1}{2}$ "ground" state of the laboratory nucleon. The possible $T = \frac{3}{2}$ state was also exhibited at that time.

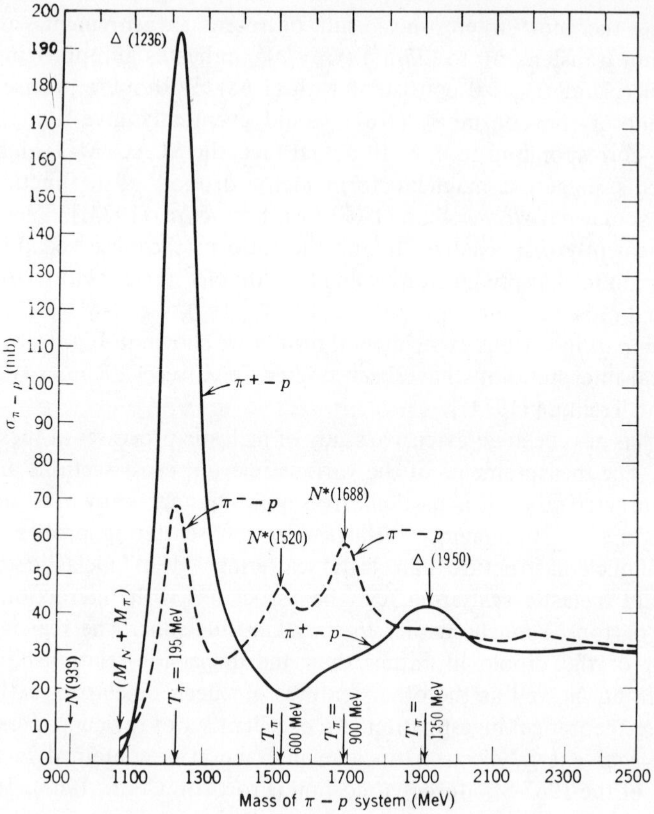

FIG. 1.5. Total cross sections for π^+–p and π^-–p scattering [taken from Gasiorowicz (1966)].

This state is, in fact, observed as a strong pion–nucleon elastic scattering resonance and might be considered to constitute the "first excited" state of the nucleon, denoted by $\Delta(1236)$. Figure 1.5 shows the observed total cross sections in $\pi^{\pm}$–p scattering. Higher resonances than the one at 1236 MeV can be taken to correspond to higher excited states.[1] For a bewildering array of such baryon resonances, classifications, energies, widths, etc., see Particle Data Group (1973). An energy level diagram showing some of the more prominent pion–proton resonances with hypercharge $Y = 1$ is given in Fig. 1.6. An excellent review of the pion–nucleon interaction is given by Moorhouse (1969) and also Gasiorowicz (1966, Chapter 19). We shall not dwell on the theoretical considerations, the most successful of which use a cut-off radius and

[1] We shall examine a more precise cataloging of "excited" states in the section discussing the quark model.

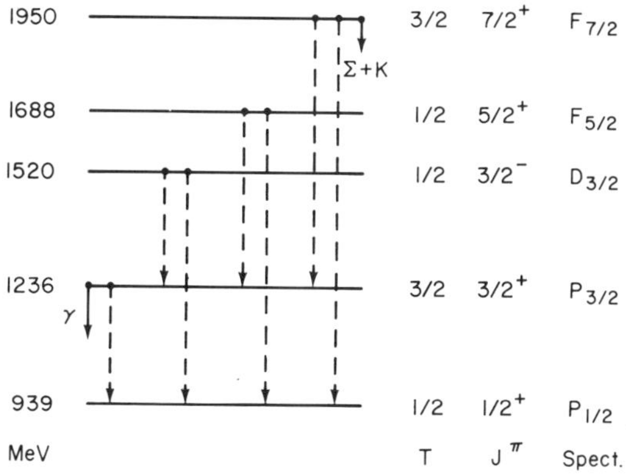

FIG. 1.6. Prominent low-energy levels or resonances of the nucleon with assignment of quantum numbers. Arrows indicate π-emission decay modes. Energies are taken from Particle Data Group (1973).

effective coupling constant. One simple zero-range approximation estimate for low-energy pion p-wave scattering without a cutoff requires $f^2/\hbar c \approx 0.2$ and approximately agrees with the value for low-energy singlet nucleon–nucleon scattering also with no cutoff. However, remarkably good agreement with data results from perturbation calculations using a cut-off radius of 0.2 fm, which can also be expressed in terms of a pion cut-off momentum or energy, viz., $E_{\max} = (\hbar^2 c^2 k^2_{\pi\max} + m_\pi{}^2 c^4)^{1/2}$. For $k_{\pi\max} = -k_{\mathrm{p,n,max}} \approx 6 m_\pi c/\hbar$ (i.e., $r_{\min} \approx \frac{1}{6}\hbar/m_\pi \mathrm{c} = 1.42/6 \approx 0.2$ fm) we have $E_{\max} \approx 800$ MeV. Best agreement with data results in using such cut-off values with an effective or renormalized coupling constant 0.085 ± 0.003 (although a value for the pseudoscalar pion–nucleon coupling constant of 0.081 is a better value when all experiments

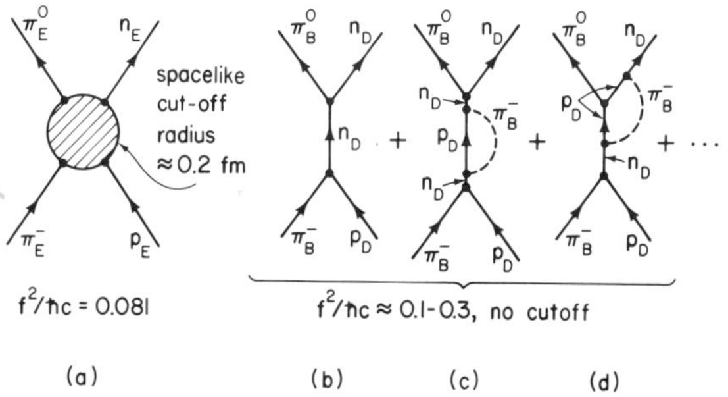

FIG. 1.7. A simple representation of charge-exchange pion–nucleon scattering (a) with a single vertex for distant collisions with an effective coupling constant and (b)–(d) showing a few of the many diagrams needed without cutoff. Nucleon–antinucleon vertices involving $G^2/\hbar c$, etc., are not shown.

relating to this coupling constant are averaged). The "bare" particle approach and the cut-off approach are illustrated schematically in Fig. 1.7. The subscript E stands for physical or laboratory particle while B and D stand for bare particles. Figure 1.7b–d shows but a very few of the simpler diagrams that would have to be summed to give the proper scattering cross section.

The current understanding of the pion–nucleon interaction also permits some success in the development of pion–nucleus optical potentials to account for the scattering of pions by nuclei. These potentials are derived from a multiple-scattering expansion summed over the nucleons in the nucleus. For example, one rather satisfactory fit to $^{12}C(\pi, \pi)$ scattering at $E = 120$–280 MeV is given by Kisslinger and Tabakin (1974).

Rather simple considerations based on conservation principles and the assignment of quantum numbers lead to a ready understanding of the $\Delta(1236)$ resonance. Since at this energy ($E_{\pi,\mathrm{LAB}} \approx 200$ MeV) elastic scattering with or without charge exchange is the only energetically allowed process, we restrict ourselves to considering

$$\pi^+ + \mathrm{p} \rightarrow \pi^+ + \mathrm{p} \tag{1-69a}$$

$$\pi^- + \mathrm{p} \rightarrow \pi^- + \mathrm{p} \tag{1-69b}$$

$$\pi^- + \mathrm{p} \rightarrow \pi^0 + \mathrm{n}. \tag{1-69c}$$

Referring to (1-57) or directly from Appendix A, we can write the various initial and final isospin states of (1-69) in terms of states $|T, T_3\rangle$, viz.:

$$\begin{aligned} |\mathrm{p}\pi^+\rangle &= |\tfrac{3}{2}, \tfrac{3}{2}\rangle \\ |\mathrm{p}\pi^-\rangle &= \sqrt{\tfrac{1}{3}}\,|\tfrac{3}{2}, -\tfrac{1}{2}\rangle - \sqrt{\tfrac{2}{3}}\,|\tfrac{1}{2}, -\tfrac{1}{2}\rangle \\ |\mathrm{n}\pi^0\rangle &= \sqrt{\tfrac{2}{3}}\,|\tfrac{3}{2}, -\tfrac{1}{2}\rangle + \sqrt{\tfrac{1}{3}}\,|\tfrac{1}{2}, -\tfrac{1}{2}\rangle. \end{aligned} \tag{1-70}$$

The reaction (1-69a) corresponds to simple elastic scattering in which an incident plane wave $e^{ikz}|\mathrm{p}\pi\rangle$ of an interacting proton and positive pion is modified by the interaction in its outgoing wave component. The reactions (1-69a) and (1-69b) involve an incident plane wave $e^{ikz}|\mathrm{p}\pi^-\rangle$ of an interacting proton and negative pion giving rise to two possible outgoing waves in the states $|\mathrm{p}\pi^-\rangle$ and $|\mathrm{n}\pi^0\rangle$. If the resonant interactions are isospin-conserving and charge independent, we can describe these processes in terms of scattering amplitudes a_3 and a_1 defined as matrix elements of the interaction operator a, viz.

$$\begin{aligned} &\langle\tfrac{3}{2}, \text{any } T_3|\mathsf{a}|\tfrac{3}{2}, \text{any } T_3\rangle = a_3, \qquad \langle\tfrac{1}{2}, \text{any } T_3|\mathsf{a}|\tfrac{1}{2}, \text{any } T_3\rangle = a_1 \\ &\langle\tfrac{3}{2}, \text{any } T_3|\mathsf{a}|\tfrac{1}{2}, \text{any } T_3\rangle = 0, \qquad \langle\tfrac{1}{2}, \text{any } T_3|\mathsf{a}|\tfrac{3}{2}, \text{any } T_3\rangle = 0. \end{aligned} \tag{1-71}$$

Thus, we have transition matrix elements

$$\langle p\pi^+|a|p\pi^+\rangle = \langle \tfrac{3}{2},\tfrac{3}{2}|a|\tfrac{3}{2},\tfrac{3}{2}\rangle = a_3$$

$$\langle p\pi^-|a|p\pi^-\rangle = \left(\sqrt{\tfrac{1}{3}}\langle \tfrac{3}{2},-\tfrac{1}{2}| - \sqrt{\tfrac{2}{3}}\langle \tfrac{1}{2},-\tfrac{1}{2}|\right)|a|\left(\sqrt{\tfrac{1}{3}}|\tfrac{3}{2},-\tfrac{1}{2}\rangle - \sqrt{\tfrac{2}{3}}|\tfrac{1}{2},-\tfrac{1}{2}\rangle\right)$$

$$= \tfrac{1}{3}a_3 + \tfrac{2}{3}a_1 \tag{1-72}$$

$$\langle p\pi^-|a|n\pi^0\rangle = \left(\sqrt{\tfrac{1}{3}}\langle \tfrac{3}{2},-\tfrac{1}{2}| - \sqrt{\tfrac{2}{3}}\langle \tfrac{1}{2},-\tfrac{1}{2}|\right)|a|\left(\sqrt{\tfrac{2}{3}}|\tfrac{3}{2},-\tfrac{1}{2}\rangle + \sqrt{\tfrac{1}{3}}|\tfrac{1}{2},-\tfrac{1}{2}\rangle\right)$$

$$= \tfrac{1}{3}\sqrt{2}\,a_3 - \tfrac{1}{3}\sqrt{2}\,a_1.$$

The various cross sections are proportional to the absolute value squared of the transition matrix elements; thus, in obvious notation, we write

$$\frac{\sigma_-}{\sigma_0} = \frac{|\tfrac{1}{3}a_3 + \tfrac{2}{3}a_1|^2}{|\tfrac{1}{3}\sqrt{2}\,a_3 - \tfrac{1}{3}\sqrt{2}\,a_1|^2} \tag{1-73}$$

and

$$\frac{\sigma_+}{\sigma_- + \sigma_0} = \frac{|a_3|^2}{|\tfrac{1}{3}a_3 + \tfrac{2}{3}a_1|^2 + |\tfrac{1}{3}\sqrt{2}\,a_3 - \tfrac{1}{3}\sqrt{2}\,a_1|^2}.$$

If the $\Delta(1236)$ resonance is a pure $T = \tfrac{3}{2}$ state, $a_1 = 0$ and $\sigma_-/\sigma_0 = \tfrac{1}{2}$, and $\sigma_+/(\sigma_- + \sigma_0) = 3$ or $\sigma_+ : \sigma_- : \sigma_0 = 9:1:2$. We note from Fig. 1.5 that the *total* cross sections are $\sigma_+/(\sigma_- + \sigma_0) = 195/68$, which ratio compares favorably with the assignment of $T = \tfrac{3}{2}$ to this resonance. Since the maximum total cross section for a resonance reaction characterized by angular momentum J is

$$\sigma_{\max} = \frac{4\pi}{k^2}\frac{2J+1}{(2s_1+1)(2s_2+1)}$$

(s_1 and s_2 are the spins of the interacting particles), we predict maximum cross sections

$$\sigma_+ = 8\pi/k^2, \qquad \sigma_- = 8\pi/9k^2, \qquad \sigma_0 = 16\pi/9k^2, \tag{1-74}$$

which also are in good agreement with the measured values.

Although rather good microscopic theories exist for the pion–nucleon interaction with renormalized coupling constants hopefully leading to believable perturbation calculations, the main phenomenological approach to elementary particle physics is generally concerned with establishing cross-section ratios or branching ratios, selection rules, etc., based on conservation principles, symmetry principles, and quantum number assignment schemes.

Some remarks are in order for techniques used to eliminate unpleasant diagrams which are totalized by using laboratory-measured intrinsic properties even for obviously composite particles and by using renormalized coupling constants for distant interactions. Consider the essentially simple "three-body" atomic problem consisting of the proton, the electron, and the electromagnetic field. One approach could be to try to treat the problem by a perturbation calculation involving the *entire* electromagnetic interaction (including the portion leading to the Coulomb interaction between the electron and proton) as a perturbation on freely propagating electron and proton plane wave states.

A calculation essentially in this spirit has been performed by Wigner (1954), who took the unperturbed state to be an electron in a large, spherical box. The entire Coulomb interaction with the proton at the center of the box was treated as a perturbation. To second order in the Rayleigh–Schrödinger perturbation theory, a finite binding energy resulted but was numerically $\frac{1}{10}$ to $\frac{1}{5}$ the required value. The convergence of the theory to a nonzero finite value was entirely dependent on the potential being of the form λr^{-n} with $n = 1$. Either $n > 1$ or $n < 1$ gives, to second order, zero binding in the first case and infinite binding in the second case. A virtually meaningless wave function was generated. Thus this approach is hardly the recommended procedure for even so simple a system as the hydrogen atom, even when a powerfully converging perturbation theory is used.

A second (and of course far more successful) approach is to include in H_0 the static portion of the electromagnetic interaction, which leads to the stationary-state solutions for the hydrogen atom bound states (as well as the distorted wave unbound states of the continuum). The *remainder* of the electromagnetic interaction can then be used to calculate radiative transition probabilities in a rapidly converging perturbation calculation. Such an approach is always possible for linear fields. In this latter case, should it develop that exact analytic eigenstates for H_0 cannot be determined due to mathematical difficulties, quantum number characterizations and symmetry properties might be deduced nonetheless for the various states suggested by experimental evidence. Even in this event, ratios of transition rates, selection rules, and the like for physically observed states might have been calculated. In a crude sense the first approach resembles efforts to use bare Dirac nucleons and the entire meson field complex, while the second more nearly resembles the cut-off calculation with the nucleon already surrounded by its complement of static field mesons resulting in an interaction with other mesons through a reduced coupling constant for manageable perturbation calculations.

7. Quark Model

Perhaps one of the more exciting, albeit bizarre, conjectures to account for the observed symmetries existing among the vast number of "elementary" particles is the quark model. While a detailed discussion is completely outside the scope of this text, a brief review insofar as it affects nucleon substructure is in order. There are many fine works to consult for a more thorough treatment of the quark model; for example, Gasiorowicz (1966, Chapters 17 and 18), a collection of important papers up to 1964 (Gell-Mann and Ne'eman, 1964), a work by Kokkedee (1969), the lecture articles by Kienzle (Zichichi, 1969, Part B., p. 372 ff.) and by Morpurgo (Zichichi, 1969, Part A., p. 84 ff.) and Morpurgo (1970). One should also cite "Quarks for Pedestrians" by Lipkin (1973).

An earlier model, the Sakata model [(1956), and Gasiorowicz (1966, Chapter 17)], noted that all the known particles at the time (1956) could be viewed as composites of three baryons ($B = 1$) p, n, and Λ and their antiparticles. [The Λ is isoscalar, with $t = 0$, intrinsic mechanical spin $\frac{1}{2}\hbar$, even parity, strangeness $S = -1$, and mass $m_\Lambda c^2 = (1115.59 \pm 0.05)$ MeV (Particle Data Group, 1973).] For example, we have already noted that the pseudoscalar isovector pion gives rise to a superstrong force between nucleon and antinucleon. From a consideration of the addition of the relevant quantum numbers, it is clear that the pion charge state π^+ in this model can be taken as[1]

$$|\pi^+\rangle = |(\mathrm{p}, \bar{\mathrm{n}})\,{}^1\mathrm{S}_0\rangle. \tag{1-75}$$

A large binding energy is assumed to account for the observed mass balance, viz., BE $\approx (m_\mathrm{p} + m_\mathrm{n})\,c^2 - m_\pi c^2 = 2 \times 939 - 140 = 1738$ MeV.

The reasons for the subsequent failure of the Sakata model need not concern us except to note that the quark model or the *eightfold way* proposed by Gell-Mann and Zweig as a replacement retains some of the features involved in the Sakata model plus a whole array of additional successes. The basic "building blocks" are the three quarks and their antiparticles. The quarks are supposedly the particlelike physical embodiment of the triplet representation *3* of the operators obeying $SU(3)$ algebra (and the antiquarks those of *3**). This algebra relates only to the quantum numbers associated with hypercharge Y and isospin t. Further quantum numbers can be assigned in addition to Y and t (with some degree of arbitrariness) to generate the "particle" properties listed in Table 1-1.

A brief discussion of search efforts to establish the existence of quarks is given by Morpurgo (Zichichi, 1969, Part A, p. 188 ff.) and Morpurgo (1970). A search for particles of anomalous charge and large mass produced by interactions of 300-GeV protons set upper limits on production cross sections of 10^{-35} cm^2 for particles with charges $e/3$ and $2e/3$, and also 5×10^{-31} cm^2 for a charge $-4e/3$, according to Leipuner *et al.* (1973). To date (1974) there is no direct experimental evidence for the physical quark. The possibility exists that an as yet unknown selection rule prevents the laboratory existence

[1] The notation implies the coupling of a proton and antineutron with $L = 0$ relative orbital angular momentum and spins coupled to the singlet state, $S = 0$.

TABLE 1-1

THE QUARK QUANTUM NUMBERS

Quark	Charge Q	Baryon number B	Hypercharge Y	Strangeness S	Mechanical spin	t_3
q_p	$\frac{2}{3}e$	$\frac{1}{3}$	$\frac{1}{3}$	0	$\frac{1}{2}$	$+\frac{1}{2}$
q_n	$-\frac{1}{3}e$	$\frac{1}{3}$	$\frac{1}{3}$	0	$\frac{1}{2}$	$-\frac{1}{2}$
q_λ	$-\frac{1}{3}e$	$\frac{1}{3}$	$-\frac{2}{3}$	-1	$\frac{1}{2}$	0

of single quarks. For some other recent comments see Rahm and Louttit (1970), Hazen (1971), Clark *et al.* (1971), and Casher *et al.* (1973).

The notation $q_{p,n,\lambda}$ is taken in analogy to the Sakata model in view of resemblances to the properties of p, n, and Λ (i.e., strangeness, t_3, and mechanical spin). The quarks also have even intrinsic parity. The Gell-Mann–Nishijima formula $Q = (t_3 + Y/2)e$ is obeyed, resulting in intrinsic charges $\frac{2}{3}e$ and $-\frac{1}{3}e$! The antiparticles have the negative of these quantum numbers and odd parity. The two nonstrange quarks form an isospin doublet $t = \frac{1}{2}$, and the strange quark an isospin singlet $t = 0$.

There may be in fact a fourth type of quark, q_D, with the properties $Q = \frac{2}{3}e$, $B = \frac{1}{3}$, $Y = \frac{4}{3}$, and $t_3 = 0$ (see Carlson and Freund, 1972). This type of quark is postulated in order to give a more satisfactory picture of hadrons when weak interactions are considered. By this addition the triplet model discussed here is extended to a quartet model. More complex substructure theories have also been conjectured, for example, a dual quark–gluon model, see Schwarz (1971).

The mesons are imagined to be made up of $q\bar{q}$ states and the baryons of qqq states.

Group-theoretic arguments show that $q\bar{q}$ states can be grouped into an "octet" (the eightfold way), namely, an isospin doublet with $Y = 1$, an isospin doublet with $Y = -1$, an isospin triplet with $Y = 0$, and an isospin singlet with $Y = 0$. In addition, there is a "singlet" group with $Y = 0$. These groups are characterized by the property of having elements that transform into each other within a group under the general unitary transformation $\xi' = U\xi$. When all mass degeneracies are removed, nine identifiable particles result, referred to as the "nonet" group. In addition to the angular momentum due to the intrinsic mechanical spin of the $(q\bar{q})$ meson structure, orbital angular momentum may be present, giving rise to so-called L-excitation with the resulting mesons having intrinsic parity $(-1)^{L+1}$. Further, additional radial oscillations leading to "n-excitations" of the $(q\bar{q})$ system may also be present. For example, with $n = 1$ (lowest radial excitation) and $L = 0$,we get a nonet of states for *each* of the $^3S_1, J^\pi = 1^-$ (vector) and $^1S_0, J^\pi = 0^-$ (pseudoscalar) possibilities in coupling the intrinsic quark spins. The $^1S_0$ coupling includes the π- and η-mesons and the $^3S_1$ coupling includes the ρ-, ω-, and ϕ-mesons already discussed. The nonets referred to result when K-mesons and additional mesons are also included. The K-mesons involve q_λ (or $\bar{q}_\lambda$) only singly in $(q\bar{q})$ and therefore give rise to the only strange mesons (in the form of a quadruplet). A simplified sketch of the situation is given in Fig. 1.8. The isospin splitting for the lower states is also shown specifically exhibiting the nonet structure.

The meson wave functions of a nonet can be written in terms of the quark states involved. Thus for the pseudoscalar mesons that comprise the lowest

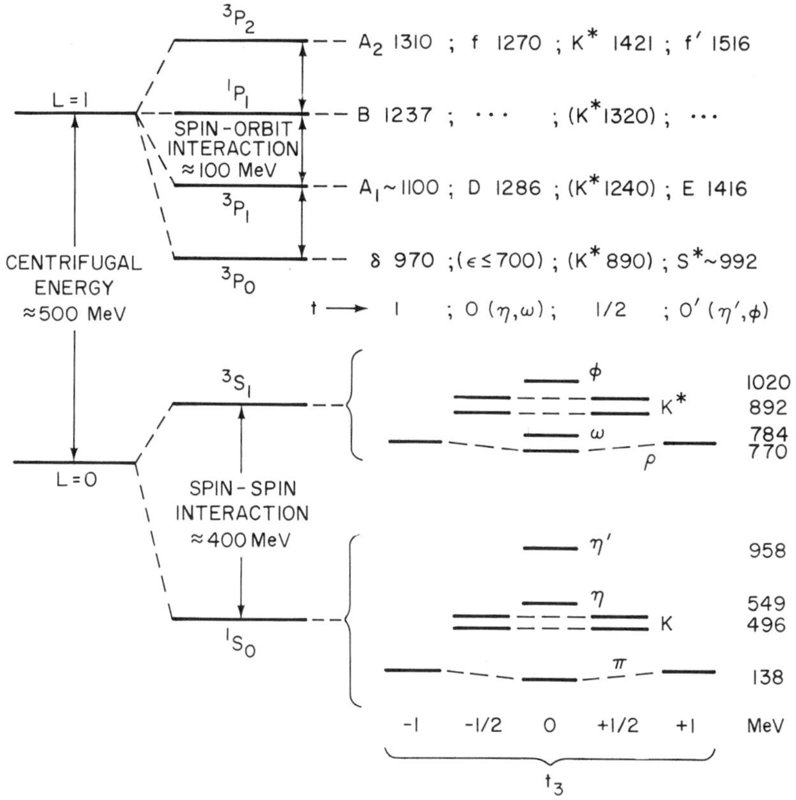

FIG. 1.8. A schematic representation of the mass spectrum of mesons according to the quark model. The energy scale is not to be taken literally and is scaled to reveal generic relationships. Doubtful entries in parentheses. The $^3P_0$-meson labeled ε was formerly also referred to as the σ-meson.

mass nonet ($^1S_0$), we have, in the notation *meson*: (Y, t_3) and the strangeness S:

$$\left.\begin{array}{ll}\left.\begin{array}{ll}\pi^-: & (0,-1) = q_n\bar{q}_p,\\ \pi^+: & (0,+1) = q_p\bar{q}_n,\\ \pi^0: & (0,0) = (1/\sqrt{2})(q_p\bar{q}_p - q_n\bar{q}_n),\end{array}\right\} S = 0 & \\ \left.\begin{array}{ll}K^+: & (1,+\tfrac{1}{2}) = q_p\bar{q}_\lambda,\\ K^0: & (1,-\tfrac{1}{2}) = q_n\bar{q}_\lambda,\end{array}\right\} S = +1 & \\ \left.\begin{array}{ll}\overline{K}^0: & (-1,+\tfrac{1}{2}) = q_\lambda\bar{q}_n,\\ K^-: & (-1,-\tfrac{1}{2}) = q_\lambda\bar{q}_p,\end{array}\right\} S = -1 & \\ \eta_8: \quad (0,0) = (1/\sqrt{6})(q_p\bar{q}_p + q_n\bar{q}_n - 2q_\lambda\bar{q}_\lambda), & S = 0\end{array}\right\} \text{(octet)}$$

$$\eta_1: \quad (0,0) = (1/\sqrt{3})(q_p\bar{q}_p + q_n\bar{q}_n + q_\lambda\bar{q}_\lambda), \qquad S = 0 \quad \text{(singlet).} \tag{1-76a}$$

The states (1-76a) represent the unmixed $SU(3)$ symmetry groups. In the presence of $SU(3)$ symmetry-breaking interactions, the last two states may be coupled

$$\begin{aligned}\eta: \quad (0,0) &= \eta_1 \cos\alpha_p + \eta_8 \sin\alpha_p\\ \eta': \quad (0,0) &= \eta_1 \sin\alpha_p - \eta_8 \cos\alpha_p,\end{aligned} \tag{1-76b}$$

with α_p referred to as the pseudoscalar *mixing angle*. To account for the observed η and η' masses and certain reaction cross sections, α_p can be estimated to be $\alpha_p \approx 90° \pm 11°$ with the result that $\eta \approx \eta_8$ and $\eta' \approx \eta_1$.[1]

The ρ- and K*-mesons of the $^3S_1$ nonet have identical quark state functions to their corresponding number in the $^1S_0$ nonet. The situation for the isoscalar mesons ω and ϕ is, however, quite different. Mass values and reaction data now suggest a vector mixing angle $\alpha_v = 37.1° \pm 1.1°$ in an expression of the form of (1-76b) for the ω and ϕ. This value is very close to the so-called *ideal mixing angle* $\tan\alpha = 1/\sqrt{2}$, $\alpha = 35.3°$. For the ideal mixing angle we have

$$\omega(0,0) \approx (1/\sqrt{2})(q_p\bar{q}_p + q_n\bar{q}_n), \qquad \phi(0,0) \approx q_\lambda\bar{q}_\lambda. \tag{1-76c}$$

Small "impurities" are, however, definitely required in (1-76c) to allow, for example, for the observed weak decay $\phi \to \rho + \pi$. The characterization of the f′-meson ($^3P_2$ nonet) appears also to require a mixing angle close to 35.3°, since the decay $f' \to 2\pi$ is weak.

[1] It is not absolutely certain that the η'-meson is correctly identified with the particle mass 958 MeV. For a complete discussion of the η and η' see DeFranceschi *et al.* (1971), and for a recent paper see Han (1973).

For a discussion of the mixing angles, including those for the $^3P_0$ scalar-meson nonet (see Fig. 1.8) as deduced from nucleon–nucleon and hyperon–nucleon scattering, refer to Nagels *et al.* (1973) [see also Capps (1972)].

▶**Exercise 1-14** Verify that (1-75) and (1-76) involve the proper combination of quantum numbers. ◀

The mass of the quark M can be estimated from the pure or "unsplit" positions of the orbital energies for the $L = 0$ and $L = 1$ excitations (see Fig. 1.8). As we shall see in Chapter IV, the energy eigenvalues for an infinitely deep, spherically symmetric well are generated by the zeros of the spherical Bessel function $j_L([(2\mu R^2/\hbar^2)T]^{1/2})$, where T is the kinetic energy of the particle measured from the bottom of the well, μ the reduced mass, and R the well radius. The solutions in increasing order are

$$T_{n,L} = (\hbar^2/2\mu R^2)\zeta^2_{n,L}, \tag{1-77}$$

with $\zeta_{10} = \pi$, $\zeta_{11} = 1.430\pi$, $\zeta_{12} = 1.835\pi$, $\zeta_{20} = 2\pi, \ldots$. Since the appropriately weighted position of the pure $L = 0$ state is ~ 800 MeV while that for the $L = 1$ state is ~ 1300 MeV, we conclude that

$$\frac{\pi^2}{2}\left(\frac{\hbar c}{R}\right)^2 \frac{1}{\mu c^2} \approx 0.7 \text{ GeV}. \tag{1-78}$$

If we take the range of the square well to be the Compton wavelength of the average meson mass with which we are dealing (namely $mc^2 \approx 1.1$ GeV) and further noting that $2\mu = M$, with M the quark mass, we get $Mc^2 \approx 17$ GeV. The use of a harmonic oscillator potential or any deep, narrow potential would give substantially the same result. We note, then, that the average binding energy is approximately $\text{BE} = 2Mc^2 - mc^2 \approx 33$ GeV. This justifies the use of (1-77) for a deep well, and, as with the nucleon–antinucleon model, ascribes the large mass difference between the composite particle (the meson) and the two substructure particles to the operation of extremely strong attractive forces. Another interesting consideration is that the internal motion is essentially nonrelativistic, since $T/Mc^2 \approx 1/30$. This leads to a situation in which interactions involving the exchange of mesons with momentum transfers in the GeV/c range must be treated relativistically, while the internal description of the meson structure in its own center-of-mass system obeys nonrelativistic quantum mechanics. Mass laws giving the observed masses within a nonet supermultiplet have been very successful and suggest that the mass of the quark q_λ with strangeness $S = -1$ is some 125 MeV more massive than the nonstrange quarks q_p and q_n. The mass excess of the q_D quark, if it exists, is estimated to be >600 MeV above the q_p and q_n quarks.

The quark state wave functions for the baryons are considerably more complicated than for the mesons since three quarks are required, qqq. The quark characteristics of Table 1-1 require the three quarks for the proton state, for example, to be q_p, q_p, and q_n and those for the neutron state to be q_n, q_n, and q_p. The nucleons fall into the *56* representation of $SU(6)$ and for the proton with "spin up," for example, the resulting wave function is

$$|p, s_z = +\tfrac{1}{2}\rangle = (1/\sqrt{18})\{2|q_p\uparrow q_n\downarrow q_p\uparrow\rangle + 2|q_p\uparrow q_p\uparrow q_n\downarrow\rangle + 2|q_n\downarrow q_p\uparrow q_p\uparrow\rangle$$
$$- |q_p\uparrow q_p\downarrow q_n\uparrow\rangle - |q_p\uparrow q_n\uparrow q_p\downarrow\rangle - |q_p\downarrow q_n\uparrow q_p\uparrow\rangle$$
$$- |q_n\uparrow q_p\downarrow q_p\uparrow\rangle - |q_n\uparrow q_p\uparrow q_p\downarrow\rangle - |q_p\downarrow q_p\uparrow q_n\uparrow\rangle\},$$

in obvious notation. As is evident, this wave function is symmetric in the interchange of spin and isospin of any two particles, as indeed is true for all members of the *56* representation. If the quarks, possessing mechanical spin $\frac{1}{2}$, obey Fermi statistics, the above spin and isospin function would have to be multipled by a space-dependent function $\chi(r_1, r_2, r_3)$ antisymmetric on space exchange of any two coordinates. Ample evidence suggests that the low-mass baryons, including the nucleons, are in an $L = 0$ spatial state. Thus $\chi_{L=0}(r_1, r_2, r_3)$ would have to be *antisymmetric* in space coordinates. A simple independent quark model (or shell model) for the nucleon states $(1s)^2(1s)^1$ with the two identical quarks coupled to $^3S_1$ would be impossible. If, however, quarks either have an additional three-valued degree of freedom or obey para-Fermi statistics of order three, such states could be allowed. Either case leads to a temptingly simple model referred to as the *symmetric quark model*. Thus a simple spectroscopic designation for the proton and neutron states would be

$$|p\rangle = |(q_p)^2\,{}^3S_1, q_n;\, J = \tfrac{1}{2}\rangle, \qquad |n\rangle = |(q_n)^2\,{}^3S_1, q_p;\, J = \tfrac{1}{2}\rangle.$$

For para-Fermi statistics the $^3S_1$ designation is the standard spectroscopic state; for Fermi statistics one requires an antisymmetric $L = 0$ state coupled to a triplet spin state giving $J_1 = 1$. In either case $J_1 = 1$ is coupled to the remaining quark spin $J_2 = \frac{1}{2}$ to give the total nucleon spin $J = \frac{1}{2}$.

In the quark model (as for *any* model in which continuous macroscopic densities are to be described using pointlike substructure particles) the charge and current operators are taken as

$$\begin{aligned} \rho_c(\mathbf{r}) &= \sum_k e_k\, \delta(\mathbf{r} - \mathbf{r}_k) \\ \mathbf{j}(\mathbf{r}) &= \sum_k \tfrac{1}{2} e_k\, \delta[\dot{\mathbf{r}}\, \delta(\mathbf{r} - \mathbf{r}_k) + \delta(\mathbf{r} - \mathbf{r}_k)\dot{\mathbf{r}}] + \sum_k \mu_k \nabla_k \times [\boldsymbol{\sigma}_k\, \delta(\mathbf{r} - \mathbf{r}_k)], \end{aligned} \qquad (1\text{-}79)$$

with $\mu_k = \mu_p e_k/e = 2.79 e_k/e$ nuclear magnetons. The magnetic moment can be calculated with the detailed, completely symmetrized wave function and these operators. Alternately, however, the magnetic moment of the proton and neutron can be determined from the simple application of the vector model. In view of the above, no orbital motion is involved, and thus for $|n\rangle$ two q_n quarks couple to give a state ($J_1 = 1$) that couples with the q_p quark ($J_2 = \frac{1}{2}$) in the "antiparallel" configuration giving $J = \frac{1}{2}$. Therefore, we have (to include $|p\rangle$ as well)[1]

$$\mu = g_1 \frac{J(J_1+1)}{J+1} - g_2 \frac{J_2 J}{J+1}, \tag{1-80}$$

with $g_k = 2\mu_p e_k/e$. This gives, noting the above,

$$\begin{aligned} \langle n|\mu_z|n\rangle &= 2\mu_p\left[-\frac{1}{3}\frac{(\frac{1}{2})(2)}{\frac{3}{2}} - \frac{2}{3}\frac{(\frac{1}{2})(\frac{1}{2})}{\frac{3}{2}}\right] = -\frac{2}{3}\mu_p \\ \langle p|\mu_z|p\rangle &= 2\mu_p\left[\frac{2}{3}\frac{(\frac{1}{2})(2)}{\frac{3}{2}} + \frac{1}{3}\frac{(\frac{1}{2})(\frac{1}{2})}{\frac{3}{2}}\right] = \mu_p. \end{aligned} \tag{1-81}$$

This result is close to being correct since the observed ratio is $\mu_n/\mu_p \approx -\frac{2}{3}$. Similar considerations predict $\mu_\Lambda = \frac{1}{2}\mu_n$ and $\mu_{\Sigma^+} = \mu_p$. Observed values for μ_Λ and μ_{Σ^+} are (-0.67 ± 0.06) and (2.59 ± 0.46) nuclear magnetons, respectively. Thus this simple model involving additivity in (1-79) gives surprisingly good results.

The model using (1-79) directly predicts the form factors G_{pE}, G_{pM}, G_{nM} to be those of (1-68) and $G_{nE} = 0$. Another success includes the photon–hadron coupling to give ρ-dominance with a decomposition 75% ρ-meson, 8% ω-meson, and 17% ϕ-meson, in excellent agreement with the Orsay experimental data quoted earlier.

As previously stated, the qqq baryon states are more complex than the $q\bar{q}$ meson states. Group-theoretic considerations now lead to many more independent representations than for the mesons. The lowest mass octet includes the well-known baryons: the nucleons n, p; the lambda particle Λ; the sigma particles $\Sigma^0, \Sigma^\pm$; and the cascade particles $\Xi^{0,-}$. There are also known decuplet groups as well as higher mass octets. Again, successful mass laws for members of these groups exist. Figure 1.9 exhibits the lowest mass baryon octet group and the pion decay branching ratios.

The baryon groups can be classified in a manner very similar to the shell model designations used for complex nuclei based on the quantum numbers for the three-dimensional harmonic oscillator. These nuclear states will concern us in detail later. Thus the mass octet of Fig. 1.9 belongs to the configuration $(1s)^3$, $L = 0$, and they are all characterized by $J^\pi = \frac{1}{2}^+$. It includes the $P_{1/2}(939)$ ground state of the nucleons.[2] The $P_{3/2}(1236)$ level

[1] This expression is correct for either Fermi or para-Fermi statistics.

[2] The spectroscopic baryon designations referenced here, such as $P_{1/2}$ for the nucleon ground state, are those of Fig. 1.8 and refer to the pion–nucleon configurations.

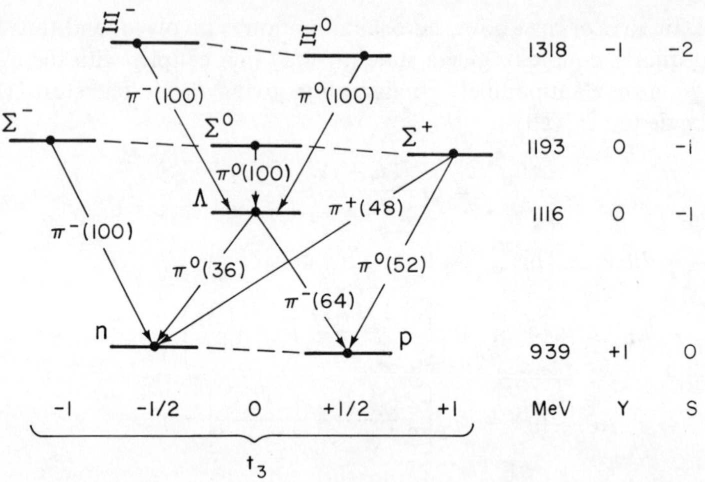

FIG. 1.9. The low-mass baryon ($B = 1$) octet $J^\pi = \frac{1}{2}^+$. The average masses mc^2, the hypercharge Y, and the strangeness S are indicated. The charge is given by the Gell-Mann–Nishijima formula $Q = (t_3 + Y/2)e$. The electromagnetic mass splitting is related by the Coleman–Glashow formula $(m_n - m_p) + (m_{\Xi^-} - m_{\Xi^0}) = (m_{\Sigma^-} - m_{\Sigma^+})$ or $1.293 + (6.4 \pm 0.6) \approx (7.94 \pm 0.09)$ MeV. The principal decay modes (percent) are also shown.

of Fig. 1.6, which we previously referred to as the "first excited state" of the nucleon, is in fact one of the $t = \frac{3}{2}$, $Y = 1$ members of the lowest mass decuplet ($J^\pi = \frac{3}{2}^+$), hence the $\Delta(1236)$ designation. This group also belongs to the $(1s)^3$ $L = 0$, configuration, but with different spin coupling.

The configuration $(1s)^3$, $L = 0$, leads to a total of 18 orthogonal states of three-quark wave functions. The states with three identical quarks, $q_p q_p q_p$, $q_n q_n q_n$, and $q_\lambda q_\lambda q_\lambda$, only appear with $J = \frac{3}{2}$. The states with all three quarks different $q_p q_n q_\lambda$, form eight magnetic substates $J_z = \frac{3}{2}, 3 \times \frac{1}{2}, 3 \times -\frac{1}{2}, -\frac{3}{2}$. The three with $J_z = +\frac{1}{2}$ form two linear combinations with $J = \frac{1}{2}$ and one with $J = \frac{3}{2}$; likewise for $J_z = -\frac{1}{2}$. These eight magnetic substates then result from two states with $J = \frac{1}{2}$ and one with $J = \frac{3}{2}$. The six states with two quarks the same each appear once with $J = \frac{3}{2}$ and once with $J = \frac{1}{2}$. Thus there are eight states with $J = \frac{1}{2}$ and ten states with $J = \frac{3}{2}$.

The lowest excited octet member corresponding to the nucleon is actually the $N^*(1470)$, $J^\pi = \frac{1}{2}^+$, corresponding to the $(1s)^2(2s)^1$ configuration, not shown in Fig. 1.6, which is only for the prominent pion resonances. The prominent $D_{3/2}$ resonance of Fig. 1.6 is in the next higher excited octet group for a member corresponding to the nucleons, designated $N^*(1520)$, $J^\pi = \frac{3}{2}^-$. It corresponds to the $(1s)^2(1p)^1$, $L = 1$, configuration. The $F_{5/2}$ and $F_{7/2}$ levels shown in Fig. 1.6 belong to the $L = 2$ mixed configurations $(1s)^2(1d)^1$ and $(1s)^1(1p)^2$.

Recent critical reviews of the experimental status of the elementary particle classifications we have been discussing are contained in the Particle Data Group review (1973) and the review by Samios *et al.* (1974).

A very remarkable conjecture has been recently put forward by a number of physicists, notably by Pati and Salam (1974) giving a unified model for strong, weak, and electromagnetic interactions of leptons and hadrons. In the theory of these authors, it is suggested that the quark quartet (i.e., q_p, q_n, q_λ, q_D) is endowed with another quantum number, the *color* quantum number. There are four color numbers in all; three involve the heavy quarks, each type of quark appearing as red, white, or blue. A fourth color (lilac) represents the lepton number L (see Chapter VIII). This last color identifies the light quarks taken to be the four known leptons, relating the q_p with the electronic neutrino ν_e, q_n with e^-, q_λ with μ^-, and q_D with the muonic neutrino ν_μ. In all, this schema leads to a sixteenfold fermionic multiplet (with Fermi number $F = B + L = 1$). The theory has a remarkable number of successes in correlating most of the elementary particle characteristics and interactions into a universal schema of great elegance. For a comprehensive, although quite advanced, review of gauge theories, as they are called, see Weinberg (1974).

In all, a very impressive number of predictions of the quark model agree with experiment. Many applications and details of the model are given in Zichichi (1969) and (*Proc. CERN School Phys.*, 1971, p. 41 ff.). There are of course many open questions still to be answered. Why $q\bar{q}$ and qqq states are strongly bound but apparently higher configurations are not, the so-called saturation problem, is still to be completely illuminated. Another difficulty has to do with the large electromagnetic radius required for the quark, when in fact the average quark–quark separation for either mesons or baryons is some 4–5 times smaller. The suggestion that the 10–20 GeV mass is only an effective mass as a means out of this difficulty requires further consideration. Yet another problem has to do with whether quarks do or do not satisfy Fermi statistics. Finally, are quarks real, perhaps the partons appearing in "deep" electron nucleon scattering, or just simply a useful model for computing observables? A rather advanced but very interesting review paper on this and related matters is by DeRújula *et al.* (1974).

C. SEMIEMPIRICAL CONSIDERATIONS

It should be suspected, from the discussion up to this point, that the nucleon–nucleon interaction is quite likely very complicated simply because the composite object called the nucleon is itself so complex. This indeed is the case, as we shall see. In addition, when many nucleons closely interact, further complications arise. It is then perhaps somewhat surprising that for low-energy nuclear phenomena, say below the pion threshold, rather simple "effective" interactions and models account for many of the main features observed. Of course, models that begin, "The nucleus behaves as if it were a ...," already contain (hopefully valid) condensations of many complexities. A central problem is to relate such behavior in as transparent a way as possible to the basic general interactions.

The two-nucleon problems at low energies, for example, can be adequately treated with a relatively modest number of simple parameters. Even the characteristics of complex nuclei where low-momentum-transfer interactions between nucleons are dominant can be similarly parameterized.

1. Classification of Nucleon–Nucleon Forces

We have seen that parity-conserving Hamiltonians must involve only true scalars. Consider first only the two-body nucleon–nucleon problem *without* velocity-dependent terms in the potential (as we shall see, this implies the potential is local). Of the quantities $\mathbf{r}$ ($\mathbf{r}_{12}$), $\boldsymbol{\sigma}_1$, and $\boldsymbol{\sigma}_2$, only the following terms are true scalars: $V(r)$, $\boldsymbol{\sigma}_1 \cdot \boldsymbol{\sigma}_2$, $(\boldsymbol{\sigma}_1 \cdot \mathbf{r})(\boldsymbol{\sigma}_2 \cdot \mathbf{r})$, $(\boldsymbol{\sigma}_1 \times \mathbf{r}) \cdot (\boldsymbol{\sigma}_2 \times \mathbf{r})$, and their products. Any imagined higher powers of $\boldsymbol{\sigma}_1$ and $\boldsymbol{\sigma}_2$ can be reduced to the above by repeated use of the commutation relationships $\sigma_x{}^2 = 1$, $\sigma_y \sigma_z + \sigma_z \sigma_y = 0$, $\sigma_y \sigma_z - \sigma_z \sigma_y = 2i\sigma_x$, and their cyclic permutations. The last term also reduces to simpler components, viz., $(\boldsymbol{\sigma}_1 \times \mathbf{r}) \cdot (\boldsymbol{\sigma}_2 \times \mathbf{r}) = r^2 \boldsymbol{\sigma}_1 \cdot \boldsymbol{\sigma}_2 - (\boldsymbol{\sigma}_1 \cdot \mathbf{r})(\boldsymbol{\sigma}_2 \cdot \mathbf{r})$.

▶**Exercise 1-15** Verify the above statements. ◀

It is convenient to modify the noncentral term $(\boldsymbol{\sigma}_1 \cdot \mathbf{r})(\boldsymbol{\sigma}_2 \cdot \mathbf{r})$ to yield a new noncentral operator S_{12}, which has a vanishing value for the average over all *directions* of $\mathbf{r}$. Since $\overline{(\mathbf{A} \cdot \mathbf{r})(\mathbf{B} \cdot \mathbf{r})} = \frac{1}{3} r^2 \mathbf{A} \cdot \mathbf{B}$, we define

$$S_{12} = (3/r^2)(\boldsymbol{\sigma}_1 \cdot \mathbf{r})(\boldsymbol{\sigma}_2 \cdot \mathbf{r}) - \boldsymbol{\sigma}_1 \cdot \boldsymbol{\sigma}_2. \tag{1-82}$$

▶**Exercise 1-16** Using spherical coordinates, show that

$$(1/4\pi) \int\!\!\int (\mathbf{A} \cdot \mathbf{r})(\mathbf{B} \cdot \mathbf{r}) \sin\theta \, d\theta \, d\varphi = \tfrac{1}{3} r^2 \mathbf{A} \cdot \mathbf{B}.$$

Derive the same result by the simpler route of using Cartesian coordinates and noting that

$$\overline{xy} = \overline{yz} = \overline{xz} = 0 \qquad \text{and} \qquad \overline{x^2} = \overline{y^2} = \overline{z^2} = \tfrac{1}{3} r^2.$$ ◀

From the discussion so far the most general local potential would be

$$V(\mathbf{r}, \boldsymbol{\sigma}) = V_{\mathrm{R}}(r) + V_\sigma(r)\, \boldsymbol{\sigma}_1 \cdot \boldsymbol{\sigma}_2 + V_{\mathrm{T}}(r) S_{12}. \tag{1-83}$$

This must now be extended to include isospin. We again repeat the comment that terms such as $\mathbf{r} \cdot \boldsymbol{\tau}$ or $\boldsymbol{\sigma} \cdot \boldsymbol{\tau}$ have no physical reality and are thus excluded.

By specifically introducing the isospin quantum numbers, we are treating nucleons as *identical* particles and hence must require that $H(1,2) = H(2,1)$ [i.e., all terms must be invariant to the exchange of *all* the coordinates ($\mathbf{r}, \boldsymbol{\sigma}$, and $\boldsymbol{\tau}$) of the two particles]. We will assume this implicitly in what follows without further comment. We also wish the Hamiltonian to be charge-conserving and hence want $[H, Q] = 0$. Since $Q = \frac{1}{2}e(\tau_{3,1} + \tau_{3,2} + 2)$ and since $[\tau_3, \tau_{2 \text{ or } 1}] \neq 0$, the only terms that can appear satisfying all the above requirements are $\boldsymbol{\tau}_1 \cdot \boldsymbol{\tau}_2$, $(\tau_{3,1})(\tau_{3,2})$, or $\tau_{3,1} + \tau_{3,2}$. Expanding on Exercise 1-6, we find the eigenvalues of these operators for the various possible symmetric and antisymmetric two-nucleon states (see Table 1-2). Clearly, $\boldsymbol{\tau}_1 \cdot \boldsymbol{\tau}_2$ leads to a charge-independent interaction, $(\tau_{3,1})(\tau_{3,2})$ leads to a charge symmetric interaction, while the presence of $(\tau_{3,1} + \tau_{3,2})$ breaks charge symmetry altogether. If we require charge independence as low-energy experimental evidence suggests, we may only use $\boldsymbol{\tau}_1 \cdot \boldsymbol{\tau}_2$.

TABLE 1-2

EIGENVALUES OF VARIOUS ISOSPIN OPERATORS

	Isospin triplet $T = 1$			Isopin singlet $T = 0$
	$T_3 = -1$ (n, n)	$T_3 = 0$ (n, p)	$T_3 = +1$ (p, p)	$T_3 = 0$ (n, p)
$\boldsymbol{\tau}_1 \cdot \boldsymbol{\tau}_2$	1	1	1	−3
$(\tau_{3,1})(\tau_{3,2})$	1	−1	1	−1
$\tau_{3,1} + \tau_{3,2}$	−2	0	2	0

Finally we write in place of (1-83)

$$V(\mathbf{r}, \boldsymbol{\sigma}, \boldsymbol{\tau}) = V_{\mathrm{R}}(r) + V_\sigma(r)\boldsymbol{\sigma}_1 \cdot \boldsymbol{\sigma}_2 + V_\tau(r)\boldsymbol{\tau}_1 \cdot \boldsymbol{\tau}_2 + V_{\sigma\tau}(r)(\boldsymbol{\sigma}_1 \cdot \boldsymbol{\sigma}_2)(\boldsymbol{\tau}_1 \cdot \boldsymbol{\tau}_2)$$
$$+ S_{12}[V_{\mathrm{T}}(r) + V_{\mathrm{T}\tau}(r)(\boldsymbol{\tau}_1 \cdot \boldsymbol{\tau}_2)], \qquad (1\text{-}84)$$

as the most general local potential. The classic paper on the subject is by Eisenbud and Wigner (1941). The first four terms of (1-84) lead to a central potential (i.e., independent of θ and φ), while the last two contribute a tensor interaction. A direct comparison with (1-45) shows that the pseudoscalar, isovector pion model is in fact of the form (1-84).

When velocity-dependent forces are included, three scalar quantities can be formed from $\mathbf{r}$ and $\mathbf{p}$. These are $p^2 = \mathbf{p} \cdot \mathbf{p}$, $r^2 = \mathbf{r} \cdot \mathbf{r}$, and $\mathbf{p} \cdot \mathbf{r}$ [this last of course would lead to a non-Hermitian operator, and we generally require writing $\frac{1}{2}(\mathbf{p} \cdot \mathbf{r} + \mathbf{r} \cdot \mathbf{p})$ in its place]. These can be included by writing in place of

$V(r)$ the general function $V(r,p,L)$. Here we have also noted that the scalar

$$\mathbf{L}\cdot\mathbf{L} = \mathbf{L}^2 = (\mathbf{r}\times\mathbf{p})\cdot(\mathbf{r}\times\mathbf{p}) = r^2p^2 - \tfrac{1}{4}r^2[\hat{\mathbf{r}}\cdot\mathbf{p}+\mathbf{p}\cdot\hat{\mathbf{r}}]^2.$$

When the spins $\boldsymbol{\sigma}_1$ and $\boldsymbol{\sigma}_2$ are included in our consideration, additional possible terms appear. One of the more important ones is $(\mathbf{r}\times\mathbf{p})\cdot(\boldsymbol{\sigma}_1+\boldsymbol{\sigma}_2) = \mathbf{L}\cdot\mathbf{S}$, the spin–orbit coupling. The static approximation used for the treatment of the pseudoscalar meson theory cannot generate the spin–orbit interaction term. When all terms that are true scalars are generated, the most general nonrelativistic nucleon potential becomes

$$V = V_c + V_{LS}\,\mathbf{L}\cdot\mathbf{S} + V_\sigma\,\boldsymbol{\sigma}_1\cdot\boldsymbol{\sigma}_2 + V_T S_{12} + V_{\sigma p}(\boldsymbol{\sigma}_1\cdot\mathbf{p})(\boldsymbol{\sigma}_2\cdot\mathbf{p}) + V_{\sigma L}(\boldsymbol{\sigma}_1\cdot\mathbf{L})(\boldsymbol{\sigma}_2\cdot\mathbf{L}), \tag{1-85a}$$

where each coefficient V_i is of the form

$$V_i = V_i^0(r,p,L) + V_i^\tau(r,p,L)\,\boldsymbol{\tau}_1\cdot\boldsymbol{\tau}_2. \tag{1-85b}$$

An alternate way of characterizing the central part of the potential of (1-84) or (1-85a, b) is in terms of exchange operators. In some sense this is more in keeping with the meson exchange model. Four central exchange potentials are required:

(a) *Ordinary or Wigner potential.* The ordinary potential, written to exhibit formally its nature as an operator, is $\tilde{V}(r) = V_W(r)\tilde{P}^W$, with $\tilde{P}^W = I$, the identity operator. Thus the Schrödinger equation with $m_n \approx m_p = m$ (and reduced mass $\mu = m/2$ for the two-nucleon problem) can be written as

$$[(\hbar^2/m)\nabla^2 + E]\,\phi(\mathbf{r})\,\chi_\sigma(1,2)\,\zeta_\tau(1,2) = V_W(r)\,\phi(\mathbf{r})\,\chi_\sigma(1,2)\,\zeta_\tau(1,2). \tag{1-86}$$

(b) *Bartlett or spin exchange potential.* We introduce the operator $\tilde{P}^B = \tilde{P}^\sigma$, which has the effect of interchanging the spin coordinates of the two nucleons, viz., $\tilde{P}^\sigma\chi_\sigma(1,2) = \chi_\sigma(2,1)$. Clearly if $\chi_\sigma(1,2)$ is symmetric (i.e., a spin triplet state), $\chi_\sigma(2,1) = \chi_\sigma(1,2)$; hence $\tilde{P}^\sigma\chi_1^{0,\pm1}(1,2) = +\chi_1^{0,\pm1}(1,2)$. Also, if $\chi_\sigma(1,2)$ is antisymmetric (i.e., the spin singlet state), $\chi_\sigma(2,1) = -\chi_\sigma(1,2)$; hence in this case $\tilde{P}^\sigma\chi_0{}^0(1,2) = -\chi_0{}^0(1,2)$. Recalling from Exercise 1-2 that $\boldsymbol{\sigma}_1\cdot\boldsymbol{\sigma}_2 = +1$ for the triplet state and $\boldsymbol{\sigma}_1\cdot\boldsymbol{\sigma}_2 = -3$ for the singlet state, we may write

$$\tilde{P}^\sigma = \tfrac{1}{2}(1+\tilde{\boldsymbol{\sigma}}_1\cdot\tilde{\boldsymbol{\sigma}}_2) \quad\text{and}\quad \tilde{V}(r) = V_B(r)\tilde{P}^B = V_B(r)\tilde{P}^\sigma. \tag{1-87}$$

The Schrödinger equation becomes

$$\begin{aligned}[(\hbar^2/m)\nabla^2 E]\,\phi(\mathbf{r})\,\chi_\sigma(1,2)\,\zeta_\tau(1,2) &= V_B(r)\,\phi(\mathbf{r})\,\zeta_\tau(1,2)\,\tilde{P}^\sigma\chi_\sigma(1,2)\\ &= (-1)^{S+1}V_B(r)\,\phi(\mathbf{r})\,\chi_\sigma(1,2)\,\zeta_\tau(1,2).\end{aligned} \tag{1-88}$$

If $V_B(r)$ is attractive in the triplet spin state ($S = 1$), then it is repulsive in the singlet spin state ($S = 0$), etc.

(c) *Heisenberg or isospin exchange potential.* We introduce the operator $\tilde{P}^H = -\tilde{P}^\tau$ (the unfortunate minus sign is of historical origin and leads to requiring some care with signs). The operator $\tilde{P}^\tau$ exchanges the isospin variables in $\zeta_\tau(1,2)$ in an exactly analogous way to the operation of $\tilde{P}^\sigma$. Thus we write

$$-\tilde{P}^H = \tilde{P}^\tau = \tfrac{1}{2}(1+\tilde{\boldsymbol{\tau}}_1\cdot\tilde{\boldsymbol{\tau}}_2) \qquad \text{and} \qquad \tilde{V}(r) = V_H(r)\tilde{P}^H = -V_H(r)\tilde{P}^\tau. \tag{1-89}$$

The Schrödinger equation becomes

$$\begin{aligned}[(\hbar^2/m)\nabla^2+E]\,\phi(r)\,\chi_\sigma(1,2)\,\zeta_\tau(1,2) &= -V_H(r)\,\phi(\mathbf{r})\,\chi_\sigma(1,2)\tilde{P}^\tau\zeta_\tau(1,2)\\ &= (-1)^T V_H(r)\,\phi(\mathbf{r})\,\chi_\sigma(1,2)\,\zeta_\tau(1,2).\end{aligned} \tag{1-90}$$

(d) *Majorana or space exchange potential.* The operator exchanging the space coordinates of the two nucleons is $\tilde{P}^M = \tilde{P}^r$, with $\tilde{P}^r\phi(\mathbf{r}) = \phi(-\mathbf{r})$ since $\mathbf{r} = \mathbf{r}_2-\mathbf{r}_1$. For the *special case* of *two identical fermions* (the two nucleons in isospin convention), we are required to have

$$\tilde{P}^\sigma\tilde{P}^\tau\tilde{P}^r[\phi(\mathbf{r})\,\chi_\sigma(1,2)\,\zeta_\tau(1,2)] = -\phi(\mathbf{r})\,\chi_\sigma(1,2)\,\zeta_\tau(1,2); \tag{1-91}$$

therefore

$$\tilde{P}^\sigma\tilde{P}^\tau\tilde{P}^r = -I. \tag{1-92}$$

From (1-92) we can write

$$\tilde{P}^M = \tilde{P}^r = -\tfrac{1}{4}(1+\tilde{\boldsymbol{\sigma}}_1\cdot\tilde{\boldsymbol{\sigma}}_2)(1+\tilde{\boldsymbol{\tau}}_1\cdot\tilde{\boldsymbol{\tau}}_2) \tag{1-93}$$

and

$$\tilde{V}(r) = V_M(r)\tilde{P}^M = V_M(r)\tilde{P}^r.$$

It was essentially to eliminate the minus sign in (1-93) that the historical convention on $\tilde{P}^H$ resulted, namely $\tilde{P}^r = -\tilde{P}^\sigma\tilde{P}^\tau$, while $\tilde{P}^M = +\tilde{P}^B\tilde{P}^H$. The Schrödinger equation in the case of a space exchange potential becomes

$$\begin{aligned}[(\hbar^2/m)\nabla^2+E]\,\phi(\mathbf{r})\,\chi_\sigma(1,2)\,\zeta_\tau(1,2) &= V_M(r)\,\chi_\sigma(1,2)\,\zeta_\tau(1,2)\tilde{P}^r\phi(\mathbf{r})\\ &= (-1)^{(1-\pi)/2}V_M(r)\,\phi(\mathbf{r})\,\chi_\sigma(1,2)\,\zeta_\tau(1,2)\\ &= (-1)^L V_M(r)\,\phi(\mathbf{r})\,\chi_\sigma(1,2)\,\zeta_\tau(1,2)\\ &= (-1)^{S+T+1}V_M(r)\,\phi(\mathbf{r})\,\chi_\sigma(1,2)\,\zeta_\tau(1,2).\end{aligned} \tag{1-94}$$

Since the nucleon–nucleon potential is short-ranged, it is often useful for purposes of a low-order approximation to take in fact the r dependence of the potentials $V_W(r)$, $V_B(r)$, $V_H(r)$, and $V_M(r)$ appearing in (1-86)–(1-94) as simple δ-functions. We now note that

$$P^r\,\delta(\mathbf{r}) = P^r\,\delta(\mathbf{r}_1-\mathbf{r}_2) = \delta(\mathbf{r}_2-\mathbf{r}_1) = \delta(\mathbf{r}_1-\mathbf{r}_2),$$

and also

$$P^\sigma\,\delta(\mathbf{r}_1-\mathbf{r}_2) = -P^\tau P^r\,\delta(\mathbf{r}_1-\mathbf{r}_2) = -P^\tau\,\delta(\mathbf{r}_1-\mathbf{r}_2).$$

Thus in the short-range limit, Majorana potentials reduce to ordinary Wigner potentials, while at the same time Bartlett and Heisenberg potentials become equivalent.

▶**Exercise 1-17** Noting that $(P^i)^2 = I$ for any i, $[P^i, P^j] = 0$, and using additional considerations, verify (1-93) and the various forms of (1-94). The quantum number π in (1-94) is the parity, $\pi = \pm 1$. ◀

Another convention that is occasionally used in the literature employs the triplet and singlet projection operators, defined as

$$\tilde{\mathscr{T}}(\xi) = \tfrac{1}{2}(1+\tilde{P}^\xi) \qquad \text{and} \qquad \tilde{\mathscr{S}}(\xi) = \tfrac{1}{2}(1-\tilde{P}^\xi) \tag{1-95}$$

where $\xi \equiv$ either σ or τ. Thus we have

$$\mathscr{S}(\sigma)\mathscr{T}(\tau) = \tfrac{1}{4}(1-P^\sigma+P^\tau+P^r), \qquad \mathscr{T}(\sigma)\mathscr{S}(\tau) = \tfrac{1}{4}(1+P^\sigma-P^\tau+P^r)$$
$$\mathscr{S}(\sigma)\mathscr{S}(\tau) = \tfrac{1}{4}(1-P^\sigma-P^\tau-P^r), \qquad \mathscr{T}(\sigma)\mathscr{T}(\tau) = \tfrac{1}{4}(1+P^\sigma+P^\tau-P^r). \tag{1-96}$$

Generally, the spatial dependence of the terms in any of the various representations of the central interaction potentials is similar. While for the two-nucleon problem to be treated in detail later we shall take into account the differences, it is useful to apply our results to the case where $V_W(r) \approx \mathscr{a}V_0(r)$, $V_B(r) \approx \mathscr{b}V_0(r)$, $V_H(r) \approx \mathscr{c}V_0(r)$, and $V_M(r) \approx \mathscr{d}V_0(r)$. Clearly $\mathscr{a}$, $\mathscr{b}$, $\mathscr{c}$, and $\mathscr{d}$ give the relative strengths and $V_0(r)$ describes a universal potential shape. The three useful characterizations of the central potential in this approximation can be written as

$$V(r) = V_0(r)[\mathscr{a}+\mathscr{b}P^B+\mathscr{c}P^H+\mathscr{d}P^M]$$
$$V(r) = V_0(r)[A+B\boldsymbol{\sigma}_1\cdot\boldsymbol{\sigma}_2+C\boldsymbol{\tau}_1\cdot\boldsymbol{\tau}_2+D(\boldsymbol{\sigma}_1\cdot\boldsymbol{\sigma}_2)(\boldsymbol{\tau}_1\cdot\boldsymbol{\tau}_2)] \tag{1-97}$$
$$V(r) = {}^{31}V(r)[\mathscr{T}(\sigma)\mathscr{S}(\tau)+\alpha\mathscr{S}(\sigma)\mathscr{T}(\tau)+\beta\mathscr{T}(\sigma)\mathscr{T}(\tau)+\gamma\mathscr{S}(\sigma)\mathscr{S}(\tau)],$$

where $A = \mathscr{a}+\tfrac{1}{2}\mathscr{b}-\tfrac{1}{2}\mathscr{c}-\tfrac{1}{4}\mathscr{d}$, $B = \tfrac{1}{2}\mathscr{b}-\tfrac{1}{4}\mathscr{d}$, $C = -\tfrac{1}{2}\mathscr{c}-\tfrac{1}{4}\mathscr{d}$, $D = -\tfrac{1}{4}\mathscr{d}$, and the notation ${}^{(2S+1)(2T+1)}V(r)$ is used.[1]

▶**Exercise 1-18** (a) Verify (1-95) and derive (1-96).

[1] The literature is divided between the notation ${}^{(2S+1)(2T+1)}V(r)$ versus ${}^{(2T+1)(2S+1)}V(r)$. The reader should be wary.

(b) Verify the relationships (1-97) and the following:

$$^{31}V(r) = V_0(r)[A+B-3C-3D] = V_0(r)[\mathscr{a}+\mathscr{b}+\mathscr{c}+\mathscr{d}]$$

$$^{13}V(r) = \alpha\,^{31}V(r) = V_0(r)[A-3B+C-3D] = V_0(r)[\mathscr{a}-\mathscr{b}-\mathscr{c}+\mathscr{d}]$$

$$^{33}V(r) = \beta\,^{31}V(r) = V_0(r)[A+B+C+D] = V_0(r)[\mathscr{a}+\mathscr{b}-\mathscr{c}-\mathscr{d}]$$

$$^{11}V(r) = \gamma\,^{31}V(r) = V_0(r)[A-3B-3C+9D] = V_0(r)[\mathscr{a}-\mathscr{b}+\mathscr{c}-\mathscr{d}].$$

◀

As a simple example of the above, let us consider the two-nucleon problem at low energy (implying $L = 0$), where the interaction is largely central. For the (p, p) or (n, n) system $T = 1$ and hence $S = 0$. Thus $V(r) = {}^{13}V(r)$. The (n, p) system occurs with either $T = 1$, $S = 0$ or $T = 0$, $S = 1$; hence $V(r) = {}^{31}V(r)\mathscr{T}(\sigma)\mathscr{S}(\tau) + {}^{13}V(r)\mathscr{S}(\sigma)\mathscr{T}(\tau)$. Actually, for $L = 0$, only one label is needed for all the two-nucleon possibilities, namely,

$$V(r) = V_t(r)\mathscr{T}(\sigma) + V_s(r)\mathscr{S}(\sigma). \tag{1-98}$$

Here $V_t(r)$ and $V_s(r)$ are simply a shorthand notation for ${}^{31}V(r)$ and ${}^{13}V(r)$, respectively, and we can further write

$$V_t(r) = {}^{31}V(r) = V_0(r)[\mathscr{a}+\mathscr{b}+\mathscr{c}+\mathscr{d}] = V_0(r)[A+B-3C-3D]$$

$$V_s(r) = {}^{13}V(r) = \alpha^{31}V(r) = V_0(r)[\mathscr{a}-\mathscr{b}-\mathscr{c}+\mathscr{d}] = V_0(r)[A-3B+C-3D]. \tag{1-99}$$

Empirically $\alpha = V_s/V_t \approx 0.6$ gives approximately the observed low-energy scattering cross sections and the deuteron binding energy.

Sometimes a fourth variant of (1-97) is used that simulates the symmetric meson-theoretic interaction with a central potential, namely,

$$V(r) = -V_{0s}(r)\,\boldsymbol{\tau}_1\cdot\boldsymbol{\tau}_2[1+\lambda\boldsymbol{\sigma}_1\cdot\boldsymbol{\sigma}_2]. \tag{1-100}$$

Here $V_{0s}(r) = {}^{31}V(r)/3(\lambda+1)$, $\alpha = (3\lambda-1)/(3\lambda+3)$, $\beta = -\frac{1}{3}$, and $\gamma = -(3\lambda-1)/(\lambda+1)$. The so-called "Rosenfeld mixture" $\lambda \approx 2.3$ gives $\alpha \approx 0.6$.

2. The Nomenclature of Low-Energy Scattering

In order to establish the necessary nomenclature for the convenient parameterization of low-energy nucleon–nucleon scattering, we give a deliberately brief although essentially accurate treatment of the process. The burden of a more comprehensive analysis is more properly undertaken in texts discussing general reaction theory. To avoid unnecessary complication, we begin by taking nucleons to be chargeless and spinless particles. We approximate the true interaction with a simple three-dimensional, real, square well of depth V_0

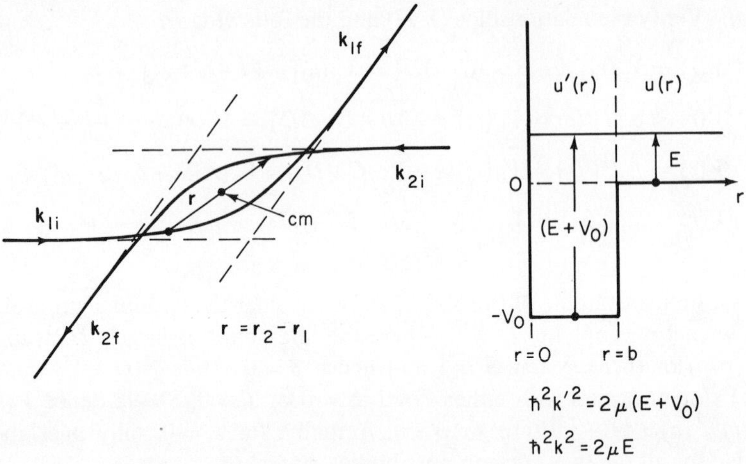

FIG. 1.10. Two-nucleon elastic scattering problem approximated by an attractive square-well potential in center-of-mass coordinates.

and well radius b (Fig. 1.10). Using a negative potential implies attractive nuclear forces. The problem is considered in center-of-mass coordinates with $E = E_{\mathrm{LAB}}/2$ and reduced mass $\mu = m/2$ ($m_{\mathrm{p}} \approx m_{\mathrm{n}} \equiv m$). Primed quantities are for the interior region $0 < r < b$ and unprimed quantities are for $r > b$. For laboratory energies below 20 MeV, we need consider only $l = 0$ or S-wave scattering and, by ignoring the relatively small capture cross section, we assume that only elastic scattering is energetically possible. Then in terms of the interior wave number $\hbar^2 k'^2 = 2\mu(E+V_0)$ and the exterior wave number $\hbar^2 k^2 = 2\mu E$, we write as acceptable general solutions to the Schrödinger time-independent equation, with $u = r\psi$,

$$\begin{aligned} u' &= A \sin k'r, \qquad && 0 < r < b \\ u &= B \sin(kr + \delta_0), \qquad && r > b. \end{aligned} \tag{1-101}$$

Matching logarithmic derivatives at $r = b$ gives

$$k \tan k'b = k' \tan(kb + \delta_0), \tag{1-102}$$

from which the *phase shift* δ_0 can be determined as a function of E, V_0, and b. With only elastic scattering possible the phase shift δ_0 is real and the subscript denotes $l = 0$.

The total wave function (1-101) in the external region $r > b$ represents a linear combination of both the incident wave and the scattered, spherically expanding wave. The incident wave can conveniently be taken to be a plane wave propagating in the direction of the polar axis:

$$\psi_{\text{inc}} = Ce^{ikz} = C\sum_{l=0}^{\infty} (2l+1)\, i^l j_l(kr)\, P_l(\cos\theta), \qquad (1\text{-}103)$$

where we have used the well-known expansion in terms of the spherical Bessel functions $j_l(kr)$ and the Legendre polynomials $P_l(\cos\theta)$. Taking only the $l = 0$ component (*partial wave*) of (1-103), we therefore write

$$\psi_0 = C[e^{ikz}]_{l=0} + D(e^{ikr}/r), \qquad r > b. \qquad (1\text{-}104)$$

We have from (1-101)

$$u = \{B[\exp(-i\delta_0)]/2i\}\,[\exp(2i\delta_0)\exp(ikr) - \exp(-ikr)], \qquad r > b$$

while from (1-104), using $j_0(kr) = [\sin(kr)]/kr$ in (1-103), we have

$$u = r\psi_0 = (C/2ik)(e^{ikr} - e^{-ikr}) + De^{ikr}.$$

Clearly $C = Bk\exp(-i\delta_0)$ and $D = B\sin\delta_0$; hence the scattered wave portion of (1-104) is

$$\psi_{\text{sc}} = D[\exp(ikr)]/r = C\{[\exp(i\delta_0)](\sin\delta_0)/kr\}\exp(ikr). \qquad (1\text{-}105)$$

The scattering cross section is defined as the flux of ψ_{sc} per unit incident flux. The incident flux per unit area from (1-103) is $I_{\text{inc}} = v|\psi_{\text{inc}}|^2 = v|C|^2$, where v is the velocity in the c.m. frame and $v = \hbar k/\mu$. The scattered flux can be evaluated by calculating the total flux through a spherical surface of radius R, where R is of the order of the macroscopic laboratory dimensions or $R \gg b$. From (1-105) we have

$$I_{\text{sc}} = 2\pi v R^2 \int_0^{2\pi} |\psi_{\text{sc}}(r = R)|^2 \sin\theta\, d\theta = 4\pi v|C|^2(\sin^2\delta_0)/k^2. \qquad (1\text{-}106)$$

Finally,

$$\sigma_{\text{sc}} = I_{\text{sc}}/I_{\text{inc}} = 4\pi(\sin^2\delta_0)/k^2. \qquad (1\text{-}107)$$

We should note that while (1-102) and the interior solution in (1-101) clearly depend on the assumed form of the potential, the form of the exterior solution at distances that are large compared to the range of the potential is correctly given in (1-101) (assuming that the true nuclear potential, while possibly differing from the square well used here, nonetheless is one possessing a well-defined range). Thus (1-105) and (1-107) are correct even for a more general potential with, however, δ_0 suitably redefined.

It is particularly instructive to examine the problem considered above in the low-energy limit $E \to 0$ (hence $k \to 0$ and $\hbar^2 k'^2 \to 2\mu V_0$). Chemical binding effects, although important at very low energies, will be ignored, it being assumed that all experimental quantities have been corrected for such effects.

It is evident from (1-107) that in this limit δ_0 must in general become proportional to k, or else the cross section becomes divergent. Indeed, for the square well

$$\lim_{k\to 0} \delta_0 \to kb\{[(\tan k'b)/k'b]-1\}, \tag{1-108}$$

verifying this conjecture. For the general potential, we therefore define the *Fermi scattering length* a by the relationships

$$\lim_{k\to 0} \delta_0 = -ak \tag{1-109}$$

and

$$\lim_{k\to 0} \sigma_0 = 4\pi a^2. \tag{1-110}$$

Actually, since $\sin^2 \delta_0$ appears in (1-107) and a^2 in (1-110), there is a sign ambiguity in (1-109); the minus sign is arbitrarily selected since this corresponds to hard-sphere scattering (i.e., the square well with $V_0 \to \pm\infty$ gives $\delta_0 \to -kb$; hence in this case $a = b$, the potential radius, and $\sigma_0 = 4\pi b^2$).

In the external region $r > b$, (1-101) becomes

$$\lim_{k\to 0} u = \lim_{k\to 0}\{[C\exp(i\delta_0)]/k\} \sin(kr+\delta_0) = C(r-a). \tag{1-111}$$

Further, we have in general from the Schrödinger equation for $l = 0$

$$\frac{1}{u}\frac{d^2u}{dr^2} = -\frac{2\mu}{\hbar^2}[E-V(r)]$$

which in the limit becomes

$$\lim_{k\to 0}\left(\frac{1}{u}\frac{d^2u}{dr^2}\right) = \frac{2\mu}{\hbar^2} V(r). \tag{1-112}$$

Thus for attractive potentials [i.e., $V(r)$ negative] we have at low energy

$$(1/u)\, d^2u/dr^2 < 0, \tag{1-112a}$$

while for repulsive potentials [i.e., $V(r)$ positive] we have

$$(1/u)\, d^2u/dr^2 > 0. \tag{1-112b}$$

Figure 1.11 summarizes some interesting cases of (1-111) with the conditions (1-112) operating. It is apparent from (1-111) that the scattering length a appears in Fig. 1.11 as the r intercept of the *external* solution [i.e., $r > b$ for the square well, or $r > b'$ for the general case—where b' is roughly defined by $V(r > b') \approx 0$]. These r intercepts have been labeled a_1, a_2, and a_3 for the three cases. It is evident from Fig. 1.11 that beginning with a vanishing small attractive potential, $a \approx 0$, and as the attractive potential becomes stronger,

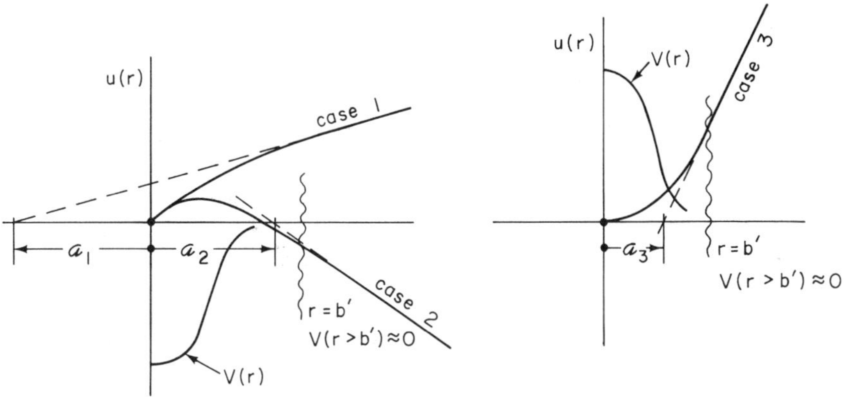

FIG. 1.11. Behavior of the wave function in the low-energy limit with: case 1, a very weak attractive potential possessing no bound-state solution; case 2, a stronger attractive potential possessing one bound state; case 3, a repulsive potential.

a takes on *negative* values, progressively going from $a = 0$ to $a = -\infty$. At the critical potential leading to $a = \pm\infty$, the first "bound" state possibility with zero binding energy appears (alternately, a zero-energy "resonance"). If the potential is made still stronger, a begins to *decrease* from $+\infty$ toward $a = 0$ again, etc. For the case of the repulsive potential, a starts out $a \approx 0$ for a very weak potential and progressively *increases* as the potential gets stronger.

It is also convenient to generalize the Fermi scattering length at $k = 0$ to a scattering length which is a function of k, viz., $a \to a(k)$ such that $a(0) = a(\text{Fermi})$. A most suitable choice for an arbitrary potential is

$$-a(k) = k^{-1}\tan\delta_0(k) \qquad \text{with} \qquad -a(0) = \lim_{k\to 0}[k^{-1}\tan\delta_0(k)]. \tag{1-113}$$

Of course, for the general case without going to the limit $k \to 0$, the phase shift δ_0 has its usual interpretation. One notes that (1-101) becomes a single equation $u = B\sin kr$ when the potential is zero and $B\sin(kr+\delta_0)$ in the distant region when it is $V(r)$. Thus δ_0 represents the *phase shift* of the sinusoidal behavior in the distant region due to the presence of the potential.

The cross section (1-107) can be rewritten in terms of the generalized scattering length, viz.,[1]

$$\sigma_{sc} = \frac{4\pi}{k^2}\sin^2\delta_0 = \frac{4\pi}{k^2}\frac{1}{1+\cot^2\delta_0} = \frac{4\pi}{1/a^2(k)+k^2}. \tag{1-114}$$

[1] With only elastic scattering energetically possible (almost true for the nucleon–nucleon case; see Section VII, F) the phase shift δ_0 and the scattering length $a(k)$ are real. When other reaction channels are open both δ_0 and $a(k)$ are complex and (1-114) becomes $\sigma_{sc} = 4\pi/|1/a(k)+ik|^2$.

3. Effective Range Approximation

It was realized rather soon after accurate cross-section measurements became available that at low energies where $\lambda\!\!\!^{-} \gg$ range of nuclear force, the behavior of scattering was determined largely by two potential parameters, a "strength" and a "range," rather than the detailed geometric shape of the potential under consideration. The earliest detailed theory resulting in the effective range approximation is due to Schwinger (1947) and Bethe (1949). We now give the "standard" derivation, although subtle mathematical problems do exist and the more careful commentary by Bethe (1949) is very instructive.

Again only the S-wave ($l = 0$) interaction is considered for spinless and chargeless particles. The Schrödinger equation is written for two different energies E_a and E_b (with $u = r\psi$) as

$$\frac{d^2 u_a}{dr^2} + \frac{2\mu}{\hbar^2}[E_a - V(r)]\,u_a = 0, \qquad \text{and} \qquad \frac{d^2 u_b}{dr^2} + \frac{2\mu}{\hbar^2}[E_b - V(r)]\,u_b = 0. \tag{1-115}$$

Multiplying the first of these equations by u_b, the second by u_a, subtracting the two resulting equations, and integrating gives

$$(u_b u_a' - u_a u_b')\big|_0^R = (k_b{}^2 - k_a{}^2)\int_0^R u_a u_b\,dr, \tag{1-116}$$

where the primes denote differentiation with respect to r and $\hbar^2 k_{a,b}^2 = 2\mu E_{a,b}$.

Two comparison functions v_a and v_b are introduced, which are solutions to (1-115) with $V(r) = 0$ *and* which asymptotically behave like u_a and u_b. Clearly, from this latter condition,

$$v_a = \mathscr{A}_a \sin(k_a r + \delta_a) \qquad \text{and} \qquad v_b = \mathscr{A}_b \sin(k_b r + \delta_b). \tag{1-117}$$

The subscripts a and b relate to the two energies E_a and E_b. Only $l = 0$ is considered and is to be understood implicitly. A convenient normalization is to have $v_{a,b}(r = 0) = 1$; hence

$$v_a = \frac{\sin(k_a r + \delta_a)}{\sin\delta_a} \qquad \text{and} \qquad v_b = \frac{\sin(k_b r + \delta_b)}{\sin\delta_b}. \tag{1-118}$$

This of course also fixes the normalization of $u_{a,b}$. Figure 1.12 shows the behavior of $u(r)$ and $v(r)$ for two energies, one simply at an energy E small compared to V_0, the other as $E \to 0$. The Fermi scattering length for the potential $a(0)$ is also indicated on the figure.

Proceeding in the same manner that for $u_{a,b}$ resulted in (1-116), we now obtain for $v_{a,b}$

$$(v_b v_a' - v_a v_b')\big|_0^R = (k_b{}^2 - k_a{}^2)\int_0^R v_a v_b\,dr. \tag{1-119}$$

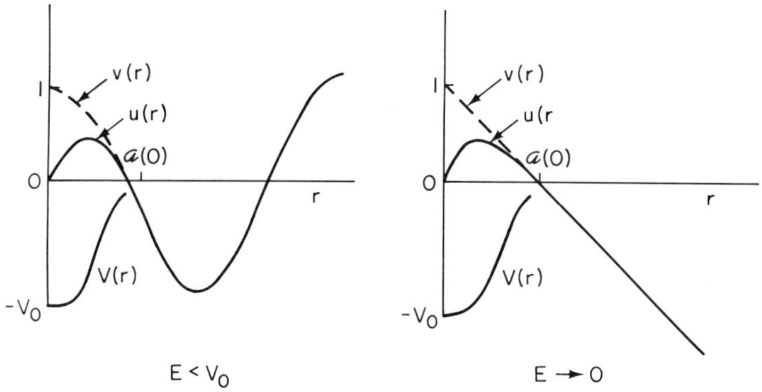

FIG. 1.12. The radial wave function $u(r)$ for the potential $V(r)$ and the comparison function $v(r)$ at two energies, one as $E \to 0$, and the other at E simply less than V_0.

Subtracting (1-116) from (1-119) gives

$$(v_b v_a' - v_a v_b' - u_b u_a' + u_a u_b')|_0^R = (k_b^2 - k_a^2) \int_0^R (v_a v_b - u_a u_b)\, dr. \qquad (1\text{-}120)$$

Allowing $R \to \infty$ in (1-116) simply leads to the well-known orthogonality for the eigenfunctions of two different energies E_a and E_b. The selection of a $v(r)$ satisfying the inhomogeneous boundary condition $v(0) = 1$ does *not* result in the corresponding orthogonality for v_a and v_b. Nonetheless, suitable limits exist for (1-120) when $R \to \infty$ due to the asymptotic equality of $u_{a,b}$ with the corresponding $v_{a,b}$ [see Bethe (1949)]. Thus, noting that $u_a(0) = 0$, $u_b(0) = 0$, $v_a(0) = 1$, $v_b(0) = 1$, $v_a'(0) = k_a \cot\delta_a$, and $v_b'(0) = k_b \cot\delta_b$, we have

$$k_b \cot\delta_b - k_a \cot\delta_a = (k_b^2 - k_a^2) \int_0^\infty (v_a v_b - u_a u_b)\, dr. \qquad (1\text{-}121)$$

Letting $E_a = 0$ ($k_a = 0$) and $E_b = E$ (dropping the subscript, $k_b = k$) and recalling the definition of the generalized scattering length $a(k)$ [from (1-113)], we have

$$-\frac{1}{a(k)} + \frac{1}{a(0)} = k^2 \int_0^\infty (v_0 v - u_0 u)\, dr. \qquad (1\text{-}122)$$

Finally, defining

$$\tfrac{1}{2}\rho(E_a, E_b) = \int_0^\infty (v_a v_b - u_a u_b)\, dr, \qquad (1\text{-}123)$$

we get

$$1/a(k) = 1/a(0) - \tfrac{1}{2}k^2 \rho(0, E) \qquad (1\text{-}124)$$

for the special case $E_a = 0$ and $E_b = E$ in (1-123). It should be noted that (1-122) and (1-124) are *exact* and give an analytic expression for the generalized scattering length $a(k)$ in terms of the Fermi scattering length $a(0)$ and the integral $\rho(0, E)$.

From the definition of $\rho(E_a, E_b)$ in (1-123) it is evident that major contributions arise mainly from the region *inside* the range of $V(r)$ where $u(r)$ and $v(r)$ differ most (see Fig. 1.12). At low energies (i.e., $E \ll V_0$), however, the imposition of the boundary conditions, the dominant influence of $V(r)$, and the near-cancellation effects setting in beyond $r = a(0)$ all conspire to give $u \approx u_0$ and $v \approx v_0$. Thus to a good approximation for $E \ll V_0$,

$$\tfrac{1}{2}\rho(0, E) \approx \tfrac{1}{2}\rho(0,0) = \int_0^\infty (v_0{}^2 - u_0{}^2)\,dr = \tfrac{1}{2}r_0. \qquad (1\text{-}125)$$

This defines the *effective range* r_0 and leads to the approximation

$$1/a(k) \approx 1/a(0) - \tfrac{1}{2}k^2 r_0. \qquad (1\text{-}126)$$

To the extent that (1-126) is a good approximation, it is also *shape independent* in that all potentials $V(r)$ that give the same $\rho(0,0)$ and $a(0)$ would generate the same cross section through the use of (1-126) in (1-114). To higher orders, shape-sensitive parameters P and Q can be introduced [see Blatt and Jackson (1949)], leading to an expression of higher accuracy,

$$1/a(k) = 1/a(0) - \tfrac{1}{2}k^2 r_0 + Pk^4 r_0{}^3 - Qk^6 r_0{}^5 + \cdots. \qquad (1\text{-}127a)$$

Diffuse potentials (including the square-well and Gaussian shapes) give small negative values for P (i.e., the square well gives $P_{sq} \approx -0.033$), while concentrated potentials such as the Yukawa and exponential shapes give small positive values (i.e., the Yukawa potential gives $P_Y \approx +0.065$).

It was soon discovered that P and Q were both required with in fact a strong correlation between their selection demanded by the data. A detailed OPE (one-pion exchange) correction suggests in place of (1-127a) the expression

$$\frac{1}{a(k)} = \frac{1}{a(0)} - \frac{1}{2}k^2 r_0 + \frac{pk^4}{1+qk^2}, \qquad (1\text{-}127b)$$

where the constants p and q are relatively complicated functions of the pion mass and coupling constant. For some additional commentary and interesting extensions of the above points see Cini *et al.* (1959), Sher *et al.* (1970), Noyes and Lipinski (1971), Razavy and Krebes (1972), and the references cited therein.

The treatment leading to (1-126) or (1-127) was for spinless and chargeless particles. It is useful at this point to give an abbreviated discussion of the effects of nucleon spin. The inclusion of nucleon spin is readily accomplished for the $l = 0$, unpolarized spin case. Neutron–proton scattering, for example, can proceed either through the singlet or the triplet spin state. With zero orbital angular momentum $J = S$ and $J_z = S_z$. Since the scattering process must conserve J and J_z, scattering from a particular initial spin substate can occur only to the same final spin substate. Due to the isotropy of space, the three spin substates of the triplet state contribute equally to the scattering. With the possibility of spin-dependent nuclear forces, we distinguish two separate equations of the form (1-126) or (1-127), one for the triplet spin state (with the subscripts t attached) and one for the singlet spin state (with the subscripts s). Thus (1-114) becomes for the case of $l = 0$ neutron–proton elastic scattering

$$\sigma_{sc} = 4\pi\left(\frac{\frac{3}{4}}{[1/a_t(0)-\frac{1}{2}k^2r_{0t}+\cdots]^2+k^2}+\frac{\frac{1}{4}}{[1/a_s(0)-\frac{1}{2}k^2r_{0s}+\cdots]^2+k^2}\right). \tag{1-128}$$

The factors $\frac{3}{4}$ and $\frac{1}{4}$ result from the appropriate averaging over initial spin states and summing over final spin states for the unpolarized case. The lack of interference terms results from the angular momentum considerations discussed above. The case of neutron–neutron or proton–proton scattering would of course only involve the singlet spin state for S-wave elastic scattering.

The proton–proton scattering case must, however, involve the Coulomb effect. The procedure in this case, although more complicated, essentially follows that leading to (1-122) for the chargeless case. The reader is referred to Jackson and Blatt (1950) for the details. The results give the exact expression

$$C_0{}^2k\cot\delta_0+\frac{h(\eta)}{R}=-\frac{1}{a_c(0)}+k^2\int_0^\infty(\phi\phi_0-uu_0)\,dr, \tag{1-129}$$

which is equivalent to (1-122). The various quantities appearing in (1-129) are: $\eta = 1/2kR$ (Coulomb factor), $R = \hbar^2/m_p e^2 = 28.8$ fm (characteristic Coulomb length), $C_0{}^2 = 2\pi\eta/(e^{2\pi\eta}-1)$ (Coulomb penetration factor), $h(\eta) = \eta^2\sum_{n=1}^{\infty}[1/n(n^2+\eta^2)]-\ln\eta-\gamma$, and $\gamma = 0.5772...$ (Euler's constant). The functions ϕ and ϕ_0 are the solutions to the Schrödinger equation with only the Coulomb potential, and in the asymptotic region are taken equal to the physical solutions u and u_0 with $V(r)$ present. The phase factor δ_0 is the nuclear phase shift produced by $V(r)$ with the Coulomb interaction additionally present. The scattering length $a_c(0)$ is the limit of the left-hand side of (1-129) with $k \to 0$,

$$-\frac{1}{a_c(0)}=\lim_{k\to 0}\left(C_0{}^2k\cot\delta_0+\frac{h(\eta)}{R}\right), \tag{1-130}$$

and is the generalization of (1-113). Indeed, with $e = 0$ (i.e., no charge) we have $C_0^2 = 1$, $h(\eta) = 0$, and $1/R = 0$, and hence (1-130) clearly reduces to (1-113); also (1-129) reduces to (1-122), and ϕ becomes v. It has been shown by Jackson and Blatt (1950) that to a good approximation (up to a slight nuclear potential shape effect) the scattering length $a_p(0)$ effective in proton–proton scattering if the Coulomb field could be "switched off" is related to $a_c(0)$ of (1-130) by

$$\frac{1}{a_p(0)} = \frac{1}{a_c(0)} + \frac{1}{R}\ln\left(\frac{R}{r_{0c}}\right) - \frac{1}{3R}. \tag{1-131}$$

Here r_{0c} is the effective range deduced from the integral in (1-129) in the approximation $\phi \approx \phi_0$ and $u \approx u_0$, which in analogous fashion to (1-125) gives

$$\int_0^\infty (\phi_0{}^2 - u_0{}^2)\, dr = \tfrac{1}{2} r_{0c}. \tag{1-132}$$

Two different sets of values for $a(0)$ and r_0 are usually quoted in the literature. Before one can deduce quantities from proton–proton elastic scattering cross-section data to determine the characteristics of the "specifically nuclear forces" operating between the protons, small corrections in addition to the Coulomb force must be taken into account. The most important of these is the effect of vacuum polarization (i.e., the exchange of an electron–positron pair between the interacting protons). Other "purely" electromagnetic effects involve various electromagnetic moment interactions. When *all* these effects arc included in the Hamiltonian and the observed cross section is interpreted as resulting from a combined nuclear and *complete electromagnetic* phase shift, the nuclear portion is labeled δ_{0e}. An effective nuclear phase shift, which can be used in connection with (1-129) and (1-130), can be obtained using $\delta_{0c} = \delta_{0e} + \tau_0 - \Delta_0$, where τ_0 shifts the reference of δ_{0e} from the complete electromagnetic functions to pure Coulomb functions and Δ_0 is a nuclear potential-dependent correction for the non-Coulomb electromagnetic effects. The resulting scattering length expansions are in terms of $a_c(0)$ and r_{0c}. Alternately, (1-129) and (1-130) can be rewritten to include small correction terms so that δ_0 is taken to refer to δ_{0e} directly. The relevant quantities are then labeled $a_e(0)$ and r_{0e}. For a complete discussion of these points refer to, e.g., Heller (1967), Sher *et al.* (1970), Noyes and Lipinski (1971), and Noyes (1972).

The result (1-120) is of course perfectly general and applies to bound-state solutions as well as to scattering states. The triplet spin neutron–proton interaction is strongly enough attractive to possess a single bound state, namely the ground state of the deuteron. If we again take $E_a = 0$ and take $E_b = -E_B$ (E_B the binding energy of the deuteron), suitable comparison functions are

$$v_a = \lim_{k\to 0} \frac{\sin(kr+\delta)}{\sin\delta}, \qquad v_b = e^{-\alpha r}, \tag{1-133}$$

with $-\hbar^2 k_b^{\,2} = \hbar^2\alpha^2 = 2\mu E_B$. The boundary condition $v_{a,b}(0) = 1$ has again been applied. The wave function u_a is the exact solution to the Schrödinger equation at zero energy and u_b is the exact deuteron wave function, each appropriately normalized to match (1-133) in the asymptotic region. Thus (1-120) becomes

$$(v_a v_b' - v_b v_a')_{r=0} = (k_b^{\,2} - k_a^{\,2}) \int_0^\infty (v_a v_b - u_a u_b)\, dr,$$

and when (1-133) is applied

$$-\alpha - \lim_{k\to 0}(k \cot\delta) = -\tfrac{1}{2}\alpha^2 \rho(0, -E_B),$$

or, finally,

$$\alpha = 1/a(0) + \tfrac{1}{2}\alpha^2 \rho(0, -E_B). \tag{1-134}$$

This equation is exact.

Since $E_B/V_0 \ll 1$, the effective range approximation can be employed, namely $\rho(0, -E_B) \approx \rho(0,0) = r_0$. This leads to an interesting relationship among the deuteron binding energy, the triplet scattering length, and the triplet effective range:

$$1/a_t(0) \approx \alpha - \tfrac{1}{2}\alpha^2 r_{0t}. \tag{1-135}$$

D. PARAMETERIZATION OF EXPERIMENTAL DATA

The two-nucleon interaction at low energies is of particular interest in understanding many properties of complex nuclei. It is therefore fortunate that a rather simple parameterization of the two-nucleon experimental data suffices even to energies of 20 MeV in the laboratory system. We thus consider the low-energy data separately from the general interaction data, which will be reviewed up to an energy of some 350 MeV.

1. Low-Energy Data

At low energy, where S-wave scattering dominates, the elastic scattering of the proton–proton system involves only the singlet spin interaction potential.

The expression for the cross section is rather more complicated than (1-107). In terms of the nuclear phase shift δ_0 and the quantities introduced in connection with (1-129), when only the Coulomb potential is used to give the electromagnetic interaction, the cross section becomes[1]

$$\frac{d\sigma}{d\Omega} = \left(\frac{e^2}{2\mu v^2}\right)^2 \left\{\left[\csc^4\frac{\theta}{2} + \sec^4\frac{\theta}{2} - \csc^2\frac{\theta}{2}\sec^2\frac{\theta}{2}\cos\left(2\eta \ln \tan\frac{\theta}{2}\right)\right] \right.$$
$$- \frac{2}{\eta}(\sin\delta_0)\left[\csc^2\frac{\theta}{2}\cos\left(\delta_0 + 2\eta\ln\sin\frac{\theta}{2}\right)\right.$$
$$\left.\left. + \sec^2\frac{\theta}{2}\cos\left(\delta_0 + 2\eta\ln\cos\frac{\theta}{2}\right)\right] + \frac{4}{\eta^2}\sin^2\delta_0\right\}. \qquad \text{(1-136)}$$

Fitting the experimental data taken at different energies with (1-136) determines δ_0 as a function of energy. It is convenient to plot the left-hand side of (1-129), using this experimental value of δ_0, as a function of k^2 or equivalently

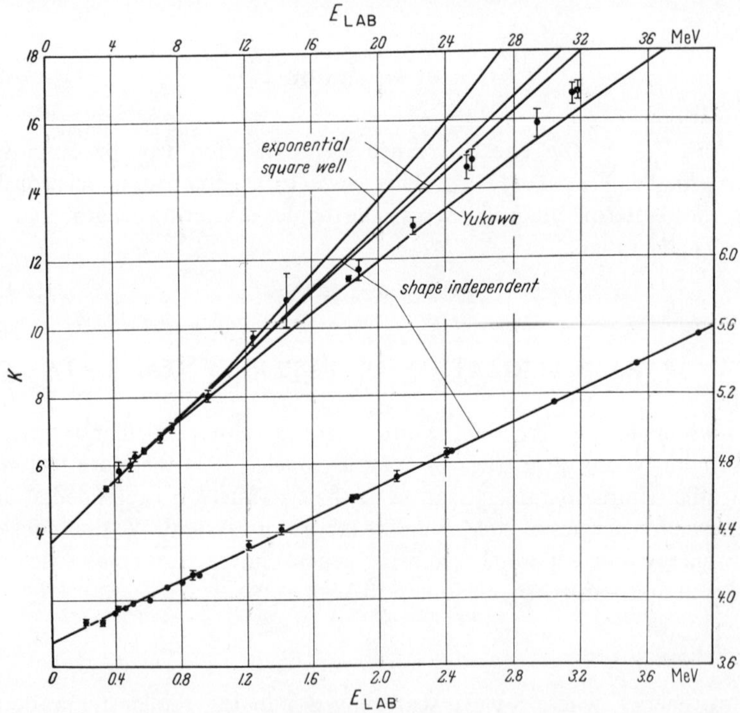

FIG. 1.13. The quantity $K = C_0^2 kR \cot\delta_0 + h(\eta)$ plotted against E_{LAB}, the laboratory energy; $E_{LAB} = \mu v^2 = \hbar^2 k^2/\mu$ [taken from Hulthen and Sugawara (1957)].

[1] The contribution from the first square-bracketed term is referred to as Mott (or pure Coulomb) scattering; the contribution from the very last term gives the pure nuclear scattering; and the second square-bracketed term is the interference term between the two.

as a function of $E_{\text{LAB}} = \hbar^2 k^2/\mu$. A typical result is shown in Fig. 1.13. (The effects of the small corrections we have mentioned would not be perceptible on the scale of this figure.) It is clear from (1-129) and (1-132) that the K intercept (i.e., the value of K for $E_{\text{LAB}} = 0$) determines $-R/a_c(0)$ and the slope at low energies determines r_{0c}. The scattering length thus is seen to be negative.

A recent analysis by Noyes and Lipinski (1971) of data up to a proton laboratory energy of 10 MeV yields $a_c(0) = (-7.823 \pm 0.01)$ fm and $r_{0c} = (2.794 \pm 0.015)$ fm for the shape-independent values. When the meson-exchange-derived shape parameters of (1-127b) are included, these values shift slightly to give $a_c(0) = -7.8275$ and $r_{0c} = 2.7937$ along with the parameters $p = 0.64788$ and $q = 3.4619$; these values of p and q are calculated for a pion coupling constant $G^2/\hbar c = 14.0$. Treating p and q as freely adjustable gives slightly different values from the above and corresponds to a pion coupling constant $G^2/\hbar c = 15.3 \pm 2.4$. In the (1-127a) parametrization, $P \approx 0.051$ and $Q \approx 0.028$. Shape-independent values of $a_e(0)$ and r_{0e} are $a_e(0) = -7.8146$ and $r_{0e} = 2.7950$. See also Sher *et al.* (1970) and Noyes (1972). When (1-131) and small nuclear potential-dependent corrections are applied to the shape-independent value of $a_c(0)$, the value $a_p(0) = -(17.1\text{–}17.3)$ fm results [see, e.g., Sher *et al.* (1970) and Henley and Keliher (1972)].

▶**Exercise 1-19** Use the quoted values for $a_c(0)$ and r_{0c} in conjunction with (1-131) to calculate the corresponding value of $a_p(0)$. ◀

The neutron–proton scattering at low energy involves both the singlet and triplet spin interaction potentials in the form (1-128). Figure 1.14 shows a plot of the observed cross section as a function of the laboratory energy in the low-energy region.

Recent values for the parameters in the shape-independent approximation of (1-128) are $a_t(0) = (5.418 \pm 0.005)$ fm, $r_{0t} = (1.742 \pm 0.006)$ fm, $a_s(0) = (-23.715 \pm 0.015)$ fm, and $r_{0s} = (2.73 \pm 0.05)$ fm [see Noyes (1972)].[1] The considerably larger error in r_{0s} results from the relatively small contribution of the singlet scattering at the moderately large values of k^2 where this parameter is important. This effect is illustrated in Fig. 1.14, where curves for two values, $r_{0s} = 0$ and $r_{0s} = 3$ fm, are exhibited. In approximate agreement with proton–proton scattering, the singlet scattering length is seen to be negative

[1] At this time there appears to be a nontrivial systematic error in the coherent scattering length determined by low-energy neutron scattering from molecular hydrogen. The values quoted here simply average over this inconsistency [see Noyes (1972)]. Should the inconsistency be resolved, resulting in the elimination of one of the troublesome values, *all* the quoted (n, p) parameters would change by an amount considerably greater than the indicated errors.

Present accuracy permits quoting $\rho_t(0,0) = r_{0t}$ and $\rho_t(0, -E_B)$ separately. The value for the latter quantity is $\rho_t(0, -E_B) = (1.755 \pm 0.005)$ fm. Slightly different values are quoted by Lomon and Wilson (1974).

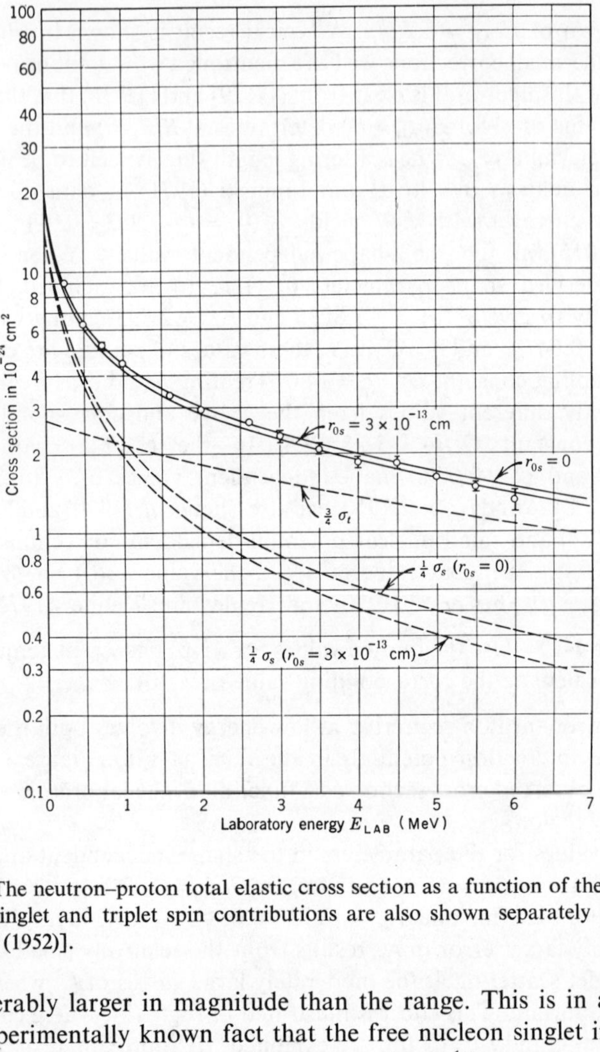

FIG. 1.14. The neutron–proton total elastic cross section as a function of the laboratory energy. The singlet and triplet spin contributions are also shown separately [from Blatt and Weiskopf (1952)].

and considerably larger in magnitude than the range. This is in agreement with the experimentally known fact that the free nucleon singlet interaction is slightly too weak to have a bound state. The virtual $^1S_0$ state is some 90 keV unbound. The triplet scattering length, on the other hand, is positive, although larger in magnitude than the range. This suggests the existence of a weakly bound discrete state. The ground state of the deuteron, which is largely a $^3S_1$ state and has a binding energy of (2.22461 ± 0.00007) MeV is, of course, this bound state.

▶**Exercise 1-20** In actual fact the triplet effective range is most accurately determined by using (1-135) with the binding energy of the deuteron and the triplet scattering length, rather than shape-fitting the data as illustrated in

Fig. 1.14. Using $E_B = (2.22461 \pm 0.00007)$ MeV and the above quoted value for $a_t(0)$, calculate r_{0t} with (1-135) and verify the value stated. ◀

Again complications arise in attempting to include terms to order $Pk^4 r_0{}^3$ [see Noyes (1963)]; however, a value of $P_t \approx +0.025$ for the triplet shape parameter at the lower energies may be assigned. This is again in agreement with the various meson models, and presumably the form (1-127b) applies. Insufficient accuracy exists to determine the corresponding shape parameter for the singlet state [see Serduke and Afnan (1971)]. In determining the shape parameters, it has been assumed that contributions to the cross section from other than S-wave scattering have been corrected for. For guidance we quote the results calculated (Noyes, 1963) for the $l \geqslant 1$ contributions to the (n, p) cross section at data laboratory energy values of 14.1 and 19.7 MeV. At $E_{LAB} = 14.1$ MeV the total (n, p) cross section measured is (689 ± 5) mb,[1] of which 5.8 mb is calculated to correspond to $l \geqslant 1$ contributions, while at $E_{LAB} = 19.7$ MeV the measured cross section is (494.2 ± 2.5) mb, of which by now 8.6 mb corresponds to $l \geqslant 1$ contributions.

Present evidence on the neutron–neutron interaction comes exclusively from experiments involving final-state interactions where three particles appear in the final state, two of them neutrons. Examples are: $D(\pi^-, \gamma)2n$; $T(n, d)2n$; $T(t, \alpha)2n$; $D(d, 2p)2n$; $D(n, p)2n$; and $T(d, {}^3He)2n$.[2] By observing the energies and angular distributions of the emerging particles the nature of the interaction between the two neutrons can be deduced. Complicating the interpretation is the need for the use of rather involved theoretical calculations. Occasionally a complementary reaction involving two protons or a proton and neutron is available for study when additional checks on both the experimental and theoretical uncertainties are possible, using the accurately known (p, p) and (p, n) interactions. Such cases occur with $T(d, {}^3He)2n$ and ${}^3He(d, t)2p$; with ${}^3He({}^3He, \alpha)2p$, ${}^3He(t, \alpha)np$, and $T(t, \alpha)2n$; and with $D(n, p)2n$ and $D(p, n)2p$, for example. For typical papers discussing such intercomparisons see, e.g., Phillips (1964), Gross *et al.* (1970), Noyes (1972), and Davis *et al.* (1973). For example, the last paper determines $a_e(0) = (-7.8 \pm 1)$ fm from the $D(p, n)2p$ reaction by analysis techniques generally used to determine $a_n(0)$. A strong case is made by all of the above-cited papers that current methods for determining $a_n(0)$ are no better than ± 1 fm.

Some typical values deduced from recent measurements are given in Table 1-3. This selection is as much to detail different approaches as it is to give the

[1] The millibarn (mb) is a unit of area equal to 10^{-27} cm^2.

[2] The nuclear reaction notation A(a, b)B is "shorthand" for an initial state A + a going over to the final state B + b, viz., $A + a \rightarrow B + b$. The initial state is always a two-body state. The final state can comprise two or more bodies by having either b or B stand for an unbound composite as in the present cases.

TABLE 1-3

LOW-ENERGY NEUTRON–NEUTRON SCATTERING PARAMETERS

Reaction	$a_n(0)$ (fm)	r_{0n} (fm)[a]	Reference
D(n, p)2n	-16.7 ± 2.8	—	Slobodrian *et al.* (1968)
D(n, p)2n	$\begin{cases} -16.0 \pm 1.2^b \\ -16.8 \pm 1.3^b \end{cases}$	(2.86)	Breunlich *et al.* (1974)
D(n, p)2n	-17.1 ± 0.8	(2.84)	McNaughton *et al.* (Int. Conf., 1972, p. 108)
D(n, p)2n	-17.1 ± 0.8	(2.86)	Zeitnitz *et al.* (Int. Conf., 1972, p. 117; 1972)
D(n, p)2n	-21.7 ± 1.2	(2.60)	Stricker *et al.* (1972)
D(n, p)2n	-25.0 ± 3.0	(2.75)	Saukov *et al.* (1972)
D(n, p)2n	-18.31 ± 0.22	(2.84)	Shirato *et al.* (1973)
$D(\pi^-, \gamma)2n$	-16.4 ± 1.6	—	Salter *et al.* (Int. Conf., 1972, p. 112)
$T(t, \alpha)2n$	-17.0 ± 0.5	2.75 ± 0.35	Gross *et al.* (1970)
$T(t, \alpha)2n$	-15.0 ± 1.0	(2.7)	Kühn *et al.* (1972)
$T(d, {}^3He)2n$	-16.45 ± 1.0	3.1 ± 0.5	Slobodrian *et al.* (Proc. Int. Conf., 1969, paper 5.3)
$T(d, {}^3He)2n$	-16.0 ± 1.0	(2.67)	Grotzschel *et al.* (1971)
D(d, 2p)2n	-15.5 ± 1.1	(2.5)	Assimakopoulas *et al.* (1970)

[a] A value of r_{0n} enclosed in parentheses means that the value was assumed in the analysis and held fixed. The analyses are relatively insensitive to r_{0n}.

[b] The $a_n(0) = -16.0$ value is based on using the Amado model, while the $a_n(0) = -16.8$ value is obtained from the Watson–Migdal approach.

more accurate values. For an earlier review see Slaus (Int. Conf., 1967, p. 575 ff.) and for a recent review see Shirato *et al.* (1973).

An uncritical average taking these somewhat contradictory values more or less at face value suggests $a_n(0) = (-17.3 \pm 0.8)$ fm and $r_{0n} = (2.8 \pm 0.3)$ fm.[1]

The results on the singlet spin two-nucleon interactions at low energy finally can be compared, viz.,

$$\begin{aligned} a_p(0) &= (-17.2 \pm 0.1)\ \text{fm} \quad &&\text{from (p, p)} \\ a_n(0) &= (-17.3 \pm 0.8)\ \text{fm} \quad &&\text{from (n, n)} \\ a_s(0) &= (-23.72 \pm 0.02)\ \text{fm} \quad &&\text{from (n, p).} \end{aligned} \tag{1-137}$$

This contrast would suggest that the nucleon–nucleon interactions are charge

[1] It might be more prudent to follow Noyes (1972) and quote $a_n(0) = (-17 \pm 1)$ fm and the theoretical value $r_{0n} = (2.84 \pm 0.03)$ fm.

symmetric, since $a_p(0) \approx a_n(0)$, but not charge independent since $a_s(0)$ is notably larger in magnitude than either $a_p(0)$ or $a_n(0)$. However, certain corrections should be considered before charge independence in the *fundamental* hadron or strong interaction is abandoned. The reader is referred to Henley and Morrison (1966) and Biswas *et al.* (1966) for a discussion of various theoretically derived differences for the $a_n(0) - a_s(0)$ anomaly for which the experimental value is $\Delta = a_n(0) - a_s(0) \approx (6.4 \pm 1.0)$ fm. The following effects can be distinguished and the approximate magnitude of the theoretical corrections estimated:

(i)	magnetic moment interaction	$\Delta \approx 0.3$ fm
(ii)	electromagnetic mass splitting of $\pi^{\pm} - \pi^0$	$\Delta \approx 2.5$–4.5 fm
(iii)	n–p electromagnetic mass difference	$\Delta \approx 0.2$ fm.[1]

A combined effect of $\Delta_{tot} \approx (3.0$–$5.0)$ fm, in which the pion mass splitting is the most important contribution, largely accounts for the experimentally observed anomaly. Similar meson-theoretic considerations as above predict $a_p(0) \approx a_n(0) + 0.3$ fm. Another meson effect (not included above) involves the charge-symmetry-violating effects that follow from the known mixing of isoscalar and isovector mesons. This phenomenon involves both the vector mesons (ρ, ω, ϕ) and the psuedoscalar mesons (η, π). These effects give $a_p(0) \approx a_n(0) + 0.8$ fm [although the sign of this correction is somewhat uncertain; with the opposite sign the appropriate correction is $a_p(0) \approx a_n(0) - 2.1$ fm].

The possibility thus exists for accounting for the observed values of the singlet scattering lengths (1-137) on the basis of the charge-independent meson theory, albeit requiring some rather sophisticated corrections. These include the "outer" electromagnetic effects such as the Coulomb effect as well as "inner" mesonic corrections. If one may speak of the laboratory nucleons as "clothed," then these latter effects only involve the "wardrobe." For additional comments see Fox and Robson (1969, p. 17 ff.), Int. Conf. (1967, p. 584 ff.), Int. Conf. (1972, p. 191 ff., p. 221 ff.), Henley and Keilehr (1972), and Arnold and Seyler (1973). The possible consequences of charge-independence violation for isobaric analog states in light nuclei is discussed by Lovitch (1965), and we defer a discussion of such points to later.

We have been discussing the largely shape-independent parameterization of the low-energy nucleon–nucleon interaction. Nonetheless it is perhaps instructive to inquire about some typical potential shapes that would yield the observed scattering lengths and effective ranges. The simplest potential shape is the square well. Values that specifically yield the observed (n, p) interaction

[1] For a general discussion of the n–p mass difference see Zee (1972).

[and to rather good approximation the (n, n) and nuclear part of the (p, p) interactions as well] are

$$V_{0t} = 38.5\text{ MeV},\quad b_t = 1.93\text{ fm};\qquad V_{0s} = 15.1\text{ MeV},\quad b_s = 2.50\text{ fm}. \tag{1-138}$$

Other useful potential shapes are the Gaussian, exponential, and Yukawa. A careful examination of equivalences has been performed in the classic paper by Blatt and Jackson (1949). Their results on this point can be summarized by giving the following parameters that yield potentials indistinguishable in the shape-independent approximation[1]:

Square well:

$$V = -V_0,\quad r < b;\qquad V = 0,\quad r > b$$

Gaussian well:

$$V = -V_G \exp(-r^2/b_G{}^2);\qquad V_G = 2.24V_0,\quad b_G = 0.697b$$

(1-139)

Exponential well:

$$V = -V_E \exp(-r/b_E);\qquad V_E = 7.36V_0,\quad b_E = 0.282b$$

Yukawa well:

$$V = -(V_Y b_Y/r)\exp(-r/b_Y);\qquad V_Y = 3.05V_0,\quad b_Y = 0.472b.$$

While considerable variations in the magnitude of the ranges and potential depths are involved in the equivalences of (1-139), the quantity $V_i b_i{}^2$ or *strength* of the potential varies far less from shape to shape. These potentials are illustrated in Fig. 1.15.

▶**Exercise 1-21** Show that the square-well potentials of (1-138) give the observed singlet and triplet scattering lengths. ◀

▶**Exercise 1-22** Calculate $d\alpha(0)/dV_0$ for a square-well potential. Using the values of (1-138) for the singlet case, evaluate this derivative in fm/MeV. What percentage change in the depth of the potential V_0 is represented by the (n, n) and (n, p) scattering length anomaly of $\Delta \approx (6.4 \pm 1.0)$ fm referred to in the text? ◀

[1] These values are for the *well depth parameter* $s = 1$ (Blatt and Jackson, 1949). Actually this is for the case of precisely zero binding. For the triplet interaction $s > 1$, while for the singlet interaction $s < 1$. Thus the values quoted in (1-139) are for rough guidance only; for exact details see Blatt and Jackson (1949).

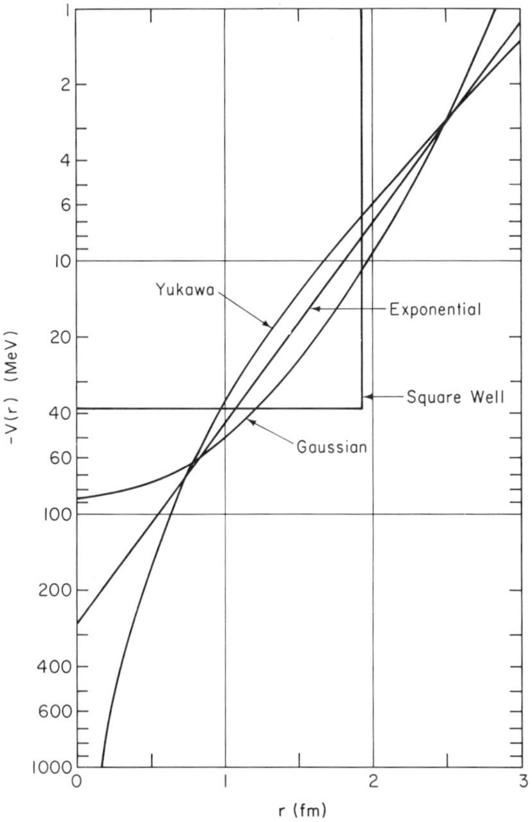

FIG. 1.15. Potentials that are equivalent in the shape-independent approximation in yielding the observed triplet spin interaction for the (n, p) system at low energy [adapted from Preston (1962)].

2. High-Energy Data

Our discussion of nucleon–nucleon scattering will be limited to a laboratory energy below 350 MeV in order to exclude the region where complex inelastic processes can occur.[1] A partial wave analysis at these energies, although involving many terms, is still manageable. A rough estimate for associating the significant onset of a particular partial wave with laboratory energy and the range of the potential is given by the semiclassical estimate for the

[1] Although the meson production threshold is $E_{LAB} \approx 280$ MeV, the production cross section above threshold is at first relatively small and to present accuracy can be ignored to $E_{LAB} \lesssim 350$ MeV.

minimum impact parameter

$$b_l \approx \frac{1}{k}\left(l+\frac{1}{2}\right) = \left(l+\frac{1}{2}\right)\left(\frac{83}{E_{\mathrm{LAB}}}\right)^{1/2} \quad \mathrm{fm}, \tag{1-140}$$

where E_{LAB} is in units of MeV. For example, at $E_{\mathrm{LAB}} = 350$ MeV, (1-140) gives $b_{1,2,3,4,5,\ldots} = 0.7, 1.2, 1.7, 2.2, 2.7, \ldots$ fm. Thus, P-waves would be expected to be sensitive to the potential shape down to dimensions well below 1 fm, while partial waves as high as H-waves ($l = 5$) would be expected to have nonzero phase shifts due to the tail of the potential extending beyond 2.5 fm or so. In the energy region up to 350 MeV, we would therefore expect the interior core region of the nucleon potential to manifest itself in modifying the S and P phase shifts and require the general phase shift analysis to be extended to include H-waves.

A more precise yardstick for considerations of this sort is a relationship for the phase shift δ_l given in the first Born approximation [see (11.85) in Merzbacher (1970)], namely,

$$\tan\delta_l \approx -(2\mu k/\hbar^2)\int_0^\infty [j_l(kr')]^2 V(r')r'^2\,dr'. \tag{1-141}$$

It is then clear from the behavior of $j_l(kr)$ [which for $l \geqslant 1$ has $j_l(0) = 0$ and reaches its first maximum near $kr \approx l+\frac{1}{2}$] that $\delta_l \approx 0$ if the range of $V(r)$ is much less than $(l+\frac{1}{2})/k$, as would be the case at low energies, and that substantial contributions to δ_l occur at elevated energies as k approaches $(l+\frac{1}{2})/b'$ (with b' the range of the potential). These considerations are of course just those associated with (1-140). Further, for S-waves ($l = 0$), $j_0(kr) = [\sin(kr)]/kr$ and thus when $kb' < \pi/4$, as would be the case at low energy, $j_0(kr) \approx 1$ so that the integral in (1-141) largely measures the average value of $V(r)$ weighted by simply r'^2 and is independent of k. It readily follows that δ_0 is proportional to k and insensitive to the detailed geometric nature of $V(r)$. This is seen to be in agreement with our earlier discussions for S-wave scattering and the effective range approximation. If $V(r)$ had an inner structure for $r < r_c$ markedly different from that for $r_c < r < b'$, its effect at very low energy on δ_0 would, of course, appear in the weighted average (1-141) represents, but would become rather dominant for higher energies where $kr_c > \pi/4$. Thus at elevated energies S-wave scattering does begin to sense preferentially the inner portions of $V(r)$, even though at very low energies only the volume integral of $V(r)$ matters.

For $l+\frac{1}{2} > kb'$, using the leading term in the power series expansion of $j_l(kr)$ in (1-141) gives

$$\tan\delta_l \approx -\frac{2^{2l+1}(l!)^2\mu k^{2l+1}}{\hbar^2[(2l+1)!]^2}\int_0^\infty V(r')\,r'^{(2l+2)}\,dr'. \tag{1-142}$$

Thus δ_l increases from the value $\delta_l = 0$ when $k = 0$, proportionate to k^{2l+1} as k increases from zero, but remains small.

In principle the scattering of two spin-$\frac{1}{2}$ particles is completely determined in spin representation by a 4×4 scattering matrix, with each element energy and angle dependent. Although these elements are not independent, a considerable amount of data is required for their empirical determination. Guidance from the general form that the nonrelativistic potential can assume [(1-85a), (1-85b)] leads to specific experimental arrangements. In particular, five suitably chosen measurements at each angle suffice to determine the spin-dependent scattering matrix. Due to the experimental difficulties associated with using polarized targets, the following arrangements or measurements are generally made:

(i) Unpolarized differential cross section, measured in single scattering, $d\sigma(\theta)/d\Omega$;

(ii) Polarization upon single scattering determined by a subsequent second scattering, $P(\theta)$;

(iii) Depolarization produced by a second scatterer in a coplanar triple scattering, $D(\theta)$;

(iv) Rotation of polarization in triple scattering using successive scattering planes at right angles, $R(\theta)$;

(v) Longitudinal polarization in triple scattering where transverse polarization resulting from the first scattering is rotated to longitudinal polarization using a magnetic field prior to second scattering, $A(\theta)$.

A good general discussion of these points as well as the entire two-nucleon problem can be found, for example, in a monograph by Moravcsik (1963). An additional good original reference is Wolfenstein (1956). The excellent review papers presented at the International Conference on the Nucleon–Nucleon Interaction (Int. Conf., 1967) give a good accounting of the intricate experimental and theoretical details involved.

The reduction of all the two-nucleon scattering data to phase shift contributions has been carried out notably by the Yale group [Breit *et al.* (1960, 1962), Hull *et al.* (1961, 1962), and most recently Seamon *et al.* (1968)], and also recently by the Livermore group (MacGregor *et al.*, 1968). Such reductions of the data lead to ambiguities in the form of several sets of phase shifts which must be resolved by theoretical considerations. To date no evidence has been found for deviations from charge independence in scattering data above 20 MeV [e.g., Hulthen and Sugawara (1957)]. Hence phase shifts can be simply categorized by the designations ${}^{2S+1}L_J$; treating the nucleons as identical in isospin formalism automatically determines T, viz., $(-1)^{L+S+T} = -1$, and hence T need not be specified. Figure 1.16 shows the almost identical phase shift results of Seamon *et al.* (1968) and MacGregor *et al.* (1968).

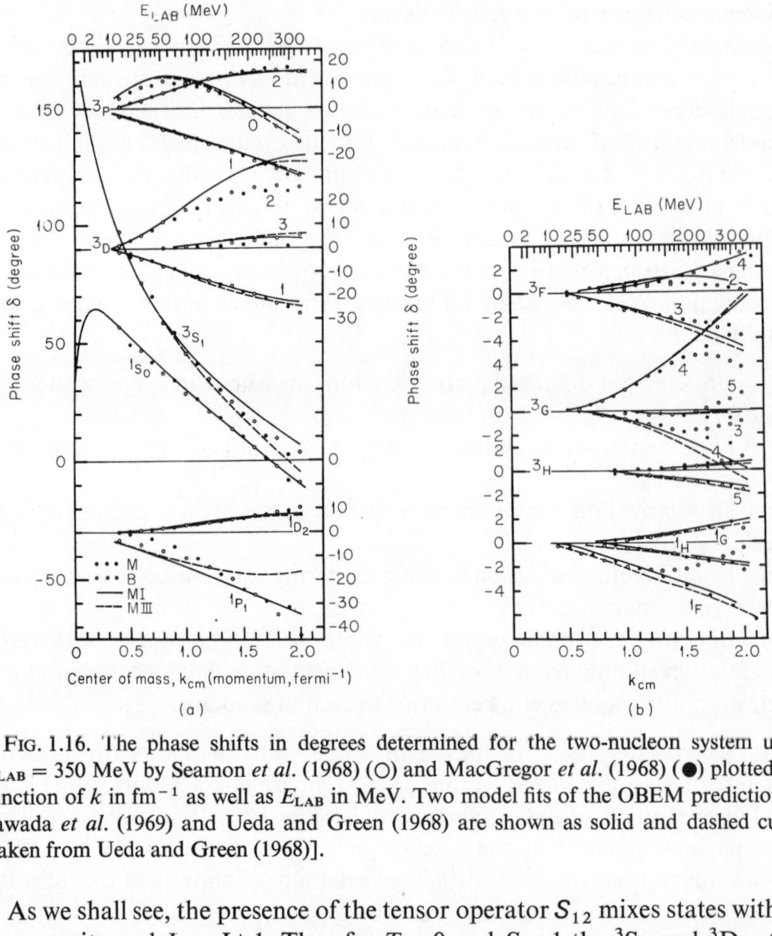

FIG. 1.16. The phase shifts in degrees determined for the two-nucleon system up to $E_{LAB} = 350$ MeV by Seamon *et al.* (1968) (○) and MacGregor *et al.* (1968) (●) plotted as a function of k in fm^{-1} as well as E_{LAB} in MeV. Two model fits of the OBEM predictions of Sawada *et al.* (1969) and Ueda and Green (1968) are shown as solid and dashed curves [taken from Ueda and Green (1968)].

As we shall see, the presence of the tensor operator S_{12} mixes states with the same parity and $L = J \pm 1$. Thus for $T = 0$ and $S = 1$ the $^3S_1$ and $^3D_1$ states are mixed (this fact will also be important in our discussion of the deuteron problem) as are the $^3P_2$ states with $^3F_2$ states, $^3D_3$ with $^3G_3$, etc. These partial waves produce coherent interference effects and hence *mixing parameters* must also be specified in addition to the phase shifts plotted in Fig. 1.16 [see e.g., Hulthén and Sugawara (1957)].

The behavior of the phase shifts in Fig. 1.16 suggests certain features that any potential designed to generate them would have to possess. In this discussion it is useful to review the singlet spin ($S = 0$) configurations first and then the triplet spin ($S = 1$) states, since both the operators S_{12} and $\mathbf{L}\cdot\mathbf{S}$ give zero when applied to a spin singlet state, thereby revealing more clearly the nature of the central potential in this case.

For the singlet spin states, the L-even (^{1}S, ^{1}D, and ^{1}G) phase shifts are positive (at least at first), while for L odd (^{1}P, ^{1}F, and ^{1}H) the phase shifts are negative. This strongly suggests a central potential of the form $V(r) = (\boldsymbol{\tau}_1 \cdot \boldsymbol{\tau}_2)(\boldsymbol{\sigma}_1 \cdot \boldsymbol{\sigma}_2) f(r)$ [precisely the form that (1-45) for the meson-theoretic potential takes for the singlet case] since the alternation of sign is guaranteed by $(\boldsymbol{\tau}_1 \cdot \boldsymbol{\tau}_2)$ as L varies from even to odd, recalling that $(-1)^{L+S+T} = -1$. The effect is to produce attractive potentials for L-even states and repulsive potentials for L-odd states at least for $r \gtrsim 1$ fm, the presumed single-pion tail of the potential. We note, however, that the ^{1}S phase shift decreases with increase in energy above a few MeV and eventually becomes negative beyond ~250 MeV. With central velocity-independent potentials this effect requires the operation of a short-range repulsion strong enough to offset the known strong attraction operating at lower energies. Recalling that potential *strength* for S-waves is more appropriately measured by $V_i b_i^2$, this offsetting core repulsion must be associated with a large positive V_0 leading in a simplified form to an infinite repulsive core (or hard core) with a short range. Since $d\delta(^1S_0)/dk \approx -0.6$ fm around $E_{LAB} \approx 250$ MeV, the core radius must be of the order 0.6 fm. These points were noted by Jastrow as early as 1950 (Jastrow, 1951). Potentials involving velocity-dependent terms in k^2 can also carry $\delta(^1S)$ negative at elevated energies.

▶**Exercise 1-23** The general decrease in the $L = 0$ phase shifts to zero (but not to negative values) is a natural consequence of the nature of even simple central potentials. The square-well potential that just binds at zero energy has a *strength* $V_0 b^2 = (\pi/2)^2 \hbar^2 / 2\mu = 102.3$ MeV·fm^2 for the two-nucleon case. Taking $b = 2.21$ fm [the average for the ^{3}S and ^{1}S potentials of (1-138)] and V_0 to just give binding, use (1-102) to plot δ_0 in degrees as a function of k in fm^{-1} up to $E_{LAB} = 350$ MeV. Take δ_0 at $k = 0$ to be $\pi/2$ or 90°. Compare with Fig. 1.16. ◀

The triplet spin S-wave phase shift $\delta(^3S_1)$ is very large (at least at first) and positive, clearly calling for the strong attractive potential known to be operating in this state from low-energy data. Although the $\mathbf{L} \cdot \mathbf{S}$ term in the potential again leaves $\delta(^3S_1)$ unaffected, the S_{12} term does produce a marked effect by coupling the $^3S_1$ and $^3D_1$ partial waves. While the effect on $\delta(^3S_1)$ is not immediately evident, the S_{12} term is required to be attractive for the triplet spin, L-even scattering states. This is also required by the positive quadrupole moment of the deuteron, as we shall see later. For triplet spin states, strong $\mathbf{L} \cdot \mathbf{S}$ and S_{12} effects (or at least velocity-dependent effects) are evident almost at once in splitting the ^{3}P and ^{3}D states. The average of the ^{3}P phases $\langle \delta(^3P) \rangle = \frac{1}{9}[5\delta(^3P_2) + 3\delta(^3P_1) + \delta(^3P_0)]$ remains small at all energies. Since such an average reveals the approximate behavior of the central interaction, this is seen to be weak. The central ^{3}P interaction required is considerably weaker than the central ^{1}P interaction. The splitting $\delta(^3P_2) > \delta(^3P_0) > \delta(^3P_1)$ does not agree with either a pure **L·S** or a pure S_{12} interaction added to the dominant central term. Thus a combination of all three interaction terms is required in the potential. The behavior of the $\delta(^3P)$ phase shifts is consistent with the **L·S** part of the potential being attractive in the $^3P_2$ state and for the S_{12} portion of the potential being repulsive in the $^3P_1$ state and attractive in the $^3P_0$ state, where the **L·S** term, however, is repulsive.[1] The anomalous behavior of $\delta(^3P_0)$ is then the result of these two oppositely tending noncentral potential terms. When in addition similar considerations are invoked for the phase shifts $\delta(^3D)$, $\delta(^3F)$, and $\delta(^3G)$, the sign of the tensor operator term in the potential is found to be negative for L even and positive for L odd. This is in agreement with (1-45), the alternation in sign again resulting from the $\boldsymbol{\tau}_1 \cdot \boldsymbol{\tau}_2$ factor. Similarly, the sign of the **L·S** term in the potential is found to be negative for L odd and positive for L even. The range of the tensor potential is larger than that of the spin–orbit potential and begins to influence the scattering at energies as low as 10 MeV and also plays an important role in the deuteron problem. There is correspondingly very little evidence for the spin–orbit interaction below 50 MeV or so.

While the meson theory in static approximation does not give rise to an **L·S** interaction, higher-order corrections do predict a generally weak term having a range $\hbar/2m_\pi c$ [see Breit (1958) and Sugawara and Okubo (1966)]. The major contribution is thus just outside the hard-core region, since $\hbar/2m_\pi c \approx 0.7$ fm. An alternate view is therefore possible that dispenses with the spin–orbit term and other velocity-dependent terms and specifies instead the boundary conditions for the wave function at the hard-core radius $r_c \approx 0.5$ fm and uses one- and two-pion exchange potentials for $r > r_c$ [see Feshbach and Lomon (1956) and more recently Int. Conf. (1967, p. 611 ff.)].

In summary, required central potentials for $r \gtrsim 1$ fm are generally in agreement with a simple meson-theoretic form such as (1-45) or (1-100). In the language of (1-97), and normalizing to ^{31}V (taken to be attractive), ^{13}V is somewhat less attractive, i.e., $\alpha \approx +0.6$; also ^{33}V is weak, i.e., $|\beta| \approx 0$, and ^{11}V is strongly repulsive, i.e., $\gamma \lesssim -1$. All require strong repulsive cores with $0.4 < r_c < 0.5$ fm in the velocity-independent approximation, although a strong case is only possible for the $^1S_0$ scattering state.

A tensor interaction of the form (1-45) is a strong requirement even at

[1] The operator **L·S** has for simultaneous eigenstates of $\mathbf{J}^2$, $\mathbf{L}^2$, and $\mathbf{S}^2$ expectation values $2\langle \mathbf{L}\cdot\mathbf{S}\rangle = J(J+1) - L(L+1) - S(S+1)$; hence for the ^{3}P states we have for $^3P_2, +1$; $^3P_1, -1$; and $^3P_0, -2$. Thus a negative sign for the **L·S** potential for case of the ^{3}P states would lead to attraction in the $^3P_2$ state and repulsion in the $^3P_0$ state. The corresponding expectation values for the tensor operator S_{12} are $^3P_2, -2/5$; $^3P_1, +2$; and $^3P_0, -4$. Thus a positive sign for the S_{12} potential for the case of the ^{3}P states would lead to repulsion for the $^3P_1$ state and attraction for the $^3P_0$ state.

relatively low energies. The situation for the spin–orbit interaction and other velocity-dependent terms is less clear, at least as far as direct compelling evidence from the phase shift curves of Fig. 1.16 is concerned.

Detailed efforts to derive potentials using theoretical guidance that would generate the experimental phase shifts have been vigorously pursued for almost and decades. Numerous texts and monographs chronicle these developments two the reader is referred to these for details, e.g., Hulthén and Sugawara (1957), Preston (1962), and Moravcsik (1963). Earlier efforts were hampered by a lack of experimental nucleon–nucleon scattering data as well as a lack of adequate experimental knowledge concerning the mesonic components of the elementary particle mass spectrum. While some encouraging results using meson-theoretic potentials were achieved, e.g., Gartenhaus (1955), generally the tendency was to incorporate more and more terms of the form (1-85a), (1-85b) on an ad hoc basis. Thus phenomenological potentials possessing as many as 30–50 parameters were derived.

Two phenomenological potential models that are rather successful in fitting the phase shift data of Fig. 1.16 as well as the other two nucleon-interaction parameters are the so-called Hamada–Johnston and the Reid potentials.

The Hamada–Johnston (1962) (HJ) potential takes strong guidance from meson-theoretic considerations. It employs a potential of the form

$$V = V_c + V_T S_{12} + V_{LS}(\mathbf{L}\cdot\mathbf{S}) + V_{LL} L_{12}, \tag{1-143}$$

with $L_{12} = [\delta_{LJ}+(\boldsymbol{\sigma}_1\cdot\boldsymbol{\sigma}_2)]\mathbf{L}^2-(\mathbf{L}\cdot\mathbf{S})^2$. The potentials V_i are taken to be

$$\begin{aligned} V_c &= V_0(\boldsymbol{\tau}_1\cdot\boldsymbol{\tau}_2)(\boldsymbol{\sigma}_1\cdot\boldsymbol{\sigma}_2)Y(x)[1+a_c Y(x)+b_c Y^2(x)] \\ V_T &= V_0(\boldsymbol{\tau}_1\cdot\boldsymbol{\tau}_2)Z(x)[1+a_T Y(x)+b_T Y^2(x)] \\ V_{LS} &= g_{LS} V_0 Y^2(x)[1+b_{LS} Y(x)] \\ V_{LL} &= g_{LL} x^{-2} Z(x)[1+a_{LL} Y(x)+b_{LL} Y^2(x)] \end{aligned} \tag{1-144}$$

with

$$V_0 = (f^2/3\hbar c)m_\pi c^2, \qquad Y(x) = e^{-x}/x, \qquad Z(x) = [1+(3/x)+(3/x^2)]Y(x)$$

$$x = (m_\pi c/\hbar)r = r/1.42 \quad (r \text{ in fm}) \quad \text{(all states)}$$

and a hard-core $r_c = 0.485$ fm or $x_c = 0.343$.

This form, using for $Y(x)$ the Yukawa potential function appropriate for a pion, simulates one-, two-, and three-pion exchange effects by using $Y(x)$, $Y^2(x)$, and $Y^3(x)$. The hard core supposedly accounts for the heavier meson contributions (although see later comments) and also requires the use of a renormalized pion coupling constant $f^2/\hbar c = 0.080$. The value of V_0 is taken to be determined a priori to be $V_0 = 3.72$ MeV, consistent with this bias.

Clearly one recognizes in (1-144) the appropriate pseudoscalar potential terms of (1-45). The 28 constants a_i, b_i are empirically determined to fit the data of Fig. 1.16. The results are given in Table 1-4. Using these values, the observed properties of the deuteron are predicted to experimental accuracy, as are the approximate low-energy effective range parameters for (n, p) and (p, p) scattering. The predicted shape parameter for the (n, p) singlet case turns out to be $P = +0.016$, even in the presence of the large core radius $r_c \approx 0.5$ fm.

TABLE 1-4

THE HAMADA–JOHNSTON POTENTIAL COEFFICIENTS WITH A HARD-CORE RADIUS $r_c = 0.485$ fm[a]

	Singlet even	Triplet even	Singlet odd	Triplet odd
a_c	8.7	6.0	−8.0	−9.07
b_c	10.6	−1.0	12.0	3.48
a_T	—	−0.5	—	−1.29
b_T	—	0.2	—	0.55
g_{LS}	—	2.79	—	7.36
b_{LS}	—	−0.1	—	−7.12
g_{LL}	−0.0334	0.101	−0.101	−0.0334
a_{LL}	0.2	1.8	2.0	−7.26
b_{LL}	−0.2	−0.4	6.0	6.92

[a] Hamada and Johnston (1962).

The Reid potentials (1968) are similar to the HJ potentials particularly for $r \gtrsim 1$ fm, but notably they include a variant in which the hard core is replaced by repulsive Yukawa-form terms. For example, the soft-core version of the potential operating in the ^{1}S state includes a repulsive component having a behavior e^{-7x}/x and a rather large relative strength. Soft core implies small but nonvanishing wavefunctions within the core region. A somewhat larger number of parameters is used than for the HJ potential, and no attempt is made to associate x with any particular meson mass. Again there is good success in predicting the deuteron parameters and the effective range parameters. The shape factor for the ^{1}S state is predicted to be $P \approx +0.020$ (the average for the various potentials). The corresponding ^{3}S shape parameter varies from $P = -0.011$ (hard core) to $P = -0.027$ (soft core). Figure 1.17 gives a comparison of the HJ potential and the Reid soft-core and hard-core potentials for the ^{1}S state as an example. The similarity of the potentials for $r \gtrsim 1$ fm is evident.

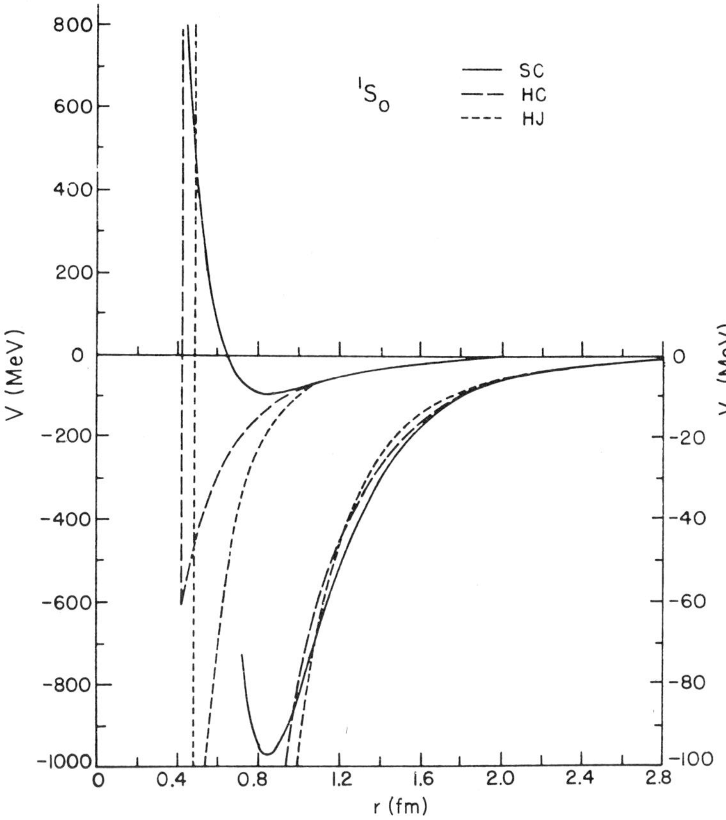

FIG. 1.17. The Reid soft-core (SC) and hard-core (HC) and the Hamada–Johnston (HJ) potentials for the $^1S_0$ state, following Reid (1968).

E. THE BOSON-EXCHANGE MODEL

Great strides have been made recently in fitting the nucleon–nucleon interaction with meson-theoretic models having a rather small number of adjustable parameters. Even many of these free parameters are circumscribed by known mesonic characteristics, and in principle the treatment suggests an eventual zero-free parameter fit. We shall begin by considering the simplest exchange model first, the so-called *OBE model* or *OBEM* (i.e., one-boson-exchange model). As the name suggests, there is no explicit inclusion of multi-pion exchanges in this approach. This model is a generalization of the linear scalar and pseudoscalar meson theories we have discussed earlier, including however, the potential contributions of a number of other mesons as well. The initial successes of the OBE model in the decade following 1960 were largely confined to the two-nucleon center-of-mass energies below 350 MeV. Near the end of this period various Ansatz constructions were introduced to allow for some two-pion exchange effects within the OBE framework and the theory was also extended into the GeV relativistic range. Most recently two-pion exchange effects involving intermediate Δ and N* nuclear "resonances" have been found to be important and we shall have a few words to say about this development as well. Recent progress has also been made in developing a formal relativistic Dirac approach to the many-body nuclear problem employing meson-field-derived interactions.

Most OBE potentials attempt to account for multiple pion exchange contributions by including the heavy mesons corresponding to the principal resonant multipion interactions. Thus, the two-pion exchange effect is primarily ascribed to a one-ρ-meson-exchange term. Indeed the low-energy pion–pion interaction is dominated by the strong resonance associated with the ρ-meson. Other resonances and nonresonant two-pion states, of course, also should be included, being careful, however, to avoid double counting. In the same manner, the dominant three-pion exchange effect is contributed by the inclusion of the one-ω-meson-exchange term. An earlier procedure was to ignore nonresonant multipion states and employ most of the bosons of Fig. 1.8 up to $mc^2 \approx 1.2$ GeV in a OBE schema, thereby hoping to include all the major multipion (and indeed multimeson) exchange effects.[1]

The OBE model then usually proceeds by involving pseudoscalar, vector, and scalar bosons of the mass spectrum of Fig. 1.8 up to and including some of the $^3P_0$-mesons. A cut-off or hard-core radius of 0.1 to 0.2 fm corresponding to the Compton wavelength of a 1.2-GeV-mass particle is required to account for the still heavier axial vector mesons, etc., that have been left out of consideration and to eliminate convergence problems in the S-wave scattering states. Alternately, "well-regulated" potentials can be used. Early investigations in which the experimental two-pion nonresonant isoscalar interactions, observed as a broad continuum in addition to the more pronounced resonances, were simulated by a hypothetical scalar meson ($T = 0$, $J^\pi = 0^+$, $mc^2 \approx 400$ MeV) met with some success, particularly in providing the needed strong

[1] Only the nonstrange members of the nonets of Fig. 1.8 are involved in the first-order or OBE strong interactions between nucleons. The strange meson members of the lowest mass nonet $J^\pi = 0^-$ form the quartet ($K^+, K^0, \bar{K}^0, K^-$): (K^+, K^0) with $Y = +1$, $B = 0$, $S = +1$ (by definition), and ($K^-, \bar{K}^0$) with $Y = -1$, $B = 0$, and $S = -1$. The conservation of strangeness in strong interactions does not allow the virtual transition of a nucleon to a state with one nucleon and one kaon. *Associated production* leading to states with two kaons could occur, e.g., $p \to p + K^+ + K^-$ or $p \to n + K^+ + \bar{K}^0$. [In hypernuclei or hyperfragments, nuclei in which a nucleon is replaced by a hyperon such as $^4_\Lambda$He in which a neutron is replaced by the Λ with $E_B \approx 2.4$ MeV, one-kaon exchange could play a role through intermediate virtual states such as $p \to K^+ + \Lambda$ (Jackson, 1958, p. 79 ff.)]. The kaons decay with relatively long lifetimes through strangeness-violating weak interactions, e.g., $K^+ \to \pi^+ + \pi^0$ with a *partial* mean life of 5.9×10^{-8} sec.

attraction in the S states at intermediate distances. Some of the criticism of such an approach was answered by allowing for a continuous distribution for the coupling constant of this hypothetical broad mass meson state appropriate for a meson spanning the range $2m_\pi c^2 = 280$ MeV $< mc^2 \lesssim 1$ GeV.

Before proceeding further, we now give a brief introduction to the generalized OBE model following the excellent treatment given by Green *et al.* (1968, Section 7.4). For many interesting details that space limitations bar us from considering, the reader may profitably consult this basic reference. For the most general form of the OBE model for the two-nucleon system, one begins with the appropriate time-independent relativistic Dirac equation

$$[c\boldsymbol{\alpha}_1 \cdot \mathbf{p}_1 + c\boldsymbol{\alpha}_2 \cdot \mathbf{p}_2 + (\beta_1 + \beta_2) m_N c^2 + V^D]\psi = E\psi \qquad (1\text{-}145)$$

with $E = W + 2m_N c^2$ and $m_1 = m_2 = m_N$. Here ψ is a direct-product wave function of the two single-nucleon Dirac spinors. When the Dirac spinors are partitioned into "large" and "small" Pauli spinor components, (1-145) can be written as a set of four coupled Pauli equations. If V^D, the Dirac OBE potential, is taken to be a sum of pseudoscalar, vector, and scalar terms, and if the reduction to the classical limit is performed, (1-145) becomes the equivalent Schrödinger–Pauli equation in the reduced mass $m_N/2$ having a potential V^P given by

$$V^P = V_c + V_\sigma(\boldsymbol{\sigma}_1 \cdot \boldsymbol{\sigma}_2) + V_{LS}(\mathbf{L} \cdot \mathbf{S}) + V_T S_{12} + V_a \nabla^2 + V_b(\mathbf{r} \cdot \nabla). \qquad (1\text{-}146)$$

Table 1-5 catalogs the contributions to (1-146) by various components of V^D.

TABLE 1-5

CONTRIBUTIONS TO THE SCHRÖDINGER POTENTIALS BY THE RELATIVISTIC PSEUDOSCALAR *P*, SCALAR *S*, AND POLAR VECTOR *V* INTERACTIONS[a]

	P	S	V_{VV}	V_{VT}
V_c	—	$-J - \frac{1}{4}\lambda^2\mu^2\tilde{J}$	J	$(F\lambda^2/2G)\mu^2\tilde{J}$
V_σ	$\frac{1}{12}\lambda^2\mu^2\tilde{J}$	—	$\frac{1}{6}\lambda^2\mu^2\tilde{J}$	$[2(F/G)+(F^2/G^2)](\lambda^2/6)\mu^2\tilde{J}$
V_{LS}	—	$\frac{1}{2}\lambda^2 J_1$	$\frac{3}{2}\lambda^2 J_1$	$(2F\lambda^2/G)J_1$
V_T	$\frac{1}{12}\lambda^2 r^2 J_2$	—	$-\frac{1}{12}\lambda^2 r^2 J_2$	$-[2(F/G)+(F^2/G^2)](\lambda^2/12)r^2 J_2$
V_a	—	$-\lambda^2 J$	$-\lambda^2 J$	—
V_b	—	$-\lambda^2 J_1$	$-\lambda^2 J_1$	—

[a] The vector interaction includes V–V coupling as well as V–T or derivative coupling, the relative strength of the latter to the former being F/G. The characteristic length λ is the Compton wavelength of the nucleon $\lambda = \hbar/m_N c$. As before $\mu = mc/\hbar$, the inverse Compton wavelength of the appropriate meson. $J = J(r)$ is the spatial behavior of the OBE potential V^D, which for the ordinary Yukawa interaction is $J(r) = (G^2/r)e^{-\mu r}$. The quantity $\mu^2\tilde{J}$ stands for the average value $\langle \nabla^2 J \rangle$; for the simple Yukawa form $\tilde{J} \equiv J$. J_1 and J_2 are defined as $J_1 = (1/r)(d/dr)J$ and $J_2 = (1/r)(d/dr)J_1$. The values listed in the table are for isoscalar fields; for isovector fields all terms are multiplied by $\boldsymbol{\tau}_1 \cdot \boldsymbol{\tau}_2$. Small additional complicated tensor terms of order $2(\lambda^2/r^2)J$ and $\lambda^2 J_1$ are also present.

The pseudoscalar component of the V^D interaction contributes in the classical limit only to V_σ and V_T (refer to Table 1-5), with the coupling constant

$$\left(\frac{m_\pi}{2m_N}\right)^2 \frac{G^2}{\hbar c} = \frac{f^2}{\hbar c},$$

in agreement with our previous discussion of (1-46)–(1-52). The expressions (1-46)–(1-52) were for the pion field, and the isovector character of this field requires that the additional multiplicative factor $\boldsymbol{\tau}_1 \cdot \boldsymbol{\tau}_2$ be included. The vector and scalar terms in the V^D potential both generate contributions to V^P that include the spin–orbit action V_{LS}. An approach emphasizing the lowest mass meson components would add only the effects of the vector bosons $J^\pi = 1^-$ (ρ, ω, ϕ) necessary to generate V_{LS} to the effects of the above pseudoscalar bosons $J^\pi = 0^-$ (π, η, η'). However, the vector interaction also gives rise to a V_c term with the coupling constant $G^2/\hbar c$ [without the factor $(m/2m_N)^2$; see Table 1-5]. When G is adjusted to give the proper spin–orbit interaction an unrealistically large repulsive central static term results. Thus in the OBE model one must also include the scalar bosons $J^\pi = 0^+$ (at least some of the $^3P_0$ population) since in this case a similar situation pertains, but with the central term having the opposite sign; see Table 1-5. A careful balancing of the vector and scalar contributions is then possible leading to a manageable V_c term. The vector and scalar terms of V^D also generate the terms V_a and V_b, or velocity-dependent terms, at the same time! In general for each of the interactions in V^D both derivative and nonderivative terms are possible (DeBenedetti, 1964, Section 7.12). While in the classical limit for the pseudoscalar interaction equivalence holds for the two forms, the vector interaction differs in basic form for the two cases and both are required to give the correct ρ-meson contribution to the magnetic form factor for the nucleons; see Scotti and Wong (1965) and Moravcsik (1963, Chapter 6). Thus derivative coupling is usually used for the ρ-meson but not for any of the others. For a detailed discussion of the above general points refer to Int. Conf. (1967, p. 594 ff.), Green *et al.* (1968), Ueda and Green (1968), and Ingber (1968).

In the foregoing it is important to note that terms in V^D that generate the spin–orbit interaction of necessity introduce the velocity-dependent Schrödinger potentials V_a and V_b. A demonstration of this behavior is present even in a simple attempt to derive the Thomas spin–orbit interaction in the one-body Dirac equation. If the free-particle Dirac equation is modified to include two simple potential terms V_1 and V_2, we have

$$[(c\boldsymbol{\alpha}\cdot\mathbf{p} + V_1) + \beta(mc^2 + V_2)]\psi = E\psi. \tag{1-147}$$

If "large" and "small" components ψ_l and ψ_s are introduced, we get the coupled equations

$$c(\boldsymbol{\sigma}^{\mathrm{P}}\cdot\mathbf{p})\psi_s + (mc^2 + V_1 + V_2)\psi_l = E\psi_l$$
$$c(\boldsymbol{\sigma}^{\mathrm{P}}\cdot\mathbf{p})\psi_l - (mc^2 - V_1 + V_2)\psi_s = E\psi_s. \tag{1-148}$$

In the nonrelativistic limit, we have for the large-component Pauli spinor ψ_l

$$\left\{\frac{p^2}{2m} + V_1 + V_2 + \frac{\hbar}{4m^2c^2}\boldsymbol{\sigma}^{\mathrm{P}}\cdot[\nabla(V_1 - V_2)\times\mathbf{p}] - \frac{i\hbar}{4m^2c^2}\nabla(V_1 - V_2)\cdot\mathbf{p}\right\}\psi_l$$
$$= (E - mc^2)\psi_l. \tag{1-149}$$

If V_1 and V_2 are spherically symmetric, $\nabla(V_1 - V_2) = \hat{r}\, d(V_1 - V_2)/dr$ and the first relativistic correction term, the Thomas term, is just equal to

$$\frac{\hbar^2}{2m^2c^2}\frac{1}{r}\frac{d(V_1 - V_2)}{dr}\mathbf{L}\cdot\mathbf{S}.$$

However, inescapably there is also the second, or Darwin, nonlocal term as well!

▶**Exercise 1-24** Verify that $(\boldsymbol{\sigma}^{\mathrm{P}}\cdot\mathbf{A})(\boldsymbol{\sigma}^{\mathrm{P}}\cdot\mathbf{B}) = \mathbf{A}\cdot\mathbf{B} + i\boldsymbol{\sigma}^{\mathrm{P}}\cdot(\mathbf{A}\times\mathbf{B})$ and using this identity, derive (1-149). ◀

Most recent OBE models try to eliminate the singularities that are inherent in the point-source, static nuclear source approximation.[1] Green [Green *et al.* (1968, Section 7.4b) and (Int. Conf., 1967, p. 594 ff.)] uses a generalized quadratic Lagrangian density which generates the "well-regulated" Yukawa potential,

$$V(r) = -\theta_1\theta_2 J = -\theta_1\theta_2\frac{G^2}{r}\left(e^{-\mu r} - \frac{U^2 - \mu^2}{U^2 - \Lambda^2}e^{-\Lambda r} + \frac{\Lambda^2 - \mu^2}{U^2 - \Lambda^2}e^{-Ur}\right), \tag{1-150}$$

where $\theta_1\theta_2$ are appropriate Dirac matrices depending on the character of the interaction, μ the inverse Compton wavelength of the corresponding meson, and U and Λ inner and outer characteristic lengths.[2] Clearly, (1-150) is nonsingular at $r = 0$. When $U \to \infty$, we have a weakly regulated function $\sim(1/r)(e^{-\mu r} - e^{-\Lambda r})$, which further reduces to the ordinary Yukawa function

[1] Recall the discussion concerning (1-45).

[2] The quantity $\mu^2\tilde{J} \equiv \langle\nabla^2 J\rangle$ appearing in Table 1-5 for the well-regulated potential function (1-150) becomes

$$\langle\nabla^2 J\rangle \equiv \mu^2\tilde{J} = \mu^2\left(e^{-\mu r} - \frac{U^2 - \mu^2}{U^2 - \Lambda^2}\left(\frac{\Lambda}{\mu}\right)^2 e^{-\Lambda r} + \frac{\Lambda^2 - \mu^2}{U^2 - \Lambda^2}\left(\frac{U}{\mu}\right)^2 e^{-Ur}\right).$$

when $\Lambda \to \infty$. Earlier results by Green and co-workers used $U \approx 20 m_N c/\hbar$ and $\Lambda = m_\Lambda c^2/\hbar c$ with $m_\Lambda c^2 \approx$ 1–2.5 GeV. It was shown [see Ueda and Green (1968)] that if the pointlike interaction for the nucleon–meson vertex was replaced by a more reasonable form factor $F(k^2) = \Lambda^2/(\Lambda^2 + \kappa^2)$ corresponding to finite size and recoil effects, the potential would assume the form

$$V(r) = -\theta_1 \theta_2 J = -\theta_1 \theta_2 \frac{G^2}{r}\left(\frac{\Lambda^2}{\Lambda^2 - \mu^2}\right)^2 \left[e^{-\mu r} - e^{-\Lambda r}\left(1 + \frac{\Lambda^2 - \mu^2}{2\Lambda} r\right)\right]. \tag{1-151}$$

This is just (1-150) when $U \to \Lambda$. Thus "regulated" Yukawa functions of the type (1-150) or (1-151) correspond to distributed or nonlocal potentials which are not only more physical than pointlike results, but also afford very well-behaved functions at $r \to 0$. Since empirical adjustment of Λ to fit observed phase shifts involves a corresponding mass of $\sim(10\text{–}20) m_\pi$, both (1-150) and (1-151) are essentially the ordinary Yukawa function for $r \gtrsim 0.5$ fm when the regularized functions pertain to the pion. Regularization also corresponds to smearing out of the nucleon in the spirit of the cut-off concept introduced earlier, albeit with a rather soft core. Thus, at least for the pion, one would expect G_π^2 to correspond to the renormalized coupling constant determined from the pion–nucleon scattering experiments, viz.,[1]

$$\frac{G_\pi^2}{\hbar c} = \left(\frac{2m_N}{m_\pi}\right)^2 \frac{f^2}{\hbar c} = \left(\frac{2 \times 939}{139}\right)^2 \times 0.081 = 14.8.$$

The regularized functions are also important in recovering the effective range approximation in the presence of the important velocity-dependent terms in the OBE potential [see Sawada *et al.* (1969)].

The procedure for fitting the phase shifts of Fig. 1.16 is to select appropriate bosons, determine their contributions to (1-146) using Table 1-5, and using either regularized Yukawa functions or the hard-core cut-off concept, solve for the phase shifts generated by this potential in the Schrödinger–Pauli equation in terms of the constants μ_i, G_i, and Λ of (1-151) [see Green *et al.* (1968, Section 7.4c)]. These constants are then either assigned when known from prior conditions or varied to obtain a best fit. Set III of Ueda and Green (1968) is a typical good fit, using regularized functions (refer to Fig. 1.16a, b). This fit also gives rather acceptable values for the various two-nucleon low-energy parameters involved in the deuteron and the scattering states.

The resulting parameters for Set III are given in Table 1-6.

[1] In both (1-150) and (1-151) the usual nucleon–meson coupling constant G' now appears as $(G')^2 = G^2[\Lambda^2/(\Lambda^2 - \mu^2)]^2$; for the pion $G_\pi' \approx G_\pi$; the effect, however, is nonnegligible for the heavier mesons.

TABLE 1-6

OBEM Coupling Constants of Ueda and Green, Set III[a]

Meson	mc^2 (MeV)	$\mu^{-1} = \hbar/mc$ (fm)	J^{π}, T	$G^2/\hbar c$	$m_{\Lambda}c^2 = \hbar c\Lambda$ (MeV)
π	138.7	1.42	$0^-, 1$	14.61	1299
ρ	763.0	0.26	$1^-, 1$	0.65[b]	1299
ω	782.8	0.25	$1^-, 0$	9.68	1299
σ_1	763.0	0.26	$0^+, 1$	1.01	1299
σ_0	782.8	0.25	$0^+, 0$	7.32	1299
σ_c[c]	416.1	0.47	$0^+, 0$	1.52	1299

[a] Ueda and Green (1968).

[b] $G_{\rho}^2/\hbar c = 0.65$ corresponds to the vector–vector coupling; the ratio of vector–tensor (derivative term) to the vector–vector term (direct term) is $F/G = 5.06$.

[c] This hypothetical meson is to allow for the nonresonant two-pion contributions cited earlier.

The Set I fit of these authors includes the η-meson. The coupling constant for the η-meson used was $G_{\eta}^2/\hbar c = 2.73$. The two scalar resonance masses were also taken differently than for Set III. The coupling constants and Λ that resulted for the various components were somewhat different. The fit (refer to Fig. 1.16) is about as good for Set I as Set III, indicating no particular need for the η-meson or for having exact masses for the scalar mesons, at least for energies below 350 MeV.

The value $G_{\pi}^2/\hbar c = 14.6$ is in good agreement with the values determined for nucleon–pion interactions. The coupling constants for the ρ-meson are also in good agreement with the electromagnetic form factor information (e.g., Meson Spectroscopy, 1968, p. 47 ff.). We also note near-cancellation of the V_c term by the approximate balancing of the (ω, σ_0) and (ρ, σ_1) coupling constants. While the assumed σ_0-meson of Set III agrees reasonably well with the ε-meson mass of Fig. 1.8, the σ_1-meson mass of 763 MeV is not in good agreement with 962 MeV for the known δ-meson having the same $(0^+, T = 1)$ properties. The similar results of the Set I fit, where the $(0^+, T = 1)$-meson was taken to have a mass of 1070 MeV, indicates that this problem for Set III is not important. The regulator mass $m_{\Lambda}c^2 \approx 1.3$ GeV is somewhat larger than that required by the pion production experiments in the GeV energy region. In a real sense, the Set III parameterization is accomplished with five free parameters, a significant philosophical improvement over the usual phenomenological fits.

Another noteworthy OBE model fit of the same vintage is that by Bryan and Scott (1969), who obtain rather similar results to Ueda and Green, discussed above, with a somewhat smaller pion coupling constant, i.e.,

$G_\pi/\hbar c = 12.6$. Additional earlier references of possible historical interest are Moravcsik (1963), Int. Conf. (1967, Session D), Bryon and Scott (1967), Sawada *et al.* (1969), and Proc. Int. Conf. (1969, Session 5).

A more recent fit of Stagat *et al.* (1971) uses, in addition to the π-, η-, ω-, ρ-, and δ-mesons of Fig. 1.8, the $(0^+, T=0)$- or $\sigma_0(\varepsilon)$-meson with a very broad mass spectrum from the threshold mass $2m_\pi$ up to and including the resonance near $m_\sigma c^2 \approx 750$ MeV. This treatment of the isoscalar–scalar $(0^+, T=0)$ contribution combines the effects of $\sigma_0(\varepsilon)$ and σ_c of the Set III parameters quoted. While the resulting parameters for most mesons are similar to those of Sets I and III and Table 1-7, the rather larger value of $G^2_{\sigma0(\varepsilon)}/\hbar c$ is required.

An interesting fit of good accuracy, involving the OBE model with π-, ρ-, ω-, and η-mesons and a two-pion exchange term, is successfully demonstrated by Lomon and Feshbach (Int. Conf., 1967, p. 611 ff.), who use a boundary condition at $r_0 = 0.51$ fm to account for inner contributions. The principle difference between this fit and Sets I and III above is definitely to require the η-meson with $G_\eta{}^2/\hbar c \approx 1.0$ and a considerably smaller $G_\omega{}^2/\hbar c \approx 1.3$ to control the strong central repulsion of the ω-meson present even for $r \gtrsim 0.5$ fm. The two-pion exchange term varies roughly proportionate to $e^{-2\mu r}$, $\mu = m_\pi c/\hbar$.

In Table 1-7 we report the range of coupling constants appearing in the literature for recent attempts to fit the nucleon–nucleon phase shifts. An

TABLE 1-7

PHENOMENOLOGICAL BOSON COUPLING CONSTANTS

	Meson		$G^2/\hbar c$		F/G	
	J^π, T	mc^2 (MeV)	Range	Typical	Range	Typical
π	$0^-,1$	138	11.7–14.9	14.6		
η	$0^-,0$	549	0–12	~2.5		
η'	$0^-,0$	958	(not used)	—		
ρ	$1^-,1$	765	0.3–3.0	0.75	+(2.5–5.5)	+4.0
ω	$1^-,0$	784	5–20	10.5	−(0.0–0.2)	−0.1
ϕ	$1^-,0$	1020	0–3	small	~ −0.1	—
$\sigma, (\delta)$	$0^+,1$	700–1100	0–6	~2		
σ_0[a]	$0^+,0$	400–1100	3–15	8.1 } (ε) : 14.0		
σ_c[b]	$0^+,0$	~500	1.5–2.0	1.8 }		

[a] Most OBEM fits require a 0^+, $T=0$ meson. We employ the notation σ_0 for this meson regardless of the mass assumed in the particular instance, leaving open its exact identification of ε, S*, or other.

[b] Hypothetical scalar meson to account for the two-pion nonresonant continuum, mostly 0^+, $T=0$ with perhaps some (~20%) 0^+, $T=1$.

interesting review giving more details as well as another successful OBEM fit is contained in a paper by Erkelenz *et al.* (1969). A good recent review relating the meson coupling constants required by the nucleon–nucleon scattering data to the known elementary particle status of the relevant mesons is by Nagels *et al.* (1973). A principal conclusion of this latter survey concerns the importance of the ε-meson in the nucleon–nucleon potential, giving a large coupling constant $G_\varepsilon^2/\hbar c = 16.2$, with a mass $m_\varepsilon c^2 = 670$ MeV and a decay width $\Gamma \approx 500$ MeV. The ε–S^* mixing angle was deduced to be small, $\theta_\varepsilon \approx 8.6°$, indicating that the ε-meson is dominantly a unitary singlet. The η'-meson (referred to as the X^0 in this paper) apparently enters with $G_{\eta'}^2/\hbar c = 7.4$. For an earlier critique of coupling constants see Int. Conf. (1967, p. 640 ff.). For an extensive review of elementary particle coupling constants see Ebel *et al.* (1971). Returning to Table 1-7, which summarizes some recently used values appearing in the literature, a word of caution is in order. The constants are determined by optimizing the statistical fit, which couples the adjustable constants particularly with respect to the meson types included (and the masses used) and the method employed to account for the nonresonant two-pion contributions. There is wide variation in the treatment of these points. The column labeled "typical" is an attempt to quote a more or less self-consistent set of values rather than a simple average.

Values similar to those in Table 1-7 fit proton–antiproton data rather well (Int. Conf., 1967, p. 681 ff.).

Significant participation of all the nonstrange mesons of mass ~3/4 GeV or lower (i.e., π, η, ρ, ω, ε, and the 2π continuum simulator σ_c) is clearly indicated in Table 1-7. The experimental information on the required mesons having mass ~1 GeV or higher, relating as it does to the behavior of the potential at internucleon distances of ~0.2 fm or less, is less certain. Of the next three heavier mesons, η', δ, and φ, only the δ seems unambiguously required. Theoretical estimates have been made (e.g., Baltay and Rosenfield 1968, pp. 47–94), leading to the expectation that $G_\rho : G_\omega : G_\phi = 1:9\text{–}13.8:4.5\text{–}6.8$, which is seen to be consistent with the entries in Table 1-7, although the value for $G_\phi/\hbar c$ is rather uncertain, if indeed required at all.

Most of the OBE results preceding 1970 were for the "static" case (although Bryan and Scott retained terms of order p^2/m_N^2). Gersten *et al.* (1971) improved the usual OBE fit to the data below 400 MeV using a relativistic treatment employing π-, η-, ε-, δ-, ω-, and ρ-mesons. The χ^2 fit was only slightly improved; however, the coupling constants were altered generally in a direction to better agree with pure elementary particle values. For example, the pion coupling constant was raised from a nonrelativistic value of 13.0 in this study to a relativistic value of 14.2. The effective mass for the troublesome ε-meson was found to be ~570 MeV. The use of purely local potentials is found to be unreliable in general.

Another study relegated to below 400 MeV, including, however, "minimal" relativity, is by Holinde *et al.* (1972). By working in momentum representation they are able easily to include retardation effects and off-energy-shell effects. These effects are found to be very important and would appear to indicate an advantage to working in momentum space. They include the π-, η-, ρ-, ω-, δ-, $\sigma_0(\varepsilon)$-, and ϕ-mesons in their calculations, and use meson form factors equivalent to the regularization we have discussed. Best fit is obtained with $m_\varepsilon c^2 \approx 500$ MeV in a single-pole (resonance) approximation. The other meson properties more or less agree with other studies.

A recent attempt to extend the OBEM to energies up to 3 GeV (Ueda, 1971) using dispersion theory including absorption effects has met with some success. A form factor regulator of the type (1-151) with $m_\Lambda c^2 \approx 1$–3 GeV (depending on the particular meson) was employed. The π-, ρ-, ω-mesons, the hypothetical σ_c isoscalar meson, and two heavy mesons, the η_V-meson ($0^+, T=0, mc^2 = 1070$ MeV) and the f-meson ($2^+, T=0, mc^2 = 1250$ MeV), were used. Coupling constants for the π-, ω-, ρ-, and σ_c-mesons close to those of Table 1-7 were found satisfactory, viz., $G_\pi{}^2/\hbar c = 15.1$; $G_\omega{}^2/\hbar c = 9.0$, $F/G = 0$; $G_\rho{}^2/\hbar c = 0.844$, $F/G = 5.85$; $G_{\sigma_c}^2/\hbar c = 1.50$. The heavier mesons required coupling constants $G_{\eta_V}^2/\hbar c = 14.4$ and $G_f{}^2/\hbar c = 10.0$. In this fit the large central repulsion of the ω-meson is in part offset by the f-meson, although still allowing enough repulsion at higher energies to change the sign of $\delta(^1S_0)$.

An alternate recent approach to extending the OBEM into the GeV range makes use of a complex potential [see Ueda *et al.* (1973)]. This approach employs the π-, ρ-, ω-, η-, and δ-mesons with characteristics fixed at the known elementary particle values, although the ρ mass is shifted to an effective value $[m_\rho c^2]_{\text{eff}} = 840$ MeV to simulate its broad width ($\Gamma_\rho = 150$ MeV). The troublesome isoscalar meson situation is treated by a two-pole approximation with a low-mass σ_c-meson ($m_{\sigma_c} c^2 \approx 350$ MeV) and a higher-mass meson tentatively identified as the S* ($m_{S*} c^2 \approx 790$ MeV). They do not appear to require the f-meson. Above 450 MeV a phenomenological imaginary potential is used involving the exchange of both $T=0$ and $T=1$ scalar mesons. Two mass parameterizations for these scalar contributions are required: the low-mass components provide the necessary attractive long-range potential and the heavier-mass components give the necessary short-range potential. The isoscalar contributions predominate. With these assumptions the elastic scattering phase shifts, inelastic cross sections, and the deuteron properties are fairly well accounted for.

It should be apparent that while the role of the π-, ρ-, η-, ω-, and possibly the δ-mesons in the nucleon–nucleon interaction is satisfactorily determined, a major problem exists for both the resonant and nonresonant scalar meson (or two-pion S-wave interaction) components. At a minimum the scalar–isoscalar interaction in the OBEM requires a broad mass spectrum for a

σ_c-meson (central mass $\sim$450 MeV) and a higher mass pole at $mc^2 \approx 1$ GeV. The known ε- and S*-mesons appear only partially to resolve the problem. We conclude the discussion of OBEM by returning to the survey of Nagels *et al.* and showing a table of their values (Table 1-8) for the nucleon–meson coupling constants that represents the best *realistic* single-meson-exchange contributions.

TABLE 1-8

OBEM Coupling Constants of Nagels, Rijken, and de Swart[a]

Designation	J^π, T	Mass (MeV)	$G^2/\hbar c$	F/G
π	$0^-,1$	138	13.53	
η	$0^-,0$	549	2.25	
X^0 or η'	$0^-,0$	958	7.37	
ρ	$1^-,1$	765 ($\Gamma = 125$)	0.588	5.45
ϕ	$1^-,0$	1020	0.554	0.647
ω	$1^-,0$	784	11.25	0.637[b]
δ } $^3P_0$	$0^+,1$	960	1.25	
S* } $^3P_0$	$0^+,0$	980	0.052	
ε } $^3P_0$	$0^+,0$	670 ($\Gamma = 500$)	16.23	

[a] Nagels *et al.* (1973).
[b] Generally $(F/G)_\omega$ is taken closer to zero or slightly negative.

The resolution of the scalar meson problem of the OBEM is probably to be sought in the more physical two-pion-exchange model (TPEM). For recent papers on TPEM refer to Partovi and Lomon (1970), Riska and Brown (1970), Green and Haapakoski (1974), and the numerous references cited therein. The points are made that in the long-range portion of the nucleon–nucleon potential (i.e., $r \gtrsim 1.5$ fm) the OBEM including its σ-mesons (as per Table 1-8, for example) can be reliably viewed as the dominant description; that in the intermediate range (i.e., $0.7 \lesssim r \lesssim 1.5$ fm, where the major σ-meson difficulties arise) the $^3P_0$-mesons [σ_c, δ, and $\sigma_0(\varepsilon, S^*)$] should be replaced by a complete TPEM calculation; and that in the core region (i.e., $r \lesssim 0.7$ fm) the potential is severely nonlocal, rendering a potential description essentially impractical. In the intermediate range or "dynamical" range the highly correlated composite meson states such as the ρ, ω, and ϕ are included in the spirit of the OBEM. The new feature then is to investigate a more satisfactory treatment of the $^3P_0$-mesons in a TPEM calculation. Some of the interaction diagrams involved in TPEM are shown in Fig. 1.18. The contributions shown in Fig. 1.18a, b represent the standard pion-exchange processes, while the contributions shown in Fig. 1.18c, d represent cases in which one or both intermediate states involve an excited configuration of the nucleons such as

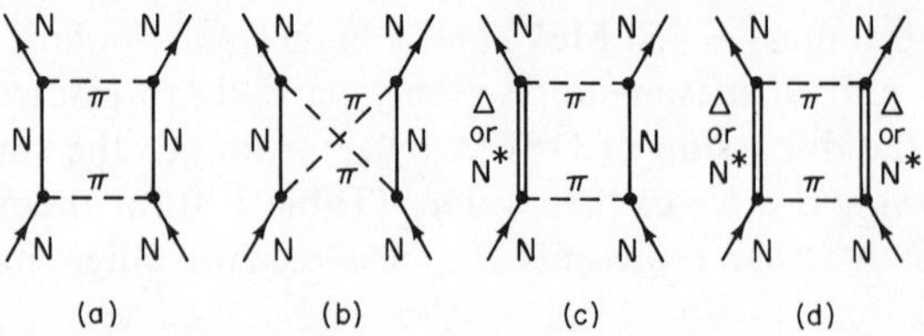

FIG. 1.18. Feynman diagrams involving two pions in the nucleon–nucleon interaction. (a, b) Standard exchange processes. (c, d) Possible resonant or excited states of the nucleon such as the Δ or N*.

the $\Delta(\frac{3}{2},\frac{3}{2})$ resonance (shown in Fig. 1.5).[1] It is successfully demonstrated that in the intermediate range the $^3P_0$-mesons can be dispensed with and that the resulting TPEM agrees rather accurately with the Hamada–Johnston and Reid potential parameterizations. The $\Delta(\frac{3}{2},\frac{3}{2})$ intermediate states play an important role in this success. A single boson of mass $700 \lesssim mc^2 \lesssim 900$ MeV simulates the TPEM behavior rather successfully down to distances $r \approx 1.0$ fm in the $^1S_0$ N–N state.

On the other hand, in a substantially zero-parameter fit employing a simple two-pion-exchange term as well as the presence of an ε-meson (derived mass $m_\varepsilon c^2 = 715$ MeV, $\Gamma = 317$ MeV) Binstock and Bryan (1971) are successfully able to fit the N–N scattering data up to 425 MeV lab energy. They estimate the ε-meson coupling constant $G_\varepsilon{}^2/\hbar c = 6 \pm 1$. It would thus appear that the $^3P_0$-meson contributions must be handled "artfully," carefully including all effects without double counting. Whether in such considerations the ε-meson (in view of its large decay width) should be treated as a discrete state is an open question [see, e.g., Casher and Susskind (1973) and the interesting discussion by Brown (1970)].

Finally, there has also been gratifying progress in applying a relativistic treatment to the many-body problem based on meson-theoretic interaction terms by both the Brueckner approach and the shell model Hartree–Fock treatment (see Section III, D, 2 for references).

F. THE DUAL MODEL

A recent theoretical development of the *dual model* prescribes certain reactions or decays to be allowed or forbidden by constructing some simple quark trajectory diagrams [Harari (1969), Rosner (1969), Mandula *et al.* (1970) and *Proc. CERN School Phys.* (1971, p. 285 ff.)]. The systematics of

[1] Graphs such as Fig. 1.18d are also possible in which the π pair is replaced by a kaon pair, one Δ (or N*) is replaced by a Λ (or Σ) and the other Δ (or N*) is replaced by a Z-particle. Such contributions would of course apply with a shorter range since the kaons are some 3.5 times as massive as the pions. The vertex coupling constants are also smaller, e.g., the NΛK vertex has the coupling constant $G^2_{N\Lambda K}/\hbar c = 6 \pm 2$ [see Ebel *et al.* (1971)].

elementary particle interactions that these diagrams represent, when properly interpreted, will greatly simplify the understanding of the relatively complex nature of the nucleon–nucleon interaction.

Specifically, *the duality diagrams* employ some simple rules:

(a) Directed lines are drawn for the three types of quarks q_n, q_p, and q_λ;
(b) These lines do not change their identity;
(c) Every free baryon has three qqq lines running in the same direction corresponding to the baryon trajectory;
(d) Every free meson has two quark lines, one in the direction of the trajectory (the quark) and one in running in the opposite direction (the (antiquark);
(e) Disconnected graphs in which quark lines of one particle are not exchanged with another are excluded (i.e., forbidden transitions);
(f) Connected graphs are *planar* if the quark lines do not cross and *non-planar* or *illegal* if they do cross (some planar graphs may also be illegal);
(g) *Legal* graphs involve only $q\bar{q}$ and qqq particles, and illegal graphs involve other combinations called *exotic* channels.

We give some illustrations of such graphs in Fig. 1.19. It can be seen from Fig. 1.19a, b that if the f′-meson is considered to be a $q_\lambda \bar{q}_\lambda$ state, its decay into two pions is forbidden, while the decay into two kaons is allowed. Figure 1.19c depicts the allowed pion–nucleon field coupling, while Fig. 1.19d shows that the ϕ-meson (if $q_\lambda \bar{q}_\lambda$) will not couple to the nucleon field. Figure 1.19e shows a legal planar graph; note that section AA' counts only qqq states while BB' intersects only $q\bar{q}$ states. Figure 1.18f, which is the equivalent of the Feynman graph in Fig. 1.18g for charge exchange, is nonplanar and includes the presence of an exotic state. *All* baryon–baryon diagrams are nonplanar and the planar baryon–antibaryon diagrams are also illegal. Thus all baryon–baryon (or antibaryon) interactions involve exotic states, the complete significance of which is not clear at this time. The exotic resonances implied have the effect of rapidly decreasing both the elastic and inelastic nucleon–nucleon cross sections with increasing energy. Forward to backward peaking in cross sections is also influenced by the presence of only exotic states. A feature of legal planar diagrams is the existence of the physical particlelike manifestation of states intersected by cuts such as AA' and BB' in Fig. 1.19e. To date very few exotic states appearing in some important nonlegal diagrams are experimentally known.

The extent to which the ϕ-meson can be characterized as $q_\lambda \bar{q}_\lambda$ leads to $G_\phi{}^2/\hbar c = 0$ in Tables 1-7 and 1-8. This also implies vanishing ϕ-meson contributions to the nucleon form factors of (1-66) and subsequent equations. Since the ϕ-meson is not a pure $q_\lambda \bar{q}_\lambda$ configuration, small contributions are not ruled out in either consideration.

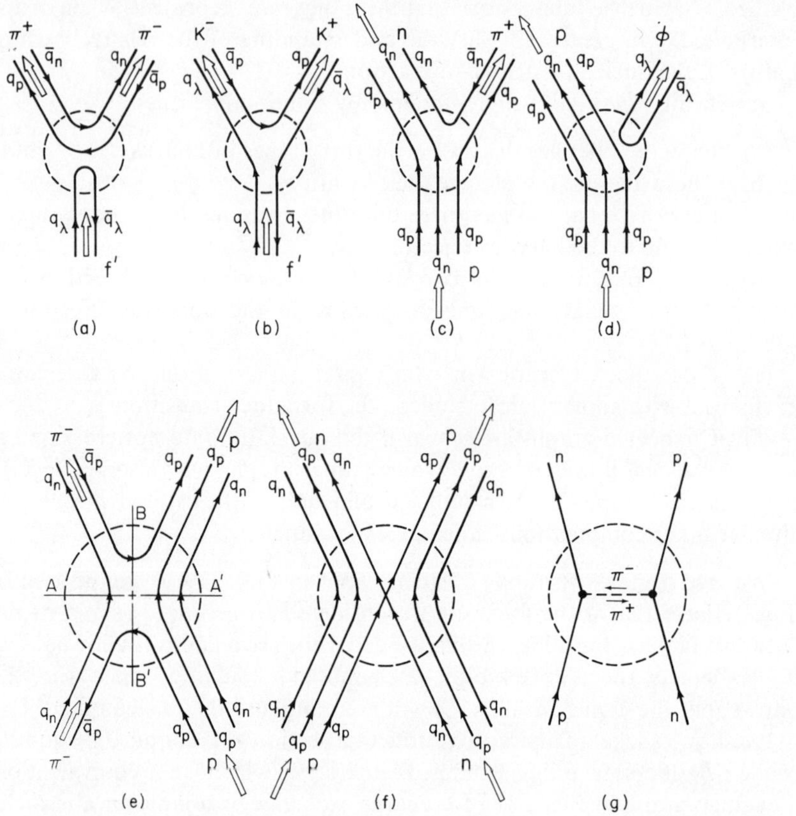

FIG. 1.19. Quark duality diagrams: (a) an uncoupled or forbidden diagram for f′-decay into two pions, (b) a legal planar graph for the decay $f' \to K^+ + K^-$, (c) a legal pion–nucleon vertex interaction, (d) an uncoupled or forbidden vertex interaction of the nucleon and ϕ-meson, (e) a legal planar diagram for the nucleon–pion elastic scattering, (f) a nonplanar baryon–baryon interaction, (g) the Feynman graph equivalent to (f).

REFERENCES

Adelberger, E. G., Swanson, H. E., Tape, J. W., Cooper, M. D., and Trainor, T. A. (1974), *Bull. A. P. S.* **19**, 478.

Alberi, J. L., Wilson, R., and Schröder, I. G. (1972), *Phys. Rev. Lett.* **29**, 518.

Arnold, L. G., and Seyler, R. G. (1973), *Phys. Rev.* **C7**, 574.

Assimakopoulos, P., Beardsworth, E., Boyd, D., and Donovan, P. F. (1970). *Bull. A. P. S.* **15**, 22.

Baltay, C., and Rosenfield, A., eds. (1968). "Meson Spectroscopy." Benjamin, New York.

Bars, I., Halpern, M. B., and Yoshimura, M. (1972). *Phys. Rev. Lett.* **29**, 969.

Bartel, W., *et al.* (1972). *Phys. Lett.* **39B**, 407.

Bergia, S., Stranghellini, A., Fubini, S., and Villi, C. (1961). *Phys. Rev. Lett.* **6**, 367.

Bertsch, G. F., and Mekjian, A. (1972). *Ann. Rev. Nuclear Sci.* **22,** 25.
Bethe, H. A. (1949). *Phys. Rev.* **76,** 38.
Binstock, J., and Bryan, R. (1971). *Phys. Rev.* **D4,** 1341.
Biswas, S. N., Kumar, A., and Saxena, R. P. (1966). *Phys. Rev.* **142,** 1141.
Blatt, J. M., and Jackson, J. D. (1949). *Phys. Rev.* **76,** 18.
Blatt, J. M., and Weiskopf, V. F. (1952). "Theoretical Nuclear Physics." Wiley, New York.
Bloom, E. D., *et al.* (1973). *Phys. Rev. Lett.* **30,** 1186.
Breit, G. (1958). *Phys. Rev.* **111,** 652.
Breit, G., Hull, M. H., Lassila, K. E., and Pyatt, K. D. (1960). *Phys. Rev.* **120,** 2227.
Breit, G., Hull, M. H., Lassila, K. E., Pyatt, K. D., and Ruppel, H. M. (1962). *Phys. Rev.* **128,** 826.
Breunlich, W. H., Tagesen, S., Bertl, W., and Chalupka, A. (1974). *Nuclear Phys.* **A221,** 269.
Brink, D. M. (1965). "Nuclear Forces." Pergamon, Oxford.
Brown, G. E. (1970). *Comments Mod. Phys.* **A4,** 140.
Bryan, R. A., and Scott, B. L. (1967). *Phys. Rev.* **164,** 1215.
Bryan, R. A., and Scott, B. L. (1969). *Phys. Rev.* **177,** 1435.
Capps, R. H. (1972). *Phys. Rev. Lett.* **29,** 820.
Carlson, C. E., and Freund, P. G. O. (1972). *Phys. Lett.* **39B,** 349.
Casher, A., and Susskind, L. (1973). *Phys. Lett.* **44B,** 171.
Casher, A., Kogut, J., and Susskind, L. (1973). *Phys. Rev. Lett.* **31,** 792.
Chan, L. H., Chen, K. W., Dunning, J. R., Ramsey, N. F., Walker, J. K., and Wilson, R. (1966). *Phys. Rev.* **141,** 1298.
Cini, M., Fubini, S., and Stranghellini, A. (1959). *Phys. Rev. Lett.* **3,** 191.
Clark, A. F., Ernst, R. D., Finn, H. F., Griffin, C. G., Hansen, N. E., Smith, D. E., and Powell, W. M. (1971). *Phys. Rev. Lett.* **27,** 51.
Davis, J. C., Anderson, J. D., Grimes, S. M., and Wong, C. (1973). *Phys. Rev.* **C8,** 863.
DeBenedetti, S. (1964). "Nuclear Interactions." Wiley, New York.
DeFranceschi, G., Reale, A., and Salvini, G. (1971). *Ann. Rev. Nuclear Sci.* **21,** 1.
DeRújula, A. (1974). *Phys. Rev. Lett.* **32,** 1143.
DeRújula, A., Georgi, H., Glashow, S. L., and Quinn, H. R. (1974). *Rev. Mod. Phys.* **46,** 391.
Drell, S. D., and Zachariasen, F. (1961). "Electromagnetic Structure of Nucleons." Oxford Univ. Press, London and New York.
Ebel, G., *et al.* (1971). *Nuclear Phys.* **B33,** 317.
Eisenbud, L., and Wigner, E. (1941). *Proc. Nat. Acad. Sci. U. S.* **27,** 281.
Eman, B., and Tadić, D. (1971). *Phys. Rev.* **C4,** 661.
Erkelenz, K., Holinde, K., and Bleuler, K. (1969). *Nucl. Phys.* **A139,** 308.
Fermi, E. (1951). "Elementary Particles." Yale Univ. Press, New Haven, Connecticut.
Feld, B. T. (1969). "Models of Elementary Particles." Ginn (Blaisdell), Boston, Massachusets.
Feshbach, H., and Lomon, E., (1956). *Phys. Rev.* **102,** 891.
Feynman, R. P., and Speisman, G. (1954). *Phys. Rev.* **94,** 500.
Fischbach, E., and Tadić, D. (1973). *Phys. Rep.* **6C,** 123.
Fox, J. D., and Robson, D., eds. (1969). "Isobaric-Spin in Nuclear Physics." Academic Press, New York.
Fox. J. D., *et al.* (1972). *Phys. Rev. Lett.* **29,** 193.
Frazer, W. R. (1966). "Elementary Particles," Prentice-Hall, Englewood Cliffs, New Jersey.
Frerejacque, D., Benaksas, D., and Drickey, D. (1966). *Phys. Rev.* **141,** 1308.
Friedman, J. I., and Kendall, H. W. (1972). *Ann. Rev. Nuclear Sci.* **22,** 203.
Gari, M. (1973). *Phys. Rep.* **6C,** 317.

Gartenhaus, S. (1955). *Phys. Rev.* **100,** 900.
Gasiorowicz, S. (1966). "Elementary Particle Physics." Wiley, New York.
Gell-Mann, M. and Ne'eman, Y. (1964). "The Eightfold Way." Benjamin, New York.
Gersten, A., Thompson, R. H., and Green, A. E. S. (1971). *Phys. Rev.* **D3,** 2069, 2076.
Green, A. E. S., Sawada, T., and Saxon, D. S. (1968). "The Nuclear Independent Particle Model." Academic Press, New York.
Green, A. M., and Haapakoski, P. (1974). *Nuclear Phys.* **A221,** 429.
Greensberg, O. W., and Yodh, G. B. (1974). *Phys. Rev. Lett.* **32,** 1473.
Grötzschel, R., Kühn, B., Möller, K., Mösner, J., and Schmidt, G. (1971). *Nucl. Phys.* **A176,** 261.
Gross, D. J., and Treiman, S. B. (1974). *Phys. Rev. Lett.* **32,** 1145.
Gross, E. E., Hungerford, E. V., Malanify, J. J., and Woods, R. (1970). *Phys. Rev.* **C1,** 1365.
Hättig, H., Hünchen, K., and Wäffler, H. (1970). *Phys. Rev. Lett.* **25,** 941.
Hamada, T., and Johnston, I. D. (1962). *Nuclear Phys.* **34,** 382.
Han, S. J. (1973). *Phys. Lett.* **47B,** 169.
Hand, L. N., Miller, D. G., and Wilson, R. (1963). *Rev. Modern Phys.* **35,** 335.
Harari, H. (1969). *Phys. Rev. Lett.* **22,** 562.
Hazen, W. E. (1971). *Phys. Rev. Lett.* **26,** 582.
Heisenberg, W. (1932). *Z. Phys.* **77,** 1.
Heller, L. (1967). *Rev. Modern Phys.* **30,** 584.
Henley, E. M. (1973). *Phys. Rev.* **C7,** 1344.
Henley, E. M., and Keliher, T. E. (1972). *Nuclear Phys.* **A189,** 632.
Henley, E. M., and Morrison, L. K. (1966). *Phys. Rev.* **141,** 1489.
Hofstadter, R. (1963). "Nuclear and Nucleon Structure." Benjamin, New York.
Holinde, K., Erkelenz, K., and Alzetta, R. (1972). *Nuclear Phys.* **A194,** 161.
Holladay, W. G., and Sachs, R. G. (1954). *Phys. Rev.* **96,** 810.
Hull, M. H., Lassila, K. E., Ruppel, H. M., McDonald, F. A., and Breit, G. (1961). *Phys. Rev.* **122,** 1606.
Hull, M. H., Lassila, K. E., Ruppel, H. M., McDonald, F. A., and Breit, G. (1962). *Phys. Rev.* **128,** 830.
Hulthén, L., and Sugawara, M. (1957). "Handbuch der Physik," Vol. 39. Springer, Berlin.
Ingber, L. (1968). *Phys. Rev.* **174,** 1250.
Int. Conf. Nucleon-Nucleon Interactions (1967). *Rev. Mod. Phys.* **39,** 495–717.
Int. Conf. Few Particle Probl. (1972). Univ. of California. North-Holland Publ., Amsterdam.
Italian Phys. Soc. (1969). *Proc. Nuclear Structure Nuclear Reactions, Varenna.* Academic Press, New York.
Jackson, J. D. (1958). "The Physics of Elementary Particles." Princeton Univ. Press, Princeton, New Jersey.
Jackson, J. D., and Blatt, J. M. (1950). *Rev. Modern Phys.* **22,** 77.
Jastrow, R. (1951). *Phys. Rev.* **81,** 165.
Jones, C. M., Ford, J. L. C., Obenshain, F. E., and Reeves, M. (1970). *Phys. Rev.* **C2,** 2113.
Kirk, P. N., *et al.* (1973). *Phys. Rev.* **D8,** 63.
Kisslinger, L. S., and Tabakin, F. (1974). *Phys. Rev.* **C9,** 188.
Kokkedee, J. J. J. (1969). "The Quark Model." Benjamin, New York.
Krane, K. S., Olsen, C. E., Sites, J. R., and Steyert, W. A. (1971). *Phys. Rev.* **C4,** 1906, 1942.
Krane, K. S., Olsen, C. E., and Steyert, W. A. (1972). *Nuclear Phys.* **A197,** 352.
Kühn, B., Kumpf, H., Parzhitsky, S., and Tesch, S. (1972). *Nuclear Phys.* **A183,** 640.
Leipuner, L. B., Larsen, R. C., Sessoms, A. L., Smith, L. W., Williams, H. H., Kellogg, R., Kasha, H., and Adair, R. K. (1973). *Phys. Rev. Lett.* **31,** 1226.
Lipkin, H. J. (1973). *Phys. Rep.* **C8,** 175.

Littauer, R. M., Schopper, H. F., and Wilson, R. R. (1961). *Phys. Rev. Lett.* **7,** 141, 144.
Lobashov, V. M., Nazarenko, V. A., Saenko, L. F., Smotritskii, L. M., Kharkevich, G. I., and Knyaz'kov, V. A. (1971). *Soviet J. Nuclear Phys.* **13,** 313.
Lobashov, V. M., Kaminker, D. M., Kharkevich, G. I., Kniazkov, V. A., Lozovoy, N. A., Nazarenko, V. A., Sayenko, L. F., Smotritsky, L. M., and Yegorov, A. I. (1972). *Nuclear Phys.* **A197,** 241.
Lomon, E., and Wilson, R. (1974). *Phys. Rev.* **C9,** 1329.
Lovitch, L. (1965). *Nuclear Phys.* **62,** 653.
MacGregor, M. H., Arndt, R. A., and Wright, R. M. (1968). *Phys. Rev.* **173,** 1272.
Mandula, J., Weyers, J., and Zweig, G. (1970). *Ann. Rev. Nuclear Sci.* **20,** 289.
Mattuck, R. D. (1967). "A Guide to Feynman Diagrams in the Many-Body Problem." McGraw-Hill, New York.
Merzbacher, E. (1970). "Quantum Mechanics." Wiley, New York.
Messiah, A. (1966). "Quantum Mechanics," Vol. 1 and II. Wiley, New York.
Moorhouse, R. G. (1969). *Ann. Rev. Nuclear Sci.* **19,** 301.
Moravcsik, M. J. (1963). "The Two-Nucleon Interaction." Oxford Univ. Press, London and New York.
Morpurgo, G. (1970). *Ann. Rev. Nuclear Sci.* **20,** 105.
Nagels, M. M., Rijken, T. A., and deSwart, J. J. (1973). *Phys. Rev. Lett.* **31,** 569.
Neubeck, K., Schober, H., and Wäffler, H. (1974). *Phys. Rev.* **C10,** 320.
Nishijima, K. (1964). "Fundamental Particles." Benjamin, New York.
Noyes, H. P. (1963). *Phys. Rev.* **130,** 2025.
Noyes, H. P. (1972). *Ann. Rev. Nuclear Sci.* **22,** 465.
Noyes, H. P., and Lipinski, H. M. (1971). *Phys. Rev.* **C4,** 995.
Particle Data Group (1973). *Rev. Modern Phys.* **45,** S1.
Partovi, M. H., and Lomon, E. L. (1970). *Phys. Rev.* **D2,** 1999.
Pati, J. C., and Salam, A. (1974). *Phys. Rev.* **D 10,** 275.
Perkins, D. H. (1972). "Introduction to High Energy Physics." Addison-Wesley, Reading, Massachusetts.
Phillips, R. J. N. (1964). *Nuclear Phys.* **53,** 650.
Preston, M. A. (1962). "Physics of the Nucleus." Addison-Wesley, Reading, Massachusetts.
Proc. CERN School Phys. (1971). Finland Lectures, Geneva.
Proc. Int. Conf. Properties Nuclear States (1969). Univ. Montreal Press, Montreal.
Proc. Int. Symp. Electron Photon Interactions High Energy (1967). Stanford U. S. AEC.
Rahm, D. C., and Louttit, R. I. (1970). *Phys. Rev. Lett.* **24,** 279.
Razavy, M., and Krebes, E. S. (1972). *Nuclear Phys.* **A184,** 533.
Reid, R. V. (1968). *Ann. Phys.* **50,** 411.
Riska, D. O., and Brown, G. E. (1970). *Nuclear Phys.* **A153,** 8.
Rosner, J. L. (1969). *Phys. Rev. Lett.* **22,** 689.
Sachs. R. G. (1952). *Phys. Rev.* **87,** 1100.
Sachs, R. G. (1954). *Phys. Rev.* **95,** 1065.
Sakata, S. (1956). *Progr. Theoret. Phys.* **16,** 686.
Samios, N. P., Goldberg, M., and Meadows, B. T. (1974). *Rev. Modern Phys.* **46,** 49.
Saukov, A. I., Chernukhin, Y. I., Oparin, E. M., and Gornitsyn, G. A. (1972). *Soviet J. Nuclear Phys.* **14,** 157.
Sawada, T., Dainis, A., and Green, A. E. S. (1969). *Phys. Rev.* **177,** 1541.
Schwarz, J. H. (1971). *Phys. Lett.* **37B,** 315.
Schweber, S. S., Bethe, H. A., and De Hoffmann, F. (1955). "Mesons and Fields," Vol. I. Row and Peterson, Evanston, Illinois.
Schwinger, J. S. (1947). *Phys. Rev.* **72,** 742.

Scotti, A., and Wong, D. (1965). *Phys. Rev.* **138,** B145.

Seamon, R. E., Friedman, K. A., Breit, G., Haracz, R. D., Holt, J. M., and Prakash, A. (1968). *Phys. Rev.* **165,** 1579.

Serduke, F. J. D., and Afnan, I. R. (1971). *Phys. Rev.* **C4,** 1002.

Shei, S., and Tow, D. M. (1971). *Phys. Rev. Lett.* **26,** 470.

Sher, M. S., Signell, P., and Heller, L. (1970). *Ann. Phys.* **58,** 1.

Shirato, S., Saitoh, K., Koori, N., and Cahill, R. T. (1973). *Nuclear Phys.* **A215,** 277.

Slobodrian, R. J., Conzett, H. E., and Resmini, F. G. (1968). *Phys. Lett.* **27B,** 405.

Smith, L. C. H. (1970). Ref. TH. 1188-CERN, July 1.

Stagat, R. W., Rieve, F., Green, A. E. S. (1971). *Phys. Rev.* **C3,** 552.

Stricker, A., *et al.* (1972). *Nuclear Phys.* **A190,** 284.

Sugawara, M., and Okubo, S. (1966). *Phys. Rev.* **117,** 605, 611.

Ueda, T. (1971). *Phys. Rev. Lett.* **26,** 588.

Ueda, T., and Green, A. E. S. (1968). *Phys. Rev.* **174,** 1304.

Ueda, T., Nack, M. L., and Green, A. E. S. (1973). *Phys. Rev.* **C8,** 2061.

Weinberg, S. (1972). *Phys. Rev. Lett.* **29,** 388.

Weinberg, S. (1974). *Rev. Med. Phys.* **46,** 255.

West, G. B. (1973). *Phys. Rev. Lett.* **31,** 798.

Wigner, E. P. (1954). *Phys. Rev.* **94,** 77.

Wilson, K. G. (1971). *Phys. Rev. Lett.* **27,** 690.

Wilson, R. R., and Levinger, J. S. (1964). *Ann. Rev. Nuclear Sci.* **14,** 135.

Wolfenstein, L. (1956). *Ann. Rev. Nuclear Sci.* **6,** 43.

Yukawa, H. (1935). *Proc. Phys. – Math. Soc. Japan* **17,** 48.

Zee, A. (1972). *Phys. Rep.* **3C,** 127.

Zeitnitz, B., Maschuw, R., Suhr, P., and Ebenhöh, W. (1972). *Phys. Rev. Lett.* **28,** 1656.

Zichichi, A., ed. (1969). "Theory and Phenomenology in Particle Physics," Parts A, B. Academic Press, New York.

Chapter II

NUCLEAR SHAPE AND NUCLEAR MOMENTS

A. INTRODUCTION

The size, shape, and electromagnetic moments of the nuclear species, while exhibiting occasional marked local variations, do follow discernible systematic patterns. In this chapter we describe the results of some of the experiments providing information concerning these nuclear features that do not specifically involve nuclear reactions. Virtually all nuclear reactions, in fact, do involve such gross features of the interacting nuclei, but these are discussed best elsewhere in the context of reaction models.

The "deuteron problem" could properly have been treated in the context of the earlier two-nucleon interaction discussion. However, since the deuteron structure, size, shape, and electromagnetic moments offer an exceptionally clear view of the origin of these features, we give that discussion in this chapter.

No attempt will be made to be complete in the following discussions since a comprehensive and detailed treatment is not appropriate for this text. The interested reader is referred to the many general and specialized texts, e.g., Hill (1957), Kopfermann (1958), Elton (1961), and Schopper (1967).

B. HIGH-ENERGY ELECTRON SCATTERING

High-energy elastic scattering of electrons from hydrogen and deuterium has resulted in form factor information concerning the neutron and proton, as discussed in Chapter I. Elastic scattering of electrons from many complex nuclei also has been carried out at numerous energies. The analysis of the data is, however, more complicated than for the case of the nucleons, since the Born approximation used to obtain (1-61) is valid only for the lightest nuclei, where the modified fine structure constant $\gamma = Ze^2/\hbar c = Z/137 \ll 1$. While the general trend suggested by (1-61) holds for $Z \lesssim 10$, large inaccuracies occur in the vicinity of scattering angles where diffraction minima appear. Nonetheless, certain features contained in (1-61) can be used for general guidance. For the extreme relativistic case the factor appearing in that equation,

$$(\hbar q/2m_\alpha c)^2 \lesssim (E/Am_\mathrm{N} c^2)^2, \tag{2-1}$$

governs the importance of the magnetic moment interaction relative to the charge density interaction (A is the mass number of the target nucleus). For the lighter nuclei, up to ^{14}N, information on the magnetic parameters is becoming available. A compendium of such parameters is contained in Schopper (1967) along with key references. In general, for $E \lesssim 1$ GeV the factor (2-1) might be quite small, in which event the main information gained in the experiment concerns the charge density distribution of the scattering nucleus. As Exercise 1-12 shows, the charge form factor can be expanded in powers of q^2 ($\hbar^2 q^2$ representing the momentum transfer) giving for spherical nuclei the result

$$F(\mathbf{q}) = F(q^2) = Z[1-(1/6)\langle r^2\rangle q^2+(1/120)\langle r^4\rangle q^4+\cdots]. \tag{2-2}$$

Thus the quantity most readily deduced from scattering data, particularly at the lower energies, 20–50 MeV, is $\langle r^2\rangle$. Elevated energies for which $\lambda\!\!\!^- < [\langle r^2\rangle]^{1/2}$ are required to determine the higher moments and thereby to reveal more information concerning the charge distribution.

The procedure for assessing the scattering data for the various nuclei is to employ exact phase shift calculations for a variety of assumed charge distributions and to compare the predicted cross sections to the data. An excellent reference for the general subject of electron scattering is the collection of reprints by Hofstadter (1963). Figure 2.1 shows the calculated cross section, using an accurate phase shift computation for ^{12}C and ^{16}O at $E = 420$ MeV based on a harmonic oscillator shell model, and the comparison to the experimental data. Figure 2.1a also shows the Born approximation for the same oscillator model charge distribution. The failure of the Born curve near the diffraction minimum is evident. We shall have more to say about these nuclear models later.

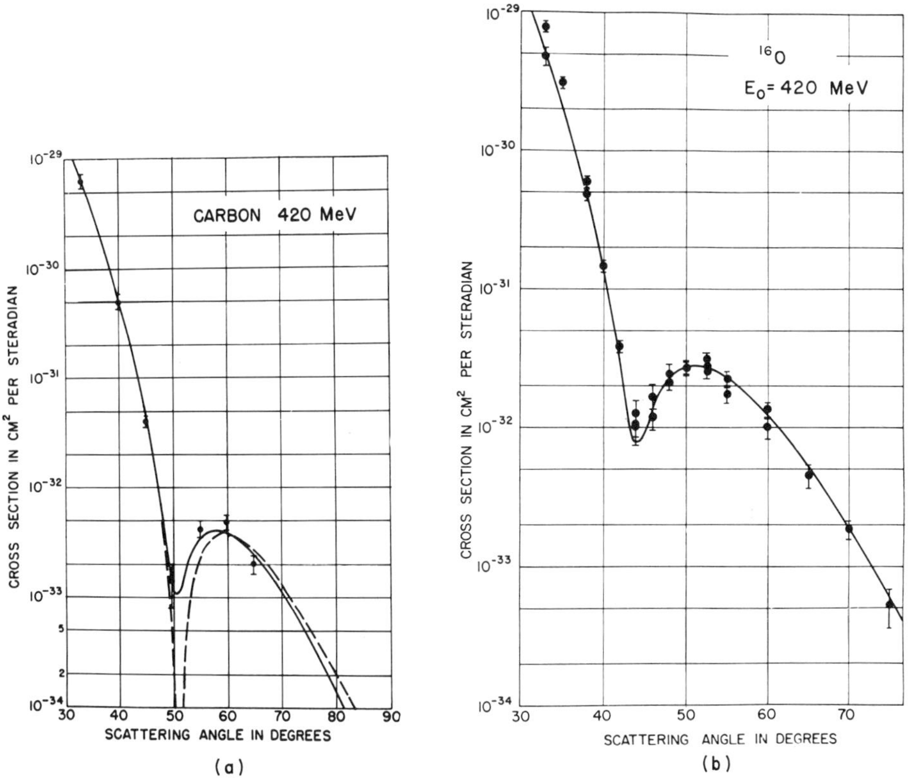

FIG. 2.1. Calculated and observed differential electron elastic scattering cross sections at 420 MeV for (a) carbon and (b) oxygen. Calculations using exact phase shift computations for the harmonic oscillator model are given by solid curves; the Born approximation calculation for the same model in the case of carbon is also shown by a dashed curve [curves taken from Hofstadter (1963)].

A rather simple spherically symmetric charge distribution, the so-called Fermi distribution, has been found to fit a vast amount of experimental data with rather good accuracy. This distribution is given by

$$\rho(r) = \frac{\rho_0}{1+e^{(r-c)/a}}, \qquad \text{with} \quad \rho_0 = \frac{3Ze}{4\pi c^3}\left(1 + \frac{\pi^2 a^2}{c^2}\right)^{-1} \tag{2-3}$$

corresponding to the normalization $4\pi \int_0^\infty \rho(r) r^2\, dr = Ze$. Figure 2.2 shows this function for the realistic case $c \gg a$, which generates for most nuclei an interior region of relatively constant density as well as a distinct surface region. The surface thickness s is defined to be the change in radius required to reduce $\rho(r)/\rho_0$ from 0.9 to 0.1. For the Fermi distribution (2-3) we have $s = 4a \ln 3 \approx 4.40a$.

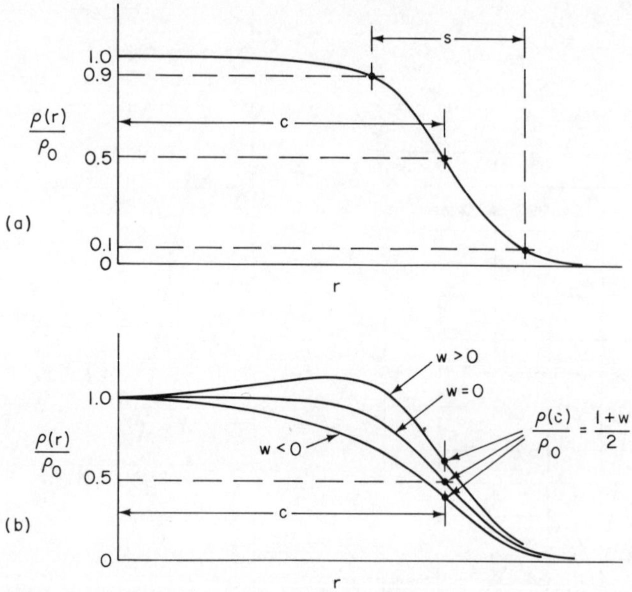

FIG. 2.2. (a) The Fermi density distribution function $\rho(r)/\rho_0 = [1+e^{(r-c)/a}]^{-1}$ and (b) the modified Fermi function $\rho(r)/\rho_0 = [1+(wr^2/c^2)][1+e^{(r-c)/a}]^{-1}$.

When the scattering data are fitted with the Fermi density function the variation of the half-density radius c and the surface or skin thickness s with the nuclear mass number A is found to be that shown in Fig. 2.3. The best $A^{1/3}$ dependence of c is $c(A) = (1.18A^{1/3} - 0.48)$ fm [see Schopper (1967)]. Earlier results (Hofstader, 1963) indicating a value $c(A) = 1.08A^{1/3}$ fm are also in reasonable accord with the data, particularly if the lightest nuclei are excluded. The surface thickness parameter s is seen to exhibit local variations correlated with the so-called magic numbers of protons, superimposed on a slight trend for s to increase with A. An approximately constant value $s \approx (2.4 \pm 0.3)$ fm fits the data reasonably well and is also in agreement with earlier findings.

Generally, a slight depression in the central region tends to give a better fit to the empirical data. Modifying (2-3) to

$$\rho(r)/\rho_0 = [1+(wr^2/c^2)][1+e^{(r-c)/a}]^n \tag{2-4}$$

yields a form that has been found to yield empirical values $-0.09 < w < +0.63$ for $n = -1$ (Schopper, 1967). The form with $n = -2$ is referred to as the Bethe distribution. For realistic values of the parameters, the form (2-4) allows greater freedom in independently setting both $\langle r^2 \rangle$ and the combination c and s to fit the wide variety of experimental data now available [see Bethe

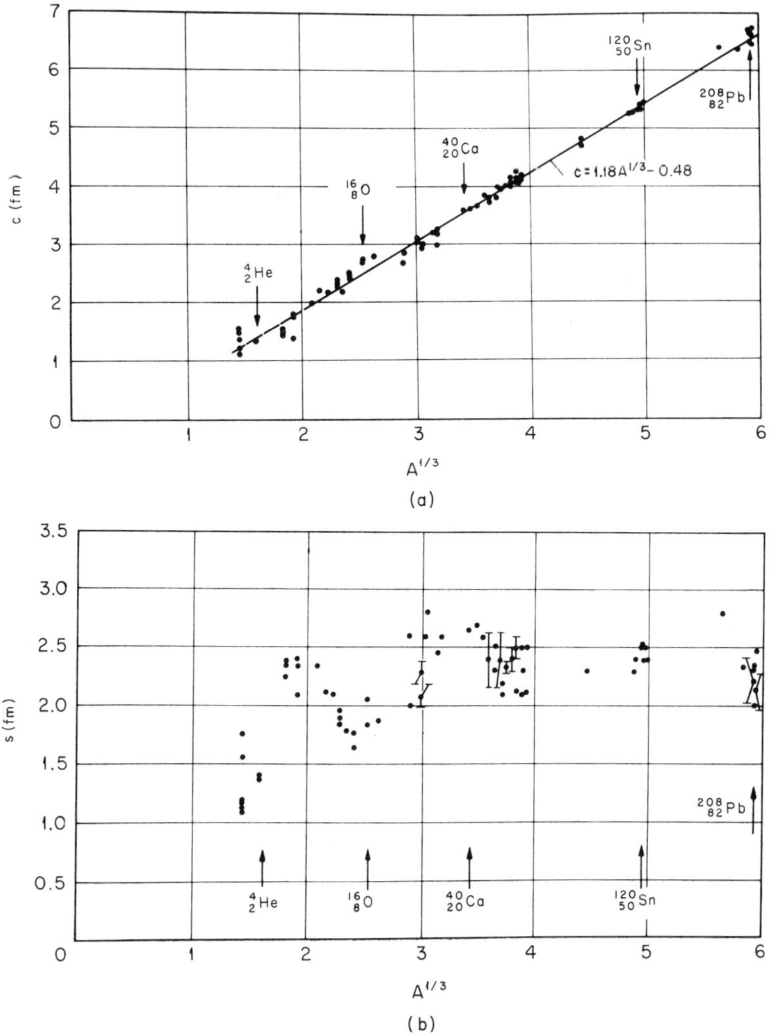

FIG. 2.3. (a) The variation of the half-density radius c versus $A^{1/3}$, determined from the charge distribution function (2-3) fitted to experimental data. (b) The surface thickness parameters versus $A^{1/3}$ [curves taken from Schopper (1967)].

and Elton (1968)]. Other variants of (2-4) to facilitate such fitting attempts have also been employed. Shell model calculations for charge distributions show slight undulations in the region where the phenomenological distribution (2-4) is relatively flat [e.g., Green *et al.* (1968, p. 79 ff.), Hill (1957, p. 235), Bethe and Elton (1968), and Negêle (1970)]. Accurate data are now becoming available bearing on such fine structure in the density function.

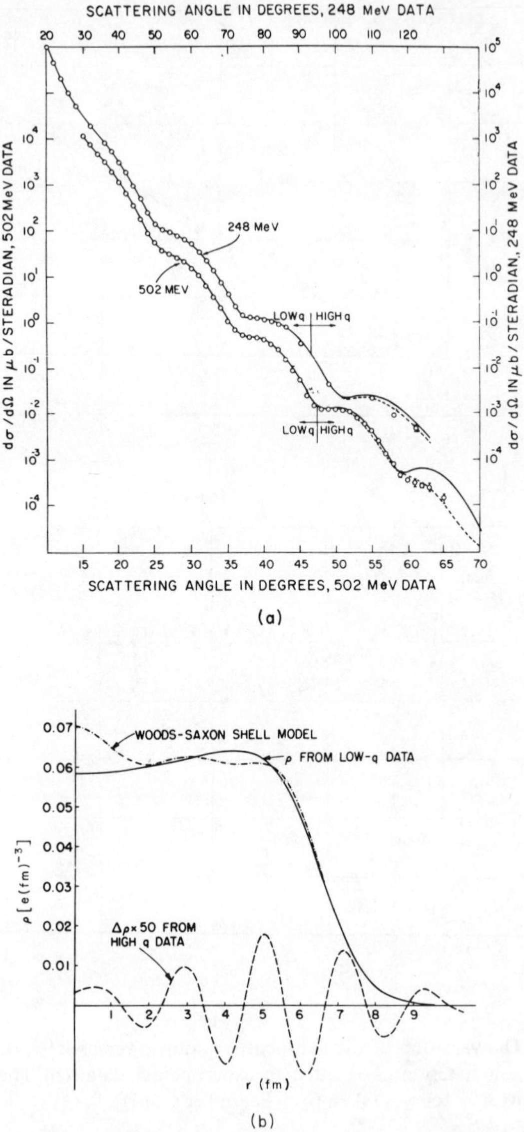

FIG. 2.4. (a) Experimental differential elastic electron scattering cross sections at 248 and 502 MeV for ^{208}Pb. The solid curve is based on the density distribution shown by the solid curve in (b). The dashed curve for high q is obtained when the undulations in density depicted by the dashed curve in (b) are added. (b) The density distribution for ^{208}Pb fitting the low-q data at *both* energies of (a) shown by the solid curve. The dashed curve shows the density modulation required to match the high-q data. The dash-dot curve is a typical shell model prediction [taken from Heisenberg *et al.* (1969)].

A paper discussing the data for ^{208}Pb at 248 and 502 MeV concludes that a central depression $1-[\rho(0)/\rho_{max}] \approx 7\%$ is required or $\bar{w} \approx 0.32$ with $n = -1.5$ in (2-4) [see Heisenberg *et al.* (1969)]. In addition, the high-q portions of the data suggest a slight undulation in the charge density. Figure 2.4 shows these data and the deduced charge distribution as well as a shell model prediction. These authors have also factored in muonic X-ray data parameters to obtain a simultaneous fit of greater reliability. The remaining parameters for (2-4) with $n = -1.5$ were found to be $c = 6.750$ fm and $a = 0.700$ fm, which give $\langle r^2 \rangle^{1/2} = 5.501$ fm. While the match between the shell model prediction and the empirical distribution in the surface region is excellent, and even the "phasing" of the undulations in both more or less agree, the disagreement in the relative magnitude of the density in the central region is marked. Possible nonlocal effects in the shell model potentials, if included in the theoretical calculations, might reconcile this difference since they are known to decrease the central density and enhance the surface density.

The charge distributions of the isotone pair ^{31}P and ^{32}S have also been examined through high-energy electron scattering experiments [see Sinha *et al.* (1972)]. Again a small central density depression and short-range undulations can be inferred from fitting the data.

Friar and Negêle (1973), in a more recent paper reviewing available electron scattering and muonic data and particularly addressing oscillatory charge density effects, cast some doubt on the reality of the central depression discussed above for ^{208}Pb.

On the other hand, there is a class of nuclei, the so-called "bubble" nuclei (^{36}Ar, ^{68}Se, ^{84}Se, ^{100}Sn, ^{116}Ce, ^{138}Ce, and ^{200}Hg, for example), that may have a substantial depression of the central density. Davis *et al.* (1973) calculate a Hartree–Fock central density that in some cases (e.g., ^{36}Ar, ^{138}Ce, and ^{200}Hg) is approximately one-half of the maximum density reached in a region just preceding the onset of the surface falloff. [See also Beiner and Lombard (1973), Campi and Sprung (1973), and Krishnan and Pu (1973).] For possible toroidal-shaped nuclei, see Wong (1972) and Nilsson *et al.* (1974).

As stated earlier, the most accurately determined aspect of the charge distribution is the rms radius or $\langle r^2 \rangle^{1/2}$. For the Fermi distribution (2-3) we find, correct up to fourth order,

$$\langle r^2 \rangle = \tfrac{3}{5}c^2[1 + (7\pi^2 a^2/3c^2)]. \qquad \text{(2-5)}$$

Figure 2.5 shows relatively recent values for the rms radius $\langle r^2 \rangle^{1/2}$. For $A^{1/3} \geqslant 2$, the expression $\langle r^2 \rangle^{1/2} \approx (0.82A^{1/3} + 0.58)$ fm accurately represents the experimental data. The results for the light elements in Fig. 2.5b show no simple systematic trend.

The rms radii for the magnetic moment density distributions are known with less accuracy [generally $\pm 10\%$, see Schopper (1967)] and within experimental error agree with the charge density values shown in Fig. 2.5b.

The fact that density distributions and their moments show little evidence of energy dependence up to $E \approx \frac{3}{4}$ GeV suggests that what is being measured is the *static* density, and that *dispersion* corrections involving virtual excitation and deexcitation of the nuclear levels by the scattering process may be small.

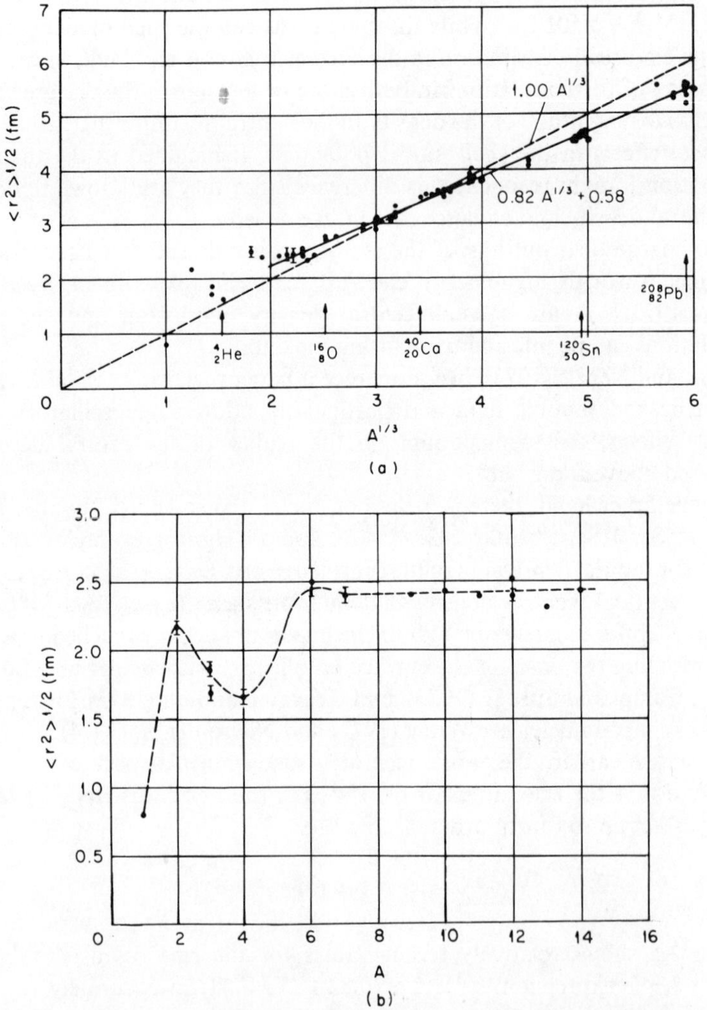

FIG. 2.5. (a) The rms radius of the charge density as determined by elastic electron scattering as a function of $A^{1/3}$. The solid line is a least-squares fit to the data $A \geqslant 8$. (b) The rms radius of the charge density for the light elements [taken from Schopper (1967)].

As we shall see, there is ample evidence for believing that nuclei can depart markedly from the simple spherically symmetric shape. For nuclei with permanent quadrupole deformations maintaining a symmetry axis (i.e., involving only $Y_2^{\,0}(\theta)$ and hence no φ dependence), the normalized density can be written in terms of a deformation parameter β in the form[1]

$$\rho\{r\} = \rho\{r[1-(\beta^2/4\pi)+\beta Y_2^{\,0}(\theta)]^{-1}\}. \tag{2-6}$$

Evaluating the mean-square radius for such a charge distribution gives

$$\langle r^2\rangle = [1+(5/4\pi)\beta^2]\langle r^2\rangle_0, \tag{2-7}$$

where $\langle r^2\rangle_0$ is the mean-square radius for the symmetric distribution. Thus deformed nuclei are expected to give enhanced values of $\langle r^2\rangle$. In addition, averaging the spatial orientation of unaligned deformed nuclei has the effect of increasing the apparent surface thickness and "washing out" the diffraction minima in the differential cross section. Indeed, all these effects are observed in electron scattering. For example, the unaligned deformed nucleus ^{181}Ta ($\beta \approx 0.24$) has $\langle r^2\rangle^{1/2} = 7.10$ fm and $s = 2.8$ fm, while the spherical nucleus ^{208}Pb has $\langle r^2\rangle^{1/2} = 6.96$ fm and $s = 2.31$ fm, even though ^{208}Pb has the larger A value.

In a recent experiment with aligned and randomly oriented ^{165}Ho nuclei ($\beta \approx 0.32$), approximately 10% differences in the scattering cross section for these cases were observed (Uhrhane *et al.*, 1971). Model calculations for the aligned case do not fit the data well. The models used attributed the deformations to localized surface effects and required some version of the Born approximation to calculate the effect on the cross section. The source of the difficulty at present is not understood.

▶**Exercise 2-1** Two very simple charge density distributions are the constant-density and the trapezoidal models. These are shown in Fig. 2.6.

(a) Show that the normalization $4\pi\int \rho(r)\, r^2\, dr = Ze$ gives

$$\rho_{01} = 3Ze/4\pi R_1^{\,3} \qquad \text{and} \qquad \rho_{02} = 3Ze/4\pi R_2^{\,3}(1+\delta^2).$$

(b) Show that the mean-square value $\langle r^2\rangle$ is

$$\langle r^2\rangle_1 = \tfrac{3}{5}R_1^{\,2}, \qquad \langle r^2\rangle_2 = \tfrac{3}{5}R_2^{\,2}(1+\tfrac{1}{3}\delta^2)(1+3\delta^2)/(1+\delta^2).$$

These should be contrasted to (2-5). ◀

[1] The assumption is made that surfaces of equal density can be written $r = r_0[1-(\beta^2/4\pi)+\beta Y_2^{\,0}(\theta)]$. The term in β^2 arises from the assumption of approximate incompressibility of nuclear matter. For a more complete discussion, see Chapter VI, particularly Exercise 6-8.

Interestingly enough, if the zero point energy of a vibrating *spherical* nucleus is considered, the value of $\langle r^2\rangle$ is again given by (2-7) with $\beta^2 \to 5\hbar\omega/2C$, where ω is the quadrupole vibrator frequency and C the effective spring constant.

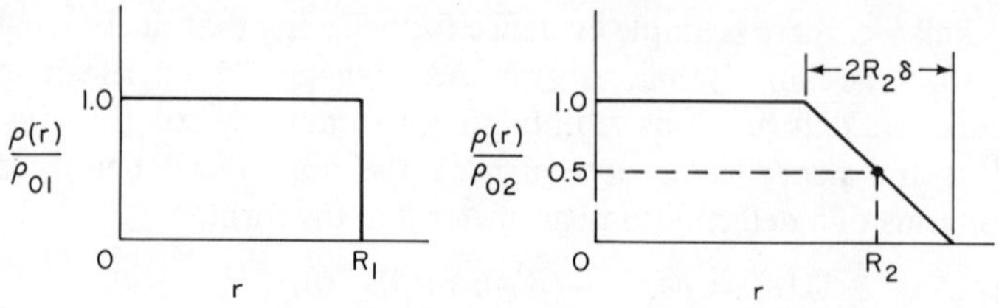

FIG. 2.6. Constant-density and trapezoidal-density functions.

▶**Exercise 2-2** If we assume that the mass density and charge density follow the same geometric shape, the mass density is obtained from the charge density by scaling with the factor A/Ze. Thus for the Fermi distribution, the central density of (2-3) becomes

$$\rho_0 = \frac{3A}{4\pi c^3}\left(1 + \frac{\pi^2 a^2}{c^2}\right)^{-1}.$$

If the central density is taken to be $\rho_0 = 0.170\ \mathrm{fm}^{-3}$ and the surface thickness $s = 4a \ln 3 = 2.30$ fm, with both assumed constant, the result is

$$\langle r^2\rangle^{1/2} = (0.87A^{1/3} + 1.56A^{-1/3} - 1.17A^{-1} + \cdots)\ \mathrm{fm}. \tag{2-8a}$$

Show that the trapezoidal distribution with $\rho_{02} = 0.170\ \mathrm{fm}^{-3}$ and $0.8(2R_2\delta) = 2.30$ fm, both taken constant, gives

$$\langle r^2\rangle^{1/2} = (0.87A^{1/3} + 1.19A^{-1/3} + \cdots)\ \mathrm{fm}. \tag{2-8b}$$

Show that $\langle r^2\rangle^{1/2}$ versus $A^{1/3}$, in the range $2 \leqslant A^{1/3} \leqslant 6$, for (2-8a), (2-8b), and the empirical fit to Fig. 2.5a, namely $\langle r^2\rangle^{1/2} = (0.82A^{1/3} + 0.58)$ fm, differ by amounts less than the experimental error, which is usually 1–2%. It should be noted that the expansions of smooth, gently varying functions in powers of $A^{1/3}$ generally lead to a strong coupling of the expansion coefficients. Thus rather different looking expansions can give similar values for the series sum. This is a recurring problem in fitting semiempirical expressions to bulk nuclear properties. For an interesting discussion of the geometrical properties of *leptodermous*[1] distributions see Myers (1973).

A useful expression is the approximation for the moment of r^n for the Fermi distribution (2-3),

$$\langle r^n\rangle = \frac{\int \rho(r)\, r^n\, d^3r}{\int \rho(r)\, d^3r} = \frac{3}{n+3}c^n\left[1 + \frac{n(n+5)}{6}\frac{\pi^2 a^2}{c^2} + \cdots\right]. \quad ◀ \tag{2-9}$$

[1] Leptodermous, having a thin skin.

C. ATOMIC EFFECTS

Several atomic effects reveal information concerning the nuclear size and shape. Generally such information relates to various moments of the nuclear charge distribution.

Fine structure in X-ray spectra can be measured to very high precision, permitting the observation of effects that result from relativistic quantum electrodynamics as well as the finite nuclear size. These latter effects are most pronounced for lines involving the $s_{1/2}$ and $p_{1/2}$ atomic electrons since these have significant electronic density in the vicinity of the nucleus (most other states are only negligibly affected). The energy shift in these atomic levels due to the finite size of the nucleus can be shown to be proportional to $\langle r^{2\sigma} \rangle$ for the nuclear charge distribution with $\sigma^2 = 1+(Z/137)^2$. While the nuclear and relativistic effects are difficult to separate, an equivalent constant-density radius (see Exercise 2-1) $R_1 \approx 1.2A^{1/3}$ fm is deduced [see Hill (1957) and Elton (1961)].

A related phenomenon is the isotope shift observed in optical and X-ray spectra. In light elements the line shift in the optical spectra of isotopes has to do with the variation of the reduced mass of the electron from isotope to isotope. In heavy elements this effect is small, and the major contribution now comes from the change in nuclear size with the isotopic species. Relativistic effects that depend largely on Z are almost constant for a particular family of isotopes and thus, in principle at least, can be separated from the nuclear size effects.

As before, the effect is greatest for spectral terms involving the $1s_{1/2}$ electron, and the shift in the spectral line largely measures this level shift. For experimental details see Kopfermann (1958, Sections 34–37). The observed isotope shift for $\delta A = \delta N$ measures the difference

$$\delta \langle r^{2\sigma} \rangle = \langle r^{2\sigma} \rangle_{Z,N} - \langle r^{2\sigma} \rangle_{Z,N-\delta N} = \left(\frac{2\sigma}{3}\right)\frac{\delta A}{A}\langle r^{2\sigma} \rangle \approx \frac{2}{3}\frac{\delta A}{A}\langle r^2 \rangle, \tag{2-10}$$

which follows if we assume that $\langle r^2 \rangle^{1/2} \approx a_1 A^{1/3}$ with a_1 a constant and take $\sigma \approx 1$. We can refer to (2-10) as defining the standard isotope shift and measure the observed shift in terms of this quantity. A summary of relevant data is given in Brix and Kopfermann (1958), Schopper (1967), and Bhattacherjee *et al.* (1969). Figure 2.7 shows a plot directly related to (2-10) as a function of the neutron number N. The figure uses $\langle r^2 \rangle \approx \frac{3}{5}(1.20A^{1/3})^2$, which from Exercise 2-2 is roughly correct for $A \approx 125$. Reference to Fig. 2.7 clearly shows large deviations from unity, indicating local departure from the behavior $\langle r^2 \rangle^{1/2} = a_1 A^{1/3}$. Some of these variations can be understood in

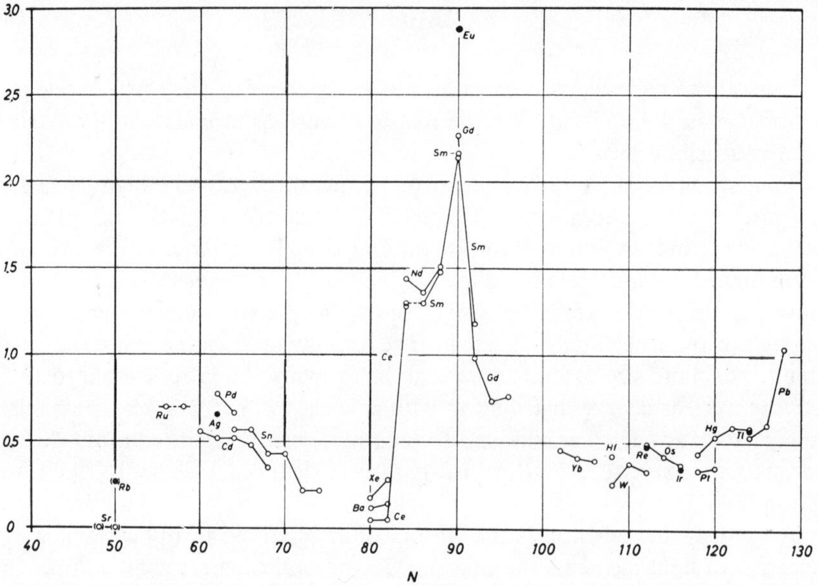

FIG. 2.7. The isotope shift (observed/standard) for even–even (A even, Z even) nuclei with the neutron number changing from $N-2$ to N for connected points, open circles. The solid points are for Z odd. The value of $a_1 = 0.93$ fm was used [taken from Brix and Kopermann (1958)].

terms of the Coulomb effect and nuclear compressibility in the neighborhood of the equilibrium radius. The compressional energy is defined as $E_0'' = [R^2\, d^2 E_v/dR^2]_{R=R_1}$, where, anticipating the discussion of Chapter III, the total nuclear energy E is written as $E = E_v + E_s$, with E_v the volume energy and E_s the surface energy. There is reason to believe that the compressional energy E_0'' is proportional to the mass number A; thus $E_0'' = KA$ with a reasonable empirical value for K of $K \approx (45 \pm 15)$ MeV.[1] Due to the additional Coulomb repulsion when a proton is added, the radius increase should be greater than when a neutron is added. When this consideration is coupled with the compressibility effect cited, it is found that (2-10) (appropriate for $\delta A = \delta N$) should be multiplied by the factor 0.65 ± 0.10.[2] If a similar isobaric

[1] While the value quoted here of $K \approx 45$ MeV leads to estimated isotope shifts in agreement with Fig. 2.7, more realistic values derived from models for "nuclear matter" suggest $K \approx 130$–190 MeV. These larger values, however, are for densities pertaining to the nuclear interior; see Section III,D,2. The nuclear surface might be more relevant in the present instance, where indeed the stiffness can be substantially smaller.

[2] This often repeated computation in the literature ignores the important modifying effects of the strong interaction of the added proton with the neutrons constituting the $N-Z$ neutron excess (see later discussion).

effect could be determined in terms of $\delta A = \delta Z$, the multiplicative factor 1.78 ± 0.35 would be required. A finite compressional energy would require a multiplicative factor less than unity even if the added nucleons δA consisted of protons and neutrons in proportion to Z and N of the initial nucleus. The relatively incompressible model for nuclear matter implied by $K \approx 45$ MeV results in an estimate of 0.95 ± 0.02 for this factor. This latter result is also consistent with the fact that $a_1 = \langle r^2 \rangle^{1/2} A^{-1/3} \approx 0.82 + 0.58 A^{-1/3}$ [or similar values obtained from (2-8a) or (2-8b)] shows only a slow decrease with increasing A.

The factor 0.65 ± 0.10 for the isotope shift with $\delta A = \delta N$ is seen to be reasonable for $55 \lesssim N \lesssim 70$ and $100 \lesssim N \lesssim 120$. The marked deviations near the magic numbers[1] $N = 50$, 82, 126 and in the rare earth region are to be associated with the larger than average value of K for the magic number nuclei and the large nuclear deformations that rapidly set in above the magic number $N = 82$ and the subsequent variation of β^2 for the rare earth region. Using (2-7) and (2-10), we have with $\delta A = \delta N = 2$,

$$\delta \langle r^2 \rangle / \delta \langle r^2 \rangle_0 \approx (15A/16\pi)\, \delta(\beta^2). \tag{2-11}$$

The change from $N = 88$ to 90 for the isotopes of Sm, Nd, Eu, and Gd produces a particularly large change in β^2 of $\delta(\beta^2) \approx 0.05$. These so-called "transitional nuclei" change rather abruptly from spherical to deformed shapes as the neutron number increases in the cited range [for numerical values see, e.g., Hill (1957, p. 227), noting that $\delta A = 2$]. Thus (2-11) gives a factor of ~ 2.2 for $N = 90$ in good agreement with Fig. 2.7. Recently the corresponding effect suddenly changing the nuclear shape from a highly deformed shape to one more or less spherical has been observed in the mercury isotopes as N increases from 105 to 107. Bonn *et al.* (1972) found $\delta(\beta^2) \approx 0.05$–$0.06$ for this corresponding change by extending the isotope shift measurements in mercury to include isotopes from ^{183}Hg ($N = 103$) to ^{201}Hg ($N = 121$). On the other hand, the anomalous behavior of ^{183,185}Hg when contrasted to that of $^{187-201}$Hg can also be explained by considering ^{183,185}Hg to be examples of bubble nuclei rather than largely quadrupole deformed nuclei [see Beiner and Lombard (1973)]. To a satisfactory degree the deviations from (2-10) observed in Fig. 2.7 can be accounted for by the introduction of finite compressibility of nuclear matter and the presence of surface deformations. The deviations due to deformations also correlate with the shell effects of Fig. 2.3b in an obvious manner.

While it is beyond the scope of this text to review all the recent applications of isotopic-like effects of the type discussed above, we should mention one interesting example relating to the change in nuclear radius with nuclear

[1] Magic numbers are defined in Section IV, C.

excitation. When the isomeric excited state in ^{57}Fe with half-life $t_{1/2} = 99.3 \times 10^{-9}$ sec and excitation energy 14.4 keV decays by M1 radiation to the ground state, it is found, by using the very sensitive recoil-free emission and resonant absorption Mössbauer technique, that the level energy is slightly shifted by the chemical binding effects of the atomic lattice containing the iron. This effect, called the *isomer shift*, depends on both the change in the atomic electron density at the nucleus and the difference in charge radius of the nucleus for the excited and ground states [e.g., see Walker *et al.* (1961)]. When both the above energy change and the change in decay lifetime are determined, the rather difficult to estimate change in electron density effects can be removed, leading to a reliable determination of change in radius. By this means Rüegsegger and Kündig (1972) were able to determine $\Delta R/R = -(3.1 \pm 0.6) \times 10^{-4}$ (i.e., the ground-state radius being *larger*) in ^{57}Fe. This value can be compared in magnitude to the *isotope shift* (i.e., the effect of adding one neutron) of $\Delta R/R = \frac{1}{3}\,\Delta A/A = 5.8 \times 10^{-3}$. Refer also to Boolchand *et al.* (1972). Such isomer "shrinkage" effects can be expected from collective nuclear behavior [see Mantri and Sood (1974)].

An effect closely related to the X-ray fine structure phenomenon discussed earlier is observed in *muonic atoms*. These are atoms in which a μ-meson has replaced an electron in the normal orbital complement of electrons accompanying the nucleus. The muon, being distinguishable from the electron, participates in atomic transitions independently of normal electronic transitions in that muonic transitions are not inhibited by the operation of the Pauli principle. In other respects the muon in this context behaves simply as a heavy electron. [The μ-meson has a mass $m_\mu c^2 - (105.6595 \pm 0.0003)$ MeV (Particle Data Group, 1973) and is thus 206.8 times as massive as the ordinary electron; it has spin $\frac{1}{2}$, a gyromagnetic ratio close to that of the electron, and two charge states $\pm e$, and it is a fermion obeying the Dirac equation. The muon–nucleon interaction is nonhadronic and is entirely electromagnetic in nature.] The length parameter in atoms, namely the Bohr radius $a = 4\pi\hbar^2/me^2$, varies inversely as the reduced mass and is therefore some 200 times smaller for the muon than for the ordinary electron. The level energies and transition energies involving the muon are correspondingly much larger. For example, for ^{208}Pb the $1S_{1/2}$ atomic level is 10.48 MeV and the $2P_{3/2}$–$1S_{1/2}$ transition energy is 5.96 MeV. These are to be contrasted to the corresponding electronic values of 87.8 and 74.8 keV. When the muon is in the 1S level in the heavier atoms the probability of the muon being in the nuclear interior can be as high as 50% or more; thus X-ray transitions involving the 1S level would have a very marked effect due to the finite nuclear size.

Transitions as high as $7I \rightarrow 6H$ have been observed, but only those between the 1S, 2P, and 3D levels give useful information on the nuclear size. The large body of data is tabulated in Schopper (1967), Wu and Wilets (1969), and Kim

(1971); the latter two references also give a thorough review of muonic atoms. Again the quantity determined, as for the X-ray fine structure, is $\langle r^{2\sigma} \rangle$. Values obtained for $\langle r^2 \rangle$ are generally accurate to $\sim 1\%$ and disagree with the electron scattering data by 2–3%. The muonic data give, for the Fermi shape (2-3), the constant central mass density 0.158 fm^{-3} and constant surface thickness $s = 2.21$ fm, which in turn result in

$$\langle r^2 \rangle^{1/2} = (0.89A^{1/3} + 1.38A^{-1/3} - 0.90A^{-1})\ \text{fm}. \tag{2-12}$$

The use of the modified Fermi or Bethe distribution (2-4) permits the reconciliation of the muonic and electron scattering data. The advent of very precise experimental data has revealed some problems in the understanding of the theory of muonic atoms [e.g., see Rinker and Rich (1972)].

The muonic atoms also exhibit the isotope shift discussed earlier for optical and X-ray spectra. The sparser data available are generally consistent with those cited earlier [see Schopper (1967), Nolen and Schiffer (1969), and Kim (1971)].

Another very interesting phenomenon involving muonic atoms is the *muonic isomer shift*. This phenomenon is not unlike the normal atomic isomer shift previously discussed. The effect occurs when the muon reaches its atomic ground state while the nucleus is in an excited state. Generally the nucleus will undergo deexcitation before the muon decays or is captured. The perturbing effect of the muon presence in altering the nuclear transition energy is referred to as the muonic isomer shift. For example, the normal nuclear transition energy in the $\frac{3}{2}^-$ to $\frac{1}{2}^-$ decay, second excited state to ground-state deexcitation, in ^{207}Pb is 897 keV; the corresponding isomer shift is almost 2 keV. The observed isomer shift in the nuclei ^{209}Bi and ^{207}Pb has been interpreted by Rinker (1971) in terms of a nuclear model coupling a single particle or hole to a collective vibrational doubly magic ^{208}Pb core. His analysis suggests the possibility that vibrational excitations are not volume-conserving.

Recently experimental data and analyses have become available on π-mesonic atoms. The pion mass is similar to that of the muon, but a very important difference in the corresponding atoms results from the strong pion–nucleon interaction. The pion–nucleon interaction can be represented by an optical potential including an absorptive term (Anderson *et al.*, 1970). Parameters to fit X-ray data from ^{10}B to ^{209}Bi require distinguishing between the neutron and proton densities. The quantity $\Delta = \langle r_n{}^2 \rangle^{1/2} - \langle r_p{}^2 \rangle^{1/2}$ is rather accurately determined. The value is found to be $\Delta = -(0.01 \pm 0.16)$ fm (assumed independent of A); thus rms radii appear to be essentially the same for both proton and neutron distributions, and only the magnitude of the densities need be distinguished. This might be taken to imply a simple scaling Z/A or N/A of the matter density distribution to get the proton or neutron density. However, various theoretical shell model calculations show differences in Δ from zero. Generally, they predict $\Delta \approx +(0.1\text{–}0.3)$ fm and indicate other slight differences in the neutron and proton density distributions (see later). For good general reviews of π-mesonic atoms, refer to Backenstoss (1970), Kim (1971), and Ponomarev (1973).

Of late, K^- atoms have been studied in a manner similar to that used for the π^- atoms. These results are more difficult to interpret; see Wiegand (1969), Bethe (1969), Aslam and Rook (1970), Bardeen and Torigoe (1971), Kim (1971), and Seki (1972).

Nuclear scattering, absorption, or capture of $\pi^{\pm}$ can also be used to determine such nuclear parameters as Δ and nucleon momentum distributions; e.g., see Kossler *et al.* (1971), Rost and Edwards (1971), Sternhein and Auerbach (1971), Batty and Friedman (1972), and Allardyce *et al.* (1973).

A good deal of information relating to proton and neutron density distributions arises from nuclear reaction studies and is best discussed in that context. To cite but one example, Greenlees *et al.* (1970), using optical potentials derived from nucleon density distributions and the nucleon–nucleon force in analyzing proton elastic scattering at 30 MeV, conclude that $\Delta \approx 0$ for medium-weight nuclei and $\Delta \approx (0.1\text{–}0.2)$ fm for heavy elements. See also Schery *et al.* (1974). For a thorough discussion concerning proton and neutron density distributions, see Wilets (1968) [also Nolen and Schiffer (1969) and Myers and Swiatecki (1969)].

D. COULOMB RADII

The classical calculation for the Coulomb energy of a spherically symmetric charge distribution is straightforward. For the constant-density model of Exercise 2-1,

$$E_{c1} = \tfrac{3}{5} Z^2 e^2 / R_1, \tag{2-13a}$$

while for the trapezoidal model,

$$E_{c2} = \frac{3}{5}\frac{Z^2 e^2}{R_2}\,\frac{1 + \frac{5}{6}\delta^2 + \frac{1}{2}\delta^3 + \frac{1}{6}\delta^4 - (1/42)\delta^5}{(1+\delta^2)^2}. \tag{2-13b}$$

The Coulomb energy for the Fermi distribution (2-3) to order a^2/c^2 is

$$E_c = \frac{1}{2}\int \frac{\rho(\mathbf{r}_1)\,\rho(\mathbf{r}_2)}{|\mathbf{r}_1 - \mathbf{r}_2|}\, d^3r_1\, d^3r_2 = \frac{3}{5}\frac{Z^2 e^2}{c}\left(1 - \frac{7}{6}\frac{\pi^2 a^2}{c^2} + \cdots\right). \tag{2-13c}$$

►**Exercise 2-3** Derive (2-13a) and (2-13b). ◄

Noting that the electrostatic energy of individual protons is included in their rest mass, the factor Z^2 in (2-13) should be replaced by $Z(Z-1)$ since each proton interacts with only $Z-1$ other protons.

We now turn from the above, purely classical derivation to at least a partial quantum-mechanical treatment based on an individual particle model in which the protons are imagined to be in individual orbitals given by wave functions $\psi_i(\mathbf{r}_i)$. For an antisymmetric total wavefunction[1] for the protons of a nucleus, we write

$$E_c = \tfrac{1}{2}\sum_{i,j} \iint [\psi_i^*(1)\,\psi_j^*(2)\,(e^2/r_{12})\,\psi_i(1)\,\psi_j(2) - \psi_i^*(1)\,\psi_j^*(2)\,(e^2/r_{12})\,\psi_j(1)\,\psi_i(2)]\; d^3r_1\, d^3r_2. \tag{2-14}$$

The first term is the direct Coulomb energy and corresponds to the classical expression when $\psi_i^*(1)\,e\psi_i(1)$ is considered the contribution of particle 1 to the charge density $\rho(\mathbf{r}_1)$. The second term is the exchange energy and represents the effect of the exclusion principle in the spatial correlation of the protons. The sum can be extended over i and j independently since when $i=j$ (the self-energy contribution) the direct and exchange terms exactly cancel.[2]

In terms of the density

$$\rho(\mathbf{r}) = e\sum_{i=1}^{Z} \psi_i^*(\mathbf{r})\,\psi_i(\mathbf{r}) \tag{2-15}$$

and the integrals

$$2e^2Z^2J = \int [\rho(\mathbf{r}_1)\,\rho(\mathbf{r}_2)/r_{12}]\; d^3r_1\, d^3r_2 \tag{2-16}$$

and

$$K = (1/J)\int [\rho(\mathbf{r})]^{4/3}\, d^3r, \tag{2-17}$$

E_c can be written [see Peaslee (1954)]

$$E_c = Z^2e^2J\{1-[C_0/(2Z^2)^{1/3}]K\}. \tag{2-18}$$

Here C_0 and K are constants of the order unity depending on the individual proton charge distribution and the form assumed for $\rho(\mathbf{r})$. For example, in the case of the constant-density model and using plane waves for ψ_i, the result is

$$E_c = \tfrac{3}{5}(Z^2e^2/R_1)[1-\tfrac{5}{4}(3/2\pi)^{1/3}Z^{-2/3}] \tag{2-19a}$$

or

$$E_c = \tfrac{3}{5}(Z^2e^2/R_1)[1-0.767Z^{-2/3}]. \tag{2-19b}$$

[1] The so-called Slater determinant that generates such functions will be discussed later. The expression (2-14) is a simple use of the general equation (3-17) developed in Chapter III.

[2] When *only* the direct energy is calculated, as in the classical case, the terms with $i=j$ must be excluded.

We note that in place of the classical factor $Z(Z-1)$ there is now $Z[Z-0.767Z^{1/3}]$. For $Z>2$ [one would hardly expect that (2-19) would make sense for $Z=1$] the second term is greater than unity, resulting in a smaller energy than the classical calculation, consistent with the requirement of the exclusion principle.

For density shapes not too dissimilar from the constant-density model, the ratio of the exchange to direct terms might be taken approximately the same as that in (2-19), in which event the Fermi density distribution would yield

$$E_c \approx \tfrac{3}{5}(Z^2e^2/c)[1-0.767Z^{-2/3}][1-\tfrac{7}{6}(\pi^2a^2/c^2)+\cdots]. \tag{2-20}$$

Similarly, the trapezoidal model from (2-13b) could be used to modify (2-19) suitably.

Pairs of nuclei exist having levels that correspond to individual particle states for their nucleons, which are identical except that a proton of one is replaced by a neutron in the other, all other quantum numbers except obviously T_3 being the same. Such states are referred to as isobaric analog or isoanalog states. Ground states of so-called *mirror* nuclei, one with $Z_1=\frac{1}{2}(A+1)$, $N_1=\frac{1}{2}(A-1)$ and the other with $Z_2=\frac{1}{2}(A-1)$, $N_2=\frac{1}{2}(A+1)$, such as

$$({}^7\text{Li},\ {}^7\text{Be};\ J^\pi=\tfrac{3}{2}^-,\ T=\tfrac{1}{2},\ T_3=\pm\tfrac{1}{2})$$

and

$$({}^{13}\text{C},\ {}^{13}\text{N};\ J^\pi=\tfrac{1}{2}^-,\ T=\tfrac{1}{2},\ T_3=\pm\tfrac{1}{2})$$

are typical examples. An isobaric triad such as the one involving the ground states of ^{14}C and ^{14}O and the first excited state of ^{14}N forms a similar grouping ($J^\pi=0^+$, $T=1$, and $T_3=\pm1, 0$). A number of reactions also identify analog states; for example, (p, n) reactions reveal isoanalog states of the target nucleus ground state in the excited residual nucleus. A thorough discussion of analog states and the broader topic of isospin is contained in the Tallahassee Conference of 1966 on isospin (Fox and Robson, 1966). The energy displacement of such member levels within a doublet or triad[1] gives information on the change in Coulomb energy resulting from the change in Z for a particular constant A. For example, when one member of a mirror pair β-decays to the other member, the β-spectrum end-point energy corrected for the neutron–proton mass difference directly measures the Coulomb energy difference, viz., for β^+-decay,

$$\Delta E_{\mathrm{c,exp}} = (E_{\beta\mathrm{max}} + 1.805)\ \mathrm{MeV}. \tag{2-21}$$

In such a case the nuclear charge changes from $Z+1=\frac{1}{2}(A+1)$ to $Z=\frac{1}{2}(A-1)$. Using (2-18), we find

$$\Delta E_c \approx 2e^2J[(Z+\tfrac{1}{2})-\tfrac{3}{2}^{2/3}(Z+\tfrac{1}{2})^{1/3}C_0K]$$

[1] By now quartets and quintets have also been discovered.

or

$$\Delta E_c \approx e^2 J A^{1/3} [A^{2/3} - (\tfrac{2}{3})^{4/3} C_0 K]. \tag{2-22}$$

Since R_1, R_2, and c in (2-13) vary with $A^{1/3}$ to first approximation, $JA^{1/3}$ in (2-22) might be expected to be relatively constant. Thus one would expect $\Delta E_{c,\exp}$ to show a straight-line dependence as a function of $A^{2/3}$, viz., $\Delta E_c \approx aA^{2/3} - b$. Data on mirror nuclei for $A \leqslant 41$ [e.g.,Peaslee (1954) and Fox and Robson (1966)], while showing such a linear dependence, indicate variations in b as one proceeds through the various nuclear shells, with b approximately constant within a shell and abruptly changing its value at the closed-shell nuclei ^{4}He, ^{16}O, and ^{40}Ca. Such shell effects on $C_0 K$ are to be expected [see Nolen and Schiffer (1969)]. An average for all the data ignoring such shell effects suggests $a = 0.73$ MeV and $b = 1.11$ MeV. Setting $a = e^2 J A^{1/3} = \frac{3}{5}(e^2/R_1) A^{1/3}$ gives

$$R_1 = 1.18 A^{1/3} \text{ fm}, \tag{2-23}$$

a value that is somewhat too small.

A general form for the Coulomb energy difference between analog states of nuclei $Z+1$, A, and Z,A given by a semiclassical expression suggested by Sengupta (1960) is similar to (2-22), viz.,

$$\Delta E_c(Z+1, Z) = (e^2/R_1 A^{1/3})[\tfrac{3}{5}(2Z+1) - 0.613 Z^{1/3} - (-1)^Z 0.30]. \tag{2-24}$$

This expression differs from (2-22) mainly in the last term, due to the change in pairing effects of proton spins on the exchange energy. Anderson *et al.* (1965) have shown that an even simpler expression for $T = \frac{1}{2}$ and 1 nuclei holds (although somewhat better results are found when the "pairing corrections" are included, particularly for the lighter mirror nuclei), namely,

$$\Delta E_c(Z+1, Z) = E_1 [(Z + \tfrac{1}{2})/A^{1/3}] + E_2 \tag{2-25}$$

with $E_1 = (1.443 \pm 0.011)$ MeV and $E_2 = (-1.13 \pm 0.11)$ MeV. These empirical values, of tested validity up to $A \approx 160$, agree well with the mirror nuclei values, which they also embrace with $E_1 = 2a$ and $E_2 = -b$, albeit they are now differently interpreted due to the pairing effect. Figure 2.8a shows the interpreted equivalent constant-density radius as a function of Z from the three principal experimental sources. The various results are seen to be in reasonable accord on the average.

Figure 2.8b, however, shows that systematic deviations from (2-25) exist correlated to magic number or shell effects. An important fact, so far ignored in our discussion, has to do with the validity of extracting ΔE_c from an expression derived for the average behavior of protons within a nucleus. In β-decay,

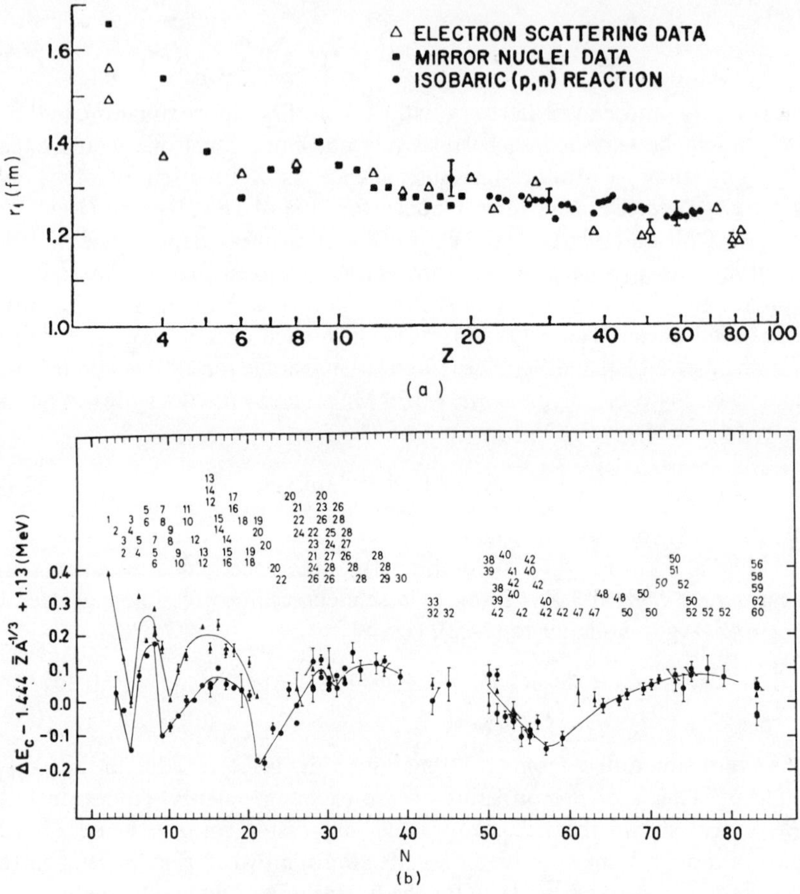

FIG. 2.8. (a) The equivalent uniform radius factor $r_1 = R_1 A^{-1/3}$ as a function of Z [taken from Anderson *et al.* (1965)]. (b) A plot of the experimental value ΔE_c minus the empirical value (2-25) versus the neutron number N. The numbers in the field represent the proton (or atomic) number Z. Even values of Z are represented by circles, odd values by triangles [taken from Harchol *et al.* (1967)].

(p, n) analog reactions, and in fact for whatever source is used in determining ΔE_c, protons predominantly in a specific orbital state (shell) are involved. Thus changes in Coulomb energy should be calculated for such protons alone rather than the average proton state. The pairing effect discussed above is only one such effect that enters. Another important difference arises from the model sensitivity of the treatment of the exchange energy contribution to ΔE_c for isobaric analogs. Shell model calculations indicate, for example, that the exchange contribution to the total change ΔE_c is 7.0% in (^{13}N, ^{13}C), while the statistical model used here gives 11.8%.[1] In heavier nuclei this effect diminishes in importance; for example, for $A = 120$, the shell model gives 2.2%, while the statistical model gives 3.3%.

Finally, an empirical expression for Coulomb displacement energies, with an accuracy of order 15 keV in the range $48 < A < 208$, has been developed by Seitz *et al.* (1972). Consideration has been given to magic and isotopic effects as well as nuclear deformations. The resulting expression is

$$\Delta E_c(Z+1, Z) = \frac{E_1 Z}{A^{1/3}}\left(1 - \frac{4}{45}\delta^2\right) + E_2 + [1-(-1)^Z]\frac{60}{N-Z} - \frac{86Z}{At} \tag{2-25a}$$

with $E_1 = (1.3941 \pm 0.003)$ MeV and $E_2 = (-0.416 \pm 0.040)$ MeV. The deformation parameter δ is defined by $\delta^2 = (45/16)\sum_\mu \langle|a_{2\mu}|^2\rangle$ (see Chapter VI) and $t = \langle\!\langle \boldsymbol{l}\cdot\boldsymbol{\sigma}\rangle\!\rangle$ is the average over the $2T+1$ neutron orbitals that constitute the analog state in a shell model description [we have $\langle \boldsymbol{l}\cdot\boldsymbol{\sigma}\rangle = l$ for $j > l$, and $\langle \boldsymbol{l}\cdot\boldsymbol{\sigma}\rangle = -(l+1)$ for $j < l$]. The third term results from the Coulomb pairing interaction and is zero except for odd-Z nuclei. The last term is a Coulomb spin–orbit interaction and is generally $\leqslant 40$ keV. The deformation parameter reaches a peak value $\delta \approx 0.30$ in the mid-rare earth region.

Coulomb energies also yield information on the difference $\Delta = \langle r_n{}^2\rangle^{1/2} - \langle r_p{}^2\rangle^{1/2}$ discussed earlier. While the inferred results are model sensitive, the results suggest $\Delta \approx +(0.05\text{–}0.15)$ fm (Nolen and Schiffer, 1969).

Excellent discussions of the many important considerations that arise are contained in papers by Carlson and Talmi (1954), Sood and Green (1957), Sengupta (1960), and Jänecke (1960). Additionally papers by Wilkinson (p. 30 ff.) and Jänecke (p. 60 ff. in Fox and Robson, 1966), Harchol *et al.* (1967), and Nolen and Schiffer (1969) are valuable commentaries and reviews.

E. SUMMARY

In all cases discussed so far in Sections II, B to II, D the current accuracy in the experimental results requires various theoretical refinements in interpretation before the bulk proton (and to some extent the bulk neutron) density distributions can be extracted. However, certain gross features can be easily discerned that are consistent with all the evidence.

A density distribution function for both protons and neutrons of the type (2-4) holds rather well, at least for $A > 20$, with a simple scaling proportional to Z or N. The simple value $n = -1$ and w ranging from ~ -0.1 to $\sim +0.4$ is indicated. The surface thickness parameter $s = 4.40a$ increases from $s \approx 2.0$

[1] The values quoted relate to $\Delta E_c(\text{exchange})/\Delta E_c(\text{direct})$, with $\Delta E_c = \Delta E_c(\text{exchange}) + \Delta E_c(\text{direct})$.

fm at $A \approx 10$ to $s = 2.5$ fm at $A \approx 200$ and the radius parameter is given by $c = (1.18A^{1/3} - 0.48)$ fm. Shell effects and deformation effects are discernible in s and small undulations in $\rho(\mathbf{r})$ also associated with shell effects may be present. Defining $\frac{3}{5}R_1^{\,2} = \langle r^2 \rangle$ and $R_1 = r_1 A^{1/3}$, an approximate value $r_1 \approx (1.06 + 0.75A^{-1/3})$ fm is consistent with all the data.[1]

The quantity $\Delta = \langle r_n^{\,2} \rangle - \langle r_p^{\,2} \rangle$, while close to zero (i.e., $\Delta_{\text{exp}} \approx +0.1$ fm), does not imply that the shell-model-dependent undulations in $\rho(\mathbf{r})$ are the same for protons and neutrons. Perhaps the principal differences arise from the contributions of the neutrons constituting the $N - Z$ neutron excess. In experiments with π^- and K^- "atomic" X-rays, the strong absorption of these mesons in nuclear matter predominantly senses the surface densities. Only a slightly larger neutron rms radius translates to a very considerable neutron excess density in the surface (and extreme surface, i.e., tail) region. For example, a shell model calculation by Nolen and Schiffer (1969) for ^{208}Pb yields $\langle r_n^{\,2} \rangle^{1/2} = 5.622$ fm and $\langle r_p^{\,2} \rangle^{1/2} = 5.507$ fm (therefore, $\Delta = +0.115$) and maintains the ratio $\rho_n/\rho_p = 1.4$ out to $r = 6.0$ fm. But by $r = 8.0$ fm this ratio is 2.1 and at $r = 10.0$ fm it is 5.9, while at $r = 12.0$ fm it rises to 21.7. Also see the results in Kömer and Schiffer (1971).

F. NUCLEAR ELECTRIC MOMENTS

Information concerning the nuclear electric moments can be obtained by measuring the hyperfine structure in atomic transitions produced by the interaction energy of the nuclear charge distribution with the field in the vicinity of the nucleus generated by the atomic electrons. It is profitable to begin the discussion of this interaction energy with a semiclassical treatment. Consider the relatively localized nuclear charge distribution $\rho(x, y, z)$ present in the electric field potential $\phi(x, y, z)$ produced by the atomic electrons; then the interaction energy is

$$W = \int_v \rho\phi \, d\tau. \tag{2-26}$$

Of interest is the level shift for the initial and final atomic states due to perturbations of the form W above. The quantity ϕ in each instance relates to the field at the nucleus appropriate for the relevant atomic state.

Expanding $\phi(x, y, z)$ about the origin in a Taylor series

$$\phi(x, y, z) = \phi_0 + \sum_j \left.\frac{\partial \phi}{\partial x_j}\right|_0 x_j + \frac{1}{2} \sum_{j,k} \left.\frac{\partial^2 \phi}{\partial x_j \, \partial x_k}\right|_0 x_j x_k + \cdots, \tag{2-27}$$

[1] Note that $\frac{3}{5}r_1^{\,2} = a_1^{\,2}$; the parameter a_1 was introduced in connection with (2-10).

with x_j and x_k standing for x, y, z, and substituting into (2-26) gives

$$W = \phi_0 \int_v \rho \, d\tau + \sum_j \frac{\partial \phi}{\partial x_j}\bigg|_0 \int_v \rho x_j \, d\tau + \frac{1}{2} \sum_{j,k} \frac{\partial^2 \phi}{\partial x_j \, \partial x_k}\bigg|_0 \int_v \rho x_j x_k \, d\tau + \cdots. \tag{2-28}$$

The terms appearing in this series involving the moments of the nuclear charge distribution can be classified as:

(a) $\int_v \rho \, d\tau = q$, a scalar, the nuclear charge q;
(b) $\int_v \rho x_j \, d\tau = p_j$, a polar vector, the nuclear electric dipole moment $\mathbf{p}$;
(c) $\int_v \rho x_j x_k \, d\tau = eQ_{jk}$, a second-rank tensor, the nuclear electric quadrupole moment tensor;

and
(d) higher terms.

This definition of Q_{jk} gives the tensor elements the units of an area. The contributions to the interaction energy W of these electric moments consist of various direct products with the corresponding field derivatives. The contribution of the scalar charge, for example, is just $W_0 = \phi_0 q$, while that of the dipole moment is $W_1 = -\mathbf{p} \cdot \mathscr{E}$ with $\mathscr{E} = -\nabla\phi|_0$. The quadrupole interaction is the direct product of the two second-rank tensors $\phi_{jk} = (\partial^2\phi/\partial x_j \, \partial x_k)|_0$ and Q_{jk} multiplied by $e/2$. Higher moments are also occasionally of interest.

The ordinary Coulomb interaction is, of course, already incorporated in the generating of the atomic level scheme. Nuclear states with well-defined parity exhibit zero electric dipole moments, since expectation values for polar vector operators under such circumstances must vanish. Thus the lowest-order nuclear electric moment that need concern us is the quadrupole term.

For the quadrupole interaction, the product of the two tensors simplifies when we take note of the cylindrical symmetry present in the field produced by the atomic electrons. Let the z axis be taken along this symmetry axis. Then, writing $\phi_{jk} = (\partial^2\phi/\partial x_j \, \partial x_k)|_0$, only $\phi_{xx} = \phi_{yy}$ and ϕ_{zz} are nonzero, and therefore

$$W_2 = \tfrac{1}{2}e\{Q_{xx}\phi_{xx} + Q_{yy}\phi_{yy} + Q_{zz}\phi_{zz}\}. \tag{2-29a}$$

Further, excluding the interaction with $s_{1/2}$ and $p_{1/2}$ electrons so that Poisson's equation gives $\nabla^2\phi = 0$,[1] we have $\phi_{zz} = -2\phi_{xx} = -2\phi_{yy}$. Since $r^2 = x^2 + y^2 + z^2$, we can define the quantity $Q_{rr} = Q_{xx} + Q_{yy} + Q_{zz}$ and obtain

$$W_2 = \tfrac{1}{4}e\phi_{zz}(3Q_{zz} - Q_{rr}). \tag{2-29b}$$

[1] Only the $s_{1/2}$ and $p_{1/2}$ electronic states have significantly nonzero wave functions (i.e., charge densities) within the nuclear volume.

It is customary to call the quantity $Q = 3Q_{zz} - Q_{rr}$ simply the "quadrupole moment." Clearly, we have

$$eQ = \int_v (3z^2 - r^2)\rho \, d\tau$$

and also

$$eQ = \int_v (3\cos^2\theta - 1)\rho r^2 \, d\tau = 2\int P_2(\cos\theta)\rho r^2 \, d\tau, \qquad (2\text{-}30)$$

with $W_2 = \frac{1}{4}eQ\phi_{zz}$.

So far nothing has been said concerning any possible symmetry of $\rho(x, y, z)$. If indeed $\rho(x, y, z)$ also possesses axial symmetry in a body-centered coordinate system x', y', z', with z' the axis of symmetry, then only $Q_{x'x'} = Q_{y'y'}$ and $Q_{z'z'}$ are nonzero when the tensor elements are evaluated in the primed coordinate system. (While we do not wish to imply that nuclei are rotating rigid bodies, in some aspects they possess collective characteristics reminiscent of such behavior.) If the z' axis is inclined at an angle ζ to the z axis, then

$$eQ = \int_v (3\cos^2\theta - 1)r^2\rho(x, y, z)\, d\tau = \tfrac{1}{2}(3\cos^2\zeta - 1)eQ'$$

with

$$eQ' = \int_v (3\cos^2\theta' - 1)r'^2\rho(x', y', z')\, d\tau'. \qquad (2\text{-}31)$$

▶**Exercise 2-4** Consider Fig. 2.9 to give the relative geometry of the (x, y, z) and (x', y', z') coordinate systems. Show that (2-31) follows.

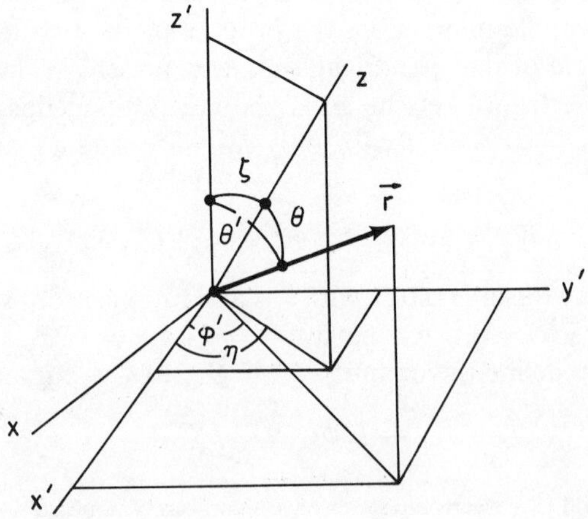

FIG. 2.9. The relationship of the (x, y, z) and (x', y', z') coordinate systems.

Hint: Use the relationships $\cos\theta = \cos\theta' \cos\zeta + \sin\theta' \sin\zeta \cos(\varphi' - \eta)$ and $\rho\, d\tau = \rho'\, d\tau'$. Also note that since $\rho' = \rho(r', \theta')$ is independent of φ', integration on φ' gives $\overline{\cos(\varphi' - \eta)} = 0$ and $\overline{\cos^2(\varphi' - \eta)} = \frac{1}{2}$, and hence the resulting trigonometric factors.

Alternately, the various elements of the quadrupole tensor Q_{st} can be calculated for any arbitrary rotation of the axes corresponding to the rotation matrix R_{sl} by computing $Q_{st} = R_{sl} Q'_{lm} R^{\dagger}_{mt}$. From this, $Q = 3Q_{zz} - Q_{rr}$ can be calculated if $Q' = 3Q_{z'z'} - Q_{r'r'}$ is known (of course, $Q_{rr} = Q_{r'r'}$). We should note that while Q'_{lm} is diagonal Q_{st} is not, but (2-29) and the following are consequences of the ϕ_{jk} tensor being diagonal, not Q_{jk}. ◀

It is instructive to note some classical values for the intrinsic (i.e., body-centered) quadrupole moments in special cases. A charge $+e$ in a circular orbit of radius r_0 in the $x'y'$ plane rotating about an otherwise spherically symmetric charge distribution gives $Q' = -e{r_0}^2$, using (2-30). Splitting the charge into two halves and placing them at $z' = \pm r_0$ would give $Q' = +2e{r_0}^2$. A pure spherically symmetric charge distribution gives $Q' = 0$, since $\overline{x'^2} = \overline{y'^2} = \overline{z'^2} = \frac{1}{3}\overline{r'^2}$. A "cigar"-shaped charge distribution $\rho(x', y', z')$ or *prolate* spheroid gives $Q' > 0$, while a "lens"-shaped or *oblate* spheroid gives $Q' < 0$.

▶**Exercise 2-5** Consider a uniformly charged ellipsoid of revolution about the z' axis with a total charge Ze. The surface equation of a typical spheroid is

$$\frac{z'^2}{b^2} + \frac{x'^2}{a^2} + \frac{y'^2}{a^2} = 1$$

with the semiaxis parallel to the z' axis equal to b and the semiaxis perpendicular to the z' axis equal to a. Show that the classical quadrupole moment in this case is

$$Q' = (8\pi/15)\,\rho a^2 b(b^2 - a^2) = \tfrac{2}{5}Z(b^2 - a^2).$$

A mean radius R and a deformation parameter δ can be introduced such that $R = (b+a)/2$, $\delta = (b-a)/R = 2(b-a)/(b+a)$, in terms of which

$$Q' = \tfrac{4}{5}(\delta Z)R^2.$$

For small δ, $b/a = 1 + \delta + \frac{1}{2}\delta^2 + \cdots$. Note also that Q' for the spheroid is evidently related to the difference in the principal moments of inertia $\frac{2}{5}Mb^2$ and $\frac{2}{5}Ma^2$. ◀

Figure 2.10 illustrates the results for general spheroids similar to the ones discussed above.

The quantum-mechanical treatment requires calculating the expectation value of the operator $e\tilde{Q} = 2r^2 P_2(\cos\theta)$ expressed in the atomic electron

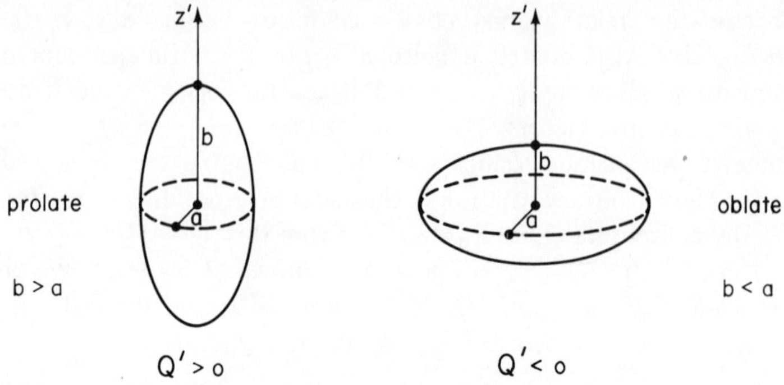

FIG. 2.10. Prolate and oblate charged spheroids and the sign of the quadrupole moment. The spheroids are figures of rotation about the z' axis.

coordinate system., i.e., the (x, y, z) system.[1] A nuclear wave function that is the simultaneous eigenstate of $\mathbf{J}^2$ and J_z in this coordinate system behaves in the semiclassical analog as a matter (and charge) distribution precessing about the z axis.[2] Under these circumstances the semiclassical value of $\cos^2\zeta = M_J{}^2/J(J+1)$ used in (2-31) would yield

$$Q = Q'[3M_J{}^2 - J(J+1)]/2J(J+1). \tag{2-32}$$

It is of interest to note that (2-32) gives zero observed quadrupole interaction when $J = \frac{1}{2}$ even though $Q' \neq 0$. While (2-32) is ambiguous for $J = 0$, the average value of $P_2(\cos\theta)$ for a space symmetric state is clearly zero. Thus, no quadrupole interaction results for $J = 0$ or $\frac{1}{2}$ even though Q' might be nonzero. This result can be seen at once by noting that $P_2(\cos\theta) = (4\pi/5)^{1/2} Y_2{}^0$ essentially represents two units of angular momentum and must satisfy the triangle condition with $\psi_J{}^*$ and ψ_J, namely $\mathbf{J} + (\mathbf{j} = \mathbf{2}) = \mathbf{J}$. The smallest value of J satisfying this condition is $J = 1$. When (2-32) is nonzero, the quoted value for the experimental quantity Q is, by convention, for the case M_J maximum or $M_J = J$.

The problem is somewhat more complicated in the actual atomic case due to the presence of the electronic angular momentum $\mathbf{I}$. The nuclear angular momentum $\mathbf{J}$ and the atomic angular momentum $\mathbf{I}$ couple to form the total angular momentum $\mathbf{F} = \mathbf{I} + \mathbf{J}$ about which each individual angular momentum

[1] More properly

$$Q = 2\sum_{i=1}^{Z} \int |\psi(\mathbf{r}_1, \mathbf{r}_2, \ldots, \mathbf{r}_A)|^2 P_2(\cos\theta_i)\, d^3r_1\, d^3r_2 \cdots d^3r_A,$$

the sum extending over all protons.

[2] It is customary in the literature on atomic fine structure to denote nuclear spins by I and the electronic angular momentum by J. Notwithstanding possible confusion, we invert this convention and refer to the nuclear angular momentum by J.

precesses in the semiclassical analog. The interaction perturbation now appropriately involves $\cos\zeta = \cos(\mathbf{I}, \mathbf{J})$, and classically

$$P_2(\cos\zeta) \approx [3(\mathbf{I}\cdot\mathbf{J})^2/2I^2J^2] - \tfrac{1}{2}. \tag{2-33}$$

The quantum-mechanical expression, due to Casimir, to which (2-33) is the classical correspondence principle limit (i.e., for large values of I and J), is

$$P_2(\cos\zeta) = \frac{3C(C+1) - 4I(I+1)J(J+1)}{2I(2I-1)J(2J-1)}, \tag{2-34}$$

with $C = F(F+1) - I(I+1) - J(J+1)$. Clearly (2-33) is the classical limit of (2-34) since $\mathbf{I}\cdot\mathbf{J} = C/2$. Details concerning the interpretation of the atomic hyperfine structure effects can be found in Kopfermann (1958).

It is instructive to calculate the quadrupole moment for a single particle of spin $\frac{1}{2}$ and orbital angular momentum l. This might represent a first-order model of a nuclear state with a very heavy, spherically symmetric core and a single proton in a simple single-particle orbital. The appropriate wave functions for the two possible values of $j = l \pm \frac{1}{2}$ and $j_z = m$ are, in obvious notation,[1]

$$|l, s=\tfrac{1}{2}, j=l+\tfrac{1}{2}, m\rangle = \frac{u(r)}{(2l+1)^{\frac{1}{2}}}\{(l+m+\tfrac{1}{2})^{\frac{1}{2}} Y_l^{m-\frac{1}{2}}\alpha + (l-m+\tfrac{1}{2})^{\frac{1}{2}} Y_l^{m+\frac{1}{2}}\beta\}$$

$$|l, s=\tfrac{1}{2}, j=l-\tfrac{1}{2}, m\rangle = \frac{u(r)}{(2l+1)^{\frac{1}{2}}}\{-(l-m+\tfrac{1}{2})^{\frac{1}{2}} Y_l^{m-\frac{1}{2}}\alpha + (l+m+\tfrac{1}{2})^{\frac{1}{2}} Y_l^{m+\frac{1}{2}}\beta\}. \tag{2-35}$$

We wish to calculate the expectation value of the quadrupole operator $\tilde{Q} = 2r^2 P_2(\cos\theta) = r^2(3\cos^2\theta - 1)$. A useful theorem that is required is

$$\langle Y_l^m | P_2(\cos\theta) | Y_l^m\rangle = [l(l+1) - 3m^2]/(2l+3)(2l-1). \tag{2-36}$$

From (2-35) and (2-36) it is readily deduced that for either case $j = l \pm \frac{1}{2}$ and for a particular value of m,

$$Q = \frac{\langle r^2\rangle}{2j(j+1)}\{j(j+1) - 3m^2\} \qquad \text{with} \quad \langle r^2\rangle = \int u^2(r)\, r^4\, dr. \tag{2-37}$$

This result is in agreement with the semiclassical result (2-32) with $Q' = -\langle r^2\rangle$.

▶**Exercise 2-6** Using the relationship

$$(\cos\theta)\, Y_l^m = \left[\frac{(l+m+1)(l-m+1)}{(2l+1)(2l+3)}\right]^{1/2} Y_{l+1}^m + \left[\frac{(l+m)(l-m)}{(2l+1)(2l-1)}\right]^{1/2} Y_{l-1}^m, \tag{2-38}$$

show that (2-36) follows. Then complete the derivation of (2-37) from (2-35) for $j = l \pm \frac{1}{2}$. ◀

[1] Refer to Appendix A.

The result (2-37) should be modified for both the recoil effects of the core and the polarizing deformation of the core induced by the single particle. The recoil effect arises from the requirement that $\tilde{Q}$ should properly be referred to the center of mass. To a first approximation, the core with charge $Z_c e$ would have a center-of-mass coordinate $\approx r_n/A_c$ and hence $(\Delta Q)_{rec} \approx (Z_c/A_c^2) Q_{sp}$ represents the recoil correction with Q_{sp} the single-particle quadrupole value. Except for the lightest nuclei, this correction would be small. A rough estimate of the polarization effect can be obtained by noting the volume eccentricity dependence on δ, namely that $\delta \approx \Delta A/A$.[1] Thus the potential that each *core* particle responds to is slightly deformed by the presence of the single particle, and noting the results of Exercise 2-5, an additional quadrupole moment for each core proton of order Q_{sp}/A results. Consequently, the core protons contribute an additional moment $(\Delta Q)_{pol} \approx (Z_c/A) Q_{sp}$. While the recoil effect is seen to be small, the polarization effect is of order Q_{sp}. A largely spherical core plus a single neutron would thus have a nonzero quadrupole moment due to the polarization of the otherwise symmetric core of $Q_n \approx (Z/A) Q_p$, with the same sign as Q_p calculated for a single proton.

The nuclei for which either Z or N equals a "magic" number[2] are very close to spherically symmetric. The addition of another proton or neutron might be expected to result in a quadrupole moment not unlike that predicted by (2-37). A magic number (closed-shell) nucleus *less* a nucleon also behaves like a single-particle *hole* and would give an opposite sign for Q. Figure 2.11 shows the observed quadrupole moments in units of $ZR^2 \times 10^{-2}$ ($R \approx 1.20 A^{1/3}$ fm), a convenience suggested by the results of Exercise 2-5. As expected, the values of Q are seen to pass through $Q = 0$ from positive to negative values as the nucleon number increases through a magic number. The magnitudes also agree reasonably well with ascribing the quadrupole moment to a single particle or hole. This behavior is clearly exhibited in the vicinity of the magic number $Z = 50$ by the nuclei ^{115}In (with $Z = 49$), for which $Q = +0.83$ b, and ^{121}Sb (with $Z = 51$), for which $Q = -0.53$ b.[3] It should be noted that $\langle r^2 \rangle \approx 0.22$ b ($A = 118$). A typical example for a neutron hole is the nucleus ^{35}S (with $N = 19$), for which $Q = +0.06$ b, where it should be noted that $(Z/A)\langle r^2 \rangle \approx 0.05$ b.

Figure 2.11 also shows that, particularly among the positive quadrupole moments, there are very large moments, many times the single-particle values of (2-37), notably in the rare earths. Such nuclei exhibit strong collective

[1] The deformation volume eccentricity ΔV can be defined as the difference between the total deformed volume $\frac{4}{3}\pi ba^2$ and the spherical central portion $\frac{4}{3}\pi a^3$, viz., $\Delta V = \frac{4}{3}\pi a^2 (b-a)$. Then $\Delta V/V = (b-a)/b \approx \delta$, and since $\Delta V/V = \Delta A/A$, the quoted result follows.

[2] The magic numbers are 2, 8, 20, (28), (40), 50, 82, (114), 126, (164), and (184). See Chapter IV.

[3] b represents the *barn*, a unit of area equal to 10^{-24} cm^2.

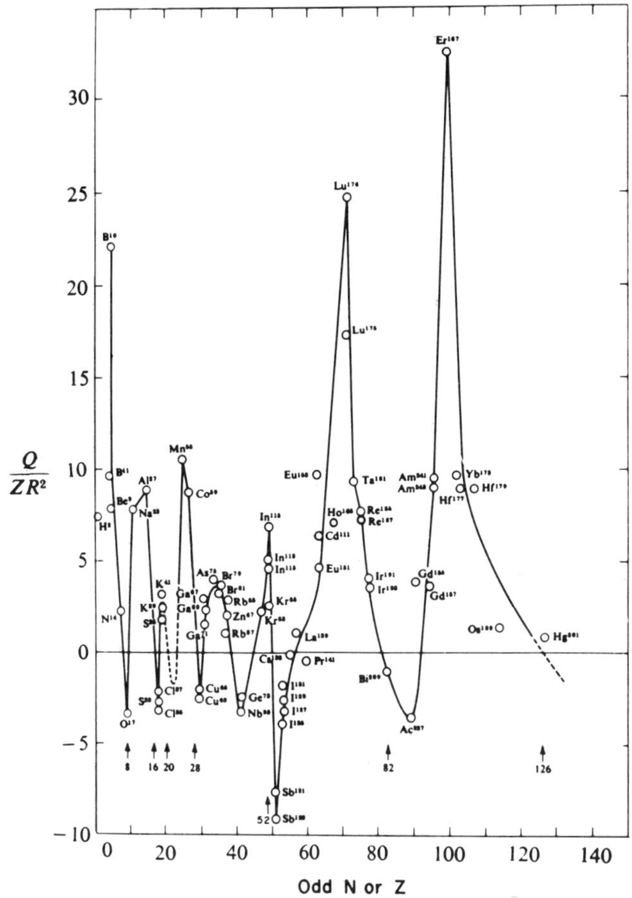

FIG. 2.11. Experimental quadrupole moments in units of $ZR^2 \times 10^{-2}$ ($R \approx 1.20A^{1/3}$ fm). Values can be found tabulated in Nuclear Data Group (1966) [figure adapted from Segrè (1964)].

behavior and have large deformations. A general discussion of these nuclei is given in Chapter VI. For present purposes we give only the briefest relevant description.

While nuclei with large intrinsic deformations do not behave as rigid rotators, rotations do arise due to circulating surface waves with moments of inertia reduced from expected rigid-body values. Body-centered axes (x', y', z') can be introduced. The nuclear wave function is assumed invariant to rotations about the z' axis, and the corresponding moment of inertia is small and usually taken as $\mathscr{I}_{z'} = 0$. The effective moments of inertia about axes perpendicular to z' are found to be $\mathscr{I}_{x'} = \mathscr{I}_{y'} \approx (0.2\text{–}0.5)\mathscr{I}_{\text{rigid}}$. A case of interest arises when a loose single nucleon couples to such a deformed core. In the relevant case of strong coupling the angular momentum **j** of the nucleon is not a constant of the motion. However, the projection of **j** on the core symmetry axis z', denoted by Ω, is a constant of the motion. The possible values of Ω are just the $2j+1$ possible projections of **j**. Finally, the total nuclear angular momentum **J** is, as shown in Fig. 2.12, characterized by a z component M_J and a projection on the z' axis Ω. As shown, the intrinsic angular momentum **R** of the deformed axially symmetric core is perpendicular to z'.

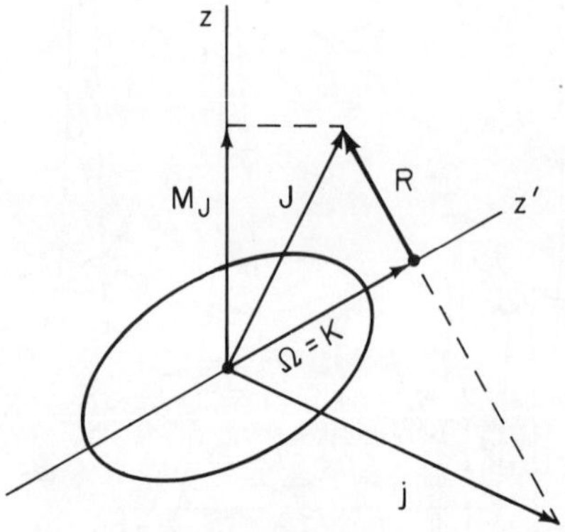

FIG. 2.12. The vector model for a loose nucleon and a deformed core with strong coupling.

The total quadrupole moment can be written as $Q = Q_j + Q_c$, composed of Q_j due to the loose nucleon and Q_c due to the core. Generally $Q_c \gg Q_j$. For this limiting model, the intrinsic core quadrupole moment Q' can be viewed as precessing about the **J** axis and in the classical limit $\cos\zeta \approx \cos(\Omega, J)$ in (2-31), so that

$$Q_c = Q'[(3\Omega^2/2J^2) - \tfrac{1}{2}]. \tag{2-39}$$

The expression (2-39) is the correspondence principle limit of the rigorous expression

$$Q_c = Q' \frac{J}{2J+3}\left[\frac{3\Omega^2}{J(J+1)} - 1\right]. \tag{2-40}$$

For the ground state we have $\Omega = J$ and Q' can be taken equal to $\frac{4}{5}\delta Z R^2$ for a deformation δ. Consider as an example ^{177}Lu with $Q_c = +5.51$ b (or $Q_c/ZR^2 = 17 \times 10^{-2}$ from Fig. 2.11) and $J = \frac{7}{2}$. These values result in a calculated deformation $\delta = 0.45$.[1] We again note that, for $J = 0$ or $\frac{1}{2}$, (2-40) gives zero.

[1] The parameter $\beta = \frac{2}{3}(4\pi/5)^{1/2}\delta \approx 1.06\delta$ is also commonly used; see (6-127).

G. NUCLEAR MAGNETIC MOMENTS

There being no magnetic monopole term in nature, the lowest-order magnetic moment of the nucleus is the dipole term. Thereafter only odd magnetic moments are nonzero for nuclear states of well-defined parity (refer to Section I, B, 2). We now turn to examining the nuclear magnetic dipole moment. It is convenient to express all contributions to the magnetic dipole moment in units of the nuclear magneton $\mu_0 = e\hbar/2m_p c = 5.0505 \times 10^{-24}$ erg/Gauss.

A fundamental assumption is made that each angular momentum-like vector quantity $\boldsymbol{\xi}$ has associated with it a magnetic moment vector $\boldsymbol{\mu}_\xi = \mu_0 g^\xi \boldsymbol{\xi}$ (the subscript on $\boldsymbol{\mu}$ signifies this causative connection).[1] The quantity g is referred to as the gyromagnetic ratio. The total nuclear magnetic moment is taken to be the vector sum of the nucleon contributions due to both orbital angular momentum and intrinsic spin-related magnetic moments. Thus

$$\boldsymbol{\mu}_J = \mu_0 g^J \mathbf{J} = \mu_0 \sum_{i=1}^{A} (g_i{}^l \boldsymbol{l}_i + g_i{}^s \mathbf{s}_i), \tag{2-41}$$

where the total nuclear magnetic moment is associated with the total nuclear angular momentum $\mathbf{J}$. Here for neutrons, $g^l = 0$ and $g^s = -2 \times 1.913$, while for protons, $g^l = +1$ and $g^s = +2 \times 2.793$; also $\boldsymbol{l} = \mathbf{r} \times \mathbf{p}/\hbar$ and $\mathbf{s} = \boldsymbol{\sigma}/2$.[2] Since the various g^l and g^s factors are not equal, g^J is not related to these in any simple way.

The nuclear magnetic moment gives rise to hyperfine structure lines in atomic spectra due to the interaction energy of this moment with the magnetic field $\mathscr{H}_0$ in the vicinity of the nucleus produced by the atomic electrons.[3] The atomic magnetic field is of course parallel to the total atomic angular momentum $\mathbf{I}$. The interaction energy is then classically

$$W_{\text{mag}} = -\boldsymbol{\mu}_J \cdot \mathscr{H}_0 = -\mu_0 g^J \mathscr{H}_0 J \cos(\mathbf{J}, \mathbf{I}). \tag{2-42}$$

If again $\mathbf{F} = \mathbf{I} + \mathbf{J}$, then $\mathbf{J} \cdot \mathbf{I} = C/2$ with C as before, $C = F(F+1) - I(I+1) - J(J+1)$; thus for the exact quantum-mechanical expression we have

$$W_{\text{mag}} = -\mu_0 g^J C \mathscr{H}_0 / 2[I(I+1)]^{1/2}. \tag{2-43}$$

[1] The generalized angular momentum $\boldsymbol{\xi}$ is in units of $\hbar$, this factor having been incorporated into μ_0.

[2] The suggestion has been put forward that mesonic-exchange currents may give anomalous effective g^l factors: $g^l(\text{p}) = 1.09 \pm 0.03$ and $g^l(\text{n}) = -0.06 \pm 0.04$ for the proton and neutron, respectively [e.g., see Nagamiya and Yamazaki (1971) and also Arima and Huang-Lin (1972)].

[3] Of course, in some experimental situations, such as that required to observe the Zeeman effect, an external magnetic field may also be present. We do not consider this case here; the reader is referred to Kopfermann (1958) for further details.

The general feature of (2-42) and (2-43), namely that a particular projection of $\boldsymbol{\mu}_J$ is required in various experimental situations, leads to a simple way of prescribing g^J. The nuclear state is taken to be the simultaneous eigenstate of $\mathbf{J}^2$ and J_z and we inquire about the z component of $\boldsymbol{\mu}_J$. The Wigner–Eckart theorem (Appendix B) states that for any vector operator $\tilde{\mathbf{A}}$

$$\langle J, M_J | A_z | J, M_J \rangle = [M_J/J(J+1)] \langle J, M_J | \tilde{\mathbf{A}} \cdot \tilde{\mathbf{J}} | J, M_J \rangle. \qquad (2\text{-}44)$$

This is essentially the basis of the semiclassical vector model to which we have occasionally referred. In this model the individual contributions $\mu_0 g_i{}^l \boldsymbol{l}_i$ and $\mu_0 g_i{}^s \mathbf{s}_i$ are considered to be precessing about $\mathbf{J}$, which in turn precesses about the z axis. Thus the z-component contributions involve two projections: one onto $\mathbf{J}$ implied by $\mathbf{A} \cdot \mathbf{J}/[J(J+1)]^{1/2}$ for the first precession and then a second onto the z axis giving the factor $M_J/[J(J+1)]^{1/2}$ for the second precession. The previous discussion of the quadrupole moment could also have been treated using the Wigner–Eckart theorem in the form applicable to tensor quantities.

Taking the form of (2-41) also to define correctly the corresponding quantum-mechanical operator, we have

$$\langle (\boldsymbol{\mu}_J)_z \rangle = \mu_0 g^J M_J = \mu_0 \int \psi^*_{J, M_J} \sum_{i=1}^{A} (g_i{}^l l_{i,z} + g_i{}^s s_{i,z}) \psi_{J, M_J} \, d\tau. \qquad (2\text{-}45)$$

By convention the quantity tabulated for various nuclei is $\langle (\boldsymbol{\mu}_J)_{z,\max} \rangle = \mu_0 g^J J$, with $M_J = J$. It is instructive to evaluate (2-45) for the extreme single-particle model in which only a single "valence" nucleon is effective in determining the entire nuclear magnetic moment. We again use the wave functions (2-35), however with $m = j$, whence

$$\begin{aligned} |l, s = \tfrac{1}{2}, j = l + \tfrac{1}{2}, m = j\rangle &= u(r) Y_l{}^l \alpha \\ |l, s = \tfrac{1}{2}, j = l - \tfrac{1}{2}, m = j\rangle &= [u(r)/(2l+1)^{1/2}][-Y_l{}^{l-1}\alpha + (2l)^{1/2} Y_l{}^l \beta]. \end{aligned} \qquad (2\text{-}46)$$

Then using the labels n and p to designate the case for either a neutron or proton, the result is

$$\begin{aligned} j = l + \frac{1}{2}: \quad & \frac{1}{\mu_0} \langle (\boldsymbol{\mu}_J)_z \rangle = g^l_{\mathrm{n,p}} \left(j - \frac{1}{2} \right) + \frac{1}{2} g^s_{\mathrm{n,p}} \\ j = l - \frac{1}{2}: \quad & \frac{1}{\mu_0} \langle (\boldsymbol{\mu}_J)_z \rangle = \frac{g^l_{\mathrm{n,p}}}{2(j+1)} \left[2j \left(j + \frac{3}{2} \right) \right] - \frac{j}{(2j+1)} g^s_{\mathrm{n,p}}. \end{aligned} \qquad (2\text{-}47)$$

▶**Exercise 2-7** (a) Carry out the derivation leading to (2-47).

(b) Show that (2-41) can be written in terms of neutron and proton projection operators to give

$$\boldsymbol{\mu}_J = \sum_{\xi=1}^{A} [\tfrac{1}{2}(1 + \tau_3{}^\xi)(\mu_{\mathrm{p}} \boldsymbol{\sigma}^\xi + \boldsymbol{l}^\xi) + \tfrac{1}{2}(1 - \tau_3{}^\xi) \mu_{\mathrm{n}} \boldsymbol{\sigma}^\xi].$$

The portion of this operator involving τ_3 (the third component of a vector) is referred to as the isovector contribution, while the remaining portion is called the isoscalar contribution. (Recall Exercise 1-1, Chapter I.) ◀

▶**Exercise 2-8** Noting that $\mathbf{J} = \mathbf{L}+\mathbf{S}$, from which $\cos(\mathbf{L},\mathbf{J})$ and $\cos(\mathbf{S},\mathbf{J})$ can be determined, use the vector model implication of the right-hand side of (2-44) to derive $\langle(\boldsymbol{\mu}_J)_{z,\max}\rangle$ for the case $J = L+S$ (the so-called parallel case) and for the case $J = |L-S|$ (the so-called antiparallel case). Particularize for the single-particle case $L = l$, $S = \frac{1}{2}$ and check (2-47).

Show that Eq. (1-80) for the quark model is correctly given and that (1-81) follows. ◀

The numerical results for (2-47) are

$$\begin{aligned} &\text{neutron} \begin{cases} j = l+\frac{1}{2}: & g^J = -1.913/j \\ j = l-\frac{1}{2}: & g^J = +1.913/(j+1) \end{cases} \\ &\text{proton} \begin{cases} j = l+\frac{1}{2}: & g^J = 1 + 2.293/j \\ j = l-\frac{1}{2}: & g^J = 1 - 2.293/(j+1). \end{cases} \end{aligned} \tag{2-48}$$

When these values of g^J are considered as functions of j, the resulting plot yields the so-called "Schüler–Schmidt lines." Figure 2.13 shows the plot of these lines and the experimental values of nuclear g^J factors. It will be observed

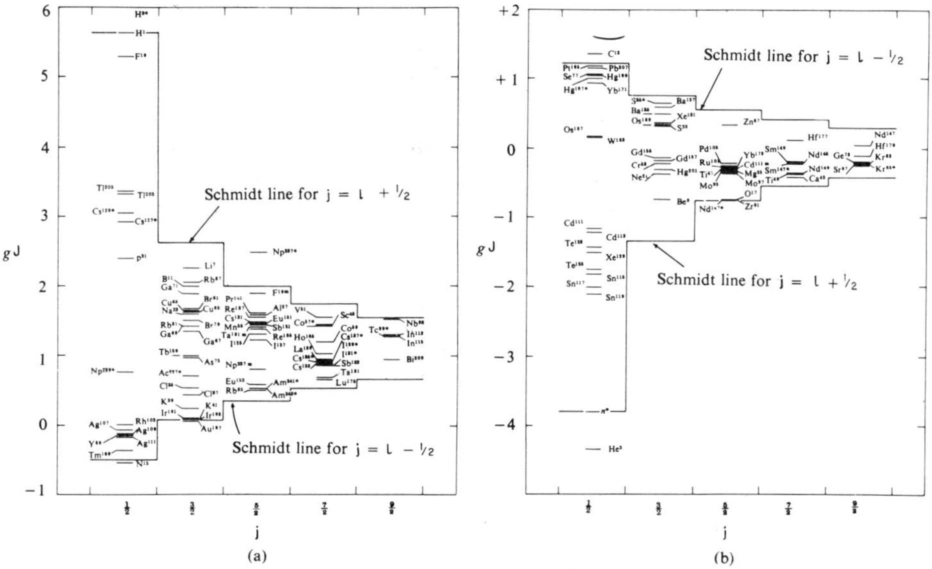

FIG. 2.13. The Schüler–Schmidt lines and the experimental nuclear g^J factors as functions of j for (a) odd-Z and (b) odd-N nuclei [taken from Segrè (1964)].

that most experimental values fall within the Schmidt limits. The nuclei with N or Z equal to a "magic" number ± 1 come close to agreeing with the Schmidt single-particle values, particularly for the lighter nuclei. The hole states have the *same signs* for g^J as that for the corresponding particles since, for the hole state, $m = -j$ rather than $m = +j$, and this *reverses* the change in sign for the charge normally associated with the hole state. For example, ^{41}Ca (odd neutron in an $f_{7/2}$ orbital) has an observed $jg^J = -1.60$, while (2-48) for $j = l + \frac{1}{2}$ predicts $jg^J = -1.91$. For ^{39}K (proton hole in a $d_{3/2}$ orbital) the observed value of the magnetic moment is $jg^J = +0.39$, while (2-48) for $j = l - \frac{1}{2}$ gives $jg^J = +0.13$.

In the case of self-conjugate nuclei (i.e., nuclei with $N = Z$ and hence $T_3 = 0$) only an isoscalar nuclear contribution to the magnetic moment is present. By averaging the moments of corresponding states in mirror nuclei (i.e., nuclear states in two isobars that have identical quantum numbers except that one has a charge T_3 while the other has charge $-T_3$) the isovector contributions are canceled, again revealing the isoscalar component, viz.,

$$\mu^S = \bar{\mu} = \tfrac{1}{2}[\mu(A, T_3) + \mu(A, -T_3)]. \qquad \text{(2-49a)}$$

In a nuclear model where the nucleons are identified as those belonging to an inert closed-shell core plus the remaining nucleons considered as valence particles all in the same j subshell [with the jj vector coupling $(\mathbf{j})^n = \mathbf{J}$ giving the total nuclear angular momentum],[1] the isoscalar magnetic moment simply becomes

$$\mu^S = \mu_0 (g^J J)^S = J\mu_0 \left(\frac{1}{2} \pm \frac{0.380}{2l+1}\right), \qquad \text{for} \quad j = l \pm \frac{1}{2}. \qquad \text{(2-49b)}$$

A comparison of the experimental moments for self-conjugate nuclei and mirror averages from (2-49a) with the predictions of (2-49b) are surprisingly close [see Van Hienen and Glandemans (1972)]. The degree of agreement with the simple j^n-configuration model, which is known to be only approximate, suggests that the calculation of the isoscalar term is rather independent of the size of the space of configurations considered (see Chapter V).

The effective magnetic moment for deformed nuclei is in general quite complicated and is governed by the strength of the coupling of loose nucleons to the core and their coupling to each other. A relatively simple description is possible for the case of a single loose nucleon strongly coupled to a deformed core, i.e., the case illustrated in Fig. 2.12. Using the vector model, we can write

$$\boldsymbol{\mu}_J = \mu_0 g^J \mathbf{J} = \mu_0 (g^\Omega \boldsymbol{\Omega} + g^c \mathbf{R}).$$

Here the gyromagnetic ratio g^c can be approximated by $g^c \approx Z_c/A$, and the

[1] For a general discussion of nucleon coupling schemes and configuration mixing refer to Chapter V.

value of g^{Ω} can be treated as an empirical constant. Then for $M_J = J$,

$$Jg^J = \frac{(\mathbf{\Omega}\cdot\mathbf{J})}{J(J+1)} Jg^{\Omega} + \frac{(\mathbf{R}\cdot\mathbf{J})}{J(J+1)} Jg^{c}. \tag{2-50}$$

Now $\mathbf{\Omega}\cdot\mathbf{J}+\mathbf{R}\cdot\mathbf{J} = \mathbf{J}\cdot\mathbf{J}$ since $\mathbf{J} = \mathbf{\Omega}+\mathbf{R}$; hence

$$g^J = g^{\Omega}\frac{\mathbf{\Omega}\cdot\mathbf{J}}{J(J+1)} + g^{c}\frac{J(J+1)-\mathbf{\Omega}\cdot\mathbf{J}}{J(J+1)}. \tag{2-51}$$

For the ground state $J = \Omega$, therefore,

$$g^J = [1/(J+1)][Jg^{\Omega}+g^{c}]. \tag{2-52}$$

If $\mathbf{j}$ is also a constant of the motion, then $g^{\Omega} \to g_S{}^J$, where $g_S{}^J$ stands for the Schmidt value for g^J. Thus, finally,

$$g^J = g_S{}^J + \frac{1}{J+1}\left(\frac{Z_c}{A} - g_S{}^J\right). \tag{2-53}$$

The net effect is to alter the Schmidt value by an amount

$$[1/(J+1)][(Z_c/A)-g_S{}^J)].$$

Two examples of permanently deformed nuclei that can be considered to have strongly coupled single-valence nucleons with j a good quantum number are ^{27}Al and ^{25}Mg. Each has $\delta \approx 0.4$; the valence proton for ^{27}Al and the valence neutron for ^{25}Mg are in $\frac{5}{2}+[202]$ Nilsson orbitals and hence $j = \frac{5}{2}$ is a good quantum number (see Section VI,E,4). Direct application of (2-53) gives calculated results $Jg_S{}^J = +3.74$ and -1.03, respectively, as compared to experimental values $Jg_S{}^J = +3.64$ and -0.86, respectively.

It is also interesting to apply (2-53) to a spherical collective vibrator such as ^{105}Pd possessing an odd neutron in a "generic" $d_{5/2}$ orbital. The value of $Jg_S{}^J$ would be -1.91; the above correction term is $+0.86$, giving an approximate theoretical value of -1.05, to be compared with the observed value $Jg^J = -0.62$. In this case j is not a good quantum number, and a more exact calculation using (2-52) gives a theoretical value of -0.81.

The effect of (2-52) or (2-53) is in general to shift the Schmidt values "inward," thus accounting for the observed values of g^J for deformed nuclei lying well inside the Schmidt limits (refer to Fig. 2.13).

H. THE TENSOR OPERATOR $\tilde{S}_{12}$

We now return to the general two-nucleon problem discussed earlier. The most general central potential was found to be (1-84), i.e.,

$$\tilde{V}_{\text{cen}} = V_R(r) + V_\sigma(r)\tilde{\boldsymbol{\sigma}}_1\cdot\tilde{\boldsymbol{\sigma}}_2 + V_\tau(r)\tilde{\boldsymbol{\tau}}_1\cdot\tilde{\boldsymbol{\tau}}_2 + V_{\sigma\tau}(r)(\tilde{\boldsymbol{\sigma}}_1\cdot\tilde{\boldsymbol{\sigma}}_2)(\tilde{\boldsymbol{\tau}}_1\cdot\tilde{\boldsymbol{\tau}}_2), \tag{2-54}$$

with $\mathbf{r} = \mathbf{r}_2 - \mathbf{r}_1$ and $r = |\mathbf{r}|$. The operators $\tilde{\mathbf{J}}^2$, $\tilde{J}_z$, $\tilde{\mathbf{L}}^2$, $\tilde{\mathbf{S}}^2$, $\tilde{L}_z$, $\tilde{S}_z$, and $\tilde{P}$ all commute with (2-54) and therefore are constants of the motion, i.e., their expectation values are independent of time. These operators do not, however, all commute among themselves, and therefore simultaneous eigenfunctions can only be constructed for various commuting subsets of these operators, for example, $\tilde{\mathbf{J}}^2$, $\tilde{J}_z$, $\tilde{\mathbf{L}}^2$, $\tilde{\mathbf{S}}^2$, and $\tilde{P}$. Various forms of notation are used for such eigenfunctions, such as $|J, M_J\rangle_{L,S}$ or its equivalent $\mathscr{Y}^{M_J}_{J,S,L}(\theta, \varphi)$. Since $\mathbf{J} = \mathbf{L} + \mathbf{S}$, the Clebsch–Gordan coefficients of Appendix A can be used to relate these functions to the simultaneous eigenfunctions of $\tilde{\mathbf{L}}^2$, $\tilde{\mathbf{S}}^2$, $\tilde{L}_z$, $\tilde{S}_z$, and $\tilde{P}$. For example,

$$\begin{aligned} \mathscr{Y}^1_{110} &= (4\pi)^{-1/2}\chi_1{}^1(\boldsymbol{\sigma}_1, \boldsymbol{\sigma}_2) = (4\pi)^{-1/2}\alpha(1)\,\alpha(2) = Y_0{}^0\chi_1{}^1 \\ \mathscr{Y}^1_{112} &= \sqrt{\tfrac{3}{5}}\,Y_2{}^2\chi_1^{-1} - \sqrt{\tfrac{3}{10}}\,Y_2{}^1\chi_1{}^0 + \sqrt{\tfrac{1}{10}}\,Y_2{}^0\chi_1{}^1. \end{aligned} \tag{2-55}$$

The two-body tensor operator $\tilde{S}_{12} = (3/r^2)(\boldsymbol{\sigma}_1 \cdot \mathbf{r})(\boldsymbol{\sigma}_2 \cdot \mathbf{r}) - \boldsymbol{\sigma}_1 \cdot \boldsymbol{\sigma}_2$ was also introduced earlier (1-82). The only property discussed at that point was the fact that S_{12} was defined in a manner to have a zero angular average. We must now examine its various commutation properties and the consequence of a potential term that includes it in the Hamiltonian.

Since S_{12} is obviously an even operator under space inversion, i.e., $\mathbf{r} \to -\mathbf{r}$, we have $[S_{12}, P] = 0$, and hence parity remains a constant of the motion. While S_{12} commutes with $\mathbf{J}^2$ and J_z, it does not with $\mathbf{L}^2$. Importantly, S_{12} does commute with $\mathbf{S}^2$. There are many ways to show this. An interesting way is to note that $[\mathbf{T}^2, S_{12}] = 0$ (trivially since they operate in different spaces), and, since $[P, S_{12}] = 0$, we use the requirement that the *two-nucleon* wave functions must obey the exclusion principle, namely, $(-1)^{\frac{1}{2}(1-\pi)+S+T} = -1$,[1] $\pi = +1$ for even parity and $\pi = -1$ for odd parity, coupled with the fact that $S = 0, 1$ only, to give $[\mathbf{S}^2, S_{12}] = 0$. Although $[\mathbf{S}^2, S_{12}] = 0$, $\mathbf{S}_z$ does not commute with S_{12} and neither does L_z. Thus with the Hamiltonian containing a term in S_{12} only $\mathbf{J}^2$, J_z, $\mathbf{S}^2$, and P are constants of the motion. Since the corresponding operators all commute with each other, we can use their simultaneous eigenvectors as a basis. These are most conveniently written as linear combinations of various states $|J, M_J\rangle_{L,S}$.

▶**Exercise 2-9** Show that $\mathbf{J}^2$ and J_z commute with S_{12}. Show that $\mathbf{L} \cdot \mathbf{S}$, $\mathbf{L}^2$, L_z, and S_z do not commute with S_{12}. ◀

As Exercise 2-9 shows, $\mathbf{L}^2$ is no longer a constant of the motion if H contains a term in S_{12}. However, since in this case the parity is still a constant of the motion, only all odd or all even values of L are mixed. Since the triangle condition $\mathbf{J} = \mathbf{L} + \mathbf{S}$ must also be satisfied, states of a given J can only have

[1] For the two-nucleon system space inversion and space exchange are equivalent and $(-1)^L = (-1)^{(1-\pi)/2}$.

$L = J-1$ and $L = J+1$ of one parity mixed, and $L = J$ alone with the other parity. Since $\mathbf{S}^2$ is also a constant of the motion, we have for basis states:

$$
\begin{array}{llllll}
\text{spin singlets S} = 0: & {}^1\mathrm{S}_0 & J = 0, & L = 0, & \pi = + \\
 & {}^1\mathrm{P}_1 & J = 1, & L = 1, & \pi = - \\
 & {}^1\mathrm{D}_2 & J = 2, & L = 2, & \pi = + \\
 & \vdots & & & \\
\text{spin triplets S} = 1: & {}^3\mathrm{P}_0 & J = 0, & L = 1, & \pi = - \\
 & \left[\begin{array}{c} {}^3\mathrm{S}_1 + {}^3\mathrm{D}_1 \\ {}^3\mathrm{P}_1 \end{array}\right] & \begin{array}{l} J = 1, \\ J = 1, \end{array} & \begin{array}{l} L = 0,2, \\ L = 1, \end{array} & \begin{array}{l} \pi = + \\ \pi = - \end{array} \\
 & \left[\begin{array}{c} {}^3\mathrm{P}_2 + {}^3\mathrm{F}_2 \\ {}^3\mathrm{D}_2 \end{array}\right] & \begin{array}{l} J = 2, \\ J = 2, \end{array} & \begin{array}{l} L = 1,3, \\ L = 2, \end{array} & \begin{array}{l} \pi = - \\ \pi = +. \end{array} \\
 & \vdots & & &
\end{array}
\tag{2-56}
$$

For various two-nucleon state computations, we also need the result for S_{12} operating on the functions $\mathscr{Y}^{M_J}_{J,S,L}$. Since for the singlet spin state the total spin operator is (omitting $\hbar$ for convenience) $\mathbf{S} = \frac{1}{2}(\boldsymbol{\sigma}_1 + \boldsymbol{\sigma}_2) = 0$, clearly $\boldsymbol{\sigma}_1 = -\boldsymbol{\sigma}_2$, and then directly from the definition of S_{12}, we have $S_{12} = -(3/r^2)(\boldsymbol{\sigma}_1 \cdot \mathbf{r})^2 + {\boldsymbol{\sigma}_1}^2 = -3+3 = 0.$[1] Thus

$$S_{12}\mathscr{Y}^{M_J}_{J,0,L} = 0. \tag{2-57}$$

The ground state of the deuteron in the presence of the tensor potential will be represented by the state that is the linear combination of ${}^3\mathrm{S}_1$ and ${}^3\mathrm{D}_1$. Of the triplet spin states $\mathscr{Y}^{M_J}_{J,1,L}$ we most particularly need $S_{12}\mathscr{Y}^{M_J}_{110}$ and $S_{12}\mathscr{Y}^{M_J}_{112}$. These results, as well as those for any other triplet states, can be calculated directly from

$$
\begin{aligned}
S_{12} = {} & 3(\sigma_{1x}\sin\theta\cos\varphi + \sigma_{1y}\sin\theta\sin\varphi + \sigma_{1z}\cos\theta) \\
& \times (\sigma_{2x}\sin\theta\cos\varphi + \sigma_{2y}\sin\theta\sin\varphi + \sigma_{2z}\cos\theta) \\
& - (\sigma_{1x}\sigma_{2x} + \sigma_{1y}\sigma_{2y} + \sigma_{1z}\sigma_{2z}).
\end{aligned}
\tag{2-58}
$$

Alternately, a short-cut method used by Blatt and Weisskopf (1952) can be employed. Consider, for example, $S_{12}\mathscr{Y}^1_{110}$. Since S_{12} and $\mathbf{J}^2, \mathbf{S}^2, J_z$ commute, using (2-56) gives

$$S_{12}\mathscr{Y}^1_{110} = A\mathscr{Y}^1_{110} + B\mathscr{Y}^1_{112}. \tag{2-59}$$

[1] This follows from ${\boldsymbol{\sigma}_1}^2 = \sigma_{1x}^2 + \sigma_{1y}^2 + \sigma_{1z}^2$, $\boldsymbol{\sigma}_1 \cdot \mathbf{r} = x\sigma_{1x} + y\sigma_{1y} + z\sigma_{1z}$, and the commutation relationships for $\sigma_i\sigma_j$.

We next average (2-59) over all directions in space (i.e., integrate over $d\Omega$). Noting that $\mathscr{Y}^1_{110}$ is independent of (θ, φ) and that both $\int S_{12}\, d\Omega = 0$ and $\int \mathscr{Y}^1_{112}\, d\Omega = 0$, clearly requires that $A = 0$. Further, since the resulting equation must hold for any (θ, φ), consider the special case $\theta = 0$ or $\hat{r} = \hat{z}$. Then with $S_{12} = 3\sigma_{1z}\sigma_{2z} - \boldsymbol{\sigma}_1 \cdot \boldsymbol{\sigma}_2$ for $\theta = 0$ and using (2-55), we have for the left-hand side of (2-59)

$$S_{12}\mathscr{Y}^1_{110} = (3\sigma_{1z}\sigma_{2z} - \boldsymbol{\sigma}_1 \cdot \boldsymbol{\sigma}_2)\,\alpha(1)\,\alpha(2)/(4\pi)^{1/2} = 2\alpha(1)\,\alpha(2)/(4\pi)^{1/2}.$$

For the right-hand side of (2-59), again using (2-55), we have

$$B\mathscr{Y}^1_{112} = B(2/\sqrt{10})(5/16\pi)^{1/2}\alpha(1)\,\alpha(2).$$

Thus, equating these, we have the result $B = \sqrt{8}$. Repeating this for $M_J = 0$ and -1 also gives $B = \sqrt{8}$; hence

$$S_{12}\mathscr{Y}^{M_J}_{110} = \sqrt{8}\,\mathscr{Y}^{M_J}_{112}. \tag{2-60}$$

The fact that B is independent of M_J follows from the more general consideration that $[S_{12}, J_\pm] = 0$, where $J_\pm = J_x \pm iJ_y$ are the raising and lowering operators. In a similar manner we find

$$S_{12}\mathscr{Y}^{M_J}_{112} = \sqrt{8}\,\mathscr{Y}^{M_J}_{110} - 2\mathscr{Y}^{M_J}_{112}. \tag{2-61}$$

▶**Exercise 2-10** (a) Noting that S_{12} is Hermitian and again taking the special case $\theta = 0$, derive (2-61) as above.

(b) Derive (2-60) from a direct general calculation, using (2-58).

(c) Show that $[S_{12}, J_\pm] = 0$.

(d) Verify that expressions such as (2-59), for a general case $S_{12}\mathscr{Y}^{M_J}_{J,S,L}$ $(S = 1)$ similarly have associated coefficients such as A and B, which are also independent of M_J. ◀

An extension of Exercise 2-10 leads to the following useful expressions for $S_{12}\mathscr{Y}^{M_J}_{J1L}$ $(S = 1)$:

$$L = J: \qquad S_{12}\mathscr{Y}^{M_J}_{J1J} = 2\mathscr{Y}^{M_J}_{J1J}$$

$$L = J-1: \quad S_{12}\mathscr{Y}^{M_J}_{J1(J-1)} = \frac{6[J(J+1)]^{1/2}}{2J+1}\mathscr{Y}^{M_J}_{J1(J+1)} - \frac{2(J-1)}{2J+1}\mathscr{Y}^{M_J}_{J1(J-1)}$$

$$L = J+1: \quad S_{12}\mathscr{Y}^{M_J}_{J1(J+1)} = -\frac{2(J+2)}{2J+1}\mathscr{Y}^{M_J}_{J1(J+1)} + \frac{6[J(J+1)]^{1/2}}{2J+1}\mathscr{Y}^{M_J}_{J1(J-1)}. \tag{2-62}$$

▶**Exercise 2-11** In discussing the phase shifts for high-energy nucleon–nucleon scattering in a previous section, we quoted expectation values for S_{12} in the $^3P_2$, $^3P_1$, and $^3P_0$ states to be $-2/5$, 2, and -4, respectively. Verify these assertions. ◀

It is sometimes convenient to describe the operator S_{12} by making use of the equivalence

$$S_{12} = (3/r^2)(\boldsymbol{\sigma}_1 \cdot \mathbf{r})(\boldsymbol{\sigma}_2 \cdot \mathbf{r}) - \boldsymbol{\sigma}_1 \cdot \boldsymbol{\sigma}_2 = (6/r^2)(\mathbf{S} \cdot \mathbf{r})^2 - 2\mathbf{S} \cdot \mathbf{S}, \quad (2\text{-}63)$$

where $\mathbf{S} = \frac{1}{2}(\boldsymbol{\sigma}_1 + \boldsymbol{\sigma}_2)$. The operator $(1/r)(\mathbf{S} \cdot \mathbf{r}) = h$ is called the spatial helicity operator and measures the component of $\mathbf{S}$ in the direction $\hat{r}$. Of course, the eigenvalues of h are 0, ± 1. Since $[S_{12}, h] = 0$ and $[\mathbf{S}^2, h] = 0$ in addition to $[S_{12}, \mathbf{S}^2] = 0$, the eigenvalue problem for S_{12} can be solved with simultaneous eigenfunctions of S_{12}, $\mathbf{S}^2$, and h. The resulting eigenvalues for the equation $S_{12}\psi = \lambda\psi$ are $\lambda = 0$ for $S = 0$; $\lambda = -4$ for $S = 1$, $h = 0$; and $\lambda = +2$ for $S = 1, h = \pm 1$. Since h does not commute with the parity operator P (i.e., h is a pseudoscalar operator), nuclear states with a well-defined parity must contain mixed $h = \pm 1$ spatial helicity states.

▶**Exercise 2-12** Prove (2-63) and verify the quoted eigenvalues λ. ◀

I. THE DEUTERON PROBLEM

The fact that the neutron–proton scattering problem has a positive Fermi scattering length in the triplet spin configuration indicates that this two-nucleon system has at least one bound state. Indeed, there does exist a bound neutron–proton system called the deuteron. The deuteron has only one bound state, its ground state with $J^\pi = 1^+$. The binding energy is known to be (2.22461 ± 0.00007) MeV. The measured magnetic moment is $(+0.857406 \pm 0.000001)\mu_0$ and the measured electric quadrupole moment is (2.875 ± 0.002) mb. The magnetic moment is close to $\mu_p + \mu_n = 2.79278 - 1.91315 = +0.87963$, and the electric quadrupole moment is rather smaller than $\langle r^2 \rangle \approx 50$ mb. A reasonable first approximation might be to consider the triplet spin potential to be central with a simple attractive well shape. The ground state then would be a pure ${}^3S_1$ state, giving $J^\pi = 1^+$, $\mu_d = \mu_p + \mu_n$, and $Q = 0$, in rough agreement with the experimental values.

The simplest central attractive potential would again be the square well of Fig. 1.10, having a depth V_0 and a radius b. We take the total energy to be $E = -E_B$ (so that the binding energy can be expressed as a positive quantity) and define[1]

$$\hbar^2 k^2 = m_N(V_0 - E_B), \qquad \hbar^2 \alpha^2 = m_N E_B. \quad (2\text{-}64)$$

The Schrödinger equation in the two regions $r < b$ and $r > b$ in terms of $u = r\psi$ becomes

$$\frac{d^2 u'}{dr^2} + k^2 u' = 0, \quad r < b; \qquad \frac{d^2 u}{dr^2} - \alpha^2 u = 0, \quad r > b. \quad (2\text{-}65)$$

[1] We have used the reduced mass equal to $m_N/2$. The nucleon mass m_N can be taken $m_N \approx m_p \approx m_n$.

Appropriate solutions can be conveniently taken to be

$$u'(r) = A \sin kr, \quad r < b; \qquad u(r) = Be^{-\alpha(r-b)}, \quad r > b. \tag{2-66}$$

Matching logarithmic derivatives at $r = b$ gives

$$k \cot kb = -\alpha, \tag{2-67}$$

while matching $u'(b) = u(b)$ also gives

$$A \sin kb = B. \tag{2-68}$$

Substituting from (2-67) into (2-68) results in

$$k^2 A^2 = (k^2 + \alpha^2) B^2. \tag{2-69}$$

The normalization $4\pi \int u^2 \, dr = 1$ in turn gives

$$\frac{A^2}{2k}(2kb - \sin 2kb) + \frac{B^2}{\alpha} = \frac{1}{2\pi}. \tag{2-70}$$

From (2-64) the known value of E_B determines $\alpha = 0.232 \text{ fm}^{-1}$. With the ground state of the deuteron just barely bound, $b \gtrsim r_{0t} = 1.73$ fm [recall that r_{0t} is determined from (1-135) when the triplet scattering length and α are known]. Actually a best value for a square well is $b = 1.93$ fm.[1] Using (2-67) and (2-64) gives $V_0 = 38.5$ MeV. We note here, borrowing a result we shall derive later, that for the square well the values of the minimum "strength" $V_0 b^2$ required to just bind the 1S, 1P, 1D, 2S, ... levels are 102, 412, 821, 903, ... MeV fm^2. The values $b = 1.93$ fm and $V_0 = 38.5$ MeV give $V_0 b^2 = 143$ MeV fm^2, hence allowing for only a bound 1S state.

From (2-69) and (2-70) we can determine A and B to rather good approximation (better than 3%) to be

$$A \approx B \approx (\alpha/2\pi)^{1/2} e^{-\alpha b/2} \tag{2-71a}$$

and hence

$$\begin{aligned} u'(r) &= (\alpha/2\pi)^{1/2} e^{-\alpha b/2} \sin kr, \qquad r < b \\ u(r) &= (\alpha/2\pi)^{1/2} e^{\alpha b/2} e^{-\alpha r}, \qquad r > b. \end{aligned} \tag{2-71b}$$

Finally, substituting numerical values into the exact expressions (2-66)–(2-70) gives

$$\begin{aligned} u'(r) &= 0.160 \sin(0.938r) \text{ fm}^{-1/2}, \qquad r < b \\ u(r) &= 0.243 e^{-0.232r} \text{ fm}^{-1/2}, \qquad r > b. \end{aligned} \tag{2-72}$$

[1] As discussed in connection with (1-138) and (1-139), a potential that just binds with zero energy has a well depth parameter $s = 1$, for which case $b = r_{0t}$ exactly. The realistic value $s \approx 1.25$ corresponding to the actual binding energy E_B would have $r_{0t}/b \approx 0.90$ for a square well, leading to $b = 1.93$ fm.

While (2-72) gives an exact solution, simultaneously giving E_B and consistent with r_{0t} and $1/a_t(0)$, the simple analytical expressions employing (2-71) are often used (recall that the assumption of a square well shape is itself an approximation).

▶**Exercise 2-13** Verify (2-68)–(2-70). Noting that (2-67) requires $kb \approx \pi/2$, with $kb = \frac{1}{2}\pi + \varepsilon$, show that $\varepsilon \approx 2\alpha b/\pi$ and hence $kb \approx \frac{1}{2}\pi + (2\alpha b/\pi)$. With $\alpha = 0.232\ \text{fm}^{-1}$ and $b = 1.93$ fm, show that this expression gives $V_0 = 40.5$ MeV. By successive approximations show that $V_0 = 38.5$ is the exact solution. Eliminating A^2 from (2-69) and (2-70), show that

$$B = \left(\frac{\alpha}{2\pi}\right)^{1/2}\left[1 + \alpha b\left(1 + \frac{\alpha^2}{k^2}\right)\left(1 + \frac{\alpha}{\pi k}\right)\right]^{-1/2} \approx \left(\frac{2\alpha}{2\pi}\right)^{1/2}\left(1 - \frac{\alpha b}{2} + \cdots\right)$$

or

$$B \approx \left(\frac{\alpha}{2\pi}\right)^{1/2} e^{-\alpha b/2}.$$

Finally, verify (2-72). ◀

Figure 2.14a shows a plot of (2-72). The figure shows that the wave function $\psi(r)$ extends well beyond $r = b$, the range of the nucleon–nucleon interaction. One qualitative indication of this feature is to calculate the radius $R_{1/2}$, which gives equal probability for finding the n–p system at separations $> R_{1/2}$ and $< R_{1/2}$. The value for $R_{1/2} = 2.50$ fm is readily determined from (2-72). Another indicator of the loosely bound nature of the deuteron is to calculate $\langle r \rangle$ from (2-72), which results in $\langle r \rangle = 3.14$ fm. The rms radius is larger still, with $\langle r^2 \rangle^{1/2} = 3.82$ fm. The radii $b = 1.93$ fm, $R_{1/2} = 2.50$ fm, and $\langle r^2 \rangle^{1/2} = 3.82$ fm are indicated on the figure.

The ground state of the deuteron for the above model gives a pure $^3S_1$ state. It is readily deduced that such a state would give rise to $\mu_d = \mu_n + \mu_p$ and $Q = 0$. The simplest improvement on the theoretical treatment given so far to obtain agreement with the observed values of μ and Q is to include the tensor interaction involving S_{12} in the potential. From (2-56) it is readily seen that S_{12} would couple the $^3S_1$ and $^3D_1$ states, giving, as we shall see, the desired effects for μ and Q. Evidence is also mounting that the D state of the deuteron must also be included in nuclear reaction calculations involving the free deuteron, as for example in (d, p) reactions [see Knutson *et al.* (1973)].

An interesting recent development concerns possible bound and unbound states of the nucleon–antinucleon system. In addition to scattering resonances, evidence exists for possible "bound" states (with annihilation widths $60 \lesssim \Gamma_a \lesssim 150$ MeV) in the $(\bar{p}, n)$ system. Specifically, theory suggests the interpretation of experimental data in terms of a bound D state with $J^\pi = 1^-$ and a mass 1855 MeV [see Bogdanova *et al.* (1972)].

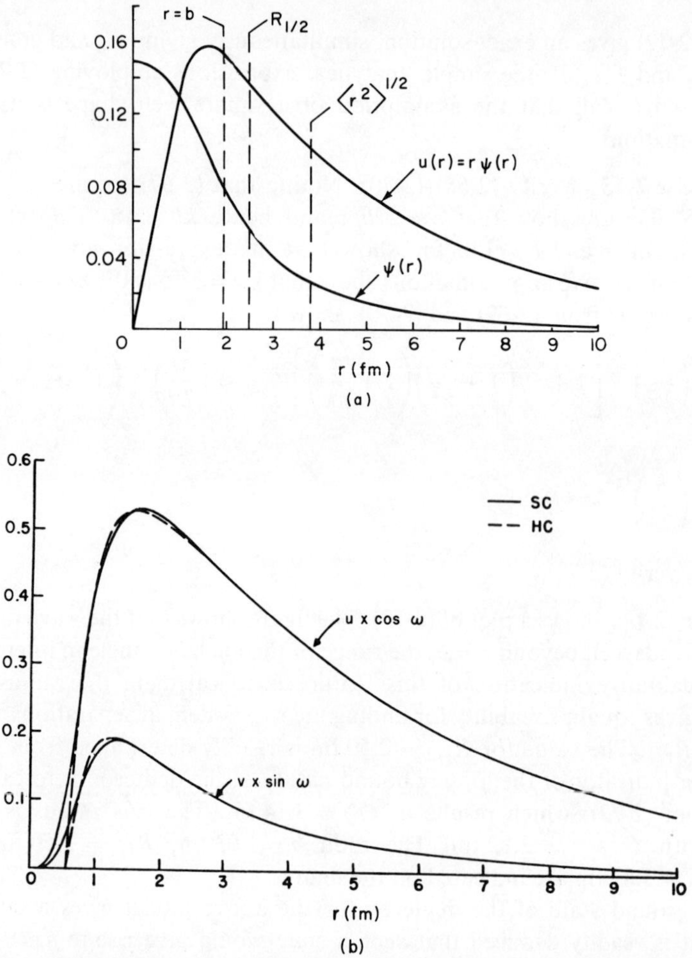

FIG. 2.14. (a) The plot of $u(r)$ and $\psi(r)$ for the square well potential approximation (2-72). For the definition of $r = b$, $R_{1/2}$, and $\langle r^2\rangle^{1/2}$ see text. (b) The plot of $u(r)\cos\omega$ and $v(r)\sin\omega$ for the hard-core (HC) and soft-core (SC) Reid potentials [taken from Reid (1968)]. The two functions $u(r)$ in (a) and (b) differ in their normalization by the factor $(4\pi)^{1/2}$.

Returning to the deuteron problem, we now write

$$(T+V)\psi(S, D) = E\psi(S, D) \tag{2-73a}$$

with

$$T = \frac{\hbar^2}{m_N}\left[-\frac{1}{r}\frac{d^2}{dr}r + \frac{l(l+1)}{r^2}\right]$$

and

$$V = {}^{31}V(r) + V_T(r)S_{12}.$$

We also introduce the wave function

$$r\psi(S, D) = (\cos\omega)\, u(r)\, \mathscr{Y}_{110}^{M_J} + (\sin\omega)\, v(r)\, \mathscr{Y}_{112}^{M_J}, \tag{2-73b}$$

where we use $\cos\omega$ and $\sin\omega$ (with $-\pi/2 \leqslant \omega \leqslant +\pi/2$) as convenient coefficients, since $\cos^2\omega + \sin^2\omega = 1$ maintains the normalization as ω is adjusted if $u(r)$ and $v(r)$ are separately normalized to unity. Direct substitution of T, V, and $r\psi$ into the Schrödinger equation and use of (2-60), (2-61), or (2-62) gives

$$\begin{aligned}
&\left(-\frac{\hbar^2}{m_N}\frac{d^2}{dr^2} + {}^{31}V\right) u\mathscr{Y}_{110}^{M_J} \cos\omega + \sqrt{8}\, V_T u \mathscr{Y}_{112}^{M_J} \cos\omega \\
&\quad - \frac{\hbar^2}{m_N}\left(\frac{d^2}{dr^2} - \frac{6}{r^2}\right) v\mathscr{Y}_{112}^{M_J} \sin\omega + ({}^{31}V - 2V_T)\, v\mathscr{Y}_{112}^{M_J} \sin\omega \\
&\quad + \sqrt{8}\, V_T v \mathscr{Y}_{110}^{M_J} \sin\omega \\
&= E(u\mathscr{Y}_{110}^{M_J} \cos\omega + v\mathscr{Y}_{112}^{M_J} \sin\omega).
\end{aligned} \tag{2-74}$$

Since (2-74) must hold for all (θ, ϕ) and the $\mathscr{Y}$'s are othogonal, (2-74) separates into the two coupled equations

$$\begin{aligned}
\left(-\frac{\hbar^2}{m_N}\frac{d^2}{dr^2} + {}^{31}V - E\right) u &= -\sqrt{8}\, V_T v \tan\omega \\
-\frac{\hbar^2}{m_N}\left(\frac{d^2}{dr^2} - \frac{6}{r^2}\right) v + ({}^{31}V - 2V_T - E) v &= -\sqrt{8}\, V_T u \cot\omega.
\end{aligned} \tag{2-75}$$

▶**Exercise 2-14** Verify (2-74) and (2-75). ◀

While the evaluation of the solutions u and v to (2-75) for even simple potential shapes requires numerical methods, their behaviors for r greater than the range of ${}^{31}V(r)$ and of $V_T(r)$ are easily found to be

$$u(r) \to Ne^{-\alpha r}, \qquad v(r) \to N'e^{-\alpha r}\left(1 + \frac{3}{\alpha r} + \frac{3}{\alpha^2 r^2}\right). \tag{2-76}$$

Solutions to (2-75) have been obtained for various potential shapes. The simple square well shape with ${}^{31}V(r) + S_{12}V_T(r) = -[1 + S_{12}]V(r)$, with $b = 2.66$ fm and $V_0 = 13.3$ MeV, was found by Guindon (1948) to give $\sin^2\omega = 0.044$ in order to match Q, μ, E_B, and the scattering parameters r_{0t} and $1/a_t(0)$. Feshbach *et al.* (1949) successfully accounted for the general low-energy nucleon–nucleon scattering problem as well as Q, μ, and E_B for the deuteron by the use of a simple Yukawa-shaped potential

$$V = -22.7\,\frac{\exp(-2.12r/b_c)}{r/b_c} - 10.9\,\frac{\exp(-2.12r/b_T)}{r/b_T}\, S_{12}$$

with $b_c = 2.47$ fm, $b_T = 3.68$ fm, and $\sin^2\omega \approx 0.033$. Here, the triplet and singlet spin interactions were taken to be different only by virtue of the term in S_{12} (which gives zero when operating on a singlet state). A potential only slightly different from this latter one also gives a satisfactory fit to the known properties of the ground state of ^{3}H [see Pease and Feshbach (1952) and also Section V,D,1].

All the recent accurate phenomenological nucleon–nucleon potential fits adequately account for all of the deuteron properties as well. A typical example is the result of Reid (1968). His computed wave functions $u(r)\cos\omega$ and $v(r)\sin\omega$ are shown in Fig. 2.14b for both the hard-core (HC) and soft-core (SC) cases. Both accurately give the observed values of E_B and Q with $\sin^2\omega \approx 0.0648$. As expected, the wave functions $u(r)$ and $v(r)$ vanish in the hard-core region $r \lesssim 0.4$ fm. A careful examination of various theoretical predictions of the deuteron wave functions for $r \lesssim 0.5$ fm, and the possibility that electron scattering polarization measurements at 6–10 GeV/c momentum transfer might distinguish these possibilities, are discussed by Moravcsik and Ghosh (1974). Another important deuteron parameter, to be discussed later, is the asymptotic $^3D_1$ to $^3S_1$ wave ratio A_D/A_S. The Reid HC and SC cases differ only slightly, and the average value is $A_D/A_S = 0.0260$.

The calculation of the magnetic moment of the deuteron, with the ground state described by a linear combination of $^3S_1$ and $^3D_1$ waves, is straightforward. In determining the contribution of the orbital angular momentum of the (n, p) system $\boldsymbol{\mu}_l = \mu_0 g^l \boldsymbol{l}$, we must determine the effective charge $\bar{q}$ and mass $\bar{m}$ to use following Exercise 1-11. Since

$$\bar{e} = \frac{m'^2 q + m^2 q'}{(m+m')^2} = \frac{m_N^2}{m_N^2}\frac{1\cdot e + 1\cdot 0}{(1+1)^2} = \frac{e}{4}$$

and

$$\bar{m} = \frac{mm'}{m+m'} = \frac{m_N}{2} \approx \frac{m_p}{2},$$

we have

$$\boldsymbol{\mu}_l = \frac{\hbar \bar{e}}{2\bar{m}c} g^l \boldsymbol{l} = \frac{\hbar e}{2m_p c}\left(2\cdot\frac{1}{4}\right)\mathbf{L} = \mu_0 \frac{\mathbf{L}}{2}. \tag{2-77}$$

For the spin contributions we have $\boldsymbol{\mu}_s = g_n{}^s \mathbf{s}_n + g_p{}^s \mathbf{s}_p$ or

$$\boldsymbol{\mu}_s = (\mu_n \boldsymbol{\sigma}_n + \mu_p \boldsymbol{\sigma}_p)\mu_0, \tag{2-78}$$

where $\mu_n = -1.9132$ and $\mu_p = +2.7928$. Hence the total magnetic moment operator is

$$\boldsymbol{\mu}/\mu_0 = \mu_n \boldsymbol{\sigma}_n + \mu_p \boldsymbol{\sigma}_p + \tfrac{1}{2}\mathbf{L}. \tag{2-79}$$

It is convenient to write (2-79) in a slightly different form, namely,

$$\boldsymbol{\mu}/\mu_0 = \tfrac{1}{2}(\mu_n+\mu_p)(\boldsymbol{\sigma}_n+\boldsymbol{\sigma}_p) + \tfrac{1}{2}(\mu_n-\mu_p)(\boldsymbol{\sigma}_n-\boldsymbol{\sigma}_p) + \tfrac{1}{2}\mathbf{L}. \tag{2-80}$$

Since the expectation value of $(\boldsymbol{\sigma}_n-\boldsymbol{\sigma}_p)$ is zero and $\tfrac{1}{2}(\boldsymbol{\sigma}_n+\boldsymbol{\sigma}_p) = \mathbf{S}$, we finally arrive at

$$\boldsymbol{\mu}/\mu_0 = (\mu_n+\mu_p)\mathbf{S} + \tfrac{1}{2}\mathbf{L}. \tag{2-81}$$

We wish to evaluate $\langle J, M_J|\,\mu_z\,|J, M_J\rangle$ by the use of the Wigner–Eckart theorem in the form (2-44). Therefore we construct

$$\mathbf{J}\cdot\boldsymbol{\mu}/\mu_0 = (\mu_n+\mu_p+\tfrac{1}{2})\tfrac{1}{2}\mathbf{J}^2 + (\mu_n+\mu_p-\tfrac{1}{2})\tfrac{1}{2}(\mathbf{S}^2-\mathbf{L}^2), \tag{2-82}$$

using (2-81) and $\mathbf{J} = \mathbf{L}+\mathbf{S}$. For wave functions consisting of linear combinations $|J, M_J\rangle_{L,S}$, *each* term contributes[1]

$$\begin{aligned}\left(\frac{\mu_z}{\mu_0}\right)(J, M_J, L, S) &= \frac{M_J}{J(J+1)}\left\{\langle J, M_J|\frac{\mathbf{J}\cdot\boldsymbol{\mu}}{\mu_0}|J, M_J\rangle\right\}_{L,S} \\ &= \frac{M_J}{2}\left(\mu_n+\mu_p+\frac{1}{2}\right)+\left(\mu_n+\mu_p-\frac{1}{2}\right) \\ &\quad\times\frac{[S(S+1)-L(L+1)]}{2J(J+1)}M_J. \end{aligned} \tag{2-83}$$

For example, when $M_J = J$ the contributions from (2-83) are

$$\begin{aligned} {}^3\mathrm{S}_1&:\ \mu_n+\mu_p \\ {}^3\mathrm{D}_1&:\ \tfrac{3}{4}-\tfrac{1}{2}(\mu_n+\mu_p) \\ {}^3\mathrm{P}_1&:\ \tfrac{1}{4}+\tfrac{1}{2}(\mu_n+\mu_p) \\ {}^1\mathrm{P}_1&:\ \tfrac{1}{2} \\ &\ \ \vdots \end{aligned} \tag{2-84}$$

Finally, using (2-73b) for the deuteron wave function,

$$\mu_d = (\mu_n+\mu_p) - \tfrac{3}{2}(\mu_n+\mu_p-\tfrac{1}{2})\sin^2\omega \tag{2-85a}$$

or

$$\mu_d = 0.8796 - 0.5694\sin^2\omega. \tag{2-85b}$$

[1] It is readily verified that the functions $|J, M_J\rangle_{L,S}$ are orthogonal in J, M_J, L, and S. Since the functions Y_l^m and $\chi_S^{m_s}$ are normalized and the Clebsch–Gordan coefficients are constructed to transform basis states while maintaining the normalization, the states $|J, M_J\rangle_{L,S}$ form an orthonormal basis.

Since the observed value of μ_d is 0.8574, use of (2-85) gives

$$\sin^2\omega = 0.0373. \tag{2-86}$$

▶**Exercise 2-15** We note that $(\boldsymbol{\sigma}_n - \boldsymbol{\sigma}_p)$ is an antisymmetric operator in spin exchange and hence $(\boldsymbol{\sigma}_n - \boldsymbol{\sigma}_p)$ converts singlet spin states into triplet spin states and vice versa. Show explicitly, for example, that

$$(\boldsymbol{\sigma}_n - \boldsymbol{\sigma}_p)_z \chi_1^{1,0,-1} = (0, 2, 0)\chi_0{}^0, \qquad (\boldsymbol{\sigma}_n - \boldsymbol{\sigma}_p)_z \chi_0{}^0 = 2\chi_1{}^0.$$

From these results show that the expectation value for $(\boldsymbol{\sigma}_n - \boldsymbol{\sigma}_p)$ is zero for a state which is a linear combination of $|J, M_J\rangle_{L,S}$ *and* for which either $S = 0$ only or $S = 1$ only.

Suppose the deuteron ground state were $J^\pi = 1^-$; show that the most general state is the one of mixed isospin ${}^1P_1 + {}^3P_1$. Evaluate the expectation value of $(\boldsymbol{\sigma}_n - \boldsymbol{\sigma}_p)_z$ for the state $(\cos\omega)\psi({}^1P_1) + (\sin\omega)\psi({}^3P_1)$ with $M_J = J$. ◀

▶**Exercise 2-16** Verify (2-82) and (2-83).

If the deuteron ground state were $J^\pi = 1^-$ with a mixed isospin state as in Exercise 2-15, calculate the expected deuteron magnetic moment μ_d in terms of ω. ◀

There are two serious problems with the result (2-86) giving a 3.7% ${}^3D_1$-state admixture to the predominantly ${}^3S_1$ character of the deuteron ground state. To begin with, since μ_d is so close to being equal to the first term of (2-85), a small change of only 1% in μ_d would lead to an approximately 40% change in $\sin\omega$. Two effects, neglected to this point, capable of generating corrections of this order of magnitude are the meson exchange current effect and a relativistic correction to the Schrödinger treatment [e.g., Hulthén and Sugawara (1957, p. 70 ff.)]. The first effect involves the effective current produced by the exchange meson momentum and the nucleon recoil and its corresponding contribution to the magnetic moment. Although this effect vanishes for the pure ${}^3S_1$ state, it is of considerable importance for the ${}^3D_1$ wave. The combined exchange and relativistic effects can be estimated to lower the predicted moment μ_d by perhaps some 2% of μ_d, although there is considerable uncertainty in this value (Hulthén and Sugawara, 1957). Thus the value of $\sin^2\omega$ can be assigned only in the rather broad range $0.02 \lesssim \sin^2\omega \lesssim 0.08$ from the consideration of the magnetic moment alone. A complete relativistic treatment for the deuteron wave function also introduces P-state wave-function components with a combined probability of approximately 0.5% [see Hornstein and Gross (1973)]. For recent calculations of meson exchange effects within the OBE model, see Horikawa *et al.* (1972).

The quadrupole moment of the deuteron for the wave function (2-73b) can be calculated when $u(r)$ and $v(r)$ are known. The quadrupole moment is given by[1]

$$Q = \int \tfrac{1}{4} r^2 (3\cos^2\theta - 1) |\psi|^2 \, d^3r \tag{2-87}$$

with ψ taken from (2-73b) and $M_J = J$. The result can be expressed as the sum of three terms[2]

$$Q = Q_{uu} \cos^2\omega + 2 \operatorname{Re} Q_{uv} \cos\omega \sin\omega + Q_{vv} \sin^2\omega \tag{2-88a}$$

with

$$\begin{aligned} Q_{uu} &= \int \tfrac{1}{4} u^2 (3\cos^2\theta - 1) |\mathscr{Y}_{110}^1|^2 r^2 \, dr \, d\Omega \\ Q_{uv} &= \int \tfrac{1}{4} uv (3\cos^2\theta - 1) \mathscr{Y}_{110}^{1*} \mathscr{Y}_{112}^1 r^2 \, dr \, d\Omega \\ Q_{vv} &= \int \tfrac{1}{4} v^2 (3\cos^2\theta - 1) |\mathscr{Y}_{112}^1|^2 r^2 \, dr \, d\Omega. \end{aligned} \tag{2-88b}$$

It is readily shown that

$$Q_{uu} = 0, \qquad Q_{uv} = (1/2\sqrt{50}) \int_0^\infty uvr^2 \, dr, \qquad Q_{vv} = -(1/20) \int_0^\infty v^2 r^2 \, dr. \tag{2-89}$$

For example, the angular dependence of the integrand in Q_{uv} is, after using (2-55),

$$(3\cos^2\theta - 1) Y_0^{0*} \chi_1^{1\dagger} \left(\sqrt{\tfrac{3}{5}} Y_2{}^2 \chi_1{}^{-1} - \sqrt{\tfrac{3}{10}} Y_2{}^1 \chi_1{}^0 + \sqrt{\tfrac{1}{10}} Y_2{}^0 \chi_1{}^1\right).$$

Orthogonality of the spin functions, noting that $(3\cos^2\theta - 1) = (16\pi/5)^{1/2} Y_2{}^0$ and $Y_0{}^0 = (4\pi)^{-1/2}$, gives

$$Q_{uv} = \left(\frac{16\pi}{5}\right)^{1/2} \frac{1}{4} \frac{1}{\sqrt{10}} \frac{1}{(4\pi)^{1/2}} \int_0^\infty uvr^2 \, dr = \frac{1}{2\sqrt{50}} \int_0^\infty uvr^2 \, dr.$$

Finally, we find

$$Q = \frac{\cos\omega \sin\omega}{\sqrt{50}} \int_0^\infty uvr^2 \, dr - \frac{\sin^2\omega}{20} \int_0^\infty v^2 r^2 \, dr. \tag{2-90}$$

Since $\sin^2\omega \ll 1$, the second term of (2-90) can be neglected. The experimental fact that $Q > 0$ clearly requires $0 < \omega < \pi/2$.

[1] Note that in the expression $Q = \sum_{i=1}^{Z} \int |\psi|^2 (3z_i{}^2 - r_i{}^2) \, d^3r_1 \, d^3r_2 \cdots d^3r_A$, we have taken $\mathbf{r}_p = \mathbf{r}/2$ and $\mathbf{r}_n = -\mathbf{r}/2$, and that $|\psi(\mathbf{r}_n, \mathbf{r}_p)|^2 \, d^3r_p \, d^3r_n = |\psi(\mathbf{r})|^2 \, d^3r$ results after integrating over the coordinates of the center of mass.

[2] The symbol Re stands for real part.

The presence of r^2 in the integral (2-90) weights the importance of $u(r)$ and $v(r)$ in the asymptotic region where (2-76) holds; thus

$$Q \approx \frac{\omega N N'}{\sqrt{50}} \int_0^{\infty} r^2 e^{-2\alpha r}\left[1+\frac{3}{\alpha r}+\frac{3}{\alpha^2 r^2}\right] dr \quad \text{or} \quad Q \approx \frac{\omega N N'}{\alpha^3 \sqrt{8}}. \tag{2-91}$$

We define the asymptotic ${}^3D_1$ to ${}^3S_1$ wave ratio to be

$$A_D/A_S = \left\{\lim_{r\to\infty}[v(r)/u(r)]\right\} \tan\omega \approx (N'/N)\omega$$

from (2-76). In the spirit of the estimate (2-91), we can normalize $u(r)$ by[1]

$$\int_0^{\infty} u^2(r)\,dr \approx N^2 \int_0^{\infty} e^{-2\alpha r}\,dr = 1 \quad \text{or} \quad N^2 \approx 2\alpha.$$

Substituting these approximations into (2-91) gives

$$Q \approx (1/\sqrt{2}\,\alpha^2) A_D/A_S. \tag{2-92}$$

Using $Q = 0.288\ \text{fm}^2$ and $\alpha = 0.232\ \text{fm}^{-1}$, we readily estimate $A_D/A_S \approx 0.022$, whereas Reid (1968) quotes 0.0260 based on an exact analysis. In any event (2-92) shows that the quadrupole moment depends rather sensitively on the asymptotic ratio A_D/A_S.

Finally, it should be noted that meson exchange current and relativistic corrections to (2-90) are now much less important than for the case of the magnetic moment in the adjustment of A_D/A_S or ω, since $Q_{uu} = 0$. Thus we can conclude by averaging the rather similar ${}^3D_1$ contribution predicted by Hamada and Johnston (1962) and Reid (1968), and quote

$$\sin^2\omega \approx 0.067, \quad \text{with} \quad 0 < \omega < \pi/2. \tag{2-93}$$

This constitutes a reasonable value essentially consistent with all the two-nucleon experimental data. However, it should be noted that the OBEM parameterization yields a somewhat smaller D-state contribution, ranging from $\sin^2\omega = 0.052$ to 0.063 (Erkelenz *et al.*, 1969).

▶**Exercise 2-17** Show that Hermiticity of the quadrupole operator leads to the second term of (2-88a) when (2-73b) is substituted into (2-87).

Verify the expressions (2-89). [*Hint:* Use (2-55) and (2-36).] ◀

[1] This normalization is the same as that of (2-71), where $N^2 = (\alpha/2\pi)e^{\alpha b}$, if we take $e^{\alpha b} \approx 1$ and note that the definition $Y_0{}^0 = (4\pi)^{-1/2}$ requires $\int_0^\infty u^2(r)\,dr = 1$ in the present instance. Reid (1968) quotes the exact computer-generated value $N\cos\omega \approx 0.879$, while from (2-72), $N\cos\omega = 0.243(4\pi)^{1/2}\cos\omega = 0.835$. For additional aspects of the two-nucleon interaction addressed by Reid's analysis, see Chapter I.

References

Allardyce, B. W., *et al.* (1973). *Nuclear Phys.* **A209,** 1.

Anderson, D. K., Jenkins, D. A., and Powers, R. J. (1970). *Phys. Rev. Lett.* **24,** 71.

Anderson, J. D., Wong, C., and McClure, J. W. (1965). *Phys. Rev.* **138,** B615.

Arima, A., and Huang-Lin, L. J. (1972). *Phys. Lett.* **41B,** 435.

Aslam, K., and Rook, J. R. (1970). *Nuclear Phys.* **B20,** 397.

Backenstoss, G. (1970). *Ann. Rev. Nuclear Sci.* **20,** 467.

Bardeen, W. A., and Torigoe, E. W. (1971). *Phys. Rev.* **C3,** 1785.

Batty, C. J., and Friedman, E. (1972). *Nuclear Phys.* **A179,** 701.

Beiner, M., and Lombard, R. J. (1973). *Phys. Lett.* **47B,** 399.

Bethe, H. A. (1969). *Nuclear Phys.* **B21,** 589.

Bethe, H. A., and Elton, L. R. B. (1968). *Phys. Rev. Lett.* **20,** 745.

Bhattacherjee, S. K., Boehm, F., and Lee, P. L. (1969). *Phys. Rev.* **188,** 1919.

Blatt, J. M., and Weisskopf, V. F. (1952). "Theoretical Nuclear Physics." Wiley, New York.

Bogdanova, L. N., Dal'korov, O. D., and Shapiro, I. S. (1972). *Phys. Rev. Lett.* **28,** 1418.

Bonn, J., Huber, G., Kluge, H. J., Kugler, L., and Otten, E. W. (1972). *Phys. Lett.* **38B,** 308.

Boolchand, P., Langhammer, D., Lin, C., Jha, S., and Peek, N. F. (1972). *Phys. Rev.* **C6,** 1093.

Brix, P., and Kopfermann, H. (1958). *Rev. Modern Phys.* **30,** 517.

Campi, X., and Sprung, D. W. L. (1973). *Phys. Lett.* **46B,** 291.

Carlson, B. C., and Talmi, I. (1954). *Phys. Rev.* **96,** 436.

Davis, K. T. R., Krieger, S. J., and Wong, C. Y. (1973). *Nuclear Phys.* **A216,** 250.

Elton, L. R. B. (1961). "Nuclear Sizes." Oxford Univ. Press, London and New York.

Erkelenz, K., Holinde, K., and Bleuler, K. (1969). *Nuclear Phys.* **A139,** 308.

Feshbach, H., Schwinger, J. S., and Harr, J. A. (1949). Effects of Tensor Range in Nuclear Two-Body Problem. Computer Lab at Harvard Univ., Cambridge, Massachusetts, Nov.

Fox, J. D., and Robson, D., eds. (1966). "Isobaric Spin in Nuclear Physics." Academic Press, New York.

Friar, J. L., and Negele, J. W. (1973). *Nuclear Physics.* **A212,** 93.

Green, A. E. S., Sawada, T., and Saxon, D. S. (1968). "The Nuclear Independent Particle Model." Academic Press, New York.

Greenlees, G. W., Makofske, W., and Pyle, G. J. (1970). *Phys. Rev.* **C1,** 1145.

Guindon, W. B. (1948). *Phys. Rev.* **145,** 145.

Hamada, T., and Johnston, I. D. (1962). *Nuclear Phys.* **34,** 382.

Harchol, M., Jaffe, A. A., Miron, J., Unna, I., and Zioni, J. (1967). *Nuclear Phys.* **90,** 459.

Heisenberg, J., Hofstadter, R., McCarthy, J. S., Sick, I., Clark, B. C., Herman, R., and Ravenhall, D. G. (1969). *Phys. Rev. Lett.* **23,** 1402.

Hill, D. H. (1957). "Handbuch der Physik," Vol. 39. Springer, Berlin.

Hofstadter, R. (1963). "Nuclear and Nucleon Structure." Benjamin, New York.

Horikawa, Y., Fujita, T., and Yazaki, K. (1972). *Phys. Lett.* **42B,** 173.

Hornstein, J., and Gross, F. (1973). *Phys. Lett.* **47B,** 205.

Hulthén, L., and Sugawara, M. (1957). "Handbuch der Physik," Vol. 39. Springer, Berlin.

Jänecke, J. (1960). *Z. Phys.* **160,** 171.

Kim, Y. N. (1971). "Mesic Atoms and Nuclear Structure." North-Holland Publ., Amsterdam.

Knutson, L. D., Stephenson, E. J., Rohrig, N., and Haeberli, W. (1973). *Phys. Rev. Lett.* **31,** 392.

Kömer, H. J., and Schiffer, J. P. (1971). *Phys. Rev. Lett.* **27,** 1457.

Kopfermann, H. (1958). "Nuclear Moments." Academic Press, New York.

Kossler, W. J., Funsten, H. O., MacDonald, B. A., and Lankford, W. F. (1971). *Phys. Rev.* **C4**, 1551.
Krishnan, R. M., and Pu, W. W. T. (1973). *Phys. Lett.* **47B**, 225.
Mantri, A. N., and Sood, P. C. (1974). *Phys. Rev.* **C9**, 2076.
Moravcsik, M. J., and Ghosh, P. (1974). *Phys. Rev. Lett.* **32**, 321.
Myers, W. D. (1973). *Nuclear Phys.* **A204**, 465.
Myers, W. D., and Swiatecki, W. J. (1969). *Ann. Phys.* **55**, 395.
Nagamiya, S., and Yamazaki, T. (1971). *Phys. Rev.* **C4**, 1961.
Negele, J. W. (1970). *Phys. Rev.* **C1**, 1260
Nilsson, S. G., Nix, J. R., Möller, P., and Ragnarsson, I. (1974). *Nuclear Phys.* **A222**, 221.
Nolen, J. A., and Schiffer, J. P. (1969). *Ann. Rev. Nuclear Sci.* **19**, 471.
Nuclear Data Group (1966). Nuclear Data Sheets, Oak Ridge Nat. Lab., Tennessee.
Particle Data Group (1973). *Rev. Modern Phys.* **45**, S1.
Pease, R. L., and Feshbach, H. (1952). *Phys. Rev.* **88**, 945.
Peaslee, D. C. (1954). *Phys. Rev.* **95**, 717.
Ponomarev, L. I. (1973). *Ann. Rev. Nuclear Sci.* **23**, 395.
Reid, R. V. (1968). *Ann. Phys.* **50**, 411.
Rinker, G. A. (1971). *Phys. Rev.* **C4**, 2150.
Rinker, G. A., and Rich, M. (1972). *Phys. Rev. Lett.* **28**, 640.
Rost, E., and Edwards, G. W. (1971). *Phys. Lett.* **37B**, 247.
Rüegsegger, R., and Kündig, W. (1972). *Phys. Lett.* **39B**, 620.
Schery, S. D., Lind, D. A., and Zafiratos, C. D. (1974). *Phys. Rev.* **C9**, 416.
Schopper, H., ed. (1967). "Nuclear Physics and Technology," Volume 2, Nuclear Radii. Springer, Berlin.
Segrè, E. (1964). "Nuclei and Particles." Benjamin, New York.
Seitz, H., Zaidi, S. A. A., and Bigler, R. (1972). *Phys. Rev. Lett.* **28**, 1465.
Seki, R. (1972). *Phys. Rev.* **C5**, 1196.
Sengupta, S. (1960). *Nuclear Phys.* **21**, 542.
Sinha, B. B. P., Peterson, G. A., Li, G. C., and Whitney, R. R. (1972). *Phys. Rev.* **C6**, 1657.
Sood, P. C., and Green, A. E. S. (1957). *Nuclear Phys.* **4**, 274.
Sternheim, M. M., and Auerbach, E. H. (1971). *Phys. Rev.* **C1**, 1805.
Uhrhane, F. J., McCarthy, J. S., and Yearin, M. R. (1971). *Phys. Rev. Lett.* **26**, 578.
Van Hienen, J. F. A., and Glaudemans, P. W. M. (1972). *Phys. Lett.* **42B**, 301.
Walker, L. R., Wertheim, G. K., and Jaccarino, V. (1961). *Phys. Rev. Lett.* **6**, 98.
Wiegand, C. E. (1969). *Phys. Rev. Lett.* **22**, 1235.
Wilets, L. (1968). *Rev. Modern Phys.* **30**, 542.
Wong, C. Y. (1972). *Phys. Lett.* **41B**, 446.
Wu, C. S., and Wilets, L. (1969). *Ann. Rev. Nuclear Sci.* **19**, 527.

Chapter III

NUCLEAR MATTER CHARACTERISTICS

A. INTRODUCTION

A very large array of nuclear models has been developed to account for the various systematic characteristics of the nuclear species. The introduction of these models is made necessary in view of the extreme difficulty associated with an exact comprehensive treatment. To date, only first steps have been taken in the direction of a universal theory.

Some models stressing specific structural features of the nucleus seem to embody assumptions that conflict with those of other models. The variants of the independent-particle shell model start from the premise that the individual nucleons in a nucleus move relatively freely in well-defined single-particle orbitals. On the other hand, variants of the collective model are based on assuming strong nucleon–nucleon interactions and resulting correlated motions that lead to pronounced cooperative nuclear features. The nucleon–nucleon force is indeed a strong, short-range, essentially two-body force, albeit of rather complex character as seen in Chapter I. The nucleons are, however, fermions and obey Fermi–Dirac statistics. The ground state of a nucleus must then to some extent resemble a Fermi gas at low temperature. Resolution of the seeming paradox involves the operation of the Pauli principle, which prevents these strong internucleon forces from causing virtual scattering of two nucleons within a nucleus into occupied states. One can then discern strong short-range correlations related to high-momentum components in the wave function for tightly bound nucleons, as well as weaker long-range correlations for nucleons near the Fermi energy (i.e., near the surface of the Fermi sphere characterized by wave number k_F) where small momentum transfers *could* scatter particles into unoccupied states.

A common starting point of many models is the well-known equivalence of either solving the Schrödinger equation

$$\tilde{H}|\Psi\rangle = E|\Psi\rangle \tag{3-1}$$

or performing the variational calculation

$$\langle\Psi|\Psi\rangle E(\Psi) = \langle\Psi|\tilde{H}|\Psi\rangle, \qquad \text{with} \qquad \delta E(\Psi) = 0, \tag{3-2}$$

where in each instance Ψ relates to the entire nuclear system. In the *independent pair* model only two-body forces are considered and the system Hamiltonian is written as

$$\tilde{H} = \sum_{i=1}^{A} \tilde{T}_i + \sum_{i>j=1}^{A} \tilde{V}_{ij} = \sum_{i=1}^{A} \tilde{T}_i + \tfrac{1}{2}\sum_{\substack{i,j\\ i\neq j}}^{A} \tilde{V}_{ij} \tag{3-3}$$

and

$$\tilde{T}_i = \tilde{p}_i^{\,2}/2m_N = -(\hbar^2/2m_N)\nabla_i^{\,2}.$$

The single-particle model results when some suitable average potential $\tilde{\bar{V}}_i$ is defined to replace $\sum_{i\neq j}\tilde{V}_{ij}$ and the auxiliary equations

$$(\tilde{T}_i + \tilde{\bar{V}}_i)|\psi_i\rangle = \varepsilon_i|\psi_i\rangle \tag{3-4}$$

are established to determine $|\psi_i\rangle$. The system energy eigenvalue E of the Hamiltonian (3-3) is not simply the sum $\sum_{i=1}^{A}\varepsilon_i$. The factor of $\frac{1}{2}$ appearing in (3-3), to avoid double-counting the pairwise potential energy interactions, imposes the relationship

$$2E = \sum_{i=1}^{A} \varepsilon_i + \sum_{i=1}^{A} \langle T_i\rangle. \tag{3-4a}$$

A further refinement requires the inclusion of a rearrangement energy contribution to (3-4a) and a more careful specification of the single-particle energies ε_i. These points and related topics are discussed in Section IV, E.

A particularly simple form results when the system wave function $|\Psi\rangle$ is taken to be just the antisymmetrized product of the $|\psi_i\rangle$'s, in which event the model is referred to as the *extreme single-particle* model. In this case no effort is made to proceed with the variational calculation implied in (3-2); rather, the simple "trial" function $|\Psi\rangle$ is assumed, and empirical verification or refinement of the form $\bar{V}_i$ is sought, thereby justifying (or perhaps improving) $|\Psi\rangle$.

Usually, a large degree of degeneracy in (3-4) results with the assignment of quantum numbers such as n, l, j. Since no attempt is generally made to relate V_{ij} and $\overline{V}_i$, $\overline{V}_i$ might be viewed to be an effective potential even in the presence of three-body forces. When a theoretical basis for $\overline{V}_i$ is sought, only two-body potentials are usually considered.

In principle, two simple extensions of the extreme single-particle model are possible among many choices. The first is to classify the nucleons outside closed shells or subshells of (3-4) as "valence" nucleons designated by n, l, j and attempt to specify *residual interactions* $\Delta V_{ij} = V_{ij} - \overline{V}_{\text{core}}$, where $\overline{V}_{\text{core}}$ is generally the average of V_{ij} operating between a particular valence nucleon and the nucleons in the closed shells or "core." The Hamiltonian (3-3) is diagonalized (usually removing some of the degeneracy of the extreme single-particle model) and new single-particle states are determined. A more ambitious program employing some form of the Hartree–Fock method might be employed. Here (3-2) can be expressed as

$$\langle\Psi|H|\Psi\rangle = \sum_{i=1}^{A} \langle\psi_i|T_i|\psi_i\rangle + \tfrac{1}{2}\sum_{\substack{i,j\\ i\neq j}} \langle\psi_i\psi_j|V_{ij}|\psi_i\psi_j\rangle \tag{3-5}$$

and $|\psi_i\rangle = \sum_\alpha C_{\alpha i}|\alpha\rangle$, where the single-particle trial function $|\psi_i\rangle$ is expanded in basis states $|\alpha\rangle$ (e.g., the shell model harmonic oscillator states) for which matrix elements of the two-particle operator of V_{ij} can be readily determined. The procedure now consists of varying the expansion coefficients $C_{\alpha i}$ to minimize (3-5) while at the same time giving a self-consistent $|\psi_i\rangle$ in (3-4) for the required $\overline{V}_i$, which can be expressed in terms of $\langle\alpha\gamma|V_{ij}|\beta\delta\rangle$.

The procedure adopted in the next section will be to construct a trial wave function directly related to the nuclear density through a rather simple, but plausible, approximation. The nuclear density will be taken to be the trapezoidal shape of Fig. 2.6, parameterized by a half-density radius R and a surface thickness $2R\delta$. The variational treatment of (3-2) will consist of varying R and δ to find the stationary value of E. The method employed for determining the functional form of E will be based on a simple Thomas–Fermi or statistical model calculation.

A rather sophisticated version of the above approach has been carried out by Brueckner *et al.* (1971). They formally relate the total nuclear energy to a density Lagrangian and express this density dependence separately for neutrons and protons. The objective of such calculations (as well as the one presented here) is to incorporate as much as possible of our knowledge of the strongly interacting fermion system and yet lead to a mathematically tractable solution. Brueckner *et al.* include density components that are consistent with infinite nuclear matter properties (see below). A variational calculation for the resulting energy-density functional leading to extremal values for the involved densities gives a satisfactory fit to the empirical data. Such calculations can also allow for deformed shapes. See also Brueckner *et al.* (1968, 1970).

In the final section general properties of the many-nucleon system are related to a more formal and detailed examination of the two-nucleon interaction in the modifying presence of the other nucleons of the system. The ideal theoretical investigation of these features treats a hypothetical, infinitely large nucleus yielding what is generally referred to as the properties of infinite nuclear matter and summarized in the prescription of an effective interaction. Current many-body treatment of the finite-nucleus problem tends to combine both statistical and effective interaction features.

B. THE STATISTICAL MODEL

The earliest historical approach to the development of the total energy equation for a nuclear system consisting of many nucleons was some variant of the statistical model. Such models generally combined the gross effects of the Pauli principle reducing the behavior of the nucleons to that of a Fermi gas, and a simplified effective central two-nucleon interaction potential of the type summarized in (1-97). Although modern calculations have far greater sophistication, much insight can be gained by examining the details of the simpler models.

We proceed in the following subsections to present the assumptions usually made or implied and to derive the energy equation of state for the nucleus. To a surprising degree, the energy equation reduces to a relatively simple expression depending on the mass number A and atomic number Z, referred to as the semiempirical mass formula. Largely for purposes of orientation, we examine the energy instability implied by such mass formulas for α- and β-decay as well as for nuclear fission.

After the introduction of the simplifying assumptions of our statistical model in the first subsection, we interrupt the development of the model to discuss some mathematical problems and examine the nature of the statistical correlations resulting from the operation of the Pauli principle for fermions.

1. Simplifying Assumptions

The simplest nuclear model would result if we ignored the internucleon potentials V_{ij} in (3-3) and simulated their average role by simply requiring the nucleons to be confined to the nuclear volume Ω. An infinitely deep square well would in effect lead to such a boundary condition. The total energy E, the eigenvalue of (3-3), would then be referenced to an arbitrary potential, which can conveniently be taken as the bottom of the well. A more realistic finite-depth square well will be considered in detail later. At this point we simply recall from (2-71) for the finite square-well approximation of the deuteron problem, that in the region $r > b$ the wave function varies as $e^{-\alpha r}$ with $\alpha^2\hbar^2 = 2\mu E_B$. It is therefore clear that a deep enough well, at least for the lower levels, i.e., E_B large, will behave essentially in the same manner as the infinitely deep well in yielding a vanishing small wave function at the well surface. We shall return to this point later.

The individual interaction-free nucleon states confined to a volume Ω would be simple plane waves characterized by the momentum wave number $\mathbf{k}_\gamma$, the spin function $\chi_\gamma(\boldsymbol{\sigma}_i)$, the isospin function $\zeta_\gamma(\boldsymbol{\tau}_i)$, and a suitable normalization N_γ, viz.,

$$\psi_\gamma(\mathbf{r}_i, \boldsymbol{\sigma}_i, \boldsymbol{\tau}_i) = N_\gamma[\exp(i\mathbf{k}_\gamma \cdot \mathbf{r}_i)]\, \chi_\gamma(\boldsymbol{\sigma}_i)\, \zeta_\gamma(\boldsymbol{\tau}_i). \tag{3-6}$$

In generating the ground-state nuclear wave function $|\Psi\rangle$, note must be taken of the fermion character of the nucleons and an antisymmetrized product wave function constructed from the single-particle states ψ_i of (3-6). This is readily achieved by writing $|\Psi\rangle$ as the *Slater determinant*,

$$\Psi(A) = |A\rangle = N_A \begin{vmatrix} \psi_1(\mathbf{r}_1, \boldsymbol{\sigma}_1, \boldsymbol{\tau}_1) & \psi_1(\mathbf{r}_2, \boldsymbol{\sigma}_2, \boldsymbol{\tau}_2) & \cdots & \psi_1(\mathbf{r}_A, \boldsymbol{\sigma}_A, \boldsymbol{\tau}_A) \\ \psi_2(\mathbf{r}_1, \boldsymbol{\sigma}_1, \boldsymbol{\tau}_1) & \ddots & & \vdots \\ \vdots & & \ddots & \\ \psi_A(\mathbf{r}_1, \boldsymbol{\sigma}_1, \boldsymbol{\tau}_1) & \cdots & & \psi_A(\mathbf{r}_A, \boldsymbol{\sigma}_A, \boldsymbol{\tau}_A) \end{vmatrix}. \tag{3-7}$$

If the normalization N_γ gives unity for (3-6) over the entire volume Ω, i.e., $N_\gamma = 1/\Omega^{1/2}$, then $N_A = (A!)^{-1/2}$. The wave function $|\Psi\rangle$ of (3-7) describes the Fermi gas model. The numerical subscripts on ψ represent the quantum numbers γ specifying $\mathbf{k}_\gamma$, $\sigma_{z\gamma}$, and $\tau_{3\gamma}$, while $\mathbf{r}_i, \boldsymbol{\sigma}_i, \boldsymbol{\tau}_i$ are the nucleon "coordinates." The "gas" is seen to consist of four distinguishable components corresponding to the four sets of quantum numbers (σ_z, τ_3) for each k_γ, namely, the neutron state with spin up $(\frac{1}{2}, -\frac{1}{2})$, the neutron state with spin down $(-\frac{1}{2}, -\frac{1}{2})$, the proton state with spin up $(\frac{1}{2}, \frac{1}{2})$, and the proton state with spin down $(-\frac{1}{2}, \frac{1}{2})$. The wave function $|\Psi\rangle$ is also seen to obey the Pauli exclusion principle, because if for two rows i and j, (σ_{zi}, τ_{3i}) and (σ_{zj}, τ_{3j}) are the same, then $\mathbf{k}_i$ must differ from $\mathbf{k}_j$ or else $|\Psi\rangle$ is identically zero. For a finite volume Ω, the boundary condition on (3-6) restricts $\mathbf{k}_\gamma$ to specific values. For example, in a cubical box with $\Omega = L^3$, we would have $k_{\gamma x} = 2\pi n_{\gamma x}/L$, $k_{\gamma y} = 2\pi n_{\gamma y}/L$, and $k_{\gamma z} = 2\pi n_{\gamma z}/L$ with $n_{\gamma x}, n_{\gamma y}, n_{\gamma z}$ being any three integers. Thus the Pauli principle in the above case would require that at least one of the three integers associated with n_γ be different for the two states occupied by the "identical" gas components. This requirement is well known to lead to the statement that *only one identical* particle can occupy each volume element

h^3 in phase space, or

$$d^2n = 4\pi p^2\,dp\,d\Omega/h^3$$

and hence

$$d\rho = d^2n/d\Omega = k^2\,dk/2\pi^2. \tag{3-8}$$

For the nuclear ground state each "lattice site" n_γ (i.e., $n_{\gamma x}, n_{\gamma y}, n_{\gamma z}$) is occupied starting with the smallest integers and progressively assigning higher ones until all of one particular type of identical particle in $d\Omega$ have been accommodated. We define

$$k_F^{\,2} = (k_x^{\,2}+k_y^{\,2}+k_z^{\,2})_{\max} = \frac{4\pi^2}{L^2}(n_x^{\,2}+n_y^{\,2}+n_z^{\,2})_{\max} = \frac{4\pi^2}{L^2}n_F^{\,2}.$$

Thus the accommodation process described above occupies all sites (n_x, n_y, n_z) within a sphere of radius n_F. Then the ground-state density is

$$\rho = (1/2\pi^2)\int_0^{k_F} k^2\,dk = k_F^{\,3}/6\pi^2. \tag{3-9}$$

A numerical quantity often quoted, for the case of nuclei with an equal number of neutrons and protons, an equal number of spin-up and spin-down states, and appropriate for the central region of the nucleus where the total matter density is ρ_0, is $\rho = \rho_0/4 = (0.170/4)\ \text{fm}^{-3}$; whence[1]

$$k_F = (\tfrac{3}{2}\pi^2\rho_0)^{1/3} = 1.36\ \text{fm}^{-1}. \tag{3-10}$$

The total energy E from (3-3), applying the constant-density model to such nuclei, is

$$E = \sum_{i=1}^{A} T_i \approx A\left(4\pi\int_0^{k_F} Tk^2\,dk\right)\Big/\tfrac{4}{3}\pi k_F^{\,3}$$

or

$$E \approx (3\hbar^2/10m_N)k_F^{\,2}A, \tag{3-11}$$

since $T = \hbar^2k^2/2m_N$. If we define $T_F = \hbar^2k_F^{\,2}/2m_N$, then $E = \frac{3}{5}T_F A$ and the average kinetic energy per nucleon is $\bar{T} = E/A = \frac{3}{5}T_F$. Using (3-10), we obtain (see previous footnote)

$$T_F = 38.3\ \text{MeV}. \tag{3-12}$$

When the interaction potentials V_{ij} in (3-3) are considered, the energy E is, of course, altered, as are the individual wave functions ψ_i. We now use as trial

[1] While it is common practice in nuclear matter studies to quote a central density $\rho_0 = 0.170\ \text{fm}^{-3}$ obtained from electron scattering data analyzed with a Fermi distribution, we should recall from Chapter II that muonic data give $\rho_0 = 0.158\ \text{fm}^{-3}$. This latter value corresponds to $k_F = 1.33\ \text{fm}^{-1}$ and $T_F = 36.6$ MeV.

wave functions

$$\psi_\gamma(\mathbf{r}_i, \boldsymbol{\sigma}_i, \boldsymbol{\tau}_i) = N_\gamma f(r_i)\{\exp[i\mathbf{k}_\gamma(r_i)\cdot\mathbf{r}_i]\}\,\chi_\gamma(\boldsymbol{\sigma}_i)\,\zeta_\gamma(\boldsymbol{\tau}_i), \tag{3-13}$$

such that $\rho_\gamma(r_i) = |\psi_\gamma(\mathbf{r}_i)|^2 = N_\gamma^{\,2} f^2(r_i)$ and $\mathbf{k}_\gamma(r_i) = (n_i/n_\mathrm{F})\,k_\mathrm{F}(r_i)$. This approximation is based on the "local uniformity assumption" [see, for example, Rodberg and Teplitz (1960) and Ayres *et al.* (1962)] and includes both an amplitude modulation and a frequency modulation of the plane waves of (3-6). The amplitude modulation through $f(r_i)$ gives a contribution to the local density $\rho_\gamma(r_i)$, which we shall assume follows the variation of the matter density $\rho(r)$ (and hence we drop the subscript γ). The frequency modulation appears through $\mathbf{k}_\gamma(r_i)$. The wave number is assumed to vary following a $[\rho(r)]^{1/3}$ dependence in the same manner as does k_F, however with the scale (and direction) $(n_x\hat{x}+n_y\hat{y}+n_z\hat{z})_\gamma/n_\mathrm{F}$. These assumptions are first of all equivalent to assuming that the four nucleon fluid types have densities simply scaling with the matter density $\rho(r)$; and second, that in the surface region where the falloff distance (~ 2.4 fm) is large compared to the range of $V_{ij}(\mathbf{r}_i - \mathbf{r}_j)$ (approximately ~ 1 fm), an essentially WKB approach is valid if the immediate region of the classical turning point is avoided.

The above assumptions are simplifications of modern local density approximations [see for example Németh and Ripka (1972), Campi and Sprung (1972), and the excellent review article by Bethe (1971)]. For a discussion of the possible singular behavior of statistical theories and classical turning point effects, see Siemens (1970). For a brief discussion of the relationship of suitable single-particle potentials to general Thomas–Fermi statistical nuclear matter models, see, e.g., Myers (1970).

Finally we write, for the trial density of the nuclear ground state,

$$\Psi^*(A)\Psi(A) = N_A^{\,2}\begin{vmatrix} u_1^*(\mathbf{r}_1,\boldsymbol{\sigma}_1,\boldsymbol{\tau}_1) & u_1^*(\mathbf{r}_2,\boldsymbol{\sigma}_2,\boldsymbol{\tau}_2) & \cdots & u_1^*(\mathbf{r}_A,\boldsymbol{\sigma}_A,\boldsymbol{\tau}_A) \\ u_2^*(\mathbf{r}_1,\boldsymbol{\sigma}_1,\boldsymbol{\tau}_1) & \ddots & & \vdots \\ \vdots & & \ddots & \\ u_A^*(\mathbf{r}_1,\boldsymbol{\sigma}_1,\boldsymbol{\tau}_1) & \cdots & & u_A^*(\mathbf{r}_A,\boldsymbol{\sigma}_A,\boldsymbol{\tau}_A) \end{vmatrix}$$
$$\times\begin{vmatrix} u_1(\mathbf{r}_1,\boldsymbol{\sigma}_1,\boldsymbol{\tau}_1) & u_1(\mathbf{r}_2,\boldsymbol{\sigma}_2,\boldsymbol{\tau}_2) & \cdots & u_1(\mathbf{r}_A,\boldsymbol{\sigma}_A,\boldsymbol{\tau}_A) \\ u_2(\mathbf{r}_1,\boldsymbol{\sigma}_1,\boldsymbol{\tau}_1) & \ddots & & \vdots \\ \vdots & & \ddots & \\ u_A(\mathbf{r}_1,\boldsymbol{\sigma}_1,\boldsymbol{\tau}_1) & \cdots & & u_A(\mathbf{r}_A,\boldsymbol{\sigma}_A,\boldsymbol{\tau}_A) \end{vmatrix}$$
$$\times\, \rho(r_1)\,\rho(r_2)\cdots\rho(r_A). \tag{3-14}$$

Here $u_\gamma(\mathbf{r}_i, \boldsymbol{\sigma}_i, \boldsymbol{\tau}_i) = \{\exp[i\mathbf{k}_\gamma(\mathbf{r}_i)\cdot\mathbf{r}_i]\}\,\chi_\gamma(\boldsymbol{\sigma}_i)\,\zeta_\gamma(\boldsymbol{\tau}_i)$ and the normalization $N_A^{\,2}$ depends on $N_\gamma^{\,2}$. If $N_\gamma^{\,2}\int\rho(r_i)\,d^3r_i = 1$, then $N_A^{\,2} = 1/A!$. Also, $u_\gamma^*(\mathbf{r}_i, \boldsymbol{\sigma}_i, \boldsymbol{\tau}_i)$

stands for $\{\exp[-i\mathbf{k}_\gamma(\mathbf{r}_i)\cdot\mathbf{r}_i]\}\,\chi_\gamma^\dagger(\boldsymbol{\sigma}_i)\zeta_\gamma^\dagger(\boldsymbol{\tau}_i)$. The potential energy operator $\tilde{V}_{ij}$ is written

$$\tilde{V}_{ij} = (e^2/r_{ij})\cdot\tfrac{1}{4}(1+\tau_{3i})(1+\tau_{3j}) + \tilde{v}_{ij} \tag{3-15a}$$

with

$$\tilde{v}_{ij} = {}^{31}V(r_{ij})\mathscr{T}(\sigma)\mathscr{S}(\tau) + {}^{13}V(r_{ij})\mathscr{S}(\sigma)\mathscr{T}(\tau) + {}^{33}V(r_{ij})\mathscr{T}(\sigma)\mathscr{T}(\tau) + {}^{11}V(r_{ij})\mathscr{S}(\sigma)\mathscr{S}(\tau). \tag{3-15b}$$

The first term of (3-15a) is the Coulomb potential operating between protons and v_{ij} is seen from (3-15b) to be the most general velocity-independent nuclear central potential.

This assumption of only a central potential is, of course, a gross oversimplification. However, the main contribution to the nuclear binding energy is contributed by the S state, for which both the tensor interaction and the spin–orbit interaction give zero contributions. A further simplification will be to take the geometric dependence of v_{ij} on r_{ij} to be a simple square well with the same well radius a for all the terms. Saturation of nuclear density will then result from the strongly repulsive singlet–singlet interaction and to some extent the weaker repulsion for the triplet–triplet interaction. These repulsive configurations correspond to the odd relative space states.

In the treatment presented here, we follow the work of Ayers *et al.* (1962), where these simplifying assumptions are made to reduce the calculational difficulties and to permit nearly analytic expressions partially revealing the role various factors play in the finite nucleus with $N \neq Z$. A more realistic approach would include a repulsive core, at least for the S-state interactions, a tensor interaction, and velocity-dependent potentials. Various more realistic potentials have been used in the literature, e.g., Tabakin (1964), Bethe (1968), Németh and Bethe (1968), Dahll and Warke (1970), Pandharipande and Prasad (1970), and Signell (1970). A particularly noteworthy treatment of the finite-nucleus case using a local-density approximation is that by Negele (1970).

In discussing suitable effective potentials, a further point extensively investigated by Coester *et al.* (1970) requires consideration. Nucleon–nucleon scattering phase shifts at all energies, and the properties of the deuteron are given equally well by a large class of unitarily equivalent Hamiltonians. The effective range approximation that was discussed in Section I,C,3 is simply the low-energy application of this equivalence. The transformation of one suitable potential into another generally involves changes in both shape and velocity dependences. Particularly due to this latter effect, these authors find the nuclear matter properties based on such equivalent potentials to vary considerably. This behavior underlines the importance of nuclear matter calculations in determining possible nucleon–nucleon potentials.

Another effect that should be considered in nuclear matter calculations relates to possible inelastic nucleon–nucleon interactions introducing N* and Δ intermediate nucleon state contributions [see Green and Schucan (1972) and the discussions pertaining to Fig. 1.18 of Chapter I and of the trinucleon β-decay in Section VIII, D].

Finally, suggestions have been put forward recently that for sufficiently dense nuclear matter an electrically neutral $\pi^{\pm}, \pi^0$ condensate might appear within the nuclear interior. In this regard recall that pions are bosons. According to calculations by Migdal (1973), the central density in the heavier nuclear species is adequate for such a condensation (nuclear matter phase change) to occur. However, relevant calculations involving the pion chemical potential require the inclusion of effects only partially understood at present, and Barshay and Brown (1973) conclude that the nuclear density ρ_0 is insufficient to produce pion condensation. They point out that, if such a condensation did occur in ^{16}O, the known excited state $T = 1$, $J^{\pi} = 0^-$ at 12.78 MeV (having the same quantum numbers as the pion), would have a considerable interaction energy shift (estimated at ~5 MeV). The empirical evidence indicates only a slight shift from an unperturbed energy of 12.42 MeV, which in fact is in the wrong direction (i.e., the interaction with the pion condensate would have lowered the energy from the unperturbed value). Pion condensation in (astronomical) neutron stars is a separate possibility.

2. Mathematical Preliminaries

We must now make a slight digression to consider certain mathematical difficulties that arise in evaluating the potential energy. It is often necessary to evaluate the expectation value for a two-body operator

$$G = \sum_{i>j=1}^{A} g_{ij} = \tfrac{1}{2} \sum_{i,\, j(i \neq j)}^{A} g_{ij}$$

for a wave function $|A\rangle$ of the type (3-7). The discussion of this problem appears in many references, notably in Condon and Shortley (1935, p. 171 ff.). The determinant $|A\rangle$ can be developed by either rows or columns, i.e., by either coordinates $(\mathbf{r}, \boldsymbol{\sigma}, \boldsymbol{\tau})$ or quantum numbers γ. We adopt the latter and for convenience run the values of γ over the indices i and j and assume the normalization $N_A{}^2 = 1/A!$, whence

$$\begin{aligned}\langle A|G|A\rangle = \frac{(A-2)!}{A!}\,\frac{A(A-1)}{2} \sum_{i>j=1}^{A} \iint d^3r_a\, d^3r_b [&\psi_i^*(a)\psi_j^*(b) g_{ij} \psi_i(a)\psi_j(b) \\ &- \psi_i^*(a)\psi_j^*(b) g_{ij} \psi_i(b)\psi_j(a) + \psi_i^*(b)\psi_j^*(a) g_{ij} \psi_i(b)\psi_j(a) \\ &- \psi_i^*(b)\psi_j^*(a) g_{ij} \psi_i(a)\psi_j(b)]. \end{aligned} \qquad (3\text{-}16)$$

It is to be emphasized that the sum $\sum_{i>j=1}^{A}$ runs over sets of quantum numbers γ of which there are also just A in total. In (3-16) a particular pair of quantum numbers $\gamma = i$ and j has been selected. There are of course $A(A-1)/2$ ways in which coordinates can be assigned to such a pair of quantum numbers, e.g., $\phi_i(1)\,\phi_j(2)$ or $\phi_i(1)\,\phi_j(8), \ldots$, and we allow a and b to stand for any such pair. Each of the four-product terms in the sum (3-16) has been integrated over the quantum numbers other than $\gamma = i$ and j. Since the ψ_γ are orthogonal, the set of $A-2$ values of γ (other than i and j) must be properly matched or the entire term vanishes. There are just $(A-2)!$ ways of permuting these matched quantum numbers. We thus arrive at (3-16).

Note that the first and third terms in (3-16) are equal and the second and fourth terms are equal, since a and b are dummy coordinates; thus

$$\langle A|\,G\,|A\rangle = \sum_{i>j=1}^{A} \iint d^3r_a\, d^3r_b [\psi_i^*(a)\,\psi_j^*(b)\,g_{ij}\,\psi_i(a)\,\psi_j(b) - \psi_i^*(a)\,\psi_j^*(b)\,g_{ij}\,\psi_i(b)\,\psi_j(a)]. \tag{3-17}$$

The first term of (3-17) is referred to as the direct integral, and the second as the exchange integral, for obvious reasons. If we introduce $\tilde{\mathbb{P}}_{ij} = \tilde{P}_{ij}^r \tilde{P}_{ij}^\sigma \tilde{P}_{ij}^\tau$, the operator that interchanges all coordinates, we can write

$$\langle A|\,G\,|A\rangle = \sum_{i>j=1}^{A} \iint d^3r_a\, d^3r_b\, \psi_i^*(a)\,\psi_j^*(b)\,g_{ij}(1-\mathbb{P}_{ij})\,\psi_i(a)\,\psi_j(b), \tag{3-18}$$

since $\mathbb{P}_{ij}\,\psi_i(a)\,\psi_j(b) = \psi_i(b)\,\psi_j(a)$. We might also note that interchanging coordinates is equivalent to interchanging quantum numbers. For antisymmetrized total wave functions, we need not exclude $i = j$, since this leads to zero contribution in (3-17) or (3-18); hence we also have $\sum_{i>j=1}^{A} = \frac{1}{2}\sum_{i,\,j(i\neq j)}^{A} = \frac{1}{2}\sum_{i,\,j}^{A}$.

▶**Exercise 3-1** Frequently it is also required to calculate the expectation value of a one-body operator of the form $F = \sum_{i=1}^{A} f_i$ (e.g., the kinetic energy operator for A particles). Show that

$$\langle A|\,F\,|A\rangle = \sum_{i=1}^{A} \int \psi_i^*(a)\, f_i\, \psi_i(a)\, d^3r_a. \tag{3-19}$$

Hint: First show that there are $(A-1)!$ permutations for matching quantum numbers γ that give nonvanishing contributions for the product of the minors of a particular $\psi_i(a)$, which itself, of course, can be selected A different ways. ◀

In evaluating the potential energy $E_\mathrm{p} = \langle A|\,v_{ij}\,|A\rangle$, it is useful to develop the expectation values for the projection operators $\mathscr{T}(\sigma)\,\mathscr{S}(\tau)$, etc. These projection operators have all been written in terms of the exchange operators

P^σ, P^τ, and P^r [see (1-96)]. It is convenient first to calculate $\langle A|P^\sigma|A\rangle$, $\langle A|P^\tau|A\rangle$, and $\langle A|P^r|A\rangle$, in terms of which $\langle A|\mathscr{T}(\sigma)\mathscr{S}(\tau)|A\rangle$, etc., can be readily evaluated. The expectation value $\langle A|P^\xi|A\rangle$ is a shorthand for $\langle A|\sum_{i>j}^A P_{ij}^\xi|A\rangle$. Since $\sum_{i>j}^A P_{ij}^\xi$ commutes with H of (3-3) when V_{ij} is given by (3-15), these quantities are constants of the motion.

▶**Exercise 3-2** Recalling (1-87), (1-89), and (1-93) and writing v_{ij} of (3-15b) in the form $V_0(r)[A+B\boldsymbol{\sigma}_i\cdot\boldsymbol{\sigma}_j+C\boldsymbol{\tau}_i\cdot\boldsymbol{\tau}_j+D(\boldsymbol{\sigma}_i\cdot\boldsymbol{\sigma}_j)(\boldsymbol{\tau}_i\cdot\boldsymbol{\tau}_j)]$, show that $\sum_{i>j}^A P_{ij}^\xi$ commutes with (3-15).

Hint: While it will be evident at once that P^ξ commutes with the kinetic energy and the central terms of the potential energy, the commutation of P^τ (and $P^r = -P^\sigma P^\tau$) with the Coulomb energy must be examined with some care. This leads to considering the commutator

$$\left[\left\{\sum_{i>j}^{A}(\tau_{3i}+\tau_{3j}+\tau_{3i}\tau_{3j})\right\},\left\{\sum_{r>s}^{A}(\tau_{1r}\tau_{1s}+\tau_{2r}\tau_{2s})\right\}\right].$$

Distinguish the three cases: (a) When all four indices $ijrs$ are different; (b) When $r=i$ and $s=j$; (c) When ijs are different. There is no loss in generality by selecting these indices such that $i>j>s$. Six different commutators must be considered *simultaneously*, viz.,

$$[E_{ij},F_{js}]+[E_{ij},F_{is}]+[E_{is},F_{ij}]+[E_{is},F_{js}]+[E_{js},F_{ij}]+[E_{js},F_{is}].$$ ◀

We now introduce a simplified *partition diagram* exhibiting the quantum numbers of the single-particle states. Under the operation of only central forces, s_{zi} is a constant of the motion, and for a particular value of $\mathbf{k}_{\gamma i}$ four combinations of s_{zi} and τ_{3i} are permitted by the Pauli exclusion principle. These states are shown in Fig. 3.1.

The partition numbers l, m, p, and q of Fig. 3.1 designate the number of "spin-up" neutrons, "spin-down" neutrons, "spin-up" protons, and "spin-down" protons, respectively. To specify completely the exact single-particle description of a nucleus (including possible excited configurations) requires additional partition numbers to identify the number of nucleons occupying levels with the same value of $\mathbf{k}_\gamma$. There exist elegant classification schemes [e.g., the Wigner supermultiplet classification, Wigner (1937)] to accomplish this in some generality. Since it is not our intention to develop such general classification schemes, we restrict ourselves to a particularly simple schema generally satisfactory for "low-energy" configurations.

A simplification, adequate for our purposes, is to consider only so-called "normal arrangements" [see Blatt and Weisskopf (1952, Chapter VI)], where for the case $N \geqslant Z$, $l \geqslant m \geqslant p \geqslant q$; q in addition to specifying the number of protons with "spin down" also represents the number of levels (designated by any occupied sequence of $\mathbf{k}_\gamma$) that have four nucleons $\mathrm{n}\uparrow, \mathrm{n}\downarrow, \mathrm{p}\uparrow, \mathrm{p}\downarrow$;

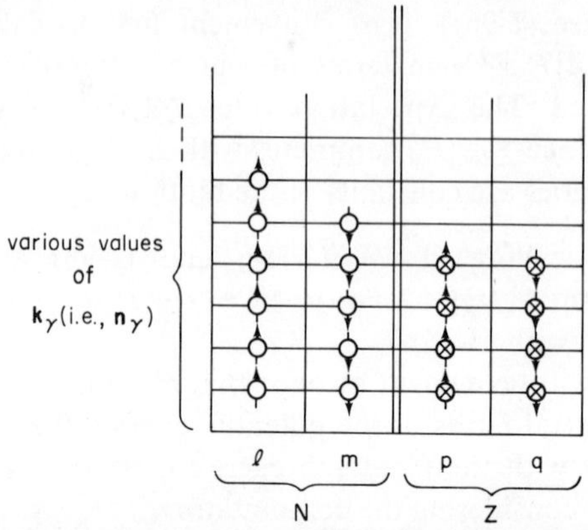

FIG. 3.1. The partition diagram for the single-particle quantum states with γ designating ($\mathbf{k}$, σ_z, τ_3). The column labeled l displays the neutrons with "spin up"; the one labeled m, the neutrons with "spin down"; and p and q refer to corresponding spin states for the protons, with $l+m = N$ and $p+q = Z$, and $N+Z = A$.

$(p-q)$ also represents the number of levels (again any sequence of $\mathbf{k}_\gamma$) that have three nucleons $n\uparrow, n\downarrow, p\uparrow$; similarly, $(m-p)$ represents the number of levels with two neutrons $n\uparrow, n\downarrow$, and $(l-m)$ the number with only one neutron, $n\uparrow$. A similar set of conditions holds for $Z \geqslant N$. While this convention precludes many possible diagrams, it does permit $(l-m) > 1$, etc., ..., and thus clearly permits some excited nuclear configurations.

Consider, for example, evaluating $\langle A|P^\tau|A\rangle$, and hence take $g_{ij} = P^\tau_{ij}$ in (3-17). Let us designate the quantum numbers γ by $(\mathbf{k}, \sigma_z, \tau_3)$. First consider the same $\mathbf{k}$ for i and j. When $i = (\mathbf{k}, \frac{1}{2}, -\frac{1}{2})$ and $j = (\mathbf{k}, -\frac{1}{2}, -\frac{1}{2})$, the direct term in (3-17) yields $+1$ and the exchange term is zero (due to the orthogonality of the spin functions), and hence the contribution to $\langle P^\tau\rangle$ is just $+1$. When $i = (\mathbf{k}, \frac{1}{2}, -\frac{1}{2})$ and $j = (\mathbf{k}, \frac{1}{2}, \frac{1}{2})$, the direct term gives zero and the exchange term -1; hence the contribution to $\langle P^\tau\rangle$ is now -1. When $i = (\mathbf{k}, \frac{1}{2}, -\frac{1}{2})$ and $j = (\mathbf{k}, -\frac{1}{2}, \frac{1}{2})$, we get zero for both the direct and exchange integrals. In a similar manner when $i = (\mathbf{k}, \frac{1}{2}, -\frac{1}{2})$ and $j = (\mathbf{k}' \neq \mathbf{k}, \pm\frac{1}{2}, -\frac{1}{2})$, we get $+1$ as the contribution to $\langle P^\tau\rangle$ and for $j = (\mathbf{k}' \neq \mathbf{k}, \pm\frac{1}{2}, \frac{1}{2})$, we get zero when orthogonality in $\mathbf{k}$ is also taken into account. The generalization of these findings leads to some simple rules.

For i = neutron with spin up and momentum $\mathbf{k}$, i.e., $i = (\mathbf{k}, \frac{1}{2}, -\frac{1}{2})$ the results are that:

(a) Charge exchange with *any* other neutron gives a contribution $+1$;

(b) Charge exchange with a spin-up proton with the same $\mathbf{k}$ gives -1 and with different $\mathbf{k}$ gives 0;

(c) Charge exchange with any spin-down proton gives 0.

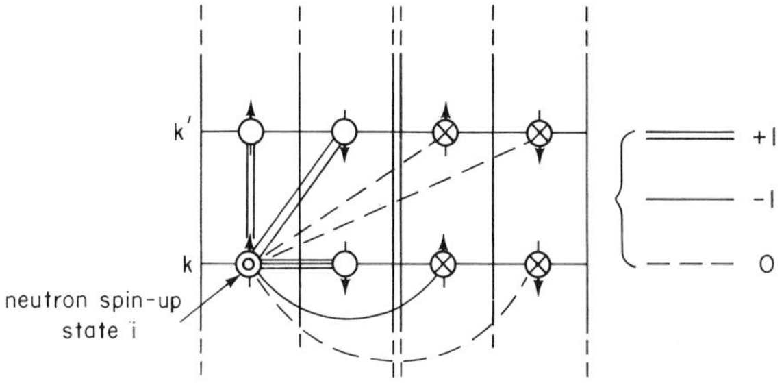

FIG. 3.2. Diagrammatic summary of charge exchange contributions to $\langle A|P^\tau|A\rangle$.

In diagram form, this is summarized in Fig. 3.2. These rules are equivalent to those given in reference Blatt and Weisskopf (1952). We note that charge exchanges between $i \equiv (\mathbf{k}, \frac{1}{2}, -\frac{1}{2})$ and those states j connected by dashed lines would result in diagrams violating the Pauli principle. We also see that these exchanges give zero contribution. Charge exchanges indicated by the double line correspond to exchanges for a two-particle "parallel" charge state and hence give a $+1$ contribution. The exchange indicated by the solid line gives -1 for the two-particle "antiparallel" charge state. We further note that both the double and single solid line exchanges are permitted by the Pauli principle. Similar rules follow for $i = (\mathbf{k}, \frac{1}{2}, \frac{1}{2}), (\mathbf{k}, -\frac{1}{2}, \frac{1}{2})$, and $(\mathbf{k}, -\frac{1}{2}, -\frac{1}{2})$.

Returning to Fig. 3.1, we then readily obtain

$$\langle|P^\tau|\rangle_{\mathrm{N}\geqslant\mathrm{Z}} = (+1)\left[\frac{N(N-1)}{2} + \frac{Z(Z-1)}{2}\right] + (-1)[2q + (p-q)]$$

$$\langle|P^\tau|\rangle_{\mathrm{N}\leqslant\mathrm{Z}} = (+1)\left[\frac{N(N-1)}{2} + \frac{Z(Z-1)}{2}\right] + (-1)[2m + (l-m)].$$

Then introducing $A = N+Z$, $I = N-Z$, $\eta = l-m$, and $\zeta = p-q$, both of these results can be combined into one simple form for any N, Z, yielding

$$\langle|P^\tau|\rangle = \tfrac{1}{4}A^2 - A + \tfrac{1}{4}I^2 + \tfrac{1}{2}|I|. \tag{3-20a}$$

In a similar manner, we find

$$\langle|P^\sigma|\rangle = \tfrac{1}{4}A^2 - A + \tfrac{1}{2}(\eta+\zeta) + \tfrac{1}{4}(\eta+\zeta)^2 \tag{3-20b}$$

$$\langle|P^r|\rangle = -\tfrac{1}{8}A^2 + 2A - \tfrac{1}{8}I^2 - |I| - \tfrac{1}{2}(\eta+\zeta) - \tfrac{1}{4}(\eta^2+\zeta^2) \tag{3-20c}$$

$$\langle|1|\rangle = \tfrac{1}{2}A(A-1). \tag{3-20d}$$

▶**Exercise 3-3** (a) Verify the considerations leading to the rules summarized in Fig. 3.2.

(b) Verify the steps leading to (3-20a).

(c) In a similar manner derive (3-20b)–(3-20d). ◀

▶**Exercise 3-4** The results (3-20) can also be derived directly from (3-18). Letting $g_{ij} = a + bP_{ij}^{\sigma} + cP_{ij}^{\tau} + dP_{ij}^{r}$ and $\mathbb{P}_{ij} = P_{ij}^{r} P_{ij}^{\sigma} P_{ij}^{\tau}$, show that $g_{ij}(1 - \mathbb{P}_{ij}) = (a - dP_{ij}^{\sigma} P_{ij}^{\tau} + bP_{ij}^{\sigma} + cP_{ij}^{\tau}) + (-aP_{ij}^{\sigma} P_{ij}^{\tau} + d - bP_{ij}^{\tau} - cP_{ij}^{\sigma}) P_{ij}^{r}$.

Consider the simple case $l = m = p = q$ (the so-called $4n$ nuclei). First consider the sums in $\sum_{i>j}^{A}$ involving the quantum numbers σ_z and τ_3.

(a) Show that there are $2^4 = 16$ combinations giving $+1$ for the identity operator (i.e., for the term involving a).

(b) Show that there are $2^3 = 8$ combinations giving $+1$ for the operator P_{ij}^{σ} (also P_{ij}^{τ}).

(c) Show that for the operator $P_{ij}^{\sigma} P_{ij}^{\tau}$ only $2^2 = 4$ combinations give $+1$.

Hence show that

$$\sum_{i,j:(\sigma_z,\tau_3)} = 4[(4a - d + 2b + 2c) + (-a + 4d - 2b - 2c) P_{ij}^{r}],$$

which we can conveniently contract to

$$\sum_{i,j:(\sigma_z,\tau_3)} = 4(D + EP_{ij}^{r}).$$

Finally,

$$\langle A|G|A\rangle = 2\sum_{i,j}^{A/4} \iint d^3r_a\, d^3r_b\, u_1^*(a)\, u_j^*(b)(D + EP_{ij}^{r})\, u_i(a)\, u_j(b),$$

where i and j now only refer to the wave number $\mathbf{k}$, or

$$\begin{aligned}
\langle A|G|A\rangle &= 2D \sum_{i,j}^{A/4} \iint d^3r_a\, d^3r_b\, u_i^*(a)\, u_j^*(b)\, u_i(a)\, u_j(b) \\
&\quad + 2E \sum_{i,j}^{A/4} \iint d^3r_a\, d^3r_b\, u_i^*(a)\, u_j^*(b)\, u_j(a)\, u_i(b) \\
&= 2D \sum_{i,j}^{A/4} 1 + 2E \sum_{i,j}^{A/4} \delta_{ij} \\
&= 2D \sum_{i,j}^{A/4} 1 + 2(E + D) \sum_{i=1}^{A/4} 1 = 2D\frac{A}{4}\left(\frac{A}{4} - 1\right) + 2(E + D)\frac{A}{4}.
\end{aligned} \qquad (3\text{-}21)$$

Substituting for D and E, show that:

(d) $\langle |1| \rangle = \frac{1}{2}A(A-1)$.

(e) $\langle |P^{\sigma}| \rangle = A(\frac{1}{4}A - 1)$.

(f) $\langle |P^{\tau}| \rangle = A(\frac{1}{4}A - 1)$.

(g) $\langle |P^{r}| \rangle = -\frac{1}{8}A^2 + 2A$.

Note that these agree with (3-20) for $I = 0$, $\eta = 0$, and $\zeta = 0$. ◀

▶**Exercise 3-5** Defining the number of pairs of nucleons in a relative spin singlet and simultaneous charge triplet state by ${}^{13}n = \langle|\mathscr{S}(\sigma)\mathscr{T}(\tau)|\rangle$, and with ${}^{31}n$, ${}^{33}n$, and ${}^{11}n$ similarly defined, show from (3-20) that

$$ {}^{13}n = \tfrac{3}{32}A^2 + \tfrac{3}{8}A - \tfrac{1}{8}|I| + \tfrac{1}{32}I^2 - \tfrac{1}{4}(\eta+\zeta) - \tfrac{1}{8}(\eta^2+\eta\zeta+\zeta^2) \quad \text{(3-22a)} $$

$$ {}^{31}n = \tfrac{3}{32}A^2 + \tfrac{3}{8}A - \tfrac{3}{8}|I| - \tfrac{3}{32}I^2 + \tfrac{1}{8}\eta\zeta \quad \text{(3-22b)} $$

$$ {}^{11}n = \tfrac{1}{32}A^2 - \tfrac{1}{8}A + \tfrac{1}{8}|I| - \tfrac{1}{32}I^2 - \tfrac{1}{8}\eta\zeta \quad \text{(3-22c)} $$

$$ {}^{33}n = \tfrac{9}{32}A^2 - \tfrac{9}{8}A + \tfrac{3}{8}|I| + \tfrac{3}{32}I^2 + \tfrac{1}{4}(\eta+\zeta) + \tfrac{1}{8}(\eta^2+\eta\zeta+\zeta^2). \quad \text{(3-22d)} $$

◀

We can also define n^+ as the number of pairs of nucleons in symmetric relative space states and n^- as the number of pairs in antisymmetric space states, giving

$$ n^+ = {}^{13}n + {}^{31}n = \tfrac{3}{16}A^2 + \tfrac{3}{4}A - \tfrac{1}{2}|I| - \tfrac{1}{16}I^2 - \tfrac{1}{4}(\eta+\zeta) - \tfrac{1}{8}(\eta^2+\zeta^2) \quad \text{(3-23a)} $$

$$ n^- = {}^{11}n + {}^{33}n = \tfrac{5}{16}A^2 - \tfrac{5}{4}A + \tfrac{1}{2}|I| + \tfrac{1}{16}I^2 + \tfrac{1}{4}(\eta+\zeta) + \tfrac{1}{8}(\eta^2+\zeta^2). \quad \text{(3-23b)} $$

Of course, $n^+ + n^- = {}^{13}n + {}^{31}n + {}^{33}n + {}^{11}n$ is seen to be just $A(A-1)/2$, as expected. When $\eta = \zeta = 0$ and $I = 0$, we have for large A

$$ {}^{13}n \to \tfrac{3}{32}A^2 \quad \text{(3-24a)} $$

$$ {}^{31}n \to \tfrac{3}{32}A^2 \quad \text{(3-24b)} $$

$$ {}^{11}n \to \tfrac{1}{32}A^2 \quad \text{(3-24c)} $$

$$ {}^{33}n \to \tfrac{9}{32}A^2. \quad \text{(3-24d)} $$

Thus we note that the ratio of the total number of spin triplet to spin singlet states is

$$ (S=1)/(S=0) = ({}^{31}n + {}^{33}n)/({}^{13}n + {}^{11}n) = (3+9)/(3+1) = 3; $$

similarly, we also have

$$ (T=1)/(T=0) = ({}^{13}n + {}^{33}n)/({}^{31}n + {}^{11}n) = (3+9)/(3+1) = 3. $$

Further, among the isospin triplet states, the ratio of spin triplet states to spin singlet states is

$$ (T=1){:}[(S=1)/(S=0)] = {}^{33}n/{}^{13}n = 9/3 = 3. $$

Similar results pertain for $T = 0$, etc. These are just the ratios expected. However, unexpectedly the ratio of antisymmetric space states to symmetric space states is $n^-/n^+ = 5/3$. This interesting result is a consequence of the Pauli exclusion principle.

Having established the expectation values ^{13}n, ^{31}n, ^{11}n, and ^{33}n of the various projection operators, we now return to calculating $\langle A|G|A\rangle$ when $G = \sum_{i>j=1}^{A} v_{ij}$, with v_{ij} given by (3-15b). Consider, for example, the term $^{31}v_{ij} = {}^{31}V(r_{ij})\mathscr{T}(\sigma)\mathscr{S}(\tau)$ substituted into (3-17). After interchanging the sum and integration, we find

$$\langle A|^{31}G|A\rangle = \iint d^3r_a\, d^3r_b \times \sum_{i>j=1}^{l+m+p+q} \{\psi_i^*(a)\psi_j^*(b)\, {}^{31}V(r_{ij})\mathscr{T}(\sigma)\mathscr{S}(\tau)\psi_i(a)\psi_j(b) - \psi_i^*(a)\psi_j^*(b)\, {}^{31}V(r_{ij})\mathscr{T}(\sigma)\mathscr{S}(\tau)\psi_i(b)\psi_j(a)\},$$

or

$$\begin{aligned}\langle A|^{31}G|A\rangle = \iint & d^3r_a\, d^3r_b\, \rho(r_a)\rho(r_b)\, {}^{31}V(|\mathbf{r}_b-\mathbf{r}_a|) \\ & \times \sum_{i>j=1}^{l+m+p+q} ([\chi_i^\dagger(\boldsymbol{\sigma}_a)\chi_j^\dagger(\boldsymbol{\sigma}_b)\mathscr{T}(\sigma)\chi_i(\boldsymbol{\sigma}_a)\chi_j(\boldsymbol{\sigma}_b)] \\ & \times [\zeta_i^\dagger(\boldsymbol{\tau}_a)\zeta_j^\dagger(\boldsymbol{\tau}_b)\mathscr{S}(\tau)\zeta_i(\boldsymbol{\tau}_a)\zeta_j(\boldsymbol{\tau}_b)] \\ & - \{\exp[i\mathbf{k}_i\cdot(\mathbf{r}_b-\mathbf{r}_a)]\exp[i\mathbf{k}_j\cdot(\mathbf{r}_a-\mathbf{r}_b)]\} \\ & \times [\chi_i^\dagger(\boldsymbol{\sigma}_a)\chi_j^\dagger(\boldsymbol{\sigma}_b)\mathscr{T}(\sigma)\chi_i(\boldsymbol{\sigma}_b)\chi_j(\boldsymbol{\sigma}_a)] \\ & \times [\zeta_i^\dagger(\boldsymbol{\tau}_a)\zeta_j^\dagger(\boldsymbol{\tau}_b)\mathscr{S}(\tau)\zeta_i(\boldsymbol{\tau}_b)\zeta_j(\boldsymbol{\tau}_a)]).\end{aligned} \tag{3-25}$$

Examining the mechanical spin products involving $\mathscr{T}(\sigma)$, we note that the direct term is identical with the exchange term for any $(\sigma_{zi}, \sigma_{zj})$ pair, while for the isospin products involving $\mathscr{S}(\tau)$, the exchange terms are just the negative of the direct counterpart for any (τ_{3i}, τ_{3j}) pair. Thus the combined mechanical spin–isospin products for the case $^{31}v_{ij}$ give exchange terms that are just the negative of the direct terms. We must, of course, eventually take note of the additional minus sign in front of the entire exchange term. Looking ahead, we note that this is also the case for $^{13}v_{ij}$ involving the projection operators $\mathscr{S}(\sigma)\mathscr{T}(\tau)$. However, for $^{11}v_{ij}$ and $^{33}v_{ij}$, the exchange terms have the same sign and are identical with the direct terms. These conclusions are readily verified by referring to Exercise 3-6.

▶**Exercise 3-6** Note that using (1-12), we have

$$\alpha(a)\alpha(b) = \chi_1{}^1(a,b), \qquad \alpha(a)\beta(b) = (1/\sqrt{2})[\chi_1{}^0(a,b)+\chi_0{}^0(a,b)]$$
$$\beta(a)\alpha(b) = (1/\sqrt{2})[\chi_1{}^0(a,b)-\chi_0{}^0(a,b)], \qquad \beta(a)\beta(b) = \chi_1^{-1}(a,b). \tag{3-26}$$

Define

$$\mathscr{T}_{ij}^{\text{D}}(\sigma) = \chi_i^{\dagger}(a)\,\chi_j^{\dagger}(b)\,\mathscr{T}(\sigma)\,\chi_i(a)\,\chi_j(b)$$
$$\mathscr{T}_{ij}^{\text{E}}(\sigma) = \chi_i^{\dagger}(a)\,\chi_j^{\dagger}(b)\,\mathscr{T}(\sigma)\,\chi_i(b)\,\chi_j(a) \tag{3-27}$$

and similar quantities $\mathscr{S}_{ij}^{\text{D}}(\sigma)$ and $\mathscr{S}_{ij}^{\text{E}}(\sigma)$. As an example, observe that for $i = +\frac{1}{2}$ and $j = -\frac{1}{2}$, we have

$$\begin{aligned}\mathscr{T}_{1/2,-1/2}^{\text{E}}(\sigma) &= \alpha^{\dagger}(a)\,\beta^{\dagger}(b)\,\mathscr{T}(\sigma)\,\alpha(b)\,\beta(a)\\ &= \tfrac{1}{2}[\chi_1^{0\dagger}(a,b)+\chi_0^{0\dagger}(a,b)]\,\mathscr{T}(\sigma)\,[\chi_1{}^0(a,b)-\chi_0{}^0(a,b)]\\ &= +\tfrac{1}{2}.\end{aligned} \tag{3-28}$$

In a like manner verify the results in Table 3-1.

TABLE 3-1

i, j	$\mathscr{T}_{ij}^{\text{D}}(\sigma)$	$\mathscr{S}_{ij}^{\text{D}}(\sigma)$	$\mathscr{T}_{ij}^{\text{E}}(\sigma)$	$\mathscr{S}_{ij}^{\text{E}}(\sigma)$
$+\frac{1}{2}, +\frac{1}{2}$	$+1$	0	$+1$	0
$+\frac{1}{2}, -\frac{1}{2}$	$+\frac{1}{2}$	$+\frac{1}{2}$	$+\frac{1}{2}$	$-\frac{1}{2}$
$-\frac{1}{2}, +\frac{1}{2}$	$+\frac{1}{2}$	$+\frac{1}{2}$	$+\frac{1}{2}$	$-\frac{1}{2}$
$-\frac{1}{2}, -\frac{1}{2}$	$+1$	0	$+1$	0

The expectation values for the projection operators ${}^{31}n$, ${}^{13}n$, ${}^{11}n$, and ${}^{33}n$ could also be calculated in this manner by now continuing with the evaluation of the sums required. The route selected earlier is perhaps more instructive. ◀

3. Statistical Correlations

The presence of the exponential factors in the exchange term of (3-25) generates a very important effect that we wish to examine in some detail. Returning to (3-25), let us carry out the double sum for the exchange term first. This double sum can be grouped into terms with i and j each confined to range over a particular pair of choices in the domains l, m, p, or q. For example, consider i in the domain l, and j in the domain p (i.e., spin-up neutron states coupled to spin-up proton states). The mechanical spin–isospin products are the same for any such pair of states and therefore simply represent a common factor [for the case ${}^{31}v_{ij}$ of (3-25) this product is equal to $-\frac{1}{2}$]. As part of such a double sum, first take the sum on i from 1 to l. This partial sum contribution is the so-called mixed density $\rho_l(r_a, r_b)$, with

$$\rho_l(r_a, r_b) = \sum_{i=1}^{l} \exp[i\mathbf{k}_i\cdot(\mathbf{r}_b-\mathbf{r}_a)] \approx \frac{3l}{4\pi k_{\text{F}}{}^3}\int_0^{k_{\text{F}}} \exp(i\mathbf{k}\cdot\mathbf{r})\,d^3k, \tag{3-29}$$

with $\mathbf{r} = \mathbf{r}_b - \mathbf{r}_a$. The Fermi momentum k_{F} refers to the particle type l (i.e., neutrons with spin up) and more properly should be written $k_{F,l}$. Since the exchange integral is smaller than the direct integral, we make the simplifying approximation of using the central density for this integral, whence $k_{F,l}(r_i) = k_{F,l}(0)$, independent of r_i [i.e., ignoring the surface effect in the frequency modulation of $u_\gamma(r_i)$]. The amplitude modulation effect of the surface is, of course, still present in the joint density factor $\rho(r_a)\,\rho(r_b)$. No problem arises in the larger direct term since the frequency modulation is not present. Various degrees of approximation to handle the surface effects appear in the literature, e.g., Rodberg and Teplitz (1960), Bethe (1968), Németh (1968), Dahll and Warke (1970), and Pandharipande and Prasad (1970). For example, Rodberg and Teplitz (1960) have shown that the effect of the frequency modulation on the potential energy is relatively small, i.e., $(\delta V)_f \approx 0.03A$ MeV.

Introducing a polar coordinate system in **k** space with the polar axis in the direction of **r**,

$$\begin{aligned}\rho_l(r_a, r_b) &= \frac{3l}{2k_F{}^3}\int_0^\pi\int_0^{k_F} \sin\theta\, d\theta\, k^2\, dk\, e^{ikr\cos\theta} \\ &= \frac{3l}{k_F{}^3 r}\int_0^{k_F} k\, dk\, \sin kr = \frac{3l}{k_F{}^3 r^3}(\sin k_F r - k_F r\cos k_F r).\end{aligned} \tag{3-30}$$

This function has a pronounced maximum at $r = 0$, with

$$\rho_l(r_a, r_b)|_{\max} = l. \tag{3-31}$$

An additional approximation is now made by noting that for the ground state $l = m$ or $l = m+1$, and thus for large A we have $l \approx m \approx N/2$. Introduce the convenient notation $k_N \equiv k_{F,N/2} \approx k_{F,l} \approx k_{F,m}$ [i.e., $k_N \approx (3\pi^2 N/\Omega)^{1/3} \approx (\frac{3}{2}\pi^2\rho_0)^{1/3}(2N/A)^{1/3} = 1.36(2N/A)^{1/3}$ fm^{-1}]. Then

$$\rho_l(r_a, r_b) = \rho_m(r_a, r_b) = \tfrac{1}{2}\rho_N(r_a, r_b) = \tfrac{3}{2}N j_1(k_N r)/k_N r, \tag{3-32a}$$

where we have noted that the spherical Bessel function is given by $j_1(\omega) = (\sin\omega - \omega\cos\omega)/\omega^2$. When i is summed on the protons, the function $\rho_Z(r_a, r_b)$ is analogously developed. We assume $p \approx q$ and introduce the Fermi momentum $k_Z \approx (3\pi^2 Z/\Omega)^{1/3}$, with $k_Z \approx k_{F,Z/2} \approx k_{F,p} \approx k_{F,q}$. We thus write

$$\rho_p(r_a, r_b) = \rho_q(r_a, r_b) = \tfrac{1}{2}\rho_Z(r_a, r_b) = \tfrac{3}{2}Z j_1(k_Z r)/k_Z r. \tag{3-32b}$$

Using (3-32) and the results of Exercise 3-6, we have for (3-25) when the sums are carried out over all the domains l, m, p, and q, and explicitly writing $r \equiv r_{ab}$,

$$\langle {}^{31}v\rangle = \frac{3}{8}NZ\iint d^3r_a\, d^3r_b\, \rho(r_a)\,\rho(r_b)\, {}^{31}V(r_{ab})\left[1 + \frac{9 j_1(k_N r_{ab})\, j_1(k_Z r_{ab})}{k_N k_Z r_{ab}^2}\right]. \tag{3-33a}$$

Similarly,

$$\langle {}^{13}v \rangle = \frac{1}{8} \int\int d^3r_a\, d^3r_b\, \rho(r_a)\rho(r_b)\, {}^{13}V(r_{ab}) \Bigg\{ (N^2 + NZ + Z^2)$$

$$+ 9 \left[\frac{N^2 j_1{}^2(k_N r_{ab})}{k_N{}^2 r_{ab}^2} + \frac{NZ j_1(k_N r_{ab}) j_1(k_Z r_{ab})}{k_N k_Z r_{ab}^2} + \frac{Z^2 j_1{}^2(k_Z r_{ab})}{k_Z{}^2 r_{ab}^2} \right] \Bigg\}, \tag{3-33b}$$

and there are equivalent expressions for $\langle {}^{33}v \rangle$ and $\langle {}^{11}v \rangle$.

▶**Exercise 3-7** (a) Verify (3-33).

(b) As an interesting exercise, (3-33) can be used to evaluate ${}^{31}n$ and ${}^{13}n$ for the case $l = m$ and $p = q$ (i.e., $\eta = \zeta = 0$). For this purpose we simply take ${}^{31}V(r_{ab})$ and ${}^{13}V(r_{ab})$ identically equal to $+1$ for all values of r_{ab}. To avoid the complication of integrating over surface effects, we can take the uniform density model with $\rho(r_a)$ and $\rho(r_b)$ constant out to a radius R. We also observe that integrating on the variable r_b while keeping r_a fixed is the same as integrating on r_{ab} with r_a fixed. Thus $\int\int d^3r_a\, d^3r_b = \int\int d^3r_a\, d^3r_{ab}$. Since $j_1(k_{N,Z} r_{ab})$ falls off rapidly with r_{ab}, this integration can be taken between the limits of 0 and ∞. This is equivalent to ignoring surface effects and the integration on r_{ab} is thus independent of r_a.

Using our normalization $\int_0^R \rho(r_a)\, d^3r_a = 1$, and taking $\rho(r_b) \approx 1/\Omega = 3/4\pi R^3$, show that (3-33) reduces to (3-22) with $\eta = 0$ and $\zeta = 0$. Note that $\int_0^\infty j_1(at) j_1(bt)\, dt = \frac{1}{6}\pi a/b^2$, with $b > a$.

It is noteworthy to observe that the quadratic terms in A and I come from the direct contributions and the linear terms in A and I come from the exchange contributions. ◀

As reference to (3-33) and Exercise 3-7 will indicate, the A and I dependences of the potential energy terms do not appear as simple multiplicative factors involving ${}^{31}n$, ${}^{13}n$, ${}^{11}n$, and ${}^{33}n$, except in the special case $l = m = p = q = A/4$. It is nonetheless useful to define effective interactions that allow a factorization in terms of the pair counts ${}^{31}n$, ${}^{13}n$, etc. An approximation correct to order $1/A$ and I^2/A, in which the direct integral is also included, is

$$\begin{aligned} E_p = {}&{}^{31}n \int\int \rho(r_a)\rho(r_b) g_0^{(+)}(r_{ab})\, {}^{31}V(r_{ab})\, d^3r_a\, d^3r_b \\ &+ {}^{13}n \int\int \rho(r_a)\rho(r_b) g_0^{(+)}(r_{ab})\, {}^{13}V(r_{ab})\, d^3r_a\, d^3r_b \\ &+ {}^{11}n \int\int \rho(r_a)\rho(r_b) g_0^{(-)}(r_{ab})\, {}^{11}V(r_{ab})\, d^3r_a\, d^3r_b \\ &+ {}^{33}n \int\int \rho(r_a)\rho(r_b) g_0^{(-)}(r_{ab})\, {}^{33}V(r_{ab})\, d^3r_a\, d^3r_b. \end{aligned} \tag{3-34a}$$

Here we have defined

$$g_0^{(+)} = 1 + \eta^2(r_{ab}) = 1 + 9\left[\frac{Z}{A}\frac{j_1(k_Z r_{ab})}{k_Z r_{ab}} + \frac{N}{A}\frac{j_1(k_N r_{ab})}{k_N r_{ab}}\right]^2 \qquad \text{(3-34b)}$$

$$g_0^{(-)} = 1 - \eta^2(r_{ab}) = 1 - 9\left[\frac{Z}{A}\frac{j_1(k_Z r_{ab})}{k_Z r_{ab}} + \frac{N}{A}\frac{j_1(k_N r_{ab})}{k_N r_{ab}}\right]^2. \qquad \text{(3-34c)}$$

Developing the square of the bracketed terms in (3-34b) and (3-34c) reflects the relative number of neutron–neutron, proton–proton, and neutron–proton pairs, ignoring the precise pair selection character of the projection operators appearing in ${}^{31}v$, ${}^{13}v$, etc. However, the exchange symmetry of the pair space states is properly accounted for in $g_0^{(+)}$ and $g_0^{(-)}$. The numerical results, discussed later in connection with (3-49), support the above simplifications.[1] The alternation in sign for $g_0^{(\pm)}$ has a simple interpretation in the exchange symmetry of the space states corresponding to the spin–isospin states selected by the projection operators $\mathscr{T}(\sigma)\mathscr{S}(\tau)$, etc. For the pairs ${}^{31}n$ and ${}^{13}n$, $\mathscr{T}(\sigma)\mathscr{S}(\tau)$ and $\mathscr{S}(\sigma)\mathscr{T}(\tau)$ require symmetric space states, hence $g_0^{(+)}$, while for ${}^{11}n$ and ${}^{33}n$, $\mathscr{S}(\sigma)\mathscr{S}(\tau)$ and $\mathscr{T}(\sigma)\mathscr{T}(\tau)$ require antisymmetric space states, hence $g_0^{(-)}$.

The expressions (3-34b) and (3-34c) are referred to as the Wigner–Seitz factors (Wigner and Seitz, 1933) or *statistical correlation functions*. The function $j_1(\omega)/\omega$ has the value 1/3 as $\omega \to 0$, and oscillates as ω increases, with, however, a diminishing amplitude that approaches zero as $1/\omega^2$ for large ω. Thus we have $[g_0^{(+)}(r_{ab}=0)] = 2$ and $[g_0^{(-)}(r_{ab}=0)] = 0$, while $[g_0^{(\pm)}(k_{Z,N}r_{ab} \gg 1)] \approx 1$. Hence two nucleons in space symmetric states are twice as likely to be found at a separation $r_{ab} = 0$, each within unit volumes d^3r_a and d^3r_b, than when they are at separations large compared to $(k_F)^{-1}$. Two nucleons in antisymmetric space states, on the other hand, have zero probability of being at a separation $r_{ab} = 0$. These short-range correlations, deviating from the uniform probability expected for distinguishable particles, are consequences of the Pauli exclusion principle operating for identical particles. For example, when $\boldsymbol{\sigma}_a = \boldsymbol{\sigma}_b$ and $\boldsymbol{\tau}_a = \boldsymbol{\tau}_b$ in (3-7) (i.e., relative spin triplet, isospin triplet states), we note that for $\mathbf{r}_a = \mathbf{r}_b$ two columns then become equal and hence the Slater determinant vanishes. At large separations compared to $(k_F)^{-1}$, these correlations go over into the uniform probabilities

[1] Note that $j_1(\omega)/\omega = \frac{1}{3} - (\omega^2/30) + (\omega^4/840) + \cdots$, and that $\omega_Z/\omega_N = k_Z/k_N = (Z/N)^{1/3} = 1 - \frac{2}{3}(I/A) + \cdots$. Thus for $r_{ab} \leqslant a$ (i.e., a separation less than the interaction well radius), $\eta^2(r_{ab})$ of (3-34b) differs from the corresponding correct factor of (3-33a) and (3-33b) by less than 5% even for $A \approx 250$ and $I \approx 55$ (a deviation only reached at $r_{ab} \approx a \approx 2.5$ fm, when η^2 is already quite small). In fact, even if the integral containing (3-34b) is evaluated for $a \to \infty$ in a manner suggested in Exercise 3-7, the error in using this approximate expression is simply to introduce a factor $[1 + (4/A) - (2I/A^2)]$, which for all practical purposes can be taken as unity.

obeyed by classically distinguishable particle pairs. Although the factor $g_0^{(\pm)}$ arose in connection with evaluating G for the sum on v_{ij}, only the symmetry behavior of the product states of ψ_γ was involved. These correlation factors would therefore apply even to noninteracting particles.

A simple consideration contrasting the behavior of a pair of noninteracting particles for the cases when they are distinguishable and when they are indistinguishable is perhaps worthwhile. Consider two space functions: $u_A(\mathbf{r})$, which is large and more or less constant when $\mathbf{r}$ is in a given region A and very small outside A, and $u_B(\mathbf{r})$, which is likewise only large and more or less constant when $\mathbf{r}$ is in the region B. Let regions A and B overlap in a small intersection as shown in Fig. 3.3. Let u_A and u_B form the spatial portion of the

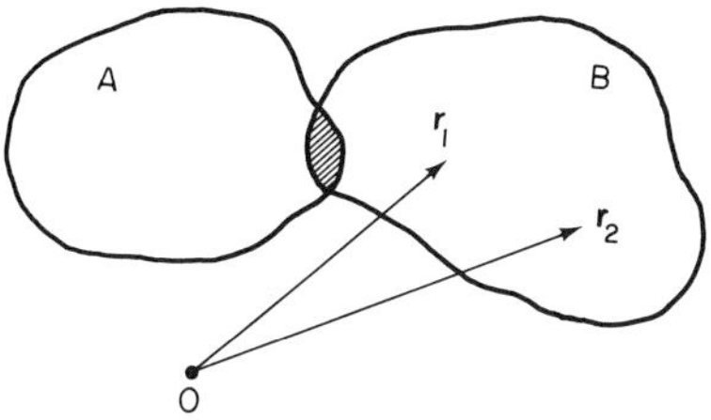

FIG. 3.3. An illustration of the regions A and B and the intersection region shown cross-hatched. The origin O for the vectors $\mathbf{r}_1$ and $\mathbf{r}_2$ is also indicated.

eigenfunction for two noninteracting distinguishable particles each having spin $\frac{1}{2}$ and write

$$\Psi_C(1,2) = u_A(\mathbf{r}_1)\,u_B(\mathbf{r}_2)\,\chi(1,2). \tag{3-35}$$

The behavior imposed on u_A and u_B for the regions A and B allows for the convenience of distinguishing the cases when r_{12} is large and when it is small. Ψ_C essentially vanishes unless $\mathbf{r}_1$ is in A and $\mathbf{r}_2$ is in B. When both are in the intersection (which is a small region by construction), $r_{12} \approx 0$. When $\mathbf{r}_1$ is in A and $\mathbf{r}_2$ is in B but not *both* simultaneously in the intersection, r_{12} is considered large.

For two noninteracting but identical fermions obeying the same Hamiltonian as above, we must have either

$$\Psi_A(1,2) = (1/\sqrt{2})\,\{u_A(\mathbf{r}_1)\,u_B(\mathbf{r}_2) - u_B(\mathbf{r}_1)\,u_A(\mathbf{r}_2)\}\,\chi_S(1,2) \tag{3-36a}$$

or

$$\Psi_S(1,2) = (1/\sqrt{2})\,\{u_A(\mathbf{r}_1)\,u_B(\mathbf{r}_2) + u_B(\mathbf{r}_1)\,u_A(\mathbf{r}_2)\}\,\chi_A(1,2). \tag{3-36b}$$

The subscripts on $\Psi(1,2)$ refer to space exchange symmetry, and the subscripts on the spin functions $\chi(1,2)$ refer to spin exchange symmetry. Both wave functions are, of course, antisymmetric on simultaneous spin and space exchange. Now consider $|\Psi|^2$, assuming the spin functions are suitably normalized; then

$$|\Psi_C|^2 = |u_A(\mathbf{r}_1)|^2 \cdot |u_B(\mathbf{r}_1)|^2 \tag{3-37a}$$

and

$$|\Psi_{\substack{S\\A}}|^2 = \tfrac{1}{2}[|u_A(\mathbf{r}_1)|^2 \cdot |u_B(\mathbf{r}_2)|^2 + |u_B(\mathbf{r}_1)|^2 \cdot |u_A(\mathbf{r}_2)|^2]$$
$$\pm \tfrac{1}{2}[u_A{}^*(\mathbf{r}_1)\, u_B{}^*(\mathbf{r}_2)\, u_B(\mathbf{r}_1)\, u_A(\mathbf{r}_2) + u_B{}^*(\mathbf{r}_1)\, u_A{}^*(\mathbf{r}_2)\, u_A(\mathbf{r}_1)\, u_B(\mathbf{r}_2)]. \tag{3-37b}$$

Thus, when $r_{12} \approx 0$ (i.e., both $\mathbf{r}_1$ and $\mathbf{r}_2$ are in the intersection)

$$|\Psi_A|^2 \approx 0, \qquad |\Psi_S|^2 \approx 2|u_A(\mathbf{r}_1)|^2 \cdot |u_B(\mathbf{r}_2)|^2 = 2|\Psi_C|^2. \tag{3-38}$$

When r_{12} is large, $\mathbf{r}_1$ is in A or B and $\mathbf{r}_2$ is in B or A (but not both in the intersection). Examining the exchange bracket in (3-37b) shows that two of the four product states in each term are small while for the direct bracket all four product states are large; hence for r_{12} large the exchange bracket can be dropped. Further, if indeed the two particles are indistinguishable, the Hamiltonian must be exchange symmetric and $H(1,2) = H(2,1)$; therefore $|u_A(\mathbf{r}_1)|^2 \cdot |u_B(\mathbf{r}_2)|^2 = |u_B(\mathbf{r}_1)|^2 \cdot |u_A(\mathbf{r}_2)|^2$. Thus for large r_{12}

$$|\Psi_A|^2 = |\Psi_S|^2 = |\Psi_C|^2. \tag{3-39}$$

For any arbitrary r_{12}, clearly,

$$|\Psi_A|^2 + |\Psi_S|^2 = 2|\Psi_C|^2. \tag{3-40}$$

Hence distinguishability manifests itself in short-range correlations such as (3-38) and the behavior of $g_0^{(\pm)}(r_{ab} = 0)$ discussed earlier, while at large separations no special correlations are evident due to distinguishability.

Figure 3.4 shows the statistical correlation factors $g_0^{(\pm)}$ as a function of the nucleon separation and the Fermi momentum $k_F = k_N = k_Z$, viz., $k_F|\mathbf{r}_b - \mathbf{r}_a| = k_F r_{ab}$. The behavior characteristics corresponding to (3-38)–(3-40) are clearly evident. Bethe (1968), Németh and Bethe (1968), and Lin [referenced in Bethe (1968)] have examined the validity of the Wigner–Seitz expression in detail. The suggestion is made that k_F should be evaluated for $\rho(\mathbf{R})$, with $\mathbf{R} = \frac{1}{2}(\mathbf{r}_b + \mathbf{r}_a)$. The suggestion is also made that $\rho_{\text{eff}} \approx [\rho(\mathbf{r}_a)\, \rho(\mathbf{r}_b)]^{1/2}$ should be used for the direct term, as we have done.

In addition to the statistical correlations and local density considerations, other correlations can also be distinguished. One consideration would relate to possible α-cluster substructure within nuclei. In view of the successes of the α-cluster model in light nuclei (see Chapter V), α-particle clustering correlations in the nuclear interior in general might also be expected. An extreme form of such correlations could give rise to density oscillations suggesting a

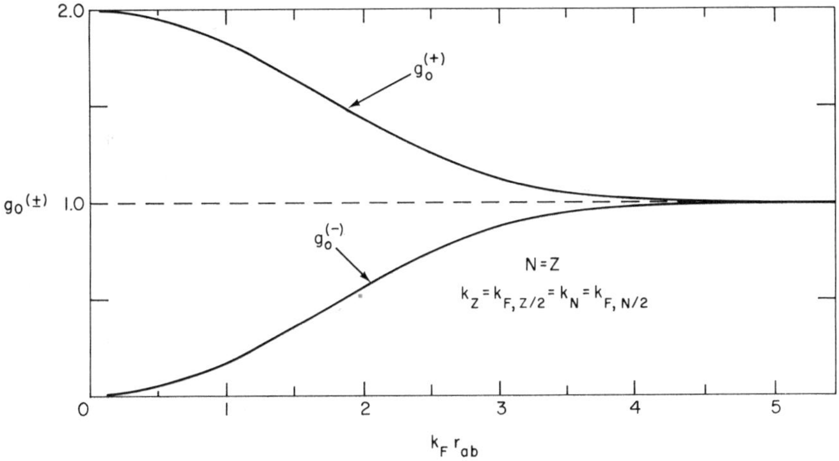

FIG. 3.4. The Wigner–Seitz statistical correlation factors $g_0^{(\pm)}$ defined by (3-34) for the case $N = Z$. Appropriate curves for ^{208}Pb with $N/A = 0.605$ and $Z/A = 0.395$ would differ only imperceptibly from those of this figure. The Fermi momenta appropriate for the central region of the nucleus are $1.2 < k_Z = k_N < 1.6\,\text{fm}^{-1}$.

periodic distribution of α-clusters similar to the structures of crystals. Theoretical studies [e.g., Brink and Castro (1973)] show, however, that using individual nucleon plane-wave states leads to greater binding in infinite nuclear matter than is found using α-cluster states of four highly correlated nucleons, at densities corresponding to the interior of heavy nuclei (i.e., $k_F \approx 1.3$–$1.5\,\text{fm}^{-1}$). On the other hand, at densities less than $\frac{1}{3}$ the central region density (i.e., $k_F \lesssim 1.0\,\text{fm}^{-1}$) α-clustering could occur. One therefore expects α-clustering effects to require attention in the surface region of heavy nuclei and throughout most of the interior of very light nuclei. We shall not consider such effects beyond the already included strong binding implied in the partitioned states of Fig. 3.1 for the four nucleons with identical space wave functions. For a more detailed discussion of the *quartet model* (i.e., four strongly interacting nucleons—two protons and two neutrons) as opposed to the more specific α-cluster model, see Arima and Gillet (1971). For effects introduced by the internal structure of the α-particle on nuclear matter and related problems, see, e.g., Turner and Jackson (1972).

4. The Energy Equation

Returning to the evaluation of (3-34), we formally integrate over all coordinates except $\mathbf{r}_{ab}$ and define

$$P(r) = \int\!\!\int \rho(r_a)\,\rho(r_b)\, d^3R_{ab}\, d\Omega \tag{3-41}$$

with $\mathbf{R}_{ab} = \frac{1}{2}(\mathbf{r}_b+\mathbf{r}_a)$.[1] When the densities $\rho(r_a)$ and $\rho(r_b)$ are taken to be trapezoidal with a central density (see Exercise 2-1)

$$\rho_{0,a,b} = \rho_0/A = 3/4\pi R^3(1+\delta^2) \tag{3-42}$$

and we recall that $\int \rho(r_a)\, d^3r_a = 1$, $P(r)$ becomes, after some manipulations (Ayres *et al.*, 1962, p. 218 ff.),

$$P(r) = 3[4\pi R^3(1+\delta^2)^2]^{-1}[C_0 - C_2 r^2 + C_3 r^3 + C_4 r^4 - C_5 r^5] \tag{3-43}$$

with

$$C_0 = (1-\delta+\delta^2-\tfrac{1}{5}\delta^3), \qquad C_2 = (4\delta R^2)^{-1}(1+\tfrac{1}{3}\delta^2)$$

$$C_3 = (32\delta^2 R^3)^{-1}(1+\delta^2), \qquad C_4 = (40\delta R^4)^{-1}, \qquad C_5 = (240\delta^2 R^5)^{-1}.$$

$P(r)$ can be expressed in polynomial forms that depend on the range of r. The polynomial form quoted here is for $0 < r < 2R\delta$. Substituting (3-41) into (3-34), noting the definition of $\eta(r)$, and recalling that the various potentials are all square-well potentials with a range a,

$$E_p = 4\pi\, {}^{31}V_0 \left\{ {}^{31}n \int_0^a P(r)[1+\eta^2(r)]\, r^2\, dr + \alpha\, {}^{13}n \int_0^a P(r)[1+\eta^2(r)]\, r^2\, dr \right.$$
$$\left. + \beta\, {}^{33}n \int_0^a P(r)[1-\eta^2(r)]\, r^2\, dr + \gamma\, {}^{11}n \int_0^a P(r)[1-\eta^2(r)]\, r^2\, dr \right\}. \tag{3-44}$$

Here the relative potential depths are ${}^{13}V_0/{}^{31}V_0 = \alpha$, ${}^{33}V_0/{}^{31}V_0 = \beta$, and ${}^{11}V_0/{}^{31}V_0 = \gamma$ as in Exercise 1-18. Equation (3-44) leads to

$$E_p = {}^{31}V_0\{({}^{31}n+\alpha\, {}^{13}n)(G+J) + (\beta\, {}^{33}n+\gamma\, {}^{11}n)(G-J)\} \tag{3-45a}$$

with

$$G = 4\pi \int_0^a P(r)\, r^2\, dr = 3[R^3(1+\delta^2)^2]^{-1}[\tfrac{1}{3}C_0 a^3 - \tfrac{1}{5}C_2 a^5 + \tfrac{1}{6}C_3 a^6 + \cdots] \tag{3-45b}$$

and

$$J = 4\pi \int_0^a P(r)\,\eta^2(r)\, r^2\, dr = 3[R^3(1+\delta^2)^2]^{-1}[C_0 D_0 a^3 - C_2 D_2 a^5 + \cdots] \tag{3-45c}$$

with $a^n D_{n-3} = \int_0^a \eta^2(r)\, r^{n-1}\, dr$ and provided that $a < 2R\delta$ and $\delta < \frac{1}{2}$. The functions D are relatively complicated functions of a parameter $\kappa =$

[1] While risking some confusion it is convenient in this section to use the quantity $\frac{1}{2}\mathbf{r}_{ab}$.

$(9\pi/4)^{1/3} a/R_0(1+\delta)$ (with $R = R_0 A^{1/3}$):

$$D_0 = (288/\kappa^6)\{\tfrac{1}{3}[(\cos\kappa - 1) + \kappa\sin\kappa + \tfrac{1}{4}\kappa^2(\cos\kappa - 3) + \tfrac{1}{4}\kappa^3 \operatorname{Si}\kappa] - (1/36)(I^2/A^2)[\kappa^2(\cos\kappa - 1) + \tfrac{1}{4}\kappa^3\sin\kappa + \tfrac{1}{4}\kappa^4]\} \qquad (3\text{-}46a)$$

(where $\operatorname{Si}\kappa$ is the sine-integral of κ) and

$$D_2 = (288/\kappa^6)\{[(\cos\kappa - 1) + \tfrac{1}{4}\kappa\sin\kappa + \tfrac{1}{4}\kappa^2] + (5/72)(I^2/A^2)[\kappa\sin\kappa - \tfrac{3}{5}\kappa^2(\cos\kappa + \tfrac{2}{3}) - \tfrac{1}{10}\kappa^3\sin\kappa - \tfrac{1}{30}\kappa^4]\}. \qquad (3\text{-}46b)$$

Evidently the integral J can be written in the form

$$J = J_0 + (I^2/A^2)J_1. \qquad (3\text{-}47)$$

The integral G has a rather simple interpretation, namely the probability of finding two nucleons at a separation less than a when both have individual densities (probabilities) that are trapezoidal with half-density radius R and surface thickness $2R\delta$. G can be written

$$G = (a/R)^3(1+\delta^2)^{-2}[C_0 - \tfrac{3}{5}C_2 a^2 + \tfrac{1}{2}C_3 a^3 + \cdots]. \qquad (3\text{-}48a)$$

As we shall see, $\delta = 0.280 \pm 0.010$ will result from the present theory for all nuclei $A \geqslant 40$. For this value of δ and for $a/R \lesssim \tfrac{3}{5}$, we have

$$G_\delta = (a/R)^3[0.684 - 0.473(a/R)^2 + 0.185(a/R)^3 + \cdots]. \qquad (3\text{-}48b)$$

Blatt and Weisskopf (1952, p. 123) have shown that this probability for the uniform model ($\delta = 0$) is, for $a/R < 2$,

$$G_{\delta=0} = (a/R)^3[1 - (9/16)(a/R) + (1/32)(a/R)^3]. \qquad (3\text{-}48c)$$

When $a/R \ll 1$, the behavior of (3-48c) is just the ratio of two volumes with radii a and R, as expected. Also, $G_{\delta=0}$ becomes unity when $a/R \geqslant 2$. The presence of the surface region, however, requires $a/R \geqslant 2(1+\delta)$ before the probability is unity. Even for small values of a/R, G_δ is considerably less than $G_{\delta=0}$ due to the reduced probability contribution of the surface region. For $\delta = 0.280$, $G_\delta/G_{\delta=0} = 0.684$ for small a/R. If, however, the comparison is made keeping the central density ρ_0 the same in both cases, Exercise 2-1 shows that a larger effective uniform density radius is required, in which event $G_\delta/G_{\delta=0} = 0.737$ for the cited case. In any event the surface effect is seen to be quite pronounced.

It is instructive at this point to examine the quantities G, J, J_0, and J_1 as functions of A and of $I = N - Z$, for values of δ, R_0, and a, which will be found to pertain when eventually a numerical analysis is performed. The values obtained are given in Table 3-2.

TABLE 3-2

	$A=56, Z=26$	$A=127, Z=53$	$A=267, Z=103$
δ	0.285	0.275	0.272
G	0.0914	0.0439	0.0218
J_0	0.0657	0.0312	0.0154
J_1	−0.0143	−0.0070	−0.0035
I^2/A^2	0.0051	0.0273	0.0522

The analysis referred to also gives $R_0 = 1.25$ fm, $a = 2.60$ fm, $\delta \approx 0.280$, and $\kappa \approx \pi$. Approximate numerical expansions of J_0 and J_1 as functions of A (through a/R and $R = R_0 A^{1/3}$) are

$$J_0 = (a/R)^3 [0.505 - 0.331 (a/R)^2 + \cdots], \tag{3-49a}$$

and

$$J_1 I^2/A^2 \approx -0.085 (a/R)^3 I^2/A^2. \tag{3-49b}$$

Thus $J_1 I^2/A^2 \ll J_0$ for all A,[1] and also the integral factors $(G+J)$ and $(G-J)$ appearing in (3-45a) are relatively independent of A, with

$$\frac{G+J}{G-J} \approx \frac{1+J_0/G}{1-J_0/G} = \frac{1.71}{0.29}. \tag{3-50}$$

The Coulomb energy for the protons E_c can be evaluated in a like manner to E_p, using the two-body Coulomb operator of (3-15a). The number of pairs of protons in various spin and isospin states are readily shown to be

$${}^{13}n_p = \langle |\mathscr{S}(\sigma)\mathscr{T}(\tau)| \rangle_{\text{protons}} = \tfrac{1}{8}Z^2 + \tfrac{1}{4}Z - \tfrac{1}{8}\zeta^2 - \tfrac{1}{4}\zeta \tag{3-51a}$$

$${}^{33}n_p = \langle |\mathscr{T}(\sigma)\mathscr{T}(\tau)| \rangle_{\text{protons}} = \tfrac{3}{8}Z^2 - \tfrac{3}{4}Z + \tfrac{1}{8}\zeta^2 + \tfrac{1}{4}\zeta \tag{3-51b}$$

$${}^{11}n_p = {}^{31}n_p = 0. \tag{3-51c}$$

Again we note that ${}^{13}n_p + {}^{31}n_p + {}^{11}n_p + {}^{33}n_p = Z(Z-1)/2$. Analogous to (3-44), we now have

$$E_c = 4\pi e^2 \left\{ {}^{13}n_p \int_0^\infty P(r)[1+\eta^2(r)]\, r\, dr + {}^{33}n_p \int_0^\infty P(r)[1-\eta^2(r)]\, r\, dr \right\}. \tag{3-52}$$

Writing this in terms of direct and exchange integrals,

$$E_c = 4\pi e^2 \left\{ ({}^{13}n_p + {}^{33}n_p) \int_0^\infty P(r)\, r\, dr + ({}^{13}n_p - {}^{33}n_p) \int_0^\infty P(r)\eta^2(r)\, r\, dr \right\}$$
$$= \tfrac{1}{2}Z(Z-1)e^2 G_c - \tfrac{1}{4}(Z^2 - 4Z + \zeta^2 - 2\zeta) e^2 J_c. \tag{3-53}$$

[1] The approximation of replacing the exact expressions (3-33) with those of (3-34) is valid to the extent that the term in J_1 is negligible compared to the term in J_0.

▶**Exercise 3-8** Verify (3-51) by direct calculation and also by taking $N = \eta = 0$ in (3-24). ◀

Rather than proceed with the evaluation of G_c and J_c, a simpler procedure will be followed to facilitate comparison with results in the literature. As discussed in connection with (2-20), we use

$$E_c \approx \frac{3}{5}\frac{e^2}{R}[Z^2 - 0.767Z^{4/3}]\frac{[1+\frac{5}{6}\delta^2+\frac{1}{2}\delta^3+\frac{1}{6}\delta^4-(1/42)\delta^5]}{(1+\delta^2)^2}. \quad (3\text{-}54a)$$

This can be conveniently written as

$$E_c = a_c A^{-1/3}[Z^2 - 0.767Z^{4/3}]. \quad (3\text{-}54b)$$

For the values $\delta = 0.280$ and $R_0 = 1.25$ fm we have

$$E_c = 0.642A^{-1/3}[Z^2 - 0.767Z^{4/3}]. \quad (3\text{-}54c)$$

It should be noted that for present purposes the quantity of interest is the bulk Coulomb energy rather than the change in Coulomb energy for near neighbors among the nuclear species that was under discussion earlier. In the present context, shell effects are much less important. A typical result is that of Swamy and Green (1958) (see also references in Section II,D), who have shown by direct shell model calculations that writing

$$E_c = \tfrac{3}{5}(e^2/R)[Z^2 - C(N,Z)Z^{4/3}]$$

gives an approximate behavior for $C(N,Z) \approx 0.767 - e^{-0.38Z}$. For $Z \gtrsim 7$, $C(N,Z)$ is within 10% of the value 0.767, and provides some justification for using (3-54).

The remaining quantity that must be calculated is the kinetic energy term of (3-3). The variation of the density in the surface region results in both the amplitude and frequency modulations discussed in connection with our trial wave functions (3-13). Rodberg and Teplitz (1960) have shown that the kinetic energy can be written

$$\langle|T|\rangle = T_0 + (\delta T)_f + (\delta T)_a \quad (3\text{-}55a)$$

with

$$T_0 = \frac{3\hbar^2}{10m_N}\int d^3r\, k_F{}^2(r)\,\rho(r) \quad (3\text{-}55b)$$

$$(\delta T)_f = \frac{3\hbar^2}{10m_N}\int d^3r\, k_F{}^2(r)\left[\frac{4r}{9}\frac{d\rho}{dr} + \frac{r^2}{27\rho}\left(\frac{d\rho}{dr}\right)^2\right] \quad (3\text{-}55c)$$

$$(\delta T)_a = \frac{\hbar^2}{8m_N}\int d^3r \frac{1}{\rho}\left(\frac{d\rho}{dr}\right)^2. \quad (3\text{-}55d)$$

The first term is just the local density approximation appropriate for the Thomas–Fermi model expressed in (3-11). The first of the two derivative terms results from the frequency modulation accompanying the change in Fermi momentum, while the second arises from the amplitude modulation. For realistic density shapes, this last term is known to overestimate the effect. Rodberg and Teplitz have also shown that variational calculations with the trial function (3-13) give $(\delta T)_a \approx +2.7A$ MeV and $(\delta T)_f \approx -3.0A$ MeV, thus tending to cancel the two effects. See also Siemens (1970).

Although an unknown and possibly serious error may result, we drop $(\delta T)_f$ and $(\delta T)_a$ and simply calculate T_0 for the four-component Fermi gas model. Using (3-9) and $\rho_l(r) = (l/A)\rho(r)$, with similar expressions for the m, p, and q components, we have

$$T_0 = \frac{3\hbar^2}{10m_N}(6\pi^2)^{2/3}\left[\left(\frac{l}{A}\right)^{5/3} + \left(\frac{m}{A}\right)^{5/3} + \left(\frac{p}{A}\right)^{5/3} + \left(\frac{q}{A}\right)^{5/3}\right]\int \rho^{5/3}(r)\, d^3r. \tag{3-56}$$

It is readily established that

$$4\pi\int \rho^{5/3}(r) r^2\, dr = \frac{4\pi}{3}\rho_0^{5/3}R^3\left(1 - \frac{3}{4}\delta + \frac{21}{22}\delta^2 - \frac{47}{308}\delta^3\right) \tag{3-57}$$

and

$$\left[\left(\frac{l}{A}\right)^{5/3} + \left(\frac{m}{A}\right)^{5/3} + \left(\frac{p}{A}\right)^{5/3} + \left(\frac{q}{A}\right)^{5/3}\right] = \left(\frac{1}{4}\right)^{2/3}\left[1 + \frac{5}{9}\frac{I^2}{A^2} + \frac{10}{9}\frac{\eta^2+\zeta^2}{A^2} + \cdots\right]. \tag{3-58}$$

Therefore, with $R = R_0 A^{1/3}$, we obtain

$$T_0 = \frac{3\hbar^2}{10m_N {R_0}^2}\left(\frac{9\pi}{8}\right)^{2/3} A\left[1 + \frac{5}{9}\frac{I^2}{A^2} + \frac{10}{9}\frac{\eta^2+\zeta^2}{A^2} + \cdots\right] \times \left(1 - \frac{3}{4}\delta + \frac{21}{22}\delta^2 - \frac{47}{308}\delta^3\right)(1+\delta^2)^{-5/3}. \tag{3-59a}$$

In obvious notation we can write

$$T_0 = FA\left[1 + \frac{5}{9}\frac{I^2}{A^2} + \frac{10}{9}\frac{\eta^2+\zeta^2}{A^2} + \cdots\right]. \tag{3-59b}$$

For the values $\delta = 0.280$ and $R_0 = 1.25$ fm, this gives

$$T_0 = 14.05A\left[1 + \frac{5}{9}\frac{I^2}{A^2} + \frac{10}{9}\frac{\eta^2+\zeta^2}{A^2} + \cdots\right]. \tag{3-59c}$$

Although the average kinetic energy given by (3-59c) is distinctly lower than the usual theoretical values [see Bethe (1971)], it is in reasonable agreement with a calculation relating the average kinetic energy per proton with the experimentally observed kinematic spectrum of protons from (p, 2p) experiments [see Koltun (1972)]. Relativistic Hartree–Fock calculations by Miller (1972) also give lower than usually assumed average kinetic energies.

▶**Exercise 3-9** Verify (3-57)–(3-59). ◀

Two very striking features of nuclei are the saturation of nuclear binding energies and the saturation of nuclear densities. The first feature relates to the observed fact that, to first order, nuclear binding energies[1]

$$E_B = [NM_n + ZM_H - M(A, I)]c^2$$

vary linearly with A rather than with A^2. The second feature relates to the fact that nuclear radii vary approximately with $A^{1/3}$, resulting in a relatively constant central matter density.

Saturation of nuclear binding energy, in the present framework, follows directly from the short-range nature of nuclear forces. If $R > a/2$, then roughly speaking only nearby neighboring nucleons interact rather than all the pairs simultaneously, and this would result in E varying with nA rather than with the total number of pairs $A(A-1)/2 \approx A^2/2$. The quantity n involves the precise nature of the nuclear interaction and the operation of the Pauli principle. In expressing $E = -E_B(A)$ as a function of A, the leading term is therefore found to be $-a_v A$, where a_v is referred to as the volume energy coefficient. A simple estimate of a_v is readily obtained. We first estimate the leading term in E_p for A large (i.e., R large). From (3-48b) and (3-49a), we find

$$G + J_0 = (a/R)^3[1.189 - 0.804(a/R)^2 + \cdots] \quad \text{(3-60a)}$$

and

$$G - J_0 = (a/R)^3[0.179 - 0.142(a/R)^2 + \cdots]. \quad \text{(3-60b)}$$

Thus, from (3-45a) and using (3-24), the result is

$$E_{p>} = 0.282\,{}^{31}V_0 A[3.567(1+\alpha) + 0.179(9\beta + \gamma)]. \quad \text{(3-61)}$$

As we shall find, ${}^{31}V_0 = -20.3$ (for $a = 2.60$ fm), $\alpha = 0.754$, $\beta = -1/3$, and $\gamma = -3.34$; hence

$$E_{p>} = -29.3A \text{ MeV}. \quad \text{(3-62)}$$

From (3-59c), we obtain

$$T_{0>} = 14.1A \text{ MeV}; \quad \text{(3-63)}$$

[1] This definition gives $E_B > 0$ and corresponds to $E = -E_B$. The quantities M_n and M_H refer to the neutron and neutral hydrogen atomic mass, respectively, and $M(A, I)$ is the neutral atomic mass of the nuclide in question.

hence

$$E = -a_v A = (-29.3+14.1)A \qquad \text{or} \qquad a_v = 15.2 \text{ MeV}. \tag{3-64}$$

This resulting realistic value was, of course, one of the restraints in arriving at the quoted values of ${}^{31}V_0$, α, β, and γ, but (3-64) does show the rather simple nature of this restraint, while (3-61) along with (3-59) clearly demonstrates the saturation effect on nuclear binding energies.

The individual contributions from the terms appearing in (3-61) to the total potential energy are: ${}^{31}E_p = -20.4$, ${}^{13}E_p = -15.4$, ${}^{33}E_p = +3.1$, and ${}^{11}E_p = +3.4$ MeV. These values can be compared to the accurate and complete calculations now available for nuclear matter based on exact phenomenological nucleon–nucleon interactions. For example, using the Reid soft-core potential with $k_F = 1.36 \text{ fm}^{-1}$, Kallio and Day (1969) obtain ${}^{31}E_p = -17.9$, ${}^{13}E_p = -18.1$, ${}^{33}E_p = -1.0$, and ${}^{11}E_p = +2.4$ MeV. The main contributions to the l-even, two-nucleon states arise from the ${}^{31}S$ and ${}^{13}S$ configurations. Note that $|{}^{31}E_p| < |{}^{13}E_p|$ (true also for the S-state contributions by themselves). The calculations also indicate that $|E_p({}^{31}S)| > |E_p({}^{13}S)|$ for $k_F \lesssim 1.3 \text{ fm}^{-1}$, agreeing with the expectation for the free two-nucleon case; however, for $k_F > 1.3 \text{ fm}^{-1}$, $|E_p({}^{13}S)| > |E_p({}^{31}S)|$ (i.e., $\alpha > 1.00$). Even for the present simplified treatment, we require $\alpha = 0.754$, a value greater than $\alpha = 0.656$ for the free two-nucleon problem, in conformity with the above trend. As expected, it is found that the softer (i.e., weaker) the short-range repulsion, the greater the binding, particularly in the ${}^{31}S$ state. The presence of the Reid soft-core repulsion reduces the binding energy per particle about 2.0 MeV. The present treatment is without a short-range repulsion, and relies on saturation from stronger repulsion in the l-odd states. As pointed out earlier, the usual nuclear matter calculations employ a larger kinetic energy term; thus Kallio and Day underestimate a_v, which they give as $a_v = 11.2$ MeV. In comparing the present results extrapolated to $A \gg 1$ with the exact calculations for infinite nuclear matter, the essentially nonvanishing surface effect must be kept in mind.

A number of other simplified nuclear matter calculations with special effective interactions have appeared in the literature to shed physical insight on various aspects of an otherwise very complicated problem; refer to Skyrme (1959), Kuo and Brown (1966), Moszkowski (1970), Ehlers and Moszkowski (1972), Vautherin and Brink (1972), Koltun (1973), Mueller (1973), Sharp and Zamick (1973), and Zamick (1973).

▶**Exercise 3-10** Define the total number of effective pair interactions n within a range a by the expression

$$nA = n^+(G+J) + n^-(G-J).$$

Note that for classically *distinguishable* particles [i.e., $\eta^2(r) = 0$] and $R(1+\delta) < a/2$, $G = 1$ and $J = 0$; hence $nA = n^+ + n^- = A(A-1)/2$ and all pairs of nucleons interact.

Show that:

(a) For A large, $\delta = 0$, distinguishable particles, $a = 2.60$ fm, and $R_0 = 1.25$ fm, we find $n = 4.50$;

(b) For conditions in (a) except $\delta = 0.280$, we find $n = 3.08$;

(c) For conditions in (b) but allowing for the appropriate statistical correlation required for *indistinguishable* fermions, we obtain $n = 2.51$. ◀

The understanding of the saturation of nuclear densities requires examining the behavior of E_p and T_0 as functions of R for a fixed A. This in turn requires noting the behavior of $G(R)$ and $J(R)$. Clearly $G(R) \approx 1$ for $0 < R(1+\delta) < a/2$ (or approximately $R \lesssim a/2$). For large R, we see from (3-48) that $G \approx (a/R)^3$. The defining equation (3-45c) for J involves $\eta^2(k_F r)$ of (3-34b) and (3-34c). For $N \approx Z$, $k_F = (9\pi/8)^{1/3}A^{1/3}/R$, if we take the nuclear volume as $\Omega \approx (4\pi/3)R^3$. Thus for $R \approx 0$, k_F is very large and $\eta^2(k_F r) \approx 0$ in the range $0 < r < a$, resulting in $J(R \approx 0) \approx 0$. As R increases from zero, k_F decreases and eventually reaches a value $k_F a \approx 1$. Figure 3.4 then shows that $\eta^2(k_F r) \approx 1$ in the entire range $0 < r < a$ and hence $J(R) \approx G(R)$. The condition $k_F a \approx 1$ implies that for $R \gtrsim (9\pi/8)^{1/3}A^{1/3}a$ we have $J(R) \approx G(R)$. The behavior of $G(R)$ and $J(R)$ is schematically represented in Fig. 3.5a.

If we take A large enough to have (3-24) pertain, then (3-45a) becomes

$$E_p = ({}^{31}V_0 A^2/32)[3(1+\alpha)(G+J) + (9\beta+\gamma)(G-J)]. \qquad (3\text{-}65)$$

We note that if $9\beta+\gamma < -3(1+\alpha)$ (a condition satisfied by $\alpha = 0.754$, $\beta = -1/3$, and $\gamma = -3.34$), and if ${}^{31}V_0 < 0$ (i.e., attractive), then $E_p > 0$ (repulsive) for $R \approx 0$. As R increases, however, the bracketed term changes sign and E_p becomes attractive and eventually goes to zero as $E_p \approx (3/16)\,{}^{31}V_0 A^2(1+\alpha)(a/R)^3$. The kinetic energy (3-59) can be written for large enough A as

$$T_0 = \frac{3\hbar^2}{10m_N}\left(\frac{9\pi}{8}\right)^{2/3}\frac{A^{5/3}}{R^2}f(\delta). \qquad (3\text{-}66)$$

If we ignore the relatively small Coulomb energy, then $E \approx T_0 + E_p$. Figure 3.5b shows the schematic behavior of $E_p(R)$ for only attractive forces, i.e., $\alpha, \beta, \gamma > 0$ and ${}^{31}V_0 < 0$, and also for the case $9\beta+\gamma < -3(1+\alpha)$ with the same α and ${}^{31}V_0$ as before. Also shown are $T_0(R)$ and the two resulting total energy functions $E(R)$.

We observe from Fig. 3.5b that with only attractive forces present (the curve labeled V_2), $E_2(R)$ has a minimum near $R \approx a/2$, a situation allowing all pairs to interact strongly. As a result all nuclei would have the radius $\sim a/2$

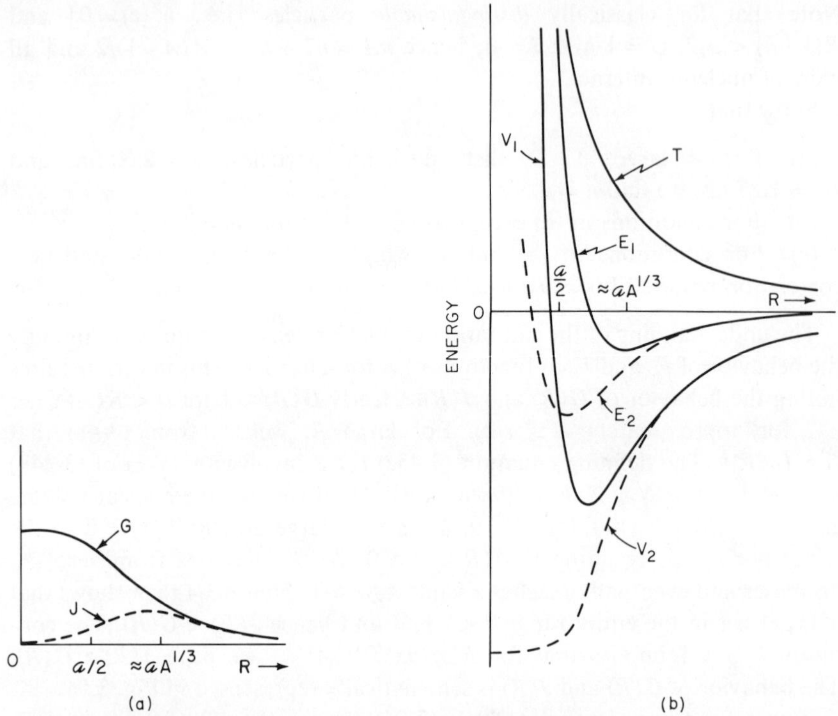

FIG. 3.5. (a) Schematic representation of the functions $G(R)$ and $J(R)$ for a fixed value of A, with $A \gg 1$. (b) Schematic behavior of the potential, kinetic, and total energies as a function of R for a fixed value of A, with $A \gg 1$. V_1 and the resulting E_1 are for $9\beta+\gamma < -3(1+\alpha)$, while V_2 and E_2 are for $\alpha, \beta, \gamma > 0$.

independent of A. The nucleus in this case would be in a degenerate collapsed state. With strong enough repulsion (as for the case labeled V_1) the minimum in the total energy $E_1(R)$ shifts to $R \sim aA^{1/3}$, thus giving a behavior of the type implied by $R = R_0 A^{1/3}$ with R_0 of the order of a. For a more detailed discussion of the saturation condition requirements on α, β, γ, and ${}^{31}V_0$ as well as the effects of the inclusion of the tensor force, see, for example, Blatt and Weiskopf (1952, p. 149 ff.) and the more recent paper of Negele (1970). Another reference on the sufficient conditions for saturation is by Calogero *et al.* (1972).

5. Determination of Numerical Parameters

We now return to the variational calculation originally proposed. The total binding energy can be written, using (3-22), (3-45), (3-47), (3-54b), and (3-59b), as

$$\begin{aligned} E_B = -E = &-\{\tfrac{1}{32}{}^{31}V_0[3(1+\alpha)A(G+J_0)+(9\beta+\gamma)A(G-J_0)]+F\}A \\ &-\tfrac{1}{8}{}^{31}V_0[3(1+\alpha)A(G+J_0)-(9\beta+\gamma)A(G-J_0)] \\ &+\tfrac{1}{8}{}^{31}V_0[(3+\alpha)A(G+J_0)-(3\beta+\gamma)A(G-J_0)]|I|A^{-1} \\ &+\{\tfrac{1}{32}{}^{31}V_0[(3-\alpha)A(G+J_0)-(3\beta-\gamma)A(G-J_0) \\ &-(3+\alpha-9\beta-\gamma)AJ_1]-\tfrac{5}{9}F\}I^2A^{-1} \\ &-a_cA^{-1/3}[Z^2-0.767Z^{4/3}] \\ &-\tfrac{1}{8}{}^{31}V_0[(1-\alpha)A(G+J_0)+(\beta-\gamma)A(G-J_0)]\eta\zeta A^{-1} \\ &-\tfrac{1}{4}{}^{31}V_0[-\alpha A(G+J_0)+\beta A(G-J_0)](\eta+\zeta)A^{-1} \\ &-\{\tfrac{1}{8}{}^{31}V_0[-\alpha A(G+J_0)+\beta A(G-J_0)]+\tfrac{10}{9}F\}(\eta^2+\zeta^2)A^{-1}. \end{aligned} \quad (3\text{-}67)$$

This equation contains the parameters that specify the four central potential strengths through ${}^{31}V_0$, α, β, γ, and a well range a, and the shape through the half-density radius R and the shell thickness $s = 0.8t = 0.8(2\delta R) = 1.60\delta R$. The procedure would be to require these parameters to yield correctly the three energy conditions $E = E_B$(observed), $(\partial E/\partial s)_{E_B} = 0$, and $(\partial E/\partial R)_{E_B} = 0$. In addition the β-stability condition $[\partial M(Z,A)/\partial Z]_A = 0$, to yield the observed regional value of I_0, would be used as a further constraint. As a first step this calculation is performed for the spherical closed-shell nucleus ${}^{208}Pb$.

The three energy conditions applied to (3-67) give three simultaneous equations [see Ayres *et al.* (1962)],

$$BP + CQ + DW = 0, \qquad B'P + C'Q + D'W = 0,$$

and

$$B''P + C''Q + D''W = 0, \quad (3\text{-}68)$$

with $P = a + b\alpha$, $Q = c\beta + d\gamma$, and $1/W = {}^{31}V_0$, and where a, b, c, and d depend on A and I and statistical weight factors and are constants for a particular nucleus; also, the quantities B, C, and D are complicated functions of a, R, and s. The condition for a solution in terms of P, Q, and W, namely that the determinant vanish, requires that $|\Delta(a, R, s)| = 0$. Thus not any combination of a, R, and s will do. It is found that in the physically acceptable regions (somewhat arbitrarily defined) $1.5 < a < 3$ fm, $1.0 < R_0 < 1.5$ fm, and $1.6 < s < 4$ fm there is no solution to $|\Delta(a, R, s)| = 0$. Had a solution been found, P, Q, and W could have been determined, giving ${}^{31}V_0$, and $c\beta + d\gamma$. The β-stability condition would then have resolved Q into β and γ.

If the condition $(\partial E/\partial R)_{E_B} = 0$ is relaxed and preferred values selected for a, R, s, and α, then solutions to (3-67) consistent with the remaining constraints

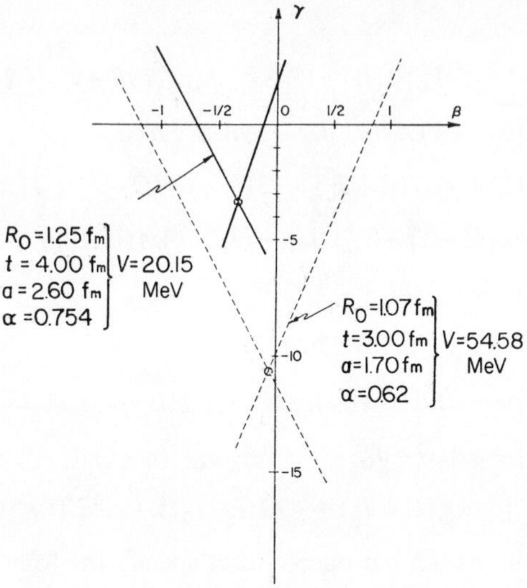

FIG. 3.6. β–γ plane representation of normalization solutions for ^{208}Pb. For each set of assumed parameters the left-hand lines (with negative slopes) are for constant E_B and $(\partial E/\partial s)_{E_B} = 0$, while the right-hand lines (with positive slopes) are for $[\partial M(Z,A)/\partial Z]_A = 0$. The quantity V is equal to $-{}^{31}V_0$. [Figures 3.6–3.14 taken from Ayres *et al.* (1962).]

can be obtained. One can view this procedure as an effort to discover whether reasonable values can be found for $^{31}V_0$, β, and γ. Figure 3.6 shows typical solutions for $^{31}V_0$, β, and γ. An "acceptable" solution for ^{208}Pb shown by the solid lines is for the selection $a = 2.60$ fm, $R_0 = 1.25$ fm, $s = 3.20$ fm, and $\alpha = 0.754$; these yield $\beta = -\frac{1}{3}$ and $\gamma = -3.34$, and $^{31}V_0 = -20.15$ MeV.

These values correspond to the magnitudes $^{31}V_0 a^2 = 136$ MeV fm^2 and $^{13}V_0 a^2 = {}^{31}V_0 \alpha a^2 = 103$ MeV fm^2. These model potentials are to be contrasted to the *free space* values quoted in Chapter I, Eq. (1-138), namely, $^{31}V_0 a_3{}^2 = 143$ MeV fm^2 and $^{13}V_0 a_1{}^2 = 94$ MeV fm^2. We note that the model potential for the two-nucleon interaction in the presence of other nuclear matter is different from the free space values. In fact a bound S-state results with a square well if $Va^2 \geqslant 102$; hence in nuclear matter in the present model the $^1S_0$ "deuteron" state is bound in addition to the $^3S_1$ state. These interaction potentials can be represented in the various forms (1-97) as square wells each with a radius $a = 2.60$ fm, and for $r < a$,

$$\begin{aligned} V &= {}^{31}V_0[\mathscr{T}(\sigma)\mathscr{S}(\tau) + \alpha\mathscr{S}(\sigma)\mathscr{T}(\tau) + \beta\mathscr{T}(\sigma)\mathscr{T}(\tau) + \gamma\mathscr{S}(\sigma)\mathscr{S}(\tau)] \\ &= V_0[A + B\boldsymbol{\sigma}_1\cdot\boldsymbol{\sigma}_2 + C\boldsymbol{\tau}_1\cdot\boldsymbol{\tau}_2 + D(\boldsymbol{\sigma}_1\cdot\boldsymbol{\sigma}_2)(\boldsymbol{\tau}_1\cdot\boldsymbol{\tau}_2)] \\ &= V_0[a + bP^{\mathrm{B}} + cP^{\mathrm{H}} + dP^{\mathrm{M}}] \\ &= V_0[a + bP^{\sigma} - cP^{\tau} + dP^{r}]. \end{aligned} \tag{3-69}$$

Setting ${}^{31}V = V_0 = -20.15$ MeV,

$$\alpha = 0.754, \qquad \beta = -1/3, \qquad \gamma = -3.342,$$
$$a = -0.480, \qquad b = 0.814, \qquad c = 1.357, \qquad d = -0.691,$$
$$A = -0.0675, \qquad B = 0.0675, \qquad C = 0.0060, \qquad D = -0.3394. \tag{3-70}$$

The above empirical values for ${}^{31}V_0 a^2$ and α, β, and γ are in general agreement with effective interactions required in Hartree–Fock and other shell model calculations. For example, Chandra and Rustgi (1973), using only the four central exchange interactions (no repulsive core or tensor interactions) calculated ${}^{31}V_0 a^2 = 139$ MeV fm^2 (square-well equivalent) and $\alpha = 0.733$, $\beta = -0.414$, and $\gamma = -0.872$. For a more detailed review of effective interactions, particularly in the ^{208}Pb region, refer to Anantaraman and Schiffer (1971). Recall also the discussion of the results of Kallio and Day relevant to (3-61).

The last line of (3-70) is not unlike the pseudoscalar meson-theoretic expression (1-45), where $A = B = C = 0$ and $V = +\frac{1}{3}|V_0|(\boldsymbol{\tau}_1 \cdot \boldsymbol{\tau}_2)(\boldsymbol{\sigma}_1 \cdot \boldsymbol{\sigma}_2) f(r)$ when applied to the dominant S states. The values $A = B = C = 0$ and $D = -\frac{1}{3}$ would, of course, give $\alpha = 1$, $\beta = -\frac{1}{3}$, and $\gamma = -3$ rather than the values obtained. For the symmetric meson theory (1-100) we have

$$V = -V_{0s}(\boldsymbol{\tau}_1 \cdot \boldsymbol{\tau}_2)[1 + \lambda \boldsymbol{\sigma}_1 \cdot \boldsymbol{\sigma}_2].$$

The "Rosenfeld mixture" often used in the literature with $\lambda = 2.33$ gives $\alpha = \frac{3}{5}$, $\beta = -\frac{1}{3}$, and $\gamma = -1.80$, while a somewhat larger value of $\lambda = 4.42$ gives $\alpha = 0.754$, $\beta = -\frac{1}{3}$, and $\gamma = -2.26$. Charge exchange scattering experiments can be used to infer λ directly [see Anderson (1966, p. 530 ff.)]. The (p, n) scattering data for $A \approx 18$ suggest $0.4 < \lambda < 0.9$, while for $A \approx 50$, the value obtained is $1 < \lambda < 3$. The reason for these wide variations is unknown. Finally, the values $\alpha = 0.754$, $\beta = -\frac{1}{3}$, and $\gamma = -3.342$ do not contradict the general conclusions discussed in Chapter I for the behavior of the central potentials in free nucleon–nucleon scattering.

Nuclei other than ^{208}Pb are treated by assuming α, β, γ, a, and R_0 to be fixed at the quoted values, by using $(\partial E/\partial s)_{E_B} = 0$ to determine s, and adjusting ${}^{31}V_0$ to give the observed E_B. These results are shown in Figs. 3.7–3.9. Figure 3.7 shows the variation of E_B/A with t ($t = 1.25s$), giving rather broad maxima. The required values of ${}^{31}V_0$ vary from ~ -17.6 MeV at $A = 20$ to ~ -20.3 MeV at $A = 267$. Although the high-energy electron scattering data do suggest that the surface thickness s increases with A (Fig. 2.3b), the variation implied by $\delta \approx 0.280$ shown in Fig. 3.9 leading to $s \approx 0.56A^{1/3}$ fm is somewhat too extreme. The neglected corrections to the kinetic energy in the surface

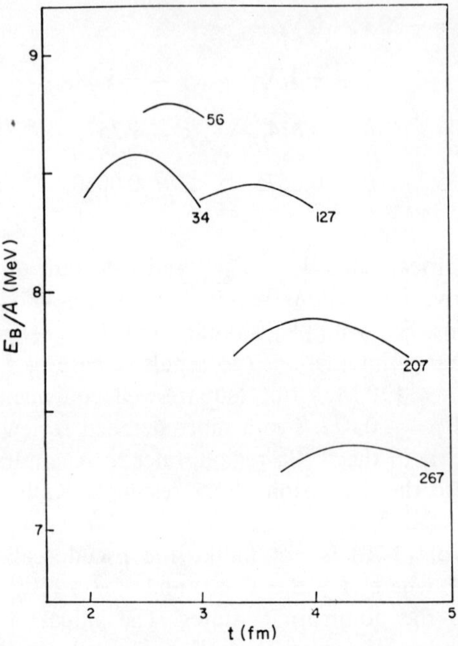

FIG. 3.7. Binding energy per nucleon E_B/A as a function of t, where $t = 1.25s$, for various values of A. In each case the maximum in E_B/A has been adjusted to occur at the observed value of E_B/A.

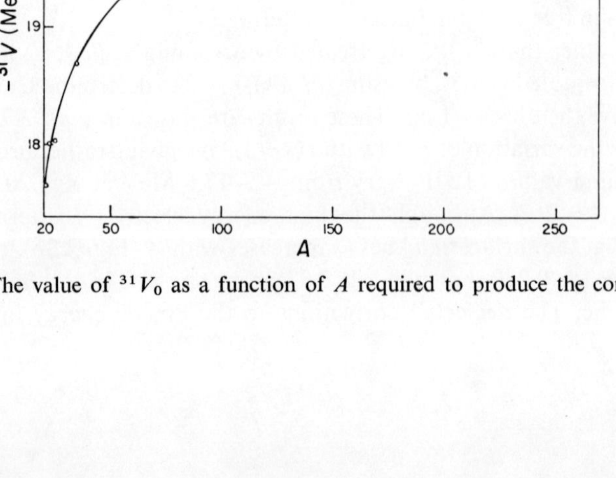

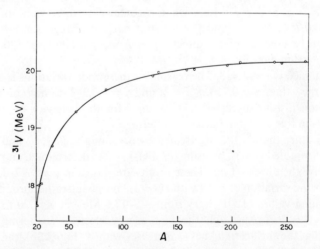

FIG. 3.8. The value of $^{31}V_0$ as a function of A required to produce the conditions of Fig. 3.7.

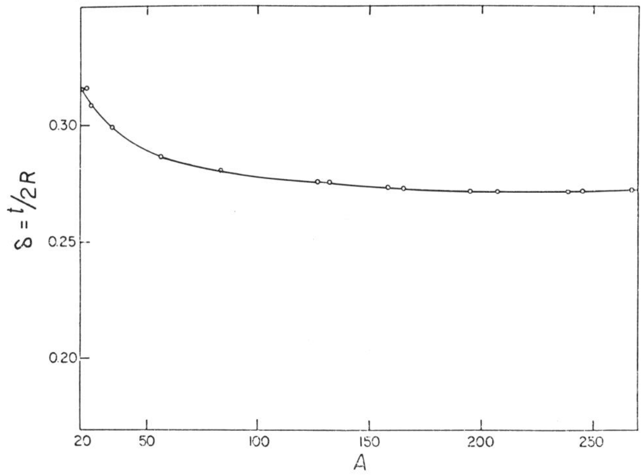

FIG. 3.9. $\delta = t/2R = t/2R_0A^{1/3}$ as a function of A.

region, particularly near the classical turning point [i.e., refer to (3-55)] are contributors to this anomalous behavior.

The energy-density variational calculations of Brueckner *et al.* (1968) cited earlier are consistent with a more or less constant value for t, although a gentle increase with A may be present, i.e., an increase from $t = 2.11$ fm for ^{16}O to $t = 2.69$ fm for ^{140}Ce.

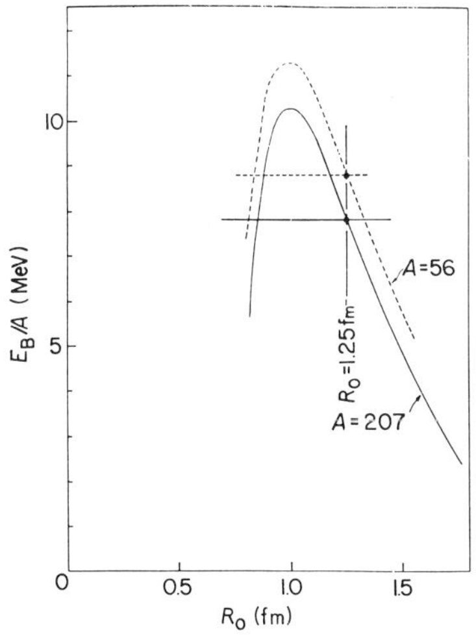

FIG. 3.10. The variation of E_B/A as a function of R_0. The observed binding energy occurs at $R_0 = 1.25$ fm in each case. The lowest energy for nuclei occurs at a lower R_0 and greater binding.

Using $\delta \approx 0.280$ and the results of Exercise 2-1 with a constant central density of $\rho_0 = 0.170\ \text{fm}^{-3}$ gives the trapezoidal half-density radius $R = 1.14A^{1/3}$ fm rather than the adopted value $R = 1.25A^{1/3}$ fm. Alternately, the value $R = 1.25A^{1/3}$ fm coupled with a constant value of $\delta \approx 0.280$ gives a central density $\rho_0 = 0.113\ \text{fm}^{-3}$ for all nuclides. The Fermi kinetic energy in the central region is thus considerably depressed from the value cited in (3-12) to $T_F = 29.2$ MeV. Although the condition $(\partial E/\partial R)_{E_B} = 0$ had to be relaxed in the present calculation, it is nonetheless instructive to calculate E_B as a function of R_0. Figure 3.10 shows the resulting behavior for $A = 56$ and $A = 207$. Saturation is seen to occur with $R_0 \approx 1.00$ fm. The resulting half-density radius $R = 1.00A^{1/3}$ fm is somewhat too small and the binding energy is approximately 2.5 MeV per nucleon too large.

C. SEMIEMPIRICAL MASS FORMULA

Although the expression for the nuclear binding energy (3-67) would appear to require an extensive numerical calculation for each particular case of interest, a good deal of simplification is possible. The resulting simplified expressions are quite useful in a number of ways, as we shall see.

Reference to (3-48b) and (3-49) shows that factors such as AG, AJ_0, and AJ_1 are largely independent of A. Since F and a_c are also only gentle functions of A (through δ), the various bracketed terms in (3-67) are largely independent of A. Then (3-67) can be written in the obvious form

$$E_B = \Phi(A) - a_i\frac{|I|}{A} - a_\tau\frac{I^2}{A} - \frac{a_c}{A^{1/3}}[Z^2 - 0.767Z^{4/3}] - \frac{a_\pi(\eta,\zeta)}{A}, \quad (3\text{-}71)$$

where the coefficients a_i, a_τ, a_c, and a_π are gentle, almost constant functions of A. The behavior of $\Phi(A)$ requires special attention. Figure 3.11 shows $\Phi(A)/A$ plotted as a function of $A^{-1/3}$. This curve can be parameterized in a number of ways. One most readily conforming to practice is

$$\Phi(A) = a_v A - a_s A^{2/3} \quad (3\text{-}72)$$

in which $a_s \equiv a_s(A)$ becomes a function of A. The "most nearly constant fit," shown by the dashed line, gives $a_v = 14.75$ MeV and $\bar{a}_s = 15.25$ MeV. Only a modest variation of $a_s(A)$ from its mean value of $\bar{a}_s$ is required by the data points. Figure 3.12 shows the required values of $a_s(A)$, as well as the calculated values of $a_\tau(A)$ and $a_c(A)$. Figure 3.13 shows $a_i(A)$ and $a_\pi(\eta,\zeta\!:\!A)$ for the cases $\eta = \zeta = 1$; $\eta = 1$, $\zeta = 0$; and $\eta = 0$, $\zeta = 1$ [the reference value is $a_\pi(0,0) = 0$].

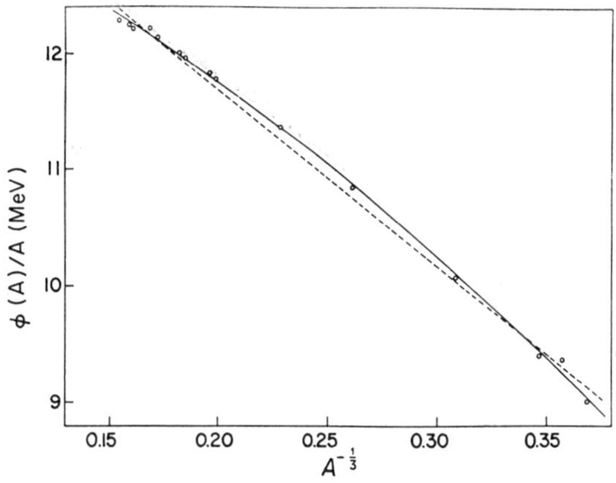

FIG. 3.11. The function $\Phi(A)/A$ versus $A^{-1/3}$. The dashed line represents the behavior $a_v - a_s A^{-1/3}$. The solid curve can be represented either as $a_v - a_s(A) A^{-1/3}$ or $a_V - a_W A^{-1/3} - a_X A^{-2/3} - a_Y A^{-1} + \cdots$ (see text).

An alternate procedure for giving a simple analytic form for $\Phi(A)$ is to expand this quantity in a descending power series in $A^{1/3}$. We write

$$\Phi(A) = a_V A - a_W A^{2/3} - a_X A^{1/3} - a_Y + \cdots. \tag{3-73}$$

The resulting least-squares fit gives (in MeV)

$$a_V = 13.811, \qquad a_W = 7.115, \qquad a_X = 15.864, \qquad a_Y \approx \text{negligible}. \tag{3-74}$$

While both (3-72) and (3-73) are equally valid, we note that a_v and a_V differ by almost 1 MeV in the two cases. Smooth and nearly constant functions

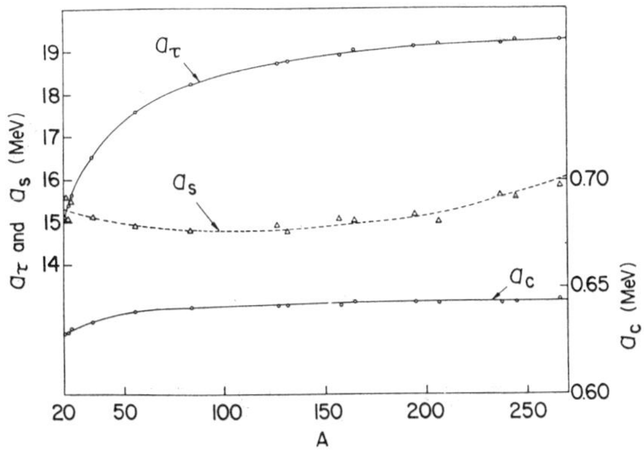

FIG. 3.12. Calculated values of a_τ, a_s, and a_c as a function of A.

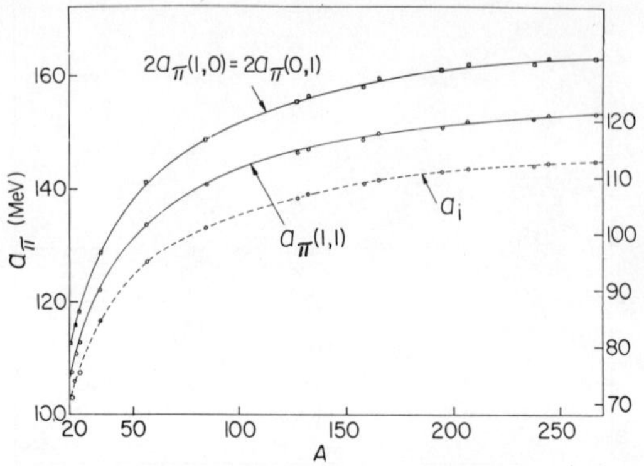

FIG. 3.13. Calculated coefficients for the pairing energies $a_\pi(\eta,\zeta)/A$ and $a_i|I|/A$ showing their variation with A.

such as $\Phi(A)/A$ cannot, in fact, be separated unambiguously into "volume" and "surface" terms that differ by only the multiplicative factor $A^{1/3}$ when the available range for $A^{1/3}$ is as small as nature provides. Additional criteria are required for a sharp delineation of these two effects. For a recent effort to treat the function $\Phi(A)/A$, see Muthukrishnan (1971). As a general conclusion the liquid drop volume coefficient is probably most correctly quoted as $a_v = (15 \pm 1)$ MeV and the surface coefficient as $a_s = (16.5 \pm 3)$ MeV, with, however, a high degree of correlation between the *pair* of values selected. The following discussion uses (3-72).

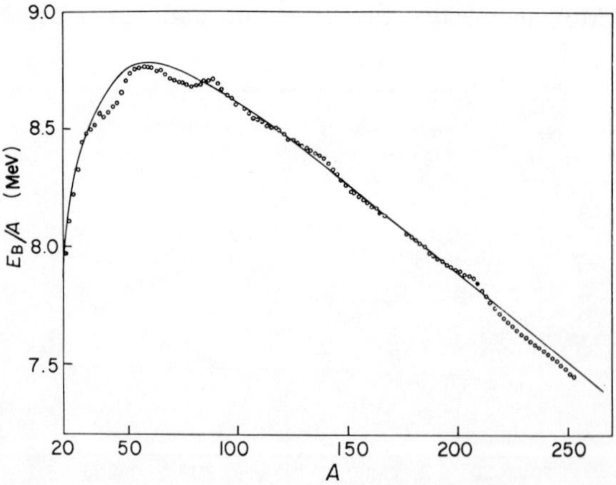

FIG. 3.14. Experimental values of E_B/A for β-stable odd-A nuclei (∘) and the calculated curve using (3-67) and (3-71).

The final fit to the experimental binding energy, when the results depicted in Figs. 3.6–3.13 are used, is shown in Fig. 3.14. A simple expression of the type (3-71) with "most nearly constant" coefficients is[1]

$$E_B \approx 14.75A - 15.25A^{2/3} - 105\frac{|I|}{A} - 18.90\frac{I^2}{A} - \frac{0.641}{A^{1/3}}[Z^2 - 0.767Z^{4/3}]$$

$$+\Delta\begin{cases} 0 & \eta = 0,\ \zeta = 0 \\ -\dfrac{80}{A} & \eta = 1,\ \zeta = 0 \quad \text{or} \quad \eta = 0,\ \zeta = 1 \\ -\dfrac{145}{A} & \eta = 1,\ \zeta = 1. \end{cases} \tag{3-75}$$

Expression (3-75) is one version of the well-known Bethe–Weizsäcker semi-empirical mass formula [von Weizsäcker (1935) and Bethe and Bacher (1936)]. Table 3-3 lists some typical values that have appeared in the literature for the simplest version of the mass formula.

TABLE 3-3

SEMIEMPIRICAL MASS FORMULA COEFFICIENTS (IN MEV)

$$E_B = a_v A - a_s A^{2/3} - a_\tau (I^2/A) - (a_c/A^{1/3})Z^2$$

a_v	a_s	a_τ	a_c	Reference
13.86	13.20	19.5	0.580	Bethe and Bacher (1936)
14.00	13.03	19.3	0.584	Fermi (1949)
15.58	17.23	23.3	0.698	Green and Engler (1953)

Since 1960 numerous and more complicated formulas have come into vogue, some with as many as 20 adjustable parameters. A rather successful mass formula that includes a surface correction to the isotopic term is by Myers and Swiatecki (1966), namely,[2]

$$E_B = 15.68A - 18.56A^{2/3} - 28.1\frac{I^2}{A}\left(1 - \frac{1.18}{A^{1/3}}\right) - 0.717\frac{Z^2}{A^{1/3}}\left(1 - \frac{1.69}{A^{2/3}}\right)$$

$$+ \text{pairing term} + \text{shell effects.}$$

[1] The literature generally takes as its pairing energy reference $\Delta(1,0) = \Delta(0,1) = 0$ with $\Delta(0,0) = +\underline{\Delta}$ and $\Delta(1,1) = -\underline{\Delta}$.

[2] The exchange term in the Coulomb energy is essentially written as $0.767Z^{4/3} = 0.767Z^2(A/2)^{-2/3}$. A more recent version by these authors quoted in Brueckner *et al.* (1971) includes an $|I|/A$ isotopic term and uses $a_v = 15.642$, $a_s = 19.232$, and $a_c = 0.727$ MeV.

Another notable mass formula is the earlier result by Seeger (1961). Like (3-75), it includes both I^2/A and $|I|/A$ isotopic terms. For a recent mass formula, see Ludwig *et al.* (1973), and for a discussion of generalized nuclidic mass relationships, see Jänecke and Behrens (1974).

Empirical values for a_v postdating 1960 are higher than those shown in Table 3-3, and generally fall in the range $15.6 \lesssim a_v \lesssim 17.1$ MeV. The corresponding surface coefficients a_s fall in the range $18.5 \lesssim a_s \lesssim 33.6$ MeV. As stated earlier, it is prudent to quote only $a_v = 15 \pm 1$ MeV and $a_s = (16.5 \pm 3)$ MeV as reasonable liquid drop values.

The representation illustrated in Fig. 3.1 and used to evaluate ^{31}n, ^{13}n, ^{33}n, and ^{11}n is correct only if the spin and isospin of the individual particle states are good quantum numbers. In addition, spherically symmetric nuclear shapes were assumed. While these conditions may prevail near closed-shell or magic-number nuclei, for nuclei with partially filled shells these assumptions are certainly in error. Thus "shell" corrections and deformation corrections are certainly required, particularly for a_π and a_i.[1]

Shell and deformation corrections have received considerable attention in the literature. A rather thorough semiempirical investigation of these effects is contained in the work of Myers and Swiatecki (1966). More recently the theoretical foundation for such empirical corrections has been given by Bunatian *et al.* (1972). The interested reader is referred to these papers and the references contained therein. An early effort to correct for shell and deformation effects with adequate detail for our present purposes was carried out by Mozer (1959). The binding energy was written

$$E_B = a_v A - a_s A^{2/3} - a_\tau I^2 A^{-1} - a_c A^{-1/3} Z(Z-1) - S(Z, A) - D(Z, A). \tag{3-76}$$

A semiempirical evaluation of $S(Z, A)$ and $D(Z, A)$ leads to the values shown in Fig. 3.15. The function $D(Z, A)$ was adjusted to be consistent with the observed nuclear deformations. The deformations are large between magic numbers, and corresponding nuclei have *greater* binding when deformed than when spherical; hence $D(Z, A) < 0$. Closed-shell nuclei are more bound than those having valence nucleons in partially filled shells, just as for the corresponding atomic cases. As defined in (3-76), $S(Z, A) \approx 0$ for doubly magic nuclei, and is notably greater than zero particularly when both N and Z are midway between magic numbers. Thus, both $D(Z, A)$ and $S(Z, A)$ correlate strongly with the magic numbers.

The coefficients in (3-76) were found to be $a_v = 15.28$, $a_s = 15.76$, $a_\tau = 22.76$, and $a_c = 0.690$ MeV, when $D(Z, A)$ and $S(Z, A)$ of Fig. 3.15 are also included.

[1] Nuclear shell structure is discussed in Chapter IV.

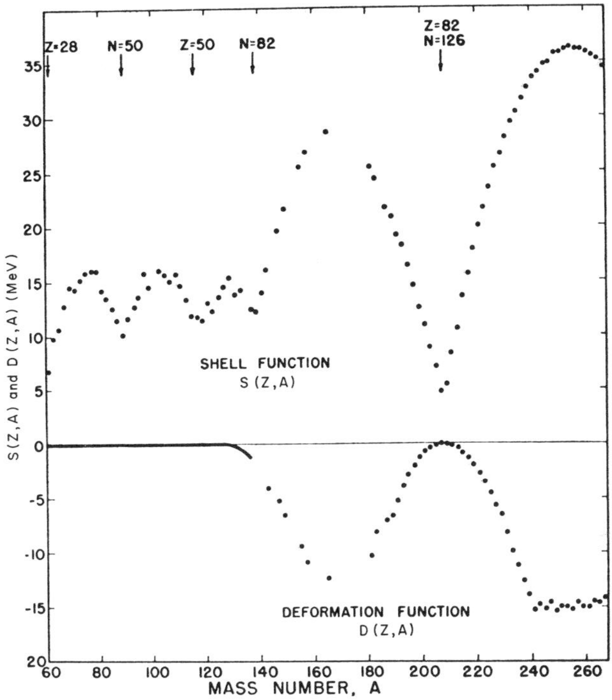

FIG. 3.15. The semiempirical shell and deformation corrections as given by Mozer (1959).

▶**Exercise 3-11** Using the semiempirical mass formula (3-75), evaluate the binding energy per particle E_B/A for ${}^{34}_{16}S$, ${}^{158}_{64}Gd$, ${}^{195}_{78}Pt$, and ${}^{256}_{100}Fm$. Consider including the corrections $S(Z,A)$ and $D(Z,A)$ from Fig. 3.15. Compare with observed masses using recent values such as those listed in Mattauch *et al.* (1965), Lederer *et al.* (1968), and Wapstra and Gove (1971). ◀

The pairing energy correction $\Delta = a_\pi(\eta,\zeta)/A$ shown in Fig. 3.13, essentially exhibiting the dependence of $a_\pi(\eta,\zeta)$ on A while giving approximately the right order of magnitude for Δ, has too strong a dependence on A. Many simple expressions for Δ have been suggested. Most have $\Delta_n = a_\pi(1,0)/A$ equal to $\Delta_p = a_\pi(0,1)/A$, and $\Delta_{np} = a_\pi(1,1)/A$ approximately twice Δ_n or Δ_p, in agreement with the present model. The origin of the pairing energy term is evident from the examination of (3-23). It will be observed that with η or ζ nonzero, n^+ is reduced and n^- increased additionally from any change in I. Since symmetric space states have attractive forces and antisymmetric states have repulsive forces, *lower* binding results when η or ζ is nonzero. The lowering of the binding energy when both η and ζ are nonzero is not quite twice as great due to the extra (n, p) interaction. This additional attractive interaction $\Delta_n + \Delta_p - \Delta_{np}$ is approximately (10–15)/A MeV (see Fig. 3.13). In the presence of short-range attractive spin–orbit forces the pairing energy can be shown to be proportional to $(2j+1)/A$ [e.g., see Mayer and Jensen (1955, p. 239 ff.)].

Numerical expressions that have appeared in the literature are

$$\Delta_n = \Delta_p = 135/A \text{ MeV} \quad \text{(Blatt and Weisskopf, 1952)} \tag{3-77a}$$

$$\Delta_n = \Delta_p = 36/A^{3/4} \text{ MeV} \quad \text{(Fermi, 1949)} \tag{3-77b}$$

$$\Delta_n = \Delta_p = 12/A^{1/2} \text{ MeV} \quad \text{(Green and Edwards, 1953)} \tag{3-77c}$$

$$\Delta_n, \Delta_p = 12.5(2j_{n,p}+1)/A \quad \text{(Mayer and Jensen, 1955).} \tag{3-77d}$$

Reference to Fig. 3.13 shows $a_\pi(\eta, \zeta)$ to be a slightly increasing function of A. An approximate power-law expression in analogy with (3-77) is possible, yielding

$$\Delta_n = \Delta_p \approx 43/A^{0.88} \text{ MeV}. \tag{3-77e}$$

Of the above possibilities (3-77c) gives the best overall agreement with empirical values. Figure 3.16 shows an empirical evaluation of Δ_n and Δ_p obtained from measured nuclear masses and also exhibits the function (3-77c) [see Zeldes *et al.* (1967)]. Rather striking variations from smooth behavior are evident, with a tendency for larger pairing energies at magic numbers for either neutrons or protons. While $\Delta_p \approx \Delta_n$, there is some tendency for Δ_p to exceed Δ_n.

The terms in (3-71) involving $I = N - Z = -2T_3$ and (η, ζ) all arise from the reduction of the binding energy for nucleon configurations involving partition diagrams with other than $l = m = p = q$. In this connection it should be recalled that (3-20) and hence (3-22) are generally valid even for $\eta > 1$ and $\zeta > 1$, and for filling the levels illustrated in the partition diagram Fig. 3.1 in a sequence not necessarily generating the lowest energy (i.e., for excited states). When n^+ is associated with an attractive potential and n^- with a repulsive potential (or at least a less attractive potential), the nuclear ground state will always result for a given N, Z when η and ζ take on their smallest possible values.

The division of the total energy into terms in T_3, the "symmetry energy," and those in (η, ζ), the "pairing energy," is somewhat arbitrary. Taking the three terms together we have,[1]

$$E_I = a_i \frac{|I|}{A} + a_\tau \frac{I^2}{A} + a_\pi \frac{(\eta, \zeta)}{A}, \tag{3-78}$$

[1] For moderately heavy nuclei

$$E_I \approx 105\frac{|I|}{A} + 18.90\frac{I^2}{A} + 41\frac{(\eta+\zeta)}{A} + 37\frac{(\eta^2+\zeta^2)}{A} - 17\frac{\eta\zeta}{A}.$$

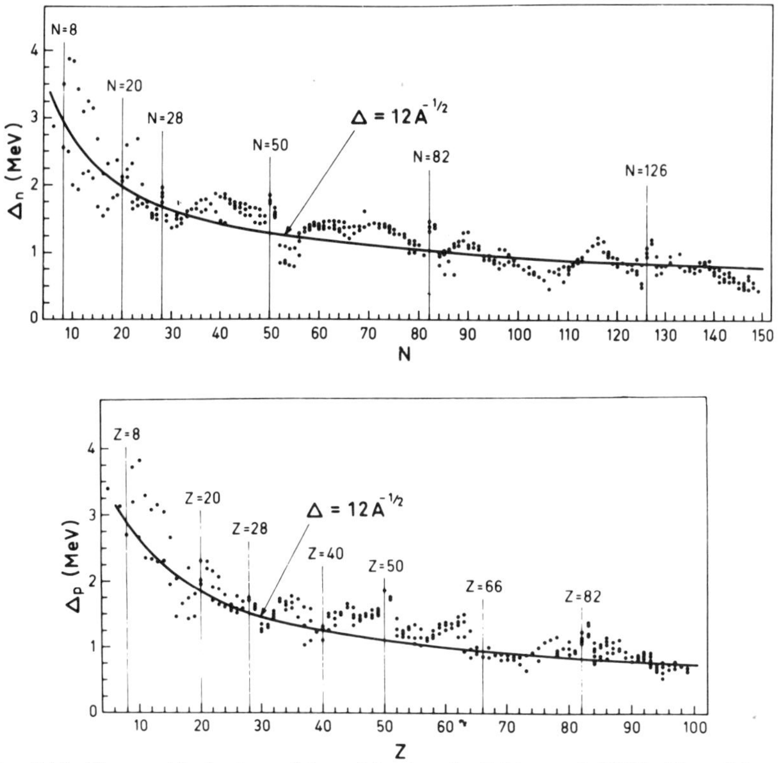

FIG. 3.16. The empirical values of Δ_n and Δ_p given by Zeldes *et al.* (1967). The solid curve is a plot of (3-77c).

which clearly leads to greatest binding when $I = 0$ and $\eta = \zeta = 0$ if no Coulomb energy were present. The repulsive Coulomb effect, on the other hand, results in strong binding for $Z = (A - I)/2$ small (i.e., I large). In the presence of both effects, optimum binding occurs for some values of $I > 0$. Thus, for each value of A, there is a unique value of I for which (3-71) results in maximum binding.

1. α- and β-Decay Instability

Experimentally the most directly determined relevant datum for the equilibrium nuclear equation of state based on (3-75) is the isobar of lowest mass that is also consequently β-stable. The mass of any nuclear type (for the present consideration each being in its ground state) can be written

$$Mc^2 = (M_H Z + M_n N)c^2 - a_v A + a_s A^{2/3} + a_i |I| A^{-1} + a_\tau I^2 A^{-1} + a_c A^{-1/3}[Z^2 - 0.767 Z^{4/3}] + \Delta(\eta, \zeta) \qquad \text{(3-79a)}$$

or

$$M(A,I)c^2 = \tfrac{1}{2}(M_H+M_n)Ac^2 + \tfrac{1}{2}(M_n-M_H)Ic^2 - a_v A + a_s A^{2/3} + a_i|I|A^{-1}$$
$$+ a_\tau I^2 A^{-1} + \tfrac{1}{4}a_c A^{-1/3}[(A-I)^2 - 0.767 \times 2^{2/3}(A-I)^{4/3}]$$
$$+ \Delta(\eta,\zeta). \qquad \text{(3-79b)}$$

In (3-79) the mass $M(A,I)$ refers to the *atomic mass*, hence the quantity $M_H A$ in terms of the atomic mass for hydrogen M_H is used to include the mass of the electrons. In a crude sense this also accounts for a portion of the electronic binding energy, which must also be considered. Present uncertainties in E_B are great enough not to warrant any further refinement of this consideration. Since $M_n \neq M_H$, the lowest mass isobars occur for a value $I = I_0$, which is not quite the value giving maximum binding. Setting $\partial M(A,I)c^2/\partial I = 0$ leads to

$$I_0 = (A^{5/3} - 2.03A - 325)/(117.7 + A^{2/3}), \qquad \text{(3-80)}$$

where we have used $(M_n - M_H)c^2 = 0.783$, $a_i = 105$, $a_\tau = 18.90$, and $a_c = 0.641$ MeV. Using the Taylor series expansion of $M(A,I)c^2$ about the value $M(A,I_0)c^2$ gives to second order

$$M(A,I)c^2 = M(A,I_0)c^2 + \lambda_I(I-I_0)^2 + \Delta(\eta,\zeta) \qquad \text{(3-81a)}$$

or

$$M(A,Z)c^2 = M(A,Z_0)c^2 + \lambda_Z(Z-Z_0)^2 + \Delta(\eta,\zeta). \qquad \text{(3-81b)}$$

The atomic number Z_0 corresponds to $(A-I_0)/2$. In a straightforward evaluation of $[\partial^2 M(A,I)c^2/\partial I]|_{I_0}$ we get for the above constants

$$\lambda_Z = 4\lambda_I = \frac{75.6}{A}\left(1 + \frac{A^{2/3}}{118}\right) \text{MeV}. \qquad \text{(3-82)}$$

▶**Exercise 3-12** Derive (3-80) and (3-82), by expressing $(A-I)^{4/3}$ in a power series in I/A and using suitable approximations when called for. ◀

Figure 3.17 shows a plot of the β-stable isobars and $I_0(A)$ determined by an exact calculation based on (3-67), which does not lead to significantly different results from the use of (3-80). Deviations in the empirical values from the smooth trend are evident. If (3-79) had also included the terms $+S(Z,A)$ $+D(Z,A)$ of Fig. 3.15, a somewhat better agreement between the observed isobars and the predicted curve would have resulted. A test of the approximate expression (3-81) with the auxiliary expressions (3-80) and (3-82) is most reasonably performed for large A, where the statistical model might be expected to be most valid. To reduce the uncertainties involved with $S(Z,A)$ and $D(Z,A)$, the region $190 < A < 200$ is selected. Nuclei in this region have

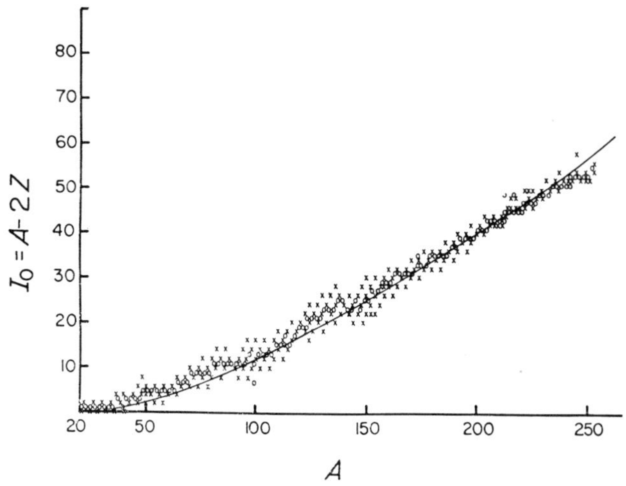

FIG. 3.17. The calculated variation of I_0 with A and the experimental points for β-stable nuclei, both A_{odd} (○) and A_{even} (×) [after Ayres *et al.* (1962)].

relatively small deformations and the discontinuities at the magic number $Z = 82$ and $N = 126$ are avoided by staying on the low side of the "cusp" at $A = 208$. The presence of the pairing energy term $\Delta(\eta, \zeta)$ in (3-81) requires distinguishing between A odd and A even. If $\Delta(0,1) = \Delta(1,0)$, the quantity $\Delta(\eta, \zeta)$ remains a simple additive constant to $M(A,Z)c^2$ when A is odd. When, however, A is even, $\Delta(\eta, \zeta)$ is alternately $\Delta(0,0)$ and $\Delta(1,1)$ as Z is alternately even or odd. Thus a double parabolic family results for A even. Figure 3.18 shows the resulting "mass parabolas" for $A = 197$, 198, and 199. The dashed line interpolation of $M(A, Z_0)$ from the two A-odd curves to the A-even curve shows that $\Delta(1,1) \approx 2\Delta(1,0)$ or $2\Delta(0,1)$.

Beta-decay with e^- emission will occur when $M(Z,A) > M(Z+1, A)$ with a maximum kinetic energy release $E_{\beta^-,\max} = [M(Z,A) - M(Z+1, A)]c^2$. Beta-decay with e^+ emission or electron capture will occur when $M(Z,A)c^2 > M(Z-1, A)c^2 + 2m_e c^2$. When $M(Z-1, A)c^2 < M(Z, A)c^2 < M(Z-1, A)c^2 + 2m_e c^2$, only electron capture is possible. The apparent asymmetry in the electron rest mass energy $m_e c^2$ is due to referring to atomic rather than nuclear masses. A detailed discussion of β-decay is given in Chapter VIII. Reference to Fig. 3.18 shows that for $A = 197$, ${}^{197}_{79}$Au is β-stable, ${}^{197}_{78}$Pt β^--decays to ${}^{197}_{79}$Au, and ${}^{197}_{80}$Hg K-captures to ${}^{197}_{79}$Au since in this case the atomic mass difference is only 0.42 MeV, well below $2m_e c^2 = 1.02$ MeV. For $A = 198$, ${}^{198}_{78}$Pt and ${}^{198}_{80}$Hg are β-stable with ${}^{198}_{79}$Au decaying by both K-capture and β^--decay to ${}^{198}_{78}$Pt and ${}^{198}_{80}$Hg, respectively. ${}^{198}_{77}$Ir β^--decays to ${}^{198}_{78}$Pt, and ${}^{198}_{81}$Tl decays by both β^+-emission and K-capture to ${}^{198}_{80}$Hg since here the

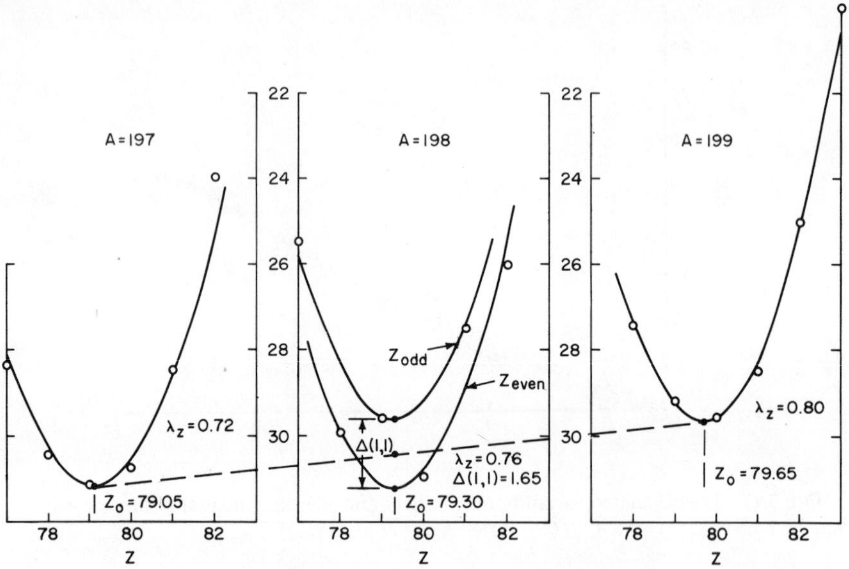

FIG. 3.18. The "locally" determined mass parabolas for $A = 197$, 198, and 199. The values of Z_0, λ_z, and $\Delta(1, 1)$ of (3-81) are determined in each case to give the best fit to the data points shown.

atomic mass difference is ~3.5 MeV. The "stability" of ${}^{198}_{78}Pt$, with a mass greater than ${}^{198}_{80}Hg$, results from the very long lifetime to be associated with the required *double* β-decay to ${}^{198}_{80}Hg$. The single mass parabola for $A = 199$ is readily interpreted with ${}^{199}_{80}Hg$ as the stable isotope.

The "local" empirical values for Z_0, λ_Z, and $\Delta(1, 1)$ for $190 < A < 200$ are given in Table 3-4. The expression (3-80) for Z_0 rather closely agrees with these empirical values, in most instances to within ±0.05. The required values for $\Delta(1, 1)$ also agree reasonably well with the estimate $\Delta(1, 1) \approx 24/A^{1/2}$ MeV. However, λ_Z as calculated from (3-82) is 0.51 at $A = 191$ and slowly *decreases* thereafter to 0.47 at $A = 201$, while the empirical value *increases* from 0.55 at $A = 191$ to 0.90 at $A = 201$. Thus while the empirical value for λ_Z is in reasonable agreement with (3-82) for $A \leqslant 190$, it increases to approximately twice

TABLE 3-4

A	191	192	193	194	195	196	197	198	199	200	201
Z_0	76.85	77.20	77.50	77.95	78.30	78.65	79.05	79.30	79.65	79.90	80.25
λ_Z	0.55	0.57	0.58	0.61	0.63	0.67	0.72	0.76	0.80	0.84	0.90
$\Delta(1,1)$	—	1.85	—	1.80	—	1.72	—	1.65	—	1.55	—

the calculated value as A approaches $A = 208$. A further feature in agreement with this behavior is the tendency for the data points for a given A to exhibit a steeper rise from $Z > Z_0$ than for $Z < Z_0$ in the region $Z_0 < 82$. The decrease in $S(Z, A) + D(Z, A)$ toward the cusp at $A = 208$ adds a significant positive contribution to $\partial^2 M(Z, A)/\partial Z^2$ and hence to λ_Z. The major contribution in the present case is from $S(Z, A)$. The approximate theoretical function for $S(Z, A)$ is generally taken to be quadratic, resulting in a behavior similar to that in Fig. 3.15, e.g.,

$$S_{ij}(N, Z) = \alpha_i(N - N_i)^2 + \alpha_j(Z - Z_j)^2 + K_{ij} \tag{3-83}$$

with N_i and Z_j representing the nearest magic numbers (in the present instance $N_i = 126$ and $Z_j = 82$). Current empirical evidence suggests that $S(Z, A) + D(Z, A)$ has an increasing curvature as A approaches the doubly magic nuclide with $A = 208$.

▶**Exercise 3-13** The region $100 < A < 110$ is a comparable region with small deformations and distinctly to the low side of $Z = 50$. Using recent empirical mass values and β-decay energies, such as those found in Wapstra and Gove (1971), Lederer *et al.* (1968), and Mattauch *et al.* (1965), determine local empirical values for Z_0, λ_Z, and $\Delta(1, 1)$. Compare these to the values calculated from (3-80), (3-82), and (3-77c) and comment. ◀

▶**Exercise 3-14** (a) Show that the most favorable case for Z odd, N odd being β-stable is $\Delta(1, 1) \leqslant \lambda_Z$. Using (3-82) and (3-77c), estimate the *largest* value of A for which Z odd, N odd might be stable. The only known such nuclei are ^{2}H, ^{6}Li, ^{10}B, and ^{14}N.

(b) Show that the most favorable case for 3 stable isobars is $\Delta(1, 1) \geqslant 3\lambda_Z$. Using (3-82) and (3-77c), determine the *smallest* value of A for which 3 stable isobars might appear. If the K-capture half-life of $^{96}_{40}Zr$, $t_{1/2} > 3.6 \times 10^{17}$ yr, is considered long enough to pronounce $^{96}_{40}Zr$ "stable," then $A = 96$ is the lowest observed A for 3 stable isobars. The lowest A for 3 stable isobars to the order requiring double β-decay is $A = 124$ with $^{124}_{50}Sn$, $^{124}_{52}Te$, and $^{124}_{54}Xe$ all stable.

(c) Show that $\Delta(1, 1) \leqslant 3\lambda_Z$ defines the *largest* value of A-even for which only one stable isobar is possible. The largest A-even, for one β-stable isobar, is $A = 200$ with only $^{200}_{80}Hg$ stable.

(d) The observed β-stable ($t_{1/2} > 10^{12}$ yr) nuclides with $A \leqslant 202$ are grouped as follows:

	Z even	*Z* odd
N even	163	48
N odd	55	4

including the three anomalous A-odd cases with two "β-stable" isobars. These are for $A = 113, 115$ with $\frac{9}{2}^+ \leftrightarrows \frac{1}{2}^+$ spin and parity-changing β^--decays having $t_{1/2} \gtrsim 10^{15}$ yr, and for $A = 123$ with $\frac{1}{2}^+ \rightarrow \frac{7}{2}^+$ low-energy electron capture having $t_{1/2} \approx 1.2 \times 10^{13}$ yr. Also, masses $A = 5$ and 8 have been excluded. In view of parts (a)–(c) of this exercise, give a simplified numerical accounting for the nuclide frequencies tabulated above. ◀

▶**Exercise 3-15** Using (3-75) and the binding energy of ^{4}He equal to 28.30 MeV, show that spontaneous α-particle emission occurs with a kinetic energy release of

$$E_\alpha = 28.30 - 4a_v + \frac{8}{3}\frac{a_s}{A^{1/3}} + \frac{4a_c Z}{A^{1/3}}\left[1 - \frac{1}{3A}(Z - 0.767Z^{1/3}) - \frac{0.511}{Z^{2/3}}\right]$$
$$- \frac{4a_i}{A}\left(1 - \frac{2Z}{A}\right) - 4a_\tau\left(1 - \frac{2Z}{A}\right)^2 \text{ MeV.} \qquad (3\text{-}84)$$

Using the suggested constants in (3-75), determine the smallest A for a β-stable nuclide that can undergo α-particle emission. In each instance use $Z = Z_0$ as determined from (3-80). The lightest known β-stable α-particle emitter is $^{144}_{60}$Nd with a half-life $t_{1/2} = 2.4 \times 10^{15}$ yr, although $^{142}_{58}$Ce might also be α-particle unstable with a comparable lifetime. In each instance the decay is to a daughter with $N = 82$. Discuss the expected effect of $N = 82$ for predictions based on (3-84) in the context of shell closure effects (see Chapter IV). ◀

It is very useful to develop a "local" isotopic effect mass law that relates isobaric analog state masses under the operation of the most general isospin-conservation-breaking interaction. The isospin-splitting interaction is considered in a perturbation approximation. A zero-order charge *independent* Hamiltonian H_0 is assumed yielding the $2T+1$ degenerate states $|\alpha T T_3\rangle$, where α is a convenient representation of all the other required quantum numbers, viz.,

$$H_0 |\alpha T T_3\rangle = E^0_{\alpha, T} |\alpha T T_3\rangle. \qquad (3\text{-}85)$$

The most general isospin-independence-breaking two-body interaction that still conserves total charge [see the discussion in Chapter I concerning (1-84)] can be written

$$H_1 = \sum_{i>j=1}^{A} \{V_0(r_{ij}) + V_1(r_{ij})(t_{3i} + t_{3j}) + V_2(r_{ij})\, t_{3i} t_{3j}\}. \qquad (3\text{-}86)$$

An equivalent form more clearly exhibiting the tensor character of (3-86) is

$$H_1 = \sum_{i>j=1}^{A} [V_0(r_{ij}) + \tfrac{1}{3}V_2(r_{ij})\mathbf{t}_1 \cdot \mathbf{t}_2] + \sum_{i>j=1}^{A} [V_1(r_{ij})(t_{3i} + t_{3j})]$$
$$+ \sum_{i>j=1}^{A} [\tfrac{1}{3}V_2(r_{ij})(3t_{3i}t_{3j} - \mathbf{t}_1 \cdot \mathbf{t}_3)]. \qquad (3\text{-}87)$$

►**Exercise 3-16** Show that the Coulomb interaction of (3-15a),

$$H_c = \sum_{i>j=1}^{A} (e^2/r_{ij})(\tfrac{1}{2}+t_{3i})(\tfrac{1}{2}+t_{3j}),$$

for example, leads to precisely the form (3-87). ◄

The first term of (3-87) is an isoscalar, the second an isovector, and the third an isotensor of second rank. Treating H_1 as a perturbation, we have

$$E_{\alpha,T,T_3} = E^0_{\alpha,T} + \langle \alpha TT_3 | H_1 | \alpha TT_3 \rangle$$

with

$$\langle \alpha TT_3 | H_1 | \alpha TT_3 \rangle = \langle \alpha T || H_1{}^1 || \alpha T \rangle + \frac{T_3}{[T(T+1)]^{1/2}} \langle \alpha T || H_1{}^2 || \alpha T \rangle$$
$$+ \frac{3T_3{}^2 - T(T+1)}{[T(T+1)(2T+3)(2T-1)]^{1/2}} \langle \alpha T || H_1{}^3 || \alpha T \rangle; \quad (3\text{-}88)$$

here we have made use of the Wigner–Eckart theorem (see Appendix C) and $H_1^{1,2,3}$ stands for the isoscalar, isovector, and isotensor terms, respectively, in (3-87). Clearly (3-88) yields a mass law of the form

$$M(\alpha, T, T_3) \text{ MeV} = a(\alpha, T) + b(\alpha, T)T_3 + c(\alpha, T)T_3{}^2, \quad (3\text{-}89)$$

and $T_3 = \frac{1}{2}(Z-N) = -\frac{1}{2}I$, as per (1-19) and the definition employed in (3-20).

A typical example of the success of (3-89) is the case for the $T = \frac{3}{2}$ levels in $A = 13$. The observed mass defects[1] are given in Table 3-5. These lead to $a = 19.257$, $b = 2.180$, and $c = 0.256$ MeV. It is important to distinguish the difference between the mass parabola (3-81), which relates to the nuclear ground states, designated in Table 3-5 by (gd), and the mass parabola (3-89), which relates to analog isospin states, such as the entries in Table 3-5 for

TABLE 3-5

MASS DEFECTS (IN MEV)

		$J^\pi = \frac{1}{2}^-$, $T = \frac{1}{2}$	$J^\pi = \frac{3}{2}^-$, $T = \frac{3}{2}$
$T_3 = +\frac{3}{2}$	$^{13}_{8}O$	—	23.11 ± 0.04 (gd)
$T_3 = +\frac{1}{2}$	$^{13}_{7}N$	5.345 ± 0.003 (gd)	20.411 ± 0.004
$T_3 = -\frac{1}{2}$	$^{13}_{6}C$	3.125 ± 0.003 (gd)	18.231 ± 0.003
$T_3 = -\frac{3}{2}$	$^{13}_{5}B$	—	16.562 ± 0.004 (gd)

[1] The mass defect is defined as $[M(Z, N) - A]c^2$.

$T = \frac{3}{2}$ (for which the quoted constants a, b, and c pertain). The quadratic term coefficient from (3-82), assuming it still to be valid for such light nuclei as $A = 13$, is $\lambda_T = \lambda_Z \approx 6.3$ MeV, which is seen to be considerably larger than $c = 0.256$ MeV. It will be noted that the $T_3 = \pm\frac{1}{2}$ mass difference for $T = \frac{1}{2}$ is (2.220 ± 0.004) MeV, while for the $T = \frac{3}{2}$ levels it is (2.180 ± 0.005) MeV. The constants $b(\alpha, T)$ and $c(\alpha, T)$ are, in general, largely dependent only on A and, as in the present case, do not depend very sensitively on α and T. The coefficient $c(A)$ slowly decreases from 0.270 for $A = 7$ to 0.172 for $A = 37$, while $b(A)$ increases from 1.318 for $A = 9$ to 6.178 for $A = 37$ in an approximately linear fashion.

In the case $A = 13$, Eq. (3-89) with the quoted constants gives a rather good fit (i.e., $\chi^2 \approx 0.01$). However for $A = 9$, which is the most accurately measured multiplet, (3-89) allows only a fit with $\chi^2 \approx 4$. The addition of a cubic term leads to a marked improvement for $A = 9$. The significance of the required cubic term may indicate the need for higher order perturbation terms, second-order Coulomb terms, or the presence of charge-dependent three-body forces. Second-order Coulomb effects are found to give a ${T_3}^3$ coefficient of order 1–3 keV [see Auerbach and Lev (1972)], while the experimental value of (7.6 ± 1.7) keV for $A = 9$ is considerably larger (Kashy *et al.*, 1974). Why out of over a dozen known examples only the $A = 9$ case requires the addition of a cubic term to (3-89) also remains to be explained. An interesting confirmation of the $A = 9$, 13, 21, and 33 coefficients is obtained in the precision experimental determination of the mass excesses for ^{9}C, ^{13}O, ^{21}Mg, and ^{33}Ar using the (^{3}He, ^{6}He) reactions at 70 MeV [see Treutelman *et al.* (1971) and Nann *et al.* (1974)]. With the successful production of ^{8}C and ^{20}Mg (two nuclides with a proton excess of 4, hence $T_3 = +2$), using the (α, ^{8}He) reaction at 156 MeV (Robertson *et al.*, 1974), the expression (3-89) has been verified for some members of the mass quintets corresponding to $T = 2$ for $A = 8$ and 20.

An excellent discussion of many additional applications of such nuclear mass relations with isospin are given by Garvey (1969) and Jänecke (1965, 1966, 1968), and Jänecke and Behrens (1974).

The second term $a_s A^{2/3}$ in the mass formulas (3-72) and (3-75) is referred to as the surface energy since the surface area of a nucleus with radius $R_0 A^{1/3}$ would vary with mass number as $A^{2/3}$, just as this term does. In the language of the liquid drop model, as the statistical model developed here is sometimes called, one could speak of a *surface tension* σ and write

$$4\pi R^2 \sigma = a_s A^{2/3} \qquad \text{or} \qquad \sigma = a_s / 4\pi {R_0}^2. \tag{3-90}$$

2. Fission Instability

An interesting application of the liquid drop concept is to the phenomenon of nuclear fission. In the fission process a relatively heavy nucleus splits into two large fragments with an energy release of ~ 150–200 MeV.[1] Fission can be either spontaneous or reaction-induced. The mass distribution spectrum of the fragments has been the object of extensive study. Characteristically, it shows that for low internal excitation of the fissioning nucleus the two fragments most probably have unequal masses, a situation referred to as asymmetric fission. When asymmetric fission is the rule, the mass spectrum has two well-defined, broad peaks at mass numbers approximately summing to the initial mass of the fissioning nucleus as shown in Fig. 3.19.

Generally, increasing the internal excitation develops a broad, single peak in the mass spectrum, indicating symmetric fission as the most probable occurrence.

The fission fragments are usually very neutron rich and initiate a succession of β-decays leading to a chain of isobars decaying in sequence one into the other. The fragments initially produced are also accompanied by the emission of several prompt evaporation neutrons. The emission of delayed neutrons can also occur when in a particular β-decay chain a neutron-unstable isobar is formed at some point, thereby exhibiting a radioactive lifetime characteristic of the preceding β-decays. The presence of the delayed neutrons, even though they represent a small fraction of the fission neutrons, are essential in permitting the control of reactor operating dynamics. A typical example of slow-neutron-induced asymmetric fission is

$$^{235}_{92}\mathrm{U} + \mathrm{n} \rightarrow {}^{92}_{36}\mathrm{Kr} + {}^{142}_{56}\mathrm{Ba} + 2\mathrm{n}.$$

The $^{92}_{36}$Kr fragment then initiates a mass $A = 92$ β-decay chain ending in β-decay-stable $^{92}_{40}$Zr,

$$^{92}_{36}\mathrm{Kr} \xrightarrow{3.0\ \mathrm{sec}} {}^{92}_{37}\mathrm{Rb} \xrightarrow{5.3\ \mathrm{sec}} {}^{92}_{38}\mathrm{Sr} \xrightarrow{2.71\ \mathrm{hr}} {}^{92}_{39}\mathrm{Y} \xrightarrow{3.53\ \mathrm{hr}} {}^{92}_{40}\mathrm{Zr}\ (\beta\text{-stable}).$$

Spontaneous fission of heavy nuclei with fission lifetimes shorter than 10^{12} yr is observed when $A \gtrsim 240$. However, spontaneous fission is actually a negligible mode of decay compared to α-particle decay until $A \gtrsim 250$. An example of a case where spontaneous fission actually becomes the main decay channel is $^{254}_{98}$Cf with $t_{1/2} = 60.5$ days. Fission induced by various projectiles (or γ-rays) on fissile nuclei can have substantial cross sections for considerably lighter target nuclei than even $A \approx 240$. The well-known slow-neutron-induced fission of $^{235}_{92}$U (thermal neutron fission cross section $\sigma_f \approx 580$ b) previously cited is but one example of such reaction-induced fission.

It is instructive to consider the energy release (and hence excitation energy) in the possible symmetric spontaneous fission of a heavy even–even nucleus

[1] Three-fragment or *ternary fission* is also known; for a recent review, see Halpern (1971) and Diehl and Greiner (1973).

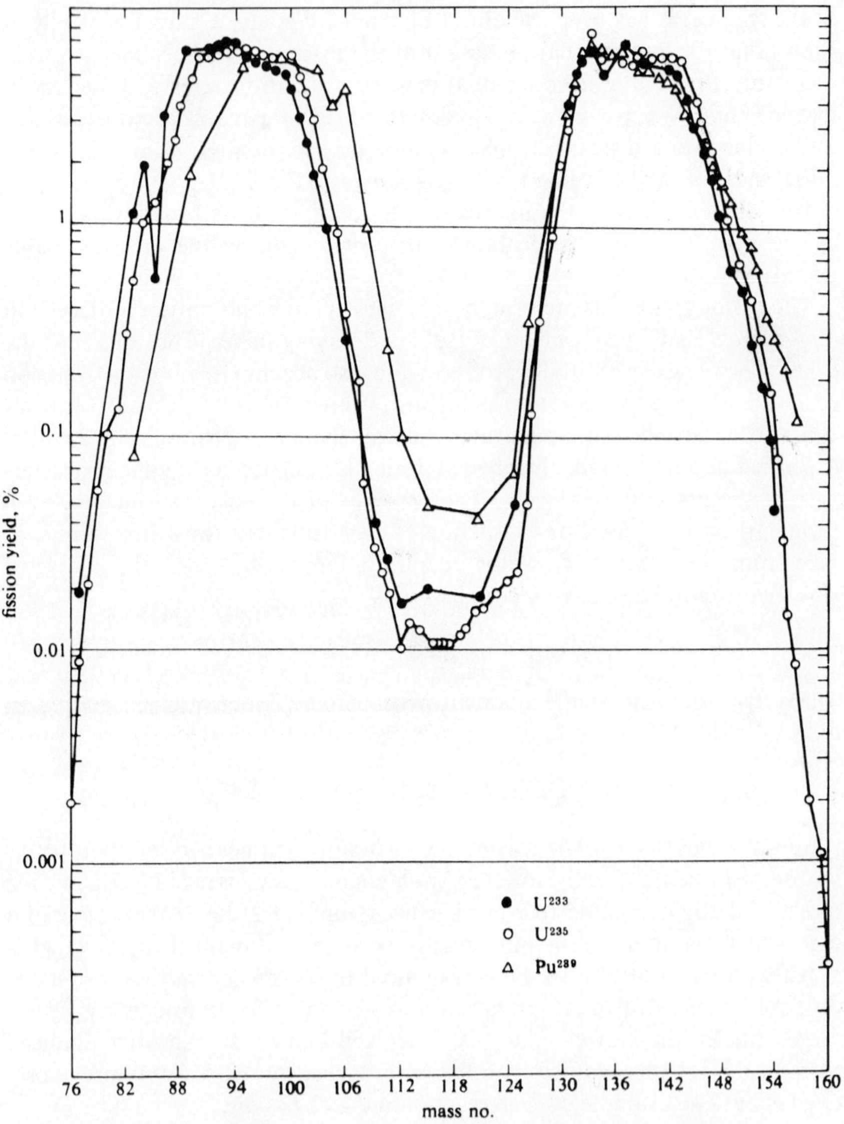

FIG. 3.19. Fission fragment mass spectrum in the slow-neutron-induced fission of ^{233}U, ^{235}U, and ^{239}Pu [taken from Weinberg and Wigner (1959)].

such as ${}^{238}_{92}U$. Using (3-75), it is readily shown that

$$\begin{aligned} Q &= M(Z, A)c^2 - 2M(Z/2, A/2)c^2 \\ &= (1-2^{-2/3})a_c Z^2 A^{-1/3} - (2^{1/3}-1)a_s A^{2/3} - a_i I A^{-1} - 2\Delta(Z/2, N/2) \\ &= 0.237Z^2 A^{-1/3} - 3.97A^{2/3} - 105IA^{-1} - 2\Delta(Z/2, N/2)\ \text{MeV}. \qquad (3\text{-}91) \end{aligned}$$

Except for the small isotopic and pairing energy effects, the Q value results from the difference between a reduced Coulomb energy and an increased surface energy when the nucleus splits into two fragments. The numerical value for Q in the case of ${}^{238}U$ calculated from (3-91) is $Q = 146$ MeV. When the two fission fragments are well separated and considered spherical in shape, the Coulomb potential energy is $V_c = e^2Z^2/4r$, where r is the separation of the centers of the fragments. If this expression held even at the separation $r = 2R(Z/2, A/2)$, with $R(Z/2, A/2)$ being the fragment radius, i.e., when the fragments assumed spherical in shape are just in contact, then

$$V_c(\text{max}) = \frac{2^{1/3}}{8}\frac{5}{3}a_c\frac{Z^2}{A^{1/3}} = 0.168\frac{Z^2}{A^{1/3}}\ \text{MeV}. \qquad (3\text{-}92)$$

The numerical value for $V_c(\text{max})$ for ${}^{238}U$ is $V_c(\text{max}) = 229$ MeV. The situation is illustrated in Fig. 3.20. The "realistic" curve of potential energy is shown as a solid line and presents a substantially lower barrier for fission

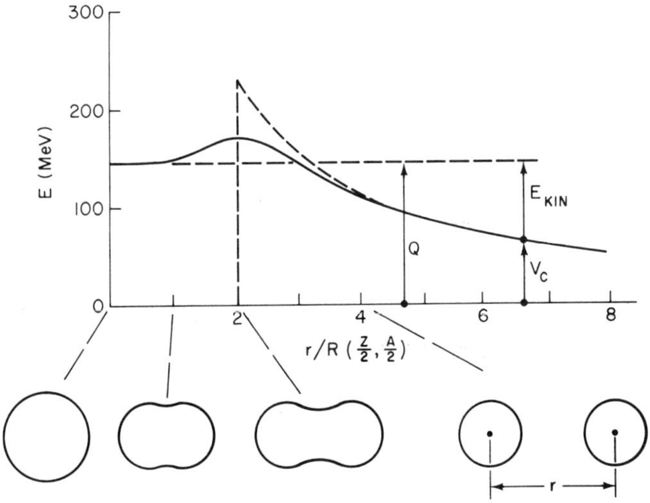

FIG. 3.20. A schematic representation of the energy considerations in the symmetric spontaneous fission of ${}^{238}_{92}U$. The separation of the centers of the fragments is in units of the fragment radius $R(Z/2, A/2)$. For interesting and more realistic diagrammatic energy–distance representations of the fission process, refer to Swiatecki and Bjørnholm (1972).

than the estimate based on $V_c(\max)$ shown by the dashed line. This reduction is largely due to the additional surface energy as the droplet becomes strongly elongated just prior to rupturing into two distinct fragments. The prompt evaporation neutrons which are observed to be preferentially emitted along a line connecting the fragments are not shown.

Clearly, spontaneous fission would occur rapidly if $Q \geqslant V_c(\max)$, and from (3-91) and (3-92) this condition leads to

$$Z^2/A \geqslant 67, \tag{3-93}$$

when the constants of (3-75) are used. The result (3-93) greatly overestimates the required value of Z^2/A for spontaneous fission. A more realistic value can be obtained by inquiring about the stability of a liquid drop under quadrupole oscillations. Suppose an essentially incompressible spherical liquid drop of radius R is deformed into an ellipsoid with a semimajor axis b and a semiminor axis a. Define a parameter ε such that

$$b = R(1+\varepsilon), \qquad a = R(1+\varepsilon)^{-1/2}, \tag{3-94}$$

which maintains the volume $4\pi ba^2/3$ constant. To terms of order ε^2, the surface energy of the ellipsoidal liquid drop having a surface tension σ is

$$E_s = 4\pi R^2\sigma(1+\tfrac{2}{5}\varepsilon^2+\cdots) = a_s A^{2/3}(1+\tfrac{2}{5}\varepsilon^2+\cdots), \tag{3-95}$$

where we have noted (3-90). The Coulomb energy for a uniform charge density can be shown to be, in the same approximation,

$$E_c = \frac{3e^2Z^2}{5R}\left(1-\frac{\varepsilon^2}{5}+\cdots\right) = \frac{a_c Z^2}{A^{1/3}}\left(1-\frac{\varepsilon^2}{5}+\cdots\right). \tag{3-96}$$

The change in energy under deformation due to these two effects is[1]

$$\Delta E = \left(\frac{2a_s A^{2/3}}{5} - \frac{a_c Z^2}{5A^{1/3}}\right)\varepsilon^2. \tag{3-97}$$

If $\Delta E > 0$, the spherical shape is stable under ellipsoidal oscillations. On the other hand, if $\Delta E < 0$, then the nucleus is unstable, giving the criterion for spontaneous fission

$$Z^2/A \geqslant 2a_s/a_c \qquad \text{or} \qquad Z^2/A \geqslant 48, \tag{3-98}$$

if we use the constants of (3-75). The experimental situation verifies the importance of the factor Z^2/A, referred to as the *fissionability*, as a general trend indicator in determining the half-life for spontaneous fission. Figure 3.21 shows the half-life for various fissile isotopes versus the parameter Z^2/A. It will be observed that fractional year lifetimes for Z^2/A occur well below even the critical value of $Z^2/A = 48$. The rapid decrease in lifetime as Z^2/A

[1] For a more detailed discussion of deformation energies, refer to Section VI,B.

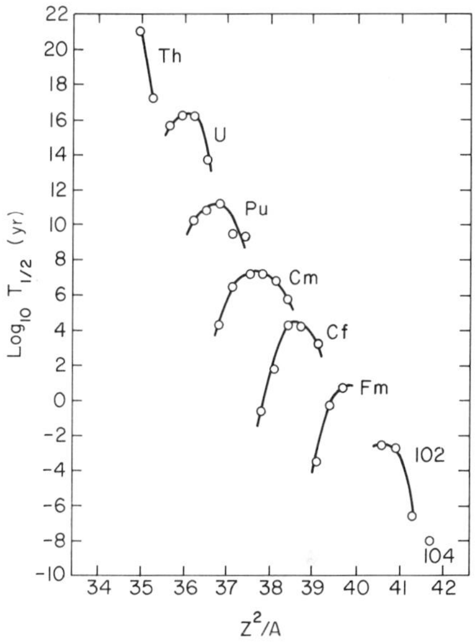

FIG. 3.21. The half-life for spontaneous fission of some even–even nuclei versus the parameter Z^2/A. [See also Hulet *et al.* (1971) and Randrup *et al.* (1974).]

increases toward the critical value is, of course, the result of the rapidly increasing "tunneling" probability or barrier penetration as Q approaches V_c(max) or its more realistic value such as that indicated in Fig. 3.20. There is some evidence that nuclei with $N > 158$ are becoming "catastrophically" unstable with respect to spontaneous fission, as evidenced by the very short half-life of (380 ± 60) μsec determined for ^{258}Fm (not shown in Fig. 3.21) [see Hulet *et al.* (1971)]. For additional discussion, see Randrup *et al.* (1974).

The classic paper of Bohr and Wheeler (1939) and those of Swiatecki (1956) should be consulted for further details concerning elementary liquid drop models. These primitive models do not account for the striking feature of asymmetric fission. It has been proposed by Griffin (1968) that collective non-equilibrium motion may carry over the characteristics of the reflection symmetry of individual particle states to the actual point of separation, *scission*. Fission potential energy trajectories as a function of deformation then lead to asymmetric fission as most probable, particularly for nuclei with large fissionability.

An interesting application of these considerations to the study of the fissioning mechanism for the postulated *superheavy* nucleus $^{298}_{114}$X based on an asymmetric two-center model is given by Mustafa and Schmitt (1973). For a general discussion of superheavy nuclei, see, e.g., Köhler (1971), Fiset and Nix (1972), and Nix (1972).

The exact behavior of the deformation energy requires a careful examination of the variations of the single-particle energies of the individual nucleons within a deforming nuclear potential. The simple liquid drop model may be satisfactory for lighter fissioning nuclei, but shell corrections significantly effect the minimum-potential-energy path to actual scission. A thorough recent review paper including such considerations is that of Nix (1972). One interesting, recently verified implication of such studies has to do with the existence of high-lying isomeric states in some transuranium elements such as $^{240,242}Am$ that, while being relatively stable to gamma-ray decay do, in fact, decay by spontaneous fission. In these cases it is believed that shell corrections to the liquid drop model produce secondary minima in the nuclear energy surface at large deformations [see Strutinsky (1967)]. The fissioning isomers are nuclei caught in such secondary minima with a greatly reduced barrier to fission. The nuclear ground state lies at the bottom of the first and deeper minimum. For typical experiments involving this phenomenon see Back *et al.* (1972), Bjørnholm *et al.* (1967), Alexander and Bimbot (1972), James *et al.* (1972), and Migneco and Theobald (1968). For general references concerning the characteristics of such potential energy surfaces see Jägare (1969), Anderson *et al.* (1970), Britt *et al.* (1971, 1973), Mosel and Schmitt (1971), Bolsterli *et al.* (1972), Ledergerber and Pauli (1973), and Sobiczewski *et al.* (1973).

▶**Exercise 3-17** Verify (3-91) and (3-92), and then show that (3-93) follows. Use a suitable approximation for the term $I/A^{5/3}$ derived from (3-80) for a mean value of $A \approx 250$. ◀

▶**Exercise 3-18** Consider the slow-neutron-induced fission

$$ {}^{A-1}_{\ \ Z}\mathrm{U} + \mathrm{n} \rightarrow ({}^{A}_{Z}\mathrm{U}) \rightarrow {}^{\frac{1}{2}A-a}_{\frac{1}{2}Z-z}\mathrm{X} + {}^{\frac{1}{2}A+a}_{\frac{1}{2}Z+z}\mathrm{Y}. $$

(a) Using (3-79), show that the energy release in the reaction for $a = 0$ and $z = 0$, i.e., when $Q_f(z, a) = Q(0, 0)$, is

$$ Q_f(0,0) = a_v - a_\tau\left(1 - \frac{4Z^2}{A^2}\right) - a_i\left(1 - \frac{2Z}{A}\right) - 0.260 a_s A^{2/3} $$

$$ + 0.370 a_c \frac{Z^2}{A^{1/3}} + \delta\Delta. \tag{3-99} $$

(b) Expanding $M(\frac{1}{2}Z \pm z, \frac{1}{2}A \pm a)$ in a Taylor series about $M(\frac{1}{2}Z, \frac{1}{2}A)$, show that an energy ellipse for $Q_f(z, a)$ results with

$$ Q_f(z, a) = Q_f(0,0) - a^2 J + 2azK - z^2 L. \tag{3-100} $$

(c) Evaluate $Q_f(0,0)$ for $Z = 92$ and $A-1 = 235$ using the constants of (3-75) and plot the ellipse (3-100) for $Q_f(z, a) = 0.95Q_f(0,0)$. Consider only both z and a even. Observed values for $Q_f(z, a)$ with $z \approx 10$ and $a \approx 24$ are 154.7 MeV from ionization chamber measurements and 167.1 MeV using time-of-flight techniques. ◀

D. INFINITE NUCLEAR MATTER

We now turn to a general discussion of the properties of nuclear matter that emphasizes the role that the presence of other nucleons plays in influencing the wave function of a particular pair of nucleons selected for examination. A thorough investigation of this sort requires a rather accurate specification of the nucleon–nucleon interaction. Even at the elementary level proposed here, the nonlocal nature of this interaction must be included in our considerations. Hence we first give a general discussion of nonlocal potentials. Although a variety of powerful techniques have been developed for treating nuclear matter, we confine our attention to a particularly revealing approach based on the so-called Bethe–Goldstone equation.

1. Nonlocal Potentials

For a basic understanding and appreciation of all the implications involved in the presence of nonlocal potentials in the wave equation, it is necessary to be rather formal. We follow the methods elucidated in a number of standard texts on linear vector spaces [e.g., Merzbacher (1970, Chapter 14)] and employ strict Dirac notation.

Let $\tilde{x}$, a Hermitian operator, represent the observable corresponding to a particle's spatial coordinate with eigenvalues x' and corresponding eigenvectors $|x'\rangle$, i.e.,

$$\tilde{x}\,|x'\rangle = x'\,|x'\rangle. \tag{3-101}$$

If the potential energy operator $\tilde{V} \equiv V(\tilde{x})$ is a nonpathological function of only the space coordinate, and hence can be expanded in a converging power series in $\tilde{x}$, for example, then $V(\tilde{x})\,|x'\rangle = V(x')\,|x'\rangle$. The matrix elements of the potential operator are then *diagonal*, viz.,

$$\langle x''|\,V(\tilde{x})\,|x'\rangle = \langle x''|\,V(x')\,|x'\rangle = V(x')\langle x''|x'\rangle = V(x')\,\delta(x''-x'), \tag{3-102}$$

where we have assumed the normalization $\langle x''|x'\rangle = \delta(x''-x')$.

If the potential operator is a function of both the particle coordinate and the momentum ($p = \hbar k$) such that $\tilde{V} \equiv V(\tilde{x}, \tilde{k})$ is a well-behaved function, then its matrix elements become

$$\langle x''|V(\tilde{x},\tilde{k})|x'\rangle = \int_{k'}\int_{k''}\langle x''|k'\rangle\, dk'\,\langle k'|V(\tilde{x},\tilde{k})|k''\rangle\, dk''\,\langle k''|x'\rangle, \tag{3-103}$$

where we have used the closure relation $\int_{k'}|k'\rangle\, dk'\,\langle k'| = 1$ twice in padding out the initial expression. The operator $\tilde{k}$ obeys a parallel construction to (3-101). Now since $\langle k''|x'\rangle = (2\pi)^{-1}\exp(-ik''x')$, we have

$$\begin{aligned}\langle x''|V(\tilde{x},\tilde{k})|x'\rangle &= (2\pi)^{-1}\int_{k'}\int_{k''}\{\exp[i(k'x''-k''x')]\}\, dk'\, dk'' \\ &\quad\times V(x',k'')\langle k'|k''\rangle \\ &= (2\pi)^{-1}\int_{k'}\{\exp[ik'(x''-x')]\}\, V(x',k')\, dk' \\ &= V(x',x'').\end{aligned} \tag{3-104}$$

In obtaining (3-104), we have used $\langle k'|k''\rangle = \delta(k'-k'')$. If at the last step $V(\tilde{x},\tilde{k})$ is taken after all to be only a function of the coordinate, (3-102) is recovered since the integral over k' is then just an integral representation of the delta function $\delta(x''-x')$. However, if $\tilde{V}$ is indeed a function of both $\tilde{x}$ and $\tilde{k}$, (3-104) now possesses *off-diagonal* elements in general.

The full significance of such off-diagonal elements is best appreciated by deriving the Schrödinger equation with $\tilde{V} \equiv V(\tilde{x},\tilde{k})$. We seek to solve the eigenvalue problem $\tilde{H}|E\rangle = E|E\rangle$, with

$$\tilde{H} \equiv (\hbar^2/2m)\,\tilde{k}^2 + V(\tilde{x},\tilde{k}). \tag{3-105}$$

In the standard manner we define the *wave function* $\psi_E(x') = \langle x'|E\rangle$ for a particle in a state having the energy E. Thus we have

$$\begin{aligned}\left\{\frac{\hbar^2}{2m}\tilde{k}^2 + V(\tilde{x},\tilde{k})\right\}|E\rangle &= \frac{\hbar^2}{2m}\int_{x'}|x'\rangle\left(-\frac{\partial^2}{\partial x'^2}\right)\psi_E(x')\, dx' \\ &\quad + \int_{x''}V(\tilde{x},\tilde{k})|x''\rangle\,\psi_E(x'')\, dx''.\end{aligned} \tag{3-106}$$

The left-hand side of (3-106) can also be written

$$\tilde{H}|E\rangle = E\int_{x'}|x'\rangle\,\psi_E(x')\, dx'. \tag{3-107}$$

For the second term on the right-hand side of (3-106), we have

$$\begin{aligned}\int_{x''}V(\tilde{x},\tilde{k})|x''\rangle\,\psi_E(x'')\, dx'' &= \int_{x'}\int_{x''}|x'\rangle\, dx'\,\langle x'|V(\tilde{x},\tilde{k})|x''\rangle\,\psi_E(x'')\, dx'' \\ &= \int_{x'}|x'\rangle\left\{\int_{x''}V(x'',x')\,\psi_E(x'')\, dx''\right\} dx',\end{aligned} \tag{3-108}$$

using (3-104). Therefore, substituting (3-108) and (3-107) into (3-106), we obtain

$$\int_{x'} |x'\rangle\, dx' \left\{ \left(-\frac{\hbar^2}{2m} \frac{\partial^2}{\partial x'^2} - E \right) \psi_E(x') + \int_{x''} V(x'', x')\, \psi_E(x'')\, dx'' \right\} = 0. \tag{3-109}$$

Since the vectors $|x'\rangle$ are linearly independent, the expression inside the curly bracket must be zero. The resulting Schrödinger equation, expressed in three dimensions and simplifying the prime notation, finally becomes

$$-(\hbar^2/2m)\, \nabla^2 \psi_E(\mathbf{r}) + \int_{r'} V(\mathbf{r}, \mathbf{r}')\, \psi_E(\mathbf{r}')\, d^3r' = E\psi_E(\mathbf{r}) \tag{3-110a}$$

with

$$V(\mathbf{r}, \mathbf{r}') = (2\pi\hbar)^{-3} \int V(\mathbf{r}, \mathbf{p}) \exp[(i/\hbar)(\mathbf{r}-\mathbf{r}')\cdot\mathbf{p}]\, d^3p. \tag{3-110b}$$

The result (3-110) permits us to appreciate the significance of the presence of off-diagonal matrix elements for $V(\tilde{\mathbf{r}}, \tilde{\mathbf{p}})$ in generating a *nonlocal* behavior. It further shows the complete equivalence of nonlocality and "velocity dependence" in the potential.

The selection or designation of a suitable potential in the form $V(\mathbf{r}, \mathbf{p})$ or $V(\mathbf{r}, \mathbf{r}')$ will be governed by whichever gives the clearest insight into the physical situation operating in a specific case. For example, in Chapter I in connection with the source function $\eta(\mathbf{r})$, we have already discussed the natural appearance of nonlocality in meson exchange potentials when the finite "size" of the nucleons is taken into account.

In usual applications detailed knowledge of the exact form to use is lacking, and we therefore note the general properties $V(\mathbf{r}, \mathbf{r}')$ should have:

(a) Since $V(\mathbf{r}, \mathbf{p})$ was assumed Hermitian, $V(\mathbf{r}, \mathbf{r}')$ must be symmetric in $\mathbf{r}$ and $\mathbf{r}'$, if it is real;

(b) The approximate validity of model calculations with local potentials requires specifically nonlocal effects to be small;

(c) The range of the nonlocality should be of order (0.5–1.0) fm to conform with experimental information and inferences concerning the "nucleon size."

A suitable often-used single-particle potential that possesses these properties and has the important feature (on an ad hoc basis) of being separable is of the form

$$V(\mathbf{r}, \mathbf{r}') = V[(\mathbf{r}+\mathbf{r}')/2]\, \delta_\ell(\mathbf{r}-\mathbf{r}'). \tag{3-111}$$

The function $\delta_\ell(\mathbf{r}-\mathbf{r}')$ is a suitable sharply peaked function at $\mathbf{r} = \mathbf{r}'$, such as

$$\delta_\ell(\mathbf{r}-\mathbf{r}') = \pi^{-3/2}\ell^{-3}\exp[-(\mathbf{r}-\mathbf{r}')^2/\ell^2]. \qquad (3\text{-}112)$$

As the parameter ℓ in (3-112), referred to as the range of the nonlocality, approaches zero, the function $\delta_\ell(\mathbf{r}-\mathbf{r}')$ approaches the usual delta function $\delta(\mathbf{r}-\mathbf{r}')$. In the zero range limit, $V[(\mathbf{r}+\mathbf{r}')/2]$ goes over into whatever local potential was being considered. For the nonzero range case, $V(x)$ is conveniently taken at the average coordinate $(\mathbf{r}+\mathbf{r}')/2$.

▶**Exercise 3-19** Letting $\mathbf{r}' = \mathbf{r}+\ell\mathbf{s}$, show that for (3-112), $\int \delta_\ell(\mathbf{s}) b^3\, d^3s = 1$. Also show that $\int s^2\, \delta_\ell(\mathbf{s}) b^3\, d^3s = \frac{3}{2}$. ◀

When we introduce $\mathbf{r}' = \mathbf{r}+\ell\mathbf{s}$, as in Exercise 3-19, the potential term in (3-110a) becomes

$$\int V(\mathbf{r}+\tfrac{1}{2}\ell\mathbf{s})\, \psi(\mathbf{r}+\ell\mathbf{s})\, \delta_\ell(\mathbf{s})\ell^3\, d^3s. \qquad (3\text{-}113)$$

Under the assumption of small nonlocality effects, we expand $V(\mathbf{r}+\frac{1}{2}\ell\mathbf{s})$ and $\psi(\mathbf{r}+\ell\mathbf{s})$ in a Taylor series about $\ell\mathbf{s} = 0$; hence

$$V(\mathbf{r}+\tfrac{1}{2}\ell\mathbf{s}) = V(\mathbf{r}) + \tfrac{1}{2}\ell\mathbf{s}\cdot\nabla V(\mathbf{r}) + \tfrac{1}{8}\ell^2(\mathbf{s}\cdot\nabla)^2 V(\mathbf{r}) + \cdots \qquad (3\text{-}114\text{a})$$

and

$$\psi(\mathbf{r}+\ell\mathbf{s}) = \psi(\mathbf{r}) + \ell\mathbf{s}\cdot\nabla\psi(\mathbf{r}) + \tfrac{1}{2}\ell^2(\mathbf{s}\cdot\nabla)^2\psi(\mathbf{r}) + \cdots \qquad (3\text{-}114\text{b})$$

Substituting (3-114) into (3-113), we have, to order ℓ^2,

$$V(\mathbf{r})\psi(\mathbf{r}) + \int (\ell^2 s^2/24)[\nabla^2 V(\mathbf{r}) + 2\nabla\cdot V(\mathbf{r})\nabla + V(\mathbf{r})\nabla^2]\psi(\mathbf{r})\, \delta_\ell(\mathbf{s})\ell^3\, d^3s,$$

and using the results of Exercise 3-19, we obtain

$$V(\mathbf{r})\psi(\mathbf{r}) + (\ell^2/16)[\nabla^2 V(\mathbf{r}) + 2\nabla\cdot V(\mathbf{r})\nabla + V(\mathbf{r})\nabla^2]\psi(\mathbf{r}).$$

Thus (3-110a) becomes

$$\left\{-\frac{\hbar^2}{2m}\nabla^2 + \frac{\ell^2}{16}[\nabla^2 V(\mathbf{r}) + 2\nabla\cdot V(\mathbf{r})\nabla + V(\mathbf{r})\nabla^2]\right\}\psi(\mathbf{r}) + V(\mathbf{r})\psi(\mathbf{r}) = E\psi(\mathbf{r}). \qquad (3\text{-}115)$$

A typical central potential used in single-particle model studies is one with a Fermi shape,[1] which we might express as $V(r) = -V_0 f(r)$, whereupon (3-115) becomes

$$-\frac{\hbar^2}{8}\left[\nabla^2\frac{1}{m^*(r)} + 2\nabla\cdot\frac{1}{m^*(r)}\nabla + \frac{1}{m^*(r)}\nabla^2\right]\psi(\mathbf{r}) - V_0 f(r)\psi(\mathbf{r}) = E\psi(\mathbf{r}), \qquad (3\text{-}116)$$

[1] The Fermi shape when used for potentials is referred to as the Woods–Saxon potential shape, $V(r) = -V_0[1+e^{(r-R)/a}]^{-1}$; see Chapter IV.

with $m^*(r) = m/[1+\beta f(r)]$ and $\beta = b^2 mV_0/2\hbar^2$. The masslike quantity $m^*(r)$ is referred to as the *effective mass* and is seen to vary from $m^*(0) = m/(1+\beta)$ in the central region, where $f(r) \approx 1$, to $m^*(r) = m$ in the tail of the potential and beyond, where $f(r) \approx 0$. In the central region of a nearly constant potential, (3-116) reduces to

$$-(\hbar^2/2m^*)\nabla^2\psi(\mathbf{r}) - V_0\psi(\mathbf{r}) = E\psi(\mathbf{r}), \qquad (3\text{-}117)$$

with $m^* \equiv m^*(0) = m/(1+\beta)$. For purposes of orientation we note that for more or less realistic values of $V_0 = 75$ MeV and $b = 0.85$ fm, we get $\beta = \frac{2}{3}$ and $m^* = \frac{3}{5}$ m. More detailed discussions will be given later for realistic values to use in (3-116). It should also be pointed out that in the surface region where $f(r)$ is rapidly varying, complicated kinetic energy corrections pertain. The possible simulation of these effects by the use of trial functions that are frequency- and amplitude-modulated plane waves as discussed in connection with (3-55) should be recalled.

▶**Exercise 3-20** Carry out the required steps leading from (3-113) to (3-115). Show that for a central potential of the form $V(r) = -V_0 f(r)$, (3-116) results. ◀

In an approach considering the possible form of $V(\mathbf{r}, \mathbf{p})$, with the possibility of small nonlocal effects in mind, one might express the "velocity"-dependent potential in a power series in the momentum $\mathbf{p}$. Of particular interest would be the form of $V(\mathbf{r}, \mathbf{p})$ in the central region of the nuclear interior or its behavior in infinite nuclear matter. In either of these cases the potential would be independent of the nucleon coordinate and since it must be composed of terms that are true scalars,

$$V(\mathbf{p}) = -V_0 + \sum_{n=1} a_n \mathbf{p}^{2n}. \qquad (3\text{-}118)$$

If the quadratic approximation is considered sufficient, $V(\tilde{\mathbf{p}})$ can be conveniently written

$$V(\tilde{\mathbf{p}}) = -V_0 + \beta\tilde{T}, \qquad (3\text{-}119)$$

where $\tilde{T}$ is the kinetic energy operator $\tilde{\mathbf{p}}^2/2m$. The Schrödinger equation then takes the form

$$\{\tilde{T}(1+\beta) - V_0\}\psi(\mathbf{r}) = E\psi(\mathbf{r}) \qquad \text{or} \qquad \{(\hbar^2/2m^*)\nabla^2 - V_0\}\psi(\mathbf{r}) = E\psi(\mathbf{r}) \qquad (3\text{-}120)$$

with $m^* = m/(1+\beta)$, and β thus has the same significance as before. Retaining terms in (3-115) to order b^2 is equivalent to the quadratic approximation (3-119) in regions of constant nuclear density.

2. The Bethe–Goldstone Equation

The theoretical investigation of the properties of "nuclear matter" as deduced from the nature of the nucleon–nucleon potential has been diligently pursued since the pioneering papers by Bethe and Goldstone (1956), Bethe (1956), Goldstone (1957), and Brueckner and Gammel (1958). A particularly lucid presentation of the basic points is contained in a paper by Gomes *et al.* (1958), which we shall closely follow here. Other thorough review papers are by Day (1967) and Bethe (1971).

The concept of "nuclear matter" is a highly idealized generalization involving a uniform density sea of an equal number of neutrons and protons. Coulomb forces are imagined "turned off." The object of such investigations is usually to demonstrate by a self-consistent calculation, using a reasonable characterization of the nucleon–nucleon potential, that a nuclear density close to that pertaining in the central region of real nuclei gives an energy minimum and to show that the resulting equilibrium energy density is close to the leading term of the semiempirical expression (3-75), namely $E_B/A \approx$ 15 MeV.

It is convenient to take as the starting point the simple single-particle Fermi gas model we discussed earlier. The system wave function would then consist of a Slater determinant of plane waves [recall (3-6) and (3-7)]. We assume that the single-particle potential,

$$V_i = \sum_{i \neq j} v_{ij}, \tag{3-121}$$

in infinite nuclear matter is independent of position $\mathbf{r}_i$ and therefore can depend at most only on the momentum $\mathbf{p}_i$ giving rise to a velocity-dependent potential, which in quadratic approximation is

$$V(\mathbf{r}_i, \mathbf{p}_i) = -V_0 + \beta(\mathbf{p}_i^2/2m). \tag{3-122}$$

The energy eigenvalue for one of the plane wave states, characterized by the set of quantum numbers γ, is

$$\varepsilon_\gamma = (\hbar^2 k_\gamma^2/2m)(1+\beta) - V_0 = (\hbar^2 k_\gamma^2/2m^*) - V_0 \tag{3-123}$$

with $m^* = m/(1+\beta)$ defining the effective mass as in (3-120). The total energy of the ground state of a system of A such nucleons of nuclear matter is

$$E = \sum_{\gamma=1}^{A} (T_\gamma + \tfrac{1}{2}V_\gamma) = (1+\tfrac{1}{2}\beta) \sum_{\gamma=1}^{A} (\tfrac{1}{2}\hbar^2 k_\gamma^2/m) - \tfrac{1}{2} \sum_{\gamma=1}^{A} V_0$$

or

$$E = A[\tfrac{3}{5}(1+\tfrac{1}{2}\beta)T_F - \tfrac{1}{2}V_0] = A\{\tfrac{3}{10}[1+(m/m^*)]T_F - \tfrac{1}{2}V_0\} \tag{3-124}$$

with $T_F = \hbar^2 k_F^2/2m$ and $k_F = (\frac{3}{2}\pi^2\rho)^{1/3}$. The quantity $\hbar k_F$ is again the Fermi momentum and T_F the characteristic Fermi energy discussed earlier in connection with the Fermi gas model.

In order to improve on this simple model, we must specify more carefully the character of $\tilde{v}_{ij}$, the nucleon–nucleon potential operative between a pair of nucleons immersed in a Fermi sea of nucleons. The nucleon sea, of course, furnishes the average single-particle potential in which each nucleon moves. In addition, the presence of the nucleons other than a specific pair under consideration prevents the interacting pair initially in states α and β from scattering into states occupied by other nucleons, a requirement imposed by the Pauli exclusion principle. While this principle requires the two particles at large separation distances to always have the well-defined momenta $\hbar\mathbf{k}_\alpha$ and $\hbar\mathbf{k}_\beta$, it does not prevent the wave functions from being distorted by the interaction at small distances.

The ordinary wave equation for an *isolated* pair of nucleons is

$$\{(\hbar^2/2m)(\nabla_1{}^2+\nabla_2{}^2)+\varepsilon\}\psi(\mathbf{r}_1,\mathbf{r}_2) = \tilde{v}(\mathbf{r}_1,\mathbf{r}_2)\psi(\mathbf{r}_1,\mathbf{r}_2) \qquad (3\text{-}125)$$

with $\varepsilon = (\hbar^2/2m)(k_\alpha{}^2+k_\beta{}^2)$, where $\mathbf{k}_\alpha$ and $\mathbf{k}_\beta$ are the asymptotic wave vectors in the initial state. The interaction potential $\tilde{v}(\mathbf{r}_1,\mathbf{r}_2)$ may induce an elastic scattering into states with asymptotic wave vectors $\mathbf{k}_\gamma$ and $\mathbf{k}_\delta$, requiring only the conservation of energy and momentum between the initial and final states. The cross section for such scattering is, in Born approximation,

$$\sigma(\mathbf{k}_\alpha\mathbf{k}_\beta;\mathbf{k}_\gamma\mathbf{k}_\delta) = \text{const}\cdot\left|\int\{\exp[-i(\mathbf{k}_\gamma\cdot\mathbf{r}_1+\mathbf{k}_\delta\cdot\mathbf{r}_2)]\}\,\tilde{v}(\mathbf{r}_1,\mathbf{r}_2)\psi(\mathbf{r}_1,\mathbf{r}_2)\,d^3r_1\,d^3r_2\right|^2. \qquad (3\text{-}126)$$

Scattering into states γ and δ is thus seen to depend on Fourier components of the right-hand side of (3-125). The Bethe–Goldstone wave equation is the modification of (3-125) to exclude the Fourier components from the right-hand side that would allow scattering into occupied states. For this purpose a projection operator $\tilde{Q}^{\mathrm{F}}_{\alpha\beta}$ is introduced which rejects all Fourier components other than α and β that are not outside the Fermi distribution. The wave equation thus becomes

$$\{(\hbar^2/2m)(\nabla_1{}^2+\nabla_2{}^2)+\varepsilon\}\psi_{\alpha\beta}(\mathbf{r}_1,\mathbf{r}_2) = \tilde{Q}^{\mathrm{F}}_{\alpha\beta}\{\tilde{v}(\mathbf{r}_1,\mathbf{r}_2)\psi_{\alpha\beta}(\mathbf{r}_1,\mathbf{r}_2)\}. \qquad (3\text{-}127)$$

Evidently $\tilde{Q}^{\mathrm{F}}_{\alpha\beta}$ can be defined for an arbitrary function $\phi(\mathbf{r}_1,\mathbf{r}_2)$ as[1]

$$\tilde{Q}^{\mathrm{F}}_{\alpha\beta}\phi(\mathbf{r}_1,\mathbf{r}_2) = \psi_\alpha(\mathbf{r}_1)\psi_\beta(\mathbf{r}_2)\langle\alpha\beta|\phi\rangle + \sum_{\substack{k_\gamma>k_{\mathrm{F}}\\ k_\delta>k_{\mathrm{F}}}}\psi_\gamma(\mathbf{r}_1)\psi_\delta(\mathbf{r}_2)\langle\gamma\delta|\phi\rangle, \qquad (3\text{-}128a)$$

[1] This equation is for two "unequal" nucleons (i.e., differing in mechanical spin or isospin). When the two are "equal" nucleons (e.g., both neutrons, spin up), $\psi_\alpha(\mathbf{r}_1)\psi_\beta(\mathbf{r}_2)$ is replaced by $(1/\sqrt{2})\{\psi_\alpha(\mathbf{r}_1)\psi_\beta(\mathbf{r}_2)-\psi_\alpha(\mathbf{r}_2)\psi_\beta(\mathbf{r}_1)\}$.

with, as usual,

$$\langle\alpha\beta|\phi\rangle = \int \psi_\alpha^*(\mathbf{r}_1)\,\psi_\beta^*(\mathbf{r}_2)\,\phi(\mathbf{r}_1,\mathbf{r}_2)\,d^3r_1\,d^3r_2, \tag{3-128b}$$

and the individual particle wave functions are of the form

$$\psi_\alpha(\mathbf{r}) = \Omega^{-1/2}\exp(i\mathbf{k}_\alpha\cdot\mathbf{r}), \quad \text{etc.} \tag{3-128c}$$

When the right-hand side of (3-127) is used in (3-126), all matrix elements for energetically allowed final states other than the initial state vanish. Hence there is no scattered wave and the asymptotic final state is the same as for $\tilde{v} = 0$ and consequently all phase shifts are zero. Clearly

$$\lim_{|\mathbf{r}_1-\mathbf{r}_2|\to\infty} \psi_{\alpha\beta}(\mathbf{r}_1,\mathbf{r}_2) \to \psi_\alpha(\mathbf{r}_1)\,\psi_\beta(\mathbf{r}_2) \tag{3-129a}$$

for unequal nucleons, or

$$\lim_{|\mathbf{r}_1-\mathbf{r}_2|\to\infty} \psi_{\alpha\beta}(\mathbf{r}_1,\mathbf{r}_2) \to 2^{-1/2}\{\psi_\alpha(\mathbf{r}_1)\,\psi_\beta(\mathbf{r}_2)-\psi_\alpha(\mathbf{r}_2)\,\psi_\beta(\mathbf{r}_1)\} \tag{3-129b}$$

for equal nucleons. For very small distances, however, where only the high Fourier components are relevant, $\psi(\mathbf{r}_1,\mathbf{r}_2)$ will, in fact, approach the solution to the isolated wave equation (3-125). The presence of $\tilde{Q}^{\mathrm{F}}_{\alpha\beta}$ in (3-127) is thus seen to allow only a modulation of the unperturbed wave function at small distances. This modulation of the unperturbed wave function can be viewed as a "wound" that becomes negligible for distances $|\mathbf{r}_1 - \mathbf{r}_2|$ larger than the "healing distance," which is roughly equal to $\lambda\!\!\!^{-}_{\mathrm{F}} = 1/k_{\mathrm{F}}$. Numerical examples presented in Gomes *et al.* (1958) and briefly discussed later bear out these points. Accepting these observations for the moment also allows for an understanding of the validity of the independent pair approximation for nuclear matter. We note that the average separation of two nucleons in nuclear matter is

$$d = \rho^{-1/3} = (\tfrac{3}{2}\pi^2)^{1/3}\lambda\!\!\!^{-}_{\mathrm{F}} \tag{3-130a}$$

or

$$d/\lambda\!\!\!^{-}_{\mathrm{F}} = k_{\mathrm{F}}d = (\tfrac{3}{2}\pi^2)^{1/3} = 2.45. \tag{3-130b}$$

Thus, if one of the partners of a pair of nucleons singled out for consideration makes a close encounter with yet a third nucleon, the average distance between the original pair is of order d. At that separation, in view of (3-130), the original pair wave function has almost completely "healed," and each member of the original pair is in a state closely approximating that prior to their interaction. Hence, the interaction with the third nucleon takes place under conditions similar to those that would pertain if the original pair had not interacted at all.[1]

[1] Possible evidence for the presence of a three-body force in the triton is discussed by Brayshaw (1973).

Finally, combining all effects, we can take the eigenvalue problem for a pair of nucleons in a Fermi sea to be given by the wave equation

$$[-(\hbar^2/2m)(\nabla_1{}^2+\nabla_2{}^2)+V(p_1)+V(p_2)-\varepsilon_{\alpha\beta}]\,\psi_{\alpha\beta} = -Q^{\mathrm{F}}_{\alpha\beta}\,v(1,2)\,\psi_{\alpha\beta}. \tag{3-131}$$

Here $V(p_1)$ and $V(p_2)$ are the effective single-particle potentials and $v(1,2)$ is the free nucleon–nucleon potential. Using the quadratic or effective mass approximation (3-122), this becomes

$$[-(\hbar^2/2m^*(\nabla_1{}^2+\nabla_2{}^2)-2V_0-\varepsilon_{\alpha\beta}]\,\psi_{\alpha\beta} = -Q^{\mathrm{F}}_{\alpha\beta}\,v(1,2)\,\psi_{\alpha\beta}. \tag{3-132}$$

The energy eigenvalue can be written, noting that $Q^{\mathrm{F}}_{\alpha\beta}\psi_{\alpha\beta}=\psi_{\alpha\beta}$,

$$\varepsilon_{\alpha\beta} = \langle\psi_{\alpha\beta}|-(\hbar^2/2m^*)(\nabla_1{}^2+\nabla_2{}^2)-2V_0+v(1,2)\,|\psi_{\alpha\beta}\rangle. \tag{3-133}$$

The energy "correction" $V_{\alpha\beta}$ caused by the interaction is the difference between (3-133) and that when $v(1,2)=0$; hence

$$V_{\alpha\beta} = \varepsilon_{\alpha\beta} - [(\hbar^2/2m^*)(k_\alpha{}^2+k_\beta{}^2)-2V_0] \tag{3-134a}$$

or

$$V_{\alpha\beta} = \langle\psi_{\alpha\beta}|-(\hbar^2/2m^*)(\nabla_1{}^2+\nabla_2{}^2+k_\alpha{}^2+k_\beta{}^2)+v(1,2)\,|\psi_{\alpha\beta}\rangle. \tag{3-134b}$$

Thus, once we have found the solution to (3-132),[1] we can determine $V_{\alpha\beta}$ using (3-134), and from this a quantity

$$V'(p_\alpha) = \sum_{\beta\neq\alpha} V_{\alpha\beta}. \tag{3-135}$$

Unless $V(p)$ in (3-122) was correctly chosen, $V'(p_\alpha)$ of (3-135) will not be equal to $V(p_\alpha)$, as of course self-consistency would require. If an incorrect value of m^* were used, $V'(p,m^*)$ might not even be a quadratic function of p. A best parabolic fit to $V'(p,m^*)$ might be attempted, which then results in the following equation for a better value of m^*:

$$\frac{1}{m^*} = \frac{1}{m} + \left[\frac{1}{p}\frac{\partial V'(p,m^*)}{\partial p}\right]_{p=p_{\mathrm{F}}}. \tag{3-136}$$

The slope of $V'(p,m^*)$ is evaluated at p_{F} since the major perturbations to $\psi_{\alpha\beta}$ come from Fourier components just above the Fermi momentum.

Having arrived at a self-consistent solution to the above, we can calculate the total energy of a subsystem of A nucleons from (3-124) either through the determined value of m^* or directly from $V_\gamma = V(p_\gamma)$ given by (3-135) and using

[1] It will be noted that solutions to the wave equations (3-132) and (3-127) are related through simple obvious factors since the only differences are the use of the effective mass m^* in (3-132) instead of m and the shift of the energy reference by $2V_0$.

the basic equation

$$E = \sum_{\gamma=1}^{A} T_\gamma + \tfrac{1}{2} \sum_{\gamma,\delta;\, \gamma\neq\delta}^{A} v_{\gamma\delta} = \sum_{\gamma=1}^{A} (T_\gamma + \tfrac{1}{2} V_\gamma).$$

The resulting value of E is still a function of the assumed value of ρ (or k_F). The ground state is presumed to correspond to the value of ρ that gives E its minimum value. It goes without saying that the resulting equilibrium values of ρ and E also depend on the validity of the representation of the free space nucleon–nucleon potential $v(1,2)$ used in (3-132) and elsewhere.

The actual carrying out of the solution of the Bethe–Goldstone equation (3-132) or (3-127) is quite complicated. A simple solution for the approximately valid conditions of taking the momentum *of the center of mass* of the two nucleons $\mathbf{p} = \hbar(\mathbf{k}_\alpha + \mathbf{k}_\beta)$ to be zero and considering only *relative* orbital angular momentum $l = 0$ is given by Gomes *et al.* (1958). They also take $v(1,2) \equiv v(\mathbf{r})$, $\mathbf{r} = \mathbf{r}_1 - \mathbf{r}_2$, in the S state to be a square-well attractive potential with a repulsive core, viz.,

$$\begin{aligned} v(r) &= +\infty, & r &< c \\ v(r) &= -v_0, & c < r &< b + c \\ v(r) &= 0, & r &> b + c. \end{aligned} \tag{3-137}$$

Specifically, they use $c = 0.4$ fm, $b = 1.9$ fm, and a value of v_0 that just produces binding for two nucleons in free space $v_0 = \hbar^2\pi^2/4m_N b^2 = 28.3$ MeV. Figure 3.22 shows their results. Figure 3.22b contrasts the effects of the potential in free space and in nuclear matter. For small separations, $u_i(r)$ (isolated or free-space two-nucleon relative wave function) $\approx u(r)$ (embedded two-nucleon relative wave function). For $k_F r > k_F d$ and $u_0(r)$ corresponding to no interaction, i.e., $v(r) = 0$, the quantity $g(r) = u(r) - u_0(r)$ is seen to converge rapidly to zero while, of course, the quantity $g_i(r) = u_i(r) - u_0(r)$ approaches a constant-amplitude sinusoid

$$g_i(r) = \sin(kr + \delta_0) - \sin kr = 2 \sin(\tfrac{1}{2}\delta_0) \cos(kr + \tfrac{1}{2}\delta_0),$$

appropriate for the S-wave phase shift δ_0. These calculations are thus seen to confirm the statements made in connection with (3-130), and contrasting $g_i(r)$ with $g(r)$ beyond $k_F d$ indeed shows that $u(r)$ has substantially "healed" by the separation corresponding to d, the average nucleon–nucleon separation in nuclear matter.

The authors (Gomes *et al.*, 1958) concluded that the features of $v(1,2)$ producing the above behavior require that v_0 be weak (in the sense of just allowing binding, as for the deuteron) and that $b \gg c$, i.e., the attractive portion of the well be of significantly longer range than the repulsive core. The repulsive core was necessary in the first place to give saturation for other than

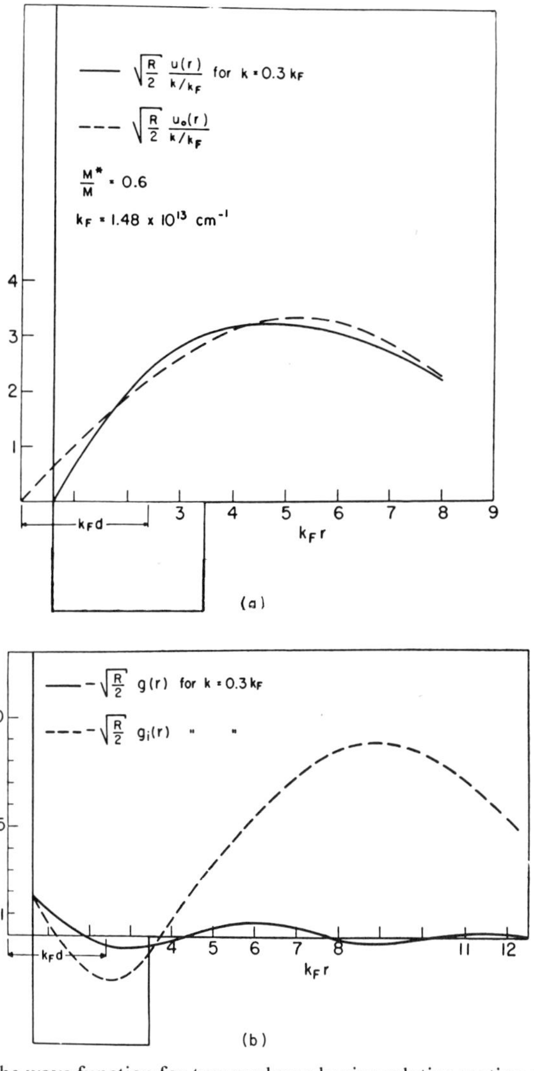

FIG. 3.22. (a) The wave function for two nucleons having relative motion with $l = 0$. The wave $u_0(r)$ corresponds to no interaction, while $u(r)$ is for the interaction (3-137) and for the pair embedded in a Fermi sea with $k_F = 1.48\ \text{fm}^{-1}$ ($\rho = 0.217\ \text{fm}^{-3}$). The relative momentum $\frac{1}{2}|\mathbf{k}_\alpha - \mathbf{k}_\beta| = k$ is $k = 0.3k_F$. The dimensionless quantity $k_F d = 2.45$ is also indicated. (b) The difference in wave functions $u(r) - u_0(r) = g(r)$ shown in (a) and also the difference $u_i(r) - u_0(r) = g_i(r)$, where $u_i(r)$ is the wave function for an isolated pair under the interaction (3-137). This latter wave function is not shown in (a). Again $k_F d$ is also shown [figures taken from Gomes *et al.* (1958)].

the collapsed state since they (Gomes *et al.*, 1958) selected a "Serber force" for $v(r)$, i.e.,

$$\tilde{v}(r)\,(\text{any } l) = -\tfrac{1}{2}(1+P^{\mathrm{M}})\,v_0, \qquad c < r < b + c, \tag{3-138}$$

which vanishes for l odd and is $-v_0$ for l even. They also found that the potential (3-138) leads to an equilibrium density very close to the realistic value $\rho = 0.170\ \mathrm{fm}^{-3}$, giving $k_{\mathrm{F}} = 1.36\ \mathrm{fm}^{-1}$. However, the resulting value of[1] $m^*/m \approx 2/3$ leads to a value of $E/A = -6.9$ MeV rather than the ~ -15 MeV given by the semiempirical mass expression.

Since the early papers of Brueckner and collaborators [e.g., Brueckner and Gammel (1958) and Brueckner *et al.* (1961)], many contributions have appeared in the literature; noteworthy for nuclear matter characteristics are the calculations of Brueckner *et al.* (1968), Kallio and Day (1969), Siemans (1970), and Sprung *et al.* (1970).

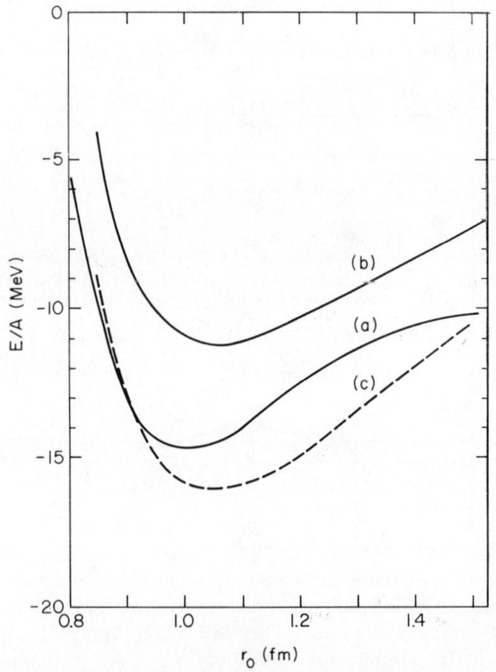

FIG. 3.23. Total energy per particle in nuclear matter as a function of r_0 (see text). (a) After Brueckner and Gammel (1958), (b) after Siemens and Sprung, quoted in Bethe (1971), (c) after Brueckner *et al.* (1968).

[1] This value of $m^*/m \approx \frac{2}{3}$ is deduced from using (3-136) evaluated at $p = p_{\mathrm{F}}$. A value $m^*/m \approx \frac{3}{5}$ results from an overall best fit.

Figure 3.23 shows the binding energy per particle in nuclear matter as a function of a radius parameter r_0 defined as the reduced radius or average volume per particle [i.e., $\Omega/A = \frac{4}{3}\pi r_0^3$, with Ω the nuclear volume, and also, clearly, $r_0 k_F = (9\pi/8)^{1/3}$]. Partly for the sake of historical interest, the earlier results of Brueckner and Gammel (1958) are compared with the later Brueckner *et al.* semiempirical results (1968) and the energy functional of Siemens and Sprung quoted in Bethe (1971). This latter result is given by a simple expression in terms of the Fermi momentum k_F, namely,

$$E(k_F) = -15.07 k_F^3 + 7.85 k_F^4. \tag{3-139a}$$

The energy functional (3-139a) is actually an empirical interpolation formula fit to a detailed microscopic calculation using an accurate nucleon–nucleon interaction. The maximum binding energy per particle is 11.25 MeV and is thus substantially lower than the expected $a_v = (15 \pm 1)$ MeV. This stationary value also yields $k_F = 1.44\ \mathrm{fm}^{-1}$ or $\rho_0 = 0.20\ \mathrm{fm}^{-3}$. Other results obtained in this calculation, which closely match those of Kallio and Day, were discussed in Section III, B, 4.

Curve (c) of Fig. 3.23 is a semiempirical energy-density functional, developed by Brueckner *et al.* (1968), based on extensive nuclear matter calculations adjusted to give an acceptable value for a_v at the extremum value of ρ (i.e., $\rho = \rho_0$). Expressed in terms of the density, the expression is

$$E(\rho) = 75.06\rho^{2/3} - 741.28\rho + 1179.89\rho^{4/3} - 467.54\rho^{5/3}. \tag{3-139b}$$

The maximum binding occurs for $\rho = \rho_0 = 0.204\ \mathrm{fm}^{-1}$ (i.e., $k_F = 1.45\ \mathrm{fm}^{-1}$ or $r_0 = 1.05$ fm), yielding $E/A = -16.57$ MeV.

There have been many attempts to calculate the nuclear energy for finite nuclei within a Brueckner or Bethe–Goldstone framework but including Coulomb and isotopic effects. For references to the work of Brueckner and his colleagues see Brueckner *et al.* (1968, 1970, 1971). These calculations consider deformed nuclei as well as the spherically symmetric nuclear examples and carry $I = N - Z$ to values $I_0 \pm 40$ (with I_0 the β-stability value). The effect of introducing $I \neq 0$ is to shift the energy minima in curves such as those shown in Fig. 3.23 to lower densities as I/A increases from zero to about 3/5. This behavior, stated in terms of the shift in the equilibrium value for k_F, is

$$k_F(I) \approx k_F(0)[1 - 0.48(I^2/A^2)].$$

In a parallel development Bethe and his colleagues use local density approximations and a Thomas–Fermi kinetic energy. Earlier references that also give useful comparisons to previous infinite nuclear matter calculations are Bethe (1968) and Németh and Bethe (1968). Realistic values are used for the nucleon–nucleon potential, although effective central potentials are introduced to simulate the behavior of neglected tensor forces. These treatments also attempt to allow for surface effects. The results conform to a central density of 0.170 fm^{-3} and a volume term in the binding energy of 15.8 MeV per particle. A surface energy term of $19.5A^{2/3}$ MeV and a surface thickness of ~ 2.2 fm are deduced, in reasonable agreement with empirical values. While the calculated density distribution is similar to the Fermi distribution discussed in Chapter II, some notable differences are found. The steepest slope occurs not at $\rho = \frac{1}{2}\rho_0$ (ρ_0 is the central density) but at the lower value $\rho \approx 0.31\rho_0$ or perhaps even lower. The resulting "asymmetric" density also approaches the central density ρ_0 more slowly than the Fermi distribution. Calculated parameters to fit the electron scattering data are nonetheless quite as successful as the values derived based on the Fermi shape.

These calculations avoid the use of the effective mass approximation because of the ensuing difficulties in the surface region discussed in connection with (3-116). Nonetheless a value of $m^*/m \approx \frac{5}{8}$ is indicated for the central region. An interesting parameter that enters these calculations is κ, which gives the probability of finding a nucleon out of its normal state inside the Fermi sea. A value of $\kappa \approx 0.15 \pm 0.02$ seems to be indicated. The "wound" in the two-nucleon wave function is intimately associated with this parameter (see Section IV, F).

The curvature of the shape of the volume energy as a function of the density in the vicinity of the equilibrium value defines the nuclear compressibility K, through $KA = [R^2\, \partial^2 E_v/\partial R^2]_{R=R_1}$, where R_1 is the effective constant-density radius. For infinite nuclear matter the total nuclear energy is just "volume" energy and generally K relates to such idealized infinite nuclear matter calculations. Coulomb and isotopic effects are usually considered separately. Literature values of compressibility range from $K \approx 130$ to 190 MeV. As discussed in Chapter II under the isotope shift, the nuclear matter "stiffness" in the surface region might be considerably smaller than this value would indicate. The total energy as a function of the density in the vicinity of the equilibrium density ρ_0, according to Bethe and co-workers, is given by the semiempirical expression

$$E/A = -15.85 + 47.5y^2 + 32.5y^3 \tag{3-139c}$$

with $y = (\rho/\rho_0)^{1/2} - 1$.

It is also found that the Thomas–Fermi expression for the kinetic energy is valid in regions where the density is greater than one-sixth the central density and that the simple mixed density expression (3-32) is also valid in the same region.

The reader interested in more details as well as a review of the general field of nuclear matter calculations is advised to refer to the excellent review article by Bethe (1971). A very informative paper discussing various modern approaches to the many-body finite nucleus problem is that by Shao and Lomon (1973).

▶**Exercise 3-21** As discussed above, the curvature of the functional for total volume energy per nucleon (i.e., E/A for infinite nuclear matter) defines the nuclear compressibility. For infinite nuclear matter we can write

$$KA = \left[k_F{}^2 \frac{d^2E}{dk_F{}^2}\right]_0 = \left[r_0{}^2 \frac{d^2E}{dr_0{}^2}\right]_0 = 9\left[\rho \frac{d^2E}{d\rho^2}\right]_0. \tag{3-140}$$

Show that $K = 134$ MeV for (3-139a). Also calculate K for (3-139b) and for (3-139c) in terms of ρ_0.

For some additional remarks on nuclear compressibility for infinite nuclear matter, see Khanna and Barhai (1973). ◀

In conclusion, something should be said about recent studies involving a relativistic OBEM treatment of complex nuclei. Infinite nuclear matter calculations have been performed using regularized OBE potentials with π, η, δ, ρ, ω, and the Ansatz σ_c mesons (also in some cases a weakly coupled ε meson) [e.g., see Holinde *et al.* (1972) and Lee and Tabakin (1972)]. It appears that satisfactory saturation and binding require adequately strong intermediate- and short-range tensor contributions and the proper inclusion of meson retardation effects. Results not very much at variance with the theories discussed earlier typically suggest $E_B/A \approx 12.4$ MeV at a saturation value of $k_F = 1.55$ fm^{-1}, a compressibility modulus $K = 170$ MeV, a defect parameter $\kappa = 0.11$, and an effective mass $m^*/m = 0.61$.

A complete Dirac relativistic Hartree–Fock calculation has been attempted for ^{16}O, ^{40}Ca, and ^{48}Ca [see Miller (1974)]. This work also confirms the importance of including meson retardation effects to prevent nuclear collapse, and reveals some ambiguities in the proper formulation of a relativistic Hartree–Fock problem.

References

Alexander, J. M., and Bimbot, R. (1972). *Phys. Rev.* **C5,** 799.
Anantaraman, N., and Schiffer, J. P. (1971). *Phys. Lett.* **37B,** 229.
Anderson, B. L. (1966). *In* "Isobaric Spin in Nuclear Physics" (J. D. Fox and D. Robson, eds.). Academic Press, New York.
Anderson, B. L., Dickmann, F., and Dietrich, K. (1970). *Nuclear Phys.* **A159,** 337.
Arima, A., and Gillet, V. (1971). *Ann. Phys.* **66,** 117.
Auerbach, N., and Lev, A. (1972). *Nuclear Phys.* **A180,** 651.
Ayres, R., Hornyak, W. F., Chan, L., and Fann, H. (1962). *Nuclear Phys.* **29,** 212.
Back, B. B., Britt, H. C., Garrett, J. D., and Hansen, O. (1972). *Phys. Rev. Lett.* **28,** 1707.
Barshay, S., and Brown, G. E. (1973). *Phys. Lett.* **47B,** 107.
Bethe, H. A. (1956). *Phys. Rev.* **103,** 1353.
Bethe, H. A. (1968). *Phys. Rev.* **167,** 879.
Bethe, H. A. (1971). *Ann. Rev. Nuclear Sci.* **21,** 93.
Bethe, H. A., and Bacher, R. F. (1936). *Rev. Modern Phys.* **8,** 82.
Bethe, H. A., and Goldstone, J. (1956). *Proc. Roy. Soc. (London).* **A238,** 551.

Bjørnholm, S., Borggreen, J., Westgaard, L., and Karnaukhov, V. A. (1967). *Nuclear Phys.* **A95,** 513.
Blatt, J. M., and Weisskopf, V. F. (1952). "Theoretical Nuclear Physics." Wiley, New York.
Bohr, N., and Wheeler, J. A. (1939). *Phys. Rev.* **56,** 426.
Bolsterli, M., Fiset, E. O., Nix, J. R., and Norton, J. L. (1972). *Phys. Rev.* **C5,** 1050.
Brayshaw, D. D. (1973). *Phys. Rev.* **C7,** 1731.
Brink, D. M., and Castro, J. J. (1973). *Nuclear Phys.* **A216,** 109.
Britt, H. C., Burnett, S. C., Erkkila, B. H., Lynn, J. E., and Stein, W. E. (1971). *Phys. Rev.* **C4,** 1444.
Britt, H. C., Bolsterli, M., Nix, J. R., and Norton, J. L. (1973). *Phys. Rev.* **C7,** 801.
Brueckner, K. A., and Gammel, J. L. (1958). *Phys. Rev.* **109,** 1023.
Brueckner, K. A., Lockett, A. M., and Rotenberg, M. (1961). *Phys. Rev.* **121,** 255.
Brueckner, K. A., Buchler, J. R., Jorna, S., and Lombard, R. J. (1968). *Phys. Rev.* **171,** 1188.
Brueckner, K. A., Clark, R. C., Lin, W., and Lombard, R. J. (1970). *Phys. Rev.* **C1,** 249.
Brueckner, K. A., Chirico, J. H., and Meldner, H. W. (1971). *Phys. Rev.* **C4,** 732.
Bunatian, G. G., Kolomietz, V. M., and Strutinsky, V. M. (1972). *Nuclear Phys.* **A188,** 225.
Calogero, F., Simonov, Y. A., and Surkov, E. L. (1972). *Phys. Rev.* **C5,** 1493.
Campi, X., and Sprung, D. W. (1972). *Nuclear Phys.* **A194,** 401.
Chandra, H., and Kustgi, M. L. (1973). *Phys. Rev.* **C7,** 180.
Coester, F., Cohen, S., Day, B., and Vincent, C. M. (1970). *Phys. Rev.* **C1,** 769.
Condon, E. U., and Shortley, G. H. (1935). "The Theory of Atomic Spectra." Cambridge Univ. Press, London and New York.
Dahll, G., and Warke, C. (1970). *Nuclear Phys.* **A147,** 94.
Day, B. D. (1967). *Rev. Modern Phys.* **39,** 719.
Diehl, H., and Greiner, W. (1973). *Phys. Lett.* **45B,** 35.
Ehlers, J. W., and Moszkowski, S. A. (1972). *Phys. Rev.* **C6,** 217.
Fermi, E. (1949). "Nuclear Physics." Univ. of Chicago Press, Chicago, Illinois.
Fiset, E. O., and Nix, J. R. (1972). *Nuclear Phys.* **A193,** 647.
Fox, J. D., and Robson, D., eds. (1966). "Isobaric Spin in Nuclear Physics." Academic Press, New York.
Garvey, G. T. (1969). *Ann. Rev. Nuclear Sci.* **19,** 433.
Goldstone, J. (1957). *Proc. Roy. Soc.* (*London*). **A239, 267**.
Gomes, L. C., Walecka, J. D., and Weisskopf, V. F. (1958). *Ann. Phys.* **3,** 241.
Green, A. E. S., and Edwards, D. F. (1953). *Phys. Rev.* **91,** 46.
Green, A. E. S., and Engler, N. A. (1953). *Phys. Rev.* **91,** 40.
Green, A. M., and Schucan, T. H. (1972). *Nuclear Phys.* **A188,** 289.
Griffin, J. J. (1968). *Phys. Rev. Lett.* **21,** 826.
Halpern, I. (1971). *Ann. Rev. Nuclear Sci.* **21,** 245.
Holinde, K., Erkelenz, K., and Alzetta, R. (1972). *Nuclear Phys.* **A198,** 598.
Hulet, E. K., Wild, J. F., Lougheed, R. W., Evans, J. E., Qualheim, B. J., Nurmia, M., and Ghiorso, A. (1971). *Phys. Rev. Lett.* **26,** 523.
Jägare, S. (1969). *Nuclear Phys.* **A137,** 241.
Jänecke, J. (1965). *Nuclear Phys.* **61,** 326.
Jänecke, J. (1966). *Phys. Rev.* **147,** 735.
Jänecke, J. (1968). *Nuclear Phys.* **A114,** 433.
Jänecke, J., and Behrens, H. (1974). *Phys. Rev.* **C9,** 1276.
James, G. D., Lynn, J. E., and Earwaker, L. G. (1972). *Nuclear Phys.* **A189,** 225.
Kallio, A., and Day, B. D. (1969). *Nuclear Phys.* **A124,** 177.
Kashy, E., Benenson, W., and Nolen, J. A. (1974). *Phys. Rev.* **C9,** 2102.
Khanna, K. M., and Barhai, P. K. (1973). *Nuclear Phys.* **A215,** 349.

Köhler, H. S. (1971). *Nuclear Phys.* **A162**, 385; **A170**, 88.
Koltun, D. S. (1972). *Phys. Rev. Lett.* **28**, 182.
Koltun, D. S. (1973). *Ann. Rev. Nuclear Sci.* **23**, 163.
Kuo, T. T. S., and Brown, G. E. (1966). *Nuclear Phys.* **85**, 40.
Lederer, C. M., Hollander, J. M., and Perlman, I. (1968). "Table of Isotopes." Wiley, New York.
Ledergerber, T., and Pauli, H. C. (1973). *Nuclear Phys.* **A207**, 1.
Lee, H. T., and Tabakin, F. (1972). *Nuclear Phys.* **A191**, 332.
Ludwig, S., von Groote, H., Hilf, E., Cameron, A. G. W., and Truran, J. (1973). *Nuclear Phys.* **A203**, 627.
Mattauch, J. H. E., Thiele, W., and Wapstra, A. H. (1965). *Nuclear Phys.* **67**, 1.
Mayer, M. G., and Jensen, J. H. (1955). "Elementary Theory of Nuclear Shell Structure." Wiley, New York.
Merzbacher, E. (1970). "Quantum Mechanics." Wiley, New York.
Migdal, A. B. (1973). *Phys. Lett.* **47B**, 96.
Migneco, E., and Theobald, J. P. (1968). *Nuclear Phys.* **A112**, 603.
Miller, L. D. (1972). *Phys. Rev. Lett.* **28**, 1281.
Miller, L. D., (1974). *Phys. Rev.* **C9**, 537.
Mosel, U., and Schmitt, H. W. (1971). *Phys. Rev.* **C4**, 2185.
Moszkowski, S. A. (1970). *Phys. Rev.* **C2**, 402.
Mozér, F. S. (1959). *Phys. Rev.* **116**, 970.
Mueller, G. P. (1973). *Nuclear Phys.* **A210**, 544.
Mustafa, M. G., and Schmitt, H. W. (1973). *Phys. Rev.* **C8**, 1924.
Muthukrishnan, R. (1971). *Phys. Rev.* **C3**, 2091.
Myers, W. D. (1970). *Nuclear Phys.* **A145**, 387.
Myers, W. D., and Swiatecki, W. J. (1966). *Nuclear Phys.* **81**, 1.
Nann, H., Benenson, W., Kashy, E., and Turek, P. (1974). *Phys. Rev.* **C9**, 1848.
Negele, J. W. (1970). *Phys. Rev.* **C1**, 1260.
Németh, J., and Bethe, H. A. (1968). *Nuclear Phys.* **A116**, 241.
Németh, J., and Ripka, G. (1972). *Nuclear Phys.* **A194**, 329.
Nix, J. R. (1972). *Ann. Rev. Nuclear Sci.* **22**, 65.
Pandharipande, V. R., and Prasad, K. G. (1970). *Nuclear Phys.* **A147**, 193.
Randrup, J., Tsang, C. F., Möller, P., Nilsson, S. G., and Larsson, S. E. (1974). *Nuclear Phys.* **A217**, 221.
Robertson, R. G. H., Martin, S., Falk, W. R., Ingham, D., and Djaloeis, A. (1974). *Phys. Rev. Lett.* **32**, 1207.
Rodberg, L. S., and Teplitz, V. L. (1960). *Phys. Rev.* **120**, 969.
Seeger, P. A. (1961). *Nuclear Phys.* **25**, 1.
Shao, J., and Lomon, E. L. (1973). *Phys. Rev.* **C8**, 53.
Sharp, R. W., and Zamick, L. (1973). *Nuclear Phys.* **A208**, 130.
Siemens, P. J. (1970). *Phys. Rev.* **C1**, 98.
Siemens, P. J. (1970). *Nuclear Phys.* **A141**, 225.
Signell, P. (1970). *Phys. Rev.* **C2**, 1171.
Skyrme, T. H. R. (1959). *Nuclear Phys.* **9**, 615.
Sobiczewski, A., Bjørnholm, S., and Pomorski, K. (1973). *Nuclear Phys.* **A202**, 274.
Sprung, D. W. L., Banerjee, P. K., Jopko, A. M., and Srivastava, M. K. (1970). *Nuclear Phys.* **A144**, 245.
Strutinsky, V. M. (1967) *Nuclear Phys.* **A95**, 420.
Swamy, N. V. V. J., and Green, A. E. S. (1958). *Phys. Rev.* **112**, 1719.
Swiatecki, W. J. (1956). *Phys. Rev.* **101**, 651; **104**, 993.

Swiatecki, W. J., and Bjørnholm, S. (1972). *Phys. Rep.* **4C**, 325.
Tabakin, F. (1964). *Ann. Phys.* **30**, 51.
Treutelman, G. F., Preedom, B. M., and Kashy, E. (1971). *Phys. Rev.* **C3**, 2205.
Turner, R. J., and Jackson, A. D. (1972). *Nuclear Phys.* **A192**, 200
Vautherin, D., and Brink, D. M. (1972). *Phys. Rev.* **C5**, 626.
von Weizsäcker, C. F. (1935). *Z. Phys.* **96**, 431.
Wapstra, A. H., and Gove, N. B. (1971). *Nucl. Data Tables* **9**, 265.
Weinberg, A. M., and Wigner, E. (1959). "The Physical Theory of Neutron Chain Reactors." Univ. of Chicago Press, Chicago, Illinois.
Wigner, E. P. (1937). *Phys. Rev.* **51**, 106, 947.
Wigner, E. P., and Seitz, F. (1933). *Phys. Rev.* **43**, 804.
Zamick, L. (1973). *Phys. Lett.* **45B**, 313.
Zeldes, N. Grill, A., and Simievic, A. (1967). *Mat. Fys. Skr. Dan. Vid. Selsk.*, **3**, No. 5.

Chapter IV

SINGLE-PARTICLE SHELL MODEL

A. INTRODUCTION

One of the more successful simple nuclear models that has provided a reasonable accounting for many observed properties of nuclides is the single-particle shell model. According to this model, the nucleons within a nucleus move in individual particle orbitals guided by a relatively simple one-body potential field extending throughout a region roughly conforming to the nuclear volume. The potential is imagined to be the result of an *averaging* procedure, accounting for all the pairwise interactions any individual nucleon makes. The specific details of any particular pair interaction are generally ignored in first approximation, leading to the so-called *extreme* single-particle model. In this form of the model a one-body potential

$$\tilde{V}(\mathbf{r}_i, \boldsymbol{\sigma}_i, \boldsymbol{\tau}_i) \equiv \left\langle \sum_{i \neq j}^{A} \tilde{v}_{ij} \right\rangle \tag{4-1}$$

is used to generate a set of eigenvalue wave equations

$$(\tilde{T}_i + \tilde{V}_i)\,|\psi_i\rangle = \varepsilon_i\,|\psi_i\rangle. \tag{4-2}$$

The set $|\psi_i\rangle$ constitutes the proper single-particle wave functions to describe the individual equations of motion having various quantum numbers associated with the constants of the motion. Even when specific pair interactions are "turned on" as perturbations, these single-particle wave functions form a useful set of basis states.

At first glance the single-particle model appears to be completely unjustifiable in view of the known "strong" character of the nucleon–nucleon forces. How can individual nucleons in any sense be in stationary states $|\psi_i\rangle$ when we realize that the average spacing $d = \rho_0^{-1/3} = 1.80$ fm in the nuclear interior is nearly the same as the effective range of nuclear forces? We have seen in Chapter III the strong inhibiting influence of the Pauli principle in suppressing the probability that a pair interaction would scatter either member out of the state $|\psi_{ij}\rangle$. The pair wave function was seen to "heal" rather effectively to $\psi_{\alpha\beta} \approx \psi_\alpha \psi_\beta$[1] when $|\mathbf{r}_1 - \mathbf{r}_2| \gtrsim d$. Hence, in spite of frequent interactions, the wave functions are very similar to those for free particles except when two nucleons are close together. Paradoxically, the condition required for healing the pair wave function at the average nuclear separation is, in fact, that the nucleon–nucleon potential be weak. In another context, we already know that the attractive portion of the internucleon potential is just strong enough to yield a single bound state for the triplet spin interaction and yields no bound state for the singlet interaction.

The concept of "interaction strength" requires further consideration. An interacting pair, at a separation less than the range of nuclear forces possessing an attractive mean depth v_0, will have an impressed wave number $k_0^2 = mv_0/\hbar^2$ (i.e., $K^2 = k^2 + k_0^2$) modulating $|\psi_i\rangle$. If the wave number k_0 is less than the Fermi wave number k_F, the attractive potential will not be very effective in inducing a scattering outside the Fermi distribution and the effect will be confined to a rather short-range modulation of $|\psi_i\rangle$. A "strength factor" $\not{f}$ can thus be introduced, defined as

$$\not{f} = \frac{k_0^2}{k_F^2} = \frac{mv_0}{\hbar^2 k_F^2} = \frac{md^2 v_0}{\hbar^2}\left(\frac{2}{3\pi^2}\right)^{2/3}. \tag{4-3}$$

For nuclear matter we have $\not{f} \approx \frac{1}{3}$. By contrast, it is instructive to consider another system of fermions in its ground state, namely a sample of solid nitrogen at zero temperature.[2] If we substitute the approximate empirical values $d \approx 10^{-8}$ cm and $v_0 \approx 7$ eV into (4-3), we get $\not{f} \approx 4 \times 10^3$. The resulting impressed long-range modulation produces a crystalline behavior for solid nitrogen rather than the behavior appropriate for a Fermi gas. Thus, in the sense of (4-3), internuclear forces might be considered "weak" when contrasted to interatomic forces.

[1] For "unequal" nucleons.

[2] The electronic ground state of the nitrogen atom is a $^4S_{3/2}$ state and the ^{14}N atom as a whole is a fermion.

In conclusion one sees that in spite of the short-range nature and strength of nuclear forces an overall smooth average potential can be used to represent the actual potential acting on a nucleon, confining the local effects to still shorter range, weak modulations of the resulting wave functions. Such smooth potentials with a variety of shapes have been used with some success.

B. SINGLE-PARTICLE POTENTIALS

The starting point of many theories involves a reference to single-particle states or orbitals. It is useful therefore to give a brief discussion of the more commonly used single-particle potential shapes and the characteristics of corresponding orbitals. We confine our attention to the harmonic oscillator, square-well, and the so-called Woods–Saxon potentials.

Important modifying effects, in addition to the potential shape, involve such factors as the spin–orbit interaction and other velocity-dependent potential terms. These and other relevant topics are discussed in subsequent sections.

1. Harmonic Oscillator Potential

We will assume familiarity with the elementary characteristics of the harmonic oscillator well, $V(r) = -V_0 + \frac{1}{2}m\omega^2 r^2$, and will review briefly only a few relevant features. The Schrödinger equation for the oscillator is separable in Cartesian, cylindrical, or spherical coordinates. In Cartesian coordinates, we have

$$-(\hbar^2/2m)\nabla^2\psi + \tfrac{1}{2}m\omega^2(x^2+y^2+z^2)\psi = (E+V_0)\psi. \tag{4-4}$$

Using the separation constants E_q, E_r, and E_s, defining

$$H_x = (\tfrac{1}{2}p_x{}^2/m) + \tfrac{1}{2}m\omega^2 x^2,$$

and similarly H_y and H_z, and writing $\psi_{qrs}(x,y,z) = \psi_q(x)\psi_r(y)\psi_s(z)$, we have

$$\begin{aligned} H_x\psi_q(x) &= E_q\psi_q(x) = (q+\tfrac{1}{2})\hbar\omega\psi_q(x) \\ H_y\psi_r(y) &= E_r\psi_r(y) = (r+\tfrac{1}{2})\hbar\omega\psi_r(y) \\ H_z\psi_s(z) &= E_s\psi_s(z) = (s+\tfrac{1}{2})\hbar\omega\psi_s(z). \end{aligned} \tag{4-5}$$

Also

$$E + V_0 = E_q + E_r + E_s \qquad \text{or} \qquad E = -V_0 + (q+r+s+\tfrac{3}{2})\hbar\omega. \tag{4-6}$$

The harmonic oscillator quantum number $n = q+r+s$ can be introduced giving $E_n = -V_0 + (n+\frac{3}{2})\hbar\omega$. The total wave function becomes

$$|q,r,s\rangle = \left(\frac{m\omega}{\hbar\pi}\right)^{3/4} \frac{\exp[-\alpha(x^2+y^2+z^2)]}{(q!\,r!\,s!\,2^n)^{1/2}} \times H_q[(2\alpha)^{1/2}x]\, H_r[(2\alpha)^{1/2}y]\, H_s[(2\alpha)^{1/2}z], \tag{4-7}$$

with $\alpha = m\omega/2\hbar$. The $H_n(\xi)$ are the Hermite polynomials, the first few of which are

$$\begin{aligned} &H_0(\xi) = 1, \qquad H_1(\xi) = 2\xi, \qquad H_2(\xi) = -2 + 4\xi^2 \\ &H_3(\xi) = -12\xi + 8\xi^3, \qquad H_4(\xi) = 12 - 48\xi^2 + 16\xi^4. \end{aligned} \tag{4-8}$$

From the general properties of the Hermite polynomials or from (4-8), we note that the parity of $|q,r,s\rangle$, given by (4-7), is

$$\pi = (-1)^{q+r+s} = (-1)^n. \tag{4-9}$$

It is readily verified that for a given value of n, and hence energy E_n, n can be partitioned among the integers q, r, and s a total of $D_n = \frac{1}{2}(n+1)(n+2)$ different ways and hence the oscillator is D_n-fold degenerate. Allowing for the two spin states for each type of nucleon, i.e., neutrons or protons, the Pauli principle will allow an occupation number $2D_n = (n+1)(n+2)$ for a level E_n for each nucleon type. The accumulated total number of a particular type of nucleon up to and including energy E_n is

$$M_n = 2\sum_{n=0} D_n = \tfrac{1}{3}(n+1)(n+2)(n+3). \tag{4-10}$$

Accordingly, closed "shells" corresponding to (4-10) are generated for a total number of each type of nucleon given by

$$M_n = 2, 8, 20, 40, 70, 112, 168, \ldots. \tag{4-11}$$

The Schrödinger equation (4-4) can also be conveniently solved in spherical coordinates by writing

$$\psi_{nlm}(r,\theta,\varphi) = R_{nl}(r)\, Y_l^m(\theta,\varphi). \tag{4-12}$$

As usual taking $u(r) = rR(r)$, the radial equation becomes

$$\left[-\frac{\hbar^2}{2m}\frac{d^2}{dr^2} + \frac{l(l+1)\hbar^2}{2mr^2} + \frac{m\omega^2}{2}r^2\right]u(r) = (E+V_0)\,u(r). \tag{4-13}$$

Equation (4-13) can be solved in terms of the associated Laguerre polynomials or equivalently in terms of the confluent hypergeometric functions.

The solution in terms of the Laguerre polynomial $L_k^{l+\frac{1}{2}}(z)$ is

$$\psi(r,\theta,\varphi) = \left\{\frac{4(2\alpha^3)^{1/2}k!}{[\Gamma(\frac{1}{2}N+\frac{1}{2}l+1)]^3}\right\}^{1/2} z^{l/2}e^{-z/2}L_k^{l+\frac{1}{2}}(z)\,Y_l^m(\theta,\varphi), \tag{4-14}$$

with $\alpha = m\omega/2\hbar$, $z = 2\alpha r^2$, $k = (N-l-1)/2$, and $N = n+1$. Occasionally, the quantum number $N = n+1$ is referred to as the harmonic oscillator number rather than n as chosen here. The solution (4-14) can also be written in terms of the confluent hypergeometric function

$${}_1F_1(a;c;z) = 1 + \frac{a}{c}z + \frac{a(a+1)}{c(c+1)}\frac{z^2}{2!} + \frac{a(a+1)(a+2)}{c(c+1)(c+2)}\frac{z^3}{3!} + \cdots \tag{4-15}$$

with $c = l+\frac{3}{2}$, $a = -k$, and again $z = 2\alpha r^2$ and $\alpha = m\omega/2\hbar$. The relationship of the hypergeometric function and the Laguerre polynomial is given by

$$L_k^{l+\frac{1}{2}}(z) = \{[\Gamma(\tfrac{1}{2}N+\tfrac{1}{2}l+1)]^2/k!\,\Gamma(l+\tfrac{3}{2})\}\,{}_1F_1(-k;l+\tfrac{3}{2};z). \tag{4-16}$$

It is evident from (4-15) that if ${}_1F_1$ is to be a terminating polynomial in z as required by normalizability, a must be a negative integer $-k$. The series then terminates with the term $(-1)^k[\Gamma(c)/\Gamma(c+k)]z^k$, and can be written

$${}_1F_1(-k;c;z) = \sum_{s=0}^{k}(-1)^s\frac{\Gamma(c)k!}{(k-s)!\,\Gamma(c+s)}\frac{z^s}{s!}. \tag{4-17a}$$

The various quantum numbers obey the conditions

$$\begin{aligned} n &= 0,1,2,\ldots; \qquad k = 0,1,2,\ldots, \quad k \leqslant n/2 \\ l &= 0,1,2,\ldots, \quad l \leqslant n; \qquad l = n-2k. \end{aligned} \tag{4-17b}$$

The radial quantum number k has the further significance of being equal to the number of intermediate nodes in $R(r)$ (i.e., excluding $r = 0$ and $r = \infty$); the number of nodes in $u(r)$ including the origin is $k+1$.

Many interchangeable notations are used for the solutions to the oscillator potential. When it is specifically required to exhibit the magnetic or projection quantum number m, the wave function or coordinate representation of the state is written $\psi_{l,k,m}(r,\theta,\varphi) = \langle r,\theta,\varphi|l,k,m\rangle$. When there is no danger of confusion, a convenient practice in the literature is simply to write $|l,k,m\rangle$ for the "wave function." A particularly useful notation is the standard spectroscopic form suppressing the projection quantum number, namely $|(k+1),l\rangle$ or simply $(k+1)l$, where in addition the symbols s, p, d, f, g, h, i, j, ... are used to stand for $l = 0,1,2,3,4,5,6,7,\ldots$, respectively. For the moment we are ignoring the nucleon spin. Thus for $n = 2$, $l = 2$, and hence $k = 0$, we speak of the 1d state or for $n = 2$, $l = 0$, and $k = 1$ of the 2s state, etc. The relationship expressed in (4-17) clearly requires l to be odd when n is odd and l even when

n is even; thus the parities of the eigenfunctions for a particular E_n are again seen to agree with (4-9), as of course they must. The degree of degeneracy also agrees with the previous findings.

Table 4-1 gives some relevant properties of the various eigenfunctions (4-14). The normalization is taken to give $\int_0^\infty R_{nl}^2(r) r^2\, dr = 1$.

TABLE 4-1

HARMONIC OSCILLATOR RADIAL WAVE FUNCTIONS[a]

n	l	k	π	Notation	$(2l+1)$	D_n	$C^{-1}R_{nl}(z)$
0	0	0	+	1s	1	1	$e^{-z/2}$
1	1	0	−	1p	3	3	$(\frac{2}{3})^{1/2} z^{1/2} e^{-z/2}$
2	2	0	+	1d	5	6	$2(15)^{-1/2} z e^{-z/2}$
	0	1	+	2s	1		$(\frac{3}{2})^{1/2}(1-\frac{2}{3}z) e^{-z/2}$
3	3	0	−	1f	7	10	$(8/105)^{1/2} z^{3/2} e^{-z/2}$
	1	1	−	2p	3		$(\frac{5}{3})^{1/2} z^{1/2}(1-\frac{2}{5}z) e^{-z/2}$
4	4	0	+	1g	9		$4(945)^{-1/2} z^2 e^{-z/2}$
	2	1	+	2d	5	15	$(14/15)^{1/2} z(1-\frac{2}{7}z) e^{-z/2}$
	0	2	+	3s	1		$(15/8)^{1/2}[1-\frac{4}{3}z+(4/15)z^2] e^{-z/2}$

[a] $C = 2(8\alpha^3/\pi)^{1/4}$; $z = 2\alpha r^2$; $\alpha = m\omega/2\hbar$.

A commonly encountered requirement is the specification of the nuclear density based on a particular model state. It is instructive to take the simple yet nontrivial case of the α-particle or ^{4}He. Consider ^{4}He to consist of four 1s nucleons, two neutrons and two protons, with each nucleon type in spin states $\sigma_z = \pm 1$. If we define the matter density $\rho(\mathbf{r})$ to be the probability that each nucleon in turn will be found within the volume element d^3r, with the normalization $\int \rho(\mathbf{r})\, d^3r = 1$, then

$$\rho(\mathbf{r}) = \int \Psi^*(A)\left[\sum_{i=1}^{A} \delta(\mathbf{r}-\mathbf{r}_i)\right]\Psi(A)\, d^3r_1\, d^3r_2 \cdots d^3r_A. \qquad (4\text{-}18)$$

When $\Psi(A)$ is taken to be a Slater determinant of the appropriate four single-particle 1s harmonic oscillator wave functions, we readily obtain

$$\rho(\mathbf{r}) = 4(1/\pi a_0{}^2)^{3/2} \exp(-r^2/a_0{}^2), \qquad (4\text{-}19)$$

with $a_0{}^2 = 1/2\alpha = \hbar/m\omega$.

▶**Exercise 4-1** Using (3-19) and the definition (4-18), show that

$$\rho(\mathbf{r}) = \sum_{i=1}^{A} |\psi_i(\mathbf{r})|^2. \qquad (4\text{-}20)$$

Then, using Table 4-1, verify (4-19). ◀

The nuclides between ^4He and ^{16}O, according to the harmonic oscillator shell model, would place nucleons, after filling the 1s shell, into the 1p shell. The nuclide ^{16}O would close this shell. The individual nucleon angular momentum characterized by the quantum number j can take the values $j = l \pm \frac{1}{2} = \frac{3}{2}, \frac{1}{2}$ for nucleons in the 1p shell when the nucleon spins are considered. As we shall see presently, there will be compelling reason for supposing that, in addition to a central potential, an effective one-body spin–orbit interaction potential must also be added. The effect of this interaction (with a suitably selected "sign") splits the twelvefold degeneracy of the 1p shell and lowers the energy of the $1p_{3/2}$ *subshell* ($j = \frac{3}{2}$), while it raises the energy of the $1p_{1/2}$ *subshell* ($j = \frac{1}{2}$). A typical single-particle wave function in the $1p_{3/2}$ subshell is (for example, for a neutron with $j = \frac{3}{2}$, $m_j = +\frac{1}{2}$), using Table 4-1 and Appendix A,

$$|1p_{3/2}\rangle = 2(4/9\pi a_0{}^6)^{1/4} z^{1/2} e^{-z/2} \{\sqrt{\tfrac{2}{3}}\, Y_1{}^0 \chi_{1/2}^{1/2} + \sqrt{\tfrac{1}{3}}\, Y_1{}^1 \chi_{1/2}^{-1/2}\} \zeta_{1/2}^{-1/2}, \qquad (4\text{-}21)$$

again with $a_0{}^2 = 1/2\alpha = \hbar/m\omega$ and $z = 2\alpha r^2 = r^2/a_0{}^2$. A nuclide with valence particles in the 1p shell would have a total wave function given by a Slater determinant of such single-particle states (including as well the four 1s states). The parity of the nuclide state, according to Exercise 1-4, would simply be $(-1)^A$.

It is readily shown, using (4-20), that for $4 \leqslant A \leqslant 16$,[1] the *average* radial density, defined by $\rho(r) = (1/4\pi)\int \rho(\mathbf{r})\, d\Omega$, is[2]

$$\rho(r) = 2(1/\pi a_0{}^2)^{3/2} \{2 + \tfrac{1}{3}(A-4)(r^2/a_0{}^2)\} \exp(-r^2/a_0{}^2). \qquad (4\text{-}22)$$

▶**Exercise 4-2** (a) Construct the remaining seven single-particle wave functions for the $1p_{3/2}$ subshell and the four additional wave functions for the $1p_{1/2}$ subshell.

(b) Using the relationship

$$\sum_{m=-l}^{+l} |Y_l{}^m|^2 = (2l+1)/4\pi \qquad (4\text{-}23)$$

show that $\rho(\mathbf{r})$ for ^{12}C (i.e., the nuclide presumably closing the $1p_{3/2}$ subshell) is *spherically symmetric*. Show that $\rho(\mathbf{r})$ for ^{16}O is also *spherically symmetric*. This very important result suggests that in the extreme single-particle shell model the nuclear density functions are spherically symmetric at closed shells and subshells.[3] ◀

[1] It is understood that we mean $2 \leqslant Z \leqslant 8$ and $2 \leqslant N \leqslant 8$ as well.

[2] Including the contributions from the 1s shell.

[3] In reality, the ground states of ^{12}C and ^{16}O are known to have configurations more complicated than the simple model suggested in this exercise. We defer to later a discussion of configuration mixing and collective effects.

The function (4-22) for ^{16}O has a decided dip in $\rho(r)$ for $r=0$. The high-energy electron scattering data in fact suggest the existence of such a depression. Before comparing density functions of the type (4-22) with such data, however, two important although canceling effects must be considered. Account must be taken of the finite size of the proton by folding in a proton charge distribution function. This distribution is generally assumed to be of the Gaussian form,

$$\rho_p(r) = (1/\pi a_p{}^2)^{3/2} \exp(-r^2/a_p{}^2) \tag{4-24}$$

with $r_p(\text{rms}) = (3/2)^{1/2} a_p \approx 0.8$ fm. The folding is achieved in the obvious manner,

$$\rho_c(r) = \int \rho_{\text{HO,c}}(r')\,\rho_p(r-r')\,dr'. \tag{4-25}$$

A second factor that must be considered is the fact that the center of mass of the entire nucleus does not coincide with the center of the shell-model potential well. When both these factors are allowed for, a fit to the electron scattering data yields values for $a_0 = (\hbar/m\omega)^{1/2}$ that increase smoothly (except for the anomalous behavior of ^6Li) from $a_0 = 1.32$ fm for ^4He to $a_0 = 1.71$ fm for ^{16}O. These results in turn yield values for $\hbar\omega = \hbar^2/ma_0{}^2$ of 23.7 MeV for ^4He and 14.1 MeV for ^{16}O. In the case of ^{16}O, the central matter density from (4-22) becomes 0.144 fm^{-3}, as compared to the more or less constant value of 0.170 fm^{-3} for heavier nuclides.

The tendency for the harmonic oscillator energy level spacing $\hbar\omega$ to decrease with increasing A as cited above is found to be quite general. A simple estimate of $\hbar\omega$ will now be given. We write the energy eigenvalue (4-6) for a particle with harmonic oscillator quantum number n as $E_n = -V_0 + \varepsilon_n$, with $\varepsilon_n = (n+\frac{3}{2})\hbar\omega$, and also define $V(r) = -V_0 + v(r)$, with $v(r) = \frac{1}{2}m\omega^2 r^2$. It can then be shown (see Exercise 4-3) that $\langle T\rangle = \langle v\rangle$ and therefore $\varepsilon_n = 2\langle v\rangle$ or

$$\langle r_n{}^2\rangle = \langle n,k,l|r^2|n,k,l\rangle = (\hbar/m\omega)(n+\tfrac{3}{2}) = a_0{}^2(n+\tfrac{3}{2}). \tag{4-26}$$

For a nucleus with mass number A we have

$$\sum_{i=1}^{A} \varepsilon_i = \sum_{i=1}^{A} m\omega^2 \langle i|r^2|i\rangle = m\omega^2 A\langle r^2\rangle, \tag{4-27}$$

where we define $\langle r^2\rangle = (1/A)\sum_{i=1}^{A}\langle i|r^2|i\rangle$. Consider a simple model with $N = Z = A/2$. Assume the nuclear ground state to have nucleons completely filling the harmonic oscillator shell with $n = \nu$ and an additional κ valence nucleons in shell $\nu+1$. Then

$$A = 4\sum_{n=0}^{\nu} D_n + \kappa, \tag{4-28}$$

and

$$(1/\hbar\omega)\sum_{i=1}^{A}\varepsilon_i = 4\sum_{n=0}^{\nu} D_n(n+\tfrac{3}{2}) + \kappa(\nu+1+\tfrac{3}{2}). \tag{4-29}$$

Combining (4-27)–(4-29) leads to

$$(m\omega A/\hbar)\langle r^2\rangle = (\tfrac{3}{2})^{4/3}\tfrac{1}{2}A^{4/3}[1-\tfrac{2}{9}(\kappa^2/A^2)+\cdots] + \tfrac{1}{2}\kappa. \tag{4-30}$$

To a rather good approximation, this gives

$$\omega\hbar \approx (\tfrac{3}{2})^{4/3}\tfrac{1}{2}A^{1/3}\hbar^2/m\langle r^2\rangle = 35.6A^{1/3}/\langle r^2\rangle \text{ MeV}, \tag{4-31}$$

and

$$\langle T\rangle \approx (\tfrac{3}{2})^{8/3}\tfrac{1}{8}A^{2/3}\hbar^2/m\langle r^2\rangle = 15.3A^{2/3}/\langle r^2\rangle \text{ MeV}, \tag{4-32}$$

where $\langle r^2\rangle$ is in units of fm^2. For the heavier nuclei $\langle r^2\rangle \approx \frac{3}{5}(1.20)^2A^{2/3}$ fm^2; hence

$$\omega\hbar \approx 41/A^{1/3} \text{ MeV} \qquad \text{and} \qquad \langle T\rangle \approx 17.7 \text{ MeV}. \tag{4-33a}$$

We note that the average kinetic energy above is independent of A and can be compared to the value of $E_k/A \approx 14.1$ MeV found in Chapter III for $I = 0$.[1] Also since $\langle\varepsilon\rangle = 2\langle T\rangle$, we can estimate V_0 to be $\gtrsim 45$ MeV. A slightly more refined estimate for $\omega\hbar$ would use the empirical value

$$\langle r^2\rangle^{1/2} = (0.82A^{1/3} + 0.58) \text{ fm}$$

found in Chapter II to give a rather good fit for nuclei with $A^{1/3} > 2$. This gives the result

$$\omega\hbar \approx 53/(A^{1/3}+1.41) \text{ MeV}. \tag{4-33b}$$

Based on (4-33b), values of $\omega\hbar$ are predicted to decrease from 13.5 MeV for ^{16}O to 7.2 MeV for ^{208}Pb.

►**Exercise 4-3** (a) Using the virial theorem $2\langle T\rangle = \langle \mathbf{r}\cdot\nabla V\rangle$ and the fact that $\langle T\rangle + \langle V\rangle = E_n$, show that for the harmonic oscillator single-particle states with $V = -V_0 + v(r) = -V_0 + \frac{1}{2}m\omega^2r^2$, we have $\langle T\rangle = \langle v\rangle$ and that the result (4-26) follows.

(b) Noting that

$$\sum_{n=0}^{\nu}(n+1)(n+2) = \tfrac{1}{3}(\nu+1)(\nu+2)(\nu+3) = \tfrac{1}{3}[(\nu+2)^3 - (\nu+2)]$$

[1] The difference in these values can be largely attributed to the somewhat larger value of $\langle r^2\rangle \approx \frac{3}{5}(1.36)^2A^{2/3}$ fm^2 used in Chapter III.

and that

$$\sum_{n=0}^{\nu} (n+1)(n+2)(n+\tfrac{3}{2}) = \tfrac{1}{4}(\nu+2)[(\nu+2)^3 - (\nu+2)],$$

derive (4-30) for the case $(\kappa/A)^2 \ll 1$.

(c) Verify (4-33a) and (4-33b). ◀

The rather simple properties of the harmonic oscillator single-particle states make them very convenient to use. Many useful characteristics of these functions are given in a paper by Talman (1970).

2. Square-Well Potential

Another simple potential well used to represent the average potential of the single-particle model is the three-dimensional square well, $V = -V_0$ for $r < R$ and $V = 0$ for $r > R$. The wave functions for the bound states, i.e., $-V_0 < E < 0$, are readily found to be

$$\psi_{nlm} = R_{nl}(r) Y_l^m(\theta, \varphi), \tag{4-34}$$

with

$$\begin{aligned} R_{nl}(r) &= Aj_l(k_1 r), \qquad r < R \\ R_{nl}(r) &= Bh_l^{(1)}(ik_2 r), \qquad r > R \end{aligned} \tag{4-35}$$

and

$$k_1{}^2\hbar^2 = 2m(E+V_0) = 2mT, \qquad k_2{}^2\hbar^2 = -2mE. \tag{4-36}$$

These radial wave functions are the spherical Bessel functions $j_l(\rho_1)$ and the Hankel functions of the first kind $h_l^{(1)}(\rho_2)$ that are solutions to the radial

TABLE 4-2

INFINITE SQUARE-WELL CHARACTERISTICS

n	l	Notation	$\zeta_{n,l}$	$D_n = 2l+1$	$M_n = 2\sum_{n=0} D_n$
1	0	1s	3.142	1	2
1	1	1p	4.493	3	8
1	2	1d	5.763	5	18
2	0	2s	6.283	1	20
1	3	1f	6.988	7	34
2	1	2p	7.725	3	40
1	4	1g	8.183	9	58
2	2	2d	9.095	5	68
1	5	1h	9.356	11	90
3	0	3s	9.425	1	92
2	3	2f	10.417	7	106
1	6	1i	10.513	13	132
3	1	3p	10.904	3	138
2	4	2g	11.705	9	156

equation

$$-\frac{\hbar^2}{2mr^2}\frac{d}{dr}\left(r^2\frac{dR(r)}{dr}\right)+\frac{\hbar^2 l(l+1)}{2mr^2}R(r) = [E-V(r)]\,R(r)$$

in the two regions $r < R$, with $V = -V_0$, and $r > R$, with $V = 0$. Matching the logarithmic derivatives of (4-35) at $r = R$ leads to a discrete spectrum of values for E. While these eigenvalues must be generated by numerical means for the general case, the case of the infinitely deep well $V_0 \to \infty$ permits a simple evaluation. For the infinitely deep well the internal solution of (4-35) must vanish at $r = R$. The energy eigenvalues can thus be determined from the roots of $j_l(k_1 R) = 0$, with the result that the eigenvalues expressed as the

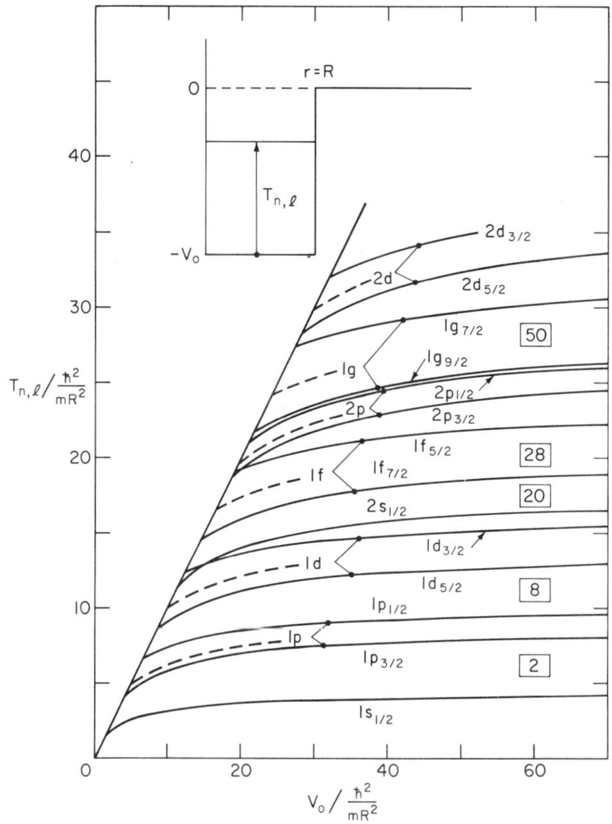

FIG. 4.1. The finite square-well eigenvalues. The spin–orbit corrections are also shown (see later in text).

kinetic energy measured from the bottom of the well are

$$T_{n,l} = (\hbar^2/2mR^2)\zeta_{n,l}^2. \tag{4-37}$$

The values of $\zeta_{n,l}$ are given in Table 4-2. For the case of the infinite square well, the normalization constant A in (4-35) to give $\int_0^R j_l^2(k_1 r) r^2\, dr = 1$ is

$$A_{n,l} = (2/R^3)^{1/2}[j_{l+1}(k_1 R)]^{-1}. \tag{4-38}$$

In order to just bind a particular level characterized by (n, l) requires $V_0 R^2$ to have at least a strength given by

$$V_0 R^2 = (\hbar^2/2m)\zeta_{n,l-1}^2, \qquad l \geqslant 1 \tag{4-39a}$$

or

$$V_0 R^2 = (\hbar^2/2m)[\tfrac{1}{2}(2n-1)\pi]^2, \qquad l = 0. \tag{4-39b}$$

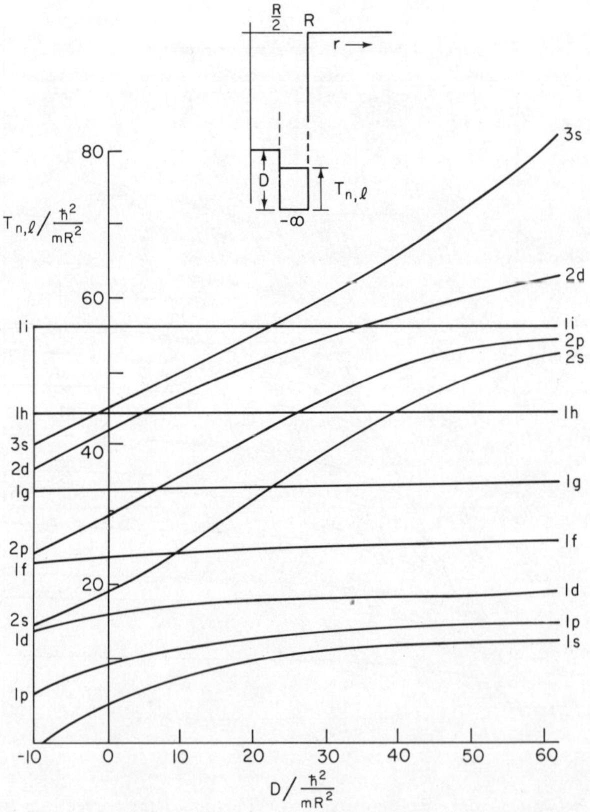

FIG. 4.2. A simplified "wine-bottle" potential and its energy eigenvalues.

Table 4-2 also shows the shell numbers M_n corresponding to the cumulative occupation numbers of degeneracies for the designated shells, allowing for two spin states. Figure 4.1 shows the values of $T_{n,l}$ as a function of V_0, each quantity expressed in units of $\hbar^2/mR^2$. The explicit expressions for the first few spherical Bessel functions and Hankel functions appearing in (4-35) are

$$\begin{aligned} j_0(z) &= \frac{\sin z}{z}, & h_0^{(1)}(z) &= -i\frac{e^{iz}}{z} \\ j_1(z) &= \frac{\sin z}{z^2} - \frac{\cos z}{z}, & h_1^{(1)}(z) &= \left(-\frac{i}{z^2} - \frac{1}{z}\right)e^{iz} \\ j_2(z) &= \left(\frac{3}{z^3} - \frac{1}{z}\right)\sin z - \frac{3}{z^2}\cos z, & h_2^{(1)}(z) &= \left(-\frac{3i}{z^3} - \frac{3}{z^2} + \frac{i}{z}\right)e^{iz}. \end{aligned} \tag{4-40}$$

Various results have suggested that the nuclear shell model potential might exhibit a slight dip in strength in the central region. A crude simulation of this effect would be the simple "wine-bottle" potential shown in Fig. 4.2. This figure shows the variation of the energy eigenvalues with a central dip in the region $r < R/2$, expressed in the dimensions of $\hbar^2/mR^2$. The behavior of these curves can be readily understood if the presence of D is viewed as a perturbation giving an energy shift $\langle\psi_{nl}|D|\psi_{nl}\rangle$ and the radial dependence of $\psi_{nl}(r)$ is kept in mind.

3. Woods–Saxon Potential

Perhaps the most realistic potential shape for the single-particle model is the Woods–Saxon shape,

$$V(r) = -V_o\{1+e^{(r-R)/a}\}^{-1}. \tag{4-41}$$

Unfortunately the solutions to the Schrödinger equation with this potential cannot be given in closed form and require numerical or computer-assisted calculations to determine the eigenfunctions and eigenvalues. For illustrative purposes Fig. 4.3 shows approximately equivalent harmonic oscillator and Woods–Saxon potentials for $A \approx 100$. The radial wave functions for $n = 4$, $l = 0$ and $l = 4$ are shown for both potentials. The Woods–Saxon potential shape will be recognized to be the same as the Fermi density shape used earlier. Detailed numerical calculations requiring self-consistency show that the potential radius R in (4-41) should correspond to a radius at which the nuclear density is one-fourth to one-third of its central value. A value of R commonly appearing in the literature is $R \approx 1.30A^{1/3}$ fm. Values for the surface diffuseness that have been used range from 0.5 to 0.8 fm and $a \approx 0.55$ fm represents a not uncommon average. The distance in which the potential drops from 0.9 to 0.1 of its central value is $s = 4a \ln 3 \approx 4.40a \approx 2.4$ fm. Typical early shell model calculations based on a local Woods–Saxon potential are given by Rost (1968). Green and co-workers (1968, Chapter II) have performed detailed

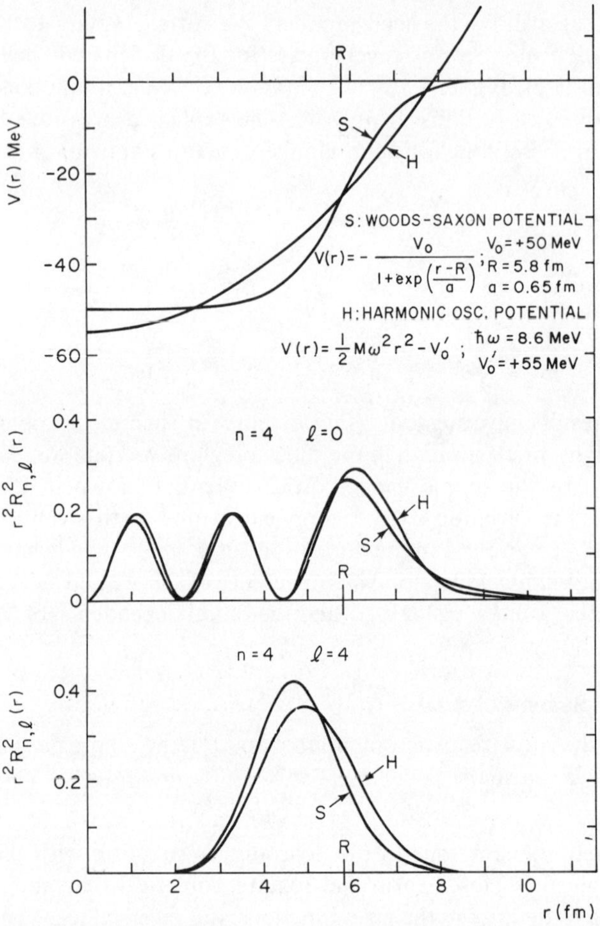

FIG. 4.3. Radial wave function solutions for the harmonic oscillator and equivalent Woods–Saxon potential for a nucleus $A \approx 100$.

numerical calculations with smooth potentials very similar in shape to the Woods–Saxon shape. For a more extensive discussion of the Woods–Saxon potential including velocity dependence, and for additional references, see Section IV, E.

C. SPIN–ORBIT INTERACTION

It was recognized at an early stage in the development of nuclear theory that the shell numbers M_n generated by any of the simple central potentials discussed up to this point do not conform to the empirically determined so-called "magic numbers" (Mayer and Jensen, 1955). Magic numbers correspond to neutron or proton number N or Z equal to $M_n = 2, 8, 20, (28), (40),$

50, 82, (114), 126, (164), (184).[1] Nuclides having N or Z equal to a magic number[2] possess zero intrinsic quadrupole moment in the ground state, suggesting a spherical charge distribution. The binding energy per nucleon is also abnormally large for such nuclides. In neighboring nuclei with N or $Z = M_n + 1$, the separation energy for the least bound nucleon is also abnormally low. These plus numerous other indications, such as the systematics of neutron capture cross sections, nuclear abundance, magnetic moments, etc., suggest that the observed magic numbers correspond to closed or filled shells for single-particle orbitals.

It was found that the addition of a one-body spin–orbit interaction of the form $V_{LS} = -2\alpha(\boldsymbol{l}\cdot\mathbf{s})$ to the simple central potentials does, in fact, generate the observed shell numbers (Mayer, 1949; Mayer and Jensen, 1955). The theoretical justification for the presence of such a potential term is not completely satisfactory, although plausible arguments were presented as early as 1955 by Blin-Stoyle (1955).[3]

As we have seen in Chapter I, the basic nucleon–nucleon potential requires a spin–orbit interaction which can be written

$$V_{LS,ij} = J(|\mathbf{r}_i - \mathbf{r}_j|)\,\{(\mathbf{r}_i - \mathbf{r}_j) \times (\mathbf{p}_i - \mathbf{p}_j)\} \cdot (\boldsymbol{\sigma}_i + \boldsymbol{\sigma}_j). \tag{4-42}$$

We define the effective spin–orbit interaction experienced by a particular nucleon at $\mathbf{r}_j$ by the average value

$$V_{LS,j} = \sum_{i \neq j}^{A} \int \Psi^* V_{LS,ij} \Psi \, d^3r \, d\sigma, \tag{4-43}$$

where Ψ is the nuclear wave function taken to be a simple Slater determinant of the individual-particle orbitals ϕ_{nlm}. If we neglect nucleon–nucleon correlation effects and ignore the exchange integrals, we have

$$V_{LS,j} = -\left\{2 \sum \int \phi_{nlm}^*(\mathbf{r}_i)\, J(|\mathbf{r}_i - \mathbf{r}_j|)\,(\mathbf{r}_i - \mathbf{r}_j)\,\phi_{nlm}(\mathbf{r}_i)\, d^3r_i\right\} \times \mathbf{p}_j \cdot \boldsymbol{\sigma}_j. \tag{4-44}$$

This result would be exact (except for the exchange term) if the particles other than the one described by $\mathbf{r}_j$ constituted a spin-saturated core with $\langle \sum \boldsymbol{\sigma}_i \rangle = 0$ and $\langle \sum \mathbf{p}_i \rangle = 0$. In the following, we shall in fact assume that these conditions

[1] The magic numbers in parentheses correspond to rather weak observed effects or are of a conjectural nature.

[2] When both N and Z are magic, the corresponding nuclides are referred to as doubly magic, e.g., ^{208}Pb with $Z = 82$ and $N = 126$.

[3] For additional remarks and an extension of the model calculation to deformed nuclei, see Mackintosh (1972). There are also nontrivial center of mass effects when relativistic two-body forces are to be included in a many body calculation [see Kim (1971)]. The spin–orbit strength parameter α used here should not be confused with the harmonic oscillator parameter α previously introduced.

hold. From the symmetry of the problem, the quantity inside the braces of (4-44) is a vector in the direction of $\mathbf{r}_j$ and hence can be written in the form $F(r_j)\mathbf{r}_j$ with

$$F(r_j) = -2\sum \int \phi^*_{nlm}(\mathbf{r}_i)\, J(|\mathbf{r}_i-\mathbf{r}_j|)\,[(\mathbf{r}_i\cdot\mathbf{r}_j/r_j^2)-1]\,\phi_{nlm}(\mathbf{r}_i)\, d^3r_i, \tag{4-45}$$

and thus

$$V_{LS,j} = F(r_j)\,(\mathbf{r}_j\times\mathbf{p}_j)\cdot\boldsymbol{\sigma}_j = F(r_j)\,\boldsymbol{l}_j\cdot\boldsymbol{\sigma}_j. \tag{4-46}$$

If we denote the mixed density, shifting the notation slightly, as

$$2\sum \phi^*_{nlm}(\mathbf{r}')\,[(\mathbf{r}'\cdot\mathbf{r}/r^2)-1]\,\phi_{nlm}(\mathbf{r}') = \rho(\mathbf{r},\mathbf{r}'), \tag{4-47}$$

we have

$$F(r) = -\int J(|\mathbf{r}-\mathbf{r}'|)\,\rho(\mathbf{r},\mathbf{r}')\, d^3r'. \tag{4-48}$$

It is convenient to rewrite $\rho(\mathbf{r},\mathbf{r}') = \rho'(\mathbf{s},\mathbf{r})$ with $\mathbf{s} = \mathbf{r}'-\mathbf{r}$, and to expand $\rho'(\mathbf{s},\mathbf{r})$ about $\mathbf{s} = 0$, giving

$$F(r) = -4\pi \int J(s)\,\rho'(0,\mathbf{r})\,s^2\, ds - \frac{4\pi}{3!}\int J(s)\,[\nabla^2\rho'(\mathbf{s},\mathbf{r})]_{s=0}\,s^4\, ds + \cdots. \tag{4-49}$$

The first term of (4-49) is zero and evaluation of the second term gives

$$F(r) = -\sum \frac{2(2l+1)}{3}\left(\int J(s)\,s^4\, ds\right)\frac{1}{r}\frac{d}{dr}[R^2_{n,l}(r)], \tag{4-50}$$

where $R_{n,l}(r)$ is the radial wave function for the state characterized by the quantum numbers (n, l). The mean density for a particle in a closed shell orbital (n, l) can be written[1]

$$\rho_{n,l}(r) = 2R^2_{n,l}(r)\,(2l+1)/4\pi \tag{4-51}$$

and hence

$$F(r) = -\sum \frac{4\pi}{3}\left(\int J(s)\,s^4\, ds\right)\frac{1}{r}\frac{d\rho_{n,l}(r)}{dr}. \tag{4-52}$$

Thus finally,

$$V_{LS} = K\frac{1}{r}\frac{d\rho(r)}{dr}\,\boldsymbol{l}\cdot\boldsymbol{\sigma} \tag{4-53}$$

with

$$K = -\tfrac{4}{3}\pi\int J(s)\,s^4\, ds.$$

▶**Exercise 4-4** Carry out the derivation from (4-44) to (4-53) providing all the detailed steps. ◀

[1] Recall (4-23) and the results (4-19) and (4-22) for the harmonic oscillator.

The form (4-53) is seen to be of the Thomas type except that the nuclear density $\rho(r)$ replaces the usual potential term. As we have indicated, a realistic single-particle potential might be expected to be very similar to $\rho(r)$, and indeed if the nucleon–nucleon potential had zero range, the two would be perforce identical in shape.

As a crude approximation, the density distribution can again be taken to be trapezoidal with a surface thickness tR. The spin–orbit splitting

$$\Delta E = E(j=l-\tfrac{1}{2}) - E(j=l+\tfrac{1}{2})$$

given by

$$\Delta E = -(2l+1)\int_0^R |R_{n,l}(r)|^2 F(r)\, r^2\, dr \tag{4-54}$$

can be readily evaluated for any assumed radial function $R_{n,l}(r)$. For example, if $R_{n,l}(r)$ is taken to be the solution for an infinite square well (of radius $R = R_0 A^{1/3}$) as given by (4-35), we write

$$\Delta E = \gamma_{n,l}(t)\, U/A^{2/3} \tag{4-55a}$$

with

$$U = (1/R_0{}^5)\int J(s)\, s^4\, ds. \tag{4-55b}$$

The quantity t is defined by Blin-Stoyle somewhat differently than in the discussion of Chapter II, leading in the present instance to $t = 2\delta/(1+\delta)$. Blin-Stoyle has calculated the values $\gamma_{n,l}$ and his results are given in Table 4-3.

For $\delta = 0.28$, and using a modified form of the Hamada–Johnston potential for the spin–orbit interaction, we obtain, to a reasonable approximation,

$$E_{nlj} = E_{nl} - (20 \pm 5) A^{-2/3}\, \boldsymbol{l} \cdot \mathbf{s} \text{ MeV}. \tag{4-56}$$

TABLE 4-3

SPIN–ORBIT ENERGY SPLITTING PARAMETER $\gamma_{n,l}(t)$[a]

State	$t = 0.3$	$t = 0.4$	$t = 0.5$
1p	2.86	4.38	5.76
1d	6.72	9.34	10.96
1f	11.99	15.17	16.22
2p	3.99	3.50	3.10
1g	18.61	21.49	21.38
2d	6.60	5.24	5.90
1h	25.28	27.28	25.70
2f	8.88	7.27	10.38
1i	32.87	33.30	30.11
3p	2.91	2.97	4.23
2g	10.54	10.19	16.14
3d	4.53	5.84	6.94

[a] After Blin-Stoyle (1955).

The energy coefficient, $(20 \pm 5 = 15\text{–}25)$ MeV, starts at its lower values for the lower orbital angular momenta and tends more or less smoothly to the larger values for the higher orbital angular momenta. We also note that for $j = l + \frac{1}{2}$, $\boldsymbol{l}\cdot\mathbf{s} = +l/2$, while for $j = l - \frac{1}{2}$, $\boldsymbol{l}\cdot\mathbf{s} = -(l+1)/2$.

▶**Exercise 4-5** Due to the short-range nature of the spin–orbit nucleon–nucleon potential, the dominant contribution to $J(s)$ would be expected to come from the relative ^{3}P state. Thus we require V_{LS} of (1-144) for the triplet-odd case.

Evaluate the integral $\int J(s) s^4 \, ds$ for

$$J(s) = g_{LS} v_0 Y^2(s)[1 + b_{LS} Y(s)] \tag{4-57}$$

with $Y(s) = [\exp(-s/s_0)]/(s/s_0)$ and $s_0 = \mu^{-1} = \hbar/m_\pi c = 1.42$ fm, $b_{LS} = -7.1$, and $v_0 = 3.72$ MeV. To avoid the effects of the hard-core repulsion, carry the integration from $s = s_0/2$ to ∞ and use $g_{LS} \approx 6.0$ rather than the 7.36 of (1-144) in order approximately to allow for the spin–orbit repulsion in the ^{3}D state.

Show that $U \approx +5.5$ MeV results if we use $R_0 \approx 1.37$ fm. [The equivalent infinite square-well radius used to generate $R_{n,l}(r)$ would be expected to be approximately the $\rho_0/3$ density radius $R_0 \approx 1.25(1 + \frac{1}{3}\delta) = 1.37$ fm.[1]] With this value of U and an average value of $2\gamma_{nl}(t)/(2l+1)$ from Table 4-3 for $t \approx 0.43$, show that (4-56) results. ◀

The spin–orbit energy shift of (4-56) can also conveniently be expressed in units of $\hbar^2/mR^2$, where $R = R_0 A^{1/3}$; using $R_0 = 1.37$ fm, we have

$$E_{nlj} = E_{nl} - (0.92 \pm 0.2)(\hbar^2/mR^2)(\boldsymbol{l}\cdot\mathbf{s}). \tag{4-58}$$

Thus the spin–orbit correction can be very simply included, in units of $\hbar^2/mR^2$, into the eigenvalue curves of Fig. 4.1, as shown.

It might be conjectured that the spin–orbit interaction should be of the relativistic Thomas type,

$$V_{LS} = -\frac{\hbar^2}{2m^2c^2}\frac{1}{r}\frac{\partial V}{\partial r}\boldsymbol{l}\cdot\mathbf{s}. \tag{4-59}$$

This form is of course quite similar to the form (4-53), as has already been remarked. For a potential of the Woods–Saxon type (4-41), the derivative becomes

$$\frac{\partial V}{\partial r} = \frac{V_0 e^{(r-R)/a}}{a[1 + e^{(r-R)/a}]^2}. \tag{4-60}$$

It is useful to note that for values near $R = 1.33A^{1/3}$ fm and $a = 0.55$ fm

[1] Extensive model calculations with realistic potentials by Green and co-workers (1968, p. 79) give the value $R_0 = 1.36$ fm for the ρ_0/e density radius.

this expression can be approximated rather closely by an equivalent Gaussian

$$\frac{\partial V}{\partial r} = \frac{V_0}{4a} \exp\left[-\left(\frac{r-R}{2.10a}\right)^2\right]. \tag{4-61}$$

For $V_0 \approx 55$–75 MeV and typical values for R and a, the value of $\langle V_{LS} \rangle$ from (4-59) is some 30–50 times less than the required empirical values expressed by (4-58). Shifting the potential origination to the electromagnetic field results in roughly the same small-magnitude spin–orbit term, in this case, however, even with the wrong sign. The anomalously strong relativistic interaction term is rather to be associated with the mesonic nature of the nucleon–nucleon force. In Chapter I (see Table 1-5 and accompanying discussion) it was pointed out that a near cancellation of the attractive and repulsive static central potential terms originating from the scalar and vector meson fields allowed for the proper balance in strength between central, spin-dependent, tensor, and spin–orbit interactions. This balance has precisely the desired effect of greatly strengthening relativistic effects relative to non-relativistic effects. For relevant papers see Miller (1972), Miller and Green (1972), and Ueda *et al.* (1973).

Using (4-61) to determine $\langle V_{LS} \rangle$ from (4-59) would indicate a spin–orbit interaction energy varying as $\sim\langle 1/ra \rangle$. If a is assumed a constant, a result such as (4-56) would be obtained, but with an A dependence varying with $A^{-1/3}$ rather than $A^{-2/3}$. If, as found in Chapter III, a is proportional to $A^{1/3}$, the $A^{-2/3}$ dependence of Blin-Stoyle's treatment is recovered. A survey by Cohen *et al.* (1962) found a magnitude for the spin–orbit energy shift roughly in accord with (4-56), with, however, an A dependence A^{-n} with $n = 0.53 \pm 0.09$. This is in agreement with the discussion in Chapter III that $a(A)$ is a slowly increasing function of A, although probably with not as strong a dependence as $A^{1/3}$. Interpreting Cohen's empirical results in terms of a shell-dependent parameter α defined by the expression

$$E_{nlj} = E_{nl} - 2\alpha \boldsymbol{l} \cdot \mathbf{s} \tag{4-62}$$

leads to the values given in Table 4-4.

TABLE 4-4

EMPIRICAL SPIN–ORBIT PARAMETER α OF (4-62)

k^a	$l=1$	$l=2$	$l=3$	$l=4$	$l=5$	$l=6$
0	—	1.15	0.93	0.76	0.46	0.44
1	0.68	0.58	0.26	0.28	—	—
2	0.39	0.19	—	—	—	—

[a] $n = l + 2k$; see (4-17b).

To cite just one example, the observed energy splitting $d_{3/2}$–$d_{5/2}$ in ^{17}O is $\Delta E = 5.083$ MeV. The value predicted by (4-56) using $-15A^{-2/3}\boldsymbol{l}\cdot\mathbf{s}$ is $\Delta E = 5.7$ MeV; the value from Table 4-3 using $U = 5.5$ MeV and $t = 0.43$ is $\Delta E = 8.3$ MeV, while Table 4-4 gives $\Delta E = 5.75$ MeV. A theoretical calculation for ^{17}O is given by Niblack and Nigam (1968), predicting $\Delta E = 5.95$ MeV, using a zeroth-order harmonic oscillator single-particle potential with a realistic two-nucleon interaction potential as a perturbation.

We should remark also that the observed spin–orbit splitting appears to be larger when the subshell $j = l + \frac{1}{2}$ is filled and the subshell $j = l - \frac{1}{2}$ is empty than for the case when both subshells are filled (Davis and McCarthy, 1971; Bethe, 1971).

In Fig. 4.4 we show the generic connections between the single-particle model energy eigenvalues and the various assumed potential characteristics.

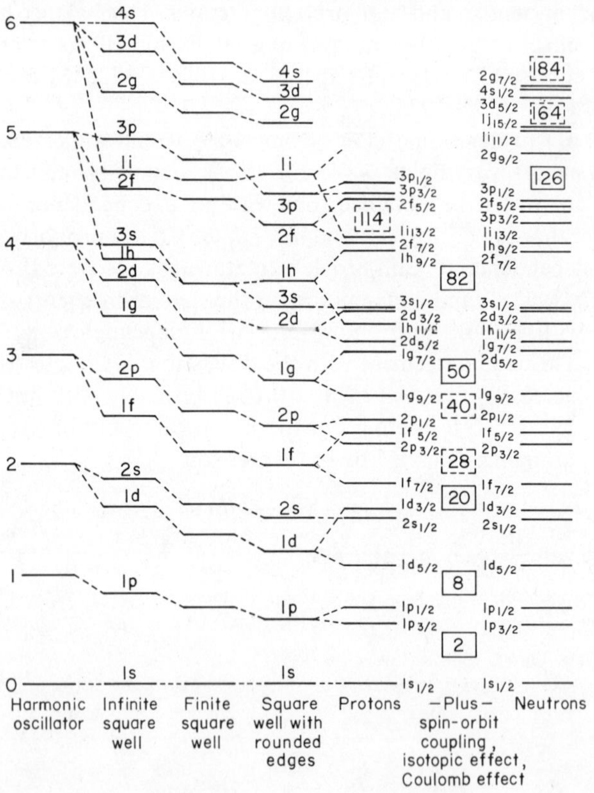

FIG. 4.4. A schematic presentation of the single-particle energy levels in the nuclear species under various potential forms. The realistic right-hand columns also indicate the energy gaps occurring at the "magic numbers."

As with Fig. 4.1, the generation of the empirical magic numbers is seen to be successfully accounted for.

At this point we should remind ourselves that for purely central potentials, i.e., $V \equiv V(|\mathbf{r}|)$, the quantities $\boldsymbol{l}^2, l_z, \mathbf{s}^2, s_z, \mathbf{j}^2 = (\boldsymbol{l}+\mathbf{s})^2, j_z = l_z + s_z$, and parity π are all constants of the motion. Suitable basis states that are simultaneous eigenfunctions of subsets of mutually commuting operators of this group can be selected in a number of ways. However, if the potential also contains a spin–orbit interaction of the type (4-53) or (4-59), then only $\boldsymbol{l}^2, \mathbf{s}^2, \mathbf{j}^2, j_z$, and π are constants of the motion (neither s_z nor l_z is separately constant). Since all the corresponding operators commute, a convenient basis consists of the states $|j, m_j\rangle_{l,s} = \mathscr{Y}^{m_j}_{j,s,l}(\theta, \varphi)$ discussed earlier in Section II,H. Explicitly, we can write these as

$$\psi_{n,l,j,m_j}(\mathbf{r}, \boldsymbol{\sigma}) = R_{n,l}(r) \sum_{m_l = m_j \pm \frac{1}{2}} (l \tfrac{1}{2} m_l, m_j - m_l | j, m_j)\, Y_l^{m_l}(\theta, \varphi)\, \chi_{1/2}^{m_j - m_l}(\sigma). \tag{4-63}$$

The wave function (4-21) encountered earlier is such a state function.

Finally, in Fig. 4.5 we compare the kinetic energy distribution of single-particle orbitals using the three-dimensional square well and the predictions of the Fermi gas model. According to the Fermi gas model with a constant density out to a radius R, the number of spin-$\frac{1}{2}$ particles (for example, neutrons) having kinetic energy less than or equal to T is given by

$$N_T = \frac{8\sqrt{2}}{9\pi} \left(\frac{T}{\hbar^2/mR^2} \right)^{3/2}. \tag{4-64}$$

When the Fermi kinetic energy T_F is reached, N_T simply becomes the total number of particles N. By actual summing of the orbital occupation numbers $(2j+1)$ we can also calculate N_T as various shells appear at the values T_{nlj} for the square-well potential model. Curves are given for both $V_0/(\hbar^2/mR^2) =$ 60 and ∞ for the case of the square-well potential shape.

In comparing these curves, it should be realized that $V_0 = \infty$ in effect imposes the boundary condition that $\psi_{nlj}(r)$ and the summed density vanish at $r = R$ in a manner generating a low-density surface region just inside $r = R$. This in effect confines the particles to a smaller volume than the constant-density gas model with radius R, thus giving a higher kinetic energy for a given number of particles. Allowing V_0 to be finite lowers the central density for the same R and hence decreases the kinetic energy, making the Fermi gas and square-well models more nearly equivalent, as seen in Fig. 4.5. In general the trend of N_T is rather accurately depicted by the Fermi gas model.

▶**Exercise 4-6** (a) Derive (4-64).

(b) Assume that a more appropriate Fermi gas model calculation to

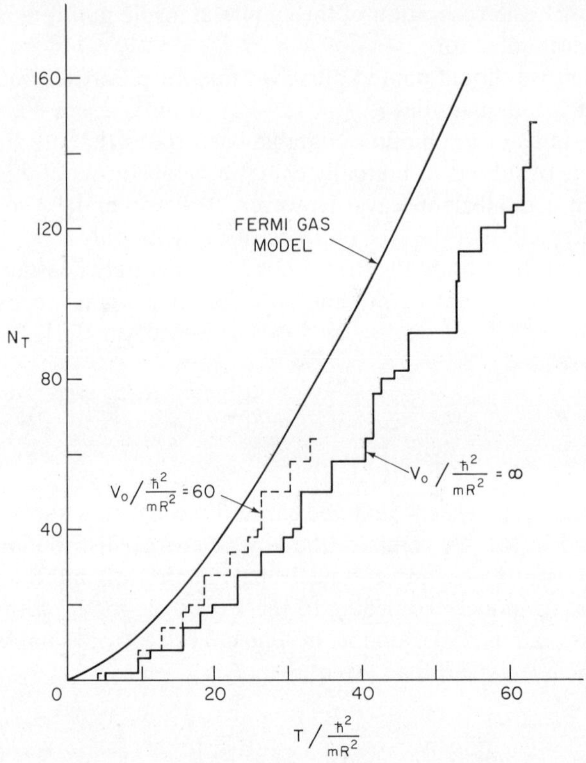

FIG. 4.5. The number of neutrons with kinetic energies less than or equal to T as a function of $T/(\hbar^2/mR^2)$ according to the Fermi gas model and two different square-well potentials.

compare to the $V_0 = \infty$ square-well case results if we use the Thomas–Fermi calculation for a trapezoidal density distribution shown in Fig. 4.6. Show that the Thomas–Fermi kinetic energy for the density distribution shown in Fig. 4.6 is greater than if the same number of particles were confined at a constant

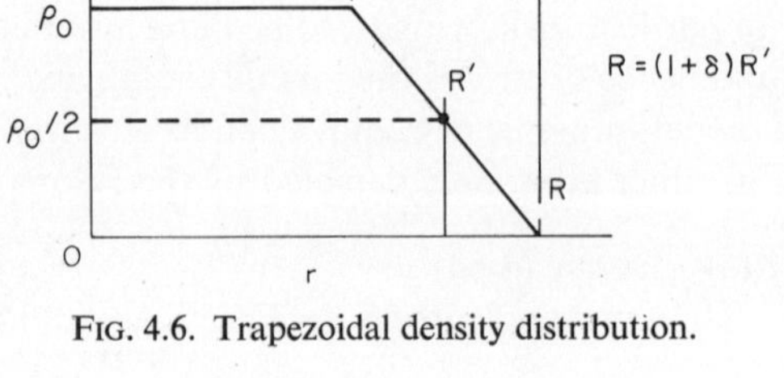

FIG. 4.6. Trapezoidal density distribution.

density to a sphere of radius R by the factor

$$(1+\delta)^2\left(1-\frac{3}{4}\delta+\frac{21}{22}\delta^2-\frac{47}{308}\delta^3\right)(1+\delta^2)^{-5/3}$$

For $\delta = 0.280$ this factor is 1.25. If for any particular N_T, we increase the Fermi gas model kinetic energy shown in Fig. 4.5 by this factor, we arrive at a curve running practically through the square-well curve for $V_0 = \infty$. ◀

D. SINGLE-PARTICLE POTENTIAL STRENGTH

Up to now we have discussed only the possible spatial and spin dependences of the single-particle potentials. To complete the discussion, we must also estimate the depth or strength of the potential experienced by the neutrons and protons in the various nuclides. We start by assuming that the potential depth for a flat-bottomed potential shape such as the square-well or the Woods–Saxon well can be written as

$$\overline{V}_Z = -V_0 - V_1\frac{I}{A} - V_2\frac{|I|}{A^2} - V_3\frac{I^2}{A^2} + \overline{V}_c \tag{4-65a}$$

$$\overline{V}_N = -V_0 + V_1\frac{I}{A} - V_2\frac{|I|}{A^2} - V_3\frac{I^2}{A^2}, \tag{4-65b}$$

with $I = N - Z$.[1]

Except for the second term in each case, the terms are readily justified by recalling the form taken by ${}^{31}n$, ${}^{13}n$, ${}^{11}n$, and ${}^{33}n$ in Chapter III. Of course for present purposes we require that these n values be divided by the number of pairs $\sim A^2$. We have also ignored the small terms in A^{-1} distinguishing $A(A-1)$ from A^2 and have ignored the pairing corrections in η and ζ. The term $\overline{V}_c$ is the average interior Coulomb potential. Since (4-65) is the average over *pairs of interactions*, the total nuclear potential energy is

$$E_p = \frac{Z}{2}\overline{V}_Z + \frac{N}{2}\overline{V}_N = -\frac{V_0}{2}A + \frac{V_1}{2}\frac{I^2}{A} - \frac{V_2}{2}\frac{|I|}{A} - \frac{V_3}{2}\frac{I^2}{A} + \frac{Z}{2}\overline{V}_c, \tag{4-66}$$

or for $N > Z$,

$$E_p = -\frac{V_0}{2}A + \frac{I^2}{2A}(V_1 - V_3) - \frac{V_2 I}{2A} + \frac{Z}{2}\overline{V}_c. \tag{4-66a}$$

[1] For either the Woods–Saxon potential shape (4-41) or the square well $V(r) = -V_0$ for $r < R$, $\overline{V}_N$ and $\overline{V}_Z$ represent modifications of V_0 to allow for the different potential strength acting on neutrons and protons and to give a crude accounting for the interior region Coulomb effect. For $N = Z$, and no Coulomb effect, $\overline{V}_N = \overline{V}_Z = -V_0$.

We note that (4-66a) is of the proper form discussed in Chapter III. The n values in fact really only give guidance for the form of (4-66a) rather than for (4-65), and we note that the second and fourth terms in (4-65) combine to give a single term of the form $I^2/2A$ in (4-66a). Thus the presence of second terms in (4-65a) and (4-65b) cannot be excluded by a simple reference to the discussions in Chapter III.

The presence of the second terms in (4-65a) and (4-65b) with their respective signs is readily justified by the arguments first presented by Lane (1962). We write the most general central nucleon–nucleon potential [see (1-97), and excluding the Coulomb interaction] as

$$v_{ij} = v_0 + (\boldsymbol{\tau}_i \cdot \boldsymbol{\tau}_j) v_1, \tag{4-67}$$

where both v_0 and v_1 are possibly spin dependent [i.e., are of the form $A_{0,1} + B_{0,1}(\boldsymbol{\sigma}_i \cdot \boldsymbol{\sigma}_j)$]. Then the potential experienced by particle j is

$$V_j = \sum_{i \neq j} v_0 + \boldsymbol{\tau}_j \cdot \sum_{i \neq j} \boldsymbol{\tau}_i v_1. \tag{4-68}$$

By introducing the suitable convenient definitions $\sum_{i \neq j} v_0 = -V_0$ and $\frac{1}{2} \sum_{i \neq j} \boldsymbol{\tau}_i v_1 = \mathbf{T} V_1/A$, with $\mathbf{T} = \frac{1}{2} \sum_{i \neq j} \boldsymbol{\tau}_i$ for the isospin of the nucleons other than the one labeled by the subscript j, we have

$$V = -V_0 + (4V_1/A)\,\mathbf{t} \cdot \mathbf{T}. \tag{4-69}$$

The nucleon under consideration with $t = \frac{1}{2}$ can couple with the combined isospin T of the $A-1$ nucleons to form states with $T' = T \pm \frac{1}{2}$. Then for

$$T' = T + \tfrac{1}{2}, \qquad \mathbf{t} \cdot \mathbf{T} = \tfrac{1}{2}[(T+\tfrac{1}{2})(T+\tfrac{3}{2}) - T(T+1) - \tfrac{3}{4}] = \tfrac{1}{2}T, \tag{4-70a}$$

and for

$$T' = T - \tfrac{1}{2}, \qquad \mathbf{t} \cdot \mathbf{T} = \tfrac{1}{2}[(T-\tfrac{1}{2})(T+\tfrac{1}{2}) - T(T+1) - \tfrac{3}{4}] = -\tfrac{1}{2}(T+1). \tag{4-70b}$$

We define the obvious quantities: for

$$T' = T + \tfrac{1}{2}, \qquad V(T, T') = -V_0 + (2T/A)V_1, \tag{4-71a}$$

and for

$$T' = T - \tfrac{1}{2}, \qquad V(T, T') = -V_0 - [2(T+1)V_1/A]. \tag{4-71b}$$

Using Appendix A, we write, in terms of the relevant Clebsch–Gordan coefficients, the average interaction

$$\bar{V} = \sum_{T' = T \pm \frac{1}{2}} (T\tfrac{1}{2}Mm|T'M')^2 V(T, T'). \tag{4-72}$$

Since the coupling of the isospins of the $A-1$ nucleons may result in a configuration that is a linear combination of several possible T values, we take this configuration to be a mixture of pure isospin states each with a weight $|a_T|^2$, properly normalized to give $\sum_T |a_T|^2 = 1$. We thus extend the definition of $\bar{V}$ to

$$\bar{V} = \sum_T |a_T|^2 \sum_{T'=T\pm\frac{1}{2}} (T\tfrac{1}{2}Mm|T'M')^2 V(T,T'). \tag{4-73}$$

Taking the nucleon under consideration to be a proton, we have $m = +\frac{1}{2}$ and $M' = M+\frac{1}{2} = \frac{1}{2}(Z-N) = -I/2$. Clearly in the sums (4-73), m, M', and M are fixed. We have, using Appendix A,

$$\begin{aligned}\bar{V}_Z &= \sum_T |a_T|^2 \left\{\frac{T+M+1}{2T+1} V\left(T, T+\frac{1}{2}\right) + \frac{T-M}{2T+1} V\left(T, T-\frac{1}{2}\right)\right\} \\ &= \sum_T |a_T|^2 \left\{\frac{T+M+1}{2T+1}\left(-V_0 + \frac{2T}{A}V_1\right) + \frac{T-M}{2T+1}\left[-V_0 - \frac{2(T+1)V_1}{A}\right]\right\} \\ &= \sum_T |a_T|^2 \left(-V_0 + \frac{2M}{A}V_1\right)\end{aligned}$$

or

$$\bar{V}_Z = -V_0 - \frac{(I+1)}{A}V_1. \tag{4-74a}$$

Note that this result does not in fact depend on the assignments a_T. In a similar manner, when the nucleon under consideration is a neutron, $m = -\frac{1}{2}$, $M' = M-\frac{1}{2} = -I/2$, we obtain

$$\bar{V}_N = -V_0 + \frac{(I-1)}{A}V_1. \tag{4-74b}$$

Thus, continuing to neglect terms in $1/A$ but not in I/A, we have

$$\begin{aligned}\bar{V}_Z &= -\left(V_0 + \frac{V_1}{A}\right) - \frac{I}{A}V_1 \approx -V_0 - \frac{I}{A}V_1 \\ \bar{V}_N &= -\left(V_0 + \frac{V_1}{A}\right) + \frac{I}{A}V_1 \approx -V_0 + \frac{I}{A}V_1,\end{aligned} \tag{4-75}$$

which give just the second terms assumed on an ad hoc basis in (4-65).

▶**Exercise 4-7** Derive (4-74b). ◀

Returning to (4-66a), we now include an estimate of the kinetic energy. Since presently we shall make use of expression (3-75) based on the discussions of Chapter III, we are required to use a consistent set of parameters. The trapezoidal density distribution of Exercise 2-1 leads to a central density

$\rho_0 = 3/4\pi R_0{}^3(1+\delta^2) = 0.113\ \text{fm}^{-3}$ for the values $R_0 = 1.25$ fm and $\delta = 0.280$ used in Chapter III. The corresponding Fermi momentum is $k_F = 1.19\ \text{fm}^{-1}$ and the Fermi kinetic energy is $T_F = (\hbar^2/2m)(\frac{3}{2}\pi^2\rho_0)^{2/3} = 29.2$ MeV, a value rather lower than the more traditional estimate $T_F = 36.6$–38.3 MeV discussed in connection with (3-12). The assumption of similar neutron and proton density distributions leads to a four-component central region total kinetic energy $E_k = \frac{3}{5}(29.2)[1+\frac{5}{9}(I^2/A^2)+\cdots]$ MeV, where $I = N-Z$. Thus we write

$$E_B = -E = -(E_p + E_k)$$

or

$$E_B = \left(\frac{V_0}{2} - 17.5\right)A - \left(\frac{V_1 - V_3}{2} + 9.7\right)\frac{I^2}{A} + \frac{V_2 I}{2A} - \frac{Z}{2}\overline{V}_c. \tag{4-76}$$

Direct comparison with (3-75), namely[1]

$$E_B = 14.75A - 15.25A^{2/3} - 105\frac{|I|}{A} - 18.90\frac{I^2}{A} - \frac{0.641}{A^{1/3}}Z(Z-1) + \Delta,$$

gives

$$V_0 = 64.6\ \text{MeV} \tag{4-77a}$$

$$V_1 - V_3 = 18.4\ \text{MeV} \tag{4-77b}$$

$$V_2 = 210\ \text{MeV} \tag{4-77c}$$

$$\overline{V}_c = (1.28/A^{1/3})(Z-1)\ \text{MeV}. \tag{4-77d}$$

As stated at the beginning of this section, we specifically seek in the above to estimate the depth of the central portion of the average single-particle potential and accept an approximate Woods–Saxon-like shape with a surface region having a thickness of some 2.4 fm. It is therefore appropriate to ignore the surface energy term in (3-75) in making the comparison with (4-76). An additional complication ignored for the present is the requirement to treat the rearrangement energy. Discussions of rearrangement energy and surface effects are deferred to the next section.

The resolution of (4-77b) into the separate contributions of V_1 and V_3 can be estimated by considering the specific case of a heavy nuclide such as ^{208}Pb (i.e., where $I = N-Z \gg 1$). The eigenvalue diagram for the least-bound neutron and proton in ^{208}Pb is illustrated in Fig. 4.7. We approximate the neutron and proton eigenvalue difference to be equal to the separation energy difference known empirically to be 0.65 MeV (with the proton more tightly

[1] We have substituted $Z(Z-1)$ for $Z^2 - 0.767Z^{4/3}$, for the sake of simplicity.

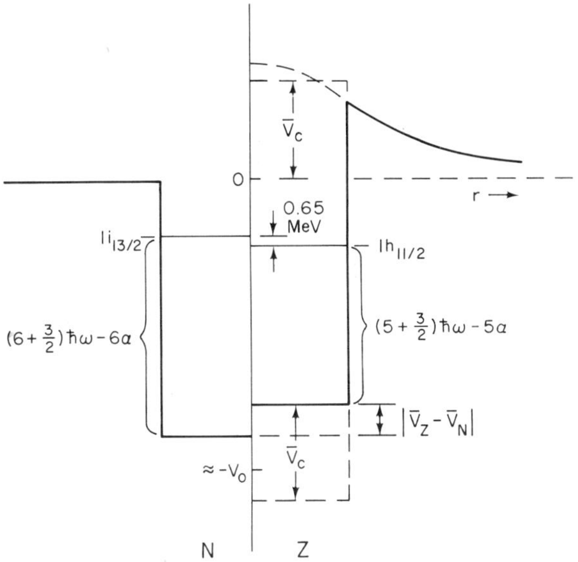

FIG. 4.7. Schematic representation of the eigenvalues and relevant energy terms for the least-bound neutron and proton in ^{208}Pb.

bound). For simplicity we use the harmonic oscillator model for the eigenvalues as modified by (4-62) to include spin–orbit effects and note that in this approximation the protons just complete the $1h_{11/2}$ subshell with $n = 5$ while the neutrons just fill the $1i_{13/2}$ subshell with $n = 6$.[1] We take the value of $\alpha = 0.45$ MeV (the average for $k = 0$, $l = 5$ and $l = 6$) from Table 4-4 and estimate $\hbar\omega \approx 7.0$ MeV from (4-33). Hence

$$(\bar{V}_Z-\bar{V}_N) + (5+\tfrac{3}{2})\,\hbar\omega - 5\alpha + 0.65 = (6+\tfrac{3}{2})\,\hbar\omega - 6\alpha, \qquad \text{(4-78a)}$$

or

$$(\bar{V}_Z-\bar{V}_N) = \bar{V}_c - 2V_1(I/A) = \hbar\omega - \alpha - 0.65. \qquad \text{(4-78b)}$$

Finally, substituting numerical values,

$$V_1 \approx 27.2 \text{ MeV}, \qquad V_3 \approx 8.8 \text{ MeV}. \qquad \text{(4-79)}$$

In conclusion, we estimate the depth of flat-bottomed effective single-particle

[1] The exact order of single-particle levels in completing the $N = 126$ and $Z = 82$ major shells need not concern us for purposes of the present estimate. Empirical evidence suggest that the $3s_{1/2}$, $2d_{3/2}$, and $1h_{11/2}$ subshells are closely spaced near $Z = 82$ and the subshells $3p_{1/2}$, $2f_{5/2}$, $3p_{3/2}$, and $1i_{13/2}$ are similarly closely spaced near $N = 126$; see Fig. 4.4. The important point is to select "generic" oscillator states with $\Delta n = 1$ in contrasting the neutron and proton states.

potentials for protons and neutrons to be given by

$$\bar{V}_Z \approx -64.6 - 27.2\frac{I}{A} - 210\frac{|I|}{A^2} - 8.8\frac{I^2}{A^2} + \frac{1.28}{A^{1/3}}(Z-1) \qquad \text{(4-80a)}$$

$$\bar{V}_N \approx -64.6 + 27.2\frac{I}{A} - 210\frac{|I|}{A^2} - 8.8\frac{I^2}{A^2}\,. \qquad \text{(4-80b)}$$

The terms in $|I|/A^2$ and I^2/A^2 are in general relatively small and can be neglected. For example, for ^{238}U with $I = 54$, these terms are only $\sim 3\%$ and $\sim 7\%$ of the term in I/A, which itself is a small correction. The major isotope effect is thus contained in the second term of (4-80), namely $\pm 27.2(I/A)$ MeV, and hence conforms quite well with the considerations discussed in connection with the derivation of (4-75).

The discussion to this point has ignored the possibility of velocity-dependent potential terms. We must now turn to consider this possibility and to consider the general question of rearrangement and surface energy effects.

E. REARRANGEMENT ENERGY AND VELOCITY-DEPENDENT POTENTIALS

An empirical datum of great importance is the relatively close equality of the separation energies for the least-bound neutron and the least-bound proton, and the average binding energy per nucleon for finite real nuclei.

▶**Exercise 4-8** (a) Using the semiempirical mass formula in the form (3-75), show that the separation energy for the least-bound neutron can be written[1]

$$E_S(N) = E_B(I, A) - E_B(I-1, A-1) = E(I-1, A-1) - E(I, A)$$

$$\approx a_v - \frac{2}{3}a_s A^{-1/3} - a_i\frac{A-I}{A^2} - a_\tau\frac{2IA - I^2 - A}{A^2}$$

$$+ \frac{a_c}{3A^{4/3}}(Z^2 - 0.767Z^{4/3}) \pm \Delta(1, 0). \qquad \text{(4-81a)}$$

Also show that the separation energy for the least-bound proton can be written

$$E_S(Z) = E_B(I, A) - E_B(I+1, A-1)$$

$$\approx a_v - \frac{2}{3}a_s A^{-1/3} + a_i\frac{A+I}{A^2} + a_\tau\frac{2IA + I^2 + A}{A^2}$$

$$- \frac{a_c Z}{A^{1/3}}\left(2 - \frac{Z}{3A} - \frac{1}{Z^{2/3}}\right) \pm \Delta(0, 1). \qquad \text{(4-81b)}$$

[1] In this convention E_B and E_S are positive quantities and the total energy $E = -E_B$ is negative as is appropriate for a bound system.

(b) Using the above results and the numerical coefficients of (3-75) with $\Delta(1,0) = \Delta(0,1) \approx 12/A^{1/2}$ MeV, evaluate $E_S(N)$, $E_S(Z)$, and E_B/A for ^{200}Hg and compare your values to the observed results $E_S(N) = 8.03$, $E_S(Z) = 7.70$, and $-\bar{E} = E_B/A = 7.91$ MeV.

(c) Note that the near equality of the three calculated quantities in part (b) results from the interplay of quite different and relatively large contributions of the isotopic and Coulomb effects. Keeping only leading terms, show that approximate equality of $E_S(N)$, $E_S(Z)$, and E_B/A is achieved only for nuclei at the bottom of the "β-stability valley" for which $I \approx I_0$ of (3-80).[1] ◀

Exercise 4-8 suggests that $E_S(N) \approx E_S(Z) \approx \bar{E}$ is an *equilibrium* property for finite nuclei and pertains therefore only to the β-stable nuclides. A more definitive illustration of this conclusion is shown in Fig. 4.8. The empirical values plotted are for the largely spherical nuclides $A = 106$ for which ^{106}Pd

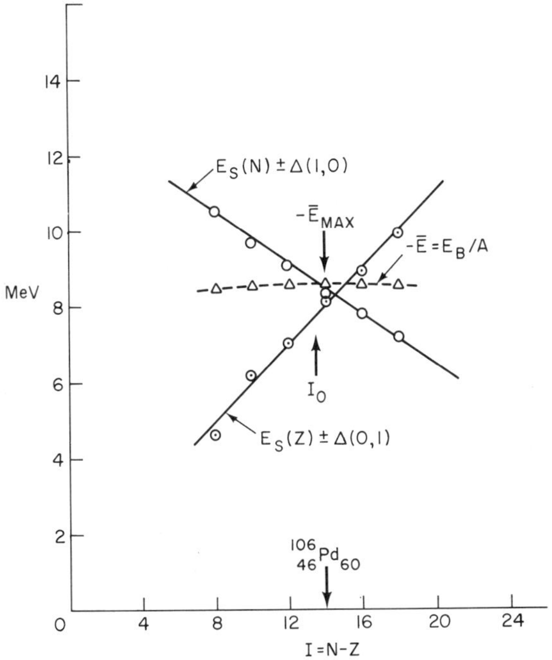

FIG. 4.8. The empirical values for the binding energy per nucleon $-\bar{E} = E_B/A$, neutron separation energy $E_S(N)$, and the proton separation energy $E_S(Z)$ for the nuclides $A = 106$. A pairing energy correction $\Delta(1,0) = \Delta(0,1) = 12/A^{1/2} = 1.20$ MeV has been applied to remove the "even–odd" effect in the observed separation energies.

[1] The leading term for I_0 is $I_0 \approx a_c A^{5/3}/4a_\tau$.

($I = 14$) has the lowest mass and greatest binding energy. Local mass parabola values give $I_0 = 13.6$ and $\Delta(1,0) = \Delta(0,1) = 1.20$ MeV, in reasonable agreement with $I_0 = 13.1$ calculated from (3-80) and $\Delta = 1.20$ MeV calculated from $\Delta = 12/A^{1/2}$ MeV. Indeed the approximate tripartite energy equality is seen to hold only for ^{106}Pd, the nuclide nearest to the physical realization of $I = I_0$. We incidentally observe that a pairing correction of 1.20 MeV leads to rather smooth curves for the separation energies.

It is readily apparent that if a model for infinite nuclear matter ($I = 0$, and no Coulomb or surface effects) results in a binding energy of the form $E_B(A) = a_v A$ for A nucleons, then perforce $-\bar{E} = E_B/A$ and $E_S(A) = \partial E_B/\partial A$ would be equal. Hugenholtz and Van Hove (1958) have shown that quite generally for the ground state of a Fermi gas of *interacting* particles confined to a volume generating an internal pressure p and having a uniform density ρ,

$$E_{S,F} = -\bar{E} + p/\rho; \tag{4-82}$$

here $\bar{E}$ is the average energy per particle and $E_{S,F}$ is the energy of the particle at the top of the Fermi sea (i.e., the least-bound particle). Appropriately, for infinite nuclear matter at equilibrium, the pressure vanishes and the expected theorem results. The extension of this theorem to finite nuclei must be done with some care. For example, the liquid drop model attributes the surface energy of a finite nucleus to the operation of surface tension $\sigma = a_s/4\pi R_0^2$ [see (3-90)]. The presence of this surface tension would be expected to give rise to an internal pressure $p = 2\sigma/R$, leading to a value for the last term of (4-82)

$$p/\rho = \tfrac{2}{3}a_s A^{-1/3}. \tag{4-83}$$

Indeed, surface effects as portrayed by the semiempirical mass formula result in appropriate contributions to $\bar{E}$ and E_s of this order (see Exercise 4-8). This discussion is primarily meant to emphasize the nontrivial difference in the equilibrium conditions between finite, real (and β-stable) nuclides and infinite nuclear matter.

Although the aforementioned investigation by Hugenholtz and Van Hove also indicated that orbitals in the sense of stationary or near stationary states can probably be ascribed only to particles near the top of the Fermi sea, it has been found useful to talk of shell model orbitals for all the nucleons of a nucleus. We shall return to the question of the nonstationary character of single-particle orbitals later. One of the physical observables associated with such orbitals is the separation energy or the necessary energy required for the removal of a specific nucleon (not necessarily the least-bound nucleon discussed up to this point). This separation energy is not equal to the energy eigenvalue of the orbital in question, due to the rearrangement energy of the remaining nucleons bringing the residual system into its lowest energy state again.

Contributions from two effects can be identified by reference to Section III, D, 2 discussing the Bethe–Goldstone equation. The removal of a particle, e.g., from the state γ, alters the operator $\tilde{Q}^{\mathrm{F}}_{\alpha\beta}$ since now either particle α or β may in fact scatter into the unoccupied state γ, which the Pauli principle originally forbade. The second effect has to do with the changes required in a self-consistent determination of the new values of ε_i due to the changes in (3-121) and (3-135) produced by the loss in interaction with the removed particle. Another aspect of the problem relates to the correctness of characterizing the nuclear state by occupied pure single-particle orbitals. For example, reference to Fig. 4.1 would indicate that the least-bound neutron in ^{41}Ca is in an $\mathrm{f}_{7/2}$ orbital, an expectation reinforced by noting that $J^{\pi} = \frac{7}{2}^{-}$ for the ground state. However, there are, in fact, considerable admixtures of higher states in the wave function. The observed separation energy for the least-bound neutron therefore does not immediately relate to the pure $\mathrm{f}_{7/2}$ neutron eigenvalue. Formal definitions embracing all these points can be found in the suggestions put forward by Brueckner and Goldman (1959) or Brandow (1966), to cite two examples.

The fact that the original system was in equilibrium tends to diminish the magnitude of these effects due to the stationary character of the total energy with respect to small parametric excursions from equilibrium. Ignoring rearrangement effects leads to the theorem, generally referred to as Koopman's theorem,[1] equating the single-particle energy eigenvalues and separation energies. However, detailed Hartree–Fock calculations [e.g., Reid *et al.* (1972) and Padjen *et al.* (1973)] suggest that rearrangement contributions to single-particle energies can be as large as 20% or more. Since the single-particle energy eigenvalue is not, in fact, an observable, rearrangement energy contributions are subject to initial definitions and interpretations [see, however, Brueckner *et al.* (1972)]. It is, for example, possible to define single-particle eigenvalues so that Koopman's theorem holds from the outset. Of course, the rearrangement energy can be further characterized, if desired, as originating from surface effects, isotopic effects, etc.

Employing Koopman's theorem without allowing for rearrangement effects does not lead to the usual estimates of the kinetic energy when empirical values are used in (3-4a), which we can rewrite

$$\overline{T} = (1/A)\left(2E - \sum_{i}^{A} \varepsilon_i\right)$$

[see, e.g., Köhler (1966) and Becker and Patterson (1971)]. A relativistic treatment of the kinetic energy [see Miller (1972) and Miller and Green (1972)]

[1] Koopman's theorem (1933) in its original form referred to ionization energies and electron energy eigenvalues for atomic systems. The approximation of using the same wave function to calculate both effects results in the cited theorem.

is found to have an important bearing on the questions under discussion by significantly reducing the expectation value of the single-particle kinetic energy.

Our purpose at present is to examine in a semiquantitative manner the consequences of the experimentally verified Hugenholtz–Van Hove energy condition, in order to obtain an improved estimate of $\bar{V}_Z$ and $\bar{V}_N$ as expressed by (4-80). No attempt will be made here to review the many recent detailed calculations appearing in the journals; rather, a simplified first approximation will be given.

We again concern ourselves with conditions pertaining in the central region of β-stable nuclei.

Sound theoretical reasons were introduced in Chapter I suggesting important velocity-dependent or nonlocal interactions in the nucleon–nucleon force. In Chapter III we also considered the many-body aspects of such forces. In the present consideration of nonlocal or velocity-dependent potential effects, the simplification will be introduced of estimating these effects only for the terms in (4-80) not involving the isotopic number I, and by taking guidance from the properties of nuclear matter. We assume that nonlocal effects can be represented by the effective mass approximation (see Section III, D, 1). Hence, for a nucleon with momentum $\hbar k$,

$$V(k) = -V_0 + \beta T(k) = -V_0 + [(m/m^*) - 1]\, T(k). \tag{4-84}$$

The average energy per nucleon is, from (3-124),

$$\bar{E} = E/A = -E_B/A = \tfrac{3}{5}(1+\tfrac{1}{2}\beta)\, T_F - \tfrac{1}{2}V_0, \tag{4-85}$$

with $T_F \equiv T(k_F)$ equal to the Fermi kinetic energy. The nucleon eigenvalue ε_k is, using (3-123),

$$\varepsilon_k = \varepsilon(k) = (\hbar^2 k^2/2m)(1+\beta) - V_0 = (1+\beta)\, T(k) - V_0. \tag{4-86}$$

We further define $V_R(k)$ to be the rearrangement energy ensuing from the removal of the nucleon from the eigenstate k; thus the separation energy $E_S(k)$ is

$$-E_S(k) = \varepsilon(k) + V_R(k). \tag{4-87}$$

Finally, the Hugenholtz–Van Hove condition for nuclear matter,

$$E_S(k_F) = -\bar{E} = a_v, \tag{4-88}$$

is applied. We assume $T_F = 29.2$ MeV and $m^*/m = 3/5$ (i.e., $\beta = \frac{2}{3}$) and use $a_v = 14.8$ MeV. These values used in (4-88) and (4-85) give

$$V_0 = 76.2 \text{ MeV}, \tag{4-89a}$$

while (4-88) used with (4-86) and (4-87) gives

$$V_R(k_F) = +12.8 \text{ MeV}. \tag{4-89b}$$

▶**Exercise 4-9** (a) Show that with both $\beta = 0$ and $V_R(k) = 0$, Eqs. (4-85)–(4-88) result in the unphysical condition $V_0 = \frac{4}{5}T_F$ (i.e., inadequate potential to bind nucleons at the top of the Fermi sea).

(b) Determine the values of β and V_0 required for consistency with $a_v = 14.8$ MeV and $V_R(k) = 0$ if T_F is either 29.2 or 38.3 MeV.

(c) Determine the value of $V_R(k_F)$ and V_0 required for consistency with $a_v = 14.8$ MeV and $\beta = 0$ if T_F is either 29.2 or 38.3 MeV.

(d) Comment on your results. ◀

We note that $V_R(k_F)$ is a positive quantity, which thus *reduces* the separation energy to a value less than $-\varepsilon_k$.[1] The variation of the rearrangement energy with k is difficult to estimate using the present crude model, although see Azziz (1970). Detailed microscopic calculations, while not always defining the rearrangement energy in identical terms, indicate that $V_R(k)$ is a positive quantity (our signs for E_S, etc.) and is a *decreasing* function of k. The difference $V_R(0) - V_R(k_F)$ is variously estimated to be from 2 to 10 MeV. A microscopic calculation by Padjen *et al.* (1973) for ^{16}O yields a larger variation in $V_R(0) - V_R(k_F)$ and they quote an average rearrangement energy per particle of 10.5 MeV [see also Reid *et al.* (1972)]. We should also mention the approximation developed by Brueckner and Goldman (1959) that $V_R \approx 240\rho^2$ MeV (with ρ in fm^{-3}). This leads to $3.0 < V_R < 7.0$ MeV for $0.113 < \rho < 0.170$ fm^{-3}, which is somewhat smaller than the value appearing in (4-89b). While the density dependence of this approximation may be questioned, the order of magnitude for V_R should be essentially correct [see also Brueckner and Meldner (1973)].

Detailed theoretical calculations [see, e.g., Köhler (1971)] suggest that single-particle eigenvalues for the highest occupied orbital (corresponding to the top of the Fermi sea) are relatively constant for the nuclides $A \geqslant 10$. This interesting observation is not unexpected, as the following reasoning shows. Separation energies are more or less constant when pairing and major shell closing effects are allowed for, as is evident from the behavior of the average binding energy for the β-stable nuclei (recall Fig. 3.14) and in view of the empirical Hugenholtz–Van Hove condition for such nuclides. The central density for $A \geqslant 10$ is essentially constant at the value ρ_0, and therefore both k_F and $V_R(k_F)$ would also be expected to be relatively constant. Thus, applying (4-87), the desired behavior follows.

In view of these various semiempirical considerations, $V_R(k)$ to a first

[1] This is in fact so for any reasonable values of β, T_F, and a_v.

approximation can be taken to be a constant equal to the value 12.8 MeV indicated by (4-89b) for $k = k_F$.

We can write as improved versions of (4-80a) and (4-80b)

$$\overline{V}_Z(k) = -76.2 + \tfrac{2}{3}T(k) - 27.2(I/A) + (1.28/A^{1/3})(Z-1) \quad \text{(4-90a)}$$

$$\overline{V}_N(k) = -76.2 + \tfrac{2}{3}T(k) + 27.2(I/A). \quad \text{(4-90b)}$$

In these expressions we have retained only the leading term of the isotopic effect and have ignored any possible momentum dependence for this term. Evaluation of the eigenvalue problem using (4-90a) and (4–90b) should employ the full nucleon mass m in the kinetic energy operator. Alternately, the effective mass approximation can be employed with the second term $\beta T(k)$ *omitted* and the nucleon mass replaced by m^* in a Hermitian version of the kinetic energy operator such as in (3-116), or more crudely as in (3-117) if the approximation is warranted. A potential shape that has been found to give very satisfactory predictions is the Woods–Saxon potential (4-41), which in the present notation can be written

$$V(r) = \overline{V}_{N\,\text{or}\,Z}\{1+e^{(r-R)/a}\}^{-1}, \quad \text{(4-91a)}$$

depending on whether the nucleon in question is a neutron or a proton. Appropriate values of R and a are

$$R = R_0 A^{1/3}, \quad \text{(4-91b)}$$

$$R_0 = 1.20\text{–}1.30 \text{ fm}, \quad \text{(4-91c)}$$

$$0.40 \leqslant a \leqslant 0.70 \text{ fm}. \quad \text{(4-91d)}$$

For rather crude, but convenient, estimates an effective square well can be used with a radius

$$R_{\text{sq}} = (1.33\text{–}1.41)A^{1/3} \text{ fm}. \quad \text{(4-91e)}$$

Spin–orbit effects can be allowed for by using one of the techniques discussed in Section IV,C; for quick estimates the central potential eigenvalues can be simply shifted using the semiempirical approximations (4-58) or (4-62). Finally, the eigenvalues and separation energies can be related using (4-87). The procedure outlined here is, of course, a single-particle semiempirical method not intended as a substitute for the more satisfying, although unfortunately far more difficult, self-consistent Hartree–Fock calculation using first principles and employing the known nucleon–nucleon force. The situation involving the above points is diagrammatically represented in Fig. 4.9.

It is perhaps instructive to use (4-90) with a simple square-well potential shape to predict some separation energies and to compare these to both more exacting theoretical calculations and experimentally determined values. For

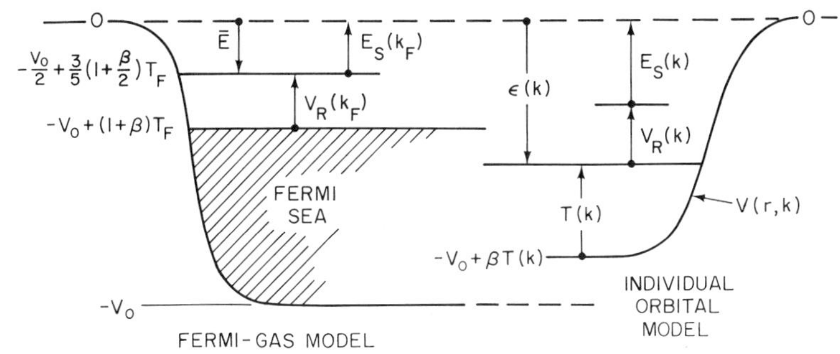

FIG. 4.9. A schematic representation of the individual nucleon eigenvalue problem. For notation see text.

this estimate we use (4-91e) with $R_0 = 1.36$ fm. The spin–orbit energy shift is estimated from (4-58). The results are given in Table 4-5 for characteristic orbitals of some spherical nuclei.

The values appearing in Table 4-5 generally agree within a few MeV with recent extensive Hartree–Fock calculations using the rather successful local density approximation [e.g., see Reid *et al.* (1972) and Campi and Sprung (1972)] and also agree with empirical values for "occupied" orbitals. Our simplified model, however, predicts rather unreliable values for the *energy gap* (the energy difference between the highest occupied level and the lowest unoccupied level eigenvalues) and in general misplaces the least-bound orbitals. The situation would be improved if the rearrangement energy were allowed to decrease some 5 MeV for the orbitals near the top of the Fermi sea from the constant value $V_R(k) = 12.8$ MeV used in arriving at the entries in Table 4-5. However, the difficulty is also traceable to the progressively increasing error in the square-well approximation, as lesser-bound orbitals are considered, due to the neglected surface behavior of more realistic potentials.

Shell model calculations in the above spirit, but using the more realistic Woods–Saxon or similar potential shapes, have been carried out by numerous investigators. The results of Ross *et al.* (1956) are typical. The velocity dependence they used was equivalent to $m^*/m = \frac{1}{2}$ in the central region, and Woods–Saxon shape parameters used were $R_0 = 1.30$ fm and $a = 0.86$ fm. The required neutron potential depth for ^{208}Pb was 69 MeV and the corresponding proton potential depth was 81 MeV (without including the Coulomb effect). These authors also point out the existence of what is usually referred to as the *continuous ambiguity* in trying to determine potential parameters to fit experimental data, namely, that the quantities most directly determined are $V_0 R_0^2$ and a/R_0. Thus a variety of potentials with the same values for these quantities will have very similar energy level spectra. For more recent calculations of the

TABLE 4-5

CALCULATED SEPARATION ENERGIES (MEV) (FOR TYPICAL ORBITALS)[a]

		^{16}O	^{40}Ca	^{90}Zr	^{208}Pb
$\hbar^2/mR^2$		3.52	1.92	1.11	0.64
$\bar{V}_N$		$-76.2+\frac{2}{3}T$	$-76.2+\frac{2}{3}T$	$-73.2+\frac{2}{3}T$	$-70.4+\frac{2}{3}T$
$\bar{V}_Z$		$-72.6+\frac{2}{3}T$	$-69.1+\frac{2}{3}T$	$-68.1+\frac{2}{3}T$	$-64.5+\frac{2}{3}T$
Neutron orbitals	$1s_{1/2}$	41.9	50.9	52.6	53.0
	$1p_{3/2}$	22.5	38.5	45.4	48.6
	$1p_{1/2}$	17.6	35.8	43.9	47.7
	$2s_{1/2}$	—	15.1	30.3	39.4
	$1g_{9/2}$	—	—	15.3	28.2
	$1g_{7/2}$	—	—	(10.7)	25.6
	$2d_{5/2}$	—	—	(1.2)	20.8
	$2d_{3/2}$	—	—	—	19.3
Proton orbitals	$1s_{1/2}$	38.7	43.8	47.6	47.1
	$1p_{3/2}$	17.6	31.7	40.5	42.8
	$1p_{1/2}$	12.7	29.0	39.0	41.9
	$2s_{1/2}$	—	9.0	25.8	33.7
	$1g_{9/2}$	—	—	(7.8)	22.6
	$1g_{7/2}$	—	—	(3.2)	20.0
	$2d_{5/2}$	—	—	—	14.8
	$2d_{3/2}$	—	—	—	13.3

[a] — indicates unbound state. ( ) indicates bound but "unoccupied" state.

above sort including higher-order momentum-dependent terms, refer to Chasman (1971), and also see Janiszewski and McCarthy (1972) and Lodhi and Waak (1972).

Suitable phenomenological nonlocal Woods–Saxon potentials fitting single-particle energies as well as electron scattering form factors have been deduced for the 1p shell [see Gamba *et al.* (1973)]. An interesting set of calculations in the mass range $4 \leqslant A \leqslant 65$ using a Woods–Saxon potential shape with a strength that is treated as local but state dependent is contained in Malaguti and Hodgson (1973) and Millener and Hodgson (1973). In addition to state-dependent effects (i.e., $|V_0(1s)| > |V_0(1p)| > |V_0(1d)$ or $V_0(2s)|$, etc.), a distinct A dependence for each shell and a substantial particle–hole difference were required to fit empirical eigenvalues.

Analyses that consider free scattering states lead to $-78 < V_0 < -71$ MeV for the depth of a suitable nonlocal potential with a range of nonlocality $0.85 < b < 0.90$ [see (3-111)–(3-117)] [refer to Perey and Buck (1962) and Schulz and Wiebicke (1966)]. The Woods–Saxon parameters found were $R_0 = 1.22$ fm and $a = 0.65$ fm.

All the preceding findings are quite consistent with the present discussion. Recent values for the energy-dependent local approximation will be cited later [(4-99) and Table 4-6].

Simplified calculations using velocity-independent potentials have also been attempted with some success, particularly when only a narrow energy range of eigenvalues is explored, as, for example, the levels near the top of the Fermi sea. Evaluating (4-90) for ^{208}Pb and $T = T_{\mathrm{F}}$ as an example, we get $V_Z(k_{\mathrm{F}}) = -62.6$ MeV (without including the Coulomb effect) and $V_N(k_{\mathrm{F}}) = -51.0$ MeV. Typical corresponding values found in the literature, e.g., Ross *et al.* (1956) and Rost (1968), are $-60 < V_Z < -57$ MeV and $-44 < V_N < -40$ MeV. These values, however, equate eigenvalues and separation energies and should therefore be shallower than the values from (4-90) by $\sim V_{\mathrm{R}}(k_{\mathrm{F}})$. The differences, of order (6 ± 3) MeV, could thus be interpreted to represent a more appropriate value of V_{R} for states just below and just above the top of the Fermi sea.

At present the most reliable empirical information on separation energies of inner orbitals is obtained from high-energy (p, 2p), (p, pn), and (e, e′p)

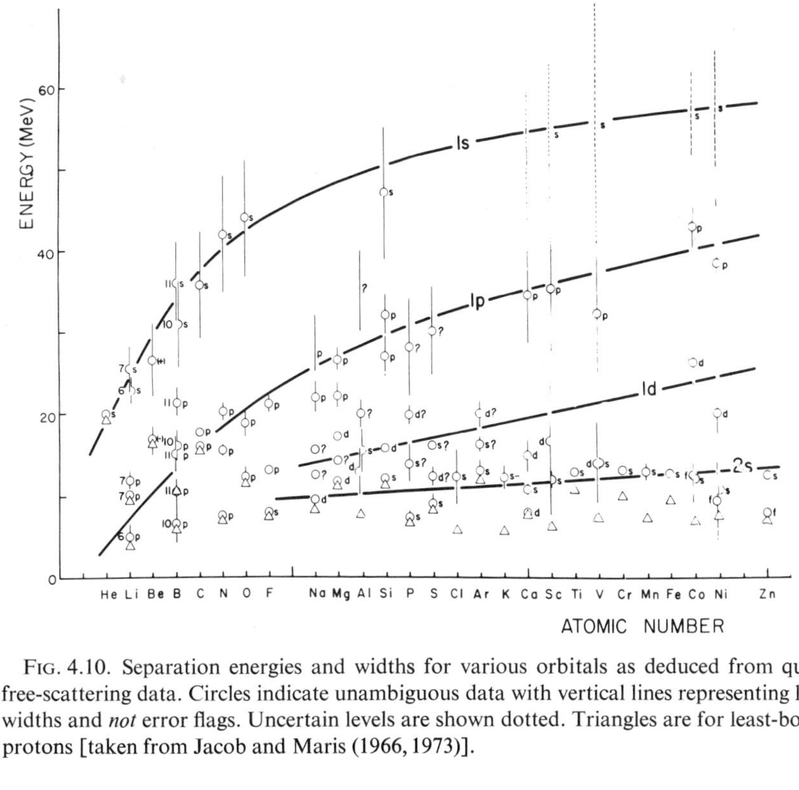

FIG. 4.10. Separation energies and widths for various orbitals as deduced from quasi-free-scattering data. Circles indicate unambiguous data with vertical lines representing level widths and *not* error flags. Uncertain levels are shown dotted. Triangles are for least-bound protons [taken from Jacob and Maris (1966, 1973)].

experiments.[1] The energy spectra of the emerging reaction particles suggest the presence of "resonant" states in the target nucleus. These can be taken as direct indications of single-particle states of the type we have been discussing. These observed particle spectra and angular distributions require kinematic corrections and a theoretical interpretation based on some form of a quasi-free-scattering model before the "hole spectroscopy" parameters are deduced. Recent values for separation energies obtained in this manner are shown in Fig. 4.10. For details refer to the review articles by Jacob and Maris (1966, 1973). The trend of the data clearly follows results shown in Table 4-5.

F. NONSTATIONARY CHARACTER OF SINGLE-PARTICLE ORBITALS

Earlier we referred to the findings of Hugenholtz and Van Hove that only states near the top of the Fermi sea could be considered stationary or near stationary. We now return to a discussion of this point.

We have seen that the removal of a nucleon from a single-particle orbital of a nucleus does not simply result in a corresponding hole state with the remaining orbitals left unaffected. In addition to the realistic modification labeled the rearrangement energy, a further, higher-order correction takes note of the fact that a hole state of a nucleus corresponds to an excited state of the residual nucleus.[2] Such excited states decay following the well-known exponential law $e^{-t/\tau}$, where τ, the *mean life*, can be related to a *level width* Γ or equivalently to an imaginary component iW_0 of the effective potential. A simple connection of these quantities can be readily established for the case of a potential largely independent of r, as, for example, for a square-well shape used to simulate the central region of a flat-bottomed potential. In this case the interior Schrödinger time-dependent equation for an attractive complex well of depth $V_0 + iW_0$ becomes

$$i\hbar\, \partial\psi/\partial t = -[(\hbar^2/2m)\nabla^2 + V_0 + iW_0]\psi. \tag{4-92a}$$

From this equation and its complex conjugate, we obtain

$$\partial\rho/\partial t = -\nabla\cdot\mathbf{j} - (2W_0/\hbar)\rho, \tag{4-92b}$$

where $\rho = \psi^*\psi$ and $\mathbf{j} = (\hbar/m)\,\mathrm{Im}(\psi^*\nabla\psi)$. If the presence of W_0 is interpreted to lead to a complex energy $E - \frac{1}{2}i\Gamma$, where E is the energy with $W_0 = 0$, we have

$$i\hbar\, \partial\psi/\partial t = (E - \tfrac{1}{2}i\Gamma)\psi, \tag{4-93a}$$

[1] For incident protons $E \geqslant 100$ MeV and for incident electrons $E \geqslant 300$ MeV.

[2] The term residual nucleus is used to refer to the nucleus that results when the particular nucleon in question is removed from the original nucleus.

and

$$\partial \rho / \partial t = -(\Gamma/\hbar)\rho. \tag{4-93b}$$

Thus, for the divergenceless case,

$$\rho = \rho_0 e^{-t/\tau}, \tag{4-94a}$$

with

$$\tau = \hbar/\Gamma \qquad \text{and} \qquad \Gamma = 2W_0. \tag{4-94b}$$

The damping of the simple hole state measured by Γ results from the coupling of this state to other degrees of freedom of the system generally involving collective effects. Currently, there is inadequate experimental resolution to detect any *fine structure* that might be present due to coupling with numerous complex modes. Ultimately, the relationship of W_0 to the energy distribution of these partial widths would be similar to that employed in discussing *strength functions* in reaction theory. This topic is best left to texts emphasizing such considerations.

Theoretical estimates [e.g., Köhler (1966)] suggest a dependence of Γ on the average momentum of the nucleon in the particular orbital in question. The functional behavior of $\Gamma(k_\alpha)$ is shown in Fig. 4.11 as evaluated by Köhler (1966). The experimental values are shown in Fig. 4.10 (the *total length* of the verticle line represents Γ).

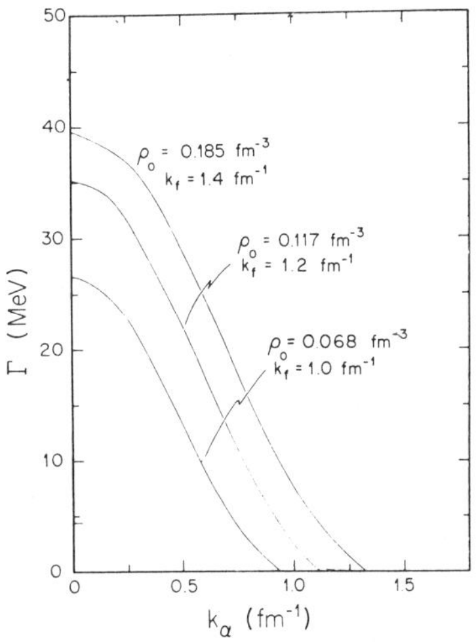

FIG. 4.11. Theoretical estimates of $\Gamma(k_\alpha)$ in terms of the average wave number k_α of a particular orbital. Curves are given for various central densities ρ_0. The quantity k_F refers to the Fermi momentum [taken from Köhler (1966)].

For example, the average momentum for a nucleon of ^{16}O in the 1s shell can be estimated to be $k_\alpha(1s) = 0.78\ fm^{-1}$. Using Fig. 4.11, we obtain $\Gamma(1s) = 10$ MeV for a value of $\rho_0 = 0.113$ adopted earlier. A more refined estimate by Köhler (1966) gives $\Gamma(1s) = 13.7$ MeV. The experimental value is $\Gamma(1s) = 14$ MeV. The value of $\Gamma(1s)$ for ^{208}Pb, referring as it does to a much smaller value of $k_\alpha = 0.37\ fm^{-1}$, is estimated to be $\Gamma(1s) \approx 27$ MeV. The validity of thinking of an "orbital" or quasistationary state for a system with such a large width is perhaps somewhat strained.

We also note from Fig. 4.11 that for $k_\alpha = k_F$, i.e., the orbital precisely at the top of the Fermi sea, $\Gamma(k_F) = 0$. The reason for this is obvious when we realize that the removal of the nucleon that is at the top of the Fermi sea (and hence the least bound) results in a hole state that is precisely the ground state of the residual nucleus.

For a related paper discussing single-particle orbitals from a nuclear matter point of view including complex energy effects, see Engelbrecht and Weidenmüller (1972).

▶**Exercise 4-10** (a) Noting that the average kinetic energy in the 1s state of a harmonic oscillator is $\langle T(1s)\rangle = \frac{3}{4}\hbar\omega$ and using (4-33) to estimate $\hbar\omega$, determine k_α for the 1s state of ^{16}O. The effective square-well value in Table 4-5 for the 1s state of ^{16}O leads to $\langle T(1s)\rangle = 12.8$ MeV. Use this value also to determine k_α and compare the result to the harmonic oscillator estimate and the quoted value in the text of $k_\alpha = 0.78\ fm^{-1}$.

(b) The effective square-well approximation for ^{208}Pb gives $\langle T(1s)\rangle =$ 2.8 MeV; estimate the corresponding value of k_α.

(c) Using (4-94b), estimate the mean life of a state with $\Gamma = 27$ MeV. Compare this to the transit time through the ^{208}Pb nucleus for a particle traveling at the velocity of light, $\Delta t = 2R/c$, and $R = 1.25A^{1/3}$ fm. Comment. ◀

A quantity closely related to the above discussion is Brandow's dimensionless "small parameter" κ (Brandow, 1966), defined as

$$\kappa = \rho \int |\zeta(r)|^2\, d^3r \tag{4-95}$$

where $\zeta(r) = \phi(r) - \phi_0(r)$ is the "wound" in the correlated Bethe–Goldstone two-body wave function discussed in Section III,D,2. Here $\phi(r)$ is the Bethe–Goldstone wave function for an average pair, while $\phi_0(r)$ is the wave function for the same pair with the interaction between them turned off. It will be recalled that $\zeta(r)$ vanishes rapidly beyond a separation corresponding to the average nucleon–nucleon separation in nuclear matter, i.e., the wave function $\phi(r)$ is said to have substantially "healed" beyond this separation. From (4-95), κ is seen to be the volume of the wound in the wave function divided by the average volume occupied by a nucleon $d^3 = 1/\rho$ [see (3-130a)]. Clearly, with no interaction, $\zeta(r)$ and, hence, κ vanish. The quantity κ in (4-95) is obtained by summing the contributions to the energy of a large class of particle–hole diagrams corresponding to various virtual excitations. The probability that an average particle line in the Fermi sea is occupied is $1-\kappa$; conversely, the probability of occupying an average hole line is κ. Hence κ permits relating the correlated nucleon pair interaction to the equivalent concept of virtual particle–hole transitions. A strong interaction producing a large wound while the interacting pair are close is equivalent to a strong virtual scattering out of the Fermi sea. In effect this reduces the occupation probability of the original particle lines and correspondingly generates a partial occupancy of hole lines during such a close encounter. Since the value $\kappa = 0.15 \pm 0.02$ appears in most of the current literature, it follows that there is an 85% probability of finding an average particle line or orbital in the Fermi sea occupied. For the effect of the renormalized fermion lines on the Hartree–Fock treatment see Reid *et al.* (1972).

G. NUCLEON ELASTIC SCATTERING POTENTIALS

It is instructive to rewrite (4-90) explicitly exhibiting the dependence of the potential on the total energy E, and to extend the region of validity to the free scattering states with $E > 0$. We shall again assume that it is appropriate to ignore the velocity dependence of the isotopic term and exclude it from the following consideration, although some investigators write the nuclear portion of the real potential depth as $V_0 = (1 + sI/A)V_1(E)$ with $s \approx \frac{1}{3}$ to $\frac{1}{2}$ [e.g., see the discussion in Sood (1966)]. We write

$$V(T) = V_0' + \beta T \tag{4-96}$$

and also

$$E = T + V = V_0' + (1+\beta)T. \tag{4-97}$$

Eliminating T from (4-96) and (4-97) gives

$$V(E) = \frac{V_0'}{1+\beta} + \frac{\beta}{1+\beta}E. \tag{4-98}$$

For the scattering states, the quantity V_0' should include the rearrangement potential $V_{\mathrm{R}}(k)$. Any reasonable estimate of $V_{\mathrm{R}}(k)$ indicates that it is a decreasing function of k. If our interest centers on the energy range $E \gtrsim 30$ MeV, we can neglect the rearrangement contribution [although see Azziz (1970)]. Thus, for neutrons we obtain, using $\beta = \frac{2}{3}$,

$$\overline{V}_N(E) = -45.7 + \tfrac{2}{5}E + 27.2I/A. \tag{4-99a}$$

For protons it is customary to quote a potential *apart* from the *usual* Coulomb potential. This requires first including $\overline{V}_c$ in V_0' and then subtracting $\overline{V}_c$ from the equivalent to (4-98); hence

$$\begin{aligned}\overline{V}_Z(E)(\text{nuclear}) &= \frac{V_0'+\overline{V}_c}{1+\beta} + \frac{\beta}{1+\beta}E - \overline{V}_c - V_\tau\frac{I}{A} \\ &= \frac{V_0'}{1+\beta} + \frac{\beta}{1+\beta}(E-\overline{V}_c) - V_\tau\frac{I}{A} \\ &= -45.7 + \frac{2}{5}(E-\overline{V}_c) - 27.2\frac{I}{A} \qquad \text{(4-99b)}\end{aligned}$$

with

$$\overline{V}_c = (1.28/A^{1/3})(Z-1). \qquad \text{(4-99c)}$$

This procedure of including $\overline{V}_c$ in (4-99b) is not above criticism and the validity of its inclusion is difficult to test experimentally due to the presence of the isotopic term, which complicates the interpretation of differences observed between neutron and proton scattering experiments. It is in general quite difficult to disentangle various proposed A- and I-dependent terms in the scattering potential. A critique on these points is given by Sood (1966) and Agrawal and Sood (1974).

Numerous analyses of elastic scattering data for neutrons and protons have resulted in expressions of the form (4-99a) and (4-99b), namely

$$\overline{V}_N(E) = -V_{00} + \gamma E + V_\tau I/A \qquad \text{(4-100a)}$$

$$\overline{V}_Z(E)(\text{nuclear}) = -V_{00} + \gamma(E-\overline{V}_c) - V_\tau I/A. \qquad \text{(4-100b)}$$

These potential depths are generally used with Woods–Saxon potential shapes (4-91) characterized by a radius parameter $R = R_0 A^{1/3}$ and a diffuseness parameter a. Some typical values are listed in Table 4-6.

TABLE 4-6

PARAMETERS FOR THE REAL PART OF THE OPTICAL POTENTIAL

V_{00} (MeV)	γ	V_τ (MeV)	R_0 (fm)	a (fm)	Reference
53.3	0.55[a]	27	1.25	0.65	Perey (1963)
47.7	0.33[a]	26.2	1.25	0.65	Rosen *et al.* (1965), Green *et al.* (1968)[b]
49.9	0.22	26.4	1.16	0.75	Fricke *et al.* (1967), Batty *et al.* (1968)
48.6	0.30	30	—	—	Azziz (1970)

[a] $\gamma\overline{V}_c$ is separately quoted as $0.4Z/A^{1/3}$.
[b] Protons only.

The value of $\bar{V}_c$ most often appearing in the literature is

$$\bar{V}_c = 1.38Z/A^{1/3}. \tag{4-100c}$$

Many studies have been undertaken to determine V_τ. The value $V_\tau = (33 \pm 5)$ MeV, with the possibility that V_τ might be complex, agrees with most findings [e.g., Sood (1966), Greenlees and Pyle (1966), Batty *et al.* (1968), and Azziz (1970)] and is consistent with the value of 27.2 MeV used here.

The energy dependence parameter $\gamma = \beta/(1+\beta)$ has been variously estimated to range from 0.21 (Slanina and McManus, 1968) to 0.55 (Perey, 1963). Recently a detailed calculation from first principles suggests $\gamma = 0.34$ [see Lerner and Redish (1972)]. The value $\gamma = \frac{2}{5}$, corresponding to $m^*/m = \frac{3}{5}$, agrees with the average value exhibited in Table 4-6.

It should not be inferred from the above that simple real potentials of the type discussed suffice for fitting data in scattering experiments. Quite important spin–orbit and imaginary potentials are also required, each of rather complicated space dependence. The strength of the spin–orbit interaction required is generally similar in both scattering and bound state problems. The reader interested in the nucleon–nucleus *optical model* is referred to the rather extensive literature [e.g., Hodgson (1967)]. For a typical discussion of the relationship of the real part of the elastic scattering optical potential and the two-nucleon force (including spin–orbit and isospin terms) refer to Greenlees *et al.* (1970).

Recently the optical model analysis for proton elastic scattering has been extended up to an incident energy of approximately 1 GeV by Van Oers and Haw (1973). They find that in the energy range $25 < E \lesssim 150$ MeV, γ for the real part of the potential decreases with increasing A, e.g., $\gamma \approx 0.45$ for ^{6}Li, the lightest nucleus studied, and $\gamma \approx 0.27$ for ^{208}Pb, the heaviest nucleus studied. In the extended energy range $E \lesssim 1$ GeV, a logarithmic energy dependence rather than a linear one appears to be called for. The real part of the potential becomes repulsive above about 500 MeV for all nuclides.

Finally, we conclude that the potential depths (4-90) agree rather well for both bound state and scattering state situations at least for incident energies below ~ 150 MeV.

H. REVIEW OF SINGLE-PARTICLE MODEL SUCCESSES

The previous sections demonstrated the usefulness of the single-particle shell model in accounting for the gross features of the nuclear system ground state. Successful experimental verification of the various separation energies that may be thought of as associated with definite shell model orbitals is reenforced by the observed angular distribution of the ejected particle, characterized in part by the l value of the appropriate orbital. A more or less satisfactory connection between microscopic models involving the shell model states, and macroscopic bulk energy parameters such as those appearing in the semiempirical mass formulas, is now essentially at hand. A review paper giving a brief generalized view of various shell model approaches and containing an outline of the shell model theory of nuclear reactions is by Mang and Weidenmüller (1968). We turn next to additional successes of the simple shell model in nuclear structure problems.

1. Angular Momentum and Parity

The shell model in its simplified form with levels either fully occupied or completely unoccupied is described in terms of *pure configurations*, which, to borrow the obvious notation from atomic physics, can be written in forms $(n_1 l_1 j_1)^{k_1}(n_2 l_2 j_2)^{k_2}(n_3 l_3 j_3)^{k_3}\cdots$, with one configuration for protons, another for neutrons. The shell quantum numbers are, of course, $n_i l_i$ and j_i; and k_i is the shell occupation number. The configuration sequence quite naturally would be written in ascending order of energy. Nuclei in the ground state would have all but k_i for the last filling shell equal to the maximum occupancy permitted by the exclusion principle, namely $k_{\max} = 2j+1$. Thus the neutron configuration for ^{17}O would be $(1s_{1/2})^2(1p_{3/2})^4(1p_{1/2})^2(1d_{5/2})^1$ (see Fig. 4.4).

In view of (1-22) the parity of the nuclear ground state would be

$$\pi = \prod_i (-1)^{l_i k_i}, \tag{4-101}$$

with k_i running over both neutron and proton numbers. This would be so whether we construed the nuclear state to be a simple product state or an antisymmetrized product state, each constructed from the individual single-particle nucleon states (see Exercise 1-5). Since $k_{\max}$ is always an even number (j is an odd half-integer), closed shells contribute $+1$ in the parity product. Hence the parity is always determined by the nucleons in the unfilled shells. The parity of the ground state of ^{17}O for the above-cited configuration would thus be even.

The fact that an important motivation for the introduction of the spin–orbit interaction into the shell model was the desire to generate the magic numbers signifying the closing of major shells has already been mentioned. The total angular momentum of nuclear ground states that consist entirely of such closed shells or those that consist of closed shells plus or minus one particle is then determined quite simply by the operation of the exclusion principle. Thus the angular momentum and parity for closed-shell nuclei such as $^{16}_{8}O$, $^{40}_{20}Ca$, $^{90}_{40}Zr$, and $^{208}_{82}Pb$ must be $J^\pi = 0^+$, while those of $^{15}_{7}N$, $^{17}_{8}O$, $^{207}_{82}Pb$, and $^{209}_{83}Bi$, for example, would be predicted to be $\frac{1}{2}^-$, $\frac{5}{2}^+$, $\frac{1}{2}^-$, and $\frac{9}{2}^-$, respectively, corresponding to an additional "valence" particle or hole present in the states $1p_{1/2}$, $1d_{5/2}$, $3p_{1/2}$, and $1h_{9/2}$, (see Fig. 4.4).

The empirical observation that the ground states of all Z-even, N-even nuclides have zero angular momentum suggests that identical nucleons (i.e., either neutrons or protons) within the same orbital couple their individual angular momenta to a zero resultant. This effect can be traced to the residual interaction between nucleons that must be considered in addition to the simple average potential felt by each nucleon in the noninteracting particle approximation. Two identical nucleons in the same orbital interact more strongly (and hence produce greater binding) than two that are in different orbitals. We have called such additional binding the *pairing effect* in Chapter III. The corollary expectation that nuclides with Z even and N odd or with Z odd and N even would have an angular momentum and parity equal to that of the odd particle, i.e., $J^\pi = j^\pi$, is in fact largely fulfilled. This model is referred to as the *extreme single-particle model.* Thus $^{67}_{30}$Zn, $^{63,65}_{29}$Cu, $^{93}_{41}$Nb, and $^{143}_{60}$Nd, for example, have ground-state angular momenta and parities $\frac{5}{2}^-$, $\frac{3}{2}^-$, $\frac{9}{2}^+$, and $\frac{7}{2}^-$, respectively (refer to Fig. 4.4).

The level sequence shown in Fig. 4.4, particularly for very closely spaced states, exhibits occasional instances of a *crossover* anomaly or level *inversion* where local microscopic considerations affect the actual ordering. Generally these interchanges of order occur between levels of markedly different l values that happen to be nearly degenerate in energy. Examples occur for the $1h_{11/2}$ and $3s_{1/2}$ states and for the $1i_{13/2}$ and $3p_{1/2}$ states. The presence of the pairing energy also requires the interpretation of the level order to be used with some discretion. For example, the three levels $1h_{11/2}$, $2d_{3/2}$, and $3s_{1/2}$ are seen to be close in energy and located just below the major shell closure at 82 nucleons. No matter which of the six permutations for the proper order of these levels is assumed, the three β-stable nuclides $^{123}_{52}$Te, $^{125}_{52}$Te, and $^{129}_{54}$Xe with $N = 71, 73$, and 75, respectively, would each have to have $J^\pi = \frac{11}{2}^-$ for the ground state; instead the observed value is $J^\pi = \frac{1}{2}^+$ for each case.[1] The values of the magnetic moment for these nuclides are $\mu/\mu_0 = -0.736$, -0.887, and -0.777, respectively. They appear plotted on Fig. 2.13 ($gJ = 2\mu/\mu_0$). These values and the parity are consistent with an $s_{1/2}$ designation for the three ground states. (However, since the magnetic moments deviate considerably from the Schmidt value of -1.913, it is evident that these states are not pure single-particle $s_{1/2}$ states of a spherical nucleus.) An explanation is readily afforded by noting that the j dependence of the pairing energy exhibited in (3-77d), namely,

$$\Delta_n, \Delta_p = 12.5(2j_{n,p} + 1)/A,$$

will favor pairing identical nucleons in states of large j over states of small j.

[1] Proof that the proper three nearly degenerate levels were considered is offered by noting that the first two excited states of these nuclei are $\frac{3}{2}^+$ and $\frac{11}{2}^-$, in that order, with the excitation energy of the $\frac{11}{2}^-$ level being less than 0.25 MeV in each case.

Actually ^{123}Te decays by electron-capture but with a half-life of 1.2×10^{13} yr.

Thus, placing an odd number of neutrons into the three closely spaced levels $1h_{11/2}$, $2d_{3/2}$, and $3s_{1/2}$ will result in the lowest system energy with pairs going into the $1h_{11/2}$ state and the remaining odd neutron always going into the $3s_{1/2}$ state. This accounts for the appearance of many more than one $J^\pi = \frac{1}{2}^+$ ground state and virtually no $J^\pi = \frac{11}{2}^-$ ground state in the region $64 < N < 82$.[1]

A similar behavior of preferentially filling larger j-value levels with pairs, rather than neighboring lower j-value levels, also occurs notably in the vicinity of the $1g_{9/2}$ and also the $1i_{13/2}$ states.

An additional problem arises for the case of three identical nucleons (or holes) in the $1d_{5/2}$, $1f_{7/2}$, or $1g_{9/2}$ shell. In these instances the total angular momentum coupling tends to generate $J = j-1$ rather than $J = j$, which would occur if two of the three nucleons (or holes) coupled to zero. Examples are ^{21}Ne and ^{23}Na with $J^\pi = \frac{3}{2}^+$ instead of the expected $\frac{5}{2}^+$; ^{47}Ti and ^{55}Mn with $J^\pi = \frac{5}{2}^-$ instead of $\frac{7}{2}^-$; and ^{79}Se with $J^\pi = \frac{7}{2}^+$ instead of $\frac{9}{2}^+$. Although the original discussion of this effect was given by Kurath (1953) on the basis of the simple shell model with plausible residual interactions, a somewhat more straightforward explanation is possible, appropriate for deformed nuclei. The nuclei cited all have very large electric quadrupole moments (see Fig. 2.11) and exhibit other characteristics suggesting that they are markedly nonspherical.

In the region of highly deformed nuclei the sequence shown in Fig. 4.4, generated for spherical potentials, could not properly be expected to apply. We leave for Chapter VI the general discussion of deformed nuclei and only give a brief accounting of a few characteristics here.

At this point it is useful to refer to Fig. 2.12 and the associated discussion in Chapter II. It was pointed out that, for a closed shell (or subshell) plus one nucleon (or hole) nucleus, the loose nucleon angular momentum j is not a constant of the motion, but only its projection Ω on the body-centered symmetry axis z' is. The ground state of such a nucleus would be expected to have $J = K = \Omega$ and further $\Omega = j$. Thus even for this case of strong deformation, $J = j$ results, giving the same prediction for the total angular momentum of the nuclear ground state that the spherical model does. In fact, this result is actually independent of the strength of the coupling between the nuclear core and the odd nucleon. Of course, other manifestations of the collective nature of the deformation, such as in its influence on the magnetic dipole moment and the electric quadrupole moment, would be discernible. The case of three identical valence nucleons, in contrast to the case of the single valence

[1] The β^--decaying nuclides $^{123}_{50}$Sn and $^{125}_{50}$Sn, each with a magic number of protons, are believed to have $J^\pi = \frac{11}{2}^-$ for their ground states and would provide the only examples of $1h_{11/2}$ ground states.

nucleon, is dependent, however, on the strength of the core-nucleon coupling relative to the mutual nucleon coupling.

When the mutual nucleon coupling is dominant (weak core coupling case), the three equal angular momenta j couple to form the lowest nucleonic state with $J' = j$ as in the simple shell model. The coupling of J' to the deformed core angular momentum R then again gives the final total angular momentum $J = j$.

However, when the core-nucleon coupling dominates (strong core coupling case), the individual projections of the nucleon angular momenta on the z' axis become the constants of the motion with $\Omega_1 = +j$, $\Omega_2 = -j$, and $\Omega_3 = j-1$, when the exclusion principle is allowed for. Hence, with $\Omega = \sum_i \Omega_i = j-1$, we now arrive at the condition $J = j-1$.

Although we have dwelled to a greater extent on the exceptional cases, the extreme single-particle model is rather successful in predicting the angular momentum and parity of nuclear ground states for the A-odd nuclides in a straightforward manner. The case N odd and Z odd is more complicated, particularly when the odd neutron and proton are in different orbitals.

A reliable prediction of the ground-state angular momentum for the odd–odd nucleus requires a careful microscopic calculation of the effect of the residual nucleon–nucleon interaction. We shall give just such a detailed calculation in Chapter V for the case of ^{6}Li. For the present we shall simply cite the results obtained from some very general considerations that are summarized in the *Nordheim rules*. If the residual nucleon–nucleon interaction is similar to the two-free-nucleon case, the lowest energy would be expected for the triplet spin state, i.e., parallel neutron and proton spin orientations, rather than for the singlet spin state.[1] We note that $j_p - l_p$ and $j_n - l_n$ are each either $\pm\frac{1}{2}$. When these differences are opposite in sign, parallel spin orientation requires that j_p and j_n have opposite orientation, leading to $J = |j_n - j_p|$. When the differences have the same sign, j_p and j_n in a parallel orientation is favored and $J = j_n + j_p$ if the $\boldsymbol{\sigma}_n \cdot \boldsymbol{\sigma}_p$ nucleon–nucleon force is strongly enough attractive. However, in general $J = |j_n - j_p|$ cannot be ruled out if the spin–spin force is weak enough; in this instance $|j_n - j_p|$ may again become the lowest energy state.

We define the *Nordheim number*,

$$\mathcal{N} = j_p - l_p + j_n - l_n, \tag{4-102}$$

and give the rules

$$\text{strong rule:} \quad \mathcal{N} = 0, \qquad J = |j_n - j_p| \tag{4-103a}$$

$$\text{weak rule:} \quad \mathcal{N} = \pm 1, \qquad J = j_n + j_p \quad \text{or} \quad |j_n - j_p|. \tag{4-103b}$$

[1] For the present we speak loosely of parallel and antiparallel spin states without resorting to precise descriptions; see (1-12) and Exercise 4-11.

These rules are discussed in great detail by DeShalit and Walecka (1961). Brennan and Bernstein (1960) have shown that even when the odd–odd case involves more than one valence neutron or proton such that n neutrons couple to an angular momentum J_n and p protons couple to an angular momentum J_p, these rules follow provided j_n is replaced by J_n and j_p is replaced by J_p. These results do, however, require that neutron and proton occupancies be either both hole- or particle-like descriptions, i.e.,

$$(2j_p+1-2p)(2j_n+1-2n) > 0.$$

No simple rule follows for the particle–hole configuration, i.e., when

$$(2j_p+1-2p)(2j_n+1-2n) < 0.$$

Similar rules extended for deformed nuclei involving the odd group model with projections Ω_p and Ω_n are given by Gallagher and Moszkowski (1958) (see also Section VI, E, 6).

▶**Exercise 4-11** (a) Consider coupling a $d_{3/2}$ nucleon with a $p_{1/2}$ nucleon to form the state with $J = 2$ and $M_J = +2$. Show by the repeated use of the Clebsch–Gordan coefficients of Appendix A that the proper wave function is

$$\begin{aligned}
&|(d_{3/2})^1 (p_{1/2})^1 : J = 2, M_J = 2\rangle \\
&\quad = \frac{1}{\sqrt{15}}\{2\sqrt{2}\,Y_2{}^2(1)\,Y_1{}^1(2)\,\chi_1^{-1}(1,2) + Y_2{}^1(1)\,Y_1{}^0(2)\,\chi_1{}^1(1,2) \\
&\qquad - [\sqrt{2}\,Y_2{}^2(1)\,Y_1{}^0(2) + Y_2{}^1(1)\,Y_1{}^1(2)]\,\chi_1{}^0(1,2) \\
&\qquad + [\sqrt{2}\,Y_2{}^2(1)\,Y_1{}^0(2) - Y_2{}^1(1)\,Y_1{}^1(2)]\,\chi_0{}^0(1,2)\}.
\end{aligned}$$

Also show that this wave function is $\frac{4}{5}$ triplet state and $\frac{1}{5}$ singlet state.

(b) Repeat the above for the case $|(d_{3/2})^1 (p_{1/2})^1 : J = 1, M_J = +1\rangle$.

(c) Assume a δ-function (i.e., very short-range) residual interaction of the form

$$v_{12} = -(1 - \xi + \xi\boldsymbol{\sigma}_1 \cdot \boldsymbol{\sigma}_2)\,v_0\,\delta(\mathbf{r}_{12})$$

and calculate $\langle v_{12}\rangle$ for the above two cases. Which coupled state has the lower energy if $0 < \xi < 1$? Does this obey the Nordheim rule? ◀

▶**Exercise 4-12** Nuclei in the vicinity of $^{90}_{40}Zr$ ($Z = 40$ and $N = 50$) are approximately spherical. Noting that the angular momentum and parity of $^{91}_{41}Nb$ and $^{91}_{40}Zr$ are $\frac{9}{2}^+$ and $\frac{5}{2}^+$, respectively, and using Fig. 4.4 and the Nordheim rules, predict J^π for $^{92}_{41}Nb$. Noting that J^π for $^{89}_{39}Y$ and $^{89}_{40}Zr$ are $\frac{1}{2}^-$ and $\frac{9}{2}^+$, respectively, predict J^π for $^{88}_{39}Y$. The most probable experimental values of J^π for $^{92}_{41}Nb$ and $^{88}_{39}Y$ are 7^+ and 4^-, respectively. ◀

2. Static Electromagnetic Moments

As a rule nuclides consisting of a magic number of neutrons and protons plus one nucleon (or hole) are fairly well described as pure single-particle spherical nuclear states. Such states constitute the most plausible examples of the extreme single-particle model. The magnetic dipole moments are close to the Schmidt limits and the electric quadrupole moments are small and appropriate for a single nucleon plus an inert spherical core. These cases were discussed in Chapter II. In addition, electromagnetic transition rates for the decay of excited states in such nuclei also conform to the single-particle estimates, as we shall see in Chapter VII. The level structure of such nuclei and nearby even–even nuclei also have characteristic features of spherical nuclei. The levels are relatively widely spaced for the magic nuclei, and the nearby even–even members show typical vibrational structure: ground state 0^+, first excited state 2^+, and nearly degenerate second, third, and fourth excited states 0^+, 2^+, and 4^+ at approximately twice the excitation of the first excited state. These features are discussed in Chapter VI.

For select cases in the region of deformed nuclei, parities and angular momenta agree with simple estimates based on Fig. 4.4 and the extreme single-particle model. These, however, are seen to be more the result of the quantified nature of these dynamical variables rather than the correctness of the model details. The nonquantized quantities such as the magnitude of the electric moments, on the other hand, depend very sensitively on the degree and nature of the collective behavior present. A satisfactory theoretical accounting for the observed values generally involves a nontrivial microscopic calculation for such deformed nuclei. Observed electromagnetic transition rates also differ from single-particle estimates, reflecting the collective nature of the states involved.

The general treatment of deformed nuclei has received a remarkable amount of attention in the literature. We defer the discussion of certain features of successful models employed in such analyses to Chapter VI. It is, however, worthwhile to cite here the characteristic hallmarks of such nuclei. Magnetic dipole moments fall to values "inward" of the Schmidt limits; electric quadrupole moments tend to be large, many times single-particle values; electromagnetic transitions are usually faster; the level structure of excited states shows well-developed rotational bands, which for the even–even nuclei follow the sequence $0^+, 2^+, 4^+, 6^+, \ldots$ with energies approximately proportional to $J(J+1)$. Strongly deformed nuclei are to be found particularly in the range $A \approx 24$, $150 < A < 190$, and $A > 230$; i.e., when neutrons as well as protons approximately half fill major shells.

References

Agrawal, D. C., and Sood, P. C. (1974). *Phys. Rev.* **C9**, 2454.
Azziz, N. (1970). *Nuclear Phys.* **A147**, 401.
Batty, C. J., Bonner, B. E., Friedman, E., Tschalar, C., Williams, L. E., Clough, A. S., and Hunt, J. B. (1968). *Nuclear Phys.* **A116**, 643.
Becker, R. L., and Patterson, M. R. (1971). *Nuclear Phys.* **A178**, 88.
Bethe, H. (1971). *Ann. Rev. Nuclear Sci.* **21**, 93.
Blin-Stoyle, R. J. (1955). *Phil. Mag.* **46**, 973.
Brandow, B. H. (1966). *Phys. Rev.* **152**, 863.
Brennan, M. H., and Bernstein, A. M. (1960). *Phys. Rev.* **120**, 927.
Brueckner, K. A., and Goldman, D. T. (1959). *Phys. Rev.* **116**, 424.
Brueckner, K. A., and Meldner, H. W. (1973). *Phys. Rev.* **C7**, 537.
Brueckner, K. A., Meldner, H. W., and Perez, J. D. (1972). *Phys. Rev.* **C6**, 773.
Campi, X., and Sprung, D. W. (1972). *Nuclear Phys.* **A194**, 401.
Chasman, R. R. (1971). *Phys. Rev.* **C3**, 1803.
Cohen, R. C., Mukherjee, P., Fulmer, R. H., and McCarthy, A. L. (1962). *Phys. Rev.* **127**, 1678.
Davis, K. T. R., and McCarthy, R. J. (1971). *Phys. Rev.* **C4**, 81.
De Shalit, A., and Walecka, J. D. (1961). *Nuclear Phys.* **22**, 184.
Engelbrecht, C. A., and Weidenmüller, H. A. (1972). *Nuclear Phys.* **A184**, 385.
Fricke, N. P., Gross, E. E., Morton, B. J., and Zucker, A. (1967). *Phys. Rev.* **156**, 1207.
Gallagher, C. J., and Moszkowski, S. A. (1958). *Phys. Rev.* **111**, 1282.
Gamba, S., Ricco, G., and Rottigni, G. (1973). *Nuclear Phys.* **A213**, 383.
Green, A. E. S., Sawada, T., and Saxon, D. S. (1968). "The Nuclear Independent Particle Model." Academic Press, New York.
Greenlees, G. W., and Pyle, G. J. (1966). *Phys. Rev.* **149**, 836.
Greenlees, G. W., Makofske, W., and Pyle, G. J. (1970). *Phys. Rev.* **C1**, 1145.
Hodgson, P. E. (1967). *Ann. Rev. Nuclear Sci.* **17**, 1.
Hugenholtz, N. M., and Van Hove, L. (1958). *Physica* **24**, 363.
Jacob, G., and Maris, T. A. J. (1966). *Rev. Modern Phys.* **38**, 121,
Jacob, G., and Maris, T. A. J. (1973). *Rev. Modern Phys.* **45**, 6.
Janiszewski, J., and McCarthy, I. E. (1972). *Nuclear Phys.* **A192**, 85.
Kim, H. (1971). *Phys. Lett.* **37B**, 347.
Köhler, H. S. (1966). *Nuclear Phys.* **88**, 529.
Köhler, H. S. (1971). *Nuclear Phys.* **A162**, 385.
Koopman, T. H. (1933). *Physica* **1**, 104.
Kurath, D. (1953). *Phys. Rev.* **91**, 1430.
Lane, A. M., (1962). *Nuclear Phys.* **35**, 676.
Lerner, G. M., and Redish, E. F. (1972). *Nuclear Phys.* **A193**, 565.
Lodhi, M. A. K., and Waak, B. T. (1972). *Phys. Rev. Lett.* **29**, 301.
Mackintosh, R. S. (1972). *Nuclear Phys.* **A192**, 80.
Malaguti, F., and Hodgson, P. E. (1973). *Nuclear Phys.* **A215**, 243.
Mang, H. J., and Weidenmüller, H. A. (1968). *Ann. Rev. Nuclear Sci.* **18**, 1.
Mayer, M. G. (1949). *Phys. Rev.* **75**, 1969.
Mayer, M. G., and Jensen, J. H. D. (1955). "Elementary Theory of Nuclear Shell Structure." Wiley, New York.
Millener, D. J., and Hodgson, P. E. (1973). *Nuclear Phys.* **A209**, 59.
Miller, L. D. (1972). *Phys. Rev. Lett.* **28**, 1281.
Miller, L. D., and Green, A. E. S. (1972). *Phys. Rev.* **C5**, 241.

Niblack, W. K., and Nigam, B. P. (1968). *Phys. Rev.* **167**, 996.
Padjen, R., Rouben, B., Le Tourneux, J., and Saunier, G. (1973). *Phys. Rev.* **C8**, 2024.
Perey, F. G. (1963). *Phys. Rev.* **131**, 745.
Perey, F. G., and Buck, B. (1962). *Nuclear Phys.* **32**, 353.
Reid, N. E., Banerjee, M. K., and Stephenson, G. J. (1972). *Phys. Rev.* **C5**, 41.
Rosen, L., Beery, J. G., Goldhaber, A. S., and Auerbach, E. H. (1965). *Ann. Phys.* (*N.Y.*) **34**, 96.
Ross, A. A., Mark, H., and Lawson, R. D. (1956). *Phys. Rev.* **102**, 1613.
Ross, A. A., Lawson, R. D., and Mark, H. (1956). *Phys. Rev.* **104**, 401.
Rost, E. (1968). *Phys. Lett.* **26B**, 184.
Schulz, H., and Wiebicke, H. (1966). *Phys. Lett.* **21**, 90.
Slanina, D., and McManus, H. (1968). *Nuclear Phys.* **A116**, 271.
Sood, P. C. (1966). *Nuclear Phys.* **89**, 553.
Talman, J. D. (1970). *Nuclear Phys.* **A141**, 273.
Ueda, T., Nack, M. L., and Green, A. E. S. (1973). *Phys. Rev.* **C8**, 2061.
Van Oers, W. T. H., and Haw, H. (1973). *Phys. Lett.* **45B**, 227.

Chapter V

INDIVIDUAL-PARTICLE MODEL

A. INTRODUCTION

In the preceding chapter we treated at length the single-particle shell model, which attempts to account for the intricate sum of forces on a particular nucleon through a simple one-body potential. Although many gross nuclear properties were satisfactorily accounted for, need did arise occasionally for considering the neglected residual interactions such as the pairing interaction. Notable deviations in the electric moments, for example, clearly signal the need for closer approximations.

The not inconsiderable successes of the single-particle shell model suggest at the very least the use of these wave functions as a suitable orthonormal set or basis Φ in terms of which the actual nuclear wave function might be determined. If the total nuclear Hamiltonian H is known, the energy matrix $\langle\Phi_i|H|\Phi_j\rangle$ can be diagonalized, yielding the energy eigenvalues E_k and eigenfunctions $\Psi_k = \sum_i \alpha_{ik}\Phi_i$. Alternately, a variational calculation requiring the expansion coefficients to yield a stationary value for the energy can be employed, at least for the ground state. The wave functions Φ_i in either case are appropriate antisymmetrized nuclear wave functions consisting of single-particle product states. These results are of course exact, and to the extent that one coefficient dominates the expansion, the single-particle shell model is recovered. The various versions of this type of approach are referred to as the individual-particle models.

The above procedure is not generally tractable and various approximations are employed to simplify the required calculations. One physically plausible approximation views the closed shells (and occasionally subshells) as constituting an inert "spectator" core, the presence of which simply provides a one-body potential for each of the remaining valence nucleons. Although models exist that further allow for coupling *core excitations* to various possible valence nucleon states, we proceed with the simpler version of the inert core model in which $J^\pi = 0^+$ for the core, and all energies refer to the total energy of the core as a base level. Thus the core state consists of closed shells in configurations $(n_1 l_1)^{2(2l_1+1)}(n_2 l_2)^{2(2l_2+1)} \cdots (n_c l_c)^{2(2l_c+1)}$, one set for protons and one for neutrons. The valence nucleons are placed in subsequent shells $(n_{c+1} l_{c+1})^{k_1}(n_{c+2} l_{c+2})^{k_2} \cdots$, with $k_1 + k_2 + \cdots = k$ the total of valence particles. When the number of possible shell configurations available for the valence nucleons is limited, one speaks of *truncating the space of configurations.* Commonly, only the closely spaced shells next in order after the closed shells are considered for occupation by the valence nucleons; for example, in the region $^4\mathrm{He} \leqslant (A, Z) \leqslant {}^{16}\mathrm{O}$ only the 1p shell is generally considered, while for $^{16}\mathrm{O} \leqslant (A, Z) \leqslant {}^{40}\mathrm{Ca}$ both the 2s and 1d shells are used. Reference to Fig. 4.4 gives a reasonable guide to the shells to be considered important in any particular case (or the equivalent Nilsson states for strongly deformed nuclei, see Chapter VI). When more than just the immediate next unfilled shell (and occasionally subshell) is considered, one speaks of *configuration mixing.*

With the advent of high-speed computers "large" shell model calculations can be performed in which up to 10,000 basis states can be included [see, e.g., Whitehead and Watt (1972)]. On the other hand, reasonable methods using much smaller truncated spaces are also possible [e.g., Wong *et al.* (1972)]. For our purposes we shall use the smallest space of configurations that still permits the dominant effect of interest to be studied without gross errors.

To be rigorously correct, the nuclear wave functions are required to be antisymmetrized with respect to the exchange of any pair of nucleons using isospin formalism (or at any rate with respect to the exchange of any pair of like nucleons). The inert core model assumes we can write $\Phi_i = \psi(\mathrm{core}) \cdot \mathscr{A}_k[\prod_{\gamma=1}^{k} \psi_\gamma]_i$, where $\mathscr{A}_k$ is the antisymmetrization operator for the k valence nucleons; thus Φ is *not* antisymmetrized with respect to the exchange of core and valence nucleons. This approximation is particularly appropriate in the absence of configuration mixing. By the expression $[\prod_{\gamma=1}^{k} \psi_\gamma]_i$ we mean a suitably constructed linear combination of product states representing a particular coupling of angular momenta, with the latter designated by the subscript i. The construction of these states and their properties will occupy our attention in subsequent sections. In the absence of residual interactions these states are highly degenerate.

In the total nuclear Hamiltonian $H = T+V$, the potential energy $V = \sum_{i<j} v_{ij}$ can be partitioned to yield $V = V_0 + \Delta V$. In the absence of configuration mixing, V_0 could be conveniently taken to be a sum of optical-model-type one-body potentials representing the interaction of the individual valence nucleons with the core. The *residual interactions* represented by ΔV would include the sum of v_{ij} over all the valence nucleons and any portions of the imagined true or best optical potential that might be omitted for convenience in defining V_0.[1] A not uncommon option, for example, is to omit the known spin–orbit interaction in the core–valence nucleon interaction in defining V_0, including it instead in ΔV.

B. COUPLING SCHEMES

The valence nucleon states $[\prod_{\gamma=1}^{k} \psi_\gamma]_i$ involve a number of possible coupling schemes to add the appropriate mechanical and spin angular momenta. Quite naturally the scheme to be preferred is the one most nearly diagonalizing the matrix elements of V. We distinguish two particularly useful schemes, *LS* or Russell–Saunders coupling and *jj* coupling.

1. LS or Russell–Saunders Coupling

Under *LS* coupling nuclear basis states are constructed that are simultaneous eigenfunctions of the mutually commuting operators $\boldsymbol{l}_i^2$, $\mathbf{s}_i^2$, $\mathbf{L}^2 = (\sum_i \boldsymbol{l}_i)^2$, $L_z = \sum_i l_{iz}$, $\mathbf{S}^2 = (\sum_i \mathbf{s}_i)^2$, and $S_z = \sum_i s_{iz}$. If both V_{0i} and ΔV are central, i.e., $V_{0i} = V_{0i}(|\mathbf{r}_i|)$ and $\Delta V = v(|\mathbf{r}_{ij}|)$, such basis vectors would be entirely appropriate. The resulting energy spectrum would possess a large degeneracy in the unspecified values of J resulting from the vector addition of L and S, viz., $\mathbf{J} = \mathbf{L} + \mathbf{S}$. The central interaction ΔV in its most general form (1-97), possessing two-body exchange terms, would split the degeneracy of the configuration depending on L and S. (Under certain circumstances the designation of additional quantum numbers might be required; see later.) It is customary to designate *LS*-coupled states in the notation borrowed from atomic physics, namely ${}^{(2S+1)}L_J$. Thus, for example, with only central forces the states ${}^3D_1$, ${}^3D_2$, and ${}^3D_3$ would be degenerate. In general there are $2S+1$

[1] This is a convenient but somewhat inappropriate use of the term "residual interaction." A strict definition conforming to the practice in atomic physics, from which the term is borrowed, would define it as the difference between an average potential $\langle A-i|\sum_j v_{ij}|A-i\rangle$ and the actual potential $\sum_j v_{ij}$ involving nucleon i.

degenerate values of J when $L \geqslant S$, or $2L+1$ values when $S \geqslant L$. While there are numerous ways of forming orthogonal subsets of these degenerate wave functions, a particularly useful choice is to specify $\mathbf{J}^2$ and J_z.

For the sake of simplicity, we restrict our consideration to two valence nucleons, one in shell $(n_1 l_1)$, the other in shell $(n_2 l_2)$. The individual nucleon wave functions are selected as the solutions to the optical potential V_{0i} without spin–orbit interaction and are written as in (4-12) or (4-34), etc., with, however, the addition of the spin function

$$\psi_{nlm_l m_s} = \phi_{nlm_l} \chi_{1/2}^{m_s} = R_{nl}(r) Y_l^{m_l}(\theta, \varphi) \chi_{1/2}^{m_s}(\sigma). \tag{5-1}$$

The spin states $\chi_{1/2}^{m_s}(\sigma)$ are, of course, either the state $\alpha = \binom{1}{0}$ for $m_s = +\frac{1}{2}$ or $\beta = \binom{0}{1}$ for $m_s = -\frac{1}{2}$. The isospin variable could also be stipulated in (5-1) by appending the multiplicative isospin function $\zeta_{1/2}^{m_t}(\tau)$.

For two LS-coupled particles we must first combine the vectors $\boldsymbol{l}_1 + \boldsymbol{l}_2 = \mathbf{L}$ and, separately, $\mathbf{s}_1 + \mathbf{s}_2 = \mathbf{S}$. For the space part of the coupled system state function, we have

$$\phi(l_1 l_2 | LM_L) = \sum_{m_{l1}} (l_1 l_2 m_{l1}, M_L - m_{l1} | LM_L) \phi_{n_1 l_1 m_{l1}}(1) \phi_{n_2 l_2, M_L - m_{l1}}(2), \tag{5-2}$$

where we have taken into account the requirement that $m_{l1} + m_{l2} = M_L$ and the arguments of the functions ϕ labeled (1) and (2) stand for r_1, θ_1, and φ_1; and r_2, θ_2, and φ_2, respectively. Similarly, we have

$$\chi(s_1 s_2 | SM_S) = \sum_{m_{s1}} (\tfrac{1}{2}\tfrac{1}{2} m_{s1}, M_S - m_{s1} | SM_S) \chi_{1/2}^{m_{s1}}(1) \chi_{1/2}^{M_S - m_{s1}}(2), \tag{5-3}$$

where the arguments (1) and (2) stand for σ_1 and σ_2. Equation (5-3) for two nucleons simply yields the well-known triplet and singlet spin functions $\chi_1^{M_S}(1,2)$ and $\chi_0^{0}(1,2)$. In both (5-2) and (5-3) the appropriate Clebsch–Gordan coefficients from Appendix A are assumed.

Finally, the total angular momentum is obtained by coupling L and S, viz., $\mathbf{J} = \mathbf{L} + \mathbf{S}$ (the name LS coupling derives from this form of the summing). The total wave function, suppressing the core contribution, then is

$$\Phi(LS|JM) = \sum_{M_L} (LSM_L, M - M_L | JM) \phi(l_1 l_2 | LM_L) \chi(s_1 s_2 | S, M - M_L), \tag{5-4}$$

where we have written simply M for M_J, recognizing that $M = M_L + M_S$. Except for any possible required antisymmetrization, (5-4) is the LS-coupled wave function for two nucleons previously designated $[\prod_{\gamma=1}^{k} \psi_\gamma]_i$, with $i \equiv (n_1, n_2, l_1, l_2, s_1, s_2, L, S, J, M)$. We note that while (5-4) is a simultaneous eigenfunction of $\mathbf{J}^2, J_z, \mathbf{L}^2, \mathbf{S}^2, \boldsymbol{l}_1^2, \boldsymbol{l}_2^2, \mathbf{s}_1^2, \mathbf{s}_2^2$, and parity $\pi = (-1)^{l_1 + l_2}$, it is not an eigenfunction of $l_{z1}, l_{z2}, s_{z1}, s_{z2}$, nor of L_z or S_z.[1] Finally, we could also append the isospin identification of the state through the multiplicative isospin function $\zeta(t_1 t_2|TM_T)$ representing the coupling of two isospin-$\frac{1}{2}$ nucleons. Perforce two neutrons or two protons would have to be in an isospin triplet state.

2. *jj* Coupling

If the optical potential V_{0i} includes a spin–orbit term $a(\boldsymbol{l}_i \cdot \mathbf{s}_i)$ in addition to any central term, we can no longer specify l_{iz} or s_{iz}, since the corresponding operators do not commute with $(\boldsymbol{l}_i \cdot \mathbf{s}_i)$. In terms of the spatial functions $\phi_{nlm_l}(\mathbf{r}) = R_{nl}(r) Y_l^{m_l}(\theta, \varphi)$ and the spin functions $\chi_{1/2}^{m_s}(\sigma)$ we write, equivalent to (4-63),

$$\psi_{nljm_j} = \sum_{m_l} (l\tfrac{1}{2}m_l, m_j - m_l | jm_j) \, \phi_{nlm_l} \chi_{1/2}^{m_j - m_l} \tag{5-5}$$

for the individual-particle wave functions. The sum in (5-5) is only over two terms at most with $m_l = m_j \pm \frac{1}{2}$.

This situation leads in a natural way to what is referred to as *jj* coupling. The convenient basis vectors to use under *jj* coupling are the eigenfunctions of the mutually commuting operators $\boldsymbol{l}_i^2$, $\mathbf{s}_i^2$, $\mathbf{j}_i^2$, $\mathbf{J}^2 = (\sum_i \mathbf{j}_i)^2$, and $J_z = \sum_i j_{iz}$.

Again for the sake of simplicity, we specialize to the case of two valence particles, one in shell $(n_1 l_1)$ with total angular momentum j_1, i.e., in subshell $(n_1 l_1 j_1)$, and the other in subshell $(n_2 l_2 j_2)$. We are now required to combine two states of the type (5-5) to obtain the coupled states with $\mathbf{J} = \mathbf{j}_1 + \mathbf{j}_2$ and $M = m_{j1} + m_{j2}$ (the name *jj* coupling derives from this form of the summing). The result quite readily is

$$\Phi(j_1 j_2|JM) = \sum_{m_{j1}} (j_1 j_2 m_{j1}, M - m_{j1}|JM) \, \psi_{n_1 l_1 j_1 m_{j1}}(1) \, \psi_{n_2 l_2 j_2, M - m_{j1}}(2), \tag{5-6}$$

where again the core portion of the wave function has been supressed. The isospin dependence could also be stipulated in (5-6) by including the function $\zeta(t_1 t_2|TM_T)$ coupling the two isospin-$\frac{1}{2}$ nucleons.

We note that (5-6) is not an eigenfunction of the operator j_{iz}. While (5-6) remains an eigenfunction of the Hamiltonian even when V includes a central potential V_0 *and* a central two-body Wigner potential ΔV, this would no longer be the case with the presence in ΔV of a general central two-body potential of the form (1-97) possessing exchange terms.

[1] Note that space inversion and the exchange of space coordinates of two nucleons is not equivalent for a system with $A > 2$. Thus the parity of the nuclear state is *not* determined by L in general, while, of course, for the deuteron states the parity is determined uniquely by L.

3. Vector Recoupling Coefficients

The two bases, *LS* coupling and *jj* coupling, form two complete orthonormal sets. Therefore it is possible, of course, to write terms of one in an expansion of the other. As we shall see, it is very useful to be able to do this to facilitate the evaluation of the matrix elements of the various operators that must be considered.

If the function $\Phi(j_1 j_2|JM)$ is given, its expansion in *LS* terms is

$$\Phi(j_1 j_2|JM) = \sum_{\substack{L,S \\ \mathbf{L}+\mathbf{S}=\mathbf{J}}} [(2L+1)(2S+1)(2j_1+1)(2j_2+1)]^{1/2} \times \mathrm{X}\begin{Bmatrix} l_1 & l_2 & L \\ s_1 & s_2 & S \\ j_1 & j_2 & J \end{Bmatrix} \Phi(LS|JM). \tag{5-7}$$

The quantity

$$\mathrm{X}\begin{Bmatrix} l_1 & l_2 & L \\ s_1 & s_2 & S \\ j_1 & j_2 & J \end{Bmatrix}$$

is referred to as the 9-*j* symbol or X-coefficient, or sometimes as the "recoupling coefficient." Alternately, if $\Phi(LS|JM)$ is given, it can be expanded in *jj* terms as follows:

$$\Phi(LS|JM) = \sum_{\substack{j_1,j_2 \\ \mathbf{j}_1+\mathbf{j}_2=\mathbf{J}}} [(2L+1)(2S+1)(2j_1+1)(2j_2+1)]^{1/2} \times \mathrm{X}\begin{Bmatrix} l_1 & s_1 & j_1 \\ l_2 & s_2 & j_2 \\ L & S & J \end{Bmatrix} \Phi(j_1 j_2|JM). \tag{5-8}$$

A general property of the 9-*j* symbols is that $\mathrm{X} = \tilde{\mathrm{X}}$, i.e., X is equal to its own transpose. Thus, as with the Clebsch–Gordan coefficients, the same table of coefficients is required for transformations as for their inverse transformations, albeit using different elements in the sums (5-7) and (5-8). These coefficients are defined to be real and normalized and are independent of the magnetic quantum numbers. The 9-*j* symbol vanishes identically unless each row and each column obeys the relevant triangle condition (i.e., $\mathbf{l}_1 + \mathbf{s}_1 = \mathbf{j}_1$, $\mathbf{l}_1 + \mathbf{l}_2 = \mathbf{L}$, etc.). Also, the sum of the nine quantum numbers,

$$Q = l_1 + l_2 + s_1 + s_2 + j_1 + j_2 + L + S + J,$$

TABLE 5-1a

THE TRANSFORMATION FROM *LS* COUPLING TO *jj* COUPLING[a, b]

j_1	j_2	$^3(J+1)$	^1J	$^3(J-1)$	^3J
$l_1+\frac{1}{2}$	$l_2+\frac{1}{2}$	$-\left\{\frac{[(J+1)^2-\Delta^2](\Sigma-J)(\Sigma-J-1)}{2(\Sigma^2-\Delta^2)(J+1)(2J+1)}\right\}^{1/2}$	$\left\{\frac{(\Sigma-J)(\Sigma+J+1)}{2(\Sigma^2-\Delta^2)}\right\}^{1/2}$	$\left\{\frac{(J^2-\Delta^2)(\Sigma+J)(\Sigma+J+1)}{2(\Sigma^2-\Delta^2)J(2J+1)}\right\}^{1/2}$	$\Delta\left\{\frac{(\Sigma-J)(\Sigma+J+1)}{2(\Sigma^2-\Delta^2)J(J+1)}\right\}^{1/2}$
$l_1+\frac{1}{2}$	$l_2-\frac{1}{2}$	$-\left\{\frac{[\Sigma^2-(J+1)^2](J-\Delta)(J-\Delta+1)}{2(\Sigma^2-\Delta^2)(J+1)(2J+1)}\right\}^{1/2}$	$\left\{\frac{(J-\Delta)(J+\Delta+1)}{2(\Sigma^2-\Delta^2)}\right\}^{1/2}$	$-\left\{\frac{(\Sigma^2-J^2)(J+\Delta)(J+\Delta+1)}{2(\Sigma^2-\Delta^2)J(2J+1)}\right\}^{1/2}$	$\Sigma\left\{\frac{(J-\Delta)(J+\Delta+1)}{2(\Sigma^2-\Delta^2)J(J+1)}\right\}^{1/2}$
$l_1-\frac{1}{2}$	$l_2+\frac{1}{2}$	$\left\{\frac{[\Sigma^2-(J+1)^2](J+\Delta)(J+\Delta+1)}{2(\Sigma^2-\Delta^2)(J+1)(2J+1)}\right\}^{1/2}$	$-\left\{\frac{(J+\Delta)(J-\Delta+1)}{2(\Sigma^2-\Delta^2)}\right\}^{1/2}$	$\left\{\frac{(\Sigma^2-J^2)(J-\Delta)(J-\Delta+1)}{2(\Sigma^2-\Delta^2)J(2J+1)}\right\}^{1/2}$	$\Sigma\left\{\frac{(J+\Delta)(J-\Delta+1)}{2(\Sigma^2-\Delta^2)J(J+1)}\right\}^{1/2}$
$l_1-\frac{1}{2}$	$l_2-\frac{1}{2}$	$\left\{\frac{[(J+1)^2-\Delta^2](\Sigma+J)(\Sigma+J+1)}{2(\Sigma^2-\Delta^2)(J+1)(2J+1)}\right\}^{1/2}$	$\left\{\frac{(\Sigma+J)(\Sigma-J-1)}{2(\Sigma^2-\Delta^2)}\right\}^{1/2}$	$-\left\{\frac{(J^2-\Delta^2)(\Sigma-J)(\Sigma-J-1)}{2(\Sigma^2-\Delta^2)J(2J+1)}\right\}^{1/2}$	$-\Delta\left\{\frac{(\Sigma+J)(\Sigma-J-1)}{2(\Sigma^2-\Delta^2)J(J+1)}\right\}^{1/2}$

[a] After Racah (1950).

[b] The quantities Σ and Δ are defined as $\Sigma = l_1+l_2+1$ and $\Delta = l_1-l_2$. The notation $^3(J+1)$ is equivalent to ^3L_J with $L = J+1$ and $2S+1 = 3$; the others follow the same obvious pattern.

must be an integer. Any two rows or any two columns can be interchanged and the 9-j symbol at most changes sign. There is no sign change for an even number of total row and column exchanges. For an odd number of total row and column exchanges, the sign change is governed by $(-1)^Q$, with Q defined as above. Thus

$$\begin{Bmatrix} a & b & c \\ d & e & f \\ g & h & k \end{Bmatrix} = (-1)^Q \begin{Bmatrix} g & h & k \\ d & e & f \\ a & b & c \end{Bmatrix} = \begin{Bmatrix} h & g & k \\ e & d & f \\ b & a & c \end{Bmatrix}. \tag{5-9}$$

For additional properties the reader is referred to numerous specialized texts [e.g., Edmonds (1957)].

Racah (1950) has evaluated the 9-j symbols appropriate for the coupling of two nucleons in shells l_1 and l_2. His results are given in Table 5-1a, using a slightly different sign convention in order to conform with the Clebsch–Gordan coefficients given in Appendix A.

A set of 9-j symbols we shall have occasion to use later for coupling two p-shell nucleons is readily obtained from Table 5-1a and is given in Table 5-1b.

Tables 5-1a and 5-1b are appropriate for any two nucleons with different radial quantum numbers $n_1 \neq n_2$, and also when $n_1 = n_2$ for the case of a neutron–proton pair only; in either event the isospin designation is not used.

When the two nucleons are in the same shell, $n_1 = n_2$ and $l_1 = l_2 = 1$ as in the present instance, then the L-even states are symmetric on space exchange

TABLE 5-1b

THE 9-j SYMBOLS FOR TWO p-SHELL NUCLEONS

	$^3D_3$
$\frac{3}{2},\frac{3}{2}$	1

	$^1D_2$	$^3D_2$	$^3P_2$
$\frac{3}{2},\frac{3}{2}$	$\sqrt{2}$	0	2
$\frac{3}{2},\frac{1}{2}$	$\sqrt{2}$	$\sqrt{3}$	-1
$\frac{1}{2},\frac{3}{2}$	$-\sqrt{2}$	$\sqrt{3}$	1

$\times (1/\sqrt{6})$

	$^3S_1$	$^3P_1$	$^1P_1$	$^3D_1$
$\frac{3}{2},\frac{3}{2}$	$2\sqrt{5}$	0	$\sqrt{30}$	-2
$\frac{3}{2},\frac{1}{2}$	-4	$3\sqrt{3}$	$\sqrt{6}$	$-\sqrt{5}$
$\frac{1}{2},\frac{3}{2}$	4	$3\sqrt{3}$	$-\sqrt{6}$	$\sqrt{5}$
$\frac{1}{2},\frac{1}{2}$	$-\sqrt{2}$	0	$2\sqrt{3}$	$2\sqrt{10}$

$\times (1/\sqrt{54})$

	$^1S_0$	$^3P_0$
$\frac{3}{2},\frac{3}{2}$	$\sqrt{2}$	-1
$\frac{1}{2},\frac{1}{2}$	1	$\sqrt{2}$

$\times (1/\sqrt{3})$

and L-odd states are antisymmetric on space exchange.[1] In addition, the LS states that are spin triplets are symmetric on spin exchange, and those that are spin singlets are antisymmetric on spin exchange. On the other hand, not all the jj states have a definite space–spin exchange symmetry even for the case $n_1 = n_2$ we are discussing. Those with $j_1 \neq j_2$ are mixed symmetry states. As we shall discuss in the next section, linear combinations of these do have a well-defined symmetry. At that juncture we shall also introduce the isospin designations.

▶**Exercise 5-1** (a) Using the Clebsch–Gordan coefficients in Appendix A, and Eq. (5-4), show that the LS-coupled states for a neutron and proton in the 1p shell with $J = 0$ are given, in the angular and spin variables, by

$$|^1\mathrm{S}_0\colon M=0\rangle = 3^{-1/2}[Y_1^{\,1}(1)Y_1^{-1}(2) - Y_1^{\,0}(1)Y_1^{\,0}(2) + Y_1^{-1}(1)Y_1^{\,1}(2)]\chi_0^{\,0}(1,2) \tag{5-9}$$

and

$$\begin{aligned}|^3\mathrm{P}_0\colon M=0\rangle = 6^{-1/2}\{&[Y_1^{\,1}(1)Y_1^{\,0}(2) - Y_1^{\,0}(1)Y_1^{\,1}(2)]\chi_1^{-1}(1,2)\\ &- [Y_1^{\,1}(1)Y_1^{-1}(2) - Y_1^{-1}(1)Y_1^{\,1}(2)]\chi_1^{\,0}(1,2)\\ &+ [Y_1^{\,0}(1)Y_1^{-1}(2) - Y_1^{-1}(1)Y_1^{\,0}(2)]\chi_1^{\,1}(1,2)\}.\end{aligned} \tag{5-10}$$

(b) Write out the complete system wave function (including the radial dependence) for LS coupling of a neutron in a $(n_1, l_1 = 1)$ state with a proton in a $(n_2, l_2 = 1)$ state to give $L = 0$, $J = 0$. Is this state space exchange symmetric (i.e., $\mathbf{r}_1 \rightleftharpoons \mathbf{r}_2$) when $n_1 \neq n_2$?

(c) Using (5-6), determine $\Phi(j_1 j_2|JM)$ for these two nucleons when $j_1 = j_2 = \frac{1}{2}$ and $J = M = 0$. Verify the entry in Table 5-1b that

$$(\tfrac{1}{2},\tfrac{1}{2}|00) = (1/\sqrt{3})(|^1\mathrm{S}_0\colon M=0\rangle + \sqrt{2}\,|^3\mathrm{P}_0\colon M=0\rangle).$$

(d) Using as a basis the jj-coupled states $(\frac{1}{2},\frac{1}{2})$ and $(\frac{3}{2},\frac{3}{2})$ each coupled to $J = 0$, evaluate the matrix elements of the *two-body* exchange operator $V_{12} = AI + B\boldsymbol{\sigma}_1\cdot\boldsymbol{\sigma}_2$ (I is the identity operator).

(*Hint:* It is useful to expand these states into LS states in order to facilitate calculating the matrix elements.) By noting that the off-diagonal elements are $\frac{4}{3}\sqrt{2}\,B$, we verify the earlier statement that the jj-coupled states are not eigenfunctions of the Hamiltonian containing the general central two-body exchange interaction, unless it is simply of the Wigner type.

(e) Using as a basis the LS-coupled states $|^1\mathrm{S}_0\rangle$ and $|^3\mathrm{P}_0\rangle$, evaluate the matrix elements of the *one-body* operator $V = a\sum_{i=1,2}\mathbf{l}_i\cdot\mathbf{s}_i$.

(*Hint:* It is useful to expand these states into jj states in order to facilitate calculating the matrix elements.) By noting that the off-diagonal elements

[1] Recall that ϕ_{nlm_l} of (5-2) *includes* the radial function $R_{nl}(r)$ and the symmetry properties quoted here therefore only follow if $n_1 = n_2$.

are $-\sqrt{2}\,\alpha$, we verify the earlier statement that the *LS*-coupled states are not eigenfunctions of the Hamiltonian containing one-body spin–orbit interaction terms. ◀

4. Symmetrization and Isospin

The recognition of the fact that nucleons are fermions imposes certain important symmetry requirements on the wave functions. The symmetry requirements imposed by the statistics for identical nucleons can be extended to all nucleons by the introduction of the isospin variable as discussed in Chapter I.[1] However, at first we shall consider the two types of nucleons as distinguishable and use the description *identical particles* to refer to all neutrons only or all protons only. We must also introduce the designation *equivalent particles* for those cases in which nucleons (either type) occupy single-particle orbitals with the same n and l quantum numbers and in jj coupling also have the same value of j. The designation *nonequivalent particles* refers to those cases in which one or more of the n or l quantum numbers (and in jj coupling the j quantum number considered as well) are different.

For two identical nucleons the Hamiltonian must have exchange symmetry, i.e., $H(1,2) = H(2,1)$, since the two nucleons are indistinguishable. Thus if $\Phi(1,2) = \phi_A(1)\,\phi_B(2)$ is a suitable nonsymmetric solution, the function $\phi_A(2)\,\phi_B(1)$ is also a solution, as are all linear combinations of these two. The required solution for two fermions must be totally antisymmetric, i.e.,

$$\Phi(1,2) = 2^{-1/2}[\phi_A(1)\,\phi_B(2) - \phi_A(2)\,\phi_B(1)]. \tag{5-11}$$

Let us first consider two identical but nonequivalent nucleons in jj coupling, with $n_1 = n_2$ and $l_1 = l_2$ but $j_1 \neq j_2$. Then, in place of the nonsymmetric solutions (5-6), we are now required to write the totally antisymmetric solution

$$\Phi_{\mathrm{A}}(j_1 j_2|JM) = 2^{-1/2}\sum_m (j_1 j_2 m, M-m|JM) \times [\phi_{j_1 m}(1)\,\phi_{j_2, M-m}(2) - \phi_{j_1 m}(2)\,\phi_{j_2, M-m}(1)], \tag{5-12}$$

where we have ϕ referring only to the angular and spin variables and have omitted the radial functions, which are identical since n and l are identical. Also, we have simply written m for m_{j1} and appended the subscript A on the system wave function to indicate total antisymmetrization. The particular case of equivalent nucleons requires further consideration. If we write $m' = M-m$ in the second half of the bracketed term, note from Appendix A that

$$(j_1 j_2 m_1 m_2|JM) = (-1)^{j_1+j_2-J}(j_2 j_1 m_2 m_1|JM), \tag{5-13}$$

[1] Recall that the nuclear force is approximately charge independent.

and take $j_1 = j_2 = j$ (i.e., take the two nucleons to be equivalent), and finally drop the prime on the dummy index m', we get

$$\Phi_A(jj|JM) = \tfrac{1}{2}[1-(-1)^{2j-J}] \sum_m (jjm, M-m|JM)\, \phi_{jm}(1)\, \phi_{j,M-m}(2). \tag{5-14}$$

Here the constant in front of the expression has been altered to maintain proper normalization. Now since j is an odd half-integer, $2j$ is odd, and hence (5-14) identically vanishes unless J is even. Thus for two identical and equivalent nucleons in a configuration $(j)^2$, only the even values of J consistent with the triangle condition are allowed. For example, for $(j=\frac{3}{2})^2$ the triangle condition alone would allow the coupling to give $J = 0$, 1, 2, and 3, but only $J = 0$ and 2 are seen to be allowed by the required antisymmetrization of (5-11) and (5-12).

Let us now adopt the attitude that all nucleons are identical, the neutron–proton difference being relegated to the specification of the three-component of the isospin. Then two equivalent nucleons in a triplet isospin state, $\zeta_1^{0,\pm 1}(1,2)$, require the space–spin part of the system wave function to be antisymmetric, as in (5-12). Hence the conclusions pertaining to (5-14) follow for any two equivalent nucleons in the $T = 1$ isospin state. Conversely, for $T = 0$, equivalent nucleons must have J odd, appropriate for the system wave function being exchange symmetric in the space–spin variables. Note that Table 5-1b shows vanishing 9-j symbol elements for just such space–spin antisymmetric LS-state admixtures in the $(j)^2$ equivalent configurations, even though triangle conditions would "allow" such elements.

Table 5-1b can be readily rewritten for pure $T = 0$ and pure $T = 1$ states for any two nucleons by noting that the LS states (i.e., columns) specify well-defined space and spin exchange symmetries. The LS state $^3S_1$ is thus necessarily associated with $T = 0$, the $^3P_1$ is associated with $T = 1$, etc.

However, care must be exercised for the nonequivalent jj states $(\frac{1}{2}, \frac{3}{2})$ and $(\frac{3}{2}, \frac{1}{2})$. In the second tableau the sum of the second and third rows generates the pure space–spin *symmetric* $^3D_2$ state, viz.,[1]

$$\Phi_A(\tfrac{3}{2},\tfrac{1}{2}|2M) = 2^{-1/2}[\Phi(\tfrac{3}{2},\tfrac{1}{2}|2M)+\Phi(\tfrac{1}{2},\tfrac{3}{2}|2M)] = |^3D_2\!:M\rangle, \tag{5-15}$$

since the $^1D_2$ and $^3P_2$ terms cancel. Paradoxically, a similar addition in the third tableau generates the pure space–spin *antisymmetric* $^3P_1$ state, viz.,

$$\Phi_A(\tfrac{3}{2},\tfrac{1}{2}|1M) = 2^{-1/2}[\Phi(\tfrac{3}{2},\tfrac{1}{2}|1M)+\Phi(\tfrac{1}{2},\tfrac{3}{2}|1M)] = |^3P_1\!:M\rangle. \tag{5-16}$$

[1] We continue to use the subscript A since including isospin exchange as well as space–spin exchange gives total antisymmetry. The symbol A should rather be read to mean "properly symmetrized."

We must, however, recall (5-13), in view of which

$$\Phi(\tfrac{1}{2},\tfrac{3}{2}|2M) = (-1)^{\frac{1}{2}+\frac{3}{2}-2}\Phi(\tfrac{3}{2},\tfrac{1}{2}|2M) = +\Phi(\tfrac{3}{2},\tfrac{1}{2}|2M) \text{ (i.e., with } 1 \rightleftharpoons 2), \tag{5-17a}$$

while

$$\Phi(\tfrac{1}{2},\tfrac{3}{2}|1M) = (-1)^{\frac{1}{2}+\frac{3}{2}-1}\Phi(\tfrac{3}{2},\tfrac{1}{2}|1M) = -\Phi(\tfrac{3}{2},\tfrac{1}{2}|1M) \text{ (i.e., with } 1 \rightleftharpoons 2). \tag{5-17b}$$

In view of these results (5-15) is indeed space–spin symmetric, while (5-16) is, in fact, space–spin antisymmetric.

Finally we arrive at Table 5-2. The entries in Table 5-2 are seen to correspond to multiplying the elements of the second row of Table 5-1a by $\sqrt{2}$ and abolishing the third row. This is generally true even for $l_1 \neq l_2$, etc. under isospin formalism.

Generating the appropriate state functions for coupling three equivalent identical particles is much more complicated. We shall simply illustrate the considerations involved by discussing the particular case of the configuration $(j=\frac{5}{2})^3$. General symmetry considerations, particularly in the LS-coupling scheme, will be discussed in the next section.

We first consider coupling two of the nucleons in jj coupling to form an intermediate state with total angular momentum J' and magnetic quantum number M'. The total angular momentum j of the third nucleon is then coupled to J' to give the desired system values of J and M. Finally a linear combination summing on J' is performed which gives the appropriate overall symmetry and normalization.

Thus we couple the nucleons with coordinate labels (1) and (2) to give[1]

$$\psi_{J',M'}(1,2) = \sum_m (jjm, M'-m|J'M')\,\psi_{jm}(1)\,\psi_{j,M'-m}(2). \tag{5-18}$$

In view of the preceding, we know that J' must be even and obey the triangle condition on (jjJ'). These considerations would allow $J' = 0, 2, 4$. A further future requirement will be the satisfying of the triangle condition on $(jJ'J)$. Thus

$$J' \equiv \text{even}, \qquad 0 \leqslant J' \leqslant 2j, \qquad |J-j| \leqslant J' \leqslant J+j. \tag{5-19}$$

Considering, as an example, the case $(j=\frac{5}{2})^3$ coupled to $J=\frac{3}{2}$ or $J=\frac{9}{2}$, we see from (5-19) that only $J' = 2$ or 4 is permitted. Only for $J=\frac{5}{2}$ are the three possible values of $J' = 0, 2, 4$ allowed. Further, values of $J=\frac{13}{2}$ and $\frac{15}{2}$ are not possible.

[1] Since all magnetic quantum numbers refer to an appropriate total angular momentum J or j, we drop the subscript on M_J and m_j for convenience and simply write M and m.

TABLE 5-2

THE 9-j SYMBOLS FOR TWO 1p-SHELL NUCLEONS IN ISOSPIN FORMALISM[a]

$T = 0$

$$\begin{array}{cc} & {}^3D_3 \\ \tfrac{3}{2},\tfrac{3}{2} & \boxed{1} \end{array} \qquad \begin{array}{cc} & {}^3D_2 \\ (1/\sqrt{2})[(\tfrac{3}{2},\tfrac{1}{2}) \oplus (\tfrac{1}{2},\tfrac{3}{2})]_A & \boxed{1} \end{array}$$

$$\begin{array}{c|ccc|} & {}^3S_1 & {}^1P_1 & {}^3D_1 \\ \tfrac{3}{2},\tfrac{3}{2} & 2\sqrt{5} & \sqrt{30} & -2 \\ (1/\sqrt{2})[(\tfrac{3}{2},\tfrac{1}{2}) \oplus (\tfrac{1}{2},\tfrac{3}{2})]_A & -4\sqrt{2} & 2\sqrt{3} & -\sqrt{10} \\ \tfrac{1}{2},\tfrac{1}{2} & -\sqrt{2} & 2\sqrt{3} & 2\sqrt{10} \end{array} \times (1/\sqrt{54})$$

$T = 1$

$$\begin{array}{cc} & {}^3P_1 \\ (1/\sqrt{2})[(\tfrac{3}{2},\tfrac{1}{2}) \ominus (\tfrac{1}{2},\tfrac{3}{2})]_A & \boxed{1} \end{array} \qquad \begin{array}{c|cc|} & {}^1S_0 & {}^3P_0 \\ \tfrac{3}{2},\tfrac{3}{2} & +\sqrt{2} & -1 \\ \tfrac{1}{2},\tfrac{1}{2} & 1 & \sqrt{2} \end{array} \times (1/\sqrt{3})$$

$$\begin{array}{c|cc|} & {}^1D_2 & {}^3P_2 \\ \tfrac{3}{2},\tfrac{3}{2} & 1 & \sqrt{2} \\ (1/\sqrt{2})[(\tfrac{3}{2},\tfrac{1}{2}) \ominus (\tfrac{1}{2},\tfrac{3}{2})]_A & \sqrt{2} & -1 \end{array} \times (1/\sqrt{3})$$

[a] The subscript A and the addition symbols $\oplus$ and $\ominus$ for the jj states imply space–spin symmetrization for $T = 0$ and space–spin antisymmetrization for $T = 1$; see (5-17a) and (5-17b).

We next couple the third nucleon to an allowed value of J' to obtain in obvious notation

$$|j^2(J')j\colon J, M\rangle = \sum_{m'} (jJ'm', M-m'|JM)\,\psi_{J', M-m'}(1,2)\,\psi_{j,m'}(3). \quad (5\text{-}20)$$

Finally, the desired state $|J, M\rangle$ is written as a suitable linear combination of states (5-20) summed on the intermediate angular momentum J', viz.,

$$|J, M\rangle = \sum_{J'} F_{J'}\,|j^2(J')j\colon J, M\rangle. \quad (5\text{-}21)$$

The coefficients $F_{J'}$ are called the *coefficients of fractional parentage* and are customarily written in a notation indicating the relevant quantities, namely, $F_{J'} \equiv [j^2(J')jJ|\}j^3J]$. Since (5-18) is already antisymmetric on the exchange (1) $\rightleftharpoons$ (2), the chosen values of $F_{J'}$ must result in antisymmetrization on the exchange (1) $\rightleftharpoons$ (3), whereupon the remaining antisymmetric behavior for the exchange (2) $\rightleftharpoons$ (3) will automatically follow. The coefficients of fractional parentage are by custom usually defined to be real and normalized. It is clear that if the conditions (5-19) were to permit only one value of J', the above symmetry requirement involving nucleon (3) could not be satisfied. Or to state it another way, values of J that produce but one value of J' in view of (5-19) are not permitted. Thus, for $(j = \frac{5}{2})^3$, values of $J = \frac{1}{2}$ and $\frac{11}{2}$ are not possible. Since it is beyond the scope of this text to develop further the properties of the coefficients of fractional parentage, the interested reader is referred to the standard works [e.g., Rose (1957)]. We close this discussion by citing in Table 5-3 the values of $F_{J'}$ for the case of identical and equivalent particles for the $(j = \frac{5}{2})^3$ configuration. We note that the nonzero entries in Table 5-3 conform to the discussion given but in addition no state with $J = \frac{7}{2}$ is possible due to an "accidental" failure to produce the required antisymmetrization.

TABLE 5-3

THE COEFFICIENTS OF FRACTIONAL PARENTAGE FOR $(j = \frac{5}{2})^3$

J	F_0	F_2	F_4
$\frac{3}{2}$	0	$(\frac{5}{7})^{1/2}$	$-(\frac{2}{7})^{1/2}$
$\frac{5}{2}$	$\sqrt{\frac{2}{3}}$	$-(\frac{5}{2})^{1/2}/3$	$-(\frac{1}{2})^{1/2}$
$\frac{9}{2}$	0	$(\frac{3}{14})^{1/2}$	$-(\frac{11}{14})^{1/2}$

The extension of the coupling scheme to more than three identical nucleons becomes quite tedious. We simply cite in Table 5-4 the results for k identical equivalent nucleons in configurations $(j)^k$. This table lists the possible J values. When both valence neutrons and protons are present in configurations

$\nu(n_1 l_1 j_1)^{k_1}$ and $\pi(n_2 l_2 j_2)^{k_2}$, these separately antisymmetrized states $|J_\nu, M_\nu\rangle$ and $|J_\pi, M_\pi\rangle$ can then be coupled to give the final system state $|J, M\rangle$. This natural "generic" procedure is not, of course, the only unique scheme and many coupling procedures are possible. Generally for $k_1 + k_2 > 3$ group-theoretic methods are employed to classify the many possible states that can be formed. An elementary discussion of some of the general considerations involved is given in the next section.

TABLE 5-4

POSSIBLE TOTAL ANGULAR MOMENTA IN THE CONFIGURATION $(j)^k$ FOR IDENTICAL PARTICLES[a]

	k	
$j = \frac{1}{2}$		
	1	$\frac{1}{2}$
$j = \frac{3}{2}$		
	1	$\frac{3}{2}$
	2	0, 2
$j = \frac{5}{2}$		
	1	$\frac{5}{2}$
	2	0, 2, 4
	3	$\frac{3}{2}, \frac{5}{2}, \frac{9}{2}$
$j = \frac{7}{2}$		
	1	$\frac{7}{2}$
	2	0, 2, 4, 6
	3	$\frac{3}{2}, \frac{5}{2}, \frac{7}{2}, \frac{9}{2}, \frac{11}{2}, \frac{15}{2}$
	4	0, 2 (twice), 4 (twice), 5, 6, 8
$j = \frac{9}{2}$		
	1	$\frac{9}{2}$
	2	0, 2, 4, 6, 8
	3	$\frac{3}{2}, \frac{5}{2}, \frac{7}{2}, \frac{9}{2}$ (twice), $\frac{11}{2}, \frac{15}{2}, \frac{17}{2}, \frac{21}{2}$
	4	0 (twice), 2 (twice), 3, 4 (3 times), 5, 6 (3 times), 7, 8, 9, 10, 12
	5	$\frac{1}{2}, \frac{3}{2}, \frac{5}{2}$ (twice), $\frac{7}{2}$ (twice), $\frac{9}{2}$ (3 times), $\frac{11}{2}$ (twice), $\frac{13}{2}$ (twice), $\frac{15}{2}$ (twice), $\frac{17}{2}$ (twice), $\frac{19}{2}, \frac{21}{2}, \frac{25}{2}$
$j = \frac{11}{2}$		
	1	$\frac{11}{2}$
	2	0, 2, 4, 6, 8, 10
	3	$\frac{3}{2}, \frac{5}{2}, \frac{7}{2}, \frac{9}{2}$ (twice), $\frac{11}{2}$ (twice), $\frac{13}{2}, \frac{15}{2}$ (twice), $\frac{17}{2}, \frac{19}{2}, \frac{21}{2}, \frac{23}{2}, \frac{27}{2}$
	4	0 (twice), 2 (3 times), 3, 4 (4 times), 5 (twice), 6 (4 times), 8 (4 times), 9 (twice), 10 (3 times), 11, 12 (twice), 13, 14, 16
	5	$\frac{1}{2}, \frac{3}{2}$ (twice), $\frac{5}{2}$ (3 times), $\frac{7}{2}$ (4 times), $\frac{9}{2}$ (4 times), $\frac{11}{2}$ (5 times), $\frac{13}{2}$ (4 times), $\frac{15}{2}$ (5 times), $\frac{17}{2}$ (4 times), $\frac{19}{2}$ (4 times), $\frac{21}{2}$ (3 times), $\frac{23}{2}$ (3 times), $\frac{25}{2}$ (twice), $\frac{27}{2}$ (twice), $\frac{29}{2}, \frac{31}{2}, \frac{35}{2}$
	6	0 (3 times), 2 (4 times), 3 (3 times), 4 (6 times), 5 (3 times), 6 (7 times), 7 (4 times), 8 (6 times), 9 (4 times), 10 (5 times), 11 (twice), 12 (4 times), 13 (twice), 14 (twice), 15, 16, 18

[a] Taken from Mayer and Jensen (1955).

Referring to Table 5-4, the maximum occupancy of any level is $k_{max} = 2j+1$. The value of J for k_{max} is, of course, $J = 0$. Table 5-4 only has entries for $k \leqslant \frac{1}{2}k_{max}$, since for $k > \frac{1}{2}k_{max}$ the results would be the same as for $k' = 2j+1-k$, with k' now representing the number of holes in the filled shell. We note that many values of J recur (several or many times) indicating the clear need for additional quantum numbers to properly characterize such more complex cases.

C. GENERAL SYMMETRY CLASSIFICATIONS, SENIORITY, AND REDUCED ISOSPIN

The coupling of three or more particles leads to spectroscopic states that cannot in general be uniquely specified by citing only the quantum numbers we have been using so far. For example, three *LS*-coupled nucleons in the p shell can generate two ^{22}P states (each with $2T+1 = 2$, $2S+1 = 2$, and $L = 1$). Similarly, four *jj*-coupled nucleons in the $p_{3/2}$ shell leads to the possible formation of two $J = 2$, $2T+1 = 1$ states. Other examples are evident in Table 5-4 for identical particles, as we have already seen. Additional quantum numbers are needed to further characterize such states.

We recall that state wave functions that are antisymmetric in the interchange of the space coordinates of two nucleons essentially vanish when this pair of nucleons are close to each other; conversely for the symmetric case the wave function exhibits a maximum amplitude when the said pair are close to each other. To the extent that nuclear forces are short-ranged, attractive, and spin independent, a system of nucleons having the highest possible space symmetry (i.e., the largest number of space symmetric pairs of nucleons) will therefore produce the greatest binding. Thus it might be expected that a specification of symmetry of the state vector with respect to the space exchange of pairs of nucleons would represent a useful classification in determining level energies under the action of residual nucleon–nucleon forces.

In the *LS* coupling scheme two nucleons will give rise to two possible values of total intrinsic spin $S = 0$ or 1 and two values of total isospin $T = 0$ or 1. Such states, of course, are either symmetric, T or $S = 1$, or antisymmetric, T or $S = 0$, under the appropriate exchange of the spin or isospin coordinates. The requirement that the entire wave function be totally antisymmetric in the exchange of *all* coordinates of the pair then requires that the spatial portion of the wave function be either symmetric (L even) or antisymmetric (L odd). Hence suitable basis states in terms of which any state can be specified are simply those exhibited in (1-13), namely

$$^{13}L(\text{even}) \quad \text{and} \quad ^{31}L(\text{even}) \tag{5-22a}$$

$$^{11}L(\text{odd}) \quad \text{and} \quad ^{33}L(\text{odd}). \tag{5-22b}$$

For three or more nucleons the construction of a suitable basis requires the inclusion of functions possessing *intermediate symmetry* under the pairwise exchanges of space coordinates. These functions must be coupled, of course, to suitable functions of intermediate symmetry in spin and isospin in order to conform to the requirements of total antisymmetrization. It is possible to classify such basis states into sets or subspaces of states that are invariant (up to a possible sign change) under any *permutation* of particle coordinates. For k nucleons, the various possible sets are designated by a partition $[\lambda] \equiv [\lambda_1, \lambda_2, \lambda_3, \ldots, \lambda_p]$. The λ_i are so defined that

$$\lambda_1 \geqslant \lambda_2 \geqslant \lambda_3 \geqslant \cdots \geqslant \lambda_p \tag{5-23a}$$

and

$$\sum_i \lambda_i = k. \tag{5-23b}$$

Clearly, λ_1 may not exceed k. When $[\lambda] = [k]$, i.e., $\lambda_1 = k$, the state in question is totally symmetric in the exchange of the space coordinates of *any* pair of nucleons. For any selection of single-particle states, there is only one member to the set $[\lambda] = [k]$ and clearly any permutation of coordinates simply reproduces the original function. The largest number of terms λ_i in the partition array occurs when $\lambda_1 = \lambda_2 = \lambda_3 = \cdots = \lambda_p = 1$, whence $p = k$. This set again only possesses a single element and is totally antisymmetric in the exchange of the space coordinates of *any* pair of nucleons. In this case the permutation of coordinates generates a factor $(-1)^\pi$ multiplying the original function, where π denotes the parity of the permutation, which in effect leaves the set invariant.[1]

Other partitions among possible λ_i arrays consistent with (5-23) yield states of intermediate symmetry.

An example will clarify these points and should also suggest the possibility that group-theoretic methods may prove very useful for the utilization of such classifications. Consider the direct product space wave function for three particles ($k = 3$) formed from among the three single-particle functions u, v, and w. When all three particles involve the same spatial function, taking u for an example, only the totally symmetric product state

$$U([\lambda] = [3]) = u(1)\,u(2)\,u(3) \tag{5-24}$$

can be formed and, of course, no state antisymmetric in the exchange of *any* pair can be constructed.

[1] All permutations can be regarded as consisting of a number of simple permutations in which $k-2$ coordinates are kept fixed and the other two exchanged. The number of such simple interchanges to yield any given permutation is uniquely odd or even. The parameter π is even when an even number of simple permutations is required, and π is odd when an odd number is required.

When two functions are the same and the remaining one is selected to be different, then in addition to product states that are totally symmetric, product states can also be formed that are *at most* antisymmetric in the exchange of a single pair. The former construction again belongs to the symmetry class $[\lambda] = [3]$, while the latter belongs to the class $[2, 1]$. It might be thought that for the choice u, u, and v, for example, three linearly independent states can be formed antisymmetric in the exchange of any one of the three pairs that can be selected from among three particles, viz.,

$$U_A([\lambda] = [2,1]) = [u(1)\,v(3) - v(1)\,u(3)]\,u(2) \tag{5-25a}$$

$$U_B([\lambda] = [2,1]) = [u(2)\,v(1) - v(2)\,u(1)]\,u(3) \tag{5-25b}$$

$$U_C([\lambda] = [2,1]) = [u(2)\,v(3) - v(2)\,u(3)]\,u(1). \tag{5-25c}$$

However, clearly only one pair of these equations is linearly independent; note, for example, that $U_C = U_A + U_B$. Moreover, linear combinations of (5-25) can be formed that are symmetric in the exchange of any one pair of coordinates. For example,

$$U_D([\lambda] = [2,1]) = U_B - U_A = 2v(1)\,u(2)\,u(3) - u(1)[u(2)\,v(3) + v(2)\,u(3)] \tag{5-25d}$$

is seen to be symmetric in the exchange $2 \rightleftharpoons 3$. Finally, for the case under discussion, the symmetry class $[2, 1]$ can be specified by a linearly independent pair of states such as U_A and U_B antisymmetric in the exchange $1 \rightleftharpoons 3$ for the one and antisymmetric in the exchange $1 \rightleftharpoons 2$ for the other, *or* the specification can be achieved by the use of a linearly independent pair of states such as U_C and U_D one of which is antisymmetric in the exchange $2 \rightleftharpoons 3$ while the other is symmetric in the *same* exchange $2 \rightleftharpoons 3$, *or* by use of a pair of states such as U_D and $U_E = -(2U_B + U_A)$, one symmetric in the exchange $2 \rightleftharpoons 3$ and the other symmetric in the exchange $1 \rightleftharpoons 3$. The *dimensionality* of the subspace $[\lambda] = [2, 1]$ thus is seen to be twofold in any event. All other states of partial symmetry are simply various linear combinations of the selected states. Any permutation of coordinates of nucleons in the selected two-member set will generate in this case a linear combination of the two, but will not involve any members of the other possible symmetry classes and in that sense leaves the group invariant. Selecting U_A and U_B for our two-dimensional representation, for example, we note that the permutation $1 \to 2$, $2 \to 3$, and $3 \to 1$ applied to U_A gives U_B, while this same permutation applied to U_B gives $-U_C = -(U_A + U_B)$, etc. It is also useful to note that the set can be completely generated starting from a single element by applying all possible coordinate permutations and selecting from the resulting terms and their linear combinations any maximal number of linearly independent terms or functions. The number of such elements will be equal to the dimensionality of the set or subspace. In the present case, starting with U_A, through permuting the coordinates, we readily generate $\pm U_A$, $\pm U_B$, and $\pm U_C$, from which we also generate U_D or U_E, etc., by taking suitable linear combinations.

Only when three different functions u, v, and w are admitted can the totally antisymmetric state $[\lambda] = [111]$ be formed, viz.,

$$U([\lambda] = [111]) = \det |u(1)\, v(2)\, w(3)|. \tag{5-26}$$

For this selection, states of symmetry $[\lambda] = [3]$ and $[2, 1]$ can also be formed, of course.

▶**Exercise 5-2** Consider three nucleons in the p shell (i.e., $k = 3, l = 1$) and using LS coupling take the various space functions U to be eigenfunctions of $\mathbf{L}^2$ and L_z. The linearly independent single-particle orbital functions that are of concern are $Y_1^{\,1}$, Y_1^{-1}, and $Y_1^{\,0}$.

Show or discuss as is appropriate, the following:

(a-i) The state $[\lambda] = [3]$ with $M = 3$ is unique, namely $U([3], M = 3) = Y_1^{\,1}(1)\, Y_1^{\,1}(2)\, Y_1^{\,1}(3)$, and must clearly correspond to $L = 3$.

(a-ii) The state $[\lambda] = [3]$ with $M = 2$ is also unique and is, in properly symmetrized form,

$$U([3], M = 2) = Y_1^{\,1}(1)\, Y_1^{\,1}(2)\, Y_1^{\,0}(3) + Y_1^{\,1}(1)\, Y_1^{\,0}(2)\, Y_1^{\,1}(3)$$
$$+ Y_1^{\,0}(1)\, Y_1^{\,1}(2)\, Y_1^{\,1}(3).$$

Since $L = 3$, $M = 2$ *must occur*, this function must be for $L = 3$. Thus $L = 2$ cannot be formed in the partition $[\lambda] = [3]$ of totally symmetric spatial functions.

(a-iii) The states $[\lambda] = [3]$ with $M = 1$ can be constructed from either of the symmetrized products

$$U_a([3], M = 1) = Y_1^{\,1}(1)\, Y_1^{\,0}(2)\, Y_1^{\,0}(3) + Y_1^{\,0}(1)\, Y_1^{\,1}(2)\, Y_1^{\,0}(3)$$
$$+ Y_1^{\,0}(1)\, Y_1^{\,0}(2)\, Y_1^{\,1}(3)$$

or

$$U_b([3], M = 1) = Y_1^{\,1}(1)\, Y_1^{\,1}(2)\, Y_1^{-1}(3) + Y_1^{\,1}(1)\, Y_1^{-1}(2)\, Y_1^{\,1}(3)$$
$$+ Y_1^{-1}(1)\, Y_1^{\,1}(2)\, Y_1^{\,1}(3).$$

Appropriate linear combinations of these correspond to $L = 3$, $M = 1$ and $L = 1$, $M = 1$ states, namely $2U_a + U_b$ for the former, and $U_a - 2U_b$ for the latter.

(a-iv) The states $[\lambda] = [3]$, $M = 0$ can be constructed two ways also, namely

$$U_a([3], M = 0) = Y_1^{\,0}(1)\, Y_1^{\,0}(2)\, Y_1^{\,0}(3)$$

and

$$U_b([3], M=0) = Y_1^{\,1}(1)\,Y_1^{\,0}(2)\,Y_1^{-1}(3) + Y_1^{\,0}(1)\,Y_1^{-1}(2)\,Y_1^{\,1}(3) + Y_1^{-1}(1)\,Y_1^{\,1}(2)\,Y_1^{\,0}(3).$$

These generate the $L=3$, $M=0$ and $L=1$, $M=0$ states.

(b-i) Clearly no state with $M=3$ can be formed with the symmetry $[\lambda]=[2,1]$, and thus no state with $L=3$ exists of this partial symmetry.

(b-ii) The states $[\lambda]=[2,1]$ and $M=2$ can be formed only from $Y_1^{\,1}$, $Y_1^{\,1}$, and $Y_1^{\,0}$. A simple choice for a basis corresponding to U_A and U_B of (5-25) is

$$U_A([2,1], M=2) = [Y_1^{\,1}(1)\,Y_1^{\,0}(3) - Y_1^{\,0}(1)\,Y_1^{\,1}(3)]\,Y_1^{\,1}(2)$$

and

$$U_B([2,1], M=2) = [Y_1^{\,1}(2)\,Y_1^{\,0}(1) - Y_1^{\,0}(2)\,Y_1^{\,1}(1)]\,Y_1^{\,1}(3).$$

These states must be for $L=2$.

(b-iii) The states $[\lambda]=[2,1]$ and $M=1$ can be formed from either the combination $Y_1^{\,1}$, $Y_1^{\,0}$, and $Y_1^{\,0}$; or $Y_1^{\,1}$, $Y_1^{\,1}$, and Y_1^{-1}. For each choice we have a two-dimensional representation, viz.,

$$U_A([2,1], M=1) = [Y_1^{\,1}(1)\,Y_1^{\,0}(3) - Y_1^{\,0}(1)\,Y_1^{\,1}(3)]\,Y_1^{\,0}(2)$$

$$U_B([2,1], M=1) = [Y_1^{\,1}(1)\,Y_1^{\,0}(2) - Y_1^{\,0}(1)\,Y_1^{\,1}(2)]\,Y_1^{\,0}(3)$$

and

$$\bar{U}_A([2,1], M=1) = [Y_1^{\,1}(1)\,Y_1^{-1}(3) - Y_1^{-1}(1)\,Y_1^{\,1}(3)]\,Y_1^{\,1}(2)$$

$$\bar{U}_B([2,1], M=1) = [Y_1^{\,1}(1)\,Y_1^{-1}(2) - Y_1^{-1}(1)\,Y_1^{\,1}(2)]\,Y_1^{\,1}(3).$$

These states are linear combinations of $L=2$ and $L=1$.

(b-iv) The states $[\lambda]=[2,1]$ and $M=0$ can be formed as four linearly independent combinations from the nine functions of the type

$$U([2,1], M=0) = [u(a)\,v(b) - v(b)\,u(a)]\,w(c),$$

in which u, v, and w stand for $Y_1^{\,1}$, Y_1^{-1}, and $Y_1^{\,0}$ and all their permutations and the coordinates a, b, and c are allowed to be permuted as well. These states correspond to $L=2$ and $L=1$.

(c) The state $[\lambda]=[111]$, $M=0$ is unique and is the totally antisymmetric function formed from the Slater determinant of $Y_1^{\,1}$, Y_1^{-1}, and $Y_1^{\,0}$. It corresponds to $L=0$. ◀

It should be apparent from the discussion to this point that the *symmetry quantum number* specified by the array $[\lambda]$ does not in fact determine a unique set of basis states, but as is appropriate specifies only their symmetry characteristics under pairwise coordinate exchanges. A convenient way of visualizing

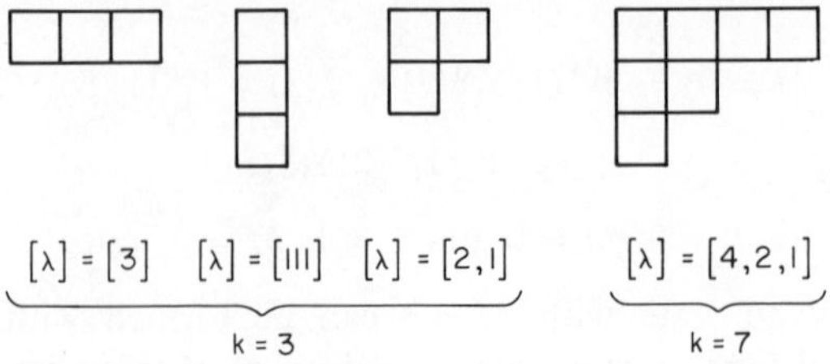

FIG. 5.1. Young tableaux for various symmetry classes $[\lambda]$.

the relevant relationships is to associate with each partition $[\lambda_1, \lambda_2, \lambda_3, \ldots, \lambda_p]$ an array of blocks arranged in a Young tableau, with λ_i blocks in the ith row. Figure 5.1 shows some examples. The length of a row determines the *largest* number of coordinates in which the set of states of partial symmetry can be taken symmetric in pairwise exchanges of coordinates. The length of a column determines the *largest* number of coordinates in which the set of states can be constructed antisymmetric in pairwise exchange of coordinates. The dimensionality of the set is determined by the number of ways the blocks can be filled in with the coordinate labels 1, 2, 3, ..., k such that "values" increase from left to right and from top to bottom.[1] Thus the tableau corresponding to $[\lambda] = [3]$ is unique and represents a totally symmetric state. The tableau $[\lambda] = [2, 1]$ is seen to be two dimensional, in agreement with the findings for the examples cited previously.

With each set $[\lambda]$ we can associate an *adjoint* or *dual* set $[\tilde{\lambda}]$ obtained by interchanging rows and columns of the tableau $[\lambda]$. Thus, for example, for $[\lambda] = [4, 2, 1]$ of Fig. 5.1, we obtain $[\tilde{\lambda}] = [3, 2, 1, 1]$. It can then be shown that the appropriate spin–isospin function to associate with the spatial function having symmetry $[\lambda]$ is in fact one of symmetry $[\tilde{\lambda}]$.

Spin–isospin functions are generated from the four linearly independent functions

$$\chi^{1/2}\zeta^{1/2}; \quad \chi^{1/2}\zeta^{-1/2}; \quad \chi^{-1/2}\zeta^{1/2}; \quad \chi^{-1/2}\zeta^{-1/2}. \tag{5-27}$$

For example, one state for $k = 2$ might be

$$\begin{aligned}\Omega &= (\chi^{1/2}\zeta^{1/2}\colon 1)(\chi^{1/2}\zeta^{-1/2}\colon 2) - (\chi^{1/2}\zeta^{-1/2}\colon 1)(\chi^{1/2}\zeta^{1/2}\colon 2) \\ &= \chi^{1/2}(1)\zeta^{1/2}(1)\chi^{1/2}(2)\zeta^{-1/2}(2) - \chi^{1/2}(1)\zeta^{-1/2}(1)\chi^{1/2}(2)\zeta^{1/2}(2) \\ &= \chi^{1/2}(1)\chi^{1/2}(2)[\zeta^{1/2}(1)\zeta^{-1/2}(2) - \zeta^{-1/2}(1)\zeta^{1/2}(2)],\end{aligned} \tag{5-28}$$

and clearly corresponds to $S = 1$, $S_z = +1$, $T = 0$, and $T_3 = 0$. It is seen to be antisymmetric on the exchange of the spin–isospin variables of the two particles and therefore has the symmetry classification $[\lambda] = [1, 1]$. The

[1] Blocks filled with coordinate labels in this manner give rise to so-called *standard arrangements*.

corresponding admissible radial function must have the symmetry $[\tilde{\lambda}] = [2]$ (i.e., must be symmetric in spatial coordinate exchange) and for two particles in the p shell can have either $L = 0$ or $L = 2$.

Quite general limitations apply to the structure of the tableaux $[\lambda]$ and $[\tilde{\lambda}]$. If $[\lambda]$ refers to the shell configuration $(l)^k$, which affords at most $2l+1$ linearly independent orbital functions, then the largest number of rows in the tableau $[\lambda]$ cannot exceed $2l+1$ since this is the largest number of particles (particle coordinates) in which antisymmetrization is possible. Similarly, since the spin–isospin functions are limited to the four of (5-27), the largest number of rows in the corresponding tableau $[\tilde{\lambda}]$ cannot exceed four. This in turn requires that the longest length of any row in $[\lambda]$ cannot exceed four. Applied to the p shell, the largest tableau $[\lambda]$ is thus $[4, 4, 4]$ and corresponds to the closed shell with 12 particles.

Finally, we note that the operators $\mathbf{L}^2 = (\sum_i \mathbf{l}_i)^2$ and $L_z = \sum_i l_{iz}$ are symmetric in particle coordinates and hence commute with any permutation operator of particle coordinates. Since the set of basic states corresponding to the symmetry $[\lambda]$ is related through permutation of particle coordinates, it is possible to select a set such that all elements are simultaneous eigenfunctions of $\mathbf{L}^2$ and L_z. Exercise 5-2 provided such examples. Similarly spin–isospin functions with definite S, S_z, T, and T_3 quantum numbers can be constructed for a particular symmetry group $[\lambda]$. Thus the proper wave function can be designated as

$$\psi([\lambda], L, S, T, L_z, S_z, T_3).$$

Table 5-5 shows such a classification for the p shell. Note that the entries for $k = 3$ are in agreement with the stated results of Exercise 5-2.

Table 5-5 need only include entries up to $k = 6$, since for $7 \leqslant k \leqslant 12$ we specify the configuration by the number of holes rather than particles. Where multiple entries exist in the last two columns, *any* combination is a permissible state. Thus for $k = 6$ and $[\lambda] = [4, 1, 1]$ four states are possible, namely, ^{11}P, ^{33}P, ^{11}F, and ^{33}F.

Table 5-5 shows that the introduction of the symmetry quantum number $[\lambda]$ did not complete the job of uniquely specifying the admissible states for the p shell with $k = 6$. For the D states with $[\lambda] = [4, 2]$, there are two states with $(2S+1, 2T+1) = (1, 3)$ and also two with $(2S+1, 2T+1) = (3, 1)$. There are also two $(2S+1, 2T+1) = (3, 3)$ states for $[\lambda] = [3, 2, 1]$ for both $L = 1$ and $L = 2$. In these cases additional quantum numbers are required.

The reader wishing further details employing the above considerations is referred to the excellent work of de-Shalit and Talmi (1963). A thorough recent discussion of nuclear symmetries particularly reviewing the realistic applicability or "goodness" of pure symmetry models for observed nuclear state characteristics is by Hecht (1973).

TABLE 5-5

LS-COUPLED WAVE FUNCTIONS FOR THE p SHELL

k	$[\lambda]$	L	$(2S+1, 2T+1)$
0	[0]	0	(1,1)
1	[1]	1	(2,2)
2	[2]	0,2	(1,3)(3,1)
	[1,1]	1	(1,1)(3,3)
3	[3]	1,3	(2,2)
	[2,1]	1,2	(2,2)(2,4)(4,2)
	[1,1,1]	0	(2,2)(4,4)
4	[4]	0,2,4	(1,1)
	[3,1]	1,2,3	(1,3)(3,1)(3,3)
	[2,2]	0,2	(1,1)(1,5)(5,1)(3,3)
	[2,1,1]	1	(1,3)(3,1)(3,3)(3,5)(5,3)
5	[4,1]	1,2,3,4	(2,2)
	[3,2]	1,2,3	(2,2)(2,4)(4,2)
	[3,1,1]	0,2	(2,2)(2,4)(4,2)(4,4)
	[2,1,1]	1	(2,2)(2,4)(4,2)(2,6)(6,2)(4,4)
6	[4,2]	$0,2^2,3,4$	(1,3)(3,1)
	[4,1,1]	1,3	(1,1)(3,3)
	[3,3]	1,3	(1,1)(3,3)
	[3,2,1]	1,2	$(1,3)(3,1)(3,3)^2(1,5)(5,1)(3,5)(5,3)$
	[2,2,2]	0	(1,3)(3,1)(3,5)(5,3)(1,7)(7,1)

Except for the lighter nuclei ($A \lesssim 10$), the low-energy states are more simply characterized in *jj* coupling than in *LS* coupling. A particularly appropriate classification of *jj*-coupled states, which otherwise remain incompletely specified when only the quantum numbers J and T are given, is the *seniority* and *reduced isospin* scheme.

It is perhaps useful first to review briefly the relationship of the ordinary isospin T to the wave function symmetry of a state having k-valence particles (both neutrons and protons) in the configuration $(nlj)^k$.[1] The state $|(j)^k JT\rangle$ can be characterized by the number of pairs $\mathscr{E}$ of particles that are in relative symmetric space and spin states and hence are in relative isospin singlet states, and $\mathscr{O} = k - 2\mathscr{E}$ remaining particles that are antisymmetrically coupled in space and spin and thus symmetrically coupled in isospin. Since the isospin singlet pairs $\mathscr{E}$ can only couple to $T(\mathscr{E}) = 0$, the final isospin of the state is

[1] The situation in which valence neutrons and protons occupy different shells, $(nlj)^k$ for neutrons and $(n'l'j')^{k'}$ for protons (important when $A \gtrsim 60$), will not be considered in what follows. The treatment of these cases is more complicated but does not require the introduction of any essentially new concepts.

completely determined by the $\mathcal{O}$ particles, and clearly $T = \mathcal{O}/2$. The situation thus can be summarized by stating that $\mathcal{O} = 2T$ gives the number of particles left after all relative isospin singlet ($T_{ij} = 0$) pairs have been decoupled from the state in question.

By a similar procedure, we might imagine decoupling the state $|(j)^k JT\rangle$ into a pair configuration coupled to $J_{12} = 0$ and $T_{12} = 1$ (i.e., a space–spin antisymmetric $J = 0$ pair), with the remaining $k-2$ particles coupled to J and isospin T' with $T-1 \leqslant T' \leqslant T+1$. Properly antisymmetrizing the resulting product configuration, we obtain

$$|(j)^k JT\rangle = \mathscr{A}_k\{|(j)^{k-2} JT'\rangle\,|(j)^2 J_{12} = 0, T_{12} = 1\rangle\}. \tag{5-29}$$

It might be possible to further continue the pairwise $J_{ij} = 0$, $T_{ij} = 1$ decoupling by next applying it to one of the possible states $|(j)^{k-2} JT'\rangle$ and so on. This procedure will eventually terminate with a principal parent $|(j)^{\mathscr{s}} J, T' = \mathscr{t}\rangle$ that cannot be further decoupled pairwise, either because applying the antisymmetrization operator $\mathscr{A}_k$ would result in the state function vanishing or because the required vector additions fail (i.e., $J > \mathscr{s}j$ and a similar appropriate condition for isospin). The quantities $\mathscr{s}$ and $\mathscr{t}$, beyond which $|(j)^{\mathscr{s}} J, T' = \mathscr{t}\rangle$ cannot be further decoupled, define the seniority $\mathscr{s}$ and the reduced isospin $T' = \mathscr{t}$.

▶**Exercise 5-3** As simple examples consider $k = 1$ and $k = 2$ for the configuration $(j = \frac{3}{2})^k$. Show or discuss as appropriate the following:

(a) For $k = 1$, clearly $\mathscr{s} = 1$ and $\mathscr{t} = \frac{1}{2}$ (also $\mathscr{E} = 0$ and $\mathcal{O} = 1$) with $J = \frac{3}{2}$, $T = \frac{1}{2}$.

(b) For $k = 2$, two values are possible for $\mathscr{s}$, namely $\mathscr{s} = 0$ or $\mathscr{s} = 2$. For the first case with $\mathscr{s} = 0$, clearly $\mathscr{t} = 0$ (also $\mathscr{E} = 0$ and $\mathcal{O} = 2$), hence $J = 0$, $T = 1$.

The case with $\mathscr{s} = 2$ could have either $\mathscr{t} = 0$ or 1. With $\mathscr{t} = 0$ (and $\mathscr{E} = 1$, $\mathcal{O} = 0$) only the J-odd states are possible, giving either $J = 1$ or 3.

The case with $\mathscr{s} = 2$, $\mathscr{t} = 1$ (and hence $\mathscr{E} = 0$, $\mathcal{O} = 2$) gives only the J-even state $J = 2$. ◀

Extending the considerations of Exercise 5-3 for the $j = \frac{3}{2}$ shell to the more complex situations of $k = 3$ and $k_{\max} = 2j+1 = 4$ yields the results summarized in Table 5-6. The entries with $T = k/2$ are clearly for identical particles and the allowed J values are seen to agree with the entries in Table 5-4 for $j = \frac{3}{2}$. The values $(\mathscr{E}, \mathcal{O})$ are redundant and are only given for completeness. Finally, we see that in this case every state is uniquely specified by $|(j)^k JT\mathscr{s}\mathscr{t}\rangle$.

It can be shown that the states $|(j)^k JT\mathscr{s}\mathscr{t}\rangle$ are diagonal not only for spin–orbit residual interactions (since we are dealing with jj representations) but also for residual central interactions having either a Wigner or Heisenberg

TABLE 5-6

SENIORITY AND REDUCED ISOSPIN CLASSIFICATION OF $(j=\frac{3}{2})^k$

k	s	t	$(\mathscr{E},\mathscr{O})$	T	J
0	0	0	—	0	0
1	1	$\frac{1}{2}$	(0, 1)	$\frac{1}{2}$	$\frac{3}{2}$
2	0	0	(0, 2)	1	0
	2	0	(1, 0)	0	1, 3
	2	1	(0, 2)	1	2
3	1	$\frac{1}{2}$	(1, 1)	$\frac{1}{2}$	$\frac{3}{2}$
	1	$\frac{1}{2}$	(0, 3)	$\frac{3}{2}$	$\frac{3}{2}$
	3	$\frac{1}{2}$	(1, 1)	$\frac{1}{2}$	$\frac{1}{2}, \frac{5}{2}, \frac{7}{2}$
4	0	0	(2, 0)	0	0
	0	0	(0, 4)	2	0
	2	0	(1, 2)	1	1, 3
	2	1	(1, 2)	1	2
	2	1	(2, 0)	0	2
	4	0	(2, 0)	0	2, 4

exchange character. While residual central interactions having a Majorana or Bartlett exchange character generally introduce nondiagonal matrix elements, these are usually small for short-range potentials. This last result follows when we note, from the discussion in Chapter I, that in the short-range limit of a delta-function interaction Majorana potentials become equivalent to Wigner potentials and Bartlett to Heisenberg.

It is fortunate that closed expressions for the interaction energy of the states $|(j)^k JTst\rangle$ can be obtained for residual central interactions of the form (1-97) for two special cases. With only ordinary Wigner forces in the short-range limit $v(r_{ij}) = V(r_{ij}) = V_0\,\delta(r_{ij})$, and k the total number of neutrons and protons, we have[1]

$$E[(j)^k JTst] \approx \{[\tfrac{1}{2}k(k-1)] + [\tfrac{3}{2}k - 2T(T+1)] + [k(j+1) - \tfrac{1}{2}g(s,t)]\}\,E_0, \tag{5-30}$$

with

$$E_0 = \frac{2j+1}{j+1}\,\frac{V_0}{16\pi}\int_0^\infty [u_{nl}(r)]^4 r^2\,dr,$$

and

$$g(s,t) = (2j+2)s - \tfrac{1}{2}s(s-1) + \tfrac{3}{2}s - 2t(t+1).$$

[1] The effective δ-function strength V_0 can be taken as $V_0 = 4\pi\int_0^\infty V(r)r^2\,dr$, and $u_{nl}(r)$ is the single-particle radial wave function normalized to give $\int_0^\infty [u_{nl}(r)]^2 r^2\,dr = 1$.

For the case with the general central interaction (1-97),

$$v(r_{ij}) = V(r_{ij})[\mathscr{a} + \mathscr{b}P^{\mathrm{B}} + \mathscr{c}P^{\mathrm{H}} + \mathscr{d}P^{\mathrm{M}}],$$

again considered in the short-range limit $V(r_{ij}) = V_0\,\delta(r_{ij})$, *and* for all k particles of the *same type* so that $T = k/2$ and $\mathscr{t} = \mathscr{s}/2$, we have

$$E[(j)^k J\mathscr{s}] \approx (\mathscr{a} + \mathscr{d} - \mathscr{c} - \mathscr{b})\{(j+1)k - \tfrac{1}{2}\mathscr{s}(2j - \mathscr{s} + 3)\}\,E_0, \qquad (5\text{-}31)$$

where E_0 is defined as before.

The expression (5-31) has some interesting consequences. Note from Exercise 1-18 and (1-99) that $V_0[\mathscr{a} + \mathscr{d} - \mathscr{c} - \mathscr{b}] = \alpha\,{}^{31}V_0$ is just the attractive $^1\mathrm{S}$(n–p) interaction. Since in addition the integral defining E_0 is greater than zero and $\mathscr{s}$ is necessarily less than $2j$, it follows from (5-31) that the levels $E[(j)^k J\mathscr{s}]$ conform to an ordering sequence giving a higher energy the larger the seniority $\mathscr{s}$.[1] For k even, the lowest value of $\mathscr{s}$ is zero, which in turn implies $J = 0$. Thus valence neutrons (or protons) alone would guarantee a ground state of $J = 0$. For k odd, the lowest value of the seniority is $\mathscr{s} = 1$, which implies $J = j$. Thus an odd number of like valence particles would give $J = j$ for the ground state. These results are, of course, in agreement with the predictions of the extreme single-particle model discussed in Section IV, H, 1. Indeed, even when both valence neutrons and protons are allowed, the cases of odd-A nuclei as well as Z-even, N-even nuclei are also expected to agree with the extreme single-particle model. The case Z odd, N odd follows the predictions already discussed under the Nordheim rules in Chapter IV.

We can also observe from (5-31), for k even and the definition of E_0, that ground-state (i.e., $\mathscr{s} = 0$) interaction energies are proportional to the factor $(2j+1)$, in agreement with the statements made in connection with the pairing energy terms (3-77).

Even under the assumption of the validity of the seniority coupling scheme, a closed-form expression for nuclear energies is not possible. However, elegant group-theoretic methods can be employed to generate a useful general parametric expression which has, at the least, proved surprisingly accurate for predicting ground-state energies of nuclei. It is shown that the barycentric energy of a group of levels for a given set of values $(\mathscr{s}, \mathscr{t})$ defined by

$$E[(j)^k \mathscr{s}\mathscr{t}] = \left\{\sum^{J} (2J+1)\,E[(j)^k JT\mathscr{s}\mathscr{t}]\right\}\Big/\sum^{J} (2J+1) \qquad (5\text{-}32)$$

can in general be written in terms of three parameters a, b, and c as

$$E[(j)^k \mathscr{s}\mathscr{t}] = \tfrac{1}{2}k(k-1)\,a + [T(T-1) - \tfrac{3}{4}k]\,b + [g(\mathscr{s}, \mathscr{t}) - 2k(j+1)]\,c. \qquad (5\text{-}33)$$

[1] The expression "seniority" is thus a somewhat unfortunate choice for $\mathscr{s}$ since ground states have low seniority rather than high seniority as the name might be thought to imply.

Not surprisingly, the three terms above are seen to be just the square-bracketed terms in (5-30) valid for k total number of neutrons and protons. Talmi and Thieberger (1956) [and Thieberger and Talmi (1956)] assumed that the ground-state binding energy for a nucleus with valence particles can be obtained relative to the binding energy of the closed-shell core by adding to (5-33) a Coulomb energy contribution E_c and a simple form of single-particle energy for each valence nucleon interacting with the core. According to the independent-particle model, this latter interaction energy can be written ka_0. Hence they write

$$E_B = E_B(\text{core}) + ka_0 + E[(j)^k s t] + E_c. \tag{5-34}$$

The four constants a_0, a, b, and c are held fixed within a shell but are allowed to vary from shell to shell. The fact that (5-32) refers to the energy of a barycentric system represents only a minor complication. For nuclei with N even, Z even, the lowest energy occurs with $(s, t) = (0, 0)$, which gives the unique value $J = 0$. For odd-A nuclei the lowest energy is for $(s, t) = (1, \frac{1}{2})$ and again we obtain the unique value $J = j$. Only for the relatively rarely occurring N-odd, Z-odd nuclei does the identification of the proper set of experimental levels involved in (5-32) pose a problem. A more recent review of the applicability of the above approach in developing nuclear binding energy relations is contained in the paper by Garvey *et al.* (1969).

The representation of the experimental binding energies by (5-34) is remarkably good for $A < 50$, yielding an rms deviation of less than 1% with notable deviations only for those few light element cases where perhaps LS coupling is more nearly appropriate. Thus, for example, using (5-34), the binding energy for ^6Li relative to ^4He is in substantial error; the observed barycentric value ($J = 1, 3$) is 2.17 MeV, while (5-34) gives 4.24 MeV.

Unfortunately, the above success of the *jj*-coupling-based seniority scheme in accounting for the observed binding energies cannot be taken as an indication of the validity of the model. Its success may be due in part to the general insensitivity of system ground-state energies to the exact details of model wave functions, as well as the not inconsiderable freedom afforded by having four free parameters to adjust. Indeed, calculations based on this scheme for the magnetic dipole moments and the electric quadrupole moments give satisfactory results only near magic number closed shells plus or minus a few nucleons. The predicted moments generally give values close to the extreme single-particle predictions and are incapable of simulating the strong collective effects observed particularly for the midregion filling of the shells; see Sections II,F and II,G.

D. INTERMEDIATE COUPLING

In previous sections we have discussed the conditions under which either *LS* coupling or *jj* coupling would be expected to prevail. The controlling factor involves the relative strength of the spin–orbit and central forces in the residual interaction. Under realistic assumptions concerning the residual interactions, both these interactions are present and the extreme limiting cases are not realized. When nuclear Hamiltonians involving such realistic residual interactions are diagonalized, they lead to the so-called *intermediate coupling model*, and considerable success has been achieved in accounting for the energy level structure of many nuclei. Of the many excellent references, we cite only the early, pioneering work of Kurath (1956, 1957) and that of Elliott and Flowers (1955), and for examples of more recent calculations, the work of Cohen and Kurath (1965), Barker (1966), Cooper *et al.* (1971), and Hauge and Maripuu (1973).

It is instructive to examine in detail the general features of such calculations for realistic (although simplified) residual interactions in the 1p shell emphasizing the simplest nontrivial cases of two nucleons $A = 6$ and two holes $A = 14$.

1. Intermediate Coupling Model for $A = 6$

The extreme single-particle shell model would suggest the possibility of treating Li^6 to a higher approximation as a stable three-body (npα) system. Four nucleons in the closed $(1s)^4$ shell constitute the α–particle, while the remaining 1p-shell neutron and proton complete the three-body system synthesis of the ^6Li nucleus. Treating such a three-body problem from first principles is very difficult and usually involves invoking some approximations. A successful solution for the $J^\pi = 1^+$, $T = 0$ ground state and the $J^\pi = 0^+$, $T = 1$ second excited state (refer to Fig. 5-3) has been given by Chuu *et al.* (1973). They use the Feshbach *et al.* n–p potential, cited in Section II, I, which gives a satisfactory accounting for the deuteron and triton problems. The α–nucleon interaction is taken as a simple S- and P-wave square well, and allowance is made for internal structure of the α-particle. The P-wave contribution of the α–nucleon interaction alone accounts for 90% of the ground-state energy. Ignoring the internal structure of the α-particle results in altering the binding energy by 6%. For similar calculations employing a separable α–nucleon potential, refer to Shanley (1969) and Ghovanlou and Lehman (1974). While such three-body de novo calculations are theoretically more satisfying, we nonetheless give a simplified shell model calculation employing shell model oscillator wave functions and effective residual interactions.

We begin by assuming that the system Hamiltonian can be written

$$H = \sum_{i}^{A} [-(\hbar^2/2m)\nabla_i^2 + \overline{V}(r_i) + \overline{U}(r_i)\boldsymbol{l}_i\cdot\mathbf{s}_i] + \sum_{i<j} v_{ij}. \tag{5-35}$$

Specializing to the case of $A = 6$, we shall assume that (5-35) has been solved in the approximation $H = H_0$, leading to a closed $(1s)^4$ core (i.e., ^{4}He) *and* two nucleons moving in an average central potential that includes only the interaction of each with the core, but not with each other, nor the spin–orbit interaction of either with the core. Thus we write

$$H = H_0 + a(k=2)\sum_{i=1}^{k=2} \boldsymbol{l}_i\cdot\mathbf{s}_i + \sum_{i<j}^{k=2} v(r_{ij}). \tag{5-36}$$

Here the spin–orbit interaction is represented as a sum of one-body potentials involving the two valence nucleons. The spin–orbit "strength" $a(k)$ includes both the interaction with the core nucleons and a suitable average of the interaction between the valence nucleons in the approximation discussed in Section IV, C. Naturally enough, $a(k)$ is allowed to vary with the number of valence particles and is not zero even for $k = 1$.[1] The potential $v(r_{ij})$ is assumed to be of the most general form for a central interaction given by (1-97). The nature of this potential would be expected to be similar to the free nucleon–nucleon potential, modified however by the fact that the nucleons are now imbedded in the nuclear matter of the core and of other valence particles when $k > 2$. In the present treatment the strengths of the various terms will be considered adjustable within limits. The Coulomb interactions involving the valence nucleons, although not exhibited in (5-36), also must be taken into account. The matrix elements of (5-36) are calculated with basis states obtained for the Hamiltonian H_0. For simplicity these will be taken to be constructed from the harmonic oscillator single-particle wave functions of Section IV, B, 1. In addition to the assumption of the validity of using a spherically symmetric core potential, we shall also assume that the ground state and low excited states are adequately described by variously coupling the valence particles within the 1p shell. These states will thus necessarily be of the so-called normal parity, which for the p shell is $\pi = (-1)^k$. Indeed, the low-lying states in the region ^{4}He to ^{16}O are of normal parity with levels of opposite parity generally confined to excitations greater than 5 MeV. The matrix elements of (5-36) would in general couple the single-particle states 1s, 1p, 1d, 2s, 1f, 2p, ..., of H_0 consistent with parity and angular momentum requirements. Thus, for $^4\text{He} + k$ valence nucleons normal parity configurations with

[1] This procedure supresses both the radial dependence and proper two-body character of the spin–orbit interaction. More recent calculations such as those of Cohen and Kurath treat the problem in terms of the proper two-body potentials. For a treatment allowing for dynamical correlations in the residual interactions, refer to Lodhi (1971).

an even number of valence particles excited into the 1d, 2s shell or an odd number into the 1f, 2p shell are possible. Such excitations are energetically unfavored in general except that the operation of strong pairing forces together with possible prolate deformations of the single-particle potential can result in a nonnegligible amount of *configuration mixing* with two p-shell particles promoted into the 1d, 2s shell. For obvious reasons such configurations are referred to as two-particle, two-hole states. This phenomenon is expected to be increasingly important as the p shell is filled and is present even for the nucleus ^{16}O at the closure of the shell. For our treatment of $A = 6$ and 14 we shall truncate the allowed configuration space to include only the 1p shell.

From Table 4-1, the appropriate radial functions for $n = 1$ and $l = 1$ are of the form $Cr\exp[-(r/r_p)^2]$, where $r_p^2 = 1/\alpha = 2a_0^2$. Since the core–valence nucleon interaction in H_0 is assumed central, the coupling of the angular momenta of the two valence particles can be conveniently selected according to either the LS or jj coupling schemes.

The isospin triplet configurations ($T = 1$) (which include the case of two identical valence nucleons), under jj coupling and following the considerations discussed in Section V, B, 4, give the five possible states

$$\left.\begin{array}{rl} (1p_{3/2})^2: & J = 0, 2 \\ (1p_{1/2})(1p_{3/2}): & J = 1, 2 \\ (1p_{1/2})^2: & J = 0 \end{array}\right\} T = 1. \tag{5-37}$$

The isospin singlet coupling ($T = 0$) yields five additional states

$$\left.\begin{array}{rl} (1p_{3/2})^2: & J = 1, 3 \\ (1p_{1/2})(1p_{3/2}): & J = 1, 2 \\ (1p_{1/2})^2: & J = 1 \end{array}\right\} T = 0. \tag{5-38}$$

Alternately, under LS coupling the following ten states result:

$$\left.\begin{array}{lll} S = 0, & L = 0, 2: & {}^1S_0, {}^1D_2 \\ S = 1, & L = 1: & {}^3P_{2,1,0} \end{array}\right\} T = 1 \tag{5-39}$$

and

$$\left.\begin{array}{lll} S = 0, & L = 1: & {}^1P_1 \\ S = 1, & L = 0, 2: & {}^3S_1, {}^3D_{1,2,3} \end{array}\right\} T = 0. \tag{5-40}$$

▶**Exercise 5-4** Verify the results (5-37)–(5-40). Note that for each coupling scheme the same set of J values appears under the isospin $T = 1$ classification and similarly another appropriate set under the isospin $T = 0$. ◀

If the valence interaction in (5-36) were zero, the above states would, of course, be totally degenerate. As an illustration of the manner in which the presence of the residual terms in (5-36) removes these degeneracies, consider the case for the five $T = 1$ states. With only the spin–orbit interaction present, the jj description leads to only diagonal nonvanishing matrix elements, viz.,

$$\langle (p_{3/2})^2\, J, T=1 |\, a(2) \sum_{i=1}^{2} \boldsymbol{l}_i \cdot \mathbf{s}_i \,| (p_{3/2})^2\, J, T=1 \rangle = +a(2)$$

$$\langle (p_{1/2})(p_{3/2})\, J, T=1 |\, a(2) \sum_{i=1}^{2} \boldsymbol{l}_i \cdot \mathbf{s}_i \,| (p_{1/2})(p_{3/2})\, J, T=1 \rangle = -a(2)/2$$

$$\langle (p_{1/2})^2\, J, T=1 |\, a(2) \sum_{i=1}^{2} \boldsymbol{l}_i \cdot \mathbf{s}_i \,| (p_{1/2})^2\, J, T=1 \rangle = -2a(2), \tag{5-41}$$

all of which are independent of J. With $a < 0$,[1] the partial removal of the fivefold degeneracy shown in Fig. 5.2a results.

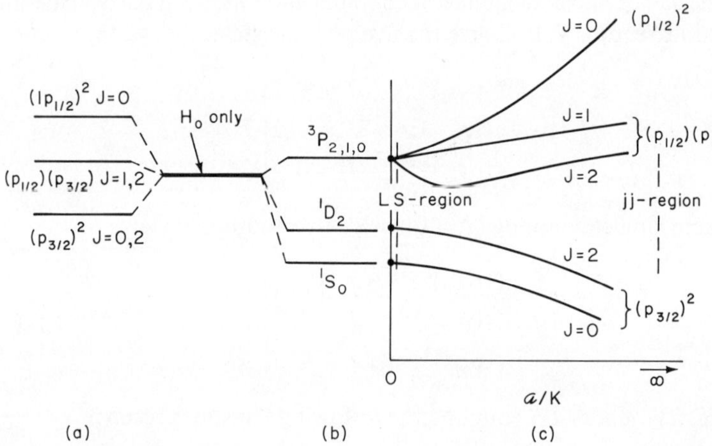

FIG. 5.2. The isospin triplet states for $A = 6$. (a) The splitting of the fivefold degeneracy under the influence of the one-body spin–orbit interaction. (b) The splitting with a realistic general central interaction. (c) Level energies as a function of the parameter a/K; see text.

▶**Exercise 5-5** Verify (5-41). ◀

On the other hand with $a = 0$ and only the central interaction $v(r_{12})$ present, it is the LS description which is diagonal. With realistic short-range, largely attractive interactions, states with the greatest probability of allowing

[1] When no confusion results, we shall write $a(2) \equiv a$, with $k = 2$ being understood.

the nucleons to be close together are favored. These are the space symmetric S and D states, favored in that order, with the antisymmetric P states the least favored.[1] The splitting of the fivefold degeneracy under the action of a realistic $v(r_{12})$ is schematically indicated in Fig. 5.2b.

Finally, Fig. 5.2c shows the "energy trajectories" for these five states when both $a < 0$ and $v(r_{12})$ are present, in terms of a parameter a/K which roughly measures the relative strength of the spin–orbit interaction to the central exchange interactions (the quantity K and other necessary parameters are discussed later). The splitting of the $T = 0$ states under the action of a realistic $v(r_{12})$ (and $a = 0$) would be expected to behave roughly as for the $T = 1$ states with the ${}^3S_1$ state being lowest in energy. For a/K small and if $v(r_{12})$ resembles the free nucleon–nucleon interaction, the lowest-energy state of all the spectroscopic possibilities would be the $T = 0, {}^3S_1$ (or ${}^{31}S_1$ state to use the notation ${}^{(2S+1)(2T+1)}L_J$ of Chapter I) when we recall the deuteron and general two-nucleon problem discussed in earlier chapters.

The wave functions to be associated with the levels of Fig. 5.2c are in general mixed LS or jj states mainly determined by the strength of the spin–orbit interaction relative to the central interaction $v(r_{12})$. Thus, for the lowest $J = 0$ level and for $a/K = 0$, we have a pure ${}^{13}S_0$ LS state. For very large values of a/K, the proper wave function asymptotically approaches the mixture $+\sqrt{\frac{2}{3}}\,{}^{13}S_0 - \sqrt{\frac{1}{3}}\,{}^{33}P_0$ which corresponds to the pure jj state $|(p_{3/2})^2 J = 0, T = 1\rangle$. Intermediate values of a/K mix the LS states ${}^{13}S_0$ and ${}^{33}P_0$ in various proportions. Among the isotriplet states the ${}^3P_1$ state is anomalous in that it happens to be exactly equivalent to the jj state $|(p_{1/2})(p_{3/2})\, J = 1, T = 1\rangle$ and hence does not change with a/K. The behavior of these states as well as the others can be readily inferred from Table 5-2.

The experimentally observed energy levels of the $A = 6$ isobaric triad with excitations below 10 MeV are shown in Fig. 5.3. The figure relates all the triad level energies to ^{6}Li by making use of β-decay energies and suitable corrections for small Coulomb energy differences and also for the n–^{1}H mass difference. It now remains to establish the energy eigenvalue E_0 of our Hamiltonian H_0 on this scale, i.e., the referencing of the ten degenerate $(1p)^2$ states relative to the true ground state of ^{6}Li. We write

$$E_0 = \{M({}^4\text{He}) + M({}^1\text{H}) + M(\text{n}) - M({}^6\text{Li})\}\, c^2 + E(\text{n–s}^4) + E(\text{p–s}^4), \tag{5-42}$$

where the quantity inside the curly bracket is simply the binding interaction energy of the true ground state of ^{6}Li relative to ^{4}He, known accurately from empirical mass values to be $+3.697$ MeV. The quantities $E(\text{n–s}^4)$ and $E(\text{p–s}^4)$

[1] More specifically the exchange nature of $v(r_{12})$ must also be considered, which raises the P-state energy still higher. For details, see later.

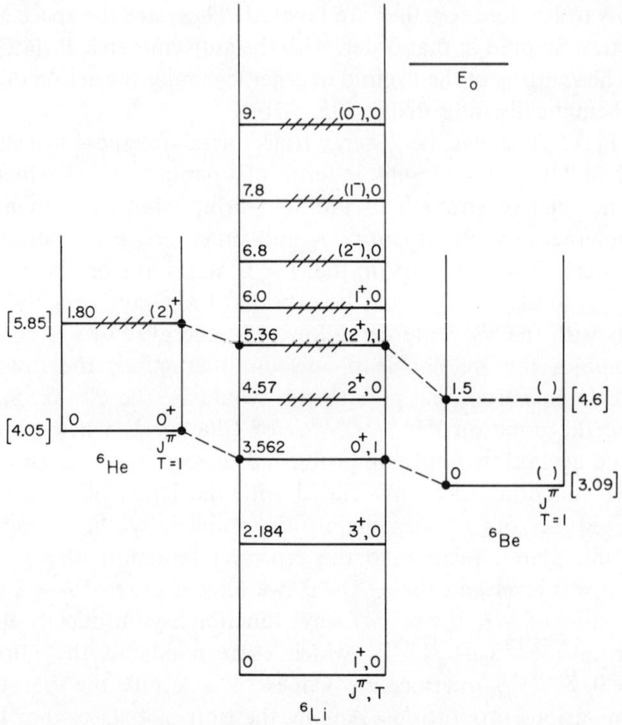

FIG. 5.3. Energy levels of the triad $A = 6$. The experimental energies are given in MeV. The nuclei ^{6}He and ^{6}Be have had their ground-state energies properly shifted relative to ^{6}Li to allow for Coulomb energy and n–^{1}H mass difference energy corrections. The ^{6}He and ^{6}Be corrected energies on the ^{6}Li scale are given in square brackets. Broad levels are indicated cross-hatched, uncertain quantum assignments are enclosed in parentheses, and the one uncertain level is shown with a dashed line. For a discussion of E_0 see text.

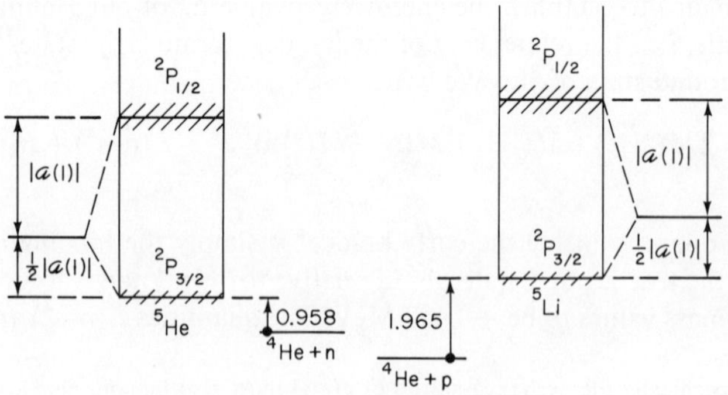

FIG. 5.4. The ^{2}P energy levels of ^{5}He and ^{5}Li.

are the neutron and proton interaction energies with the $(1s)^4$ core *excluding* the spin–orbit interaction but in the $E(p\text{–}s^4)$ case *including* the Coulomb interaction. These interaction energies can be inferred from the relevant level structure of the $A = 5$ system, ^{5}He and ^{5}Li. Figure 5.4 shows the pertinent experimental information. Nuclear reaction data, particularly elastic neutron and proton scattering from ^{4}He, can be interpreted in terms of a spin–orbit-split $^2P_{3/2,1/2}$ doublet configuration for both mirror nuclei ^{5}He and ^{5}Li. The required interaction energies are[1]

$$E(n\text{–}s^4) = 0.958 + \tfrac{1}{2}|a(1)| \text{ MeV}, \tag{5-43a}$$

$$E(p\text{–}s^4) = 1.965 + \tfrac{1}{2}|a(1)| \text{ MeV}, \tag{5-43b}$$

where $a(1) < 0$ is the appropriate spin–orbit interaction strength for $k = 1$ in the 1p shell. The width of the particle unstable $^2P_{3/2}$ ground states and particularly the very large width of the $^2P_{1/2}$ "excited states" make the experimental value of $a(1)$ uncertain. The best estimate is $a(1) \approx -(3.6 \text{ to } 4.0)$ MeV derived from the R-matrix resonance parameters obtained by Stammbach (Stammbach and Walter, 1972). This value can be compared also to $a = -4.17$ MeV observed for the ^{15}N and ^{15}O hole states in the ^{16}O core. For the case of the hole states, the spin–orbit matrix elements change sign. This results in the $^2P_{1/2}$ states being lower in energy than the $^2P_{3/2}$ states by 6.32 and 6.18 MeV, respectively; hence $\tfrac{3}{2}|a(14)| = 6.25$ MeV. Finally, we estimate $E_0 = 6.62 + |a(1)| = (10.2\text{–}10.6)$ MeV.

In evaluating the matrix elements of the residual interactions of (5-36), it will be convenient to use the LS-coupled representation for the basis states since the inherently complicated evaluation of the $v(r_{12})$ term is then somewhat simpler. For convenience we take $v(r_{12})$ of the form (1-97) with a radial dependence $V(r_{12})$ having a range r_0 and a Gaussian shape, viz.,

$$v(r_{12}) = V(r_{12})[\mathscr{T}(\sigma)\mathscr{S}(\tau) + \alpha\mathscr{S}(\sigma)\mathscr{T}(\tau) + \beta\mathscr{T}(\sigma)\mathscr{T}(\tau) + \gamma\mathscr{S}(\sigma)\mathscr{S}(\tau)], \tag{5-44}$$

with $V(r_{12}) = V_0 \exp[-(r_{12}/r_0)^2]$, and V_0 a suitable negative (i.e., attractive) strength constant. It is immediately evident that the representation $|LSJT\rangle \equiv {}^{(2S+1)(2T+1)}L_J$ is diagonal in (5-44) with each state involving only one of the terms in (5-44). For example,

$$\langle {}^{33}P_J|v(r_{12})|{}^{33}P_J\rangle = \beta V_0 \langle P_J| \exp[-(r_{12}/r_0)^2]\,|P_J\rangle.$$

A fundamental mathematical difficulty arises in evaluating matrix elements of the type encountered here, since the state vectors are functions of $\mathbf{r}_1$ and $\mathbf{r}_2$

[1] The core–nucleon interaction energies in the $A = 6$ system are not exactly those of (5-43) due to the additional polarizing effect of the other nucleon not present in the $A = 5$ system. These effects would be expected to be small, however, compared to the uncertainties in the value of $a(1)$. Also note that $a(1)\boldsymbol{l}\cdot\mathbf{s}$ is $+\tfrac{1}{2}a(1)$ for $j = \tfrac{3}{2}$, $l = 1$ and $-a(1)$ for $j = \tfrac{1}{2}$, $l = 1$.

separately, while the potential is a function of $r_{12} = |\mathbf{r}_1 - \mathbf{r}_2|$. One procedure is to expand the radial portion of the potential, $V(r_{12})$, in terms of Legendre polynomials coupled to functions depending on r_1 and r_2 separately, viz.,[1]

$$V(|\mathbf{r}_1 - \mathbf{r}_2|) = \sum_{\kappa=0}^{\infty} V_\kappa(r_1, r_2)\, P_\kappa(\cos\theta_{12}), \tag{5-45a}$$

with θ_{12} the angle between $\mathbf{r}_1$ and $\mathbf{r}_2$. Explicitly, we have

$$V_\kappa(r_1, r_2) = \tfrac{1}{2}(2\kappa+1) \int_{-1}^{+1} V(|\mathbf{r}_1 - \mathbf{r}_2|)\, P_\kappa(\cos\theta_{12})\, d(\cos\theta_{12}). \tag{5-45b}$$

We thus get

$$\begin{aligned} \langle L_J | V(r_{12}) | L_J \rangle &= \sum_\kappa \sum_{m, m'} (11m, -m | L, M = 0)(11m', -m' | L, M = 0) \\ &\times \langle Y_1^m(1) Y_1^{-m}(2)\, u_{11}(r_1)\, u_{11}(r_2) |\, P_\kappa(\cos\theta_{12}) V_\kappa(r_1, r_2) \\ &\times | Y_1^{m'}(1) Y_1^{-m'}(2)\, u_{11}(r_1)\, u_{11}(r_2) \rangle. \end{aligned} \tag{5-46}$$

When the spherical harmonic addition theorem is used to expand the Legendre polynomials, namely

$$P_\kappa(\cos\theta_{12}) = [4\pi/(2\kappa+1)] \sum_q (-1)^q Y_\kappa^{\,q}(1)\, Y_\kappa^{-q}(2), \tag{5-47}$$

(5-46) factors into angular and radial terms that are in principle straightforward to evaluate. The vector triangle condition and parity requirements limit the values of κ that may appear in these calculations to $\kappa = 0$ and $\kappa = 2$. Thus only the two radial integrals

$$F_\kappa = \int_0^\infty \int_0^\infty [u_{11}(r_1)\, u_{11}(r_2)]^2 V_\kappa(r_1, r_2)\, r_1^{\,2} r_2^{\,2}\, dr_1\, dr_2, \qquad \kappa = 0, 2, \tag{5-48a}$$

with

$$\int_0^\infty [u_{nl}(r)]^2 r^2\, dr = 1 \tag{5-48b}$$

are required in the present case. These are associated with various angular coupling factors. The results are

$$\begin{aligned} L = 0: &\qquad \langle \mathrm{S}_J | V(r_{12}) | \mathrm{S}_J \rangle = F_0 + \tfrac{2}{5} F_2 \\ L = 1: &\qquad \langle \mathrm{P}_J | V(r_{12}) | \mathrm{P}_J \rangle = F_0 - \tfrac{1}{5} F_2 \\ L = 2: &\qquad \langle \mathrm{D}_J | V(r_{12}) | \mathrm{D}_J \rangle = F_0 + \tfrac{1}{25} F_2. \end{aligned} \tag{5-49}$$

[1] This expansion is the general form of the familiar expansion $|\mathbf{r}_1 - \mathbf{r}_2|^{-1} = \sum_{\kappa=0}^{\infty} [P_\kappa(\cos\theta_{12})] r_<^{\,\kappa}/r_>^{\,\kappa+1}$ with $r_<$ the lesser of r_1 and r_2, and $r_>$ the larger of the two.

The results (5-49) are expressed by Kurath in terms of two slightly different integrals, the direct integral $\Lambda = F_0 + \frac{4}{25}F_2$ and the exchange integral $K = \frac{3}{25}F_2$. For a more complete and general treatment of the procedure briefly outlined above the reader is referred to numerous standard works [e.g., de-Shalit and Talmi (1963)]. In addition, the use of harmonic oscillator wave functions, as in the present case, leads to a particularly straightforward calculation of the otherwise troublesome radial integrals, following a method developed by Talmi (1952).

Finally, we write the matrix elements of $v(r_{12})$:

$$T = 0: \quad \begin{cases} \langle {}^{31}\mathrm{S}_1 | v(r_{12}) | {}^{31}\mathrm{S}_1 \rangle = \Lambda + 2K \\ \langle {}^{31}\mathrm{D}_J | v(r_{12}) | {}^{31}\mathrm{D}_J \rangle = \Lambda - K \\ \langle {}^{11}\mathrm{P}_1 | v(r_{12}) | {}^{11}\mathrm{P}_1 \rangle = \gamma(\Lambda - 3K) \end{cases} \tag{5-50a}$$

$$T = 1: \quad \begin{cases} \langle {}^{13}\mathrm{S}_0 | v(r_{12}) | {}^{13}\mathrm{S}_0 \rangle = \alpha(\Lambda + 2K) \\ \langle {}^{13}\mathrm{D}_2 | v(r_{12}) | {}^{13}\mathrm{D}_2 \rangle = \alpha(\Lambda - K) \\ \langle {}^{33}\mathrm{P}_J | v(r_{12}) | {}^{33}\mathrm{P}_J \rangle = \beta(\Lambda - 3K), \end{cases} \tag{5-50b}$$

which are independent of J and M_J. We now turn to the evaluation of the matrix elements of $\sum_{1,2} a(2) \boldsymbol{l}_i \cdot \mathbf{s}_i$. These are readily evaluated after transforming the LS states into their equivalent jj terms, using Table 5-2. We simply give one example by evaluating the off-diagonal element $\langle {}^{31}\mathrm{S}_1 | a \sum_{1,2} \boldsymbol{l}_i \cdot \mathbf{s}_i | {}^{11}\mathrm{P}_1 \rangle$. From Table 5-2, we have

$$\sqrt{54}\, | {}^{31}\mathrm{S}_1 \rangle = 2\sqrt{5}\,(\tfrac{3}{2}, \tfrac{3}{2}) - 4[(\tfrac{3}{2}, \tfrac{1}{2}) \oplus (\tfrac{1}{2}, \tfrac{3}{2})]_{\mathrm{A}} - \sqrt{2}\,(\tfrac{1}{2}, \tfrac{1}{2})$$

and

$$\sqrt{54}\, | {}^{11}\mathrm{P}_1 \rangle = \sqrt{30}\,(\tfrac{3}{2}, \tfrac{3}{2}) + \sqrt{6}\,[(\tfrac{3}{2}, \tfrac{1}{2}) \oplus (\tfrac{1}{2}, \tfrac{3}{2})]_{\mathrm{A}} + 2\sqrt{3}\,(\tfrac{1}{2}, \tfrac{1}{2}).$$

Recalling that

$$(\boldsymbol{l} \cdot \mathbf{s})\, | l = 1, j = \tfrac{3}{2} \rangle = +\tfrac{1}{2}\, | l = 1, j = \tfrac{3}{2} \rangle$$

and

$$(\boldsymbol{l} \cdot \mathbf{s})\, | l = 1, j = \tfrac{1}{2} \rangle = -1\, | l = 1, j = \tfrac{1}{2} \rangle$$

and the orthonormality of the $|l, j\rangle$ states, we get

$$\begin{aligned} \langle {}^{31}\mathrm{S}_1 | a \sum_{1,2} \boldsymbol{l}_i \cdot \mathbf{s}_i | {}^{11}\mathrm{P}_1 \rangle &= \frac{2\sqrt{5}\sqrt{30}}{54}\left(\frac{1}{2} + \frac{1}{2}\right) a \\ &\quad - \frac{4\sqrt{6}}{54}\left[\left(\frac{1}{2} - 1\right) + \left(-1 + \frac{1}{2}\right)\right] a \\ &\quad - \frac{2\sqrt{2}\sqrt{3}}{54}(-1 - 1)\, a = \frac{a\sqrt{6}}{3}. \end{aligned}$$

Evaluating the other spin–orbit matrix elements in a similar manner, we arrive at the complete residual matrix elements given in Table 5-7. Since the operators $\mathbf{J}^2$ and $\mathbf{T}^2$ commute with the Hamiltonian of (5-36), only states of the same isospin and J value can couple in agreement with the entries in Table 5-7. The same does not hold for $\mathbf{L}^2$ and $\mathbf{S}^2$.

▶**Exercise 5-6** (a) Verify the matrix element entries in Table 5-7.
(b) Show that the diagonal matrix elements of $a\sum_{1,2} \boldsymbol{l}_i \cdot \mathbf{s}_i$ are $\frac{1}{4}a[J(J+1)-L(L+1)-S(S+1)]$. ◀

By confining our attention to ^{6}Li we avoid the need for considering the Coulomb interaction between the valence nucleons. Anderson *et al.* (1972) have determined all relevant Coulomb matrix elements for the 1p-shell nuclei, which again can be expressed in terms of direct and exchange integrals defined analogously to Λ and K. For example, Coulomb shifts (Coulomb energy contributions to the corrections necessary to compare isobaric nuclear levels on a common scale, such as in Fig. 5.3) are rather well accounted for. Also, a reasonable accounting is given for the constants appearing in the isobaric mass formula (3-89) of Chapter III. See additional references cited in Anderson *et al.* (1972), and also refer to McCarthy and Walker (1974) and Goldhammer (1974).

The problem of diagonalizing the residual interaction energy matrix using the *LS* representation is seen from Table 5-7 to factor into the diagonalization of six submatrices. In the following numerical evaluation we shall treat as many of the parameters as possible as unknowns to be determined. While there is a preferred range of values for E_0 (i.e., $E \approx 10.2$–10.6 MeV), the large associated uncertainties lead us to allow some adjustment in its value. As we shall see, various other uncertainties and the lack of necessary experimental data will require the assignment of values to some parameters based on general considerations rather than the specifics pertaining to the $A = 6$ system (or the $A = 14$ system).

If we identify the (4.57 ± 0.03) MeV and (2.184 ± 0.002) MeV states of ^{6}Li shown in Fig. 5.3 with the $^{31}D_2$ and $^{31}D_3$ states, we have, according to Table 5-7,[1]

$$-E_0 + 4.57 = \Lambda - K - a/2 \tag{5-51a}$$

$$-E_0 + 2.184 = \Lambda - K + a. \tag{5-51b}$$

These two equations determine $a(2) = -1.59 \pm 0.02$ and in addition give the value of $\Lambda - K$,

$$\Lambda - K = 3.77 - E_0. \tag{5-52}$$

[1] In the following all energies will be in units of MeV.

TABLE 5-7

RESIDUAL MATRIX ELEMENTS FOR THE $A = 6$ STATES

$^{(2S+1)(2T+1)}L_J$	$^{31}S_1$	$^{31}D_1$	$^{11}P_1$	$^{13}S_0$	$^{33}P_0$	$^{13}D_2$	$^{33}P_2$	$^{31}D_3$	$^{31}D_2$	$^{33}P_1$
$^{31}S_1$	$\Lambda+2K$	0	$\frac{1}{3}a\sqrt{6}$	—	—	—	—	—	—	—
$^{31}D_1$	0	$\Lambda-K-\frac{3}{2}a$	$-\frac{1}{6}a\sqrt{30}$	—	—	—	—	—	—	—
$^{11}P_1$	$\frac{1}{3}a\sqrt{6}$	$-\frac{1}{6}a\sqrt{30}$	$\gamma(\Lambda-3K)$	—	—	—	—	—	—	—
$^{13}S_0$	—	—	—	$\alpha(\Lambda+2K)$	$-a\sqrt{2}$	—	—	—	—	—
$^{33}P_0$	—	—	—	$-a\sqrt{2}$	$\beta(\Lambda-3K)-a$	—	—	—	—	—
$^{13}D_2$	—	—	—	—	—	$\alpha(\Lambda-K)$	$\frac{1}{2}(2a)^{1/2}$	—	—	—
$^{33}P_2$	—	—	—	—	—	$\frac{1}{2}(2a)^{1/2}$	$\beta(\Lambda-3K)+\frac{1}{2}a$	—	—	—
$^{31}D_3$	—	—	—	—	—	—	—	$\Lambda-K+a$	—	—
$^{31}D_2$	—	—	—	—	—	—	—	—	$\Lambda-K-\frac{1}{2}a$	—
$^{33}P_1$	—	—	—	—	—	—	—	—	—	$\beta(\Lambda-3K)-\frac{1}{2}a$

In particular we note that this value of $a(2)$ is considerably smaller in magnitude than the value $a(1) \approx -(3.6 \text{ to } 4.0)$ required for the $A = 5$ system.

Treating the three $T = 0$, $J = 1$ states to first order (see Appendix D) would give eigenvalues, in order of increasing excitation

$$^{31}\mathrm{S}_1: \quad {}^1\lambda_1 = \Lambda + 2K = -E_0 \tag{5-53a}$$

$$^{31}\mathrm{D}_1: \quad {}^1\lambda_2 = \Lambda - K - \tfrac{3}{2}a = -E_0 + 6.0 \tag{5-53b}$$

$$^{11}\mathrm{P}_1: \quad {}^1\lambda_3 = \gamma(\Lambda - 3K) > 0. \tag{5-53c}$$

This last follows from the failure to observe the $^{11}\mathrm{P}_1$ state below E_0 (i.e., below ~ 10 MeV of excitation). Indeed, lack of a definitive assignment of the P states[1] in the $A = 6$ system largely prevents the determination of β and γ. In fact, we shall use the first-order values obtained from (5-52) and (5-53a), namely $a(2) = -1.59$ and $E_0 \approx 10.40$, coupled to reasonable values of β and γ to determine probable eigenvalues for the P states. Solutions under these conditions give the values $K = -1.26$ and $\Lambda = -7.89$ and, as a rough check, $^1\lambda_2(^{31}\mathrm{D}_1 \text{ state}) = -E_0 + 6.16$, in close enough agreement with (5-53b). We assume a "Rosenfeld mixture" of central forces, [see (1-100) and (3-70) with subsequent discussion in Chapter III] to be appropriate. This mixture or essentially similar ones have been very successfully used in shell model calculations. Thus with $\beta = -\frac{1}{3}$ and $\gamma = -1.80$, we obtain the first-order estimates for the excitation of the P states above the ground state, i.e.,

$$E_x = E_0 + \lambda,$$

$$E_x(^{33}\mathrm{P}_2) = E_0 + 0.57 \approx 11.0, \qquad E_x(^{33}\mathrm{P}_1) = E_0 + 2.17 \approx 12.6$$
$$E_x(^{33}\mathrm{P}_0) = E_0 + 2.96 \approx 13.4, \qquad E_x(^{11}\mathrm{P}_1) = E_0 + 7.42 \approx 17.8. \tag{5-54}$$

Given the above approximate value for $^1\lambda_3(^{11}\mathrm{P}_1 \text{ state}) \approx 7.42$, we return to treat the $T = 0$, $J = 1$ states to second order. Using Appendix D, we have, to a good approximation,

$$^2\lambda_1 = \Lambda + 2K - [\tfrac{2}{3}a^2/(^1\lambda_3 - {}^1\lambda_1)]. \tag{5-55a}$$

Estimating $^1\lambda_3 - {}^1\lambda_1 \approx 17.8$, writing $^2\lambda_1 = -E_0$, and using $a = -1.59$, we obtain

$$-E_0 = \Lambda + 2K - 0.10. \tag{5-55b}$$

Coupling (5-55b) and (5-52) gives $K = -1.22 \pm 0.05$, which is the value, along with $a = -1.59 \pm 0.02$, that we adopt for all subsequent calculations for $A = 6$. To the same approximation, we have

$$^2\lambda_2 = \Lambda - K - \tfrac{3}{2}a - [\tfrac{5}{6}a^2/(^1\lambda_3 - {}^1\lambda_2)], \tag{5-56}$$

[1] Anticipating a largely *LS*-coupling scheme to prevail for $A = 6$, we identify levels by the principal term in the eigenfunction.

which, for ${}^1\lambda_3 - {}^1\lambda_2 \approx 11.8$ and with the aid of (5-52), gives ${}^2\lambda_2 = 5.98$ to be compared to the experimental value 6.0 ± 0.2.

Finally, from the $T = 0$, $J = 1$ triad and the ${}^{31}\mathrm{D}_2$ and ${}^{31}\mathrm{D}_3$ pair, we obtain

$$a = -1.59 \pm 0.02, \qquad K = -1.22 \pm 0.05, \qquad E_0 + \Lambda = +2.55 \pm 0.05. \tag{5-57}$$

The resolution of the above into E_0 and Λ separately is not possible with the uncertainties that exist in the input quantities.

With the excitation energies of the lower member of the $T = 1, J = 0$ doublet and the $T = 1$, $J = 2$ doublet known, reasonable estimates of β for the more distant P states could in principle yield values of α, Λ, and E_0 when coupled with (5-57) since now the linear combination $E_0 + \alpha\Lambda$ appears, rather than just $E_0 + \Lambda$ as was the case for all the known energy input values leading to (5-57).

▶**Exercise 5-7** Show that by associating the (3.562 ± 0.004)-MeV level with the eigenvalue near the pure ${}^{13}\mathrm{S}_0$ state and similarly the (5.36 ± 0.015)-MeV level with the ${}^{13}\mathrm{D}_2$ state, the value $\beta = -\frac{1}{3}$ leads to

$$\alpha = 0.44 \pm 0.01, \qquad \Lambda = -5.15 \pm 0.2$$

$$\Lambda/K = 4.32, \qquad E_0 = 7.70 \pm 0.2$$

when (5-57) is used.

What is the effect of taking $\beta = 0$? ◀

Although values near those found in Exercise 5-7 are quoted in the literature, both α and Λ are anomalously low when compared to values obtained for nuclei in the remainder of the 1p shell. The difficulty may be associated with the known complexity of the broad 5.36-MeV state. We therefore prefer to choose the reasonable value $\alpha = \frac{3}{5}$ consistent with our preference for the Rosenfeld mixture and deduce Λ (and E_0) from the ${}^{13}\mathrm{S}_0$ and ${}^{33}\mathrm{P}_0$ doublet. Thus, using $\alpha = \frac{3}{5}$ and $\beta = -\frac{1}{3}$, we derive the final set of parameters

$$\alpha = \tfrac{3}{5}, \qquad \beta = -\tfrac{1}{3}, \qquad \gamma = -1.80$$

$$a = -1.59 \pm 0.02, \qquad K = -1.22 \pm 0.05, \qquad \Lambda = -7.52 \pm 0.06; \tag{5-58}$$

hence $E_0 = 10.07$, $\Lambda/K = 6.17$, and $a/K = 1.30$.

The $T = 0$, $J = 1$ triad energy matrix with the values (5-58) becomes, using the sequence of matrix elements given in Table 5-7,

$$\begin{pmatrix} -9.96 & 0 & -1.30 \\ 0 & -3.81 & +1.45 \\ -1.30 & +1.45 & +6.95 \end{pmatrix}. \tag{5-59}$$

The eigenvalues are readily found, using the techniques discussed in Appendix D, to be

$$\lambda_1 = -10.07, \qquad \lambda_2 = -4.00, \qquad \lambda_3 = +7.25. \tag{5-60}$$

Similarly, the values (5-58) allow us to determine all the remaining eigenvalues of Table 5-7 and the corresponding eigenfunctions. These are given in Table 5-8.

TABLE 5-8

RESULTS OF DIAGONALIZING THE ENERGY MATRIX OF TABLE 5-7 FOR ^{6}Li[a]

E_x(exp)	J, T	E_x(calc)	$\psi = \sum_{L,S} C_{LS} \vert ^{(2S+1)(2T+1)}L_J\rangle$
0	1, 0	0.00	$0.997\vert^{31}S_1\rangle - 0.018\vert^{31}D_1\rangle + 0.078\vert^{11}P_1\rangle$
2.184 ± 0.002	3, 0	2.18	$\vert^{31}D_3\rangle$
3.562 ± 0.004	0, 1	3.56	$0.973\vert^{13}S_0\rangle - 0.231\vert^{33}P_0\rangle$
4.57 ± 0.034	2, 0	4.57	$\vert^{31}D_2\rangle$
5.36 ± 0.015	2, 1	6.06	$0.975\vert^{13}D_2\rangle + 0.222\vert^{33}P_2\rangle$
6.0 ± 0.2	1, 0	6.07	$0.991\vert^{31}D_1\rangle + 0.028\vert^{31}S_1\rangle - 0.128\vert^{11}P_1\rangle$
—	2, 1	≈ 10.8	$0.975\vert^{33}P_2\rangle - 0.222\vert^{13}D_2\rangle$
—	1, 1	≈ 12.2	$\vert^{33}P_1\rangle$
—	0, 1	≈ 13.5	$0.973\vert^{33}P_0\rangle + 0.231\vert^{13}S_0\rangle$
—	1, 0	~ 17.3	$0.984\vert^{11}P_1\rangle - 0.075\vert^{31}S_0\rangle + 0.129\vert^{31}D_1\rangle$

[a] Based on values given by (5-58).

▶**Exercise 5-8** Verify the entries in Table 5-8. ◀

The results given in Table 5-8 show the $A = 6$ states to be rather satisfactorily accounted for by an intermediate coupling model that in this instance is not very different from pure *LS* coupling. This situation rapidly alters as similar intermediate coupling calculations are extended to the remainder of the 1p shell. For example, for $A = 14$, to which all of the matrix elements given in Table 5-7 apply if the sign of the spin–orbit terms are reversed,[1] a much larger

[1] When the energy reference is shifted to the closed 1p shell nucleus ^{16}O, matrix elements for hole states are equal to those of corresponding particle states with possible sign changes. The matrix elements for the one-body spin–orbit operator changes sign. Thus while we consider the strength of the particle spin–orbit operator to remain negative throughout the 1p shell, the matrix elements appearing in Table 5-7 when applied to the two-nucleon hole states of $A = 14$ require all the signs of a (more appropriately $\langle a \rangle$) to reverse. Alternately, some authors prefer to keep the signs in Table 5-7 and use numerical values $a > 0$ for holes and $a < 0$ for particles as, for example, Inglis (1953) (see Fig. 5.5). The matrix elements of all two-body *charge-independent* operators between hole states carry the same sign as for particle states. For a rather complete discussion of these points, see, for example, Visscher and Ferrell (1957).

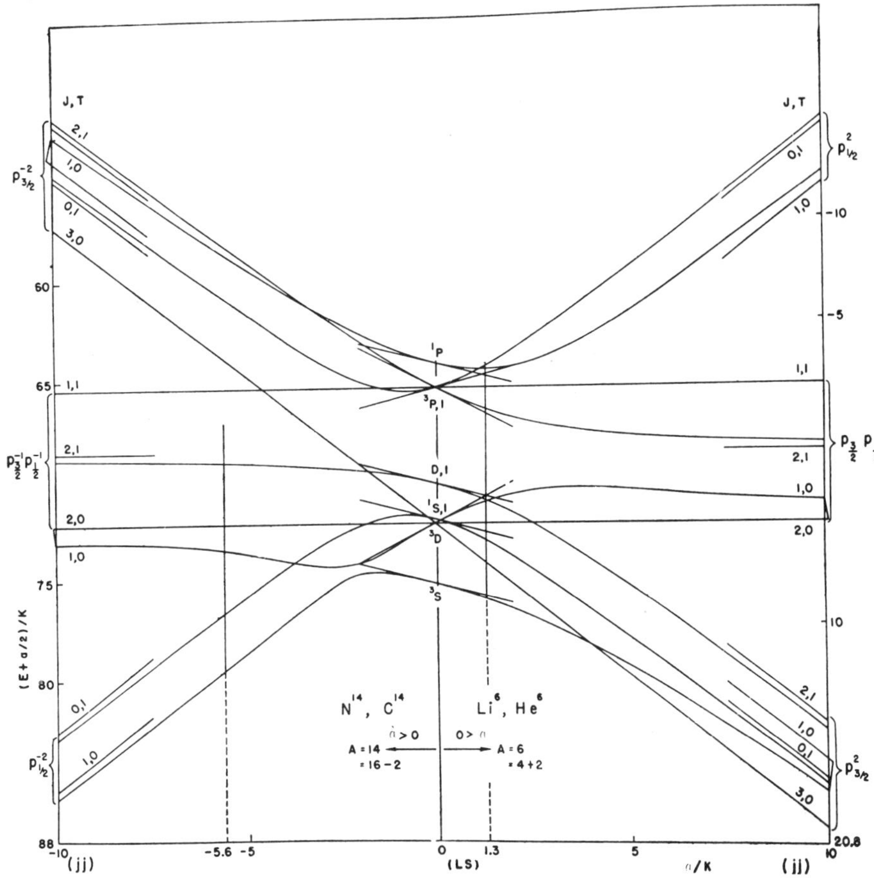

FIG. 5.5. Intermediate coupling energy eigenvalues for the $A = 6$ and $A = 14$ nuclear systems for $\Lambda/K = 6.0$, $\alpha = 0.6$, $\beta = -0.6$, and $\gamma = -1.0$ [after Inglis (1953)].

value of a/K is required to match data. Figure 5.5 shows the energy trajectories for $A = 6$ and $A = 14$ nuclear systems as functions of the parameter a/K. Values of a/K giving satisfactory energy eigenvalues are indicated in the figure.

While no predominantly P states have been definitely identified in ^6Li at about 10 MeV excitation or higher, inelastic electron scattering data suggest possible $T = 1$, $J^\pi = (0^+, 2^+)$ levels at 9.3, 14.0, and 15.8 MeV. Some of these might be associated with the levels listed in Table 5-8 at 10.8 and 13.5 MeV.

In examining the matrix elements of Table 5-7, we note that virtually all states having the same J value are coupled in fact by the interactions we have assumed. Thus, in general, ignored interactions such as the tensor interaction S_{12} would not be expected to change the output eigenvalues much if their numerical magnitudes are masked by the values of the interactions already considered. The ignored interaction matrix elements are mostly less than 0.3 MeV in magnitude (Cohen and Kurath, 1965). Two notable exceptions, however, must be considered. Realistic two-body interactions of the tensor type S_{12} and the quadratic spin–orbit type L_{12} [e.g., refer to (1-143) describing the Hamada–Johnston phenomenological interaction] couple the $^{31}S_1$ and $^{31}D_1$ states with matrix elements estimated in the range $-(0.3 \text{ to } 1.4)$ MeV. The inclusion of this matrix element is essential both to give the right sign for the quadrupole moment of ^6Li and to allow for the very slow β-decay transition $^{14}C(\beta^-)^{14}N$. The other notable case involves the question of the assumed isospin purity of the states exhibited in Table 5-7. The effects of Coulomb interactions are expected to particularly mix states with $L = J$ and $J > 0$ with matrix elements of order $-(0.03 \text{ to } 0.5)$ MeV. Such a matrix element would be expected to couple the $^{13}D_2$ and $^{31}D_2$ states. Indeed, an experimentally observed small $T = 0$ admixture in the mostly $T = 1$ state at 5.36 MeV ($J^\pi = 2^+$) in fact may result from this coupling [see Kane *et al.* (1972) and Noble (1968)]. Coupling of this level with other, as yet undetected $T = 0$ states at nearby energies cannot be ruled out and may account for the rather poor energy fit evident in Table 5-8 and in the following exercise.

▶**Exercise 5-9** Assume that in addition to the matrix elements in Table 5-7 for ^6Li we have $\langle ^{13}D_2 | V_c | ^{31}D_2 \rangle = -0.15$ MeV.[1] Using (5-58), show that the resulting triad with $J = 2$ gives for the two lowest energy levels

$$E_x = 4.55 \text{ MeV}: \qquad \psi = 0.995\,|^{31}D_2\rangle + 0.096\,|^{13}D_2\rangle + 0.017\,|^{33}P_2\rangle$$

$$E_x = 6.07 \text{ MeV}: \qquad \psi = 0.969\,|^{13}D_2\rangle + 0.221\,|^{33}P_2\rangle - 0.094\,|^{31}D_2\rangle.$$

◀

▶**Exercise 5-10** Assume that in addition to the matrix elements in Table 5-7 for ^6Li we have $\langle ^{31}S_1 | V(S_{12}) + V(L_{12}) | ^{31}D_1 \rangle = -0.35$ MeV. Using (5-58), show that while the ground-state eigenvalue is essentially unaffected, the corresponding wave function now changes the sign of the $|^{31}D_1\rangle$ component and becomes

$$\psi(^6\text{Li})_{\text{gd}} = 0.996\,|^{31}S_1\rangle + 0.039\,|^{31}D_1\rangle + 0.072\,|^{11}P_1\rangle. \quad ◀ \tag{5-61}$$

The two exercises above show that small to moderate changes in off-diagonal matrix elements, while not significantly affecting energy eigenvalues, can produce very significant changes in the wave functions. The 0.9% $T = 0$ probability admixture into the largely $T = 1$, 5.36-MeV (calculated value 6.07 MeV) state now would allow for measurable α-particle decay,

[1] This matrix element is substantially larger than afforded by a simple Mott–Schwinger interaction; for a numerical estimate pertaining to the present case, see Bray *et al.* (1973). For a general discussion of isospin impurities in nuclear states, see Bertsch and Mekjian (1972).

while the reversal in sign of the ${}^{31}D_1$ component in the ground state of ${}^6Li$ allows for the correct sign in the calculated quadrupole moment.

In an inelastic electron scattering experiment, Hutcheon *et al.* (1970) report compatible wave functions for the ${}^6Li$ ground state and the 5.36-MeV excited state [they determined the energy to be (5.38 ± 0.02) MeV and ignored any possible isospin mixing]; specifically, they quote

$$\psi(0) = 0.995|{}^{31}S_1\rangle + 0.024|{}^{31}D_1\rangle + 0.065|{}^{11}P_1\rangle$$

and

$$\psi(5.36) = 0.780|{}^{13}D_2\rangle + 0.625|{}^{33}P_2\rangle.$$

While the first of these agrees fairly well with (5-61), the latter is in striking disagreement with the values in Table 5-8. Even the set of parameters designed to give the correct energy for the 5.36-MeV state given in Exercise 5-7 would not generate enough P state, since this set yields only

$$\psi(5.36) = 0.875|{}^{13}D_2\rangle + 0.482|{}^{33}P_2\rangle.$$

Perhaps the best way to gauge the difficulty with the experimental wave function of Hutcheon *et al.* is to realize that in *jj* representation it would require a linear combination that is 92.4% $|\frac{3}{2},\frac{3}{2}\rangle$ with only a 7.6% $|\frac{3}{2},\frac{1}{2}\rangle$ admixture. This is largely incompatible with the relatively small value of a/K otherwise required for $A = 6$. For additional comments on the above isospin mixing problem, refer to DeVries *et al.* (1972) and Bray *et al.* (1973).

2. Intermediate Coupling Model for $A = 14$

Turning to the $A = 14$ case, we take the parameter set equivalent to (5-58) to be[1]

$$\alpha = \tfrac{3}{5}, \qquad \beta = -\tfrac{1}{3}, \qquad \gamma = -1.80$$

$$a = -4.80 \pm 0.05, \qquad K = -1.07 \pm 0.05, \qquad \Lambda = -6.60 \pm 0.08$$

$$E_0 = +13.66, \qquad \langle {}^{31}S_1|V(S_{12}) + V(L_{12})|{}^{31}D_1\rangle = -1.30; \tag{5-62}$$

hence $\Lambda/K = 6.17$ and $a/K = 4.50$. Diagonalization of the resulting energy matrix then gives the results shown in Table 5-9.

Table 5-9 lists only the normal parity states. The first odd parity state, a likely $(1p)^9(2s)^1$ configuration, already appears at $E_x = 4.913$ MeV. In addition normal parity states clearly requiring consideration of higher configurations than $(1p)^k$ appear by $E_x = 6.198$ MeV, labeled in the table as "interlopers." These states will mix, of course, with the truncated 1p-shell

[1] All signs of the spin–orbit matrix elements in Table 5-7 reversed.

TABLE 5-9

RESULTS OF DIAGONALIZING THE ENERGY MATRIX OF TABLE 5-7 for ^{14}N WITH THE ADDITIONAL $\langle {}^{31}S_1 | V(S_{12}) + V(L_{12}) | {}^{31}D_1 \rangle$ TERM

$E_x(\exp)^a$	J, T	$E_x(\text{calc})$	$\psi = \sum_{L,S} C_{LS} \lvert^{(2S+1)(2T+1)}L_J\rangle$
0	1,0	−0.07	$0.976\lvert^{31}D_1\rangle + 0.100\lvert^{31}S_1\rangle + 0.195\lvert^{11}P_1\rangle$
2.31281	0,1	2.37	$0.748\lvert^{13}S_0\rangle + 0.665\lvert^{33}P_0\rangle$
3.9447	1,0	3.96	$0.967\lvert^{31}S_1\rangle - 0.049\lvert^{31}D_1\rangle - 0.262\lvert^{11}P_1\rangle$
6.1976	1,0	—	Interloper
6.4436	3,0	—	Interloper
7.028	2,0	5.73	$\lvert^{31}D_2\rangle$
8.617	0,1	—	Interloper
8.963	5,0	—	Interloper
8.979	2,(0)	—	Interloper
9.172	2,1	8.94	$0.925\lvert^{13}D_2\rangle - 0.382\lvert^{33}P_2\rangle$
—	1,1	≈ 12.4	—
—	3,0	≈ 12.9	—
—	0,1	≈ 16.0	—
—	2,1	~ 18.6	—
—	1,0	~ 21.7	—

[a] Experimental error usually several parts in last place.

generic states of the same J value. In all probability only the first three states listed can in any sense be expected to be mostly 1p-shell states. Admixture of higher configurations, particularly $(1s)^4(1p)^{A-6}(2s, 1d)^2$, is expected to become more significant toward the end of the shell. The increased strength of the tensor interaction adopted in (5-62) compared to the strength used in Exercise 5-10 was partly to simulate this effect. Configuration mixing particularly relevant to $A = 14$ is discussed in detail by Lie (1972), who identifies the states labeled interlopers as predominantly $(1p)^8(2s, 1d)^2$ states. The ground state and the first excited state are each calculated to be 96% $(1p)^{10}$; on the other hand, the 3.94-MeV state appears to have a 29% $(1p)^8(2s, 1d)^2$ admixture. See also Rose *et al.* (1968).

A set of phenomenological or experimental wave functions for the ^{14}N ground state and first excited state, within the $(1s)^4(1p)^{10}$ configuration, is constructed by Ensslin *et al.* (1974) at considerable variance with Table 5-9 and previous literature values. They give

$$\psi(\text{gd}) = 0.913\lvert^{31}D_1\rangle + 0.403\lvert^{31}S_1\rangle - 0.068\lvert^{11}P_1\rangle$$

$$\psi(2.31) = 0.995\lvert^{33}P_0\rangle - 0.093\lvert^{13}S_0\rangle.$$

These ad hoc adjusted coefficients rather accurately account for the relevant observed electron scattering form factors, static magnetic dipole and electric quadrupole moments, and β and γ lifetimes. No attempt is made to reconcile the poor agreement with other determinations in terms of either the energy matrix elements involved or the nature of the effective interactions.[1] It is prudent, however, to cite this variance as a caution against too readily accepting this or that model wave function as "reality" when fits are produced with experimental data, particularly if numerous free parameters exist.

▶**Exercise 5-11** Verify the entries in Table 5-9, using the parameters of (5-62). For example, the energy matrix for the $J=1$, $T=0$ states, using the sequence of matrix elements in Table 5-7, becomes

$$\begin{pmatrix} -8.74 & -1.30 & 3.92 \\ -1.30 & -12.73 & -4.38 \\ 3.92 & -4.38 & 6.10 \end{pmatrix}.$$

Observe that the large spin–orbit effect produces the typical "level crossing" phenomenon between the eigenvalues generically related to the $a=0$, ${}^{31}\mathrm{S}_1$ and ${}^{31}\mathrm{D}_1$ states; refer also to Fig. 5.5 for $\langle a\rangle/K \approx -2$ and the discussion in Appendix D. ◀

▶**Exercise 5-12** Show that the lowest three levels of ${}^{14}\mathrm{N}$ in *jj* representation are

$$\psi_{jj}(J=1,\,T=0) = 0.913\,|\tfrac{1}{2},\tfrac{1}{2}\rangle - 0.406\,|\tfrac{1}{2},\tfrac{3}{2}\rangle - 0.059\,|\tfrac{3}{2},\tfrac{3}{2}\rangle$$

$$\psi_{jj}(J=0,\,T=1) = 0.977\,|\tfrac{1}{2},\tfrac{1}{2}\rangle + 0.228\,|\tfrac{3}{2},\tfrac{3}{2}\rangle$$

$$\psi_{jj}(J=1,\,T=0) = 0.847\,|\tfrac{3}{2},\tfrac{1}{2}\rangle - 0.406\,|\tfrac{3}{2},\tfrac{3}{2}\rangle + 0.352\,|\tfrac{1}{2},\tfrac{1}{2}\rangle.$$ ◀

▶**Exercise 5-13** (a) Show that the magnetic moment and quadrupole moment for the ${}^{6}\mathrm{Li}$ and ${}^{14}\mathrm{N}$ ground states can be written

$$\mu/\mu_0 = 0.690J - [0.190/(J+1)] \sum_{L,S} [L(L+1) - S(S+1)]\,|C_{LS}|^2$$

and

$$Q = \pm\tfrac{1}{5}\langle r^2\rangle [C_{\mathrm{P}}{}^2 - (7/10)\,C_{\mathrm{D}}{}^2 - (4/\sqrt{5})\,C_{\mathrm{S}}\,C_{\mathrm{D}}],$$

with the plus sign for the quadrupole moment of ${}^{6}\mathrm{Li}$ (two particles) and the minus sign for ${}^{14}\mathrm{N}$ (two holes). Particle and hole states produce the same magnetic moment contributions.

[1] The large ${}^{33}\mathrm{P}_0$ component in $\psi(2.31)$ and the phase difference with the ${}^{13}\mathrm{S}_0$ component leads to physically unacceptable values for at least some of the parameters of (5-62) in order to prevent the other ($J=0$, $T=1$) state (having the wave function $\psi' = 0.995\,|{}^{13}\mathrm{S}_0\rangle + 0.093\,|{}^{33}\mathrm{P}_0\rangle$) from dropping in energy *below* the selected first excited state and to give the right first excited-state energy relative to the ground state.

[*Remarks:* Note (2-83). While the expression for the quadrupole moment is most readily derived using the Wigner–Eckart theorem, it is instructive to derive the required result directly by first determining the relevant angular and isospin functions in terms of the individual coordinates (1) and (2) for the two particles or holes, viz.,

$$|{}^{31}\mathrm{S}_1, M_J = +1\rangle = \zeta_0{}^0(1,2)\,\chi_1{}^1(1,2)[3^{-1/2}Y_1{}^1(1)\,Y_1{}^{-1}(2) - 3^{-1/2}Y_1{}^0(1)\,Y_1{}^0(2) + 3^{-1/2}Y_1{}^{-1}(1)\,Y_1{}^1(2)]$$

$$|{}^{11}\mathrm{P}_1, M_J = +1\rangle = \zeta_0{}^0(1,2)\,\chi_0{}^0(1,2)[-2^{-1/2}Y_1{}^1(1)\,Y_1{}^0(2) + 2^{-1/2}Y_1{}^0(1)\,Y_1{}^1(2)]$$

$$|{}^{31}\mathrm{D}_1, M_J = +1\rangle = \zeta_0{}^0(1,2)\{10^{-1/2}\chi_1{}^1(1,2)[6^{-1/2}Y_1{}^1(1)\,Y_1{}^{-1}(2) + \sqrt{\tfrac{2}{3}}\,Y_1{}^0(1)\,Y_1{}^0(2) + 6^{-1/2}Y_1{}^{-1}(1)\,Y_1{}^1(2)] - \sqrt{\tfrac{3}{10}}\,\chi_1{}^0(1,2)[2^{-1/2}Y_1{}^1(1)\,Y_1{}^0(2) + 2^{-1/2}Y_1{}^0(1)\,Y_1{}^1(2)] + \sqrt{\tfrac{3}{5}}\,\chi_1{}^{-1}(1,2)\,Y_1{}^1(1)\,Y_1{}^1(2)\},$$

and then use (2-36).]

(b) Show that the calculated magnetic moments for the wave functions (5-61) for the ^{6}Li ground state and the entry in Table 5-9 for the ^{14}N ground state give

$$^{6}\mathrm{Li}: \quad \mu(\mathrm{calc}) = +0.876\mu_0, \qquad \mu(\mathrm{obs}) = +0.82201\mu_0$$

$$^{14}\mathrm{N}: \quad \mu(\mathrm{calc}) = +0.322\mu_0, \qquad \mu(\mathrm{obs}) = +0.40361\mu_0.$$

(c) To evaluate the quadrupole moment requires a knowledge of the mean square radius $\langle r^2\rangle$. This offers some difficulties. Values of $\langle r^2\rangle$ vary in the range $6.0 \lesssim \langle r^2\rangle \lesssim 11.0\ \mathrm{fm}^2$ depending on whether high-energy electron scattering data, Coulomb energy radii, or various single-particle well parameters are used to obtain the result. In addition, collective effects are undoubtedly present leading to enhanced values of $\langle r^2\rangle$ [see (2-7) and attendant discussion]. In the spirit of the present discussion we more or less arbitrarily adopt $\langle r^2\rangle = (8.0\pm2)\ \mathrm{fm}^2$ for both ^{6}Li and ^{14}N. This value intersects the prediction $\langle r^2\rangle = \frac{5}{4}r_\mathrm{p}{}^2$ based on a harmonic oscillator radius $r_\mathrm{p} = \sqrt{2}\,a_0 = 2.25$ fm deduced from electron scattering when corrections for the finite charge radius of the proton and center-of-mass effects are included (see Chapters II and IV).

Using the estimated value for $\langle r^2\rangle = (8.0\pm2)\ \mathrm{fm}^2$ and the above wave functions, show that the quadrupole moments for ^{6}Li and ^{14}N are

$$^{6}\mathrm{Li}: \quad Q(\mathrm{calc}) = -0.10\ \mathrm{fm}^2, \qquad Q(\mathrm{obs}) = -(0.080\pm0.008)\ \mathrm{fm}^2$$

$$^{14}\mathrm{N}: \quad Q(\mathrm{calc}) = +1.3\ \mathrm{fm}^2, \qquad Q(\mathrm{obs}) = +(1.6\pm0.7)\ \mathrm{fm}^2.$$

◀

The known location of the $^2P_{1/2,3/2}$ levels in the $A = 15$ system and the known mass values relative to ^{16}O permit an estimate of E_0 for the $A = 14$ system in a manner analogous to the $A = 6$ case. We now write

$$E_0 = \{M(^{16}O) - M(^1H) - M(n) - M(^{14}N)\}c^2 + E(n\text{–}^{16}O) + E(p\text{–}^{16}O). \tag{5-63}$$

The value of the quantity in the curly brackets is -22.96 MeV. The value of the neutron hole interaction with the closed 1p shell ^{16}O core *without* the spin–orbit interaction is obtained from the known neutron separation energy and the $^2P_{1/2,3/2}$ energy difference, viz., $E(n\text{–}^{16}O) = +15.67 + \langle a(k = 15)\rangle = 15.67 + 4.12 = 19.79$ MeV. Similarly $E(p\text{–}^{16}O) = +12.13 + \langle a'(k = 15)\rangle = 12.13 + 4.22 = 16.35$ MeV (the proton hole doublet splitting is a little larger). Thus

$$E_0 = -22.96 + 19.79 + 16.35 = +13.18 \text{ MeV}. \tag{5-64}$$

This value is in reasonable agreement with the value of 13.66 MeV required in the set (5-62). Discrepancies in such estimates for E_0 could be expected to be of the order of the energy differences observed in analog state comparisons when only simple Coulomb corrections and n–^{1}H mass difference corrections are made, e.g., in Fig. 5.3. These differences have an rms deviation of the order of 0.3 MeV in the light nuclei. The comparison of the estimated E_0 and the required E_0 for the $A = 6$ system is also consistent with this conclusion.

The small required change in K and the substantial required change in $\langle a\rangle/K$ in going from (5-58) for $A = 6$ to (5-62) for $A = 14$ is consistent with the empirical trend observed in intermediate coupling calculations within the entire 1p shell. In comparison to the more usual literature values, the present changes are somewhat smaller in magnitude and have the gratifying advantage of keeping the Rosenfeld mixture of central forces constant. The major factor accounting for these differences is the present reluctance to require an exact energy fit to the D_2 states, which more than likely require significant configuration mixing.

Literature values of K change more or less monotonically from $K \approx -1.2$ to $K \approx -0.8$ from the beginning to the end of the 1p shell while maintaining $5.0 \lesssim \Lambda/K \lesssim 6.8$. The spin–orbit strength increases somewhat irregularly from $a = -1.6$ to $a \approx -5$. The gradual transition from largely LS coupling at the beginning of the shell to largely jj coupling at the end of the shell is clearly evident. No significant closed $1p_{3/2}$ subshell effect is present for ^{12}C. In fact in the range $A = 10$–12 the ground-state wave function has an *increasing* probability of containing the configuration $(p_{3/2})^{k-2}(p_{1/2})^2$, ranging from ~10% for $A = 10$ to ~40% for $A = 12$.

In conclusion we examine the required phenomenological central potential deduced from (5-58) and (5-62). Figure 5.6 gives the relationship between

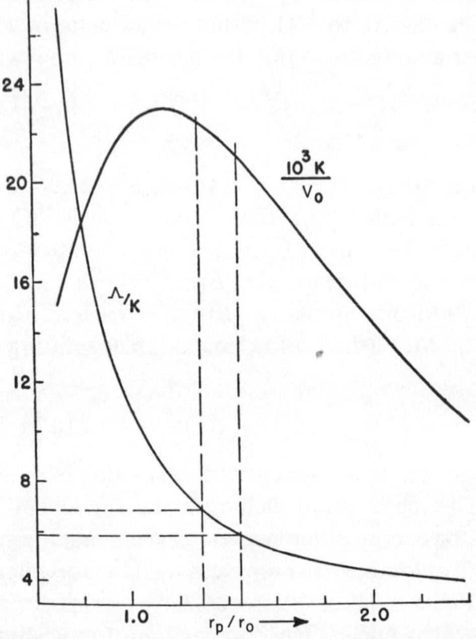

FIG. 5.6. The functional relationship between Λ, K, and V_0 of the central potential in terms of the parameter r_p/r_0. The dashed lines denote the general range of values empirically necessary to fit the 1p shell level schemes [taken from Kurath (1956)].

Λ, K, V_0, and the parameter r_p/r_0. The value of $\Lambda/K = 6.17$ for both $A = 6$ and $A = 14$ is seen to be consistent with the assumption of the same value of r_p for both. Using $r_p = 2.25$ fm and $r_p/r_0 = 1.30$ for $\Lambda/K = 6.17$, we obtain $r_0 = 1.74$ fm. From Fig. 5.6 we also obtain $10^3 K/V_0 = 21.4$.

Hence for the average value of $K = -1.15$ MeV, we obtain $V_0 = -53.8$ MeV. It is perhaps advisable to reduce these quantities for the Gaussian well to an equivalent square-well strength $V_{o,sq} b^2$, using the effective range equivalences summarized in Chapter I by (1-139). We obtain

$$V_{o,sq} b^2 = \frac{53.8}{2.24}\left(\frac{1.74}{0.697}\right)^2 = 150 \text{ MeV fm}^2 \qquad (5\text{-}65)$$

for the spin triplet interaction ^{31}V. This value is perhaps in fortuitously good agreement with the value 136 MeV fm² found in Chapter III [(3-70)] and the free-space value of 143 MeV fm².

The values $\alpha = \frac{3}{5}$, $\beta = -\frac{1}{3}$, and $\gamma = -1.80$ for the Rosenfeld mixture also compare reasonably to the values found in (3-70). Literature values for the relative central interactions generally fall in the ranges $0.45 \leqslant \alpha \leqslant 0.70$,

$-0.6 \leqslant \beta \leqslant 0.15$, and $-1.8 \leqslant \gamma \leqslant 0$. The values for

$$\langle {}^{31}\mathrm{S}_1 | V(S_{12}) + V(L_{12}) | {}^{31}\mathrm{D}_1 \rangle$$

were selected to give an acceptable prediction for the quadrupole moment of ^{6}Li and to give the cancellation required in the Gamow–Teller matrix element for the ^{14}C–^{14}N β-decay (see Chapter VIII). The values -0.35 MeV for $A = 6$ (Exercise 5-10) and -1.30 MeV for $A = 14$ [(5-62)] used for the above tensor plus quadratic spin–orbit matrix elements are within the range calculated with realistic Hamada–Johnston-type potentials and conform to the expected trend within the 1p shell. Thus we find that quite plausible interactions can give a rather satisfactory shell model accounting for the $A = 6$ and $A = 14$ 1p-shell nuclei, at least to first approximation.

Since collective effects are clearly discernible within the 1p shell, various refinements, particularly configuration mixing, must certainly be included to obtain a higher order of approximation. This is particularly true when γ-ray transition rates are to be evaluated for $\Delta L = 1$ (see Chapter VII).

3. Center-of-Mass Corrections

Allowance must be made for the fact that the central shell model potential in H_0 should be referred to the appropriate center of mass of the system rather than simply taken as a potential fixed in space. This is relatively simple to achieve in the harmonic oscillator approximation, particularly when valence particles are confined to a single (n, l) shell as in our treatment of ^{6}Li and ^{14}N for the normal parity states.

In treating these nuclei we implicitly assumed that the ten originally degenerate states of the $(1s)^4(1p)^{2,10}$ configurations were obtained from a simplified version of the Hamiltonian (5-35) with no spin–orbit or central two-body interactions and further with $\overline{V}(r_i) = \frac{1}{2}\kappa r_i^{\,2}$, where κ is an effective harmonic oscillator spring constant. We thus were using the harmonic oscillator many-body Hamiltonian

$$H = \sum_{i=1}^{A} [(1/2m)\,\mathbf{p}_i^{\,2} + \tfrac{1}{2}\kappa \mathbf{r}_i^{\,2}], \tag{5-66}$$

with a coordinate frame of reference fixed in space. The proper Hamiltonian for the internal energy of our system should have been of the translationally invariant form

$$H' = \sum_{i=1}^{A} \left[\frac{1}{2m} \left(\mathbf{p}_i - \frac{1}{A}\mathbf{P} \right)^2 + \frac{\kappa}{2} (\mathbf{r}_i - \mathbf{R})^2 \right] \tag{5-67}$$

with $\mathbf{P} = \sum_{i=1}^{A} \mathbf{p}_i$ and $\mathbf{R} = (1/A) \sum_{i=1}^{A} \mathbf{r}_i$. The Hamiltonian (5-67) contains only $A-1$ independent coordinate and momentum variables since

$\sum_{i=1}^{A}(\mathbf{r}_i-\mathbf{R})=0$ and $\sum_{i=1}^{A}[\mathbf{p}_i-(1/A)\mathbf{P}]=0$. Choosing to eliminate $\mathbf{r}_A$ and $\mathbf{p}_A$ and writing $\mathbf{r}_i'=\mathbf{r}_i-\mathbf{R}$ and $\mathbf{p}_i'=\mathbf{p}_i-(1/A)\mathbf{P}$, we obtain

$$H' = \frac{1}{2m}\left[\sum_{i=1}^{A-1}\mathbf{p}_i'^2+\left(\sum_{i=1}^{A-1}\mathbf{p}_i'\right)^2\right]+\frac{\kappa}{2}\left[\sum_{i=1}^{A-1}\mathbf{r}_i'^2+\left(\sum_{i=1}^{A-1}\mathbf{r}_i'\right)^2\right]. \tag{5-68}$$

It is possible to find a linear transformation that eliminates the cross terms contained in (5-68). In terms of these new coordinates $\mathbf{r}_i'\to\boldsymbol{\xi}_i$ and new conjugate momenta $\mathbf{p}_i'\to\boldsymbol{\pi}_i$, we obtain

$$H' = \sum_{i=1}^{A-1}[(1/2m)\boldsymbol{\pi}_i^2+\tfrac{1}{2}\kappa\boldsymbol{\xi}_i^2]. \tag{5-69}$$

For the case of the harmonic oscillator the relationship between the original fixed-center Hamiltonian and the translationally invariant Hamiltonian is readily obtained, with the result

$$H=(1/2Am)\mathbf{P}^2+\tfrac{1}{2}A\kappa\mathbf{R}^2+H' \tag{5-70a}$$

or

$$\sum_{i=1}^{A}\left(\frac{1}{2m}\mathbf{p}_i+\frac{\kappa}{2}\mathbf{r}_i^2\right)=\left(\frac{1}{2Am}\mathbf{P}^2+\frac{A\kappa}{2}\mathbf{R}^2\right)_{\text{c.m.}}+\sum_{i=1}^{A-1}\left(\frac{1}{2m}\boldsymbol{\pi}_i^2+\frac{\kappa}{2}\boldsymbol{\xi}_i^2\right). \tag{5-70b}$$

In view of (5-70), we have

$$\psi_A(\mathbf{r}_i)=\psi_{\text{c.m.}}(\mathbf{R})\,\psi_{A-1}(\boldsymbol{\xi}_i), \tag{5-70c}$$

with the oscillator angular frequency $\omega^2=\kappa/m$ and energy $\hbar\omega$ *the same for all three terms* in (5-70b).

▶**Exercise 5-14** Show that the relationships (5-66)–(5-70) follow from the definitions

$$\mathbf{R}=(1/A)\sum_{i=1}^{A}\mathbf{r}_i,\qquad M=Am,\qquad \mathbf{P}=M\dot{\mathbf{R}},$$

$$\mathbf{r}_i'=\mathbf{r}_i-\mathbf{R},\qquad \mathbf{p}_i=m\dot{\mathbf{r}}_i,\qquad \mathbf{p}_i'=m\dot{\mathbf{r}}_i'.$$ ◀

Note that if a symmetric appearance of the coordinates is desirable, $\psi_{A-1}(\boldsymbol{\xi}_i)$ in (5-70c) can be readily transformed back to $\psi_A(\mathbf{r}_i')$. Since $\psi_{\text{c.m.}}$ is symmetric in all particles, antisymmetrization (or partial antisymmetrization) of $\psi_A(\mathbf{r}_i)$ will still permit a factorization as in (5-70c), viz.

$$\begin{aligned}\psi_A'(\mathbf{r}_i)&=\mathscr{A}\psi_A(\mathbf{r}_i)=\mathscr{A}[\psi_{\text{c.m.}}(\mathbf{R})\,\psi_A(\mathbf{r}_i')]\\&=\psi_{\text{c.m.}}(\mathbf{R})\,\mathscr{A}\psi_A(\mathbf{r}_i')=\psi_{\text{c.m.}}(\mathbf{R})\,\psi_A'(\mathbf{r}_i').\end{aligned}$$

So long as $\psi_A'(\mathbf{r}_i)$ and $\psi_A'(\mathbf{r}_i')$ refer to all valence particles in the same (n, l) shell,

$$\psi_{\text{c.m.}} = (1\text{S}) = N \exp(-\alpha A R^2), \tag{5-71}$$

with $\alpha = m\omega/2\hbar$ as before. Thus the solutions to H in (5-66) include having the center of mass in a (1S) harmonic oscillator state (for $A = 6$ and 14, an ${}^{11}\text{S}_0$ state) instead of the usual simple translational state. States such as these, which have the same center-of-mass motion as the (1S) ground state, are referred to as *good states*. They generally give correct matrix elements for operators that themselves are improperly referred to the fixed coordinate frame rather than the center of mass, simply because the center of mass is in a (1S) state.

When valence particles are in several (n, l) shells, it is possible to generate states that appear different in terms of $\psi_A(\mathbf{r}_i)$ but which in fact have the same internal wave function as a lower (n, l) configuration with only the center-of-mass motion different. Such states are referred to as *spurious* states, and in general will contribute erroneous values to matrix elements. For example, the excited ${}^{16}\text{O}$ configurations $(1\text{s})^4(1\text{p})^{11}(2\text{s})$ and $(1\text{s})^4(1\text{p})^{11}(1\text{d})$ *each* could apparently give rise to a ${}^{11}\text{P}$ state of symmetry $[\lambda] = [444]$. However, a linear combination of the two turns out to be just $(1\text{s})^4(1\text{p})^{12}[4444]\,{}^{11}\text{S}_0$ multiplied by the center of mass in a (1P) state. Since the ground state of ${}^{16}\text{O}$ is $(1\text{s})^4(1\text{p})^{12}[444]\,{}^{11}\text{S}_0$ coupled to a center of mass in a (1S) state, this linear combination is a spurious state which must be discarded. However a different linear combination, namely $(\frac{5}{6})^{1/2}(1\text{s})^4(1\text{p})^{11}(2\text{s}) + (\frac{1}{6})^{1/2}(1\text{s})^4(1\text{p})^{11}(1\text{d})$, is found to be a good state with the center of mass in a (1S) state. Thus of the two possible ${}^{11}\text{P}$ states, only one corresponding to a particular linear combination survives. For other potential shapes this correction can be considerably more complicated and the "spurious states," introduced by failure to allow for proper translational invariance, pose some difficulties. Techniques are available for approximations that correct for center-of-mass effects. One method exploits the a priori known separability of the exact eigenfunction in internal and cm coordinates [see Vincent (1973)].

A more problematic effect, particularly prominent for the harmonic oscillator Hamiltonian H_0, relates to the unphysical behavior of the single-particle radial wave functions in the nuclear surface region and beyond. In the harmonic oscillator approximation *all* states are bound and the radial dependence at larger radii is dominated by the rapid damping produced by the exponential factor $\exp(-\alpha r^2)$, with $\alpha = m\omega/2\hbar = 1/r_{\text{p}}^2$ (see Section IV, B, 1). A more realistic surface and asymptotic behavior is that obtained with potentials that rapidly vanish beyond the nuclear radius. These yield a gentler exponential dependence $\exp(-k_2 r)$, with $k_2 = (2mE_{\text{B}}/\hbar^2)^{1/2}$, in the region where the potential is essentially zero (see Section IV, B, 2). The effect is to produce a more pronounced "tail" to the wave function, particularly for states that are weakly bound, i.e., E_B small. The difference in behavior for states that are unbound for realistic potentials and the bound counterpart states for infinite potentials, such as the harmonic oscillator potential, assumes a central position in one approach to reaction theory. The fundamental questions involved here are best treated under the topic of reaction theory.

E. CLUSTER MODEL

There are numerous phenomena of nuclear behavior that suggest the clustering of nucleons into groups within a nucleus. The earliest and perhaps simplest nuclear model to consider such characteristics is the α-particle model. Heavy nuclei that spontaneously decay by α-particle emission have decay rates suggesting at least a tendency for the preformation of α-particle clusters in nuclear matter. In the simplified theory of nuclear matter given earlier, the fact that four nucleons in a relative $^{11}S_0$ state could strongly interact played an important role in accounting for the binding energy of nuclei. Nuclear ground states would thus be expected to favor such quasi-α-particle configurations and also consequently exhibit large spatial symmetry. Indeed the nuclei with $N = Z$ and $A = 4n$, $n = 1, 2, 3, \ldots$, the so-called α-particle-like nuclei, have exceptionally large binding energies. In addition these binding energies are largely accounted for as multiples of the α-particle binding energy, 28.30 MeV. This suggests viewing such nuclei as consisting of n α-particle clusters with relatively weak intercluster "bond" energies. Table 5-10 gives an accounting for the binding energies of the α-particle-like nuclei on the basis of such a scheme with

$$E_B = 28.30n + Cm, \tag{5-72}$$

where C is the intercluster bond energy and m the number of bonds for the tightly packed α-clusters. The intercluster interaction is assumed to be short-ranged and of the order of the size of the α-particle. Some model of the cluster

TABLE 5-10

Intercluster Bond Energies of the Light 4*n* Self-Conjugate Nuclei

Nucleus	n	m	$E_B - 28.30n$ (MeV)	C (MeV)
^{8}Be	2	1	−0.092	—
^{12}C	3	3	7.275	2.46
^{16}O	4	6	14.44	2.41
^{20}Ne	5	8	19.17	2.40
^{24}Mg	6	12	28.48	2.37
^{28}Si	7	16	38.47	2.40
^{32}S	8	19	45.40	2.39

configuration is required to determine the number of interacting pairs, particularly for $m < n(n-1)/2$. Thus ^{20}Ne, the first nucleus for which $m \neq n(n-1)/2$, is imagined to have α-particles arranged in the form of a trigonal bipyramid. Such configuration assignments present some difficulties [e.g., see Hauge *et al.* (1971)]. When the dynamical variations of the geometry are considered for excited modes, even ^{12}C and ^{16}O pose considerable complexity [see Bertsch and Bertozzi (1971), Onishi and Sheline (1971), and De Takacsy (1972)].

Excepting ^{8}Be, which is actually unbound for decay into two α-particles, the intercluster bond energy is seen to be essentially constant for the nuclei listed in Table 5-10. These nuclei also have larger than average separation energies for the least-bound proton or neutron, suggesting the requirement of breaking up one of the α-clusters. Thus ^{8}Be, while unstable to break up into two α-particles, is nonetheless bound with respect to individual nucleon emission. The separation energies for a neutron and proton are $E_S(N) = 18.900$ MeV and $E_S(Z) = 17.256$ MeV. These values should be compared to the corresponding nucleon separation energies for the α-particle itself, which are 20.578 and 19.815 MeV, respectively. For some additional considerations of the α-particle model see McDonald *et al.* (1970), Abulaffio and Irvine (1972), Avishai (1972), and Basu (1972), and the references contained therein.

The pioneering work of Perring and Skyrme (1956) has shown that the antisymmetrized S states of maximum spatial symmetry for the shell model configurations $(1s)^4(1p)^{4(n-1)}$ can be written in a form exhibiting α-particle clustering. For the truncated 1p-shell representation of such $4n$-nuclei, these states correspond to the ground states $^{11}S_0[44\cdots]$.

The cluster model in its present form is due largely to the investigations of Wildermuth and co-workers [e.g., Wildermuth and Kanellopoulos (1958, 1959), Wildermuth and McClure (1966)] and of Neudatchin and Smirnov and their co-workers [e.g.,Neudatchin and Smirnov (1969), and Kudeyarov *et al.* (1971)]. An earlier formulation, the so-called resonating group theory of Wheeler (1937), should also be cited.

One of the simpler examples extends the equivalence mentioned above to the ground state and first two excited states of ^{8}Be. Up to a normalization constant these three-shell model states $(1s)^4(1p)^4$, $J = 0, 2, 4$ are *identical* with the antisymmetrized α-cluster wave functions

$$\Psi(^8\mathrm{Be})_J \equiv \mathscr{A}\psi(\alpha 1)\,\psi(\alpha 2)\,X_J(\mathbf{R})\,W(\mathbf{R}_{\mathrm{c.m.}}). \tag{5-73a}$$

We define

$$\mathbf{R}_{\alpha 1} = \tfrac{1}{4}\sum_{i=1}^{4}\mathbf{r}_i, \qquad \mathbf{R} = \mathbf{R}_{\alpha 1} - \mathbf{R}_{\alpha 2},$$

$$\mathbf{R}_{\alpha 2} = \tfrac{1}{4}\sum_{i=5}^{8}\mathbf{r}_i, \qquad \mathbf{R}_{\mathrm{c.m.}} = \tfrac{1}{2}(\mathbf{R}_{\alpha 1}+\mathbf{R}_{\alpha 2}) = \tfrac{1}{8}\sum_{i=1}^{8}\mathbf{r}_i.$$

The wave functions $\psi(\alpha 1)$ and $\psi(\alpha 2)$ are the internally unexcited harmonic oscillator α-particle states,

$$\psi(\alpha 1) = \exp\left(-\alpha \sum_{i=1}^{4} |\mathbf{r}_i - \mathbf{R}_{\alpha 1}|^2\right)$$

and

$$\psi(\alpha 2) = \exp\left(-\alpha \sum_{i=5}^{8} |\mathbf{r}_i - \mathbf{R}_{\alpha 2}|^2\right). \tag{5-73b}$$

$W(\mathbf{R}_{\text{c.m.}})$ is the (1S) harmonic oscillator c.m. state

$$W(\mathbf{R}_{\text{c.m.}}) = \exp(-8\alpha R_{\text{c.m.}}^2). \tag{5-73c}$$

$X_J(\mathbf{R})$ is appropriately one of the three harmonic oscillator states with excitation $4\hbar\omega$ given in Table 4-1, viz.,

$$X_{J=4}^{M}(\mathbf{R}) = |1\text{G}\colon L = 4, M\rangle = R^4[\exp(-2\alpha R^2)]\, Y_4{}^M(\hat{R})$$

$$X_{J=2}^{M}(\mathbf{R}) = |2\text{D}\colon L = 2, M\rangle = R^2\left(1 - \frac{8\alpha}{7}R^2\right)[\exp(-2\alpha R^2)]\, Y_2{}^M(\hat{R}) \tag{5-73d}$$

$$X_{J=0}^{O}(\mathbf{R}) = |3\text{S}\colon L = 0, M = 0\rangle$$

$$= \left(1 - \frac{16}{3}\alpha R^2 + \frac{64}{15}\alpha^2 R^4\right)[\exp(-2\alpha R^2)]\, Y_0{}^0(\hat{R}).$$

In *all* of the above, $\alpha = m\omega/2\hbar$, and in (5-73d) we have noted that the appropriate mass for the relative motion of the two α-particle clusters is the reduced mass $\frac{1}{2}(4m) = 2m$. For a discussion of the problems involved in performing the antisymmetrization called for in (5-73a) refer to Sünkel and Wildermuth (1972).

►**Exercise 5-15** Account for the total harmonic oscillator shell model energy of $16\hbar\omega$ for the cluster model state (5-73) including the zero-point energy. (*Hint:* Note that of the four coordinates in each of the internal α-particle states, only three are independent.) ◄

Once again the simple harmonic oscillator model would leave the three states (5-73), and indeed all the $(1s)^4(1p)^4$ states, degenerate. Numerous procedures might be employed at this point to remove the degeneracies and attempt to fit the observed energies and other level characteristics. One procedure would be to use a perturbation approach with residual interactions coupled with an appropriate energy reference, as we have in the preceding intermediate coupling calculation. Another approach would be to view the wave functions such as (5-73) as trial functions in a variational calculation, perhaps allowing the harmonic oscillator parameter α to be different in the intercluster portion of the wave function from that used for the internal α-particle states.

A simplified calculation, which, however, gives some degree of confidence in the essential correctness of (5-73) in describing the lowest three states of ^{8}Be, is presented by Wildermuth. The wave function (5-73) is used with a system Hamiltonian

$$H = (1/2m) \sum_{i=1}^{8} p_i^2 + \tfrac{1}{2} \sum_{i,j} v_{ij}, \tag{5-74a}$$

in which v_{ij} is taken to be of the form (5-44) with[1]

$$v_{ij} = V_0 \{\exp[-(r_{12}/r_0)^2]\} \times [\mathcal{T}(\sigma)\mathcal{S}(\tau) + \alpha'\mathcal{S}(\sigma)\mathcal{T}(\tau) + \beta'\mathcal{T}(\sigma)\mathcal{T}(\tau) + \gamma'\mathcal{S}(\sigma)\mathcal{S}(\tau)]. \tag{5-74b}$$

Values which in addition to describing the low-energy two-nucleon problem also adequately give a satisfactory value for the α-particle binding energy and size are

$$2\alpha = 1/a_0^2 = 0.47 \text{ fm}^{-2}, \qquad V_0 = -68.6 \text{ MeV}, \qquad r_0 = 1.55 \text{ fm},$$
$$\alpha' = 0.64, \qquad \beta' = \gamma' = 0. \tag{5-74c}$$

It should also be noted that the parameters (5-74c) are essentially equivalent to those used in our intermediate coupling calculations since the magnitude $V_0 r_0^2 = 165$ MeV fm^2 is the same for both potentials and since the low-lying states of high spatial symmetry are influenced only weakly by β and γ through coupling or mixing with states lying considerably higher in energy. The Hamiltonian (5-74), for a *completely de novo* calculation of the energy expectation value $\langle H \rangle$, gives essentially the correct binding energy for the ground state and a level spacing shown in Fig. 5.7. Similar results were also obtained by Thompson *et al.*, using a resonating-group calculation to account for the (α, α) scattering phase shifts up to 15 MeV in the c.m. system. They obtained an oscillator parameter $2\alpha = 0.514$ fm^{-2} for the α-particle cluster in (5-73b) with $V_0 = -72.98$ MeV, $r_0 = 1.48$ fm, and a near-Serber mixture with $\alpha' = 0.63$ [see Thompson *et al.* (1969)].

Since we are using harmonic oscillator wave functions, it follows that the kinetic energy is $\langle (1/2m) \sum_{i=1}^{8} p_i^2 \rangle = 8\hbar\omega$ for any state of the configuration $(1s)^4(1p)^4$. Hence all energy differences come from variations in the expectation values of the potential energy of (5-74). These variations in turn arise

[1] The Coulomb interaction is again excluded.

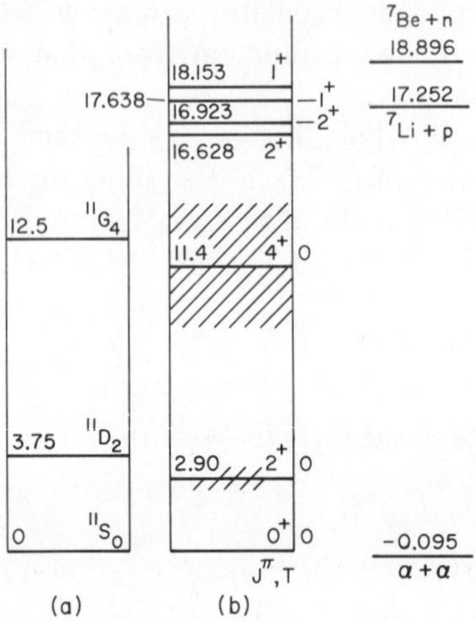

FIG. 5.7. (a) The energy levels of ^{8}Be based on the cluster model. (b) The first seven definitely established, experimentally observed states of ^{8}Be. The two 1^+ states and the upper pair of 2^+ states are strongly mixed isospin states of $T = 0, 1$. The experimental separation energies $E_S(N)$, $E_S(Z)$, and $E_S(^4\text{He})$ are also shown. Possible very broad states at ~ 6 MeV, $J^\pi = 0^+$, and ~ 10 MeV, $J^\pi = 2^+$, are not shown.

from the various ways of coupling the 1p valence nucleons. Thus the above results should be virtually identical with the intermediate coupling calculations for the lowest states of high spatial symmetry. Indeed, with $K = -1.18$ MeV, $\Lambda/K = 5.8$, and $a/K = 2.0$, Kurath (1956) finds approximately the same results for the lowest three states, which are the nearly pure $^{11}S_0$, $^{11}D_2$, and $^{11}G_4$ states.[1] In addition, however, the intermediate coupling calculations also generate numerous states in the excitation energy region $16 \lesssim E_x \lesssim 20$ MeV. These are states generically related to the pure *LS*-coupled states $^{31}P_{0,1,2}$, $^{33}P_{0,1,2}$, $^{31}D_{1,2,3}$, $^{33}D_{1,2,3}$, and $^{13}P_1$. The cluster model with internally unexcited α-particles cannot generate any of these states, and consequently they require breaking up at least one of the α-clusters. The necessarily even parity states of two interacting (spinless) α-particles can have only $J = L$ with L even. A new cluster description in terms of (^{7}Be + n) or (^{7}Li + p) is more appropriate for some of these states. For such an analysis as well as a discussion of the isospin mixing, see Marion (1965) and also Barker (1966).

It is interesting to note further that the cluster model energy ratio $[E(G) - E(S)]/[E(D) - E(S)] \approx 10/3$ is just the ratio expected for the lowest

[1] Kurath's values correspond to a two-body potential with essentially the same "strength" $V_0 r_0^2$ and $\alpha'' = 0.60$, $\beta'' = -0.60$, and $\gamma'' = -1.00$.

three states of a rigid rotator variant of the α-particle model. The energy levels of such a rotator are proportional to $J(J+1)$, J even.[1] Paradoxically, in the present instance this ratio relationship is due to potential energy differences, whereas for the rigid rotator this is a kinetic (rotational) energy effect. The resolution of this seeming paradox is through the effects of antisymmetrization. Many seemingly quite different nuclear wave functions become very similar or even identically equivalent under antisymmetrization. Thus in the present instance the lowest states of ^{8}Be are equally well accounted for (at least to first approximation) by a shell model, cluster model, or some form of collective rotational model. For a critical intercomparison of these models for ^{8}Be refer to Harvey and Jensen (1972). The question eventually arises as to what model to use. As with the selection of a suitable coordinate system in mechanics, the convenience in calculating the desired effect in question usually determines the choice to be made. It must also be said, however, that the inclusion of higher-order approximations may be more transparently manageable within one or another of approximately equivalent models.

In considering the excitation of higher shell model configurations and configuration mixing, the cluster model may give a more straightforward view of the matter. One notable example of such an instance occurs for the nucleus ^{19}F. Many of the low-lying levels of normal (even) parity involve the configuration $(1s)^4(1p)^{12}(2s, 1d)^3$, and a satisfactory accounting of these states is given by an intermediate coupling shell model calculation (Elliott and Flowers, 1955). However, the first excited state at only 0.11 MeV above the ground state is a nonnormal parity $\frac{1}{2}^-$ state. One possible description of this state would be a $(1s)^4(1p)^{12}(2s, 1d)^2(2p, 1f)^1$ shell model state involving $19\hbar\omega$ oscillator excitation quanta, $1\hbar\omega$ more than the ground state. However, a rather more satisfactory accounting of this state is given by a cluster model parentage $(^{15}N+\alpha)$ in an intercluster 5S state, also involving $19\hbar\omega$ excitation quanta. This cluster model description in equivalent shell model language is a $(1s)^4(1p)^{11}(1s, 2d)^4$ configuration. In this instance the cluster model definitely selects one of two apparently competing shell configurations. The lower normal parity states in the cluster model description are $(^{16}O+{}^{3}H)$ clusters. The $\frac{1}{2}^+$ ground state involves an intercluster 4S state, for example. Alternatively, the level structure of ^{19}F can also be described by a rotational model [see Benson and Flowers (1969), Bingham and Fortune (1972), and Rogers (1973)].

[1] The analysis of the α–α scattering experiments also suggests two very broad states in ^{8}Be, at ~ 6 MeV, $J^\pi = 0^+$, and ~ 10 MeV, $J^\pi = 2^+$. It has been conjectured that these may be collective vibrational and vibrational–rotational states [e.g., see Chang (1974)].

1. Cluster Model of ^{6}Li

Returning to the $A = 6$ nuclei that we previously discussed under intermediate coupling, we again note that, as for ^{8}Be, having $\beta \leqslant 0$ and $\gamma \leqslant 0$ results in driving the ^{3}P and ^{1}P generic states to rather higher energies than the ground state. Thus the lowest six states of ^{6}Li are all generic S and D states. A rather logical cluster model description of these states would appear to be an (α+d)-cluster. We note that coupling a deuteron in its $^{31}S_1$ ground state to an α-particle with an intercluster harmonic oscillator state of excitation $2\hbar\omega$, expected for a $(1s)^4(1p)^2$ configuration, yields all the required $T = 0$ shell model LS states for ^{6}Li, namely the states $^{31}S_1$ and $^{31}D_{1,2,3}$. The S state is associated with the intercluster 2S state while the D states are associated with the intercluster 1D state (see Table 4-1). Some of the low-lying states in ^{6}Li might be assumed to result from a similar cluster configuration involving the relatively low excited state of the deuteron.[1] Since this deuteron state is a $^{13}S_0$ state, we would thus just generate the two required $T = 1$ shell model LS states $^{13}S_0$ and $^{13}D_2$ for ^{6}Li.

It is important to realize that this internal excitation of the deuteron is *not* a harmonic oscillator excitation involving excitation quanta $\hbar\omega$. In a zeroth-order harmonic oscillator model, this excited state and the ground state are degenerate. The degeneracy is removed by the difference between the realistic ^{31}V and ^{13}V potentials for the actual deuteron. Similarly, in the cluster model description of ^{6}Li this degeneracy is again removed in any energy calculation using (5-74) by the operation of the same potential terms, albeit with re-adjusted phenomenological strengths and with matrix elements determined for the harmonic oscillator internal 1S wave function rather than for the relevant true deuteron wave functions. Interestingly enough, we note from Table 5-7 that for $a = 0$ the average energy difference between the $T = 0$ quartet and the $T = 1$ doublet is $\Delta E \approx (\alpha - 1)\Lambda = 3.0$ MeV, some 25% larger than the actual excitation energy for the real deuteron. As suggested above, this difference in energy is largely due to the difference between the internal deuteron cluster harmonic oscillator wave functions and the real deuteron wave functions, particularly for the virtual state. We thus have an example of a cluster configuration in which the excited state of one of the clusters also plays an important role and in addition also an example of the apparent approximate "additivity" of such internal excitation energies. The generalization of this situation to other cases of internally excited clusters requires careful treatment when indeed changes in the internal harmonic oscillator states are involved or even when significant configuration mixing is present in the "normal" states. (The literal treatment of the cluster model

[1] The virtual state of the deuteron is ~90 keV unbound or at an excitation of 2.315 MeV above the ground state (see Section I, D, 1).

taking into account the $^{31}D_1$ admixture in the ground state of the deuteron offers similar problems.) Since the lowest excited state of the α-particle is at 20.2 MeV, it would be generally expected to participate in cluster configurations only without internal excitation.

The first six states of ^{6}Li thus can have their wave functions written in the (α + d) representation in the general form[1]

$$\Psi(^6\text{Li}) = \mathscr{A}\psi(\alpha)\psi_{\tau,\sigma}(\text{d})\,X(\mathbf{R})\,W(\mathbf{R}_{\text{c.m.}}), \tag{5-75a}$$

where

$$\mathbf{R}_\alpha = \tfrac{1}{4}\sum_{i=1}^{4}\mathbf{r}_i, \qquad \mathbf{R} = \mathbf{R}_\alpha - \mathbf{R}_\text{d}$$

$$\mathbf{R}_\text{d} = \tfrac{1}{2}(\mathbf{r}_5+\mathbf{r}_6), \qquad \mathbf{R}_{\text{c.m.}} = \tfrac{1}{6}(4\mathbf{R}_\alpha+2\mathbf{R}_\text{d}) = \tfrac{1}{6}\sum_{i=1}^{6}\mathbf{r}_i.$$

The c.m. wave function $W(\mathbf{R}_{\text{c.m.}})$ is again a 1S harmonic oscillator state and $X(\mathbf{R})$ is one of the 2S or 1D states of Table 4-1 (with reduced mass $\frac{4}{3}m$). The internal α-particle wave function $\psi(\alpha)$ is the same as (5-73b). The internal deuteron wave function $\psi_\tau(\text{d})$ for the $T = 0$, $S = 1$, ^{6}Li states is

$$\psi_\tau(\text{d}) = \{\exp[-\alpha(|\mathbf{r}_5-\mathbf{R}_\text{d}|^2 + |\mathbf{r}_6-\mathbf{R}_\text{d}|^2)]\}\,\chi_3^{M_s}(5,6)\,\zeta_1{}^0(5,6) \tag{5-75b}$$

and for the $T = 1$, $S = 0$, ^{6}Li states, it is $\psi_\sigma(\text{d})$, with

$$\psi_\sigma(\text{d}) = \{\exp[-\alpha(|\mathbf{r}_5-\mathbf{R}_\text{d}|^2 + |\mathbf{r}_6-\mathbf{R}_\text{d}|^2)]\}\,\chi_1{}^0(5,6)\,\zeta_3{}^0(5,6). \tag{5-75c}$$

As a guide for constructing cluster model wave functions of the type (5-73) and (5-75), the harmonic oscillator noninteracting many-body shell model Hamiltonian is rewritten for q clusters each containing n_q nucleons. We have $\sum_q n_q = A$ and define

$$\mathbf{R}_q = (1/n_q)\sum_{i\in q}\mathbf{r}_i, \qquad \text{and} \qquad \mathbf{r}_i{}' = \mathbf{r}_i - \mathbf{R}_q \qquad \text{for} \quad i \in q. \tag{5-76}$$

Clearly, the internal relative coordinates $\mathbf{r}_i{}'$ satisfy the condition $\sum_{i\in q}\mathbf{r}_i{}' = 0$ and hence only $n_q - 1$ of these n_q coordinates of a particular cluster are independent. Introducing these coordinates, the Hamiltonian

$$H = \sum_{i=1}^{A}\left[(1/2m)\,\mathbf{p}_i{}^2 + \tfrac{1}{2}\kappa\mathbf{r}_i{}^2\right]$$

becomes

$$H = \sum_q H_q + (1/2m)\sum_q (1/n_q)\,\mathbf{P}_q{}^2 + \tfrac{1}{2}\kappa\sum_q n_q\mathbf{R}_q{}^2, \tag{5-77a}$$

with $\mathbf{P}_q = \sum_{i\in q}\mathbf{p}_i$. The internal cluster Hamiltonians H_q have the form

$$H_q = (1/2m)\sum_{i\in q}[\mathbf{p}_i - (1/n_q)\,\mathbf{P}_q)]^2 + \tfrac{1}{2}\kappa\sum_{i\in q}(\mathbf{r}_i - \mathbf{R}_q)^2. \tag{5-77b}$$

[1] The subscripts τ, σ refer to the triplet and singlet n–p spin states, respectively.

▶**Exercise 5-16** (a) Consider the internal cluster Hamiltonian for an α-particle. From (5-77b) with $n_q = 4$ and writing $\mathbf{p}_i' = \mathbf{p}_i - \frac{1}{4}\mathbf{P}_\alpha$ and $\mathbf{r}_i' = \mathbf{r}_i - \mathbf{R}_\alpha$, we get

$$H_\alpha = (1/2m) \sum_{i=1}^{4} \mathbf{p}_i'^2 + \tfrac{1}{2}\kappa \sum_{i=1}^{4} \mathbf{r}_i'^2.$$

Of the four relative momenta and four relative coordinates only three of each are independent. A straightforward attempt to eliminate, say, $\mathbf{r}_4'$, i.e., $\mathbf{r}_4' = -(\mathbf{r}_1' + \mathbf{r}_2' + \mathbf{r}_3')$, would give for the potential energy term

$$\tfrac{1}{2}\kappa \sum_{i=1}^{3} \mathbf{r}_i'^2 + \tfrac{1}{2}\kappa(\mathbf{r}_1' + \mathbf{r}_2' + \mathbf{r}_3')^2$$

containing numerous cross terms. A similar situation prevails for the momenta. Show that a linear transformation

$$\begin{aligned}
\boldsymbol{\xi}_1 &= -(1/\sqrt{2})(\mathbf{r}_2' + \mathbf{r}_3'), & \boldsymbol{\pi}_1 &= -(1/\sqrt{2})(\mathbf{p}_2' + \mathbf{p}_3') \\
\boldsymbol{\xi}_2 &= -(1/\sqrt{2})(\mathbf{r}_1' + \mathbf{r}_3'), & \boldsymbol{\pi}_2 &= -(1/\sqrt{2})(\mathbf{p}_1' + \mathbf{p}_3') \\
\boldsymbol{\xi}_3 &= -(1/\sqrt{2})(\mathbf{r}_1' + \mathbf{r}_2'), & \boldsymbol{\pi}_3 &= -(1/\sqrt{2})(\mathbf{p}_1' + \mathbf{p}_2')
\end{aligned}$$

results in

$$H_\alpha = (1/2m) \sum_{i=1}^{3} \boldsymbol{\pi}_i^2 + \tfrac{1}{2}\kappa \sum_{i=1}^{3} \boldsymbol{\xi}_i^2$$

with no cross terms.

Are the above $\boldsymbol{\xi}$ and $\boldsymbol{\pi}$ variables canonical (i.e., determine the relevant commutators $[\xi_i, \pi_j]$)?

(b) Show that for only two clusters n_1 and n_2, the c.m. portion of (5-77a) becomes

$$\frac{1}{2m}\left(\frac{\mathbf{P}_1^2}{n_1} + \frac{\mathbf{P}_2^2}{n_2}\right) + \frac{\kappa}{2}(n_1 \mathbf{R}_1^2 + n_2 \mathbf{R}_2^2) = \frac{1}{2Am}\mathbf{P}_{\text{c.m.}}^2 + \frac{\kappa}{2} A \mathbf{R}_{\text{c.m.}}^2 + \frac{A}{2n_1 n_2 m}\mathbf{P}^2 + \frac{\kappa}{2}\frac{n_1 n_2}{A}\mathbf{R}^2$$

with

$$\mathbf{R} = \mathbf{R}_1 - \mathbf{R}_2, \qquad \mathbf{P} = \frac{n_2}{A}\mathbf{P}_1 + \frac{n_1}{A}\mathbf{P}_2,$$

$$\mathbf{R}_{\text{c.m.}} = \frac{1}{A}(n_1 \mathbf{R}_1 + n_2 \mathbf{R}_2) = \frac{1}{A}\sum_{i=1}^{A} \mathbf{r}_i, \qquad \mathbf{P}_{\text{c.m.}} = \mathbf{P}_1 + \mathbf{P}_2 = \sum_{i=1}^{A} \mathbf{p}_i,$$

$$A = n_1 + n_2.$$

(c) If the harmonic oscillator constant κ is held fixed, are all the various oscillator frequencies the same for the terms in parts (a) and (b)? ◀

In view of the preceding, it is evident that (5-77) has a high degree of degeneracy in cluster expansion forms and that these solutions may not be completely linearly independent or orthogonal. The states of ^{6}Li discussed earlier in terms of (α + d)-cluster states are in fact a case in hand. These states can be expressed as well in terms of (^{3}H + ^{3}He)-cluster states. The internal unexcited ^{3}H and ^{3}He states would both be 1S states with spin $\frac{1}{2}$ and isospin $\frac{1}{2}$. The relative intercluster state would again contain $2\hbar\omega$ quanta of excitation and hence correspond to either 2S or 1D states. Since ^{3}H and ^{3}He are identical fermions in isospin formalism, only the cluster-coupled spin triplet–isospin singlet or spin singlet–isospin triplet states can be associated with the intercluster S and D states (i.e., antisymmetrization would result in the singlet–singlet and triplet–triplet states identically vanishing). Thus we arrive at the identical same six *LS* states as in the (α + d)-cluster expansion. Since these *LS* states are all orthogonal, the equivalence is on an exact one-to-one basis. This equivalence has been established by Wildermuth in a detailed examination of the involved wave functions. The basic reason for the above behavior can be traced to the fact that a fixed oscillator constant κ implies that intercluster separations are of the same order as the size of the individual clusters, resulting in a large degree of interpenetration of clusters. Antisymmetrization then has the further effect of largely "blurring" the physical reality of any such actual clusters.

Generally the solutions to (5-77) such as (5-73) and (5-75) should be taken in the context of trial solutions. Energy matrices for realistic Hamiltonians [i.e., versions of (5-74) with perhaps additional terms] might be computed. These could be used in variational calculations, for example, allowing the harmonic oscillator constant κ (or α) where it appears in the intercluster portion of the Hamiltonian and where it appears in the internal cluster portions to be separately adjusted to achieve an energy minimum [assuming, of course, that a potential capable of satisfying the saturation condition is used in (5-74) to prevent a collapsed state solution; see Chapter III]. It might develop then that an energy minimum establishes a smaller value of κ for the intercluster motion than for the internal cluster states, i.e., $\kappa_{\text{c.m.}} < \kappa_{\text{int}}$, thus resulting in larger intercluster separations [note $1/\alpha = 2\hbar/m\omega = 2\hbar/(\kappa m)^{1/2}$]. This has the effect of reducing the blurring produced by antisymmetrization and also reducing the overlap between various possible cluster expansions. In the asymptotic limit $\kappa_{\text{c.m.}} \ll \kappa_{\text{int}}$, the clusters become physically separated, leading to a trivial orthogonality of cluster states similar to that encountered for different exit channels of reaction theory when the channel particle wave packets reach large separations.

For example, if treating the ground state of ^{6}Li in terms of the (α + d)-cluster state should yield a value $\kappa_{\text{c.m.}} < \kappa_{\text{int}}$ (or $\omega_{\text{c.m.}} < \omega_{\text{int}}$, note that $\omega^2 = \kappa/m$), the exact transformability into a (^{3}H + ^{3}He)-cluster state would be broken. The (^{3}H + ^{3}He) overlap of such a ^{6}Li state would be expected to vary quadratically with $\Delta\omega/\omega$ to lowest order, viz.,[1]

$$\langle(^3\mathrm{H}+{}^3\mathrm{He})|{}^6\mathrm{Li}\rangle = 1 - C(\Delta\omega/\omega)^2 + \cdots. \tag{5-78}$$

The cluster state (α+d) with $\kappa_{\mathrm{c.m.}} < \kappa_{\mathrm{int}}$ when transformed back into a shell model form of six nucleons would be expected to involve various mixed shell model configurations since now mixed oscillator energies are involved. It is, of course, the presence of these states that gives the result (5-78). At the very least an exercise of the sort discussed here would then give strong suggestions of the type of configuration mixing that straightforward shell model calculations might require.

Finally, the question again arises as to which trial cluster expansion should be used to begin a variational (or other) calculation. The investigation of the case of $A = 5$ by Wildermuth provides an instructive example (Wildermuth and McClure, 1966). Two possible cluster states considered for ^{5}He were the (α+n) and (^{3}H+d) states:

$$\Psi_1(^5\mathrm{He}) = \mathscr{A}\left[\exp\left(-\alpha_4 \sum_{i=1}^{4} r_i'^2\right)\right] R[\exp(-\tfrac{4}{5}\alpha R^2)]\, Y_1^M(\hat{R}) \tag{5-79a}$$

$$\Psi_2(^5\mathrm{He}) = \mathscr{A}\left[\exp\left(-\alpha_3 \sum_{i=1}^{3} r_i'^2\right)\right] \{\exp[-\alpha_2(r_4'^2 + r_5'^2)]\}(1 - \tfrac{8}{5}\alpha R^2) \times [\exp(-\tfrac{6}{5}\alpha R^2)]\, Y_0^0(\hat{R}) \tag{5-79b}$$

$$\Psi_3(^5\mathrm{He}) = \mathscr{A}\left[\exp\left(-\alpha_3 \sum_{i=1}^{3} r_i'^2\right)\right] \{\exp[-\alpha_2(r_4'^2 + r_5'^2)]\} \times R[\exp(-\tfrac{6}{5}\alpha R^2)]\, Y_2^M(\hat{R}), \tag{5-79c}$$

in obvious notation and supressing the motion of the c.m. as a whole. For a reasonable Hamiltonian of the type (5-74), but including a two-body spin–orbit interaction, the results of Wildermuth are sketched in Fig. 5.8.[2]

Figure 5.8 shows the calculated values of the energy expectation value $\langle H \rangle$ as a function of the intercluster parameter $1/\alpha = 2\hbar/(\kappa m)^{1/2}$. As $1/\alpha$ decreases from larger to smaller values the kinetic energy continually increases as the nucleons are compressed into smaller and smaller volumes, due basically to the uncertainty principle. Due to the finite range of the nuclear forces used, the attractive potential terms become operative at a small enough value of $1/\alpha$

[1] See Amado and Noble (1971), where practicing nuclear physicists are again reminded of the implications of the nonorthogonality of cluster states. Apparently such warnings must be periodically sounded.

[2] If only α and α_2 are varied in (5-79) but α_3 and α_4 for the triton and α-particle clusters are kept fixed (adjusted to give reasonable ad hoc binding and size), the requirement that the potential in (5-74) satisfy the saturation condition can be relaxed.

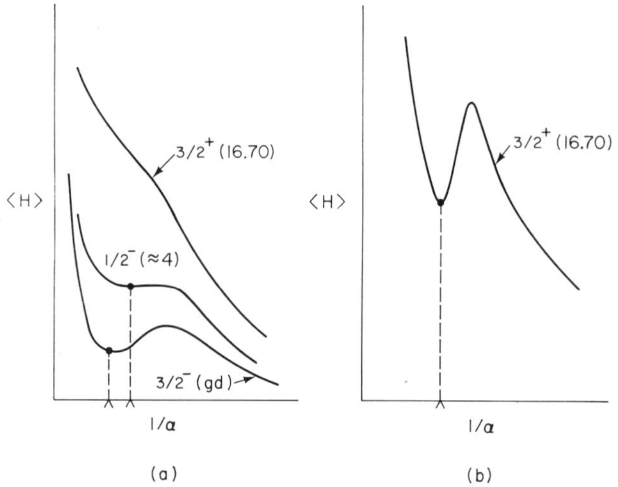

FIG. 5.8. Schematic representations of the expectation value of the energy $\langle H \rangle$ as a function of the intercluster oscillator parameter $1/\alpha = 2\hbar/m\omega$ for the states of ^{5}He based on the trial wave functions (5-79) [after Wildermuth and McClure (1966)]. The curves are labeled by $J^{\pi}(E_x$ in MeV). (a) For the trial wave function of the type (5-79a). (b) For the trial wave function of the type (5-79b).

to give significant overlap of the clusters. Finally, as $1/\alpha \to 0$, the potential energy becomes essentially constant while the continued increase in the kinetic energy produces the very sharp rise in $\langle H \rangle$ at very low values of $1/\alpha$. This behavior may or may not produce a local minimum in $\langle H \rangle$.

Figure 5-8a shows that for the $\frac{3}{2}^-$ ground state and the $\frac{1}{2}^-$ first excited states, also shown in Fig. 5.4, minima in $\langle H \rangle$ are achieved for special values of $1/\alpha$ using the trial function (5-79a). An attempt to use (5-79a) (suitably modified by replacing Y_1^M with Y_L^M) for the $\frac{3}{2}^+$ state observed at an excitation of 16.70 fails to produce a local minimum. On the other hand, the trial function (5-79b) in fact does generate a relatively deep minimum as shown in Fig. 5-8b. The trial function (5-79b) (an intercluster S state) is found to have a lower energy than (5-79c) (an intercluster D state). The calculated value of $\langle H \rangle \approx 17$ MeV results when the value of $1/\alpha$ is used that gives the minimum with (5-79b). For $\alpha_4 = \alpha_3 = \alpha_2 = \alpha$, i.e., the shell model 1p-shell equivalent, $\langle H \rangle \approx 23$ MeV. The implications are quite clearly that the $\frac{3}{2}^+$ state is best described as a modified shell model (^{3}H+d)-cluster in a 2S intercluster oscillator state coupled to unexcited ^{3}H and deuteron internal states as per (5-79b). A recent resonating-group method calculation confirms this cluster description of the $\frac{3}{2}^+$ state, and incidentally also gives an interesting insight into the operation of the Pauli exclusion principle [see Chwieroth *et al.* (1973)]. The two lower states are best described in terms of modified $(\alpha + n)$-

clusters in 1P intercluster states with the α-particle unexcited and the spin–orbit interaction added to (5-74) providing the $\frac{3}{2}$–$\frac{1}{2}$ splitting.

The depth of the minima and the "width of the barrier" exhibited in Fig. 5.8 give indications of the lifetimes of the states for decay into the relevant free cluster channels. The observed widths for the three states we have been discussing are: $\Gamma(\frac{3}{2}^-) = (0.58 \pm 0.02)$ MeV, $\Gamma(\frac{1}{2}^-) = (4 \pm 1)$ MeV, and $\Gamma(\frac{3}{2}^+) = 81$ keV, in general agreement with expected barrier penetration probabilities for the local minima depicted in Fig. 5.8.

While no general rule can be given, the best cluster descriptions for nuclear states appear to be in terms of clusters having separation energies from the ground state near the excitation energy in question. This is not to say that overlaps with other clusters are necessarily small. Thus particular reaction processes favoring some relevant cluster structure may be the dominant factor to consider. A striking example of such considerations occurs in the reaction ^{6}Li(p, ^{3}He)^{4}He at sufficiently high energies. The marked forward peaking observed in the yield of ^{3}He particles suggests a simple two-body transfer pickup process $^6\mathrm{Li} + \mathrm{p} \to (\alpha + \mathrm{d}) + \mathrm{p} \to \alpha + (\mathrm{d} + \mathrm{p}) \to \alpha + {}^3\mathrm{He}$, while the pronounced tendency for the yield of ^{3}He particles to peak also in the backward direction can be taken to indicate the presence of heavy particle stripping or triton pickup $^6\mathrm{Li} + \mathrm{p} \to ({}^3\mathrm{He} + {}^3\mathrm{H}) + \mathrm{p} \to {}^3\mathrm{He} + ({}^3\mathrm{H} + \mathrm{p}) \to {}^3\mathrm{He} + \alpha$. The resulting α-particle traveling in the forward direction leaves the recoiling ^{3}He to travel in the backward direction. The two processes, of course, occur simultaneously and produce interference effects. In any event, one process requires viewing ^{6}Li as a $(\mathrm{d} + \alpha)$-cluster while the other suggests a $({}^3\mathrm{H} + {}^3\mathrm{He})$-cluster description. For typical experimental results see Werby *et al.* (1973).

For a description of a cluster model analysis of the deuteron knockout reaction for ^{6}Li(p, pd)^{4}He at very high energies see Jain *et al.* (1970). Incidentally, the fact that this cross section is comparable to the quasi nucleon–nucleon scattering or knockout process ^{6}Li(p, 2p)^{5}He is in itself an indication of the importance of the $(\mathrm{d} + \alpha)$-cluster structure in ^{6}Li.

To give some additional results, we again return to considering the ground state of ^{6}Li in the anharmonic $(\alpha + \mathrm{d})$-cluster model. There has been notable progress in understanding the nature of the ^{6}Li wave function particularly in accounting for experimental results that are sensitive to the finer details [i.e., quadrupole moment, charge form factor in (e, e) scattering, Coulomb disintegration, etc.]. Usual shell model functions have far more difficulty in explaining the experimental results than phenomenological cluster models. Most successful are cluster functions of the general form of (5-75) which, suppressing the motion of the c.m. as a whole, can be written,

$$\Psi = \mathscr{A}\psi_\alpha(1,2,3,4)\,\psi_\mathrm{d}(5,6)\,X(\mathbf{R}). \tag{5-80}$$

The intercluster portion $X(\mathbf{R})$ in the standard harmonic oscillator 2S state would be

$$X_1(\mathbf{R}) = [1-(16/9)\alpha R^2][\exp(-\tfrac{4}{3}\alpha R^2)]\,Y_0{}^0(\hat{R}). \tag{5-81}$$

As discussed at the end of Section V, D, 3, the asymptotic behavior of a function such as (5-81) is incorrect and gives too small a tail. Consequently, the long-range part of (5-81) is usually modified to

$$X_2(R) = R^2[\exp(-c_1 R^2)+c_2\exp(-c_3 R^2)] \tag{5-82a}$$

with values of c_1, c_2, and c_3 giving energy minima in variational calculations using realistic hard-core nucleon–nucleon potentials, namely

$$c_1 = 0.18\ \text{fm}^{-2}, \qquad c_2 = 1/4, \qquad \text{and} \qquad c_3 = 0.065\ \text{fm}^{-2}. \tag{5-82b}$$

Attempts have been made to use an asymptotic behavior more appropriate for a (d,α) separation energy $E_{\mathrm{B}}(\mathrm{d},\alpha) = 1.472$ MeV matched[1] to the function (5-82a) at a matching radius R_0 $(3 < R_0 < 6\ \text{fm})$, viz.,

$$\begin{aligned} X(R) &= X_2(R), && R \leqslant R_0, \\ X(R) &= X_3(R) = (c_4/R)\exp(-c_5 R), && R \geqslant R_0, \end{aligned} \tag{5-83}$$

with $c_5{}^2 = (8m/3\hbar^2)E_{\mathrm{B}}(\mathrm{d},\alpha)$.

It has also been found that deformation of the intercluster function offers a satisfactory way of simultaneously giving a suitable charge form factor and quadrupole moment, a point that generally offers difficulties. Such deformed functions typically have the exponential form

$$\exp[-\tfrac{4}{3}\alpha R^2(1-\delta\cos^2\theta_R)]. \tag{5-84}$$

The internal deuteron wave function (5-75b) for the standard 1S harmonic oscillator form is

$$\begin{aligned} \psi_1(\mathrm{d}) &= \exp[-\alpha(|\mathbf{r}_5-\mathbf{R}_\mathrm{d}|^2+|\mathbf{r}_6-\mathbf{R}_\mathrm{d}|^2)] = \exp(-\tfrac{1}{2}\alpha|\mathbf{r}_5-\mathbf{r}_6|^2) \\ &= \exp(-\tfrac{1}{2}\alpha\rho^2). \end{aligned} \tag{5-85}$$

A satisfactory "variational" value of α in (5-85) is $\alpha = 0.33\ \text{fm}^{-2}$. Again a more suitable form would be the wave functions illustrated in Fig. 2.14 of Chapter II. Manageable analytic forms would be either that of (2-71b) or the so-called Hulthén form,

$$\psi_2(\mathrm{d}) = (1/\rho)[\exp(-b_1\rho)-B\exp(-b_2\rho)]. \tag{5-86}$$

Typical values are $B = 1$, $b_1 = 0.232\ \text{fm}^{-1}$, $b_2 \approx 7b_1$. (When only the long-range part is considered, i.e., $\rho \approx 0$ is excluded, B need not equal unity.) Again, deformed versions of (5-85) resembling (5-84) also have been used.

[1] Matching both the functions and their derivatives.

The internal α-particle wave function (5-73b) for the standard 1S harmonic oscillator form is

$$\psi_1(\alpha) = \exp(-\alpha \sum_{i=1}^{4} |\mathbf{r}_i - \mathbf{R}_\alpha|^2) \tag{5-87}$$

with a typical "variational" value of $\alpha = 0.22\ \text{fm}^{-2}$. This function also has been taken in modified forms resembling (5-82a) and with deformed shapes resembling (5-84). The value of α cited for (5-87) and the c_1 of (5-82b) should be contrasted to the "normal" shell model value $\bar{\alpha} = 0.235\ \text{fm}^{-2}$ for the 1p shell.[1]

An interesting application of the cluster model has been to the pion reaction $^6\text{Li}(\pi^-, 2\text{n})^4\text{He}$ [see Park and Rickett (1971)]. Satisfactory theoretical accounting for the experimental data suggests a deuteron oscillator coefficient $\alpha = 0.150\ \text{fm}^{-2}$ in (5-85), which leads to an rms radius of 3.16 fm for the deuteron in ^{6}Li (the rms radius for the "free" deuteron is 3.82 fm; see Chapter II). In addition, the intercluster oscillator parameter α appearing in (5-81) is determined to be $\alpha = 0.090\ \text{fm}^{-1}$.

For typical relevant references to the above general discussion of ^{6}Li see Jain *et al.* (1970), Cheon (1971), Jain (1972), Kurdyumov *et al.* (1972), Raphael (1973a, b), Grossiord *et al.* (1974), Noble (1974), and Shakin and Weiss (1974), and the references listed therein. For a recent paper emphasizing the energy level structure of ^{6}Li in a cluster model treatment, refer to Kramer and Schenzle (1973).

We have dwelt at length on the application of the cluster model to the nucleus ^{6}Li in order to better understand the various model descriptions that have been applied to nuclei by examining one case in detail. The cluster model description also has been applied successfully to many other nuclei. One striking example even suggests the presence of individual α-particle substructure contributions to the high-energy electron scattering form factor for ^{16}O. The form factor for $^{16}\text{O}(\text{e}, \text{e})$ has two minima in the range of momentum transfer $q \leqslant 3.8\ \text{fm}^{-1}$, one at $q = 1.52\ \text{fm}^{-1}$ and the other at $3.2\ \text{fm}^{-1}$. McDonald and Überall (1970) suggest the possibility that the diffraction minimum at $1.52\ \text{fm}^{-1}$ is due to the interference of the scattered waves from α-particles located at the four corners of a tetrahedron with sides of length 3.36 fm, and the minimum at $3.2\ \text{fm}^{-1}$ is due to an intrinsic individual α-particle form factor contribution. An alternate description based on the harmonic oscillator model requires the inclusion of pair correlation effects. Thus the α-particle model, at least in Born approximation, leads to a simpler and perhaps more physical picture. See also Ciofi Degli Atti and Kabachnik (1970) for both ^{16}O and ^{6}Li.

[1] Values in the literature for α range over $0.16 < \alpha < 0.28\ \text{fm}^{-2}$ in the 1p shell. There is a trend for α to decrease with increasing A; see Section IV, B, 1.

References

Abulaffio, C., and Irvine, J. M. (1972). *Phys. Lett.* **38B**, 492.
Amado, R. D., and Noble, J. V. (1971). *Phys. Rev.* **C3**, 2494.
Anderson, R. K., Wilson, M. R., and Goldhammer, P. (1972). *Nuclear Phys.* **C6**, 136.
Avishai, Y. (1972). *Phys. Rev.* **C6**, 677.
Barker, F. C. (1966). *Nuclear Phys.* **83**, 418.
Basu, M. K. (1972). *Phys. Rev.* **C6**, 476.
Benson, H. G., and Flowers, G. H. (1969). *Nuclear Phys.* **A126**, 305.
Bertsch, G. F., and Bertozzi, W. (1971). *Nuclear Phys.* **A165**, 199.
Bertsch, G. F., and Mekjian, A. (1972). *Ann. Rev. Nuclear Sci.* **22**, 25.
Bingham, H. G., and Fortune, H. T. (1972). *Phys. Rev.* **C6, 1900**.
Bray, K. H., Cameron, J. M., Fearing, H. W., Gill, D. R., and Sherif, H. S. (1973). *Phys. Rev.* **C8**, 881.
Chang, F. C. (1974). *Phys. Rev.* **C9**, 1.
Cheon, Il-T. (1971). *Phys. Rev.* **C3**, 1023.
Chuu, D. S., Han, C. S., and Lin, D. L. (1973). *Phys. Rev.* **C7**, 1329.
Chwieroth, F. S., Brown, R. E., Tang, Y. C., and Thompson, D. R. (1973). *Phys. Rev.* **C8**, 938.
Ciofi Degli Atti, C., and Kabachnik, N. M., (1970). *Phys. Rev.* **C1**, 809.
Cohen, S., and Kurath, D. (1965). *Nuclear Phys.* **73**, 1.
Cooper, B. S., Seaborn, J. B., and Willliams, S. A. (1971). *Phys. Rev.* **C4**, 1997.
de-Shalit, A., and Talmi, I. (1963). "Nuclear Shell Theory." Academic Press, New York.
De Takacsy, N. (1972). *Nuclear Phys.* **A178**, 469.
DeVries, R. M., Slaus, I., Sunier, J. W., Tombrello, T. A., and Nero, A. V. (1972). *Phys. Rev.* **C6**, 1447.
Edmonds, A. R. (1957). "Angular Momentum in Quantum Mechanics." Princeton Univ. Press, Princeton, New Jersey.
Elliott, J. P., and Flowers, B. H. (1955). *Proc. Roy. Soc. London* **A229**, 536.
Ensslin, E., *et al.* (1974). *Phys. Rev.* **C9**, 1705.
Garvey, G. T., Gerace, W. J., Jaffe, R. L., Talmi, I., and Kelson, I. (1969). *Rev. Modern Phys.* **41**, S1.
Ghovanlou, A., and Lehman, D. R. (1974). *Phys. Rev.* **C9**, 1730.
Goldhammer, P. (1974). *Phys. Rev.* **C9**, 813.
Grossiord, J. Y., Coste, C., Guichard, A., Gusakow, M., Jain, A. K., Pizzi, J. R., Bagieu, G., and de Swiniarski, R. (1974). *Phys. Rev. Lett.* **32**, 173.
Harvey, M., and Jensen, A. S. (1972). *Nuclear Phys.* **A179**, 33.
Hauge, P. S., and Maripuu, S. (1973). *Phys. Rev.* **C8**, 1609.
Hauge, P. S., Williams, S. A., and Duffey, G. H., (1971). *Phys. Rev.* **C4**, 1044.
Hecht, K. T. (1973). *Ann. Rev. Nuclear Sci.* **23**, 123.
Hutcheon, R. M., Neuhausen, R., and Eigenbrod, F. (1970). *Z. Naturforsch.* **25A**, 973.
Inglis, D. R. (1953). *Rev. Modern Phys.* **25**, 390.
Jain, A. K., Sarma, N., and Banerjee, B. (1970). *Nuclear Phys.* **A142**, 330.
Jain, B. K. (1972). *Nuclear Phys.* **A194**, 651.
Kane, R. J., Lambert, J. M., and Treado, P. A. (1972). *Nuclear Phys.* **A179**, 725.
Kramer, P., and Schenzle, D. (1973). *Nuclear Phys.* **A204**, 593.
Kudeyarov, Y. A., Kurdyumov, I. V., Neudatchin, V. G., and Smirnov, Y. F. (1971). *Nuclear Phys.* **A163**, 316.
Kurath, D. (1956). *Phys. Rev.* **101**, 216.
Kurath, D. (1957). *Phys. Rev.* **106**, 975.

Kurdyumov, I. V., Neudatchin, V. G., Smirnov, Y. F., and Korennoy, V. P. (1972). *Phys. Lett.* **40B**, 607.
Lie, S. (1972). *Nuclear Phys.* **A181**, 517.
Lodhi, M. A. K. (1971). *Phys. Rev.* **C3**, 503, 2116.
McCarthy, R. J., and Walker, G. E. (1974). *Phys. Rev.* **C9**, 809.
McDonald, L. J., and Überall, H. (1970), *Phys. Rev.* **C1**, 2156.
McDonald, L. J., *et al.* (1970). *Nuclear Phys.* **A147**, 541.
Marion, J. B. (1966). *Phys. Lett.* **14**, 315.
Mayer, M. G., and Jensen, J. H. D. (1955). "Elementary Theory of Nuclear Shell Structure." Wiley, New York.
Neudatchin, V. G., and Smirnov, Y. F. (1969). *Prog. Nuclear Phys.* **10**, 275.
Noble, J. V. (1968). *Phys. Rev.* **173**, 1034.
Noble, J. V. (1974). *Phys. Rev.* **C9**, 1209.
Onishi, N., and Sheline, R. K. (1971). *Nuclear Phys.* **A165**, 180.
Park, S. C., and Rickett, J. P. (1971). *Phys. Rev.* **C3**, 1926.
Perring, J., and Skyrme, T. H. R. (1956). *Proc. Phys. Soc. London.* **A69**, 600.
Racah, G. (1950). *Physica* **16**, 651.
Raphael, R. B. (1973a). *Phys. Rev.* **C8**, 269.
Raphael, R. B. (1973b). *Nuclear Phys.* **A201**, 621.
Rogers, D. W. O. (1973). *Nuclear Phys.* **A207**, 465.
Rose, H. J., Häusser, O., and Warburton, E. K. (1968). *Rev. Modern Phys.* **40**, 591.
Rose, M. E. (1957). "Elementary Theory of Angular Momentum." Wiley, New York.
Shakin, C. M., and Weiss, M. S. (1974). *Phys. Rev.* **C9**, 1679.
Shanley, P. E. (1969). *Phys. Rev.* **187**, 1328.
Stammbach, T., and Walter, R. L. (1972). *Nuclear Phys.* **A180**, 225.
Sünkel, W., and Wildermuth, K. (1972). *Phys. Lett.* **41B**, 439.
Talmi, I. (1952). *Helv. Phys. Acta.* **25**, 185.
Talmi, I., and Thieberger, R. (1956). *Phys. Rev.* **103**, 718.
Thieberger, R., and Talmi, I. (1956). *Phys. Rev.* **102**, 923.
Thompson, D. R., Reichstein, I., McClure, W., and Tang, Y. C. (1969). *Phys. Rev.* **185**, 1351.
Vincent, C. M. (1973). *Phys. Rev.* **C8**, 929.
Visscher, W. M., and Ferrell, R. A. (1957). *Phys. Rev.* **107**, 781.
Werby, M. F., Greenfield, M. B., Kemper, K. W., McShan, D. L., and Edwards, S. (1973). *Phys. Rev.* **C8**, 106.
Wheeler, J. A. (1937). *Phys. Rev.* **52**, 1083, 1107.
Whitehead, R. R., and Watt, A. (1972). *Phys. Lett.* **41B**, 7.
Wildermuth, K., and Kanellopoulos, T. (1958). *Nuclear Phys.* **7**, 150.
Wildermuth, K., and Kanellopoulos, T. (1959). *Nuclear Phys.* **9**, 449.
Wildermuth, K., and McClure, W. (1966). "Springer Tracts in Modern Physics," Vol. 41. Springer-Verlag, Berlin and New York.
Wong, S. K. M., Gomez, J. M. G., and Zuker, A. P. (1972). *Phys. Lett.* **42B**, 157.

Chapter VI

COLLECTIVE NUCLEAR EFFECTS

A. INTRODUCTION

In this chapter we shall examine the characteristics of nuclear levels of low excitation (i.e., E_x generally less than 5 MeV). There is an encouragingly high level of basic understanding concerning the systematic behavior of essentially all of the various level properties observed throughout the periodic table for both the spherical and deformed nuclei. However, we shall primarily content ourselves with a brief recounting of the salient features of various macroscopic models, particularly for the deformed nuclei employing collective parameters. With one exception, the Nilsson model, and a few introductory remarks, we leave the numerous microscopic theories specifically involving the individual nucleons to specialized works on the subject. Excellent general references that also include discussions of microscopic theories are Belyaev (1968), Davidson (1968), Eisenberg and Greiner (1970), Lane (1964), and Rowe (1970). Numerous journal and review articles also exist.

The starting point in the discussions carried out in Chapters III–V was the description of the nucleus as a system of noninteracting fermions moving in an average potential simulating to lowest order the effects of all the individual nucleon interactions. However, treatment of even some of the simplest observed gross nuclear features required the refinement of introducing residual interactions to some approximation. It is convenient at this point to review the effects of the residual nucleon–nucleon interaction in terms of largely short-range and also largely long-range correlation effects.

In Chapter III we saw that strong short-range, two-nucleon correlations are present in nuclear matter manifesting themselves through the production of a "wound" in the two-nucleon wave functions. Except for the presence of this wound within the relatively short collision distance of two interacting nucleons (i.e., within a range smaller than the "healing" distance) the nucleons move in a random and largely uncorrelated way. These short-range effects are largely induced by the high-momentum Fourier components in the interaction potential $v(\mathbf{r}_1, \mathbf{r}_2)$. Short-range structures in the interaction potential, such as the presence of a short-range repulsive core, play prominent roles in such behavior.

While the presence of strong short-range correlations in the nuclear wave function materially affects such bulk nuclear properties as, for example, the total nuclear binding energy, to lowest order these effects can be simulated by the introduction of an effective nucleon mass m^*, or alternately by the introduction of an effective nucleon–nucleon potential, as in Section III, B and elsewhere. Additionally, the presence of such short-range correlations does not generally imply a strong scattering of the two nucleons out of their initial states, largely due to the inhibiting effects of the Pauli principle. Thus the orbital assignment of a typical *average nucleon* is relatively free of configuration mixing and can be associated with the shell model orbitals insofar as bulk nuclear properties are concerned.

The low-momentum Fourier components in the interaction potential $v(\mathbf{r}_1, \mathbf{r}_2)$, while relatively unimportant for two relatively strongly bound nucleons, in fact do allow nucleons in occupied orbitals near the top of the Fermi sea to scatter into any of the low-lying empty orbitals. Such low-momentum interaction components imply long-range nuclear correlations. These effects, confined to a relatively small fraction of the nucleons in a nucleus, i.e., those near the top of the Fermi sea, therefore would not be expected to alter significantly bulk properties such as the total binding energy. On the other hand, nuclear excitation energies, the main subject of this chapter, generally involve relatively small energy differences of the total system. The excited-state levels, particularly at lower excitation energies, involve largely the interaction of just these nucleons near the top of the Fermi sea and therefore might be profoundly affected by the presence of any long-range correlations in these wave functions. The scattering of these nucleons into unoccupied orbitals also implies in principle the possibility of considerable configuration mixing. Figure 6.1 gives a greatly simplified pictorial representation of these effects.

The fact that the smeared-out Fermi distribution of Fig. 6.1b results in a

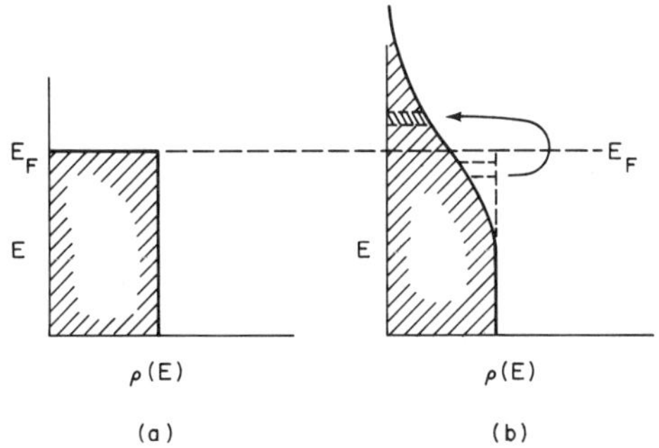

FIG. 6.1. A simplified schematic representation of the result of scattering nucleons out of occupied levels near the top of the Fermi sea into unoccupied levels through the operation of residual interactions. (a) The probability distribution $\rho(E)$ for the noninteracting Fermi sea with characteristic Fermi energy E_F. (b) The equilibrium probability distribution in the presence of residual interactions, which have promoted interacting pairs from levels below the Fermi energy to pairwise occupied levels above the Fermi energy.

lower total energy than the noninteracting distribution of Fig. 6.1a can be attributed to the operation of a strong orbital-dependent attractive pairing interaction. Such pairing effects relate to the relatively large interaction energy when two particles are in the same (n, l, j) state and differ only in the projection quantum number, being $+m$ for one member of the pair and $-m$ for the other. [In this context recall the discussion of the seniority coupling scheme implied by the result (5-31).] Up to a possible difference in phase (required to be set by convention), these two orbitals have as their classical analog two particles traversing the same orbit but one moving in a time-reversed sense with respect to the other. Thus one may adopt the point of view that the strongly interacting pair is in two relative time-reversed single-particle states.

Since the gained pairing energies must be reduced by appropriate particle–hole excitation energies before any possible depression of the total energy can be deduced, the effect will be more important in some nuclei than others. For example, empirical evidence suggests that the closed-shell gap for neutrons in spherical potential wells at the magic number 82 is some 5 MeV (i.e., the $3s_{1/2}$–$2f_{7/2}$ energy difference, refer to Fig. 4.4). The energy spread for *all the orbitals* from the neutron magic numbers 82–126 is, however, less than 4 MeV. Thus the smearing out of the Fermi distribution produced by pairing effects would be expected to increase rapidly as we go from neutron number $N = 82$ to numbers closer to the mid-major-shell values, $82 < N < 126$, due to the sharp reduction in the particle–hole excitation energy "penalty." Consequently midshell nuclei would also exhibit a large degree of configuration mixing and the level structure of any of these nuclei would be expected to deviate considerably from any simple shell model prediction. Indeed, the observed fact is that level structure resembling a simple series of *intrinsic* shell model states occurs only for nuclei very close to being doubly magic.

The problem of adding an additional nucleon to the noninteracting Fermi distribution depicted in Fig. 6.1a is rather straightforward. For the system ground state the nucleon is simply placed in the very next energy level available to the new system above the original Fermi energy E_F. The situation pertaining in the case of Fig. 6.1b is more complex since to achieve the lowest energy for the new system the added particle must have its occupation probability spread over a number of levels that are already partially occupied. It is possible by a mathematical device to attribute the resulting implied interactions (between the added new particle and those partially occupying the levels in the vicinity of E_F) to a hybrid particle or *quasi-particle*. The quasi-particle thus is "clothed" with the polarization effects of these pairing interactions. Placed in any particular angular momentum state with a specified projection quantum number, such a quasi-particle is partially in a real particlelike nucleon state (*augmenting* the original partial occupation of that part of the paired spin state with the same projection quantum number), and partly in a real holelike nucleon state (*displacing* the original matched nucleon partner in the other portion of the state with the opposite projection quantum number by essentially hindering or blocking the level promotion of the original pair).[1] In certain circumstances it is even useful to conceive of treating a combined pair of quasi-particles as a single quasi-boson entity. Two notable early papers pioneering the above considerations are by Kisslinger and Sorensen (1960) and Baranger (1960).

A concomitant feature of the pairing interaction considerations is the appearance of characteristics resembling those of a metallic superconductor. Particularly important in producing this effect is the presence of many closely spaced particle levels available in midshell nuclei, and simultaneously, a strong enough pairing energy to create an "energy gap." The landmark paper by Bohr *et al.* (1958) pointed out the empirical evidence for strong pairing interactions in midshell deformed nuclei and suggested the possible applicability of the theory of superconducting metals to the problem of the excitation spectra of such nuclei. The basic empirical observation to note is that the first intrinsic excitations (i.e., shell-model-like as distinguished from possible simple collective-mode-type excitations) in even–even nuclei are

[1] To be more precise, in the language of second quantization the creation operator for a quasi-particle in quantum state $|nljm\rangle$ is a linear combination of a real nucleon creation operator for the state $|nljm\rangle$ and a real nucleon annihilation operator for the state $|nlj, -m\rangle$. The translation from real nucleon language to quasi-particle language is a formal set of canonical transformations of creation and annihilation operators. In quasi-particle variables the nuclear system again becomes noninteracting and takes on some of the desirable features of simple single-particle shell model theory.

several times larger than expected from similar excitation energies in nearby odd-A nuclei in the midshell regions $150 < A < 180$ and $A > 230$. These nuclei have very large electric quadrupole moments and are well known to have large, nonspherical deformations. The large observed energy gap in these even–even nuclei is attributed to the presence of strong residual pairing forces between the nucleons and the correlations to which they give rise.

We should again recall that all the long-range correlation effects that we have been discussing, while important, nevertheless involve only a relatively small number of nucleons near the top of the Fermi sea. These higher orbital effects, as might be expected, can produce rather strong changes primarily in the surface region.[1] Starting with a closed-shell or magic number configuration, added nucleons progressively produce a greater smearing out of the Fermi surface, reflecting an increasing particle–hole population in the band of single particle orbitals between magic numbers. The availability of total nuclear

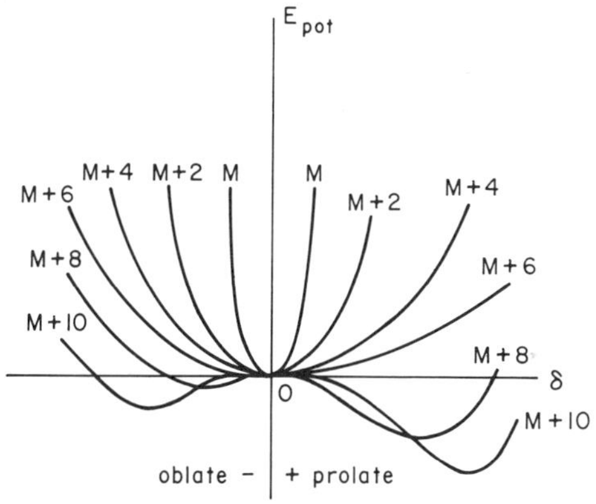

FIG. 6.2. A schematic representation of the total nuclear potential energy as a function of the surface deformation parameter δ as the number of neutrons is varied from the magic number M to $M+2$, $M+4$, etc. (with $M = 82$ as a specific example).

configurations with small energy differences coupled with the predominant surface effects leads to a reduction in the effective surface tension, i.e., a lower stiffness to surface-deforming excitations. Eventually, in fact, an energy minimum for the ground state itself develops for a nonzero deformation of the nuclear surface when the nuclear configuration departs far enough from the closed shell. Figure 6.2 illustrates this behavior in a schematic way. We note that as the neutron number is increased from the magic number M to $M+6$

[1] All realistic shell model potentials have in common the trend of extending single-particle wave functions to larger radii with increased eigenvalue energy; e.g., recall (4-26), etc.

the nuclear surface becomes easier to deform, although the potential energy minimum still occurs for a spherical shape. Then rather suddenly at neutron number $M+8$ the lower of two energy minima occurs at a substantial prolate deformation. Further increasing the neutron number increases the equilibrium deformation relatively slowly. Indeed, in the region $150 < A < 190$ the equilibrium deformation parameters δ_0 for the ground state begin at the lower mass end for the Nd and Sm isotopes with $0.20 \lesssim \delta_0 \lesssim 0.30$, peak at $\delta_0 \approx 0.35$ for Gd and Dy, and then slowly decrease back to $\delta_0 \approx 0.20$ for the W and Os isotopes.[1] Nuclei just outside the above range of A are spherical, albeit some are readily deformable in excited modes.

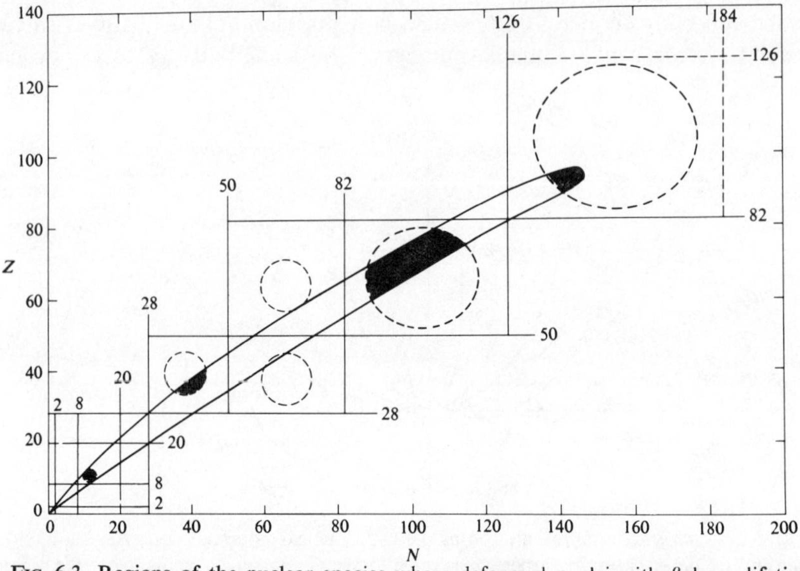

FIG. 6.3. Regions of the nuclear species where deformed nuclei with β-decay lifetimes longer than 1 min are observed (shown by shaded regions). The representation is in the N, Z plane with the magic numbers indicated.

In Fig. 6.3 we give a schematic representation of the nuclei known to have β-decay lifetimes longer than 1 min, encompassing the so-called β-stability "valley" in the N, Z plane. The various magic numbers are indicated and the regions enclosed by the dashed lines show where deformed nuclei are to be expected. The shaded subregions correspond to the observed deformed nuclei.

As we shall see, the "soft" spherical nuclei have level structures compatible

[1] For a definition of δ_0, or simply δ when the equilibrium value is clearly implied, see (6-127).

with a nuclear model exhibiting surface vibrations of a quantized irrotational liquid drop. Additionally, the nuclei with deformed stable ground states exhibit rotational excitation level band structures. Of course, complicated vibrational–rotational as well as intrinsic excitations also are observed in certain cases.

To digress for the moment into the realm of atomic physics, we note that diatomic molecules in general possess three distinct modes of excitation: the electronic orbital or intrinsic excitations generally have characteristic energies of approximately 2 eV (in the optical region); vibrational excitation modes involving the coordinate separation of the two atomic centers generally have energy level spacings of order 0.1 eV; finally, rotational excitation modes involving the rotational degree of freedom of the two-center system have level spacings typically of order 0.005 eV. The factor of the order of 20 separating these energies (and the relevant characteristic frequencies) leads to a high degree of adiabatic accommodation and generally also allows for a relatively clear identification of the three separate effects with very little coupling between them.

Since a number of different energy-related effects are involved in these considerations, it may be useful to refer to a schematic figure to sort them out more clearly. Figure 6.4 gives a schematic representation of these effects.

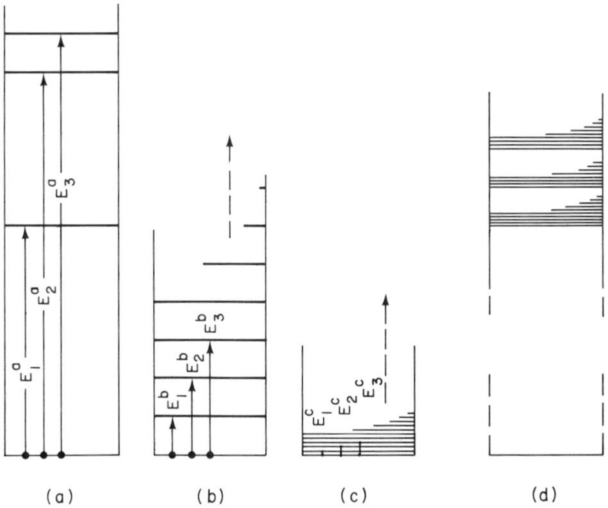

FIG. 6.4. A schematic representation of an idealized diatomic system showing: (a) electronic or intrinsic mode energy spectrum (the energies $E_n{}^a$ are referred to the bottom of the effective single-particle potential well), (b) vibrational mode spectrum, (c) rotational mode spectrum, and (d) combined molecular system spectrum built on the first excited intrinsic state.

The adiabatic approximation requires that $E_l^a \gg E_m^b \gg E_n^c$ with l, m, and n the relevant small integers. In view of this inequality, and since the time dependence involves $e^{-i\omega t} = e^{-iEt/\hbar}$, this approximation consists of noting, for example, that the relevant electrons execute many intrinsic mode periods during each fractional rotational period, and that in this sense they "track" or accommodate to the slower rotation of the diatomic nuclear centers. As illustrated in Fig. 6.4d, we clearly can speak of any particular system level as a member of a particular rotational and vibrational band built on a specific intrinsic state. However, it may happen that $E_2^a - E_1^a \approx E_m^b$ quite apart from the above adiabatic consideration. If individual molecular levels with the same constants of motion should then fall at nearby energies belonging to the two intrinsic states E_1^a and E_2^a, small coupling terms in the Hamiltonian will mix the eigenstates and the purity of the description is lost. Thus, in general, a clear identification of characteristic bands also would require little mixing and consequently $\Delta E^a > \Delta E^b > \Delta E^c$. These energy inequalities generally pertain for atomic systems.

The nuclear case is more complicated for a number of reasons. There are no corresponding heavy, massive centers of force as in the molecular case; rather it is the correlated nuclear motions that permit such analogous designations as are possible. In addition, some of the above energy considerations are met only approximately. Thus a blurring of intrinsic, vibrational, and rotational characterizations can result. Nonetheless, in some cases a relatively clear identification of a pure "rotational band," for example, can be discerned.

An additional distinction from the atomic case relates to the appropriate moment of inertia required to account for any possible observed rotational band. The diatomic moment of inertia closely conforms to the simple classical value. However, even when a permanently deformed nucleus rotates, the rotation carries relatively little nuclear matter with it in the manner of a macroscopic rigid rotator. Rather, the microscopic collective nucleon motions resemble the effective motion of a surface wave. This latter view of course would involve a lower moment of inertia than the rigid rotator analogy and is entirely appropriate for the nuclear model in which largely the higher orbital or "surface" nucleons are participants. As might be expected, the exact calculation of the proper moment of inertia, however, must take into consideration the "polarizing" effect on the inner or "core" nucleons produced by the motion of the surface nucleons. This polarizing or drag effect results in a moment of inertia intermediate between that of a rigid rotator and that appropriate for a pure hydrodynamic surface wave.

The ideal solution to the many-body nuclear problem might involve a complete self-consistent Hartree–Fock calculation, which in appropriate cases would reveal the requirement of a deformed potential and density distribution to achieve a proper energy minimum. Since such calculations are very difficult to carry out, often models attempt to consider separately the "core" nucleons and the "outer or valence" nucleons. The valence nucleons might be taken as only the few nucleons placed in the topmost orbitals of an appropriate potential or might in a more ambitious effort include an explicit accounting of all the nucleons above suitable magic number closed shells. For the permanently deformed nuclei the appropriate single-particle potential can be taken to be suitably deformed as well. The deformed potential can conveniently be viewed as one produced by a polarized core and possibly include in addition the average effect of the valence nucleons themselves.

In any event, the total nuclear Hamiltonian is considered to consist of a "collective" term relating largely to the core, an "intrinsic" term relating largely to the valence nucleons, and an appropriate interaction term coupling the intrinsic and collective degrees of freedom.

The collective Hamiltonian is prescribed in terms of collective variables relating to the description of the nuclear surface, its possible deformation, and, when appropriate, its spatial orientation. The principal economy in these models is that a relatively small number of such collective variables is used to replace the far larger number of individual nucleon variables that would appear in any complete microscopic calculation. Such models, parameterizing the nuclear surface rather than deriving its character from first principles, are thus quite properly referred to as phenomenological. Only the valence particles are described in terms of variables relating to the spin and space coordinates. It should be borne in mind, particularly when the number of nucleons explicitly accounted for is large, that a redundancy exists between individual and collective degrees of freedom. There are, for example, only $3A$ total spatial degrees of freedom available to a nucleus with A nucleons.[1] However, since collective effects are most striking for heavy nuclei and only a few valence particles are generally considered, this problem is usually ignored. Additionally, the spurious effects introduced by ignoring the coordinate redundancy do not materially affect the energy spectra.

We must distinguish two important cases for the interaction Hamiltonian coupling intrinsic and collective motions. For spherical nuclei the collective behavior consists of small surface vibrations about a spherically symmetric equilibrium shape. The intrinsic motion can be taken as the single-particle solution to some appropriate spherically symmetric potential (suitably vector-coupled single-particle solutions when more than one valence nucleon is considered). The interaction coupling in this case is weak, and a perturbation

[1] In this regard, we should recall the discussion of the center-of-mass motion in Chapter V, since in a very real sense the center-of-mass coordinate might be thought of as a collective variable. Thus, specifically exhibiting the c.m. coordinates leaves only $3(A-1)$ spatial degrees of freedom to describe internal nuclear motion.

treatment using a basis of product states of vibrational and single-particle states can be employed.

For permanently deformed nuclei both vibrational and rotational modes of collective excitation are possible. The intrinsic–collective interaction coupling clearly is strong since the permanent deformation is ultimately a result of the behavior of the valence particles both as a direct effect of their correlated motions and the resultant induced core polarization they produce. Perturbation methods for handling the intrinsic–collective coupling are thus inappropriate. Rather, intrinsic single-particle solutions in a body-centered frame of a deformed potential are sought as a generalization of the usual, simpler spherically symmetric shell model. Some arbitrariness exists in grouping the various terms in the true total Hamiltonian into collective, intrinsic, and interaction terms. This follows from the fact that the single-particle potential contains the principal nuclear deformation as a parameter.

In any event, the adiabatic approximation is usually invoked with respect to the collective and intrinsic modes of excitation for both the weak and strong coupling cases. This follows from the fact that the nucleon intrinsic orbital motion is rather rapid for the typical nuclear kinetic energies involved when compared to the period associated with the collective motions. A typical nucleon near the top of the Fermi sea might complete some hundred orbital oscillations in a single period of collective motion, allowing the view that the specific particle motions simply follow or accommodate to the slower collective motions.[1]

Since the usual product basis states for the total nuclear system involve the eigenfunctions of the collective Hamiltonian, we next turn to its discussion. A very useful and simple model is the hydrodynamic liquid drop model.

B. THE CLASSICAL LIQUID DROP

We will wish to examine aspects of the experimental evidence concerning nuclear excited states in terms of a collective model that is referred to as the hydrodynamic model. This model is closely associated with the liquid drop model of the nucleus discussed earlier. By way of introduction it is useful to derive the classical Hamiltonian for such a hydrodynamic model. For the sake of simplicity the nuclear fluid will be considered incompressible and confined within a volume possessing a sharp boundary. In addition the nuclear matter flow will be taken as irrotational and nonviscous. These latter assump-

[1] The nuclear transit time for a nucleon with, say, 30–40 MeV kinetic energy might be $t_N = 2R/(2T_k/m)^{1/2} \approx 10^{-22}$ sec. The period for a typical collective excitation mode with $\hbar\omega \approx 0.5$ MeV would be $t_c = 2\pi/\omega \approx 10^{-20}$ sec. In this respect one important energy criterion discussed in connection with Fig. 6.4 is met.

tions imply a pure hydrodynamic streaming flow free of any microscopic eddies.

We begin by a discussion of the nuclear liquid drop surface for nuclei that have equilibrium or ground states possessing spherical symmetry. Such nuclei are expected to be found in the region of the periodic system where N or Z is close to a magic number. The equilibrium surface is then simply that appropriate for a sphere, $S(\mathbf{r}) = R$, where R is the constant-density radius discussed in Chapter II. A collective excited state would involve time-varying surface deformations. An appropriate general expression for the surface can be taken as the expansion

$$S(R,\theta,\varphi,t) = R\left[1+\sum_{\lambda=0}^{\infty}\sum_{\mu=-\lambda}^{+\lambda}\alpha_{\lambda\mu}(t)Y_{\lambda}^{\mu}(\theta,\varphi)\right]. \tag{6-1}$$

For the ground state and our zero-energy reference, all $\alpha_{\lambda\mu} = 0$. Also, since S must be real and $Y_{\lambda}^{-\mu} = (-1)^{\mu}Y_{\lambda}^{\mu *}$, we must have in general

$$\alpha_{\lambda,-\mu} = (-1)^{\mu}\alpha_{\lambda\mu}^{*}. \tag{6-2}$$

It is clear that if we are to exclude "dimples" in the nuclear surface of dimensions smaller than that of the order of the internucleon separation, we should limit $\lambda \lesssim A^{1/3}$ for realistic models. Other considerations permit us to make further assumptions concerning $\alpha_{\lambda,\mu}$. Since $Y_0^{\ 0} = (4\pi)^{-1/2}$ is independent of θ and φ, the presence of a time-dependent coefficient $\alpha_{00}(t)$ would give rise to a monopole "breathing" mode oscillation expressly ruled out by the assumption of incompressibility.

The possibility exists of describing some nuclear levels such as, e.g., the first excited state in the doubly magic nucleus ^{16}O as a 0^{+} breathing mode. But the excitation energy for this state is 6.05 MeV.[1] In general the weak compressibility of nuclear matter places this mode of excitation at energies higher than that under consideration [e.g., see Zamick (1973)], and we therefore shall take $\dot{\alpha}_{00} = 0$, i.e., α_{00} is not time dependent.[2]

The $\lambda = 1$ mode, or dipole mode, corresponds to a simple oscillation of the center of mass of the entire nucleus without a change in spherical shape and shall be excluded from consideration. However, it should be remarked that a dipole mode in which protons and neutrons oscillate with *different* values of $\alpha_{1\mu}$ gives rise to a possible model for the well-known "giant dipole" resonance

[1] The 0^{+} level in ^{16}O at 6.05 MeV is a highly collective state. In a shell model basis it is a mixed 0p–0h, 2p–2h, and 4p–4h state. The state can be best described in fact as a shape isomer; see Sørensen (1973). The centroid of the breathing mode strength is probably at a higher energy, near 30 MeV, and the 6.05-MeV state has only a small portion of this total strength. In any case the arguments presented relating to breathing modes are nonetheless valid.

[2] We shall use a dot over a quantity to indicate differentiation with respect to time, thus $\dot{\alpha}_{00}(t) = \partial\alpha_{00}(t)/\partial t$.

in photonuclear reactions (the Goldhaber–Teller model). This resonance excitation energy is usually some 20 MeV, and therefore can be ignored for present purposes.[1] We shall assume in the following that protons and neutrons at all times move in an identical manner, i.e., have identical values of $\alpha_{\lambda\mu}$.

The quadrupole or $\lambda = 2$ mode thus becomes our lowest order vibration to receive detailed study. Indeed, many collective effects can be understood by involving the quadrupole mode alone.

▶**Exercise 6-1** We note that the nuclear volume can be written

$$V = \int_\Omega \int_0^S r^2\, dr\, d\Omega = \tfrac{1}{3} \int_\Omega S^3(\theta, \phi)\, d\Omega,$$

or

$$V = \tfrac{1}{3} R^3 \int_\Omega \Big[1 + \sum_{\lambda,\mu} \alpha_{\lambda\mu} Y_\lambda{}^\mu(\theta, \phi) \Big]^3 d\Omega.$$

Show that to quadratic terms in α, holding the volume constant at $V = \frac{4}{3}\pi R^3$ yields

$$\alpha_{00} = -(4\pi)^{-1/2} \sum_{\lambda,\mu} |\alpha_{\lambda\mu}|^2. \tag{6-3}$$

As we shall presently see, the quantization of the liquid drop leaves the right-hand side of (6-3) time independent. Thus, while α_{00} is required to be time independent, it may not in general be taken identically zero (without a renormalization of the ground-state volume and energy reference).

As is evident from (6-3), the quantity α_{00} is of quadratic order in $\alpha_{\lambda\mu}$ with $\lambda \geqslant 1$. Thus a sum such as that appearing on the right-hand side of (6-3) can be taken as

$$\sum_{\text{all } \lambda,\mu} |\alpha_{\lambda\mu}|^2 \approx \sum_{\lambda \geqslant 1,\mu} |\alpha_{\lambda\mu}|^2,$$

when expansions to only quadratic order in $\alpha_{\lambda\mu}$ are desired.

If constancy of the position of the center of mass is also required the deformation coefficients for $\lambda = 1$ also must be taken nonzero. These coefficients are of order $\alpha_\lambda \alpha_{\lambda+1}$ with $\lambda \geqslant 2$. Since for most purposes only $\lambda = 2$ is considered to order α^2, this correction can be ignored. ◀

We now introduce a nuclear matter velocity field $v(\mathbf{r})$. Since our model assumes no local sources or sinks, $\nabla \cdot v = 0$, and irrotational flow requires $\nabla \times v = 0$. Thus the velocity field can be derived from a scalar potential

[1] While space limitations prevent us from discussing the giant dipole resonances, they are in fact particularly important in shedding light on the relationships connecting microscopic and macroscopic collective behavior theories [e.g., see Eisenberg and Greiner (1970)].

$v(\mathbf{r}) = \nabla\Phi(\mathbf{r})$. In view of the above,

$$\nabla^2\Phi(\mathbf{r}) = 0. \tag{6-4}$$

The general solution to Laplace's equation (6-4), regular at the origin, is well known to be

$$\Phi(\mathbf{r}) = \sum_{\lambda,\mu} \xi_{\lambda\mu} r^\lambda Y_\lambda{}^\mu(\theta,\varphi). \tag{6-5}$$

The boundary condition at the surface requires that the radial fluid velocity correspond to the rate of change of the radius describing the surface in (6-1). Thus

$$(v_r)_{S(\theta,\varphi)} = \left(\frac{\partial\Phi}{\partial r}\right)_{S(\theta,\varphi)} = \frac{\partial S(\theta,\varphi,t)}{\partial t}. \tag{6-6}$$

From (6-1), (6-5), and (6-6),

$$R\sum_{\lambda,\mu} \dot{\alpha}_{\lambda\mu} Y_\lambda{}^\mu(\theta,\varphi) = \sum_{\lambda,\mu} \lambda\xi_{\lambda\mu} S^{\lambda-1} Y_\lambda{}^\mu(\theta,\varphi). \tag{6-7}$$

If for small oscillations we approximate S in (6-7) by R, we obtain the relationship between α and ξ,

$$\xi_{\lambda\mu} = \dot{\alpha}_{\lambda\mu}/\lambda R^{\lambda-2}. \tag{6-8}$$

The term in Φ for ξ_{00} just gives a constant potential and is uninteresting; hence consistent with $\dot{\alpha}_{00} = 0$, we take $\xi_{00} = 0$.

The nuclear kinetic energy with constant density ρ is

$$T = \tfrac{1}{2}\rho\int |v|^2\,d\tau = \tfrac{1}{2}\rho\int |\nabla\Phi|^2\,d\tau,$$

or

$$T = \frac{\rho}{2}\int\left[\left|\frac{\partial\Phi}{\partial r}\right|^2 + \frac{1}{r^2}\left|\frac{\partial\Phi}{\partial\theta}\right|^2 + \frac{1}{r^2\sin^2\theta}\left|\frac{\partial\Phi}{\partial\varphi}\right|^2\right]d\tau. \tag{6-9}$$

Using (6-5), we obtain

$$T = \frac{\rho}{2}\sum_{\lambda,\mu}\sum_{\lambda',\mu'} \xi_{\lambda\mu}\xi^*_{\lambda'\mu'}\int_\Omega\int_0^S r^{\lambda+\lambda'}\,dr\left[\lambda\lambda' Y_\lambda{}^\mu Y_{\lambda'}^{\mu'*} + \frac{\partial Y_\lambda{}^\mu}{\partial\theta}\frac{\partial Y_{\lambda'}^{\mu'*}}{\partial\theta} + \frac{1}{\sin^2\theta}\frac{\partial Y_\lambda{}^\mu}{\partial\varphi}\frac{\partial Y_{\lambda'}^{\mu'*}}{\partial\varphi}\right]d\Omega. \tag{6-10}$$

▶**Exercise 6-2** Using the raising and lowering operators $L_\pm = L_x \pm iL_y$, show that

$$\partial Y_\lambda{}^\mu/\partial\theta = (1/2\hbar)(e^{-i\varphi}L_+ - e^{i\varphi}L_-)Y_\lambda{}^\mu,$$

and

$$m(\cot\theta)Y_\lambda{}^\mu = -(1/2\hbar)(e^{-i\varphi}L_+ + e^{i\varphi}L_-)Y_\lambda{}^\mu.$$

Using the above, show that

$$\int_{\Omega}\left[\frac{\partial Y_{\lambda}{}^{\mu}}{\partial\theta}\frac{\partial Y_{\lambda'}^{\mu'*}}{\partial\theta}+\frac{1}{\sin^2\theta}\frac{\partial Y_{\lambda}{}^{\mu}}{\partial\varphi}\frac{\partial Y_{\lambda'}^{\mu'*}}{\partial\varphi}\right]d\Omega = \lambda(\lambda+1)\,\delta_{\mu\mu'}\,\delta_{\lambda\lambda'}. \quad (6\text{-}11)$$

[*Hint:* Collect all like terms of (6-11) before integrating.] ◀

Using result (6-11) of Exercise 6-2 and the approximation $S = R$ in the radial integral, (6-10) becomes

$$T = \tfrac{1}{2}\rho\sum_{\lambda,\mu}|\xi_{\lambda\mu}|^2\lambda R^{2\lambda+1}.$$

Defining

$$B_\lambda = \frac{\rho R^5}{\lambda} = \frac{3m_N AR^2}{4\pi\lambda} = \frac{3m_N R_0{}^2 A^{5/3}}{4\pi\lambda}$$

(since $m_N A = \frac{4}{3}\pi\rho R^3$ and $R = R_0 A^{1/3}$) and using (6-8), we obtain[1]

$$T = \tfrac{1}{2}\sum_{\lambda,\mu}B_\lambda|\dot{\alpha}_{\lambda\mu}|^2. \quad (6\text{-}13)$$

The potential energy associated with the work done against the surface tension σ of the liquid drop can be written,

$$E_\sigma = \sigma\int_{\Omega}S^2\left[1+\frac{1}{S^2}\left|\frac{\partial S}{\partial\theta}\right|^2+\frac{1}{S^2\sin^2\theta}\left|\frac{\partial S}{\partial\phi}\right|^2\right]^{1/2}d\Omega. \quad (6\text{-}14)$$

To quadratic order in α, (6-14) becomes

$$E_\sigma = 4\pi\sigma R^2 - \sigma R^2\sum_{\lambda,\mu}|\alpha_{\lambda\mu}|^2+\frac{\sigma}{2}\int_{\Omega}\left[\left|\frac{\partial S}{\partial\theta}\right|^2+\frac{1}{\sin^2\theta}\left|\frac{\partial S}{\partial\phi}\right|^2\right]d\Omega, \quad (6\text{-}15)$$

where we have used (6-1) and the constraint leading to (6-3). Direct use of (6-11) in (6-15) gives

$$E_\sigma = 4\pi\sigma R^2 + \tfrac{1}{2}\sigma R^2\sum_{\lambda,\mu}|\alpha_{\lambda\mu}|^2(\lambda-1)(\lambda+2). \quad (6\text{-}16)$$

We recognize the first term of (6-16) to be the surface energy of the undeformed ground state, $E_\sigma(0) = 4\pi\sigma R^2$, and therefore use only the second term as a contribution to the deformation potential energy.

▶**Exercise 6-3** Verify (6-14)–(6-16). ◀

The total Coulomb energy of a uniformly charged liquid drop can be

[1] The approximate $A^{5/3}$ dependence of the deformation kinetic energy in (6-13) should be distinguished clearly from the approximate simple linear A dependence for the principal term in the total ground-state kinetic energy discussed in Chapter III.

written

$$E_c = \frac{1}{2}\int\int \frac{\rho_c(\mathbf{r}_1)\,\rho_c(\mathbf{r}_2)\,d\tau_1\,d\tau_2}{r_{12}} = \frac{1}{2}\int \rho_c(\mathbf{r}_1)\,\Phi_c(\mathbf{r}_1)\,d\tau_1 \qquad \text{(6-17)}$$

with

$$\Phi_c(\mathbf{r}_1) = \int \frac{\rho_c(\mathbf{r}_2)}{r_{12}}\,d\tau_2, \qquad r_{12} = |\mathbf{r}_1 - \mathbf{r}_2|, \qquad \text{and} \qquad \rho_c(\mathbf{r}) = \frac{eZ}{m_N A}\rho(\mathbf{r})$$

connecting the charge and matter densities. We can consider the density factors in the above as constants with

$$\rho_c(\mathbf{r}_1) = \rho_c(\mathbf{r}_2) = \rho_{0c} = (eZ/m_N A)\,\rho_0 = 3eZ/4\pi R^3$$

if the relevant integrations are carried out only over the nuclear volume. Thus we have

$$\Phi_c(\mathbf{r}_1) = \rho_{0c}\int\int_0^{S(\theta_2,\varphi_2)} (r_2{}^2\,dr_2\,d\Omega_2/r_{12}), \qquad \text{(6-18a)}$$

and

$$E_c = \tfrac{1}{2}\rho_{0c}\int\int_0^{S(\theta_1,\varphi_1)} \Phi_c(\mathbf{r}_1)\,r_1{}^2\,dr_1\,d\Omega_1. \qquad \text{(6-18b)}$$

We make use of the well-known expansions

$$\begin{aligned} \frac{1}{r_{12}} &= \sum_l \frac{r_2{}^l}{r_1^{l+1}} P_l(\theta_{12}), \qquad r_2 < r_1 \\ &= \sum_l \frac{r_1{}^l}{r_2^{l+1}} P_l(\theta_{12}), \qquad r_2 > r_1, \end{aligned} \qquad \text{(6-19)}$$

and write, correct to order α^2,[1]

$$\Phi_c(\mathbf{r}_1) \approx \rho_{0c}\sum_{l=0}\left[\int\int_0^{r_1} \frac{r_2^{l+2}}{r_1^{l+1}} P_l(\theta_{12})\,dr_2\,d\Omega_2 + \int\int_{r_1}^{S(\theta_2,\varphi_2)} \frac{r_1{}^l}{r_2^{l-1}} P_l(\theta_{12})\,dr_2\,d\Omega_2\right]. \qquad \text{(6-20)}$$

[1] This expression for $\Phi_c(\mathbf{r}_1)$ is rigorously correct for $r_1 \leqslant S_{\min}$. However, in the interior region where $S_{\min} \leqslant r_1 \leqslant S_{\max}$ there occur pairs of angles θ_1, φ_1 and θ_2, φ_2 for which $r_1 = |\mathbf{r}_1 \equiv (r_1, \theta_1, \varphi_1)| > S(\theta_2, \varphi_2)$. When this occurs, the integrals of (6-20) are manifestly incorrect. Davidson (1968), using a slightly different approach to obtain E_c, requires evaluating $\Phi_c(\mathbf{r}_1)$ in the exterior region. The expression used there,

$$\Phi_c(\mathbf{r}_1) = \rho_{0c}\sum_{l=0} r_1^{-(l+1)}\int\int_0^{S(\theta_2,\varphi_2)} r_2^{l+2} P_l(\theta_{12})\,dr_2\,d\Omega_2,$$

is similarly incorrect in portions of the exterior region for which $S_{\min} \leqslant r_1 \leqslant S_{\max}$, namely when $r_1 = |\mathbf{r}_1 \equiv (r_1, \theta_1, \varphi_1)| < S(\theta_2, \varphi_2)$. The fact that E_c of (6-18) is nevertheless correctly given by (6-20) to order α^2 is due to the continuity condition on the potential and its first derivative allowing $\Phi(\mathbf{r}_1)$ to be extended into the region $S_{\min} \leqslant r_1 \leqslant S_{\max}$ to requisite order. For a rigorous treatment see Frankel and Metropolis (1947). For other approximate methods treating the surface by a more explicit calculation see Alaga (1969) and Eisenberg and Greiner (1970).

Introducing $F(\theta,\varphi) = \sum_{\lambda \geq 1,\mu} \alpha_{\lambda\mu} Y_\lambda^\mu(\theta,\varphi)$ and $\alpha_0 = \alpha_{00}(4\pi)^{-1/2}$, we have $S(\theta,\varphi) = R[1+\alpha_0+F(\theta,\varphi)]$. After performing the integration over r_2 in (6-20) and expanding the result in powers of F, we obtain to quadratic order

$$\begin{aligned}\Phi_c(\mathbf{r}_1) = \rho_{0c}\Bigg[& -\frac{2\pi r_1{}^2}{3} + R^2 \sum_{l\neq 2} \left(\frac{r_1}{R}\right)^l \frac{(1+\alpha_0)^{-l+2}}{2-l} \int P_l(\theta_{12})\, d\Omega_2 \\ & + r_1{}^2 \ln\left\{\frac{(1+\alpha_0)R}{r_1}\right\} \int P_2(\theta_{12})\, d\Omega_2 \\ & + R^2 \sum_{l=0}^{\infty} \left(\frac{r_1}{R}\right)^l (1+\alpha_0)^{-l+1} \int P_l(\theta_{12}) F(\theta_2,\varphi_2)\, d\Omega_2 \\ & + \frac{R^2}{2} \sum_{l=0}^{\infty} \left(\frac{r_1}{R}\right)^l (1-l)(1+\alpha_0)^{-l} \int P_l(\theta_{12}) F^2(\theta_2,\varphi_2)\, d\Omega_2 \Bigg].\end{aligned} \tag{6-21}$$

Finally, using the addition theorem,

$$P_l(\theta_{12}) = \frac{4\pi}{2l+1} \sum_{m=-l}^{+l} Y_l^m(\theta_1,\varphi_1)\, Y_l^{m*}(\theta_2,\varphi_2),$$

we obtain, to order α^2,

$$\begin{aligned}\Phi_c(\mathbf{r}_1) = \rho_{0c}\Bigg[& -\frac{2\pi}{3} r_1{}^2 + 2\pi R^2 (1+\alpha_0)^2 \\ & + 4\pi R^2 \sum_{\lambda \geq 1,\mu} \left(\frac{r_1}{R}\right)^\lambda \frac{(1+\alpha_0)^{-\lambda+1}}{2\lambda+1} \alpha_{\lambda\mu} Y_\lambda^\mu(\theta_1,\varphi_1) + \frac{R^2}{2} \sum_{\lambda \geq 1,\mu} |\alpha_{\lambda\mu}|^2 \Bigg].\end{aligned} \tag{6-22}$$

Evaluating E_c using (6-22) in (6-18) gives

$$E_c = \frac{3}{5}\frac{(Ze)^2}{R}\left[1 - \frac{5}{4\pi} \sum_{\lambda \geq 1,\mu} \frac{\lambda-1}{2\lambda+1} |\alpha_{\lambda\mu}|^2\right]. \tag{6-23}$$

We recognize the first term in (6-23) to be just the Coulomb energy of the undeformed ground state, $E_c(0) = \frac{3}{5}(Ze)^2/R$.

▶**Exercise 6-4** Verify (6-18)–(6-23). Note that since $F(\theta,\varphi)$ is real, $F^2 = |F|^2$ and $\sum \alpha_{\lambda\mu} Y_\lambda^\mu = \sum \alpha_{\lambda\mu}^* Y_\lambda^{\mu *}$. ◀

Combining the deformation potential energies due to surface tension and Coulomb effects gives

$$E_p = \tfrac{1}{2} \sum_{\lambda,\mu} C_\lambda |\alpha_{\lambda\mu}|^2, \tag{6-24}$$

with

$$C_\lambda = (\lambda - 1)\left[(\lambda + 2) R^2 \sigma - \frac{3}{2\pi} \frac{(eZ)^2}{(2\lambda + 1) R} \right].$$

The above expression leads to unstable surface vibrations for $\lambda = 2$ (i.e., with $C_2 \leqslant 0$) when $3(eZ)^2/10\pi R \geqslant 4\sigma R^2$. Using $R = R_0 A^{1/3}$, $a_c = \frac{3}{5}e^2/R_0$, and $a_s = 4\pi\sigma R_0{}^2$, this is equivalent to the condition $Z^2/A \geqslant 2a_s/a_c$ expressed in Chapter III in the discussion concerning spontaneous fission. A useful reference for the deformation energy of a liquid charged drop including higher order terms in α for E_p is by Swiatecki (1956). See also Leander (1974).

Recalling (6-13), the total deformation energy of the charged liquid drop is

$$H = T + E_p = \tfrac{1}{2}\sum_{\lambda,\mu} B_\lambda |\dot{\alpha}_{\lambda\mu}|^2 + \tfrac{1}{2}\sum_{\lambda,\mu} C_\lambda |\alpha_{\lambda\mu}|^2. \tag{6-25}$$

Expression (6-25) is the desired Hamiltonian for the classical liquid drop. We defer the examination of its properties to following sections employing suitable quantum mechanical considerations.

C. VIBRATIONAL STATES OF SPHERICAL NUCLEI

In the hydrodynamic model, nuclei can undergo modes of excitation appropriate for the Hamiltonian (6-25). The simplest application occurs for cases in which the deformations are kept small, and thus our attention first turns to the spherical nuclei found in the periodic system where N and Z are near magic numbers. The ground state of these nuclei would have $\alpha_{\lambda\mu} = 0$ in such a classical treatment.

As mentioned in Section VI, A, we wish to employ a Hamiltonian of the form $H = H(\text{collective}) + H(\text{single particle}) + H(\text{particle–collective coupling})$. The collective Hamiltonian for our present purposes is based on a simple adaptation of (6-25). It is useful to consider the quantized solutions to this Hamiltonian first.

1. Collective Vibrational States

We wish first to cast (6-25) into its canonical form by introducing the formal canonically conjugate momentum

$$\Pi_{\lambda\mu} = \partial T/\partial\dot{\alpha}_{\lambda\mu} = B_\lambda \dot{\alpha}^*_{\lambda\mu}, \tag{6-26}$$

to obtain

$$H = \tfrac{1}{2}\sum_{\lambda,\mu} [(1/B_\lambda)|\Pi_{\lambda\mu}|^2 + C_\lambda |\alpha_{\lambda\mu}|^2]. \tag{6-27}$$

If (6-27) is considered the Hamiltonian for the quantization of the charged liquid drop, its form is evidently that for a harmonic oscillator. In the present instance, however, the canonical "coordinates" refer to the amplitude characterization of surface deformations. The ad hoc limitation of (6-27) to a particular λ-mode yields a behavior similar to that of a $(2\lambda+1)$-dimensional isotropic oscillator. The presence of different λ terms introduces nonisotropic effects since the characteristic oscillator frequency is λ dependent, viz.,

$$\omega_\lambda = (C_\lambda/B_\lambda)^{1/2}. \tag{6-28}$$

A convenient representation of the Hamiltonian is through the introduction of some elementary second quantization formalism. The transformation to new operators

$$\begin{aligned} \alpha_{\lambda\mu} &= (\hbar/2B_\lambda \omega_\lambda)^{1/2}[b_{\lambda\mu}+(-1)^\mu b^\dagger_{\lambda,-\mu}] \\ \dot{\alpha}_{\lambda\mu} &= -i\omega_\lambda(\hbar/2B_\lambda \omega_\lambda)^{1/2}[b_{\lambda\mu}-(-1)^\mu b^\dagger_{\lambda,-\mu}] \end{aligned} \tag{6-29}$$

yields for the Hamiltonian in the $b_{\lambda\mu}$ representation

$$\tilde{H} = \tfrac{1}{2}\sum_{\lambda\mu} \hbar\omega_\lambda(b^\dagger_{\lambda\mu} b_{\lambda\mu}+b_{\lambda\mu} b^\dagger_{\lambda\mu}), \tag{6-30a}$$

or

$$\tilde{H} = \sum_\lambda \tilde{H}_\lambda, \tag{6-30b}$$

with[1]

$$\tilde{H}_\lambda = \hbar\omega_\lambda \sum_{\mu=-\lambda}^{+\lambda} (b^\dagger_{\lambda\mu} b_{\lambda\mu}+\tfrac{1}{2}) = \hbar\omega_\lambda\left[\sum_\mu b^\dagger_{\lambda\mu} b_{\lambda\mu}+\tfrac{1}{2}(2\lambda+1)\right]. \tag{6-30c}$$

We now introduce the *number operators*

$$\tilde{n}_{\lambda\mu} = b^\dagger_{\lambda\mu} b_{\lambda\mu}, \tag{6-31a}$$

and

$$\tilde{N}_\lambda = \sum_{\mu=-\lambda}^{+\lambda} \tilde{n}_{\lambda\mu} = \sum_\mu b^\dagger_{\lambda\mu} b_{\lambda\mu}. \tag{6-31b}$$

In view of these definitions,

$$\tilde{H} = \sum_\lambda \hbar\omega_\lambda[\tilde{N}_\lambda+\tfrac{1}{2}(2\lambda+1)] = \sum_\lambda \hbar\omega_\lambda\left[\sum_\mu \tilde{n}_{\lambda\mu}+\tfrac{1}{2}(2\lambda+1)\right]. \tag{6-32}$$

Evidently the operator $\tilde{n}_{\lambda\mu}$ of (6-31a) commutes with the Hamiltonian and hence $n_{\lambda\mu}$ is a good quantum number.[2] Since (6-32) involves a sum on $\tilde{n}_{\lambda\mu}$ in

[1] It is readily verified that the original commutation rule $[\alpha_{\lambda\mu}, \Pi_{\lambda'\mu'}] = i\hbar\,\delta_{\lambda\lambda'}\,\delta_{\mu\mu'}$ leads to $[b_{\lambda\mu}, b^\dagger_{\lambda'\mu'}] = \delta_{\lambda\lambda'}\,\delta_{\mu\mu'}$. Hence $b_{\lambda\mu} b^\dagger_{\lambda\mu} = b_{\lambda\mu} b^\dagger_{\lambda\mu}+1$, and (6-30c) immediately follows.

[2] In the following, particular care should be exercised to distinguish between the operator $\tilde{n}_{\lambda\mu}$ and its allowable eigenvalues $n_{\lambda\mu}$ and similarly between $\tilde{N}_\lambda$ and N_λ.

both the indices λ and μ, the most general state requires specifying a complete set of quantum numbers $n_{\lambda\mu}$. For each possible value of λ and μ, $n_{\lambda\mu}$ can take on any integral value $\geqslant 0$, i.e., $n_{\lambda\mu} = 0, 1, 2, 3, \ldots$. The energy eigenvalues in this number representation have the values

$$E = \sum_{\lambda} E_{\lambda},$$

with

$$E_{\lambda} = \sum_{\mu} E_{\lambda\mu} = \hbar\omega_{\lambda} \sum_{\mu} (n_{\lambda\mu} + \tfrac{1}{2}) = \hbar\omega_{\lambda}[N_{\lambda} + \tfrac{1}{2}(2\lambda+1)]. \qquad (6\text{-}33)$$

As before, ω_{λ} is given by (6-28).

In turning to the discussion of the eigenstates in the number representation, it is useful to review and extend some standard results appearing in a simpler context [e.g., Merzbacher (1970, Chapter 15, Section 8)]. We begin by defining the deformation ground state $|0\rangle$ as one for which $n_{\lambda\mu} = 0$ for any $\lambda\mu$. The operator $b_{\lambda\mu}^{\dagger}$ has the character of a number-raising operator or creation operator and the operator $b_{\lambda\mu}$ behaves like a number-lowering operator or annihilation operator. Essentially by definition $b_{\lambda\mu}|0\rangle = 0$ for any $\lambda\mu$. It then follows that $\tilde{n}_{\lambda\mu}|0\rangle = b_{\lambda\mu}^{\dagger} b_{\lambda\mu}|0\rangle = 0$ for any $\lambda\mu$, i.e., all $n_{\lambda\mu} = 0$. We also have the state $b_{\lambda\mu}^{\dagger}|0\rangle$ corresponding to $n_{\lambda\mu} = 1$, and the corresponding system is referred to as having only one $\lambda\mu$ mode of excitation or having one *$\lambda\mu$-phonon* present.[1] This follows since

$$\begin{aligned}\tilde{n}_{\lambda'\mu'} b_{\lambda\mu}^{\dagger}|0\rangle &= b_{\lambda'\mu'}^{\dagger} b_{\lambda'\mu'} b_{\lambda\mu}^{\dagger}|0\rangle = b_{\lambda'\mu'}^{\dagger}(b_{\lambda\mu}^{\dagger} b_{\lambda'\mu'} + \delta_{\lambda\lambda'}\delta_{\mu\mu'})|0\rangle \\ &= \delta_{\lambda\lambda'}\delta_{\mu\mu'} b_{\lambda'\mu'}^{\dagger}|0\rangle. \end{aligned} \qquad (6\text{-}34)$$

It is straightforward to show that a state $(b_{\lambda\mu}^{\dagger})^{n}|0\rangle$ corresponds to having n phonons of type $\lambda\mu$ present and hence $n_{\lambda'\mu'} = n\,\delta_{\lambda\lambda'}\,\delta_{\mu\mu'}$, i.e., $n_{\lambda\mu} = n$.

▶**Exercise 6-5** (a) Verify the above statement concerning the state $(b_{\lambda\mu}^{\dagger})^{n}|0\rangle$ and also show that $N_{\lambda'} = n\,\delta_{\lambda\lambda'}$.

(b) Verify that for the state $(b_{\lambda\mu}^{\dagger})^{n}(b_{\lambda'\mu'}^{\dagger})^{m}|0\rangle$, we have $n_{\lambda\mu} = n$, $n_{\lambda'\mu'} = m$ and $N_{\lambda} = n + m\,\delta_{\lambda\lambda'}$, $N_{\lambda'} = m + n\,\delta_{\lambda\lambda'}$. We thus see that N_{λ} determines the number of phonons present in mode λ and any possible μ. This is a useful cataloguing of phonon states since the system energy (6-33) is degenerate in the index μ. ◀

In view of the above, the reference to $b_{\lambda\mu}^{\dagger}$ as a phonon creation operator and $b_{\lambda\mu}$ as a phonon annihilation operator is quite apt. The state vector $|0\rangle$ is also graphically referred to as the phonon vacuum state. The normalization of the vacuum state is taken $\langle 0|0\rangle = 1$. The standard convention is to associate this phonon state vector with the nuclear ground state. For the present, we

[1] Sometimes the name *surfon* is used for this unit of excitation.

shall discuss only the case of even–even nuclei, leaving for later the cases with an odd number of valence nucleons. We take the complete system state, |nuclear ground state⟩ |0⟩, to have zero angular momentum, $J = 0$, and even parity, $\pi = +1$, appropriate for all even–even nuclei. We note from the above discussion that a phonon state with N_λ phonons present (of whatever combination of μ indices) has an energy $E_\lambda(N_\lambda) = \hbar\omega_\lambda[N_\lambda + \frac{1}{2}(2\lambda+1)]$. The *zero-point* energy $\hbar\omega_\lambda(2\lambda+1)/2$ is common to all these states and most conveniently is considered an internal energy for the nuclear ground state. If two different λ-modes λ and λ' are to be considered, then

$$E_{\lambda\lambda'}(N_\lambda, N_{\lambda'}) = [\hbar\omega_\lambda N_\lambda + \hbar\omega_{\lambda'} N_{\lambda'}] + [\hbar\omega_\lambda \cdot \tfrac{1}{2}(2\lambda+1) + \hbar\omega_{\lambda'} \cdot \tfrac{1}{2}(2\lambda'+1)].$$

Again the sum zero-point energy in the second bracket is attributed to the nuclear ground-state internal energy. The apparent absurdity of an infinite zero-point energy when all conceivable λ modes are considered is avoided by an ad hoc limitation of λ to some maximum value (of order $\lambda_{\max} \approx A^{1/3}$).

A suitably defined angular momentum operator in second quantization formalism has the expected properties

$$\begin{aligned} \tilde{\mathsf{L}}^2|0\rangle &= 0|0\rangle, & \tilde{L}_z|0\rangle &= 0|0\rangle \\ \tilde{\mathsf{L}}^2 b^\dagger_{\lambda\mu}|0\rangle &= \lambda(\lambda+1)\hbar^2 b^\dagger_{\lambda\mu}|0\rangle, & \tilde{L}_z b^\dagger_{\lambda\mu}|0\rangle &= \mu\hbar b^\dagger_{\lambda\mu}|0\rangle. \end{aligned} \tag{6-35}$$

Hence a $\lambda\mu$-mode phonon is seen to have an "intrinsic" angular momentum or "spin" of λ units with a z component of angular momentum equal to $\mu\hbar$. The phonons thus behave as bosons, appropriate for the integral spins. We can also ascribe an intrinsic parity to a $\lambda\mu$-mode phonon of $(-1)^\lambda$.

When two or more phonons are present, the individual angular momenta can be coupled to form eigenstates of L^2 and L_z for the entire system. Such states are designated by the notation $|N_\lambda, J, M\rangle$. The quantum number N_λ is associated with the operator N_λ of (6-31b), and J and M refer to the eigenvalues of L^2 and L_z. The vacuum state is thus $|0\rangle \equiv |0,0,0\rangle$. The state $b^\dagger_{\lambda\mu}|0\rangle$ can be written $|1, \lambda, \mu\rangle$.

As an example of a case with two phonons, consider two phonons with $\lambda = 2$ present, appropriate for the commonly considered quadrupole vibrations. One might expect $J = 0, 1, 2, 3, 4$ as possible coupled angular momenta. However, no properly normalized two-boson states are possible for $J = 1$ and 3.[1] The properly normalized states are

$$|2, J, M\rangle = 2^{-1/2} \sum_{\substack{m_1, m_2 \\ m_1 + m_2 = M}} (22m_1m_2|JM)\, b^\dagger_{2,m_1} b^\dagger_{2,m_2}|0,0,0\rangle \tag{6-36}$$

[1] The considerations here involve the same symmetry relations for the Clebsch–Gordan coefficients as in connection with (5-13) and (5-14), in this instance, however, applied to two bosons.

with $J^\pi = 0^+, 2^+, 4^+$. The value $N_2 = 2$ appearing in (6-36) results from a straightforward evaluation of the expectation value for $N_2 = \sum_{\mu=-2}^{+2} n_{2,\mu}$ of (6-31b) for the states (6-36). The energy E_2 of the state (6-36) is $(2+\frac{5}{2})\hbar\omega_2$, of which the zero-point energy $5\hbar\omega_2/2$ can again be attributed to the nuclear ground state.

Continuing the consideration of higher quadrupole excitations, the many allowable angular momentum-coupled states for any given N_2 lead to a highly degenerate system. All such states, when properly normalized, form an equally spaced level structure obeying (6-33) in the phonon numbers N_2, as depicted in Fig. 6.5.

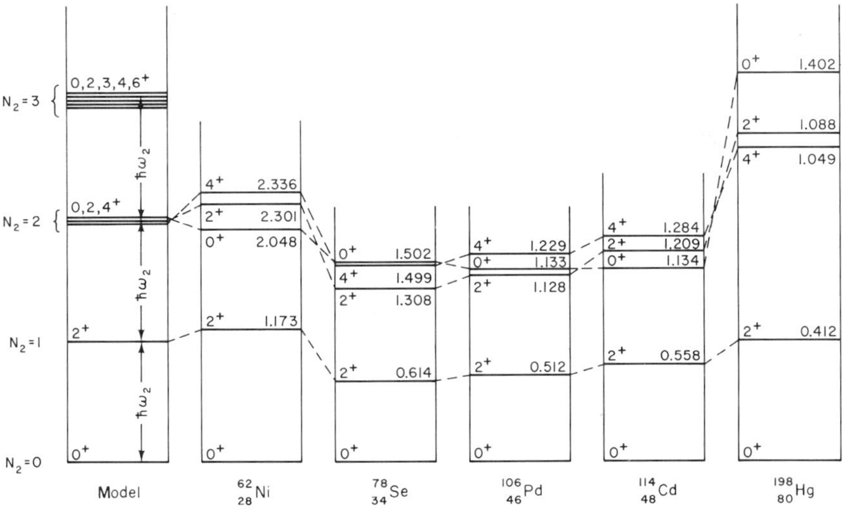

FIG. 6.5. The first four excited states of the candidate spherical nuclei ^{62}Ni, ^{78}Se, ^{106}Pd, ^{114}Cd, and ^{198}Hg. The experimental values of J^π and experimental energies are given. The energy scale has been adjusted to remove the (A, Z) dependence implied in (6-37) by plotting $A^{1/2}[1-(Z^2/A)/(Z^2/A)_f]^{-1/2}E_x$ for the observed excitation energies E_x. States presumed to be characterized by the same phonon numbers N_2 and J^π are connected by dashed lines.

▶**Exercise 6-6** (a) Show that in the terminology of Chapter III, namely $a_s = 4\pi {R_0}^2 \sigma$, $a_c = \frac{3}{5}e^2/R_0$, and $R = R_0 A^{1/3}$, we have

$$\hbar\omega_2 = \frac{2\hbar c}{R_0}\left(\frac{2a_s}{3Am_N c^2}\right)^{1/2}\left[1 - \frac{Z^2/A}{(Z^2/A)_f}\right]^{1/2}, \qquad (6\text{-}37a)$$

where $(Z^2/A)_f$ refers to the critical value for spontaneous fission, i.e., $(Z^2/A)_f = 2a_s/a_c$.

(b) Evaluate (6-37a) for the empirical liquid drop model constants of Chapter III, $a_s = 15.25$ MeV, $a_c = 0.641$ MeV, and take $R_0 \approx 1.38$ fm, to obtain

$$\hbar\omega_2 \approx \frac{29.7}{A^{1/2}}\left(1 - \frac{1}{48}\frac{Z^2}{A}\right)^{1/2} \text{ MeV.} \tag{6-37b}$$

(c) Compute the values of $\hbar\omega_2$ for ^{62}Ni, ^{78}Se, ^{106}Pd, ^{114}Cd, and ^{198}Hg. Compare these values with the experimental results shown in Fig. 6.5. ◀

Although the (A, Z) trend for $\hbar\omega_2$ contained in (6-37) is approximately verified for the nuclei shown in Fig. 6.5, the observed values of $\hbar\omega_2$ are only about one-third of the magnitudes calculated. When additional experimental data relevant to the hydrodynamic model are considered, the parameters C_λ and B_λ of (6-27) can be determined individually. The experimental values of these parameters show marked shell effects associated with the magic numbers.

The observed "mass" parameter B_λ is always substantially larger than that appropriate for the irrotational fluid assumption leading to (6-13). This deviation is most pronounced just at the magic numbers for N and Z. The increased values for the observed parameter B_λ are ascribable to the greater inertia drag due to core polarization effects. The "stiffness" parameter C_λ resulting from the opposing effects of surface tension and Coulomb repulsion also shows marked shell effects. Its behavior is consistent with the interpretation of Fig. 6.2, suggesting a larger effective surface tension (and greater stiffness) to be associated with near-magic-number values of N and Z. A useful pioneering reference in the determination of experimental values for B_λ and C_λ is Temmer and Heydenburg (1956). See also Siegbahn (1965) for systematic trends.

It should be remarked further with respect to the nuclei ^{106}Pd and ^{114}Cd, shown in Fig. 6.5, that some levels to be associated with phonon numbers $N_2 = 3$, 4, and 5 have also been identified for $1.5 < E_x < 3.0$ MeV. These two nuclei are among the best examples to date showing simple vibrational states. A recent compilation of nuclear levels identified by the phonon quantum number N_λ is that by Sakai (1972).

A principal mode of electromagnetic decay of the collective vibrational states is by emission of strongly enhanced electric quadrupole (E2) radiation.[1] The selection rules for such transitions, in addition to requiring the usual triangle condition $|J_f - J_i| \leqslant 2 \leqslant J_f + J_i$, can be shown also to require $\Delta N_2 = \pm 1$. This latter fact, that strong (E2) transitions should be found connecting consecutive vibrational bands, offers a useful additional test for attributing a simple vibrational character to nuclear levels such as those that appear to be suggested by the energy and J^π values shown in Fig. 6.5. In fact, strong (E2) transitions with $\Delta N_2 \pm 1$ are observed for these particular nuclei.

[1] See Chapter VII for nomenclature and a general discussion of electromagnetic transitions.

▶**Exercise 6-7** We note that the important term in the (E2)-transition operator of (7-77) in Chapter VII, when applied to our model of a distributed charge, is

$$\tilde{Q}_2{}^m = \sum_{\kappa=1}^{A} q_\kappa r_\kappa{}^2 Y_2^{m*}(\theta_\kappa, \varphi_\kappa) \equiv \rho_{0c} \iint_0^{S(\theta,\varphi)} r^4 Y_2^{m*}(\theta, \varphi)\, dr\, d\Omega.$$

Show, to order α, that

$$\tilde{Q}_2{}^m = \frac{3Ze}{4\pi} R^2 \alpha_{2m} = \frac{3Ze}{4\pi} R^2 \left(\frac{\hbar}{2B_2 \omega_2}\right)^{1/2} [b_{2m} + (-1)^m b^\dagger_{2,-m}]. \quad (6\text{-}38)$$

We first note that the total nuclear charge Ze is effective in inducing the transition in the collective case rather than just the charge of one proton as would be the case for a single-particle transition discussed in Chapter VII. The reference to a strong (E2) transition rate is in this context. The appearance of the annihilation and creation operators immediately gives the selection rule $\Delta N_2 = \pm 1$. See also Exercise 7-10. ◀

It is instructive to evaluate the average value of the lowest moments of the deformation parameters $\alpha_{\lambda\mu}$. In view of (6-29) and the orthogonality of the different N_λ-phonon states, we immediately conclude that the expectation value

$$\langle|\alpha_{\lambda\mu}|\rangle = 0 \quad (6\text{-}39)$$

for any phonon state. For the case of the second moment, it is useful to introduce the parameter

$$a_\lambda{}^2 \equiv \langle a_\lambda{}^2 \rangle = \left\langle \left| \sum_\mu |\alpha_{\lambda\mu}|^2 \right| \right\rangle. \quad (6\text{-}40)$$

The quantity $a_\lambda{}^2$ is referred to as the ms (mean-square) deformation parameter. If we recall that for harmonic oscillator states $\langle|T|\rangle = \langle|E_p|\rangle = E/2$, we conclude from (6-27), (6-28), and (6-33) that

$$a^2_{\lambda,N} = \frac{\hbar\omega_\lambda}{C_\lambda}\left(N_\lambda + \frac{2\lambda+1}{2}\right) = \frac{\hbar}{\omega_\lambda B_\lambda}\left(N_\lambda + \frac{2\lambda+1}{2}\right), \quad (6\text{-}41)$$

where we have included the subscript N to completely specify $a_\lambda{}^2$. Observe that $a^2_{\lambda,N}$ is nonzero even for $N_\lambda = 0$ due to the zero-point oscillation. The quantity $a^2_{\lambda=2}(N_2 = 0)$ is usually referred to as $\beta_0{}^2$ in the literature. We depart from our usual custom of employing standard literature notation in this case to avoid unnecessary confusion.

▶**Exercise 6-8** Define the mean-square charge radius as

$$R_c{}^2 = \langle|\tilde{R}_c{}^2|\rangle, \qquad \text{with} \quad \tilde{R}_c{}^2 = \rho_{0c} \iint_0^{S(\theta,\varphi)} r^4\, dr\, d\Omega. \quad (6\text{-}42)$$

Show that

$$\tilde{R}_c{}^2 = \rho_{0c} R^5 \left[(4\pi/5) + \sum_{\lambda \geq 1, \mu} |\alpha_{\lambda\mu}|^2 + \cdots \right]$$

and hence that

$$R_c{}^2 = \tfrac{3}{5} Z R^2 \left[1 + (5/4\pi) \sum_{\lambda \geq 1} a_{\lambda, N}^2 + \cdots \right]. \quad ◀ \qquad (6\text{-}43)$$

We readily find from (6-28) that neglecting the Coulomb energy gives the ratio of the characteristic octupole ($\lambda = 3$) to quadrupole ($\lambda = 2$) vibrational frequencies

$$\omega_3/\omega_2 = (15/4)^{1/2} \approx 2. \qquad (6\text{-}44)$$

We would thus expect to find one-phonon octupole $J^\pi = 3^-$ states at about the same excitation energy as two-phonon quadrupole states. Such low-lying candidate octupole states are observed, although usually at higher energies than (6-44) would predict.

Although up to this point we have been discussing a model suitable for spherical even–even nuclei found in the vicinity of N and Z magic, our examples used in Fig. 6.5 avoided nuclei with either N or Z exactly magic. As discussed earlier, the empirical evidence for the collective parameters B_λ, C_λ, and $\hbar\omega_\lambda$ shows marked shell effects. The very large values of $\hbar\omega_\lambda$ for N and Z exactly magic have the effect of comingling collective states with simpler configuration shell model or intrinsic states. For example, the first excited state of ^{208}Pb (the 3^- level at $E_x = 2.615$ MeV) is known to be highly collective (or alternately, to have a shell model description consisting of many particle–hole configurations).[1] The next two levels at $E_x = 3.198$ and 3.475 MeV appear to be rather simpler $[(2g_{9/2})^1(3p_{1/2})^{-1}]_{4^-}$ and $[(2g_{9/2})^1(3p_{1/2})^{-1}]_{5^-}$ shell model states. The next possible collective state appears to be the 2^+ state at 4.086 MeV.

When both intrinsic and collective states of the same J^π character appear at nearby unperturbed energies, perturbations produced by coupling terms in the Hamiltonian will result in mixed states with the collective strength distributed among a number of levels. Possible strong deviations from simple collective model predictions in such properties as radiative transition rates then might be expected for individual levels. Such effects raise the question of the validity of using a simple collective model for such nuclei in the first place.

[1] For present purposes we can consider a shell model description involving many particle–hole configurations for a particular level to be equivalent to its description in collective terms. It should be pointed out, however, that a central question in such equivalences in microscopic theories is the nature of the phase relationship between such configuration states due to long-range correlations. For a typical paper on such shell model calculations for ^{208}Pb, see True *et al.* (1971).

The levels of ^{208}Pb as well as those of the doubly magic nuclei ^{16}O and ^{40}Ca are shown in Fig. 6.6. Since many of these levels are more related to simple shell model configurations, the graph scale has been adjusted to remove the nuclear "size effect" for single-particle energies [see (4-33a) and (6-141)].

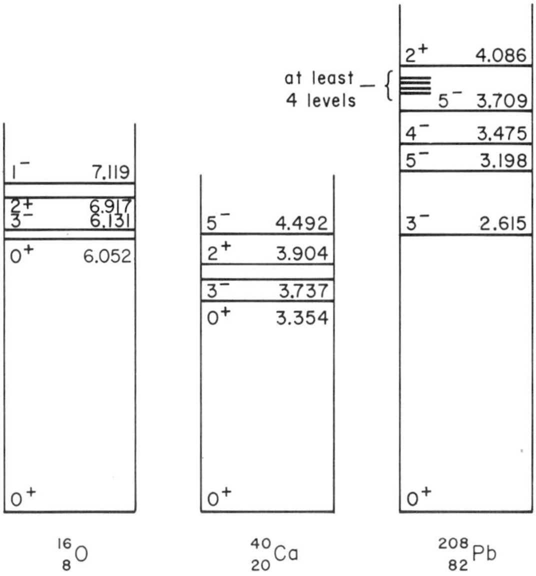

FIG. 6.6. Lowest levels of the doubly magic nuclei ^{16}O, ^{40}Ca, and ^{208}Pb. The energy scale has been adjusted to remove the A dependence implied in (4-33a) and (6-141) by plotting $A^{1/3}E_x$ for the observed excitation energies E_x.

For nuclei with Z_0, $N_0 \pm 2$ and Z_0+2, N_0 ($Z_0 = 82$, $N_0 = 128$) the first excited state in each case is a 2^+ level. The excitation energies are $E_x(2^+) = 0.803$, 0.795, and 1.180 MeV for $^{206}_{82}Pb$, $^{210}_{82}Pb$, and $^{210}_{84}Po$, respectively. Although these energies are higher than a simple hydrodynamic model extrapolation from the nuclei of Fig. 6.5 would indicate, they are, however, approximately one-fifth of that for the 2^+ level of ^{208}Pb and indicate the abruptness of the shell effects on the collective parameters. In view of this behavior, it would be reasonable to expect the valid application of a simple collective vibrational model description to even–even nuclei to commence with $N = N_0 \pm \Delta N$ and $Z = Z_0 \pm \Delta Z$ (Z_0 and N_0 magic) for ΔN and $\Delta Z \gtrsim 2$ or 4, at least insofar as freedom from mixing with simple intrinsic shell model states is concerned. Of course, where ΔN and $\Delta Z \gtrsim 8$ or 10, we more or less abruptly develop sizable permanent deformations for the nuclear states as discussed in the introduction to this chapter. We defer discussion of such cases to later. In conclusion one is led to looking for candidate vibrational states based on a spherical permanent shape in relatively narrow (N, Z) ranges near magic numbers with 2 or $4 \lesssim (\Delta N$ and $\Delta Z) \lesssim 8$ or 10; refer also to Fig. 6.3.

2. Vibrational States under Weak Coupling

In the preceding section we confined our attention to the case of spherical even–even nuclei. In turning to the case N or Z odd in nearby nuclei we assume a Hamiltonian of the form (again specializing to the case $\lambda = 2$)

$$H = H_0 + H_{\text{sp}} + H_{\text{pair}} + \tfrac{1}{2}C_2 \sum_{\mu=-2}^{+2} |\alpha_{2\mu}|^2 + \tfrac{1}{2}B_2 \sum_{\mu=-2}^{+2} |\dot{\alpha}_{2\mu}|^2 + H_{\text{int}}. \tag{6-45}$$

Here

$$H_0 + H_{\text{sp}} + H_{\text{pair}} = H_0 + \sum_{i=1}^{k} [T(r_i) + V(r_i)] + \sum_{i \neq j}^{k} v(r_{ij}) \tag{6-46}$$

is the usual shell model Hamiltonian for k valence nucleons including residual interactions. The reference Hamiltonian H_0 can be taken as a closed-shell magic configuration or, in a simplified picture, some nearby even–even nucleus assumed to be an appropriate core configuration. The index k ranges over the nucleons to be considered explicitly. The potential energy $V(r_i)$ can be taken either as a shell model potential determined by the interaction of a valence particle with the core configuration alone or can include an average potential interaction among the k nucleons themselves. In either event, the residual interaction is suitably the remaining interaction between the valence nucleons. In this connection the full range of choices discussed in Chapter V is available in determining the eigenfunctions of the Hamiltonian (6-46) and give rise to what we have been referring to as the intrinsic states.[1]

The fourth and fifth terms in (6-45) are for the phonon surface excitations discussed in the preceding section. The last term represents the interaction of the k valence particles (or holes) with the surface phonons. The interaction of a particle (or hole) with an average deformed field can always be expanded in a form

$$V(\mathbf{r}_i, \alpha_{\lambda\mu}) = V(r_i) \mp F(r_i) \sum_{\lambda,\mu} \alpha_{\lambda\mu} Y_\lambda{}^\mu(\theta_i, \varphi_i) + F'(r_i) \Big|\sum_{\lambda\mu} \alpha_{\lambda\mu} Y_\lambda{}^\mu(\theta_i, \varphi_i)\Big|^2 + \cdots, \tag{6-47}$$

where $V(r_i)$ is simply an effective spherically symmetric potential that is

[1] It is also convenient to include the zero-point energy of the quadrupole vibrational mode in the ground-state energy E_0.

already considered included in (6-46). The $\mp$ signs refer to particles $(-)$ and holes $(+)$. Thus, to first order in the particle and deformation coordinates, we have

$$H_{\text{int}} = \sum_{i=1}^{k} \left\{ \mp F(r_i) \sum_{\mu=-2}^{+2} \alpha_{2\mu} Y_2^{\mu}(\theta_i, \varphi_i) \right\}, \qquad \text{for} \quad \lambda = 2, \qquad (6\text{-}48)$$

where we have summed over the valence particles. So long as the number of explicitly treated nucleons is kept small, the redundancy appearing in the implicit connection between particle coordinate and collective coordinates can be ignored.

A rather simple model calculation can be profitably used both to verify the expansion (6-47) and to shed insight on the nature of the function $F(r_i)$. We assume, as would be appropriate for very short-range nucleon–nucleon interactions, that the average nucleon potential is proportional to the local nuclear density. Under the adiabatic assumption discussed previously, the valence nucleons execute many periods of orbital motion as this potential slowly alters its spatial aspect. At any instant of time the nucleons feel this potential essentially as if it were at rest.[1] The nuclear equipotential surface would be expected to have the same r, θ, φ dependence as the surface of constant density, viz.,

$$V(\alpha_{2\mu}, \mathbf{r}, \boldsymbol{l}, \mathbf{s}) = V\left(\frac{r}{1 + \sum_{\mu} \alpha_{2\mu} Y_2^{\mu}(\theta, \varphi)}, \boldsymbol{l}, \mathbf{s} \right), \qquad (6\text{-}49)$$

where we have noted a possible spin and orbital angular momentum dependence, and again consider only quadrupole $(\lambda = 2)$ deformations. Clearly the surface[2]

$$S_V(\alpha_{2\mu}, \theta, \varphi) = r_V \left(1 + \sum_{\mu} \alpha_{2\mu} Y_2^{\mu}(\theta, \varphi) \right)$$

is an equipotential surface of (6-49), as required.

With $\alpha_{2\mu}$ small we can use the expansion (suppressing the $\boldsymbol{l}$ and $\mathbf{s}$ dependences

[1] More rigorous treatments of coupled collective and intrinsic equations of motion in a time-dependent framework exist. An early example of historical interest is the so-called cranking model, while more recent applications include the vibrating potential model and time-dependent Hartree–Fock calculations. Refer, for example, to Rowe (1970) and Belyaev (1968).

[2] We allow for the possibility of relaxing the condition of a sharp boundary containing a constant-density fluid, and generalize the surface equation (6-1) to the constant-nuclear-matter-density "isotonic surface,"

$$S_\rho(\alpha_{2\mu}, \theta, \varphi) = r_\rho \left(1 + \sum_{\mu} \alpha_{2\mu} Y_2^{\mu}(\theta, \varphi) \right).$$

In particular, a density distribution such as (2-3) could thus be generalized to have

$$c = c_0 \left(1 + \sum_{\mu} \alpha_{2\mu} Y_2^{\mu}(\theta, \varphi) \right).$$

momentarily)

$$V(\alpha_{2\mu}, \mathbf{r}) = V(0, r) + \sum_{\mu} \left[\frac{\partial V(\alpha_{2\mu}, r)}{\partial \alpha_{2\mu}} \right]_0 \alpha_{2\mu} + \cdots$$

$$= V(0, r) + \frac{dV(0, r)}{dr} \sum_{\mu} \left[\frac{\partial}{\partial \alpha_{2\mu}} \left(\frac{r}{1 + \sum_{\mu} \alpha_{2\mu} Y_2{}^{\mu}(\theta, \varphi)} \right) \right]_0 a_{2\mu} + \cdots$$

or

$$V(\alpha_{2\mu}, r, \theta, \varphi) = V(0, r) - r \frac{dV(0, r)}{dr} \sum_{\mu} \alpha_{2\mu} Y_2{}^{\mu}(\theta, \varphi) + \cdots. \tag{6-50a}$$

Thus we identify $V(\alpha = 0, r, \mathbf{l}, \mathbf{s})$ with the effective central potential $V(r_i)$ appearing in (6-46) and (6-47) and also have

$$F(r) = r\, dV(\alpha = 0, r, \mathbf{l}, \mathbf{s})/dr. \tag{6-50b}$$

Since the derivative of the potential is largely confined to the surface region, we note that H_{int} indeed has the character of a particle–surface interaction. For the simple square well corresponding to a constant–density assumption, we have

$$F(r) = R\overline{V}_0\, \delta(r - R), \tag{6-51}$$

where R is the effective undeformed radius appearing in (6-1).

Since the effect of (6-48) is confined to the surface and involves the deformation parameters $\alpha_{2\mu}$ (assumed to be small in the present case of essentially spherical nuclei), a perturbation treatment of (6-45) in the term H_{int} would appear to be appropriate and leads to the weak coupling approximation. Suitable basis states would be the product states formed from the eigenfunctions of the spherical shell model Hamiltonian (6-46) and the phonon eigenfunctions of the preceding section. These basis states are, of course, the eigenstates of the system Hamiltonian (6-45) with $H_{\text{int}} \equiv 0$.

We designate the eigenstates of (6-46) by the state vector $|j, m\rangle$, where j and m refer to the total angular momentum and its z component of the k valence nucleons. When $k \geqslant 2$, $|j, m\rangle$ is a suitable coupled state of the individual particle (or hole) angular momenta.[1] Following the standard convention of the literature, the phonon states $|N_\lambda, J, M\rangle$ such as (6-36) are rewritten $|N_\lambda, R, M_R\rangle$, thus allowing J and $M = M_J$ to refer to the system total angular momentum and its z component.[2] The product states can be written

$$|j; N_\lambda, R; J, M\rangle = \sum_{\substack{m, M_R \\ m + M_R = M}} (jRmM_R|JM)\, |j, m\rangle\, |N_\lambda, R, M_R\rangle. \tag{6-52a}$$

[1] Additional quantum numbers may be required when $k \geqslant 2$.

[2] It is unfortunate that the symbol R thus represents both the nuclear radius and the collective total angular momentum. The distinction between the two is, however, always clear from the context.

The system energy is correspondingly

$$E = E_0 + \hbar\omega_\lambda N_\lambda + E(j), \tag{6-52b}$$

where the reference energy E_0 is that of the core plus the phonon zero-point energy. The system ground state will have $N_\lambda = 0$ and have the k valence particles in the coupled state $|j, m\rangle$ having the lowest energy $E_0(j)$.

To the extent that a perturbation approach is valid, we could expect to find at least an approximation to the simple energy additivity contained in (6-52b). Figure 6.7 shows the spectra for the nuclei near ^{106}Pd with $N = 60$ and

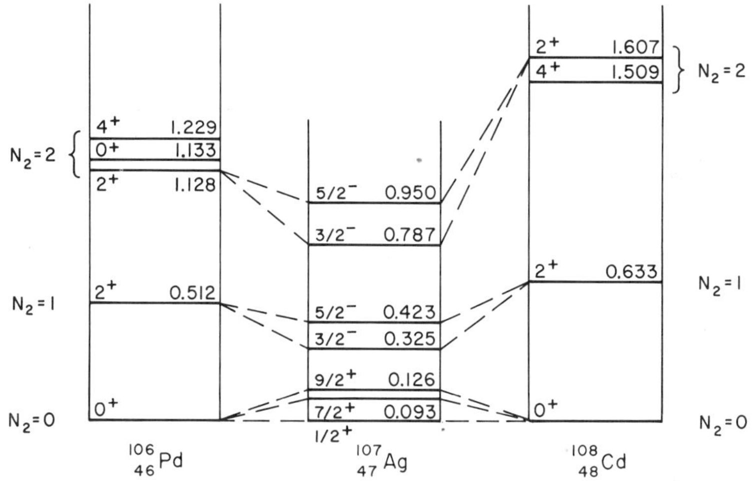

FIG. 6.7. The low-energy spectra of the spherical nuclei with $N = 60$ and $Z = 46$, 47, and 48. Possible phonon shell model coupled states are linked by dashed lines (see text).

$Z = 46$, 47, and 48. We observe from Fig. 4.4 for spherical shell model states that low-energy candidate configurations for the odd-nucleon case of $Z = 47$ would be the hole states (referred to $Z = 50$)[1]

$$|j^\pi = \tfrac{1}{2}^-: (1g_{9/2})_0^{-2}(2p_{1/2})_{1/2^-}^{-1}\rangle,$$

$$|j^\pi = \tfrac{9}{2}^+: a(1g_{9/2})_{9/2^+}^{-3} + b(1g_{9/2})_{9/2^+}^{-1}(2p_{1/2})_0^{-2}\rangle,$$

and

$$|j^\pi = \tfrac{7}{2}^+: (1g_{9/2})_{7/2^+}^{-3}\rangle.$$

[1] The fact that the indium isotopes $^{109,111,113,115}_{49}$In all have a ground state $\frac{9}{2}^+$ and a low-lying first excited state $\frac{1}{2}^-$ while the silver isotopes $^{105,107,109,111}_{47}$Ag all have a ground state $\frac{1}{2}^-$ is indicative of the close energy competition matching the $(g_{9/2})^2$ pairing energy against the considerably stronger binding of the $2p_{1/2}$ compared to the $1g_{9/2}$ orbital due to the presence of the weak magic number $Z = 40$ presumably at the closure of the $2p_{1/2}$ orbital.

The three lowest states of ${}^{107}_{47}\mathrm{Ag}$, $J^\pi = \frac{1}{2}^-, \frac{7}{2}^+$, and $\frac{9}{2}^+$, thus can be associated with these shell model states $|j^\pi, m\rangle$ coupled to the zero-phonon state $|0,0,0\rangle$. The third and fourth excited states, $J^\pi = \frac{3}{2}^-$ and $\frac{5}{2}^-$, possibly can be associated with coupling the $|j^\pi = \frac{1}{2}^-, m\rangle$ state with the one-phonon state $|1, 2, M_R\rangle$. Similarly the fifth and sixth excited states shown, again having $J^\pi = \frac{3}{2}^-$ and $\frac{5}{2}^-$, can be associated with the two-phonon state $|2, 2, M_R\rangle$. The electromagnetic transition rates for all of the ${}^{107}\mathrm{Ag}$ levels are consistent with the above identifications.

▶**Exercise 6-9** Examine possible zeroth-order states corresponding to (6-52) for the A-odd nuclei near ${}^{62}\mathrm{Ni}$, ${}^{78}\mathrm{Se}$, ${}^{114}\mathrm{Cd}$, and ${}^{198}\mathrm{Hg}$ (the even–even spherical nuclei of Fig. 6.5). ◀

A core–valence particle coupling scheme in terms of simple product states and energy additivity similar to (6-52) is, of course, of broader applicability than that for the model discussed above. Primarily, it is a feature of any problem with a separable Hamiltonian that in zeroth order has no cross terms in coordinates specifying different degrees of freedom. The examples are particularly striking when the various degrees of freedom have markedly different level density spectra as in Fig. 6.7 [$\Delta E_\lambda = \hbar\omega_\lambda \approx 0.55$ MeV and $\Delta E(j) \lesssim 0.12$ MeV]. A similar situation prevails for the triad ${}^{207,208,209}_{82}\mathrm{Pb}$ shown in Fig. 6.8. Whatever the nature of collectivity in the ground state of the

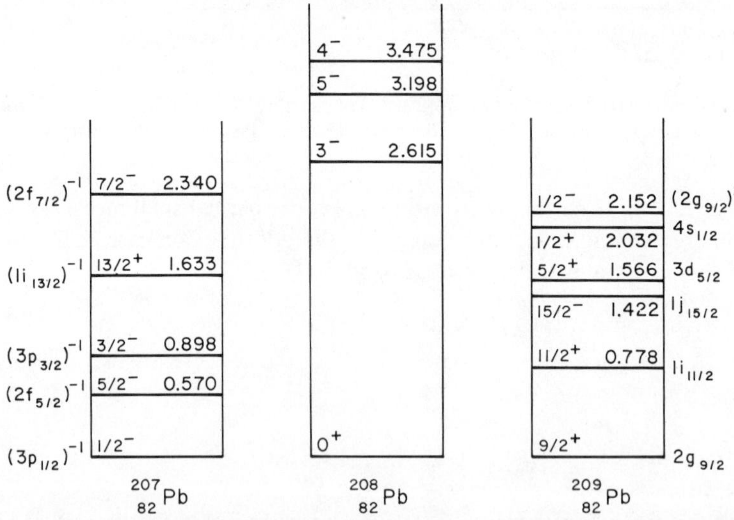

FIG. 6.8. The doubly magic core nucleus ${}^{208}\mathrm{Pb}$ and states in ${}^{207,209}\mathrm{Pb}$ that may be interpreted as simple particle or hole states coupled to the ground state of the core. The particle and hole shell model configurations are indicated.

doubly magic core nucleus ^{208}Pb, a core–particle or core–hole coupling scheme in lowest order would give the same result. Here $Z = 82$ and $N = 126$ or $N = 126 \pm 1$. Note that the first particle–hole state of any complexity occurs at 2.15 MeV in ^{209}Pb. The empirical single–particle level sequence of Fig. 4.4 is obtained in fact by reviewing data of the type given in Fig. 6.8.

When the particle–collective coupling term H_{int} is taken into account, we are required to evaluate its matrix elements for the basis states (6-52a), viz.,

$$\langle j'; N_2'R'; J, M | H_{\text{int}} | j; N_2 R; JM \rangle, \tag{6-53}$$

with H_{int} to lowest order given by (6-48).

▶**Exercise 6-10** Using (6-29) and the simple surface delta-function form of $F(r)$ given in (6-51), show that for the special case of a single valence nucleon (i.e., $k = +1$) and where $|j; N_2 R; JM\rangle \equiv |j; 00; J = j, M = m\rangle$ refers to the phonon or collective ground-state band, the matrix elements (6-53) become

$$(-1)^{j-j'+1}\left(\frac{2j'+1}{2j+1}\right)^{1/2}\left(\frac{\hbar}{2B_2\,\omega_2}\right)^{1/2}\langle j'||Y_2||j\rangle\langle n'l'|F(r)|nl\rangle$$
$$= (-1)^{j-j'+1}\left(\frac{2j'+1}{2j+1}\right)^{1/2}\frac{a_{20}}{\sqrt{5}}\langle j'||Y_2||j\rangle \overline{V}_0\, R^3 u_{n'l'}(R)\, u_{nl}(R). \tag{6-54}$$

Here $u_{nl}(r) \equiv |nl\rangle$ represents the single-particle radial function and we have noted the result (6-41). [*Hint:* Use (A-12) and (A-14) and the definition (B-25) of the Appendices A and B to arrive at a simple sum on the square of the Clebsch–Gordan coefficients

$$\sum_{\mu} (j', 2, m+\mu, -\mu | jm)^2,$$

which by normalization is just unity.]

In arriving at (6-54), it is found that only those states $|j'; N_2'R'; JM\rangle$ contribute for which

$$|j'-j| \leqslant 2 \leqslant j'+j \qquad \text{and} \qquad N_2' = 1 \quad \text{and hence} \quad R' = 2.$$

While the parities of the states $|j, m\rangle$ and $|j', m'\rangle$ have not been explicitly stated, clearly

$$\pi' = \pi \qquad \text{and} \qquad (-1)^{l'+l} = +1.$$

For other than the simple case above, the angular momentum algebra becomes complicated. Many useful explicit expressions for treating the general case can be found in Alaga (1969).

According to standard first-order perturbation theory, the new phonon or collective ground-state band becomes

$$\psi_{JM} = |j;00;J=j,M=m\rangle + \overline{V}_0 R^3 (a_{20}/\sqrt{5})$$
$$\times \sum_{j'} (-1)^{j-j'} \left(\frac{2j'+1}{2j+1}\right)^{1/2} |j';12;JM\rangle \frac{\langle j'||Y_2||j\rangle u_{n'l'}(R) u_{nl}(R)}{\hbar\omega_2 + E(j') - E(j)}. \tag{6-55}$$

◀

▶**Exercise 6-11** Repeat the above for the case of a harmonic oscillator single-particle shell model potential $V(r)=\frac{1}{2}m_N\omega^2 r^2$, whence $F(r)=-m_N\omega^2 r^2$. Examine the selection rules on n,l and n',l' and show that only the combinations $(n',l')=(n,l)$; $(n-1,l)$; $(n,l-2)$; $(n+1,l-2)$; $(n-2,l+2)$ are possible. ◀

If the above results are applied to the case of ^{107}Ag illustrated in Fig. 6.7, we see that the phonon or collective ground-state band $J^\pi = \frac{1}{2}^-, \frac{7}{2}^+$, and $\frac{9}{2}^+$ remains pure, while the two $\frac{3}{2}^-$ states are mixed as are the two $\frac{5}{2}^-$ states. The situation for the general case may be more complicated. For example, with N or Z just above the magic number 50, $N_2=0$ and $N_2=1$ phonon states coupled to both $1g_{7/2}$ and $2d_{5/2}$ shell model single-particle states would be expected to be strongly mixed.

Although the discussion of the weak coupling model to this point has been motivated by a desire to consider the case A odd, clearly the treatment given also covers the case k even. A good example of an application for the case $k=-2$ (two holes) is for ^{198}Hg. Since Hg has $Z=80$, an obvious two-hole configuration based on the closed shell $Z=82$ is suggested with possible candidate orbitals $3s_{1/2}^{-1}$, $2d_{3/2}^{-1}$, $1h_{11/2}^{-1}$, and $2d_{5/2}^{-1}$ (see Fig. 4.4 for protons). When all four possible hole states coupled to collective states with as many as three phonons of excitation were considered, the observed splitting of the unperturbed $N_2=2$, $J^\pi=0,2,4^+$ states was achieved yielding energies in approximate conformity with Fig. 6.5 [see Alaga (1969)]. In general, rather good energy spectra and γ-ray transition rates can be obtained for vibrational nuclei under interaction Hamiltonians such as (6-48).

When higher order terms in α neglected in the Hamiltonian (6-25) are considered, the effect is to introduce anharmonic effects that also mix different phonon states of the same (J,M), with the consequence that N_λ is no longer a good quantum number due to this effect as well as the mixing discussed above. It is thus rather satisfying that nuclei exist for which perhaps relatively pure phonon states can be assigned. For example, in the case of ^{114}Cd the possible mixing of the states illustrated in Fig. 6.5 would lead to the pair of 2^+ states

$$\begin{aligned} |2^+,M\rangle_1 &= a|1,2,M\rangle + (1-a^2)^{1/2}|2,2,M\rangle \\ |2^+,M\rangle_2 &= a|2,2,M\rangle - (1-a^2)^{1/2}|1,2,M\rangle. \end{aligned} \tag{6-56}$$

Satisfactory empirical electromagnetic transition rates result with $a = \pm 0.99$. Unfortunately, the small anharmonic effect this implies is not quite able to account for the intrinsic quadrupole moment measured for the first excited 2^+ state. Recently, experimental techniques using heavy ion Coulomb excitation, Doppler-shift, Mössbauer, and muonic atom effects have been able to determine intrinsic quadrupole moments for excited states of nuclei [see, e.g., Davidson (1968), deBoer and Eichler (1968), Eisenberg and Greiner (1970), and Schwalm *et al.* (1972)].[1] Rather large values are obtained for the first excited 2^+ states of a number of spherical vibrators. The value for the ^{114}Cd nucleus is (-0.7 ± 0.2) b, while calculations with the small anharmonic effects required to give reasonable level energy and electromagnetic transition rates give values of ~ -0.25 b at best [Alaga (1969)]. Discrepancies of this magnitude suggest that a significant modification in the vibrational model assumptions may be required. All things considered, however, ^{106}Pd and ^{114}Cd remain the best examples of nearly pure spherical harmonic vibrators.

D. THE SIMPLE SYMMETRIC TOP

In discussing nuclei with permanent deformations, it is useful to have the case of the canonical symmetric top before us since it offers a ready framework to build on. Therefore we digress at this point to review briefly the behavior of the quantized symmetric top or symmetric rigid rotator.

A body-fixed coordinate frame (x', y', z') is selected to correspond to the principal axes of the rigid rotator or top. We designate the principal moments of inertia in this frame by $\mathscr{I}_{x'}$, $\mathscr{I}_{y'}$, and $\mathscr{I}_{z'}$. The symmetric top case results when two of the principal moments of inertia are taken to be equal. For convenience, we assume symmetry about the z' axis and take $\mathscr{I}_{x'} = \mathscr{I}_{y'} \equiv \mathscr{I}_0$.

It is necessary to introduce angular momentum operators in terms of both the body-fixed frame (x', y', z'), namely $L_{x'}$, $L_{y'}$, and $L_{z'}$, and the laboratory frame (x, y, z), namely L_x, L_y, and L_z. Since it is not the present intention to enter into a complete detailed presentation of the transformations connecting the operators in these frames, we refer the interested reader to other texts [e.g., Eisenberg and Greiner (1970)]. We simply note that as expected $L_x{}^2 + L_y{}^2 + L_z{}^2 = L_{x'}^2 + L_{y'}^2 + L_{z'}^2 = \mathbf{L}^2$.

The Hamiltonian for the rigid rotator can be written[2]

$$H_{\text{top}} = \frac{\hbar^2}{2}\left(\frac{L_{x'}^2}{\mathscr{I}_{x'}} + \frac{L_{y'}^2}{\mathscr{I}_{y'}} + \frac{L_{z'}^2}{\mathscr{I}_{z'}}\right) = \frac{\hbar^2}{2\mathscr{I}_0}(\mathbf{L}^2 - L_{z'}^2) + \frac{\hbar^2}{2\mathscr{I}_{z'}} L_{z'}^2. \tag{6-57}$$

[1] Techniques also exist for measuring the magnetic dipole moments of excited nuclear states using ion-implantation and Mössbauer methods, etc.; e.g., see Grodzins (1972) for a general reference and Sioshansi *et al.* (1972) as a typical recent paper.

[2] Following general custom, the operators $\mathbf{L}$ and $\mathbf{L}'$ used here do not include the factor $\hbar$.

From a general consideration of the relevant commutation properties of H_{top}, $\mathbf{L}$, and $\mathbf{L}'$, the eigenstates of this Hamiltonian also can be taken as simultaneous eigenfunctions of $\mathbf{L}^2$, $L_{z'}$, and L_z, and are characterized as[1]

$$D_{MK}^L \equiv |LMK\rangle, \tag{6-58}$$

with

$$L_z|LMK\rangle = M|LMK\rangle, \qquad |M| \leqslant L \tag{6-59a}$$

$$L_{z'}|LMK\rangle = K|LMK\rangle, \qquad |K| \leqslant L \tag{6-59b}$$

$$\mathbf{L}^2|LMK\rangle = L(L+1)|LMK\rangle. \tag{6-59c}$$

The rotational kinetic energy (or total energy) is, in this notation,

$$T_{top} = \langle LMK|\frac{\hbar^2}{2}\sum_{\xi}\frac{L_\xi^{\,2}}{\mathscr{I}_\xi}|LMK\rangle, \qquad \xi = x', y', \text{and } z',$$

or

$$T_{top} = \frac{\hbar^2}{2\mathscr{I}_0}[L(L+1)-K^2] + \frac{\hbar^2K^2}{2\mathscr{I}_{z'}}. \tag{6-60}$$

The functions D_{MK}^L have considerable intrinsic interest. We cite only a few simple properties. Arbitrary rotations from a frame (x,y,z) to a frame (x',y',z') are best described in terms of rotations through the Euler angles θ_a, θ_b, and θ_c. To arrive at the frame (x',y',z'), one first rotates the original frame (x,y,z) through a right-hand rotation θ_a $(0 \leqslant \theta_a \leqslant 2\pi)$ about the z axis, thereby changing the position of the x and y axes within the original equatorial plane. Then one rotates that frame again through a right-handed rotation θ_b $(0 \leqslant \theta_b \leqslant \pi)$ about the new position of the y axis, thus changing the position of the z axis. Finally, one continues by rotating once again through a right-handed rotation θ_c $(0 \leqslant \theta_c \leqslant 2\pi)$ about this new position of the z axis.

A problem that commonly arises when two such frames are given is to specify the relationship between the representations of a state vector in the two coordinate frames. If we specify a state vector that is an eigenvector $|JK\rangle$ of the operators $\mathbf{J}^2$ and J_z, the rotated vector in frame (x',y',z') is $|JK\rangle' = R|JK\rangle$, where $R = \exp(-i\theta\hat{\mathbf{n}}\cdot\mathbf{J})$ is the rotation operator symbolically specifying the *ordered rotations* through the Eulerian angles θ_a, θ_b, and θ_c carrying (x,y,z) into (x',y',z'). It is shown in standard texts on angular momentum [e.g., Rose (1957)] that the rotation operator in this case can also be written

$$R = R(\theta_c)R(\theta_b)R(\theta_a) = \exp(-i\theta_a J_z)\exp(-i\theta_b J_y)\exp(-i\theta_c J_z).$$

[1] More properly we shall be using $D_{MK}^L(\theta_a,\theta_b,\theta_c) = \langle\theta_a\theta_b\theta_c|LMK\rangle$.

Using closure of the set $|JM\rangle$, we have

$$|JK\rangle' = R|JK\rangle = \sum_M |JM\rangle\langle JM|R|JK\rangle$$

or

$$|JK\rangle' = \sum_M D^{J*}_{MK}(\theta_a, \theta_b, \theta_c)\,|JM\rangle \tag{6-61}$$

by defining[1]

$$D^{J*}_{MK}(\theta_a, \theta_b, \theta_c) = \langle JM|\exp(-i\theta_a J_z)\exp(-i\theta_b J_y)\exp(-i\theta_c J_z)|JK\rangle. \tag{6-62a}$$

Explicitly,

$$D^{J*}_{MK}(\theta_a, \theta_b, \theta_c) = [\exp(-iM\theta_a)]\,d^J_{MK}(\theta_b)\exp(-iK\theta_c) \tag{6-62b}$$

and also

$$D^{J}_{MK}(\theta_a, \theta_b, \theta_c) = [\exp(iM\theta_a)]\,d^J_{MK}(\theta_b)\exp(iK\theta_c), \tag{6-62c}$$

with $d^J_{MK}(\theta_b)$ equal to the real function defined by

$$d^J_{MK}(\theta_b) = \sum_s \frac{(-1)^s[(J+M)!\,(J-M)!\,(J+K)!\,(J-K)!]^{1/2}}{(J-M-s)!\,(J+K-s)!\,s!\,(M-K+s)!} \times \left(\cos\frac{\theta_b}{2}\right)^{2J+K-M-2s}\left(-\sin\frac{\theta_b}{2}\right)^{2s+M-K}, \tag{6-62d}$$

where s is an integer for which the factorial arguments are zero or greater. The D^J_{MK} matrices are unitary, viz.,

$$\sum_M (D^J_{MK'})^* D^J_{MK} = \delta_{KK'} \qquad \text{and} \qquad \sum_K (D^J_{M'K})^* D^J_{MK} = \delta_{MM'},$$

from which it follows that

$$|JM\rangle = \sum_K D^J_{MK}(\theta_a, \theta_b, \theta_c)\,|JK\rangle'. \tag{6-63}$$

The functions $D^{J*}_{MK}(\theta_a, \theta_b, \theta_c)$ obey the orthonormality condition

$$\int \sin\theta_b\, d\theta_b\, d\theta_a\, d\theta_c\, D^{J*}_{MK}(\theta_a, \theta_b, \theta_c)\, D^{J'}_{M'K'}(\theta_a, \theta_b, \theta_c) = \frac{8\pi^2}{2J+1}\,\delta_{MM'}\,\delta_{KK'}\,\delta_{JJ'}.$$

Simple examples of (6-61) and (6-63) are for the spinless case $J = L$, as in

[1] Unfortunately, a profusion of phase conventions exist in the literature for the rotation matrices $D^J_{MK}(\theta_a, \theta_b, \theta_c)$. The reader must simply beware. Here we use the complex conjugate of the matrices defined by Rose (1957), viz. $D^J_{MK} = D^{J*}_{MK}(\text{Rose})$. Our convention also follows that adopted by Rowe (1970).

the present instance of the symmetric top, whence

$$Y_L{}^K(\theta', \varphi') = \sum_M D^{L*}_{MK}(\theta_a, \theta_b, \theta_c) Y_L{}^M(\theta, \varphi)$$

and

$$Y_L{}^M(\theta, \varphi) = \sum_K D^L_{MK}(\theta_a, \theta_b, \theta_c) Y_L{}^K(\theta', \varphi').$$

Returning to our discussions of the solution (6-58), we now note that the functions D^L_{MK} have important symmetry properties. A rotation through an angle ϕ about the z' axis, transforming the Euler angles into θ_a, θ_b, and $\theta_c + \phi$, gives

$$D^L_{MK}(\theta_a, \theta_b, \theta_c + \phi) = e^{iK\phi} D^L_{MK}(\theta_a, \theta_b, \theta_c). \tag{6-64}$$

Our eigenfunction for the symmetric top must be invariant to any arbitrary rotation about the symmetry axis z', hence $K = 0$. This simply reflects the physical fact that it is impossible to uniquely specify the orientation of the x' and y' axes for a system possessing complete axial symmetry (along the z' axis). Thus the angular momentum component along the z' axis must be zero, while, of course, $\mathscr{I}_{z'}$ need not be zero. The reason for retaining K up to this point is that when we finally consider coupling particle and collective modes, (6-64) will have altogether different consequences (see below).

A further important symmetry property required by our canonical symmetric top is that a rotation through π about any arbitrary axis through the center of the top perpendicular to the symmetry axis z' should leave the eigenfunction invariant. Selecting the x' axis, the Euler angles transform to $\theta_a + \pi$, $\pi - \theta_b$, and θ_c, which gives

$$D^L_{MK}(\theta_a + \pi,\ \pi - \theta_b,\ \theta_c) = e^{\pi i(L+K)} D^L_{M,-K}(\theta_a, \theta_b, \theta_c). \tag{6-65}$$

Clearly $L + K =$ even integer, or, since from (6-64) $K = 0$, we have $L =$ even integer. Since the rotation of π about the x' axis is equivalent to the parity operation for the symmetric rotator, the states with $L =$ even integer have even parity.[1]

In conclusion, we have for the canonical symmetric top or symmetric rigid rotator: the total angular momentum and parity $J^\pi = L^+$, $L = 0, 2, 4, 6, \ldots$ all even integers; $J_z = M\hbar$, $M = -L, -L+1, \ldots, 0, \ldots, L-1, L$; and $J_{z'} = 0$ and $K = 0$. The energy eigenvalues are

$$T_{\text{symm top}} = (\hbar^2/2\mathscr{I}_0) J(J+1), \tag{6-66a}$$

[1] In principle the condition (6-65) also could be considered to generate a class of anti-symmetric odd-parity states with $L =$ odd integer. We shall return to this point later when we identify the rotational states with surface phonon states and equate the lowest level with the intrinsic nuclear ground state.

and[1]

$$\psi_{J,M}(\theta_a, \theta_b, \theta_c) = \left(\frac{2J+1}{8\pi^2}\right)^{1/2} D^J_{M0}(\theta_a, \theta_b, \theta_c). \tag{6-66b}$$

Again it should be emphasized that (6-64) and (6-65) will have quite different consequences when we consider the more realistic nuclear case with coupled particle and collective modes present.

E. COLLECTIVE STATES OF DEFORMED NUCLEI

In an earlier section we considered the vibrational excitation of the spherically symmetric collective nuclear system. The classical hydrodynamic model treatment given there can be readily extended to the case of permanently deformed systems. We consider describing the permanent nuclear deformation by introducing a set of principal axes fixed in the nucleus. The orientation of these nuclear axes is described in the laboratory frame through the Eulerian angles $(\theta_a, \theta_b, \theta_c)$.

We continue to describe the nuclear surface in the laboratory frame through (6-1), which for the case of quadrupole excitations is

$$S(R, \theta, \varphi, t) = R\left[1 + \sum_\mu \alpha_{2\mu}(t) Y_2{}^\mu(\theta, \varphi)\right]. \tag{6-67}$$

This same surface can also be described in the body-centered frame

$$S(R, \theta', \varphi', t) = R\left[1 + \sum_\nu \alpha_{2\nu}(t) Y_2{}^\nu(\theta', \varphi')\right]. \tag{6-68}$$

In these surface equations we are allowing for the possibility that $\alpha_{2\mu}(t)$ can oscillate about a nonzero fixed value $\langle\alpha_{2\mu}(t)\rangle$ given by the permanent deformation.

If we consider the case of a rotating, permanently deformed nucleus, the more complicated time-dependent required set of $\alpha_{2\mu}(t)$ in the laboratory frame appears as a simple constant set of $\alpha_{2\nu}$ in the body-centered frame with the Eulerian angles displaying a time dependence appropriate to the nature of the rotation. The second description is fundamentally preferable in that it more simply describes the character of the motion. With such examples in

[1] It can be shown that $D^L_{M0}(\theta_a, \theta_b, \theta_c) = [4\pi/(2L+1)]^{1/2} Y_L^{M*}(\theta_b, \theta_a)$; recall that $0 \leqslant \theta_b \leqslant \pi$ and $0 \leqslant \theta_a \leqslant 2\pi$. While in the case $K = 0$, θ_c is a harmless redundant variable, it does appear in the normalization integral, and since $0 \leqslant \theta_c \leqslant 2\pi$, we obtain an additional factor $(2\pi)^{1/2}$. Thus we also can write $\psi_{J,M}(\theta_a, \theta_b, \theta_c = 0) = (2\pi)^{-1/2} Y_L^{M*}(\theta_b, \theta_a)$.

mind we relate the sets $\alpha_{2\mu}$ and $\alpha_{2\nu}$.[1] Since any point in space can be described by either (6-67) or (6-68), we have

$$\sum_{\mu} \alpha_{2\mu} Y_2^{\mu}(\theta, \varphi) = \sum_{\nu} \alpha_{2\nu} Y_2^{\nu}(\theta' \varphi'). \tag{6-69}$$

In view of (6-61) we thus have

$$\alpha_{2\nu} = \sum_{\mu} \alpha_{2\mu} D^2_{\mu\nu} \tag{6-70a}$$

and

$$\alpha_{2\mu} = \sum_{\nu} \alpha_{2\nu} D^{2*}_{\mu\nu}. \tag{6-70b}$$

Since the body-centered frame was selected as the principal axes, we have for the set $\alpha_{2\nu}$ in this frame $\alpha_{21} = \alpha_{2,-1} = 0$ and $\alpha_{22} = \alpha_{2,-2}$. In this situation we have no coordinate redundancies since the set of five independent $\alpha_{2\mu}$ in the laboratory frame is replaced by the two independent $\alpha_{2\nu}$, namely, α_{20} and α_{22}, and the three independent Eulerian angles θ_a, θ_b, and θ_c. It is convenient to replace the variables α_{20} and α_{22} by β and γ defined as

$$\alpha_{20} = \beta \cos\gamma, \qquad \alpha_{22} = (\beta/\sqrt{2}) \sin\gamma. \tag{6-71}$$

From (6-69) and (6-71), we have[2]

$$\sum_{\mu} |\alpha_{2\mu}|^2 = \sum_{\nu} |\alpha_{2\nu}|^2 = \sum_{\nu} \alpha_{2\nu}^2 = \beta^2. \tag{6-72}$$

In view of (6-72) the potential energy in the collective Hamiltonian (6-25) (dropping the subscript $\lambda = 2$ on C_λ and B_λ) is simply

$$V_C = \tfrac{1}{2} \sum_{\mu} C |\alpha_{2\mu}|^2 = \tfrac{1}{2} C\beta^2. \tag{6-73}$$

While β^2 is seen to determine the total mean square deformation and the collective potential energy, the significance of the γ coordinate is best interpreted by noting the surface extension at select values of (θ', φ') corresponding

[1] The literature uniformly adopts the unfortunate convention of using what normally are considered dummy summation indices μ and ν to label the reference frames, with μ relating to the laboratory frame and ν to the body-centered frame. With apologies to the reader, we follow this convention.

[2] We simply integrate the absolute value squared of the right- and left-hand sides of (6-69) over all angles and note the orthogonality of the Y_l^m.

Note that the operator $a^2_{\lambda=2}$ defined in (6-40) is identically equal to the present dynamical variable β^2. The literature convention spoken of in connection with (6-41) follows for the ground state $N_2 = 0$, hence the symbol $\beta_0{}^2$ appearing in the literature.

to the intersection of the x', y', and z' axes with the surface, viz.,

$$S_1 = S(\theta', \varphi') = S\left(\frac{\pi}{2}, 0\right) = R\left[1 + \left(\frac{5}{4\pi}\right)^{1/2} \beta \cos\left(\gamma - \frac{2\pi}{3}\right)\right]; \qquad x' \text{ axis},$$

$$S_2 = S\left(\frac{\pi}{2}, \frac{\pi}{2}\right) = R\left[1 + \left(\frac{5}{4\pi}\right)^{1/2} \beta \cos\left(\gamma - \frac{4\pi}{3}\right)\right]; \qquad y' \text{ axis},$$

$$S_3 = S(0, \varphi') = R\left[1 + \left(\frac{5}{4\pi}\right)^{1/2} \beta \cos\gamma\right]; \qquad z' \text{ axis},$$

or, in summary,

$$S_\xi = R\left[1 + \left(\frac{5}{4\pi}\right)^{1/2} \beta \cos\left(\gamma - \xi\frac{2\pi}{3}\right)\right], \qquad \xi = 1, 2, 3. \tag{6-74}$$

▶**Exercise 6-12** (a) Verify (6-74).
(b) Show that:

(i) $\gamma = 0$ yields a prolate spheroid with the z' axis as the axis of symmetry;
(ii) $\gamma = 2\pi/3$ and $4\pi/3$ yield prolate spheroids with the x' and y' axes as the axes of symmetry;
(iii) $\gamma = \pi$, $\pi/3$, and $5\pi/3$ lead to corresponding oblate spheroids;
(iv) $\gamma =$ not a multiple of $\pi/3$ yields an asymmetric figure. ◀

We now turn to calculating the classical kinetic energy for the collective excitation in the laboratory system expressed in terms of β, γ, and the Eulerian angles θ_a, θ_b, and θ_c. From (6-25) and (6-70b), we write (again for $\lambda = 2$),

$$T_C = \tfrac{1}{2}B\sum_\mu |\dot{\alpha}_{2\mu}|^2 \qquad \text{or} \qquad T_C = \tfrac{1}{2}B\sum_\mu \left|\sum_\nu (\dot{\alpha}_{2\nu} D^{2*}_{\mu\nu} + \alpha_{2\nu}\dot{D}^{2*}_{\mu\nu})\right|^2. \tag{6-75}$$

The time derivative of the rotation matrix can be evaluated most simply by decomposing the rotation $\boldsymbol{\theta}$, which carries the laboratory reference frame (x, y, z) into coincidence with the body-fixed frame (x', y', z'), into components θ_ξ along the body-fixed axes. Then with $\xi \equiv 1, 2, 3$ referring to the x', y', z' axes, respectively, and L_ξ representing the angular momentum components in this frame, and defining $\omega_\xi = \dot{\theta}_\xi$, we have, using the basic definition of the rotation matrices,

$$(\partial/\partial t)\, D^{L*}_{MK} = (\partial/\partial t)\langle L, M|\exp(-i\boldsymbol{\theta}\cdot\mathbf{L})|L, K\rangle$$

$$= -i\sum_{\xi=1}^{3} \omega_\xi \langle L, M|L_\xi \exp(-i\boldsymbol{\theta}\cdot\mathbf{L})|L, K\rangle$$

or[1]

$$(\partial/\partial t)\, D^{L*}_{MK} = -i \sum_{\xi=1}^{3} \sum_{Q=-L}^{L} \langle L, M|L_\xi|L, Q\rangle\, D^{L*}_{QK}\,\omega_\xi. \qquad (6\text{-}76)$$

Here the ω_ξ are the body-fixed components of the angular velocity of the body, which can be expressed in terms of the Eulerian angles:

$$\begin{aligned} \omega_1 &\equiv \omega_{x'} = \dot\theta_b \sin\theta_c + \dot\theta_a \sin\theta_b \cos\theta_c \\ \omega_2 &\equiv \omega_{y'} = \dot\theta_b \cos\theta_c - \dot\theta_a \sin\theta_b \sin\theta_c \\ \omega_3 &\equiv \omega_{z'} = \dot\theta_c - \dot\theta_a \cos\theta_b . \end{aligned} \qquad (6\text{-}77)$$

When (6-76) is used in (6-75) and a number of double sums are evaluated using the properties of the D^L_{KM} functions, an intermediate equation is reached,

$$T_C = \tfrac{1}{2}B\left[\sum_\nu |\dot\alpha_{2\nu}|^2 + \sum_{\xi,\nu,\sigma} (-1)^\sigma \langle 2,\nu|\, L_\xi{}^2\, |2,-\sigma\rangle\, \alpha_{2\nu}\alpha_{2\sigma}\,\omega_\xi{}^2\right]. \qquad (6\text{-}78)$$

We define the effective moments of inertia

$$\mathscr{I}_\xi = B \sum_{\nu,\sigma} (-1)^\sigma \langle 2,\nu|\, L_\xi{}^2\, |2,-\sigma\rangle\, \alpha_{2\nu}\alpha_{2\sigma} \qquad (6\text{-}79)$$

in terms of which

$$T_C = \tfrac{1}{2}B \sum_\nu |\dot\alpha_{2\nu}|^2 + \tfrac{1}{2}\sum_\xi \mathscr{I}_\xi\,\omega_\xi{}^2. \qquad (6\text{-}80)$$

The moments of inertia in terms of $(\alpha_{20}, \alpha_{22})$ or (β, γ) become[2]

$$\mathscr{I}_1 = \mathscr{I}_{x'} = B(3\alpha_{20}^2 + 2\sqrt{6}\,\alpha_{20}\alpha_{22} + 2\alpha_{22}^2) = 4B\beta^2 \sin^2(\gamma - \tfrac{2}{3}\pi) \qquad (6\text{-}81a)$$

$$\mathscr{I}_2 = \mathscr{I}_{y'} = B(3\alpha_{20}^2 - 2\sqrt{6}\,\alpha_{20}\alpha_{22} + 2\alpha_{22}^2) = 4B\beta^2 \sin^2(\gamma - \tfrac{4}{3}\pi) \qquad (6\text{-}81b)$$

$$\mathscr{I}_3 = \mathscr{I}_{z'} = 8B\alpha_{22}^2 = 4B\beta^2 \sin^2\gamma \qquad (6\text{-}81c)$$

or, in summary,

$$\mathscr{I}_\xi = 4B\beta^2 \sin^2(\gamma - \tfrac{2}{3}\xi\pi). \qquad (6\text{-}81d)$$

The sum in the first term of (6-80) is simply

$$\sum_\nu |\dot\alpha_{2\nu}|^2 = (\dot\alpha_{20})^2 + 2(\dot\alpha_{22})^2 = \dot\beta^2 + \beta^2\dot\gamma^2. \qquad (6\text{-}82)$$

[1] We use the closure condition on the set $|L, Q\rangle$, viz.,

$$\begin{aligned} \langle L, M|L_\xi \exp(-i\boldsymbol{\theta}\cdot\mathbf{L})|L, K\rangle &= \sum_Q \langle L, M|L_\xi|L, Q\rangle\langle L, Q|\exp(-i\boldsymbol{\theta}\cdot\mathbf{L})|L, K\rangle \\ &= \sum_Q \langle L, M|L_\xi|L, Q\rangle D^{L*}_{QK}. \end{aligned}$$

[2] For details of the derivation, see, for example, Eisenberg and Greiner (1970, Vol. I, Chapter 5).

Thus the total Hamiltonian becomes (for $\lambda = 2$)

$$H_C = \tfrac{1}{2}B(\dot{\beta}^2 + \beta^2\dot{\gamma}^2) + \tfrac{1}{2}C\beta^2 + \tfrac{1}{2}\sum_{\xi=1}^{3} \mathscr{I}_\xi \omega_\xi^{\,2}. \tag{6-83}$$

We recall that all the quantities in (6-83) refer to the body-centered frame with $\xi = 1 \equiv x'$, $\xi = 2 \equiv y'$, and $\xi = 3 \equiv z'$. While the derivation of (6-83) utilized the angular momentum operators to derive suitable properties of the D^L_{MK} functions, the Hamiltonian (6-83) is nonetheless classical. Although this Hamiltonian is of the form $H_C = H_C(q, \dot{q})$ with

$$(q_1, q_2, \ldots, q_5) \equiv (\beta, \gamma, \theta_a, \theta_b, \theta_c),$$

these q_i and $\dot{q}_i$ are not orthogonal coordinates and proper conjugate momenta and hence the transformation of (6-83) to a proper Schrödinger operator is not simple. In addition, of course, $\mathscr{I}_\xi$ is a function of (β, γ) producing vibrational–rotational coupling terms.

The classical Hamiltonian (6-83) is, however, useful in discussing vibrational–rotational systems in an approximate lowest order. The first two terms can be viewed as a pure vibrational Hamiltonian, while the last term can be used as an approximately pure rotational Hamiltonian.

At least formally we can write

$$H_C = H_{\beta,\gamma} + T_{\text{rot}} \tag{6-84a}$$

with

$$H_{\beta,\gamma} = \tfrac{1}{2}B_2(\dot{\beta}^2 + \beta^2\dot{\gamma}^2) + \tfrac{1}{2}C\beta^2 \tag{6-84b}$$

and

$$T_{\text{rot}} = \tfrac{1}{2}\sum_{\xi=1}^{3} \mathscr{I}_\xi \omega_\xi^{\,2} = \tfrac{1}{2}\hbar^2 \sum_{\xi=1}^{3} (L_\xi^{\,2}/\mathscr{I}_\xi). \tag{6-84c}$$

Although (6-84) was derived for an irrotational hydrodynamic model and rigid body motion is *not* irrotational, the restriction $\dot{\beta} = \dot{\gamma} = 0$, which would be appropriate for a rigid body, does lead to very similar behavior. This result readily follows from the particularly simple form of (6-84). Consider a permanently deformed rotating nucleus, with $\dot{\beta} = 0$ and $\dot{\gamma} = 0$ in the body-centered frame characterized by the average or expectation value of $\langle\beta\rangle = \beta_0$ and $\langle\gamma\rangle = \gamma_0$.[1] This would lead to effective moments of inertia $\mathscr{I}_\xi \equiv \langle\mathscr{I}_\xi\rangle$ which, while being constants, i.e., $\beta \equiv \beta_0$ and $\gamma \equiv \gamma_0$ inserted in (6-81), would be quite different in magnitude than the moments obtained for a rigid body of the same density confined within the deformed surface $S(\beta_0, \gamma_0)$ of (6-74). Nevertheless, (6-84) has the form of a Hamiltonian for an asymmetric top. The now trivial constant term $\tfrac{1}{2}C\beta_0^{\,2}$ loses its significance (more appropriately, it should be replaced by the work done to arrive at the

[1] The symbol β_0 is rather overworked in the literature. Care must be taken not to confuse the present usage with the case a_{20}^2 of (6-41), which is also written $\beta_0^{\,2}$ in the literature.

equilibrium deformed shape), and can be considered absorbed in the intrinsic or internal energy of the nuclear ground state. If in addition to $\dot{\beta} = \dot{\gamma} = 0$ we also have $\gamma_0 = 0$ or π, a symmetric rotator results with $\mathscr{I}_{z'} = 0$ and the discussion of Section VI, D pertains. As for the rigid symmetric rotator, we have $L_{z'} = 0$. Of course, for the rigid rotator, $\mathscr{I}_{z'}$ is zero only in the extreme case of all the mass concentrated along the z' axis. We defer to the next section the discussion of quantum mechanical solutions of the rotational Hamiltonian under the hydrodynamic model for both the symmetric and asymmetric cases.

If quadrupole vibrational modes are permitted about an axially symmetric, permenently deformed shape $\langle\beta\rangle = \beta_0$ and $\langle\gamma\rangle = \gamma_0 = 0$ or π, a critical parameter involves the measure of the vibrational amplitudes about the equilibrium nuclear surface $S(\beta_0, \gamma_0 = 0)$. The body-centered coordinates are written $\alpha_{20} = \beta_0 + \zeta$ and $\alpha_{22} = \alpha_{2,-2} = 0$, and we assume a *locally harmonic* potential energy $V = \frac{1}{2}C\zeta^2$. Customarily a parameter μ^2 is defined,

$$\mu^2 = 2\langle 0|\zeta^2|0\rangle/\beta_0{}^2 = \hbar/\beta_0{}^2(BC)^{1/2} \tag{6-85}$$

where $\langle 0|\zeta^2|0\rangle$ is just the square of the zero-point vibrational amplitude about the equilibrium surface (noting that body-centered β vibrations correspond to a one-dimensional oscillator). For small enough values of μ^2, the proper quantum mechanical solutions to (6-84) are rotation band states built on various β-vibrational modes with little vibrational–rotational coupling. Clearly when $\mu^2 = 0$ an exact adiabatic condition obtains. The parameter μ is referred to as *the nonadiabaticity parameter*. The rotational states for small μ resemble those of the rigid rotator with $\mathscr{I}_{x'} = \mathscr{I}_{y'} = \mathscr{I}_0 = 3B\beta_0{}^2$ and $\mathscr{I}_{z'} = 0$. Empirical evidence suggests, however, values of $\mu \approx \frac{1}{3}$ and an adiabatic approximation can be expected to have nontrivial errors (see below).

It is useful to compare the hydrodynamic moments of inertia based on (6-81), with $\beta \equiv \beta_0$ and $\gamma \equiv \gamma_0 = 0$, with the moment of inertia of a rigid sphere of radius R and the same density, viz.,

$$\mathscr{I}_{\text{rigid}} = \tfrac{2}{5} m_{\text{N}} A R^2 = (16\pi/15)B$$

to give[1]

$$\mathscr{I}_0 = (45/16\pi)\,\beta_0{}^2 \mathscr{I}_{\text{rigid}}. \tag{6-86}$$

Even for the most deformed nuclei, $\beta_0{}^2 \approx \frac{1}{10}$ and thus, as expected, the prediction of (6-86) again illustrates that the nuclear rotation actually carries very little collective mass with it. The rotational effect can be viewed as a wave coursing over the nuclear surface with an associated angular momentum.

[1] This computation is purely for guidance concerning numerical magnitudes for the moments of inertia. A perfectly spherical nucleus of course would have all three axes x', y', and z' as symmetry axes and hence $\langle \mathbf{L}^2 \rangle = 0$ and no rotational excitation is possible.

Although empirical evidence suggests that the moments of inertia $\mathscr{I}_{x'}$ and $\mathscr{I}_{y'}$ are seriously underestimated by (6-81) (generally by a factor of over four), there appears to be a validity to the prediction that $\mathscr{I}_{z'}$ in many instances is very small if not identically zero. To a large extent this can be traced to the rotational invariance about the body-centered symmetry axis of the nuclear potential in which all the nucleons move.

1. Rotational Excitations

Landmark papers in giving a detailed quantum mechanical analysis of collective nuclear behavior derived from the classical Hamiltonian (6-83) are by Davydov and co-workers (Davydov and Filippov, 1958; Davydov and Rostovsky, 1959; Davydov and Chaban, 1960) and Faessler and Greiner (1962). More recent summary reviews are contained in Eisenberg and Greiner (1970) and Rowe (1970). It will suit our purpose to use (6-83) and (6-84) in lowest order adiabatic approximation to give guidance in classifying a number of striking features in the spectroscopy of deformed nuclei. In spite of the fact that typical nonadiabaticity factors μ substantially differ from zero, such a classification is not without virtue. We shall also give a very brief description of the more exact treatments including nonadiabatic effects for completeness.

It again proves useful to consider the case of even–even nuclei first.[1] Since it is easier to excite rotational states in such nuclei than vibrational modes, we examine rotational modes first, beginning with the symmetric case $\mathscr{I}_{x'} = \mathscr{I}_{y'}$. According to (6-84c), the rotational energy term with $\mathscr{I}_{x'} = \mathscr{I}_{y'}$ would give rise to a rotational excitation band similar to (6-66a). Since the ground state of even–even nuclei are $J^\pi = 0^+$, the ground state of (6-66a) could in fact be associated with the intrinsic nuclear ground state. Thus the band would consist of the states $0^+, 2^+, 4^+\ 6^+, \ldots$ in order of increasing excitation, with

$$E_J = (\hbar^2/2\mathscr{I}_0)J(J+1), \qquad J = 0, 2, 4, \ldots. \tag{6-87}$$

A next order of approximation is possible in treating the Hamiltonian (6-84) for the hydrodynamic case. Since the vibrational frequencies ω_β are higher than the rotational frequencies, at least for low excitations, we might expect the collective elastic stretching implied in the potential energy term $\frac{1}{2}C\beta^2$ to adiabatically accommodate to some extent at least to the rotational motion, and thus give rise to a centrifugal stretching and a subsequent increase in the moment of inertia with angular momentum. This stretching accommodation would be nonoscillatory and occur with a relaxation time constant presumably of the order of $2\pi/\omega_\beta$. We write the potential energy for small

[1] An illuminating paper discussing such nuclei, which may be profitably referred to, is by Sheline (1960).

β deformations about the equilibrium value β_0 as

$$V(\beta) = E_0 + \tfrac{1}{2}C(\beta-\beta_0)^2, \tag{6-88}$$

where E_0 is the intrinsic equilibrium energy. The total energy when hydrodynamic rotational excitation modes are considered with specific allowance for (6-88) is

$$E(\beta) = E_0 + \tfrac{1}{2}C(\beta-\beta_0)^2 + (\hbar^2/6B\beta^2)J(J+1). \tag{6-89}$$

We treat the deformation coordinate β as a variational parameter in (6-89) and minimize $E(\beta)$ for each specific value of J. Taking $\partial E(\beta)/\partial\beta = 0$ gives

$$\beta \approx \beta_0\left[1+12\left(\frac{\hbar^2}{2\mathscr{I}_0}\right)^2 \frac{J(J+1)}{(\hbar\omega_\beta)^2}\right], \tag{6-90}$$

where we have written $\omega_\beta{}^2 = C/B$ for the angular frequency associated with β vibrations and $\mathscr{I}_0 = 3B\beta_0{}^2$ refers to the unperturbed moment of inertia. To this approximation we also have the stretched moment of inertia

$$\mathscr{I} \approx \mathscr{I}_0\left[1+24\left(\frac{\hbar^2}{2\mathscr{I}_0}\right)^2 \frac{J(J+1)}{(\hbar\omega_\beta)^2}\right]. \tag{6-91}$$

When (6-90) and (6-91) are substituted into (6-89), the result is

$$E_J = E_0 + E_a J(J+1) - E_b[J(J+1)]^2 \tag{6-92a}$$

with

$$E_a = \frac{\hbar^2}{2\mathscr{I}_0} = \frac{\hbar^2}{6B\beta_0{}^2} = \frac{4\pi}{9}\frac{\hbar^2}{m_N R_0{}^2}\frac{A^{-5/3}}{\beta_0{}^2} \tag{6-92b}$$

and

$$E_b = 12\left(\frac{\hbar^2}{2\mathscr{I}_0}\right)^3 \frac{1}{(\hbar\omega_\beta)^2} = \frac{12}{(\hbar\omega_\beta)^2}E_a{}^3. \tag{6-92c}$$

Thus, while the equilibrium energy for the ground state, i.e., with $J=0$, remains E_0 as is immediately clear from (6-88) and (6-89), the effective rotational energy assumes the more complicated form (6-92). The added term $E_b[J(J+1)]^2$ comes both from the change in potential energy and the increased moment of inertia. Again E_0 can be considered part of the intrinsic nuclear ground-state energy.

When we also allow for an asymmetric or γ mode of stretching, we can extend (6-88) to yield

$$V(\beta,\gamma) = E_0 + \tfrac{1}{2}C^\beta(\beta-\beta_0)^2 + \tfrac{1}{2}C^\gamma\gamma^2. \tag{6-93}$$

We introduce the stiffness and mass factors C^β, C^γ and B^β, B^γ to distinguish the two modes and facilitate any empirical fitting. This expression supposes the collective ground state to again be axial symmetric with equilibrium values $\langle\beta\rangle \equiv \beta_0$ and $\langle\gamma\rangle \equiv \gamma_0 = 0$. The analysis is now somewhat more complicated but a rather simple result follows, viz.,

$$E = E_0 + E_a J(J+1) - (E_b{}^\beta + E_b{}^\gamma)[J(J+1)]^2, \tag{6-94a}$$

with

$$E_a = \hbar^2/2\mathscr{I}_0, \tag{6-94b}$$

$$E_b{}^\beta = 12(\hbar^2/2\mathscr{I}_0)^3(\hbar\omega_\beta)^{-2}, \tag{6-94c}$$

$$E_b{}^\gamma = 4(\hbar^2/2\mathscr{I}_0)^3(\hbar\omega_\gamma)^{-2}, \tag{6-94d}$$

and where $\mathscr{I}_0 = 3B^\beta\beta_0{}^2$, $\omega_\beta{}^2 = C^\beta/B^\beta$, and $\omega_\gamma{}^2 = C^\gamma/B^\gamma$.[1]

▶**Exercise 6-13** (a) Verify (6-90)–(6-92).

(b) In a proposed variable-moment-of-inertia model [see Mariscotti *et al.* (1969), Scharff-Goldhaber and Goldhaber (1970), and Gupta (1973)] a heuristic equivalent to (6-89) is assumed of the form

$$E(\mathscr{I}) = E_0 + \tfrac{1}{2}D(\mathscr{I} - \mathscr{I}_0)^2 + (1/2\mathscr{I})J(J+1),$$

with $\mathscr{I}$ in units of $\hbar^2$. Here $\mathscr{I}_0$ and D are fixed parameters empirically fitted to any particular nucleus and represent respectively the ground-state moment of inertia and an effective "restoring force constant." Show that if $\partial E(\mathscr{I})/\partial\mathscr{I} = 0$ is used to determine the equilibrium energy for each value of J, the corresponding value of $\mathscr{I}_J$ is given by the cubic equation

$$\mathscr{I}_J{}^3 - \mathscr{I}_J{}^2\mathscr{I}_0 - [J(J+1)/2D] = 0.$$

Also show that in terms of this equilibrium value of $\mathscr{I}_J$ the rotational energy becomes

$$E_J = E_0 + \frac{J(J+1)}{2\mathscr{I}_J}\left[1 + \frac{J(J+1)}{4D\mathscr{I}_J{}^3}\right].$$

[A unified description of all phenomenological "stretching" energy corrections to the ground-state rotational band of even–even nuclei is given by Mantri and Sood (1973). See also Saethre *et al.* (1973). The variable-moment-of-inertia model considered in this exercise is just one of many proposed two- and three-parameter semiempirical models.]

(c) Show that the irrotational hydrodynamic model for $R_0 = 1.38$ fm ($R = R_0 A^{1/3}$) gives

$$E_a \approx (30.3/\beta_0{}^2)A^{-5/3}\ \text{MeV}. \quad ◀ \tag{6-95}$$

A good example illustrating the simple application of (6-94) might appear to be the highly deformed nucleus ^{170}Hf, the level scheme for which is given

[1] For a derivation of (6-94) refer, for example, to Davidson (1968, Chapter 3).

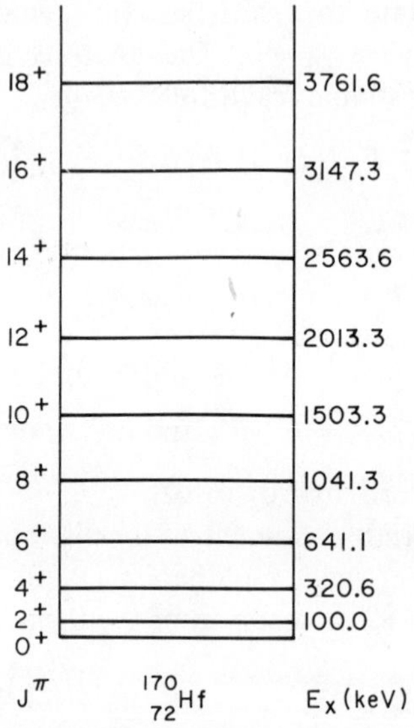

FIG. 6.9. The observed ground-state rotational band for $^{170}_{72}$Hf.

in Fig. 6.9. This figure is based on the experimental results of Stephens *et al.* (1965).[1] Rotational levels with J^π up to 14^+ were also reported for ^{164,166}Yb, ^{166,168,172}Hf, and ^{172,174,176}W. A useful tabulation of many experimental rotational band energies is given by Mariscotti *et al.* (1969).

The expressions (6-94) can be used to fit the overall data of Fig. 6.9 to yield less than a 4% error with the values $E_a = 15.93$ keV and $E_b^\beta + E_b^\gamma = 16.1$ eV. Such a fit, however, is totally inappropriate from a physical point of view. The adiabatic approximation should not be applied to the higher J values. We shall presently examine in some detail the appropriate description of the higher angular momentum states. In fact, at very high angular momenta a new phenomenon called "backbending" enters. (This behavior is discussed in Section VI, F.) Using only the levels up to $J^\pi = 4^+$, we get $E_a = 16.94$ keV and $E_b^\beta + E_b^\gamma = 45.4$ eV. These values are typical for the highly deformed rare earth and nearby nuclei with $156 \leqslant A \leqslant 184$.

[1] It should be pointed out that the experiment was specifically biased to produce very high J values by using high-energy heavy-ion-beam reactions. In the case of ^{170}Hf the reaction was ^{165}Ho(^{11}B, 6n)^{170}Hf. In addition to the high angular momentum imparted by the incident ^{11}B ions of 83 MeV, the large ground-state spins ^{165}Ho($J^\pi = \frac{7}{2}^-$) and ^{11}B($J^\pi = \frac{3}{2}^-$) also materially contributed. The six "evaporation" neutrons carried off relatively little angular momentum, leaving ^{170}Hf behind with an angular momentum up to $J \approx 20$. The cascading E2 transitions then were measured in a precision conversion electron spectrometer for the radioactive ^{170}Hf ($t_{1/2} = 12.2$ hr).

For a sample calculation in this nuclear mass range we might select $\beta_0 \approx 0.30$ and $A \approx 170$. The value calculated from (6-95) gives $E_a(\text{calc}) \approx 65$ keV and is thus some four times greater than the observed values. This implies an experimental moment of inertia four times larger than predicted by the irrotational liquid drop model but still less than the rigid body value. We can also estimate $E_b^{\beta} + E_b^{\gamma}$ of (6-94) from the observed values of E_a, $\hbar\omega_\beta$, and $\hbar\omega_\gamma$. Since empirical values of $\hbar\omega_\beta$ and $\hbar\omega_\gamma$ near $A = 170$ are $0.8 \lesssim (\hbar\omega_\beta \approx \hbar\omega_\gamma) \lesssim 1.2$ MeV, we can use $\hbar\omega_\beta \approx \hbar\omega_\gamma \approx 1$ MeV and $E_a \approx 15$ keV to calculate $E_b^{\beta} + E_b^{\gamma} \approx 55$ eV. Observed values of $E_b^{\beta} + E_b^{\gamma}$ generally range about half this predicted value.

Generally the mass range $150 \leqslant A \leqslant 194$ is considered one of the regions encompassing highly deformed nuclei. Nuclei in the subregion $156 \leqslant A \leqslant 184$ cited above follow the simple rotational rule (6-94) with relatively small values of $E_b^{\beta} + E_b^{\gamma} \lesssim 50$ eV, indicating considerable stiffness to β and γ vibrational excitation (i.e., moderately large values of $\hbar\omega_\beta$ and $\hbar\omega_\gamma$ compared to E_a). In the region $150 \leqslant A \leqslant 156$ and $184 \leqslant A \leqslant 194$, with a notable dependence on the exact value of N and Z,[1] the empirically required values for $E_b^{\beta} + E_b^{\gamma}$ rise sharply to values exceeding 100 eV, suggesting an abrupt departure from the simple $J(J+1)$ energy rule. This phenomenon can be viewed as indicative of the onset of a transitional region between the highly deformed, axially symmetric rotational nuclei and the spherical vibrational nuclei that are found just above $A \approx 194$ and below $A \approx 150$. The effect is relatively striking. For example, the β-decay-stable nuclei $^{150}_{62}\text{Sm}$ and $^{152}_{64}\text{Gd}$ both exhibit an apparent vibrational level sequence $0^+, 2^+, (0^+, 2^+, 4^+)$, while $^{152}_{62}\text{Sm}$ and $^{154}_{64}\text{Gd}$ both exhibit a rotational sequence $0^+, 2^+, 4^+, 6^+$. The first 2^+ levels in $^{150}_{62}\text{Sm}$ and $^{152}_{64}\text{Gd}$ are at 0.334 and 0.344 MeV, while the first 2^+ levels in $^{152}_{62}\text{Sm}$ and $^{154}_{64}\text{Gd}$ are at 0.122 and 0.123 MeV. The value $E_1(2^+) \approx 340$ keV can be more or less favorably compared to the systematic single vibrational phonon energies of the spherical nuclei shown in Fig. 6.5, while the value $E_1(2^+) \approx 120$ keV is closer to those of the axially symmetric rotators such as ^{170}Hf shown in Fig. 6.9. The situation at the other mass end near $^{190,192}_{76}\text{Os}$ and $^{190,192}_{78}\text{Pt}$, while similar, is less abrupt.

These transitional nuclei can also be described in terms of an asymmetric rotator model. Davydov and Filippov (1958) solved the adiabatic quantum mechanical problem corresponding to the asymmetric rigid rotator Hamiltonian (6-84c).[2] If, in addition to $\mathscr{I}_1 \neq \mathscr{I}_2 \neq \mathscr{I}_3$, one also invokes the hydrodynamic model employing nonzero static values of β_0 and γ_0 in (6-81), two free parameters result, namely $B\beta_0$ and γ_0. The quantum number K pertaining

[1] Particularly for $N \approx 90$ and $N \approx 114$.

[2] The adiabatic condition arises from the assumption that the (β, γ) vibrational energies $\hbar\omega_\beta$ and $\hbar\omega_\gamma$ are large enough compared to the low-lying rotational levels so that effects produced by their presence can be neglected.

to the z' component of angular momentum is no longer a constant of the motion. It is, however, very useful to use as a basis for the axially symmetric rigid rotator states allowing K its full range of values $|K| \leqslant J$. This procedure is immediately suggested by writing (6-84c) as

$$H_{\text{rot}} = \frac{\hbar^2}{2} \sum_{\xi=1}^{3} \frac{L_\xi^{\ 2}}{\mathscr{I}_\xi} = H_{\text{S}} + H_{\text{A}}, \tag{6-96a}$$

with

$$H_{\text{S}} = \frac{\hbar^2}{2}(\mathbf{L}^2 - L_{z'}^2)\left(\frac{1}{2\mathscr{I}_{x'}} + \frac{1}{2\mathscr{I}_{y'}}\right) + \frac{\hbar^2 L_{z'}^2}{2\mathscr{I}_{z'}} \tag{6-96b}$$

and

$$H_{\text{A}} = \frac{\hbar^2}{2}(L_{x'}^2 - L_{y'}^2)\left(\frac{1}{2\mathscr{I}_{x'}} - \frac{1}{2\mathscr{I}_{y'}}\right). \tag{6-96c}$$

Thus H_{S} is cast in the form of the symmetric rigid rotator Hamiltonian (6-57). The eigenvalue problem is solved by using the basis states of H_{S} to diagonalize H_{rot} including H_{A}. The generalized basis of H_{S} is written

$$|JMK\rangle = \left(\frac{2J+1}{16\pi^2(1+\delta_{K0})}\right)^{1/2} [D_{MK}^J(\theta_a, \theta_b, \theta_c) + (-1)^J D_{M,-K}^J(\theta_a, \theta_b, \theta_c)] \tag{6-97}$$

with $K = 0, 2, 4, 6, \ldots$ and

$$J = L = \begin{cases} K, K+1, K+2, \ldots, & K \neq 0 \\ 0, 2, 4, 6, \ldots, & K = 0. \end{cases}$$

The form (6-97) results when symmetry conditions are imposed requiring invariance under a rotation of π about both the x' and z' axes.[1]

The final solutions to (6-96a) are written

$$\psi_{J,M,i}(\theta_a, \theta_b, \theta_c) = \sum_K A_{K,i}^J(\gamma_0)\,|JMK\rangle, \tag{6-98}$$

where the index i simply enumerates the various eigenvalues possible for a given J in ascending order $1, 2, 3, \ldots$. Diagonalizing (6-96a) with (6-98) determines the coefficients $A_{K,i}^J(\gamma_0)$ and the energy. In general for each J there are as many

[1] We note that (6-66b) required invariance to any arbitrary rotation ϕ about the z' axis, not just $\phi = \pi$. This difference in requirements clearly is governed by the relationship of $\mathscr{I}_{x'}$ to $\mathscr{I}_{y'}$. We also note that for $K = 0$, hence $\delta_{K0} = 1$, (6-97) reduces to (6-66b), as it should. Also, only positive values of K are cited, since for a given (J, M), $|JMK\rangle = (-1)^J |JM, -K\rangle$. Again the invariance to a rotation of π about the x' axis is equivalent to imposing an even parity condition, thus generating only $J^\pi = J^+$ states. This is entirely appropriate for quadrupole deformations of even–even nuclei.

roots to the secular equation as coefficients A_K^J. Since there is one coefficient A_K^J for each allowed value of K for a particular J, we have only one 0^+ state, no 1^+ state, two 2^+ states, one 3^+ state, etc. There are $\frac{1}{2}J+1$ states if J is even and $\frac{1}{2}(J-1)$ states if J is odd.

Computer-generated solutions were obtained by DeMille *et al.* (1959) and the results are shown in Fig. 6.10. For $\gamma_0 = 0$ we of course just recover the

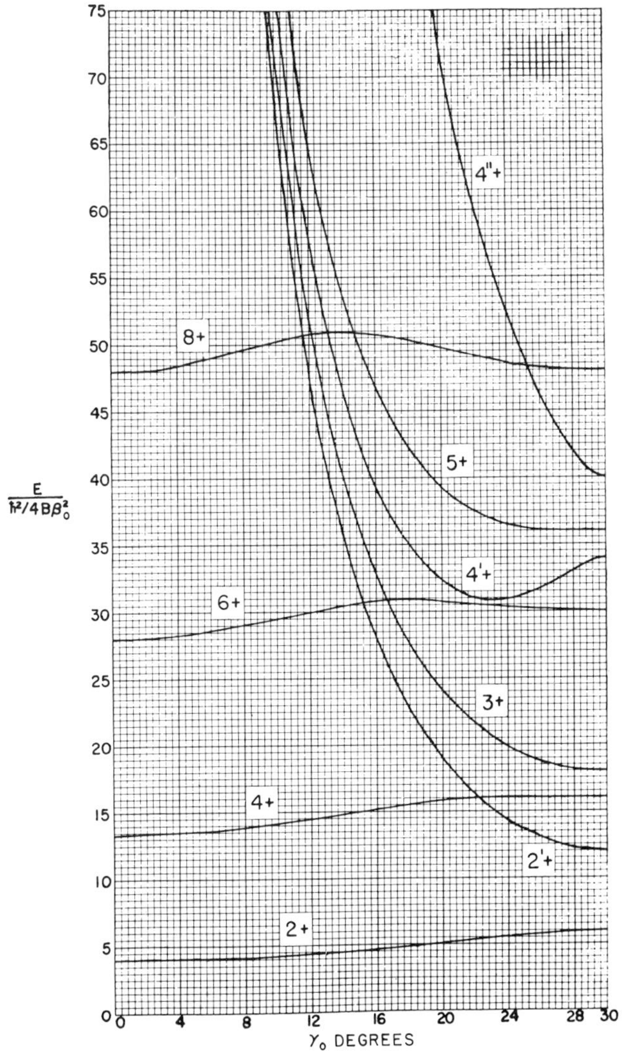

FIG. 6.10. Rotational energy levels of asymmetric, quadrupole-deformed, even–even nuclei [taken from DeMille *et al.* (1959)].

simple $J(J+1)$ rotational sequence of (6-66a). As γ_0 increases from zero, the irregular solutions (i.e., infinite energies generated by $\mathscr{I}_{z'} \to 0$ for $\gamma_0 \to 0$) decrease in energy and beyond $\gamma_0 \approx 15°$ begin to comingle with the low-lying $0^+, 2^+, 4^+, 6^+, \ldots$ sequence. It is interesting to note that for *any* γ_0, $E_1(2^+)+E_2(2^+) = E_1(3^+)$ and $4E_1(2^+)+E_2(2^+) = E_1(5^+)$. If some centrifugal stretching (so far ignored by omitting rotational–vibrational coupling) is permitted, the following rules obtain:

$$\begin{aligned} E_1(2^+) + E_2(2^+) &= E_1(3^+)(1-\Delta_3), \\ 4E_1(2^+) + E_2(2^+) &= E_1(5^+)(1-\Delta_5) \end{aligned} \tag{6-99}$$

with $\Delta_{3,5} \ll 1$. These rules are sometimes used to establish the validity of applying this simple asymmetric model. When $\gamma_0 = \pi/6$ (i.e., 30°), maximum triaxial asymmetry for the hydrodynamic model is reached with $\mathscr{I}_{x'} = 4B{\beta_0}^2$ and $\mathscr{I}_{y'} = \mathscr{I}_{z'} = B{\beta_0}^2$. The level sequence in increasing energy order becomes $0^+\ 2^+, 2^+, 4^+, 3^+, \ldots$. The energy values themselves become $E_1(2^+) = 6E_0$, $E_2(2^+) = 12E_0$, $E_1(4^+) = 16E_0$, and $E_1(3^+) = 18E_0$, with $E_0 = \hbar^2/4B{\beta_0}^2$. Thus, paradoxically, not only does the sequence resemble that for the spherical quadrupole vibrator but, by identifying $6E_0 = \hbar\omega_{\lambda=2}$, the two 2^+ and the 3^+ level energies exactly agree with (6-33) and the 4^+ level is close in energy. The conditions (6-99) in this case also trivially obtain. There is thus some ambiguity introduced in trying to determine whether a more or less equally spaced sequence $0^+, 2^+, (2^+, 4^+), 3^+$ is indicative of a maximally asymmetric rotator or a spherically symmetric vibrator! The presence of a second 0^+ state near the triplet $E_2(2^+)$, $E_1(4^+)$, and $E_1(3^+)$, the exact location of the 3^+ state energy, and carefully determined electromagnetic transition rates are required to make the distinction rather than simplistic energy considerations such as (6-99). For example, an effort to apply the energy eigenvalues of Fig. 6.10 to the presumed spherical nuclei ^{106}Pd and ^{114}Cd of Fig. 6.5 leads to an attempted assignment of $\gamma_0 \approx 28°$ and condition (6-99) gives $\Delta_3 \approx +0.05$ and $+0.02$, respectively.

In the mass range $156 \leqslant A \leqslant 184$, an effort to fit the energies with the eigenvalues of Fig. 6.10 gives values $12° \lesssim \gamma_0 \lesssim 16°$. The values required for γ_0 then rapidly rise toward $\gamma_0 \approx 30°$ in the transitional range $184 \lesssim A \lesssim 194$ and $150 \lesssim A \lesssim 156$. A rather good example for the present model would be the case of $^{192}_{78}$Pt shown in Fig. 6.11. Rule (6-99) is obeyed with $\Delta_3 = +0.0087$. The nucleus $^{192}_{76}$Os is another good example of a triaxial $\gamma_0 \approx 30°$ case, with a level scheme closely resembling Fig. 6.11.

For a very interesting review of possible nuclear potential energy surfaces within the framework of quadrupole deformations but including hypothetical cases for harmonic and anharmonic spherical vibrators with or without deformed excited states, nuclei with deformed ground states but spherical

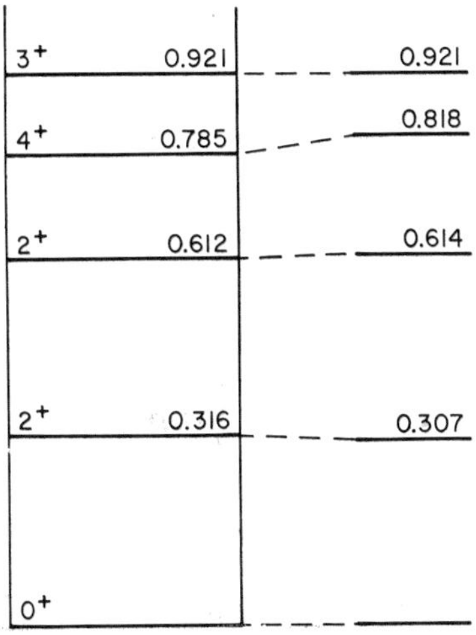

FIG. 6.11. The lowest four excited states of $^{192}_{78}$Pt (experimental, left) and the asymmetric rotator model predictions (right) with $\gamma_0 = \pi/6$ and $\hbar^2/4B\beta_0 = 0.307$ MeV.

excited states, etc., see Greiner and co-workers (Eisenberg and Greiner, 1970, Vol. I, Chapter 4, and Gneuss and Greiner, 1971).

A thorough exploration of the potential energy surface as a function of possible deformation modes should of course include consideration of $\lambda > 2$ deformations. Most of the recent model-calculated potential energy surfaces [e.g., Götz *et al.* (1972) and Pauli (1973)] use realistic Woods–Saxon-shaped single-particle potentials and include hexadecapole deformations [i.e., $\beta_{\lambda=4} Y_4{}^0(\theta, \phi)$ deformation terms] in addition to the quadrupole deformations governed by the parameters $(\beta, \gamma)_{\lambda=2}$.[1] Such calculations generally do not reveal γ-asymmetric ground-state deformations even in the transition mass regions, although exceptions with shallow minima do occur. The admixture of the hexadecapole deformations obviates the need for the γ-asymmetric deformations. In the upper transitional region near $A = 190$, the quadrupole deformation appears to change abruptly from a prolate ($A \lesssim 190$) to an oblate ($A \gtrsim 190$) shape with sizable $\beta_{\lambda=4}$ coefficients present (i.e. $|\beta_{\lambda=4}| \approx \frac{1}{4}|\beta_{\lambda=2}|$). For nuclei beyond ^{208}Pb, octupole ($\lambda = 3$) permanent deformations are also possible [see Möller *et al.* (1972)].

The more or less abrupt change in the nuclear shape in the transitional region also manifests itself in a marked change in the isotope shift discussed in Section II,C. Apparently the behavior of the relatively soft potential energy surface in this region places several possible deformation modes in

[1] Multipoles are referred to by a 2^λ-pole designation; thus for $\lambda = 1$, as dipoles; $\lambda = 2$, as quadrupoles; $\lambda = 3$, as octupoles; $\lambda = 4$, as hexadecapoles, etc.

close competition for the ground-state configuration. Microscopic calculations are necessary to distinguish between alternative possibilities. An excellent and thorough review article by Brack *et al.* (1972) explores the interrelationship between microscopic single-particle structure and bulk collective characteristics. It also includes an exhaustive investigation of nuclear potential energy surfaces employing various shape parameterizations. We shall return to a general discussion of hexadecapole deformations in Section VI, F.

Finally, we should remark that evidence suggests that in the $A = 70$–80 mass region, spherical and deformed shapes may coexist [see Hamilton *et al.* (1974)]. Level characteristics in ^{70,72}Ge and ^{72}Se can be successfully accounted for by assuming a strong coupling of coexisting rotational states, to be associated with deformed shapes, and vibrational states, normally associated with a spherical ground state.

2. Vibrational–Rotational Excitations

A more realistic model in the hydrodynamic framework is the symmetric rotator model allowing for an exact (and therefore nonadiabatic) treatment of coupled (β, γ) vibrations. We again give only a brief outline and refer the reader to the literature for details [e.g., Faessler and Greiner (1962) and Eisenberg and Greiner (1970)]. This model assumes a potential surface resembling that shown schematically in Fig. 6.12, for which we can write, using a harmonic approximation in a Taylor series expansion about the minimum at (β_0, γ_0),

$$V(\zeta, \eta) = \tfrac{1}{2}C_0\zeta^2 + C_2\eta^2, \tag{6-100}$$

where $\alpha_{20} = \beta_0 + \zeta$ and $\alpha_{22} = 0 + \eta$ in the body-centered frame. The kinetic energy is again

$$T = \frac{1}{2}B(\dot{\zeta}^2 + 2\dot{\eta}^2) + \frac{\hbar^2}{2}\sum_{\xi=1}^{3}\frac{L_\xi{}^2}{\mathscr{I}_\xi(\beta_0+\zeta, \eta)}\,. \tag{6-101}$$

When generalized orthogonal canonical coordinates and momenta $q, \dot{q}$ are introduced and a suitable Hamiltonian constructed from (6-100) and (6-101), it can be formally written $H = H_{\text{rot}} + H_{\text{vib}} + H_{\text{vib-rot}}$. The terms H_{rot} and H_{vib} are so selected that they are separable, and so that solutions of the form

$$\psi[(\theta_a, \theta_b, \theta_c), \zeta, \eta] = D^J_{MK}(\theta_a, \theta_b, \theta_c)\,\chi_{K,n_2}(\eta)\,|n_0\rangle \tag{6-102}$$

exist. The usual invariance conditions on rotating the x', y', z' frame give symmetrized solutions,

$$\begin{aligned}|JMKn_2n_0\rangle &= |JMK\rangle\,\chi_{K,n_2}(\eta)\,|n_0\rangle \\ &= \left[\frac{2J+1}{16\pi^2(1+\delta_{K0})}\right]^{1/2}[D^J_{MK}(\theta_a, \theta_b, \theta_c) \\ &\quad + (-1)^J D^J_{M,-K}(\theta_a, \theta_b, \theta_c)]\,\chi_{K,n_2}(\eta)\,|n_0\rangle,\end{aligned} \tag{6-103}$$

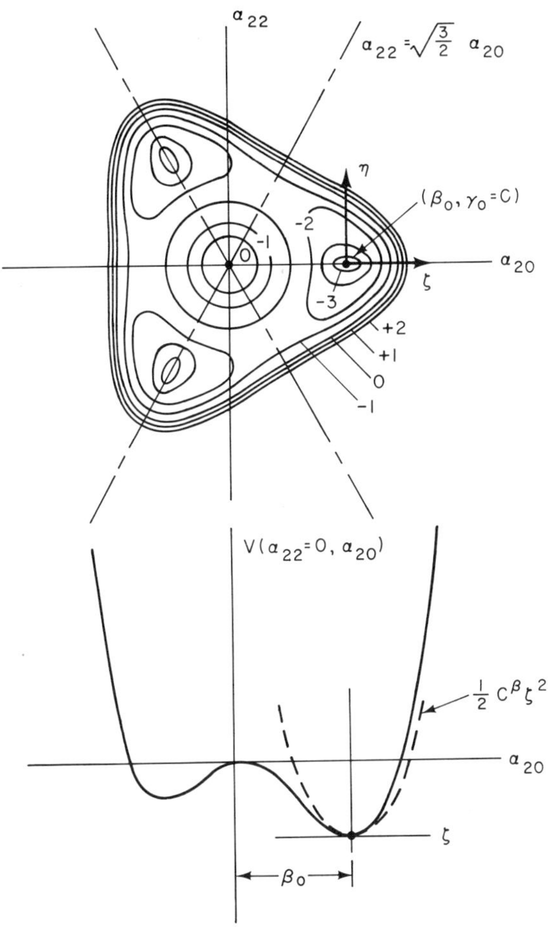

FIG. 6.12. Schematic representation of the potential surface for an axially symmetric vibrational–rotational nucleus. The quantity C^{β} may also be written as C_0 (see text).

with the same conditions on J and K as for (6-97). The β vibrations give rise to simple phonon states $|n_0\rangle$, the subscript on n_0 referring to the second subscript on α_{20}, with energy eigenvalues

$$E(n_0) = (n_0 + \tfrac{1}{2})\hbar\omega_\beta; \qquad \omega_\beta{}^2 = C_0/B. \tag{6-104}$$

The γ vibrational mode involves a centrifugal potential term leading to solutions $\chi_{K,n_2}(\eta)$ that are hypergeometric functions and have energy eigenvalues

$$E_K(n_2) = (\tfrac{1}{2}|K| + 2n_2 + 1)\hbar\omega_\gamma; \qquad \omega_\gamma{}^2 = C_2/B, \tag{6-105}$$

where the subscript on n_2 refers to the second subscript on α_{22}. Since the eigenvalues of $|JMK\rangle$ are those of (6-60), we have for the eigenenergy corresponding to (6-103), and hence for the zeroth-order energies for the complete Hamiltonian,

$$E(J,K,n_2,n_0) = (\hbar^2/2\mathscr{I}_0)[J(J+1)-K^2] + (\tfrac{1}{2}|K|+2n_2+1)\,\hbar\omega_\gamma + (n_0+\tfrac{1}{2})\,\hbar\omega_\beta, \qquad (6\text{-}106)$$

with $\mathscr{I}_0 = 3B\beta_0{}^2$.[1]

When first-order perturbation theory is used to evaluate the effect of the remaining term $H_{\text{vib-rot}}$, omitted to this point, an energy contribution of precisely the form (6-94) results, namely $\Delta E = -(E_b{}^\beta + E_b{}^\gamma)[J(J+1)]^2$, in the limit $\hbar^2/2\mathscr{I}_0 \ll \hbar\omega_{\beta,\gamma}$. For $J \gtrsim 4$, however, an exact diagonalization procedure is necessary with $H_{\text{vib-rot}}$ included and with the functions $|JMKn_2n_0\rangle$ as basis states. The application of exact diagonalization to ^{170}Hf and other nuclei in the region $156 \leqslant A \leqslant 184$ results in a fitting error of generally less than 0.4%. The deduced empirical values of $\hbar\omega_{\beta,\gamma}$ range from about 0.7 to 1.7 MeV, with $\hbar\omega_\beta$ tending to be the somewhat larger of the two. The points cited here justify the comments made earlier when the centrifugal stretching was first discussed in connection with Fig. 6.9.

The zeroth-order energy eigenvalues of (6-106) provide a very convenient way to classify the various rotational bands that can be built on the β and γ vibrational modes. These are shown schematically in Fig. 6.13. The total zero-point energy (i.e., $J=0$, $K=0$, $n_0=n_2=0$) $\hbar\omega_\gamma + \frac{1}{2}\hbar\omega_\beta$ is absorbed in the ground-state energy and not shown in the figure. An appropriate nucleus exhibiting a level structure of the type discussed here is $^{160}_{66}$Dy and is shown in Fig. 6.14. This nucleus exhibits a well-developed ground-state band with $K=0$ and $n_0=n_2=0$ as well as a clear $K=2$ and $n_0=n_2=0$ γ-vibrational band. The band head $K=4$ and $n_0=n_2=0$ γ-vibrational level is possible.

[1] It is trivial to show that in the body-centered frame with ζ, η, and γ small and with $\beta \approx \beta_0$ we have

$$\alpha_{22} = 0+\eta = (\beta/\sqrt{2})\sin\gamma \approx \beta_0\gamma/\sqrt{2}, \qquad \alpha_{20} = \beta_0+\zeta = \beta\cos\gamma \approx \beta.$$

Also,

$$V_{\text{local}} = \tfrac{1}{2}C^\beta(\beta-\beta_0)^2 + \tfrac{1}{2}C^\gamma\gamma^2 = \tfrac{1}{2}C_0\zeta^2 + C_2\eta^2$$

and

$$\text{K.E.} = \tfrac{1}{2}B^\beta\dot{\beta}^2 + \tfrac{1}{2}B^\gamma\dot{\gamma}^2 = \tfrac{1}{2}B_0\dot{\zeta}^2 + B_2\dot{\eta}^2,$$

with $C^\beta = C_0$, $C^\gamma = C_2\beta_0{}^2$, $B^\beta = B_0$, and $B^\gamma = B_2\beta_0{}^2$. Thus

$$\omega_\beta{}^2 = C^\beta/B^\beta = C_0/B_0, \qquad \omega_\gamma{}^2 = C^\gamma/B^\gamma = C_2/B_2,$$

establishing the connection between (6-100), (6-101), (6-104), and (6-105) on the one hand and (6-93) and (6-94) on the other hand if $B_2 = B_0 = B$.

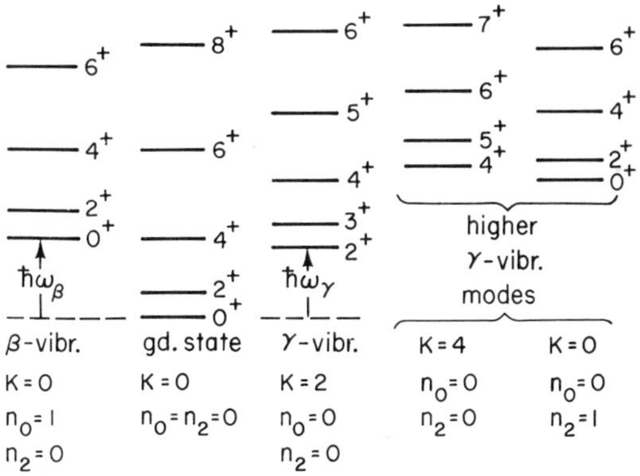

FIG. 6.13. Rotational bands with $J = K, K+1, K+2, \ldots$ for $K \neq 0$ and $J = 0, 2, 4, 6, \ldots$ for $K = 0$ built on the separate $\hbar\omega_\beta$ and $\hbar\omega_\gamma$ vibrational band heads [see (6-106) and text]. The figure refers to even–even, quadrupole-deformed, rotational nuclei.

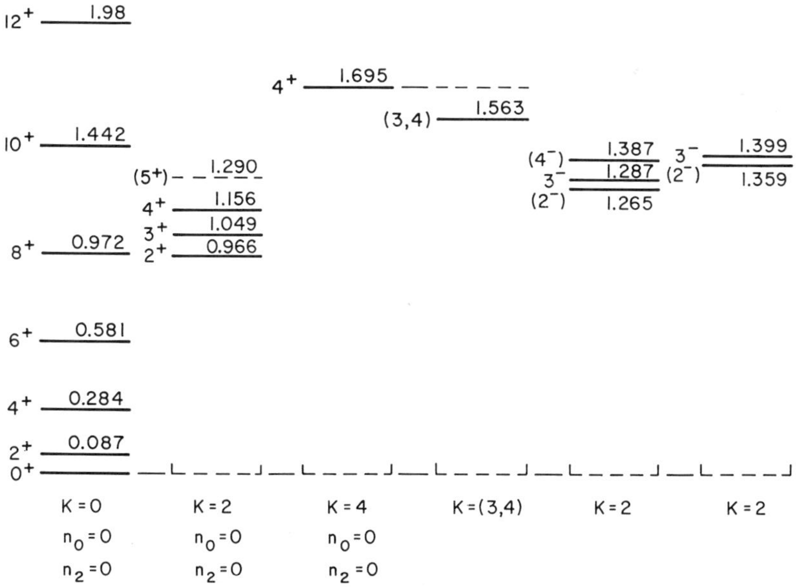

FIG. 6.14. The observed level scheme of $^{160}_{66}$Dy labeled according to possible vibrational–rotational quantum numbers.

However, we note two possible $K = 2$ *negative* parity bands. These would require at least coupling one octupole vibrational phonon (3^- character) with an appropriate quadrupole vibrational phonon of even parity (Davidson, 1968). Alternate explanations for the last three columns would involve possible intrinsic particle excitations to be discussed later.

Davydov and Chaban (1960) considered an asymmetric rotator model with possible coupled β vibrations. They assume nonzero values for both β_0 and γ_0, and in addition take $C^\gamma \gg C^\beta$ (i.e., $\hbar\omega_\gamma \gg \hbar\omega_\beta$). In these assumptions there is the implicit hope that γ vibrations of the type previously discussed are in fact reflections of asymmetric rotational effects. We are led to a simplified Hamiltonian in which γ is treated as a constant parameter γ_0. Solutions to the resulting Hamiltonian are obtained in the form

$$\Psi(\zeta; \theta_a, \theta_b, \theta_c) = \psi_{J,M,i}(\theta_a, \theta_b, \theta_c)\, \phi_J(\zeta), \tag{6-107}$$

where $\psi_{J,M,i}(\theta_a, \theta_b, \theta_c)$ are the solutions (6-98) and $\phi_J(\zeta)$ are functions of the local coordinate ζ of (6-100) and the subscript J refers back to the parameters describing $\psi_{J,M,i}$. In addition to (γ_0, β_0), another parameter that plays an important role is μ, the nonadiabaticity parameter discussed in connection with (6-85), with C of that expression particularized in the present context to C^β. Although Davydov and Chaban (1960) give a closed-form expression for the exact eigenvalues, we simply cite the approximate form valid for $\mu \leqslant \frac{1}{3}$,

$$E(J, \gamma_0, n_0, \mu) = \frac{\hbar^2}{4B{\beta_0}^2}\mathscr{E}_i(J)\left[1 + 3\mu^2\left(n_0 + \frac{1}{2}\right)\right] - D\frac{\hbar^2}{4B{\beta_0}^2}[\mathscr{E}_i(J)]^2 + \left(n_0 + \frac{1}{2}\right)\hbar\omega_\beta, \tag{6-108}$$

where $D = \frac{1}{2}\mu^4[1 + (57/4)(n_0 + \frac{1}{2})\mu^2]$ and $\mathscr{E}_i(J)$ is the dimensionless numerical factor given in Fig. 6.10 as a function of J, i, and γ_0. Clearly, as $\mu \to 0$, the simple additivity of rotational and vibrational energies appropriate for an adiabatic approximation results. For $\mu \neq 0$, (6-108) also possesses a "centrifugal" stretching term through the quantity $[\mathscr{E}_i(J)]^2$.

This model has been very successfully used to fit the empirical rotational spectra of many nuclei [see, e.g., Stephens *et al.* (1965)], giving values of $0.25 \leqslant \mu \leqslant 0.36$ for $156 \leqslant A \leqslant 184$.[1] The asymmetry parameter γ_0 for these nuclei is relatively small, namely $\gamma_0 \lesssim 12°$. The empirical fit so obtained even up to $J = 18$ is remarkably good. There is, however, some difficulty in deciding between the merits of the Faessler and Greiner model allowing γ vibrations for a symmetric rotator and the Davydov and Chaban model of only β vibrations

[1] For example, $\mu = 0.290$ is found for the spectrum of ^{170}Hf illustrated in Fig. 6.9.

for an asymmetric rotator. The ability to fit the ground-state rotational bands is more or less equally accurate. Perhaps the clearest determination, at least locally, would be the observing of the $K = 0$, $n_0 = 0$, $n_2 = 1$ band if it could be disentangled from the $K = 0$, $n_0 = 1$, $n_2 = 0$ band, both shown in Fig. 6.13, or any nonequivalent intrinsic particle excitations. The 0^+ level at 1.086 MeV in ^{188}Os in fact may be such an $n_2 = 1$ band head. For a detailed discussion of these points see Eisenberg and Greiner (1970) and Davidson (1968). The latter reference also discusses the lifting of the restriction $\dot{\gamma} = 0$.

Again the empirical values for γ_0 increase sharply for the transitional nuclei. The situation for the "spherical" nuclei is somewhat less paradoxial, having now allowed for at least β vibrational modes. The nucleus ^{114}Cd now requires $\mu = 0.54$ and $\gamma = 22.5°$.

A particularly useful way of illustrating the possible vibrational–rotational modal character of even–even nuclei is to plot the observed ratio of the level energies, particularly $E_1(4^+)/E_1(2^+)$, $E_1(6^+)/E_1(2^+)$, $E_1(8^+)/E_1(2^+)$, ... or possibly $E_2(0^+)/E_1(2^+)$, $E_2(2^+)/E_1(2^+)$, etc., where the subscript i on

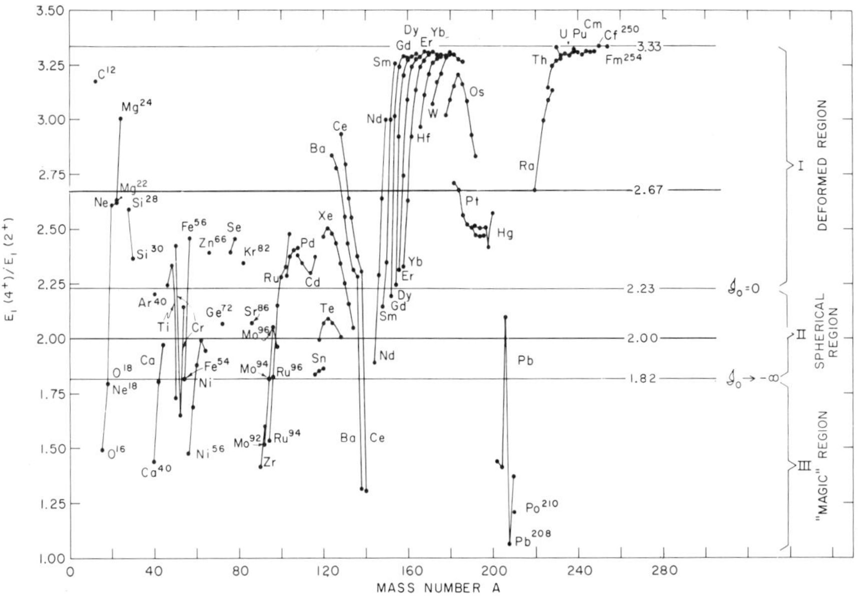

FIG. 6.15. The experimental-level energy ratios $E_1(4^+)/E_1(2^+)$ are shown as a function of A with the isotopes of a given Z connected with lines. The Davydov and Filippov model would restrict the range of this ratio to 2.67–3.33. The variable-moment-of-inertia model with the ground state moment of inertia $\mathcal{I}_0 \geqslant 0$ allows ratios between 2.23 and 3.33. An extension of this model to *negative* values of $\mathcal{I}_0$ permits ratios as low as 1.82. A pure spherical vibrator model gives exactly 2.00 for this ratio. [Figure provided in a private communication by Scharff-Goldhaber; see also Mariscotti *et al.* (1969).]

$E_i(J^\pi)$, as before, refers to the ordinal number of the particular J^π state, i.e., $i = 1$ first occurrence, $i = 2$ second occurrence, ... For pure axially symmetric rotators these ratios are $E_1(4^+)/E_1(2^+) = \frac{10}{3}$, $E_1(6^+)/E_1(2^+) = 7$, $E_1(8^+)/E_1(2^+) = 12$, etc. For pure spherical vibrational nuclei

$$E_2(0^+)/E_1(2^+) = E_2(2^+)/E_1(2^+) = E_1(4^+)/E_1(2^+) = 2, \quad \text{etc.}$$

Perhaps the most revealing of these ratios when availability of data is also considered is $E_1(4^+)/E_1(2^+)$. Figure 6.15 shows such a plot. The limiting range $\frac{8}{3} \leqslant E_1(4^+)/E_1(2^+) \leqslant \frac{10}{3}$ of the pure asymmetric rotator model of Davydov and Filippov is indicated along with the ratio $E_1(4^+)/E_1(2^+) = 2$ for the pure spherical vibrator model. Indeed, the mass ranges $150 < A < 190$, the so-called rare earth group (not too precisely conforming to $57 \leqslant Z \leqslant 71$), and $A \geqslant 220$, the so-called actinide group, show strong rotational character. The "spherical" vibrational nuclei have rather scattered energy ratios with $\frac{7}{4} \leqslant E_1(4^+)/E_1(2^+) \lesssim \frac{10}{4}$ rather than the exact value of 2. The very nearly closed-shell nuclei exhibit rather simple intrinsic particle excitations. Coupling such simple intrinsic states to pure phenomenologically collective states successfully accounts for the scatter in the vibrational data.

It is also instructive to examine the systematic trend in the excitation energy of the first 2^+ level, $E_1(2^+)$, in even–even nuclei. Figure 6.16 shows the empirical values and $0.27\hbar\omega_2$, where $\hbar\omega_2$ is the vibrator energy estimated from (6-37). The factor of 0.27 is arbitrarily introduced to facilitate comparison.

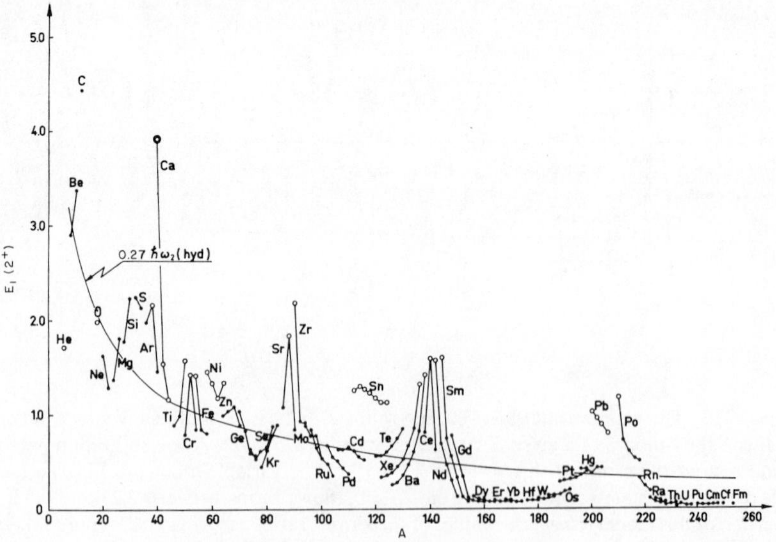

FIG. 6.16. The experimental excitation energies $E_1(2^+)$. The curve is for $0.27\hbar\omega_2$ of (6-37) [taken from Siegbahn (1965)].

In the immediate vicinity of the closed-shell magic nuclei we observe the expected elevated energies. There is a brief region nearby on either side of these nuclei where the candidate harmonic spherical vibrators are located. The strongly depressed energies pertain to the rotator nuclei $150 \leqslant A \leqslant 190$ and $A \geqslant 220$. Of course, for these nuclei estimates based on (6-37) are inappropriate.

If, for any candidate rotational even–even nucleus at least displaying in increasing energy sequence the states 2^+, 4^+, and 6^+ above the 0^+ ground state, we determine a ground-state moment of inertia $\mathscr{I}_0$ for the observed sequence, the values shown in Fig. 6.17 result. This figure includes entries for

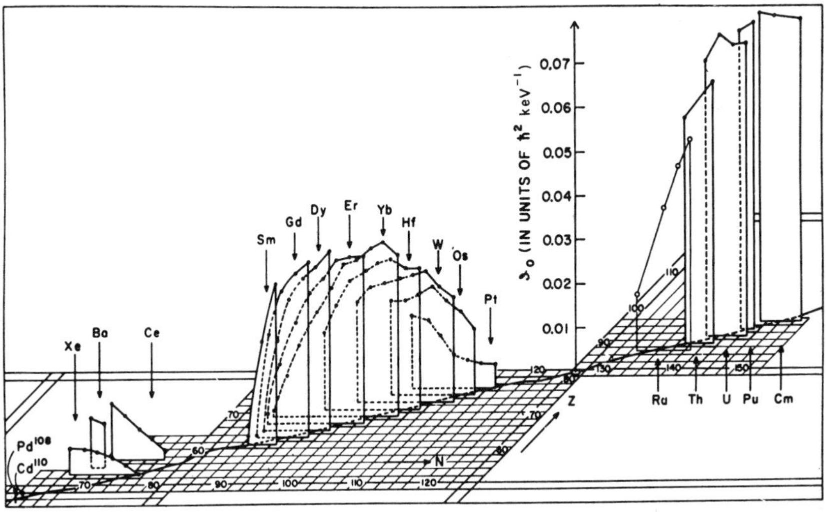

FIG. 6.17. Inferred ground state moments of inertia for candidate rotational nuclei exhibiting at least the ascending energy level sequence 2^+, 4^+, and 6^+ above the 0^+ ground state, except for the Ra isotopes. The moment of inertia $\mathscr{I}_0$ is plotted in units of $\hbar^2$ (keV)$^{-1}$ (i.e., $\mathscr{I}_0 = 0.05$ in these units corresponds to $\hbar^2/2\mathscr{I}_0 = 10$ keV) [taken from Mariscotti *et al.* (1969)].

many nuclei that are also candidate asymmetric Davydov and Filippov type rotators (such as the Pt isotopes) by using only the data for the lowest levels of any particular J^+, and ascribing departures from the $J(J+1)$ energy rule to "stretching" effects discussed earlier. Whether one considers $\mathscr{I}_0$ proportional to $\beta_0{}^2$ as in the hydrodynamic model or to depend linearly on a generalized coordinate replacing β_0 as in the variable-moment-of-inertia model [see Exercise 6-13 and Mariscotti *et al.* (1969)], the permanent deformation is seen from Fig. 6.17 to decrease essentially to zero near N and/or Z magic.

In Fig. 6.18, we use selected portions of the data shown in Fig. 6.17 for the nearly axially symmetric subgroup near the rare earth nuclei to display various theoretical possibilities. The microscopic theory based on the so-called cranking model including quasi-particle pairing effects gives rather satisfactory results as can be seen in this figure. We leave the discussion of such models to specialized texts and papers [e.g., Rowe (1970)].

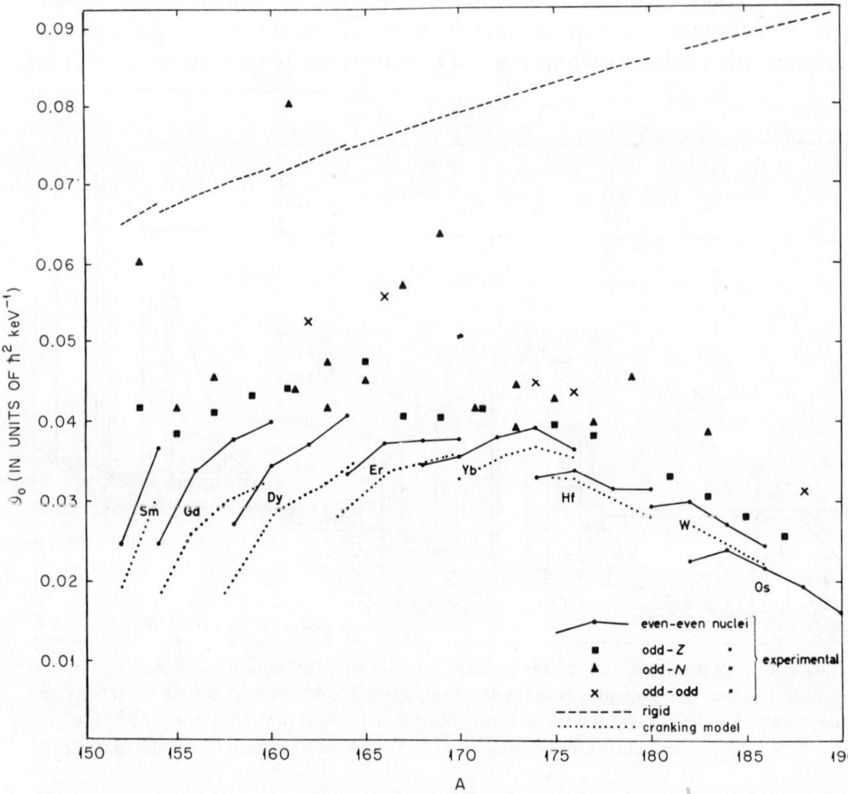

FIG. 6.18. Ground-state moments of inertia $\mathscr{I}_0$ for the "rare earth" nuclei, comparing experimental values with various theoretical possibilities. The units are the same as in Fig. 6.17 [adapted from Siegbahn (1965)].

3. Strong Coupling of Particle and Collective Motions

We now discuss the odd-A permanently deformed nuclei. Varying degrees of elaboration are possible depending on the number of extra-core nucleons explicitly considered, with the remainder relegated to an appropriate collective core. The simplest treatment for an odd-A nucleus is a collective form of the extreme single-particle model in which only the odd nucleon is considered, interacting with a collective core consisting of the appropriate adjacent even–even nucleus in the periodic table. The adiabatic approximation, relating on the one hand to the intrinsic motion of the extra-core nucleons in the potential generated by the deformed core, and on the other hand to the collective vibrational–rotational motion of the core itself, constitutes a convenient starting point. The coupling of the intrinsic and collective motions is very strong in the present instances and the perturbation approach used in Section VI, C, 2 is inappropriate. Rather, we seek to solve the intrinsic single-particle motion using a Schrödinger equation Hamiltonian

$$\tilde{H}_i \psi_n = [\tilde{T}_i + \tilde{V}(\beta, \gamma; \mathbf{r}_i', \boldsymbol{l}_i', \mathbf{s}_i')] \psi_n, \qquad (6\text{-}109)$$

in the body-centered frame (x', y', z'). The potential energy reflects the core deformations in terms of the two parameters β and γ (we again consider only the case for quadrupole deformations, $\lambda = 2$). The subscript n refers to all required relevant quantum numbers.

The total system Hamiltonian will in general consist of contributions from the extra-core particles $H_p = \sum_{i=1}^{k} H_i + \frac{1}{2} \sum_{i,j}^{k} v_{ij}$, the core H_{core}, and coupling terms H_{int}. In practice this total Hamiltonian is rearranged and reduced to the form

$$H = H_{0,p} + H_{0,\text{core}} + H'.$$

The term $H_{0,p}$ is taken to be a convenient solvable form and includes only the intrinsic or extra-core nucleon coordinates in the body-centered frame and an effective potential of the type in (6-109) with (β, γ) treated as fixed parameters (β_0, γ_0). When we particularize to the case of nuclear axial symmetry, which is assumed also to give a particle–core potential that is axially symmetric, (6-109) becomes

$$H_i \psi_n = [T_i + V(\beta_0, \gamma_0 = 0; \mathbf{r}_i', \boldsymbol{l}_i', \mathbf{s}_i')] \psi_n. \qquad (6\text{-}110)$$

The term $H_{0,\text{core}}$ is conveniently chosen and includes only the collective coordinates and core-related system constants of the motion. The remnant terms are contained in H' and are considered as perturbation terms. We defer to the next section the detailed discussion of the form of the Hamiltonian (6-110) and its solutions. For the present we suppose the solutions to exist and suitably couple the single-particle wave functions of all the extra-core nucleons according to whatever internucleon forces are assumed to operate.

We introduce the operators **J**, **R**, and **j**. The operator **J** refers to the total angular momentum of the entire system, **R** refers to the rotational angular momentum of the core, and $\mathbf{j} = \sum_{i=1}^{k} \mathbf{j}_i$ refers to the angular momentum of the k extra-core nucleons, and $\mathbf{J} = \mathbf{R} + \mathbf{j}$.

As a collective model for the core one might employ the Hamiltonian of an axially symmetric liquid drop undergoing (β, γ) vibrations discussed in connection with (6-100)–(6-106). The collective coordinates $(\zeta, \eta; \theta_a, \theta_b, \theta_c)$ would appear in the Hamiltonian $H_{\text{core}} = H_{\text{rot}} + H_{\text{vib}} + H_{\text{vib-rot}}$ with, however, the body-centered components of the angular moment L_ξ replaced by R_ξ. For example, in H_{rot},

$$\frac{\hbar^2}{2} \sum_{\xi=1}^{3} \frac{L_\xi{}^2}{\mathscr{I}_\xi(\beta_0+\zeta, \eta)} \to \frac{\hbar^2}{2} \sum_{\xi=1}^{3} \frac{(J_\xi - j_\xi)^2}{\mathscr{I}_\xi(\beta_0+\zeta, \eta)}. \tag{6-111}$$

A similar modification is called for in $H_{\text{vib-rot}}$. While in a completely general treatment we would, of course, retain $H_{\text{vib}} + H_{\text{vib-rot}}$ in our collective Hamiltonian [see, e.g., Eisenberg and Greiner (1970)] giving rise to rotational bands built on various (β, γ) vibrational excitations, it is adequate for purposes of displaying the essential behavior of the particle–core system to limit the present discussion to the core ground-state rotational band and use only the term H_{rot}. Since we have assumed axial symmetry with $\mathscr{I}_{x'} = \mathscr{I}_{y'} = \mathscr{I}_0$, we have

$$\begin{aligned} H_{\text{rot}} &= \frac{\hbar^2}{2\mathscr{I}_0}[(\mathbf{J}-\mathbf{j})^2 - (J_{z'} - j_{z'})^2] + \frac{\hbar^2}{2\mathscr{I}_{z'}}(J_{z'} - j_{z'})^2 \\ &= \frac{\hbar^2}{2\mathscr{I}_0}[\mathbf{J}^2 + \mathbf{j}^2 - (J_+ j_- + J_- j_+) - (J_{z'}^2 + j_{z'}^2)] + \frac{\hbar^2}{2\mathscr{I}_{z'}}(J_{z'} - j_{z'})^2, \end{aligned} \tag{6-112}$$

where we have used $2\mathbf{J}\cdot\mathbf{j} = J_+ j_- + J_- j_+ + 2J_{z'} j_{z'}$ with $J_\pm = J_{x'} \pm iJ_{y'}$ and $j_\pm = j_{x'} \pm ij_{y'}$. While all angular momentum operator components in (6-112) are in the body-centered frame, $J_{x',y',z'}$ only operates on the Eulerian angle variables and $j_{x',y',z'}$ only operates on the intrinsic particle coordinates. It is convenient to regroup (6-112) into three terms,

$$H_{\text{rot}} = H_{0,\text{rot}} + V_{\text{rpc}} + H'_{\text{rot}} \tag{6-113a}$$

with

$$H_{0,\text{rot}} = (\hbar^2/2\mathscr{I}_0)[\mathbf{J}^2 - (J_{z'}^2 + j_{z'}^2)] + (\hbar^2/2\mathscr{I}_{z'})(J_{z'} - j_{z'})^2 \tag{6-113b}$$

$$V_{\text{rpc}} = -(\hbar^2/2\mathscr{I}_0)(J_+ j_- + J_- j_+) \tag{6-113c}$$

$$H'_{\text{rot}} = (\hbar^2/2\mathscr{I}_0)\mathbf{j}^2. \tag{6-113d}$$

The term H'_{rot} just involves the intrinsic extra-core particle coordinates and will be considered in connection with (6-110). The term labelled rpc (rotation-particle coupling) corresponds to the classical Coriolis interaction that appears whenever a system described in a rotating frame has intrinsic angular momentum as well. If this term were omitted, the intrinsic system would experience no force originating from the rotation of its coordinate frame of reference. This situation would represent the extreme adiabatic limit of strong coupling. The extra-core particles execute motions strictly adhering to the body-centered potential as if the nucleus were not rotating at all. To the extent that the interaction V_{rpc} is operative, nonadiabaticity is introduced and the core and extra-core particle motions are thereby suitably *decoupled.* Except for one notable exception, which we shall treat later, V_{rpc} is treated as a perturbation on the term $H_{0,\mathrm{rot}}$.

In lowest order, i.e., the extreme adiabatic strong coupling model,[1] only $H_{0,\mathrm{rot}}$ is used as the collective core Hamiltonian $H_{0,\mathrm{core}}$. Since the body-centered potential in (6-110) has axial symmetry, evidently $j_{z'}$ is a constant of the motion, and we take

$$j_{z'}\chi_\Omega = \Omega\hbar\chi_\Omega, \tag{6-114}$$

where χ_Ω is the coupled extra-core nucleon wave function. However, since the potential in (6-110) is not central, $\mathbf{j}^2$ is not a constant of the motion. It is useful to expand χ_Ω in terms of solutions $\chi_{j,\Omega}$ generated by (6-110) when the non-spherically symmetric terms in the potential are omitted. We write

$$\chi_\Omega = \sum_j C_{j\Omega}\chi_{j\Omega}. \tag{6-115}$$

Identifying (6-113b) with (6-57), the normalized general solution to $H_{0,\mathrm{rot}}+H_{0,\mathrm{p}}$ is of the form

$$\psi = \left(\frac{2J+1}{8\pi^2}\right)^{1/2} D^J_{MK}\chi_\Omega = \left(\frac{2J+1}{8\pi^2}\right)^{1/2} D^J_{MK}\sum_j C_{j\Omega}\chi_{j\Omega}, \tag{6-116}$$

where $K\hbar$ is again the component of the total angular momentum along the z' axis and $M\hbar$ is the component along the laboratory z axis.

As yet, the solution (6-116) does not have the required symmetry properties. A rotation through an arbitrary angle ϕ about the z' axis gives[2]

$$D^J_{MK}(\theta_a, \theta_b, \theta_c+\phi) = e^{iK\phi}D^J_{MK}(\theta_a, \theta_b, \theta_c)$$

and

$$\chi_\Omega(\phi_c+\phi) = e^{-i\Omega\phi}\chi_\Omega(\phi_c).$$

Clearly $K-\Omega = 0$ is required. A rotation of π about the x' axis gives

$$D^J_{MK}(\theta_a+\pi, \pi-\theta_b, \theta_c) = e^{\pi i(J+K)}D^J_{M,-K}(\theta_a, \theta_b, \theta_c)$$

and

$$\chi_{j\Omega}(x'\to x', y'\to -y', z'\to -z') = e^{-\pi i(j+\Omega)}\chi_{j,-\Omega}(x', y', z').$$

[1] Also referred to as the symmetric core model.

[2] Rotations of intrinsic functions and of collective functions take opposite signs.

Thus solutions that are invariant to a rotation of π about the x' axis must be formed of linear sums of the original and rotated solutions. Finally, properly invariant and normalized solutions are, with $\Omega = K$,

$$\psi = |JM, K=\Omega, n\rangle = \left(\frac{2J+1}{16\pi^2}\right)^{1/2} \sum_j C_{j\Omega}^{(n)} [D_{MK}^J \chi_{j\Omega}^{(n)} + (-1)^{J-j} D_{M,-K}^J \chi_{j,-\Omega}^{(n)}]. \tag{6-117}$$

The energy eigenvalues corresponding to (6-117) are

$$E(JM, K=\Omega, n) = E_n(K=\Omega) + (\hbar^2/2\mathscr{I}_0)[J(J+1) - 2K^2]. \tag{6-118}$$

Since $J \geqslant K$, we have the rotational band beginning with $J = K$ $(=\Omega)$, and continuing with $K+1, K+2, \ldots$, built on the intrinsic particle state having the energy $E_n(K=\Omega)$. In writing (6-117) in its present form, the number of extra-core nucleons is imagined odd. If (6-117) is generalized to allow for an even number of nucleons as well, the case $j = 0$ (and hence $\Omega = K = 0$) just gives (6-66b).

A convenient diagrammatic representation of the above discussion for the strong coupling of extra-core nucleons to a deformed core is given in Fig. 2.12 of Chapter II. The vector $\mathbf{j}$ in that figure refers to the combined angular momenta of the loose extra-core nucleons. In general $\mathbf{j}^2$ is not a constant of the motion (however, see below).

We wish to examine the effect of including the decoupling term V_{rpc} of (6-113c) in the total Hamiltonian. The matrix elements of V_{rpc} for the basis states (6-117) are

$$\begin{aligned} -(\hbar^2/2\mathscr{I}_0)&\langle JM, K'=\Omega'|J_+ j_- + J_- j_+|JM, K=\Omega\rangle \\ &= -(\hbar^2/2\mathscr{I}_0) \sum_j C_{j\Omega'}^* C_{j\Omega} \{[\delta_{K',K-1} + (-1)^{J-j} \delta_{K',-(K-1)}] \\ &\quad \times [(J+K)(J-K+1)(j+\Omega)(j-\Omega+1)]^{1/2} \\ &\quad + [\delta_{K',K+1} + (-1)^{J-j} \delta_{K',-(K+1)}] \\ &\quad \times [(J-K)(J+K+1)(j-\Omega)(j+\Omega+1)]^{1/2}\}. \end{aligned} \tag{6-119}$$

This result readily follows from

$$J_\pm D_{MK}^J = [(J \pm K)(J \mp K+1)]^{1/2} D_{M,K\mp 1}^J \tag{6-120a}$$

and

$$j_\pm \chi_{j\Omega} = [(j \mp \Omega)(j \pm \Omega + 1)]^{1/2} \chi_{j,\Omega\pm 1}. \tag{6-120b}$$

When $K = \Omega \neq \frac{1}{2}$, only states with $K' = K \pm 1$ couple [i.e., the matrix elements (6-119) have only off-diagonal elements]. In this event, K-band mixing of the rotational states having the same J can occur. When the relevant unperturbed band members of the same J are well separated in energy, relatively little mixing occurs. If two bands K and $K+1$ should happen to closely comingle, placing two levels of the same J near each other in energy, the exact diagonalization of the required 2×2 energy matrix is easily obtained, viz.,

$$\begin{vmatrix} E(J,K=\Omega)-E & \langle J,K=\Omega|V_{\text{rpc}}|J,K'=\Omega'=K+1\rangle \\ \langle J,K'=\Omega'=K+1|V_{\text{rpc}}|J,K=\Omega\rangle & E(J,K'=\Omega'=K+1)-E \end{vmatrix} = 0, \tag{6-121}$$

with the result

$$\begin{aligned} E_{\pm} &= \tfrac{1}{2}\{E(J,\, K=\Omega)+E(J,\, K'=\Omega'=K+1) \\ &\qquad \pm \Delta E[1+4|\langle J,\, K'=\Omega'=K+1|V_{\text{rpc}}|J,\, K=\Omega\rangle/\Delta E|^2]\}^{1/2}, \end{aligned} \tag{6-122}$$

where $\Delta E = E(J,\, K=\Omega)-E(J,\, K'=\Omega'=K+1)$. The resulting two states with shifted energies (6-122) are no longer characterized by K being a good quantum number. Clearly the states approach the pure unperturbed states as ΔE becomes larger and larger.

A significant exception to the above occurs for $K=\Omega=\frac{1}{2}$. In this case (and this case only) a diagonal matrix element results for $K'=K=\frac{1}{2}$ from the factor $\delta_{K',-(K-1)} = \delta_{\frac{1}{2},-(\frac{1}{2}-1)} = \delta_{\frac{1}{2},\frac{1}{2}} = 1$. In general this produces a significant departure for the level sequence of a $K=\frac{1}{2}$ band from the simple rotational pattern of (6-118). We readily find from (6-119)

$$\begin{aligned} -\frac{\hbar^2}{2\mathscr{I}_0}&\left\langle J,\, K=\Omega=\frac{1}{2}\middle|V_{\text{rpc}}\middle|J,\, K=\Omega=\frac{1}{2}\right\rangle \\ &= -\frac{\hbar^2}{2\mathscr{I}_0}\sum_j |C_{j\frac{1}{2}}|^2(-1)^{J-j}\left(J+\frac{1}{2}\right)\left(j+\frac{1}{2}\right) \\ &= \frac{\hbar^2}{2\mathscr{I}_0}(-1)^{J+\frac{1}{2}}\left(J+\frac{1}{2}\right)a, \end{aligned} \tag{6-123a}$$

where the so-called *decoupling parameter* a is

$$a = \sum_j (-1)^{j-\frac{1}{2}}(j+\tfrac{1}{2})|C_{j\frac{1}{2}}|^2, \tag{6-123b}$$

and we have noted that $2j-1$ is an even integer. Finally for the $K=\Omega=\frac{1}{2}$ band, (6-118) becomes, when corrected for the V_{rpc} contribution, (6-123a),

$$\begin{aligned} E(JM,\, K=\Omega=\tfrac{1}{2},\, n) &= E_n(\Omega=\tfrac{1}{2}) + (\hbar^2/2\mathscr{I}_0) \\ &\qquad \times [J(J+1)+(-1)^{J+\frac{1}{2}}(J+\tfrac{1}{2})\,a-2K^2]. \end{aligned} \tag{6-124}$$

With $a<0$, the $J=\frac{3}{2},\frac{7}{2},\frac{11}{2},\ldots$ levels are depressed in energy while the $J=\frac{1}{2},\frac{5}{2},\frac{9}{2},\ldots$ levels are increased in energy; with $a>0$ the reverse is true. However, so long as $-1 \leqslant a \leqslant +1$, the normal order of the level sequence is maintained, i.e., $E(J, K=\frac{1}{2}) < E(J+1, K=\frac{1}{2})$. Outside this range, when $|a|>1$, the normal order is not preserved. For example, for $a=-2$ the sequence in increasing energy is $J=\frac{3}{2},\frac{1}{2},\frac{7}{2},\frac{5}{2},\frac{11}{2},\frac{9}{2},\ldots$.

The anomalous case for $K=\Omega=\frac{1}{2}$ was treated above from a collective point of view based on the behavior of the rotation functions D^J_{MK}. An alternate and very physical cranking-model treatment also can be given [see Inglis (1973)].

Returning to the general discussion for the case when many members of a band are observed, a centrifugal stretching $[J(J+1)]^2$ term of the type (6-94) must be added to (6-118) and (6-124) to account for the data properly. Generally, when additional perturbation effects contained in the total Hamiltonian are to be considered, it is within a framework using (6-117) as basis states. In general only (J, M) remain as good quantum numbers.

We defer giving experimental illustrations of rotational odd-A nuclei until we have examined the nature of the body-centered eigenfunctions $\chi_\Omega(\mathbf{r}')$ of the Hamiltonian (6-110) in the next section.

4. Single-Particle States of Deformed Potentials

One of the most successful models for generating realistic intrinsic single-particle states of deformed potentials is that first proposed by Nilsson (1955). This model was limited to axially symmetric quadrupole deformations. We also limit our considerations to this case. A central problem in developing a single-particle model Hamiltonian is the reduction of the potential energy term in (6-109), $V(\beta,\gamma;\mathbf{r}',\boldsymbol{l}',\mathbf{s}')$, to a mathematically tractable form. Insofar as the deformation variables β and γ are concerned, a lowest-order Taylor series expansion of the form (6-50) for a local harmonic oscillator potential of the type (6-100) gives[1]

$$V(\beta,\gamma;\mathbf{r}) = \tfrac{1}{2}m_{\mathrm{N}}\,\omega^2 r^2 - m_{\mathrm{N}}\,\omega^2 r^2\{\alpha_{20}Y_2{}^0(\theta,\varphi)+\alpha_{22}[Y_2{}^2(\theta,\varphi)+Y_2^{-2}(\theta,\varphi)]\} \tag{6-125}$$

with the collective variables $\alpha_{20}=\beta_0+\zeta$ and $\alpha_{22}=0+\eta$ for the case of an equilibrium deformation $\beta_0, \gamma_0=0$. If we again omit consideration of β and γ vibrations and thus consider only the collective vibrational ground state, we have for (6-110)

$$\begin{aligned} V(\beta_0,\gamma_0=0;\mathbf{r}) &= \tfrac{1}{2}m_{\mathrm{N}}\,\omega_0{}^2 r^2[1-2\beta_0 Y_2{}^0(\theta,\varphi)] \\ &= \tfrac{1}{2}m_{\mathrm{N}}[\omega_\perp{}^2(x^2+y^2)+\omega_z{}^2 z^2] \end{aligned} \tag{6-126a}$$

[1] We have made use of Exercise 6-11, and for convenience we drop the primes, it being understood that all coordinates are in the body-centered frame.

with[1]

$$\omega_x^2 = \omega_y^2 = \omega_\perp^2 = \omega_0^2[1+(5/4\pi)^{1/2}\beta_0], \qquad \omega_z^2 = \omega_0^2[1-2(5/4\pi)^{1/2}\beta_0]. \tag{6-126b}$$

It is relatively common practice, although by no means universal, to introduce a deformation parameter δ related to β_0 through the definition

$$\delta = \tfrac{3}{2}(5/4\pi)^{1/2}\beta_0 \approx 0.946\beta_0. \tag{6-127}$$

In this notation

$$\omega_\perp^2 = \omega_0^2(1+\tfrac{2}{3}\delta), \qquad \omega_z^2 = \omega_0^2(1-\tfrac{4}{3}\delta). \tag{6-128}$$

Although in arriving at (6-126) we have made use of (6-49), which involves assuming equipotential surfaces proportional to constant-density surfaces, it is instructive to demonstrate this explicitly for (6-126). The equation of a constant-density spheroidal surface in Cartesian coordinates is

$$(1/a^2)(x^2+y^2) + (z^2/b^2) = 1.$$

The enclosed volume $\frac{4}{3}\pi a^2 b$ is maintained a constant for all vibrations of an incompressible liquid drop. Clearly, (6-126a) has equipotential surfaces of the form $\omega_\perp^2(x^2+y^2)+\omega_z^2 z^2 = \text{const}$, and the referred to proportionality thus is evident. The condition of incompressibility requires

$$\omega_\perp^2\omega_z = \text{const} = \omega_{00}^3. \tag{6-129}$$

Thus, using (6-128),

$$\omega_0 = \omega_{00}[(1+\tfrac{2}{3}\delta)^2(1-\tfrac{4}{3}\delta)]^{-1/6} = \omega_{00}[1-\tfrac{4}{3}\delta^2-(16/27)\delta^3]^{-1/6}, \tag{6-130}$$

which to linear terms is independent of δ, as it should be.

From (6-126) we see that the principal effect of the permanently deformed surface is to convert the harmonic oscillator single-particle potential into a nonisotropic oscillator potential. In addition to the nonisotropic potential Nilsson also introduced a spin–orbit interaction $C\boldsymbol{l}\cdot\mathbf{s}$ and an interaction $D\boldsymbol{l}^2$, writing

$$V(\beta_0, \gamma_0 = 0, \mathbf{r}, \boldsymbol{l}, \mathbf{s}) = \tfrac{1}{2}m_N[\omega_\perp^2(x^2+y^2)+\omega_z^2 z^2] + C\boldsymbol{l}\cdot\mathbf{s} + D\boldsymbol{l}^2. \tag{6-131}$$

The spin–orbit interaction with $C < 0$ is necessary, of course, if (6-131) is to reduce to the required spherical potential when $\beta_0 \to 0$, discussed in Section IV,C. As per that discussion, the value of C is required to depend on the harmonic oscillator shell number n in order to match empirical evidence. The

[1] Following custom, ω of (6-125) is now written ω_0 and we have used $Y_2{}^0$ in terms of Cartesian coordinates, namely $Y_2{}^0 = (5/16\pi)^{1/2}(2z^2-x^2-y^2)/r^2$.

ad hoc term Dl^2 is necessary to reduce the unrealistic surface effect introduced by having the oscillator potential increase without limit as the coordinates increase. Reference to Fig. 4.4 shows that a realistic modification of the oscillator potential depresses the energy of the large l-value orbitals and hence a term of the form Dl^2 with $D < 0$ would appear to be reasonable. One also would expect the required value of D to vary with the harmonic oscillator shell number n.[1]

Finally, we should note the possibility of including H'_{rot} of (6-113d) in (6-131). This term can be written

$$H'_{\text{rot}} = (\hbar^2/2\mathscr{I}_0)\mathbf{j}^2 = (\hbar^2/2\mathscr{I}_0)(\boldsymbol{l}^2+\mathbf{s}^2+2\boldsymbol{l}\cdot\mathbf{s}). \qquad (6\text{-}132)$$

In the representation we presently shall introduce, $\mathbf{s}^2$ is diagonal and simply adds a constant term $3\hbar^2/8\mathscr{I}_0$ to the energy. The terms in $\boldsymbol{l}\cdot\mathbf{s}$ and $\boldsymbol{l}^2$ of (6-132) while of the form already contained in (6-131), cannot in fact be considered as the source of those terms since they are incorrect in both sign and magnitude. Instead, the possible added terms from (6-132) can be viewed as additional effects requiring an empirical adjustment of C and D appearing in (6-131).

▶**Exercise 6-14** It is instructive to consider a nonisotropic oscillator potential

$$V(\mathbf{r}) = \tfrac{1}{2}m\omega_0{}^2[e^{\sigma}(x^2+y^2)+e^{-2\sigma}z^2].$$

The exponential factors lead to a volume-conserving scale transformation carrying the harmonic oscillator problem into the anharmonic case. The equipotential surfaces are spheroids with major and minor axes proportional to e^{σ} and $e^{-2\sigma}$ enclosing a volume independent of σ. If we set $\sigma = \frac{2}{3}\delta$, the above potential corresponds to (6-126) for $\delta \ll 1$.

(a) Consulting Section IV, B, 1, show that the energy eigenvalues of the Hamiltonian containing the above potential energy is

$$E(n_1, n_2, n_3) = \hbar\omega_0[(n_1+n_2+1)e^{\sigma/2}+(n_3+\tfrac{1}{2})e^{-\sigma}].$$

Defining the principal oscillator quantum number $n = n_1+n_2+n_3$, the harmonic degeneracy with $\sigma = 0$, $D_n^{(0)} = (n+1)(n+2)$ (including spin) is now reduced through the partition involving n_3. The energy is now characterized by the quantum numbers n and n_3.

(b) Show that for a single particle in a state with (n, n_3), treating σ as a variational parameter results in an equilibrium deformation σ_{eq} for an energy

[1] The use of a term $D[l^2-\langle n|l^2|n\rangle] = D[l^2-\frac{1}{2}n(n+3)]$ reduces the required n dependence of D [see Gustafson *et al.* (1968)] by preserving the distance between the center of gravity of two adjacent oscillator shells at the value $\hbar\omega_0$. No such problem exists with the spin–orbit interaction since the energy center of gravity is left unaltered by this interaction.

minimum with

$$\sigma_{\text{eq}} = \frac{2}{3}\ln\frac{2n_3+1}{n-n_3+1} = \frac{4}{3}\left[\frac{3n_3-n}{n+n_3+2}+\frac{1}{3}\left(\frac{3n_3-n}{n+n_3+2}\right)^3+\cdots\right].$$

For k total particles in the oscillator shells (n,n_3), show that we have

$$\sigma_{\text{eq}} = \tfrac{2}{3}\ln\left[\sum_{i=1}^{k}(2n_3+1)_i \Big/ \sum_{i=1}^{k}(n-n_3+1)_i\right].$$

(c) When k is equal to the oscillator magic number M_n, the principal oscillator shells up to the one denoted by n are full, having $k = M_n = \frac{1}{3}(n+1)(n+2)(n+3)$ particles in all [see (4-10)]. Show that under these circumstances $\sigma_{\text{eq}} = 0$.

[*Hint:* You must show that $3\sum_i^{(\text{shell}\,n)}(n_3)_i = \sum_i^{(\text{shell}\,n)}(n)_i = nD_n^{(0)}$ for any of the filled shells. Note that $1+2+3+\cdots+n = \frac{1}{2}n(n+1)$ and $1^2+2^2+3^2+\cdots+n^2 = \frac{1}{6}n(n+1)(2n+1)$.]

(d) Plot the single-particle energy $E(n,n_3)$, in units of $\hbar\omega_0$, as a function of σ for the particle orbitals with principal oscillator shell numbers $n = 0$, 1, and 2, and all possible values of n_3 in each case. Allow σ to vary in the range $-\frac{3}{2} < \sigma < +\frac{3}{2}$.

(e) Consider filling the (n,n_3) subshells in a manner always giving the smallest *system* energy. Plot the resulting system energy as a function of σ with $-\frac{3}{2} < \sigma < \frac{3}{2}$, and $k = 8, 10, 12$ (i.e., just filling the $n=1$ principal oscillator shell at $M_n = 8$, and starting to fill the shell $n = 2$). In each case calculate the value of σ_{eq} for which the total system energy is a minimum. Consider carefully the consequences of the level crossing between the various (n,n_3) shells and ignore for the sake of simplicity any level crossing with $n \geqslant 3$ shells [actually the level $(n,n_3) = (3,3)$ would be very important for the $k = 12$ case]. Do these level crossing effects influence your answer to part (c)? ◀

The solution of the final Hamiltonian equivalent of (6-110),

$$H = -(\hbar^2/2m_{\text{N}})\nabla^2 + \tfrac{1}{2}m_{\text{N}}[\omega_\perp{}^2(x^2+y^2)+\omega_z{}^2z^2] + C\mathbf{l}\cdot\mathbf{s} + D\mathbf{l}^2, \tag{6-133}$$

is most readily accomplished using dimensionless coordinates

$$x_1 = (m_{\text{N}}\omega_0/\hbar)^{1/2}x, \qquad x_2 = (m_{\text{N}}\omega_0/\hbar)^{1/2}y$$
$$x_3 = (m_{\text{N}}\omega_0/\hbar)^{1/2}z, \qquad \rho^2 = \sum_{\tau=1}^{3}x_\tau{}^2. \tag{6-134}$$

We also introduce, following Nilsson, the new parameters

$$\kappa = -C/2\hbar\omega_0 \qquad \text{and} \qquad \mu = 2D/C.$$

The Hamiltonian (6-133) becomes

$$H = H_0 + H_\beta + H_{l,s} \tag{6-135}$$

with

$$H_0 = \tfrac{1}{2}\hbar\omega_0 \sum_{\tau=1}^{3} [(-\partial^2/\partial x_\tau^{\ 2}) + x_\tau^{\ 2}]$$

$$H_\beta = -\hbar\omega_0 \beta_0 \rho^2 Y_2^{\ 0}(\theta,\varphi) = -\tfrac{2}{3}(4\pi/5)^{1/2}\hbar\omega_0 \delta\rho^2 Y_2^{\ 0}(\theta,\varphi)$$

$$H_{l,s} = -\hbar\omega_0(2\kappa \boldsymbol{l}\cdot\mathbf{s} + \mu\kappa \boldsymbol{l}^2).$$

Solutions that simultaneously diagonalize H_0, $\boldsymbol{l}^2$, l_3, and s_3 can be used as a basis for arriving at the final solution of the complete energy eigenvalue problem. The associated quantum numbers for the operators H_0, $\boldsymbol{l}^2$, l_3, and s_3 are n (more commonly written N), l, Λ, and Σ. The resulting eigenvalues are[1]

$$\begin{aligned} H_0|nl\Lambda\Sigma\rangle &= \hbar\omega_0(n+\tfrac{3}{2})|nl\Lambda\Sigma\rangle, & l_3|nl\Lambda\Sigma\rangle &= \Lambda|nl\Lambda\Sigma\rangle \\ \boldsymbol{l}^2|nl\Lambda\Sigma\rangle &= l(l+1)|nl\Lambda\Sigma\rangle, & s_3|nl\Lambda\Sigma\rangle &= \Sigma|nl\Lambda\Sigma\rangle, \end{aligned} \tag{6-136}$$

with $n = 0, 1, 2, \ldots$; $l = n, n-2, n-4, \ldots, 1$ or 0; $|\Lambda| \leqslant l$; $\Sigma = \pm\tfrac{1}{2}$.

In addition, we can define the diagonal operator $j_3 = l_3 + s_3$ with eigenvalue $\Omega = \Lambda + \Sigma$. The parity $(-1)^n$ is also a good quantum number.

If we consider adding the term $H_{l,s}$ to H_0, the quantum numbers Λ and Σ no longer refer to constants of the motion, due to the presence of the spin–orbit operator. The operators $H_0, \boldsymbol{l}^2, j_3$ remain diagonal and hence n, l, and Ω are good quantum numbers. If the addition of $H_{l,s}$ were our only concern, the basis $|nlj\Omega\rangle$ would be more convenient, and indeed was our choice for the spherical nuclear potentials of Chapter IV. The two bases are connected as before through

$$|nlj\Omega\rangle = \sum_{\Lambda=\Omega\pm\frac{1}{2}} (l\tfrac{1}{2}\Lambda\Sigma|j\Omega)|nl\Lambda\Sigma\rangle. \tag{6-137}$$

When the term H_β is also added to $H_0 + H_{l,s}$, yielding the complete Hamiltonian (6-135), neither the operator $\boldsymbol{l}^2$ nor $\mathbf{j}^2$ commutes with the new Hamiltonian. Retaining the basis (6-136), the matrix elements of $\rho^2 Y_2^{\ 0}$ can be explicitly calculated, using

$$\langle n'l'\Lambda'\Sigma'|\rho^2 Y_2^{\ 0}|nl\Lambda\Sigma\rangle = \langle n'l'|\rho^2|nl\rangle\langle l'\Lambda'|Y_2^{\ 0}|l\Lambda\rangle\,\delta_{\Sigma'\Sigma}. \tag{6-138}$$

The submatrix element $\langle n'l'|\rho^2|nl\rangle$ connects states with $n' = n$ and $n' = n\pm 2$ and also $l' = l$ and $l' = l\pm 2$ in all nine combinations of $n'l'$ and nl. (The states $n' = n\pm 1$ are not coupled due to parity conservation.) The matrix

[1] The angular momenta are again in units of $\hbar$. Since $\Sigma = \pm\tfrac{1}{2}$, we also have the notation $|nl\Lambda, +\rangle$ and $|nl\Lambda, -\rangle$, respectively, for these two possibilities.

elements connecting $n' = n \pm 2$ involve mixing states differing in unperturbed energy of the order $2\hbar\omega_0$. Since this energy difference is generally much larger than the energy differences between states within a principal harmonic oscillator shell (i.e., $n' = n$), even under the perturbing influences that are present, the matrix elements with $n' = n \pm 2$ can be ignored.[1] We thus consider only the cases for $n' = n$, for which

$$\begin{aligned} \langle nl|\,\rho^2\,|nl\rangle &= n + \tfrac{3}{2} \\ \langle n,(l+2)|\,\rho^2\,|nl\rangle &= [(n-l)(n+l+3)]^{1/2} \\ \langle n,(l-2)|\,\rho^2\,|nl\rangle &= [(n-l+2)(n+l+1)]^{1/2}. \end{aligned} \tag{6-139}$$

The remaining submatrix involving the spherical harmonic is readily evaluated to be

$$\langle l'\Lambda'|\,Y_2{}^0\,|l\Lambda\rangle = [5(2l+1)/4\pi(2l'+1)]^{1/2}(l2\Lambda 0|l'\Lambda'), \tag{6-140}$$

with the added parity selection rule $(-1)^{l'+l} = +1$.

The diagonalization of the complete Hamiltonian H of (6-135), using the basis states $|n,l,\Lambda,\Sigma\rangle$, requires a numerical solution and was first performed by Nilsson. The numerical values used for $\hbar\omega_0$, κ, and μ are usually selected to give the observed level sequence and spacing for the "spherical" nuclei. Typical values for the parameters are

$$\hbar\omega_0 = 41A^{-1/3}\ \text{MeV}, \tag{6-141}$$

and for neutrons $\kappa_N \approx 0.06$ with μ_N increasing with the oscillator shell number n from $\mu_N \approx 0$ for $n = 0, 1, 2$ to $\mu_N \approx 0.45$ for $n = 6$, while for protons $\kappa_Z \approx 0.08$ and $\mu_Z \approx 0$ for $n = 0, 1, 2$ and $\kappa_Z \approx 0.06$ for $n \geqslant 3$ with μ_Z increasing with n to $\mu_Z \approx 0.60$ for $n = 5$.

When the level systematics of the deformed rare earths and the actinides are considered, Nilsson *et al.* (1969) find that along with (6-141), the parameters κ and μ also exhibit an A dependence. For $A \gtrsim 100$ they give[2]

$$\begin{aligned} &\text{for neutrons:} \quad \begin{cases} \kappa_N = 0.0641 - 0.0026(A/1000) \\ \mu_N = 0.624 - 1.234(A/1000) \end{cases} \\ &\text{for protons:} \quad \begin{cases} \kappa_Z = 0.0766 - 0.0779(A/1000) \\ \mu_Z = 0.493 + 0.649(A/1000). \end{cases} \end{aligned} \tag{6-142}$$

[1] If we introduce new coordinates $\xi_\tau = (\omega_\tau/\omega_0)^{1/2}x_\tau$ with $\omega_1 = \omega_2 = \omega_\perp$, $\omega_3 = \omega_z$, $\sigma^2 = \sum_\tau \xi_\tau{}^2$, and $\nabla_\xi{}^2 = \sum_\tau \partial^2/\partial\xi_\tau{}^2$, we can directly diagonalize $H_0 + H_\beta$. The resulting states would automatically include the interaction between shells with $\Delta n = 2$. However, in this pseudospherical coordinate system, the true angular momentum operator is not $-i\hbar\boldsymbol{\xi}\times\nabla_\xi$ and the $\boldsymbol{l}^2$ and $\boldsymbol{l}\cdot\mathbf{s}$ terms also offer some problems (Nilsson, 1955).

[2] Using $H_{l,s} = -\hbar\omega_0\{2\kappa\boldsymbol{l}\cdot\mathbf{s} + \mu\kappa[\boldsymbol{l}^2 - \tfrac{1}{2}n(n+3)]\}$, from Nilsson *et al.* (1969).

It is also useful to consider solutions to the basic Hamiltonian (6-135) in the asymptotic limit of very large deformations for which, from (6-128), we have $\omega_\perp \gg \omega_z$. Under these circumstances the oscillator number n_3 (refer to Exercise 6-14) involving the z vibrational modes becomes a good quantum number in addition to the principal oscillator number n. We may treat the $\boldsymbol{l}\cdot\mathbf{s}$ and $\boldsymbol{l}^2$ terms as small perturbations and use the asymptotic quantum numbers n, n_3, Λ, and Ω, traditionally written in the notation $\Omega\pi[nn_3\Lambda]$.[1]

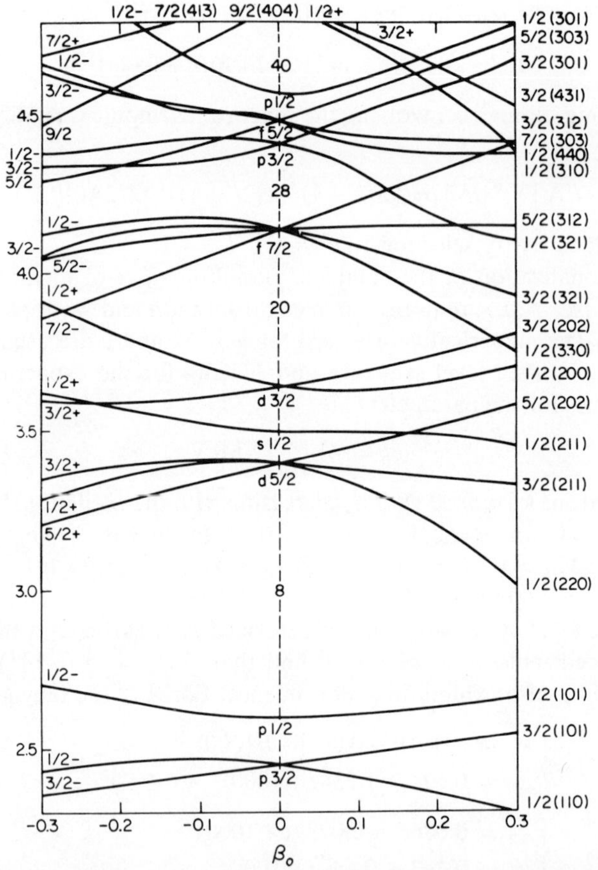

FIG. 6.19. Energy eigenvalues in units of $\hbar\omega_0$ plotted against the deformation parameter β_0 for the Nilsson model in the oscillator shells $n = 1$, 2, and 3. The spin–orbit coupling parameters μ_n for these shells are $\mu_1 = \mu_2 = 0$ and $\mu_3 = 0.35$. These energy levels are to be associated with odd-neutron or -proton nuclei in the 1p, 2s–1d, and 2p–1f shells [taken from Davidson (1968)].

[1] The parity designation $\pi = (-1)^n$ is redundant, but nonetheless is quoted traditionally for convenience.

The basis states now have the properties

$$l_3 \,|nn_3 \Lambda\Omega\rangle = \Lambda \,|nn_3 \Lambda\Omega\rangle, \qquad s_3 \,|nn_3 \Lambda\Omega\rangle = \Sigma \,|nn_3 \Lambda\Omega\rangle \tag{6-143}$$

with $\Omega = \Lambda + \Sigma$. The total energy in the asymptotic limit becomes

$$E(nn_3 \Lambda\Omega) = (n_3 + \tfrac{1}{2})\,\hbar\omega_z + (n - n_3 + 1)\,\hbar\omega_\perp + C\Lambda\Sigma$$
$$+ D[\Lambda^2 + n(2n_3 + 1) + n_3 - 2n_3{}^2]. \tag{6-144}$$

Finally, in Figs. 6.19–6.23, using the calculations of Davidson (1968), we

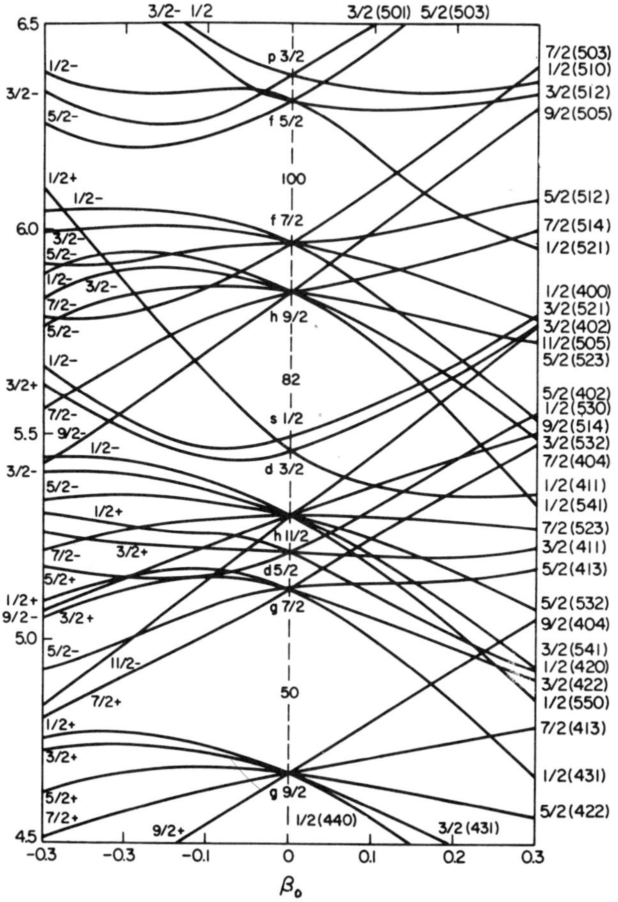

FIG. 6.20. Energy eigenvalues in units of $\hbar\omega_0$ plotted against the deformation parameter β_0 for the Nilsson model in the oscillator shells $n = 4$ and 5. The spin–orbit coupling parameters μ_n for these shells are $\mu_4 = 0.625$ and $\mu_5 = 0.630$. These energy levels are to be associated with odd-A nuclei in the range $50 < (N \text{ or } Z) < 82$ [taken from Davidson (1968)].

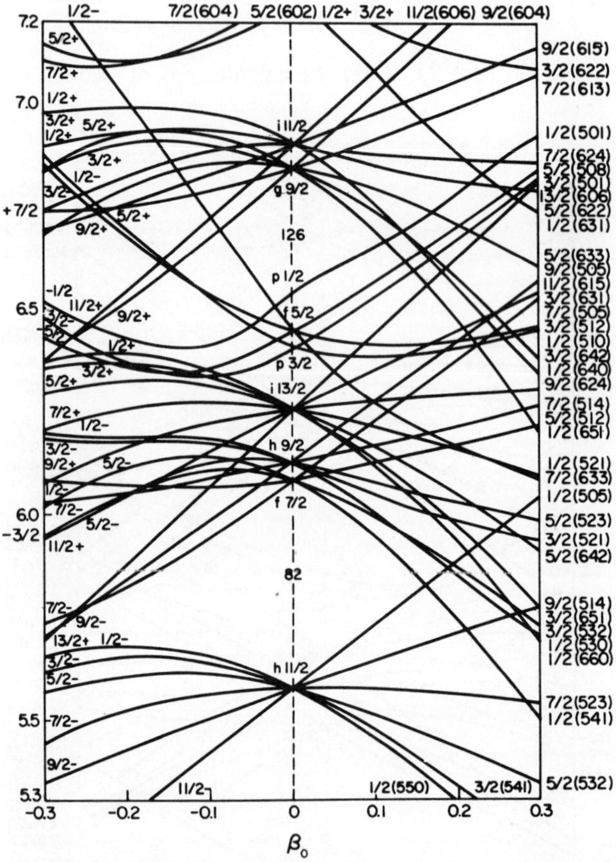

FIG. 6.21. Energy eigenvalues in units of $\hbar\omega_0$ plotted against the deformation parameters β_0 for the Nilson model in the oscillator shells $n = 5$ and 6. The spin–orbit coupling parameters μ_n for these shells are $\mu_5 = 0.450$ and $\mu_6 = 0.448$. These energy levels are to be associated with odd-neutron nuclei in the range $82 < N < 126$ [taken from Davidson (1968)].

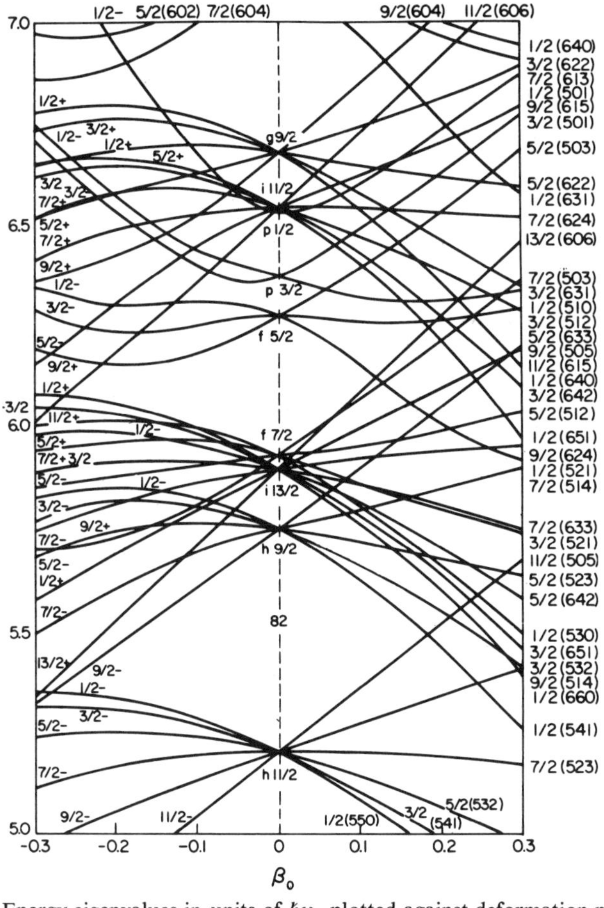

FIG. 6.22. Energy eigenvalues in units of $\hbar\omega_0$ plotted against deformation parameter β_0 for the Nilsson model in the oscillator shells $n = 5$ and 6. The spin–orbit coupling parameter μ_n for these shells are $\mu_5 = 0.700$ and $\mu_6 = 0.620$. These energy levels are to be associated with odd-proton nuclei with $Z > 82$. [taken from Davidson (1968)].

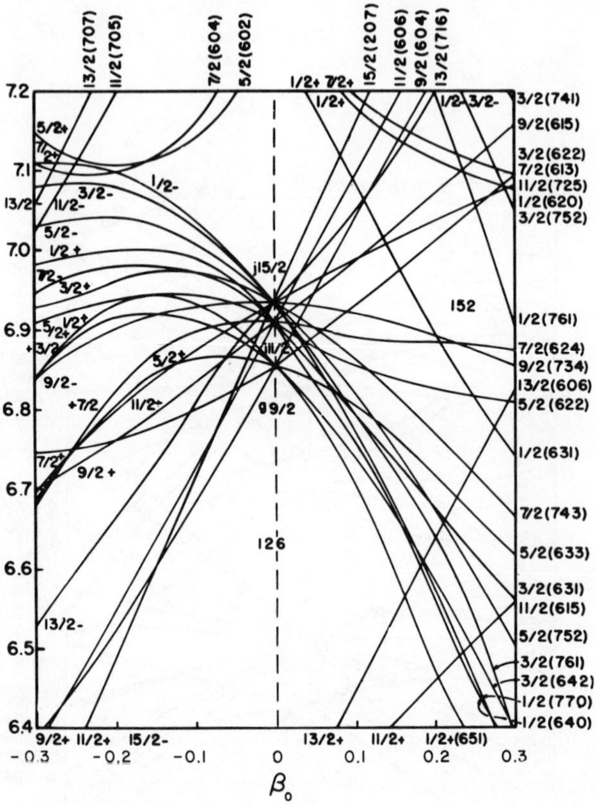

FIG. 6.23. Energy eigenvalues in units of $\hbar\omega_0$ plotted against the deformation parameter β_0 for the Nilsson model in the oscillator shells $n = 6$ and 7. The spin–orbit coupling parameters μ_n for these shells are $\mu_6 = 0.448$ and $\mu_7 = 0.434$. These energy levels are to be associated with odd-neutron nuclei with $N > 126$ [taken from Davidson (1968)].

show the results of diagonalizing the complete Hamiltonian (6-135). The energy is plotted in units of $\hbar\omega_0$ as a function of the quadrupole deformation parameter β_0. The value adopted for κ is 0.05 throughout. The values for μ in the different N, Z regions are labeled μ_n to indicate the oscillator number dependence and are given in the figure legends. The asymptotic labels $\Omega\,[nn_3\,\Lambda]$ appear along the right-hand side of the figures. In each case $\pi = (-1)^n$ is indicated for the orbitals along the left-hand side of the figure.

The magic numbers M_n (including spin–orbit effects) and the (l, j) quantum numbers are indicated for the spherical (i.e., $\beta_0 = 0$) cases. The level energy sequence is seen to follow essentially those appearing in Fig. 4.4. For nonzero values of β_0, $(2j+1)/2$ energy trajectories radiate from each such (l, j) $\beta_0 = 0$ concentration point, corresponding to the allowable values of Ω

when we take note of the energy degeneracy in Ω and $-\Omega$. For prolate deformations (i.e., $\beta_0 > 0$), these radiating trajectories always appear in order of increasing energy in the sequence $[n, n_3 = n, \Lambda = 0]$, $[n, n_3 = n-1, \Lambda = 1]$, $[n, n_3 = n-2, \Lambda = 2], \ldots, [n, n_3 = 0, \Lambda = n]$. Thus the highest energy trajectory $[n, n_3 = 0, \Lambda = n]$ corresponds to oscillation only in the direction of the minor axes. This type of behavior is characteristic of the anharmonic oscillator and Exercise 6-14 can be referred to profitably at this point.

A commonly used example to illustrate the behavior of the eigenfunctions of (6-135) as a function of the deformation β_0 is the case for $n = 5$, $\Omega = \frac{7}{2}$. It is somewhat more convenient to write these eigenfunctions using the basis $|nlj\Omega\rangle$ rather than the basis $|nl\Lambda\Sigma\rangle$. We therefore write

$$\chi_\Omega^{(n)} = |n\Omega\rangle = \sum_j C_{j\Omega} |nlj\Omega\rangle. \tag{6-145}$$

In this sum the value of j runs from $j = \Omega$ to $j = (2n+1)/2$. For each value of j the value of l is uniquely determined by $l = j \pm \frac{1}{2}$ *and* the parity condition $(-1)^n = (-1)^l$. Table 6-1 gives the values of $C_{j\Omega}$ using the results of Davidson for the relevant $\frac{7}{2}-[523]$, $\frac{7}{2}-[514]$, and $\frac{7}{2}-[503]$ asymptotic energy trajectories.

TABLE 6-1

THE COEFFICIENTS $C_{j\Omega}$ FOR THE EIGENFUNCTIONS OF THE ENERGY TRAJECTORIES LABELED $\frac{7}{2}-[523]$, $\frac{7}{2}-[514]$, AND $\frac{7}{2}-[503]$[a]

β_0 =		-0.3	-0.2	-0.1	0.0	0.1	0.2	0.3
K = 7/2-[523]								
E		5.215928	5.268329	5.295458	5.305000	5.302772	5.292658	5.277188
L	J							
3	7/2	-.403379	-.231732	-.096673	.000000	.068080	.116650	.152100
5	9/2	.095522	.088737	.050531	.000000	-.049855	-.093760	-.130681
5	11/2	.910034	.968724	.994033	1.000000	.996433	.988738	.979688
K = 7/2-[514]								
E		5.563549	5.677593	5.790853	5.855000	5.903576	5.953915	6.006885
L	J							
3	7/2	.767266	.784362	.492012	.000000	-.107764	-.131268	-.133505
5	9/2	.577215	.606072	.870581	1.000000	.992540	.985345	.979420
5	11/2	.279509	.132112	.003594	.000000	.057023	.108925	.151373
K = 7/2-[503]								
E		5.784819	5.807608	5.856453	5.972000	6.114888	6.263897	6.415632
L	J							
3	7/2	-.498586	-.575393	.865204	1.000000	.991843	.984460	.979307
5	9/2	.810986	.790445	-.489423	.000000	.111262	.142496	.153817
5	11/2	-.306127	-.210048	.109023	.000000	-.062200	-.102633	-.131524

[a] The basis states are $|n = 5, l, j, \Omega = \frac{7}{2}\rangle$ and the energies are in units of $\hbar\omega_0$. The parameters $\mu_5 = 0.63$ and $\kappa = 0.05$ are used [taken from Davidson (1968)].

The first trajectory $\frac{7}{2}-[523]$ has an energy $5.305\hbar\omega_0$ for $\beta_0 = 0$ and a corresponding eigenfunction which is a pure ($j = l+\frac{1}{2} = \frac{11}{2}$, $m_j = \frac{7}{2}$) spherical wave function now appearing in the notation $|nlj\Omega\rangle = |5, 5, \frac{11}{2}, \frac{7}{2}\rangle$. For $\beta_0 = 0$ the other two eigenfunctions at energies $5.855\hbar\omega_0$ and $5.972\hbar\omega_0$ are pure ($j = l-\frac{1}{2} = \frac{9}{2}$, $m_j = \frac{7}{2}$) and ($j = l+\frac{1}{2} = \frac{7}{2}$, $m_j = \frac{7}{2}$) spherical wave functions. When β_0 assumes nonzero values, these three spherical wave functions are mixed in varying degrees; thus, for example, when $\beta_0 = +0.2$, the energy of the $\frac{7}{2}-[523]$ trajectory has become $5.293\hbar\omega_0$ with the mixed wave function

$$|n, \Omega\rangle = 0.117\,|5, 3, \tfrac{7}{2}, \tfrac{7}{2}\rangle - 0.094\,|5, 5, \tfrac{9}{2}, \tfrac{7}{2}\rangle + 0.989\,|5, 5, \tfrac{11}{2}, \tfrac{7}{2}\rangle.$$

▶**Exercise 6-15** (a) In view of (6-137), show that

$$|5, 3, \tfrac{7}{2}, \tfrac{7}{2}\rangle = |533, +\rangle$$

$$|5, 5, \tfrac{9}{2}, \tfrac{7}{2}\rangle = -0.426\,|553, +\rangle + 0.905\,|554, -\rangle$$

$$|5, 5, \tfrac{11}{2}, \tfrac{7}{2}\rangle = +0.905\,|553, +\rangle + 0.426\,|554, -\rangle.$$

(b) Calculate the eigenfunctions of the $\frac{7}{2}-|503\rangle$ trajectory for $\beta_0 = 0$, 0.1, 0.2, and 0.3 in the $|nl\Lambda\Sigma\rangle$ basis.

(c) Why would you expect the $|554, -\rangle$ component in part (b) to approach zero as $\beta_0 \to +\infty$? ◀

Although for single-particle orbitals with $\beta_0 \neq 0$, $\mathbf{j}^2$ is clearly seen not to be a constant of the motion, there are some exceptions. These occur (ignoring the $\Delta n = 2$ oscillator shell mixing) for the cases with $|\Omega| = j = (2n+1)/2$, as, for example, for the $\Omega\pi[nn_3\Lambda]$ trajectories: $\frac{3}{2}-[101]$; $\frac{5}{2}+[202]$; $\frac{7}{2}-[303]$; $\frac{9}{2}+[404]$, etc.

For a recent critique on the single-particle model in nonspherical nuclei, refer to Ogle *et al.* (1971).

5. Odd-A Deformed Nuclei

We are now in a position to return to the discussion of the strong coupling of particle and collective motions in the A-odd nuclei. The Nilsson model provides us with the particle wave function $\chi_\Omega^{(n)} = |n\Omega\rangle$ required in (6-115)–(6-117), when only a single extra-core nucleon is considered. If several extra-core nucleons are to be considered explicitly, an additional interaction Hamiltonian $\frac{1}{2}\sum_{i,j}^{k} v_{ij}$ coupling the extra-core nucleons must be given.

For the simple case of a single extra-core nucleon, A odd, $J = K = \Omega$ ($K \neq \frac{1}{2}$) corresponds to the entire nuclear system ground-state angular momentum with energy $E_n(K = \Omega) - \hbar^2 K(K-1)/2\mathcal{I}_0$, determined by (6-118). This level serves as the band head for a rotational band with subsequent angular momenta $J = K+1, K+2, \ldots$, the energy in each case again being given by (6-118). The case $K = \Omega = \frac{1}{2}$ first requires the determination of the decoupling parameter a for the particular assumed deformation β_0, using (6-123b) and the appropriate tabulated values of $C_{j\Omega}$. The actual level sequence of angular momenta in order of increasing energy is determined by (6-124). For illustrative purposes we give Davidson's values of a in Table 6-2 for a few low-energy orbitals as a function of β_0. It can be readily seen by reference to this table that not uncommonly values of $a > 1$ and $a < -1$ occur and hence nonnormal J sequences are to be expected.

TABLE 6-2

DECOUPLING PARAMETER a AS A FUNCTION OF β_0 FOR THE ORBITALS $\Omega\pi[nn_3\Lambda]$

$\beta_0 =$	-0.3	-0.2	-0.1	0	$+0.1$	$+0.2$	$+0.3$
$\frac{1}{2}-[110]$	-0.804	-1.157	-1.698	-2.000	-1.846	-1.650	-1.512
$\frac{1}{2}-[101]$	-0.196	$+0.157$	0.698	1.000	0.846	0.650	0.512
$\frac{1}{2}+[220]$	1.562	1.775	2.264	3.000	2.624	2.252	1.997
$\frac{1}{2}+[211]$	0.929	1.236	1.391	1.000	0.837	$+0.193$	-0.090
$\frac{1}{2}+[200]$	-0.492	-1.011	-1.655	-2.000	-1.461	-0.445	$+0.093$
$\frac{1}{2}-[330]$	-2.351	-3.175	-3.795	-4.000	-3.815	-3.383	-2.926
$\frac{1}{2}-[321]$	0.097	$+0.431$	-0.446	-2.000	-0.499	$+0.495$	0.606

The first illustration of an odd-A rotational nucleus is the interesting case of $^{25}_{13}$Al shown in Fig. 6.24. This nucleus was one of the earliest examples of a definitive deformed rotational light nucleus. The odd 13th proton, counting up in two's from the lowest Nilsson orbitals shown in Fig. 6.19 for the 1p shell and higher,[1] could be placed in either of the crossing orbitals $\frac{1}{2}+[211]$ or $\frac{5}{2}+[202]$, assuming $0.1 < \beta_0 < 0.3$. The actual ground state turns out to be the $\frac{5}{2}+[202]$ orbital, with the $\frac{1}{2}+[211]$ only 0.45 MeV higher in energy. These states generate rotational bands with $K = \frac{5}{2}$ and $K = \frac{1}{2}$, respectively.

Figure 6.19 also shows three nearby closely spaced $\Omega = \frac{1}{2}$ states: $\frac{1}{2}+[211]$, $\frac{1}{2}+[200]$, and $\frac{1}{2}-[330]$. These serve as good examples of the important role of the decoupling parameter a. Table 6-2 shows that for $\beta_0 \approx 0.24$, the parameter a for three cases is ~ 0, -0.3, and -3.2, respectively. With these values of a used in (6-124) we can readily confirm the level ordering and spacing shown in Fig. 6.24 for the $K = \frac{1}{2}$ bands. With the large negative value of a, the $K = \frac{1}{2}$ odd-parity sequence actually has the J order $\frac{3}{2}, \frac{7}{2}, \frac{1}{2}, \frac{11}{2}, \frac{5}{2}, \ldots$.

The highest rotational band shown in Fig. 6.24 is based on the $\frac{3}{2}+[211]$ band head. This is presumably a hole state with a strongly coupled pair in the $\frac{5}{2}+[202]$ orbital. The analog state nucleus $^{25}_{12}$Mg has virtually the same level

[1] The 1s shell pertaining to the nuclides $A \leqslant 4$ is not shown. The occupied orbitals are filled in pairs if possible, with quantum numbers $\pm\Omega$.

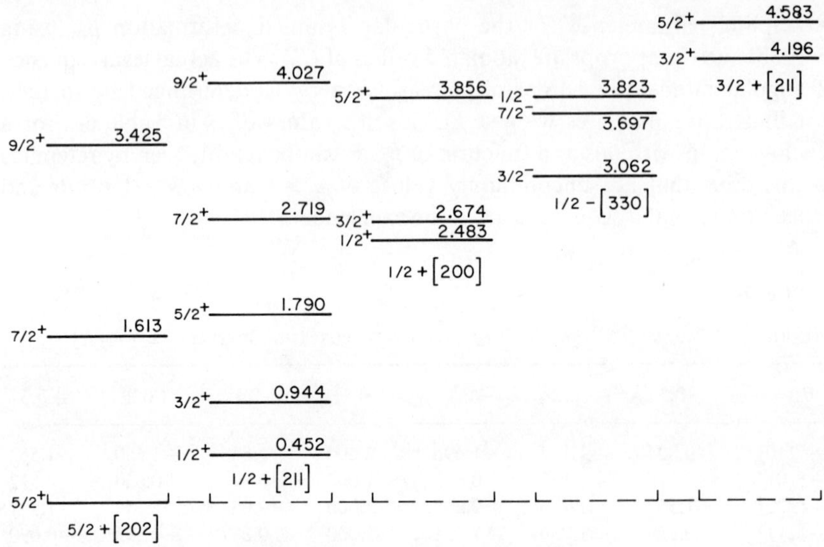

FIG. 6.24. The rotational band structure of $^{25}_{13}$Al. The Nilsson orbitals are indicated in the notation $K = \Omega$, $\pi[nn_3\Lambda]$; see text. Level energies in MeV.

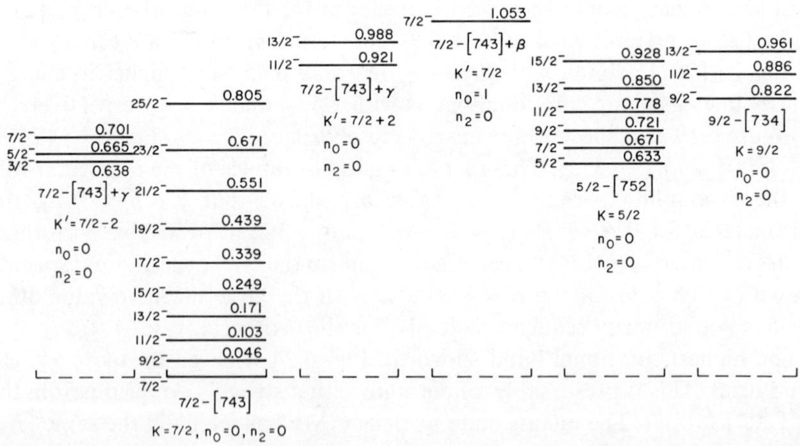

FIG. 6.25. The odd-parity rotational band structure of $^{235}_{92}$U. The Nilsson orbitals are indicated in the notation $K = \Omega$, $\pi[nn_3\Lambda]$; see text. Level energies in MeV. Even-parity bands are also known.

structure as ${}^{25}_{13}$Al with the lowest levels being $(J^\pi, E_x) = (\frac{5}{2}^+, \text{gd}), (\frac{1}{2}^+, 0.58), (\frac{3}{2}^+, 0.98), (\frac{7}{2}^+, 1.61), (\frac{5}{2}^+, 1.96), \ldots$. Other level characteristics such as electromagnetic transition rates and ground-state magnetic moment also confirm the deformed rotational character of these two nuclei.

The second example of an odd-A nucleus is the case of the actinide nucleus ${}^{235}_{92}$U shown in Fig. 6.25. Referring to Fig. 6.23, the odd 143rd neutron is seen to fall into the $\frac{7}{2}-[743]$ orbital for $0.2 < \beta_0 < 0.3$ when we allow for the strong pairing energy for two $\frac{13}{2}+[606]$ neutrons. A very well-developed rotational band based on this ground-state Nilsson orbital in fact is observed experimentally.

Although reference was made to possible core excitations in treating the strong coupled particle–collective motion theory, this possibility was suppressed in the discussion for the sake of simplicity. The present example was also selected to indicate the known presence of such core excitations. The neighboring even–even nuclei ^{234}U and ^{236}U have well-developed ground-state rotational bands $0^+, 2^+, 4^+, 6^+, 8^+$. In addition each shows a $0^+, 2^+, 4^+$ β band (i.e., $K = 0$, $n_0 = 1$, $n_2 = 0$; refer to Fig. 6.13), at band head energies 0.810 and 0.920 MeV, respectively. The fourth column of Fig. 6.25 shows a possible $\frac{7}{2}-[743]$ particle–core β-vibrational coupled mode at an energy of 1.053 MeV in ^{235}U. The nucleus ^{234}U also has a well-developed $2^+, 3^+, 4^+$ γ-vibrational band ($K = 2$, $n_0 = 0$, $n_2 = 0$) at a band head energy of 0.927 MeV. The first and third columns of Fig. 6.25 show the $\frac{7}{2}-[743]$ particle–core γ-vibrational coupled mode with $K' = \frac{7}{2} \pm 2$ with band heads at 0.638 and 0.921 MeV.

Again referring to Fig. 6.23, additional Nilsson particle intrinsic states of odd parity could involve the $\frac{9}{2}-[734]$ particle orbital and a $\frac{5}{2}-[752]$ hole state, with a presumed coupled pair in the $\frac{7}{2}-[743]$ orbital. The two odd-parity rotational bands shown in columns 5 and 6 of Fig. 6.25 for ^{235}U are to be associated with these two intrinsic particle states. Well-developed even-parity rotational bands in ^{235}U are also known: a low-lying $K = \frac{1}{2}$ band based on the $\frac{1}{2}+[631]$ orbital, a $K = \frac{3}{2}$ band, and two $K = \frac{5}{2}$ bands. These bands are not shown in Fig. 6.25. Again the comparisons of the predicted and observed electromagnetic transition rates, ground-state magnetic moments, etc., confirm the deformed rotational character of the ^{234,235,236}U nuclei.

For a thorough review of intrinsic collective states in odd-A nuclei in the region $150 < A < 190$, refer to Bunker and Reich (1971) and Ogle *et al.* (1971).

6. Odd–Odd Deformed Nuclei

We now turn to a brief description of the odd-N, odd-Z deformed nuclei. The simplest assumption is to consider only the last odd proton and neutron with each moving independently in an appropriate Nilsson orbital. The residual neutron–proton interaction can be considered as a final perturbation. The total angular momentum is the sum of the core angular momentum $\mathbf{R}$ and the coupled nucleon angular momenta $\mathbf{j}_p + \mathbf{j}_n$, or $\mathbf{J} = \mathbf{R} + \mathbf{j}_p + \mathbf{j}_n$. The formalism introduced with (6-111) and following thereafter pertains. The quantum number K is now

$$K = |\Omega_p \pm \Omega_n|. \tag{6-146}$$

This leads to two rotational bands based on the band heads $K_1 = |\Omega_p - \Omega_n|$ and $K_2 = \Omega_p + \Omega_n$, viz.,

$$\begin{aligned} J_1 &= |\Omega_p - \Omega_n|,\ |\Omega_p - \Omega_n| + 1,\ |\Omega_p - \Omega_n| + 2, \ldots \\ J_2 &= \Omega_p + \Omega_n,\ \Omega_p + \Omega_n + 1,\ \Omega_p + \Omega_n + 2, \ldots. \end{aligned} \tag{6-147}$$

The corresponding energies are given by a straightforward modification of (6-118),

$$E(J, M, K_{1,2}) = E(\Omega_p) + E(\Omega_n) + (\hbar^2/2\mathscr{I}_0)[J_{1,2}(J_{1,2}+1) - 2K_{1,2}^2]. \tag{6-148}$$

The two band head energies are in general close to being equal and the precise nature of the neutron–proton residual interaction is needed to determine the character of the ground state. Again, rules paralleling the Nordheim rules [see (4-102), (4-103), and relevant discussion in Chapter IV] can be formulated. The Gallagher and Moszkowski rule cited in Chapter IV gives

$$\begin{aligned} K_{gd} &= |\Omega_p - \Omega_n| \qquad \text{for} \quad \Omega_p = \Lambda_p \pm \tfrac{1}{2}, \quad \Omega_n = \Lambda_n \mp \tfrac{1}{2} \\ K_{gd} &= \Omega_p + \Omega_n \qquad \text{for} \quad \Omega_p = \Lambda_p \pm \tfrac{1}{2}, \quad \Omega_n = \Lambda_n \pm \tfrac{1}{2}. \end{aligned} \tag{6-149}$$

The total wave function for the odd–odd case is a somewhat more complicated version of (6-117), namely

$$\begin{aligned} |J, M, K = |\Omega_p \pm \Omega_n|, n, n'\rangle = \left(\frac{2J+1}{16\pi^2}\right)^{1/2} \sum_{j_n, j_p} C_{j_p\Omega_p}^{(n)} C_{j_n\Omega_n}^{(n')} \, [\chi_{j_p\Omega_p}^{(n)} \chi_{j_n\Omega_n}^{(n')} D_{MK}^J \\ + (-1)^{J - j_p - j_n} \chi_{j_p, -\Omega_p}^{(n)} \chi_{j_n, -\Omega_n}^{(n')} D_{M,-K}^J], \end{aligned} \tag{6-150}$$

with $K = |\Omega_p \pm \Omega_n|$.

Now the special case $K = 0$ does not reduce to the sequence $J^\pi = 0^+, 2^+, 4^+, \ldots$, as with (6-66b). The total angular momentum can take on all the integral values $J = 0, 1, 2, 3, 4, \ldots$, with a parity $(-1)^{n+n'}$, where n and n' refer to the principal oscillator shell numbers of the proton and neutron. Since the residual neutron–proton interaction is different in relative even and odd states, this angular momentum sequence splits into two rotational bands $J = 0, 2, 4, \ldots$, and $J = 1, 3, 5, \ldots$, when this interaction is included.

For a typical theoretical treatment of the parallel and antiparallel coupled odd-neutron, odd-proton states for both zero-range and finite-range nuclear forces, see Jones *et al.* (1971) and Nakagawa *et al.* (1972).

The standard example of an odd–odd nucleus is $^{166}_{67}Ho$. For the ground-state band, the odd 67th proton is in a $\frac{7}{2}-[523]$ Nilsson orbital, while the odd 99th neutron is in the $\frac{7}{2}+[633]$ orbital. Thus we have $\Omega_p = \Lambda_p + \frac{1}{2}$ as well as $\Omega_n = \Lambda_n + \frac{1}{2}$ (i.e., $\Lambda = 3$ and $\Omega = \frac{7}{2}$ in each case). The Gallagher–Moszkowski rule (6-149) would predict that of the two possible values for band head angular momenta $|\Omega_p \pm \Omega_n|$, namely $J^\pi = 0^-$ and 7^-, the ground state should be $J^\pi = 7^-$. Instead the experimental situation has the $K = 0$ band 9 keV lower in energy than the $K = 7$ band. The ground-state band splits into an even-J sequence $0^-, 2^-, 4^-, \ldots$, and an odd-J sequence $1^-, 3^-, 5^-, \ldots$, where this second sequence is displaced considerably upward in energy (i.e., the 1^- level is actually higher than the 2^- level). Additional bands based on promoting the neutron into the $\frac{1}{2}-[521]$ and $\frac{5}{2}-[512]$ orbitals exist. The experimental situation is displayed in Fig. 6.26.

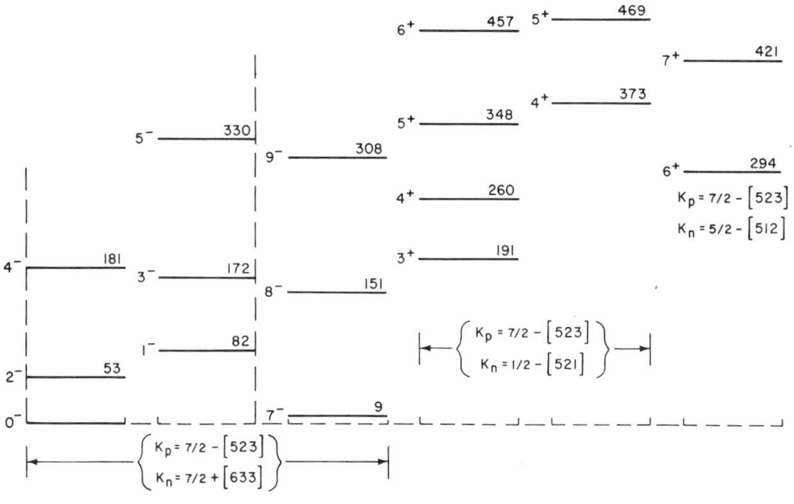

FIG. 6.26. The rotational band structure of the odd-N, odd-Z nucleus $^{166}_{67}Ho$. The Nilsson orbitals are indicated in the notation $K_{n,p} = \Omega_{n,p}, \{n[nn_3\Lambda]\}_{n,p}$; see text. Level energies in keV.

F. CONCLUDING REMARKS

Only the briefest introductory treatment has been given in this chapter to the highly successful collective model of the nucleus. The theoretical and experimental research in this area is currently proceeding vigorously and the reader is well advised to consult the recent literature.

The Nilsson model discussed in the last section has been remarkably successful in giving a description of single-particle states in deformed nuclei using only the symmetric quadrupole deformation contribution $H_\beta(\lambda=2)$ in the particle Hamiltonian. Several improvements on this simple model have been proposed. Perhaps the most important of these has been the inclusion of a hexadecapole deformation term of the form $\beta(\lambda=4)Y_4{}^0(\theta,\varphi)$ in the particle Hamiltonian. The work of Andersen (1971) gives an example of its inclusion in a basic oscillator potential framework similar to the original Nilsson approach we have discussed.

A further improvement would be the use of a basic Woods–Saxon potential shape in place of the oscillator potential. Significant early papers by Lemmer (1960) and Lemmer and Green (1960) used a particle Hamiltonian of a nonlocal character combining features of the general nonlocal form (3-116) and a deformed Woods–Saxon potential with spin–orbit interaction, viz.,

$$\left\{\frac{1}{4}\left[\frac{1}{2m^*}\mathbf{p}^2+2\mathbf{p}\cdot\frac{1}{2m^*}\cdot\mathbf{p}+\mathbf{p}^2\frac{1}{2m^*}\right]+V\left(\frac{r}{1+\sum_\mu\alpha_{2\mu}Y_2{}^\mu}\right)+C\boldsymbol{l}\cdot\mathbf{s}\right\}\psi = E\psi, \tag{6-151}$$

with m^* as defined in connection with (3-116). The effect of the additional terms in $\mathbf{p}$ was found to be nearly equivalent to that of the $\boldsymbol{l}^2$ term introduced orginally on an ad hoc basis by Nilsson. The overall results were in fact quite similar to those found by Nilsson.

We have already cited the work of Götz *et al.* (1972) in connection with the transitional nuclei in the rare earth mass region. They use a Woods–Saxon potential and include hexadecapole deformations. A notation generally used for the nuclear surface when such deformations are included is

$$\begin{aligned} S(\theta,\varphi) = R(\beta,\gamma,\beta_4)[1+\beta_2Y_2{}^0(\theta,\varphi) \\ +\beta_{22}\{Y_2{}^2(\theta,\varphi)+Y_2^{-2}(\theta,\varphi)\}+\beta_4Y_4{}^0(\theta,\varphi)], \end{aligned} \tag{6-152}$$

with $\beta_2=\beta\cos\gamma$ and $\beta_{22}=(\beta/\sqrt{2})\sin\gamma$. Nuclear incompressibility is achieved by having R depend on (β,γ) and β_4. An extensive programmed search by these authors for minima in the ground-state potential energy surfaces for the rare earth nuclei revealed very few stable, deformed, triaxial ellipsoidal shapes. On the other hand, important β_4 contributions were found for the equilibrium potential values. Their results are shown in Fig. 6.27. Attention is called to the abrupt change from prolate to oblate shapes at $A\approx190$ and the large values of β_4 for the transitional nuclei. The nucleus ^{192}Pt, previously given as a possible example of a triaxially asymmetric rotator illustrated in Fig. 6.11, and the nucleus ^{192}Os fall into the abruptly changing or discontinuous narrow range near $A=190$.

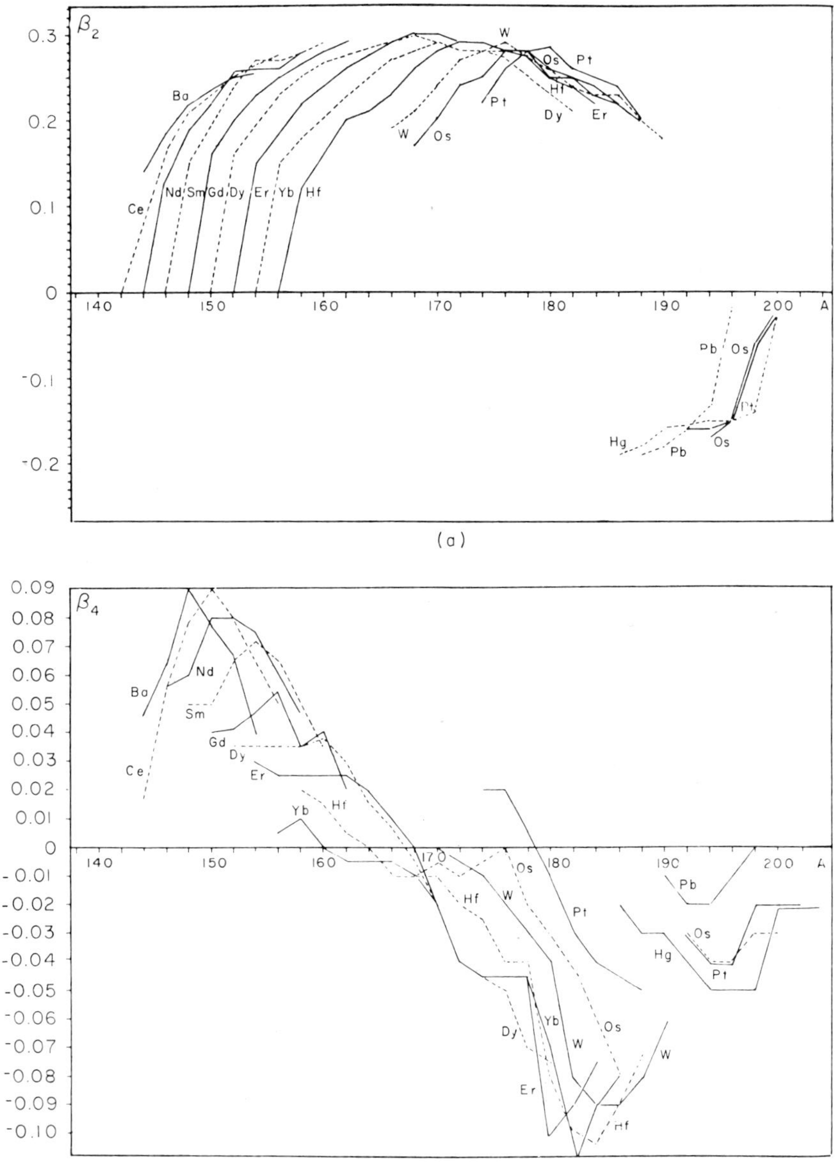

FIG. 6.27. (a) Equilibrium values of the quadrupole deformation parameter β_2; (b) equilibrium values of the hexadecapole deformation parameter β_4 found by Götz *et al.* (1972) for the rare earth nuclei.

There are numerous experimental indications of the presence of hexadecapole deformations. One interesting example is for the elastic and inelastic scattering of high-energy α-particles (i.e., $E_\alpha \geqslant 50$ MeV) from deformed nuclei. Since α-particles are highly absorbed in the nuclear interior, these reactions desirably emphasize the effects of the target nucleus surface. Such high-energy α-particles interacting with the surface of the target nucleus can impart large direct angular momentum transfers and thus excite rotational bands and allow the discovery of the presence of any possible higher nuclear multipoles. The analysis of the data using a coupled-channels technique permits a determination of β_2 and β_4 (including their relative phase). Such experiments have been performed for many nuclei. The rare earth region has been investigated by Hendrie *et al.* (1968). Their results even suggest the possibility of the presence of a $\beta(\lambda=6)Y_6{}^0(\theta,\varphi)$ deformation in addition to the clearly required $\beta(\lambda=4)Y_4{}^0(\theta,\varphi)$ term. The experimental β_4 values closely agree with those of Fig. 6.27. Other interesting examples are for the target nuclei ^{20}Ne, ^{28}Si, and ^{46,48,50}Ti reported by Rebel *et al.* (1971, 1974). They give $\beta_2 = +0.35$ and $\beta_4 = +0.11$ for prolate ^{20}Ne, and $\beta_2 = -0.32$ and $\beta_4 = +0.08$ for oblate ^{28}Si. The nucleus ^{46}Ti appears to be characterizable as a soft prolate rotator, ^{48}Ti as a β-soft, asymmetric, deformed nucleus of the triaxial Davydov–Chaban type, and ^{50}Ti as an anharmonic vibrator.

For an interesting determination of the hexadecapole moment for the rare earth nuclei Sm, Gd, Dy, and Er using precision Coulomb excitation, see Erb *et al.* (1972) and Saylor *et al.* (1972). The deduced values of β_2 and β_4 are in accord with Fig. 6.27.

Detailed knowledge of the nuclear surface deformation is also evidently very important in understanding such phenomena as fission, shape isomerism, and the possible existence of "superheavy "nuclei near $Z = 114$ and $N = 184$. In addition to models discussed up to this point, another approach also has been successful in shedding light on surface deformations, the so-called "two-center" model. In this model the single-particle equipotential surfaces generally consist of two separated spheroidal shapes smoothly joined by a transitional surface having overall axial symmetry. Asymmetric surfaces possessing two separate centers of reference also are considered. Such models are particularly useful when very large deformations are involved as nuclei approach scission stage and final fission, since the parameter referring to the separation of the two centers of reference offers a natural characterization of these phenomena. Useful recent references are Andersen *et al.* (1970), Fabian *et al.* (1972), and Albrecht (1973), and the references cited therein.

Considerable insight into the nature of the collective nuclear fluid behavior has been recently derived from the study of very large angular momentum states, particularly in rotational nuclei. In earlier sections we saw that experimental rotational energies increased less rapidly than the simple $J(J+1)$

behavior, explained somewhat simplistically as a centrifugal stretching [see (6-92a), (6-94a), or Exercise 6-13] or in the vibrational–rotational model explained as a nonadiabatic coupling perturbation [see discussion relevant to (6-106) and (6-108)]. Alternately the effect can be viewed as an increase in the moment of inertia with increasing J.

In exhibiting the variation of the moment of inertia $\mathscr{I}$ with J the custom is to plot experimental local values of $2\mathscr{I}/\hbar^2$ as a function of $(\hbar\omega)^2$, where ω is the local rotational angular velocity defined by the canonical relation

$$\hbar\omega = dE/d[J(J+1)]^{1/2}. \tag{6-153a}$$

This relationship can also be written as

$$(\hbar\omega)^2 = 4J(J+1)\{dE/d[J(J+1)]\}^2. \tag{6-153b}$$

The moment of inertia can be defined in the standard manner,

$$\mathscr{I}\omega = \hbar[J(J+1)]^{1/2}, \tag{6-154a}$$

which combined with (6-153a) gives

$$2\mathscr{I}/\hbar^2 = \{dE/d[J(J+1)]\}^{-1}. \tag{6-154b}$$

The indicated derivatives are locally determined from the energy spacing $E(J_i)-E(J_i-2)$, and $J(J+1)$ is taken as the mean value in the interval, viz.,

$$J(J+1) = \tfrac{1}{2}[J_i(J_i+1)+(J_i-2)(J_i-1)] = J_i^2 - J_i + 1.$$

The result is

$$\left.\frac{dE}{d[J(J+1)]}\right|_{J(J+1)=J_i^2-J_i+1} = \frac{E(J_i)-E(J_i-2)}{4J_i-2}. \tag{6-155}$$

When the empirical value obtained from (6-155) is substituted into (6-153b) and (6-154a), curves of the type shown in Fig. 6.28 result. Figure 6.28 shows that $\mathscr{I}$ tends to increase slowly at first, possibly due to an effective centrifugal stretching effect; however, at a more or less critical angular momentum J_c the curve breaks into a sharp S curve or *backbending* curve. Although earliest experimental results discovered the backbending phenomenon in even–even rotational nuclei [e.g., see Johnson *et al.* (1972) and Khoo *et al.* (1973)], the effect has also been observed for odd-A rotational nuclei [e.g., Grosse *et al.* (1974)] and even in "spherical" nuclei [e.g., those nuclei exhibiting quasi-rotational bands such as the Te isotopes; see Marshalek (1972)].

The backbending phenomenon was in fact predicted on theoretical grounds in 1960 by Mottelson and Valatin. Recall that the collective behavior of nuclei is intimately associated with the pairing correlations existing between a relatively small number of pairs of particles in two time-reversed single-particle states. It might be expected that relatively small disturbances of the

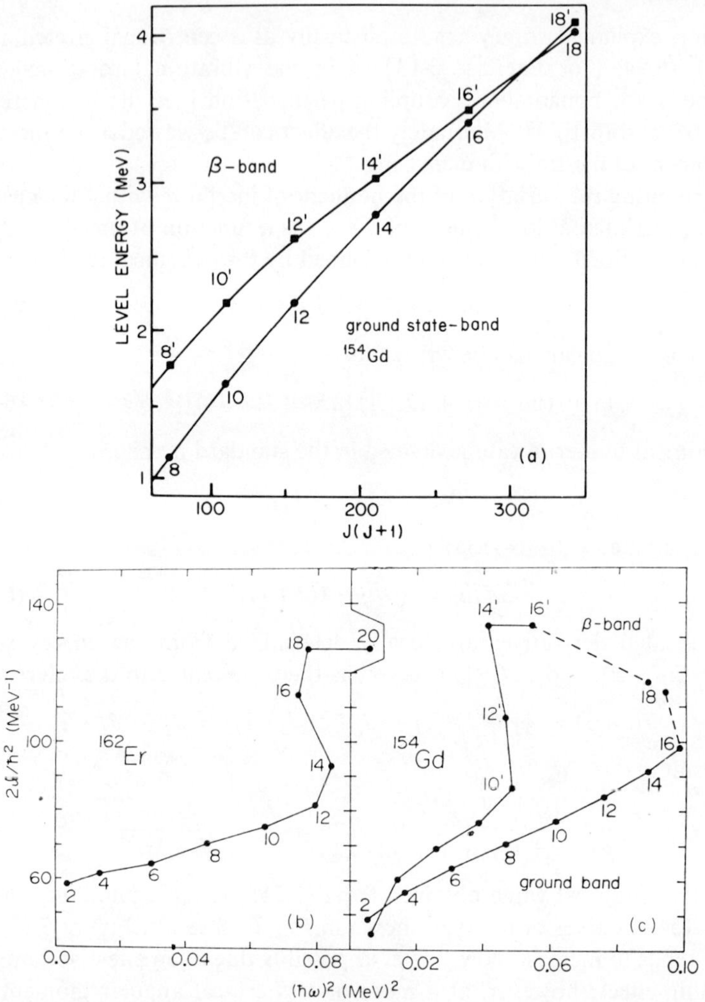

FIG. 6.28. (a) Plot of the upper energy end of the level energies for both the ground-state band and β band of ^{154}Gd. (b, c) Backbending curves plotting $2\mathscr{I}/\hbar^2$ as a function of $(\hbar\omega)^2$ for (b) the ground-state band of ^{162}Er and (c) the ground-state band and β band of ^{154}Gd. In all cases the values of J_i are indicated as parameter labels attached to the experimental points [curves taken from Khoo *et al.* (1973) and Ross and Nogami (1973)].

entire system might be sufficient to destroy these correlations. The Coriolis force acts in opposite directions on the paired particles and thus tends to decouple their pairing correlation, a behavior characterized as *Coriolis antipairing* (CAP). The main result of the above as we consider rotational states with increasing angular momentum is to at first slowly increase the effective collective moment of inertia, but then at a critical value the pairing is abruptly altered (reduced) corresponding to a phase transition not unlike the phenomenon occurring in superconductors at the critical magnetic field. Thus one speaks of the backbending as indicating a phase transition from a superfluid phase to a nonsuperfluid phase. A number of other competing theoretical explanations have been put forward [e.g., see Krumlinde and Szymański (1974) and Stephens *et al.* (1974)]. Sheline (1972) has given a careful discussion of the critical angular momentum J_c. Suggested empirical values are $J_c \approx 3$, 8–10, and 10–12 for A regions ~ 22, ~ 102, and ~ 126, respectively. Various theories predict a behavior

$$J_c(J_c+1) \approx \xi A^n, \tag{6-156}$$

with $\frac{7}{6} < n < \frac{5}{3}$. For additional theoretical comment and review papers, see Grosse *et al.* (1973), Johnson and Szymański (1973), Ross and Nogami (1973), Gorfinkiel *et al.* (1974), Goodman and Goswami (1974), Dasso and Klein (1974), and Mantri and Sood (1974).

We have mentioned several examples of permanently deformed nuclei having relatively low mass numbers. The mass region $19 \leqslant A \leqslant 31$ by now has been established clearly to consist of permanently deformed nuclei and contains the cited nuclei ^{20}Ne, ^{25}Al, and ^{28}Si. In spherical shell model language, all these nuclei are in the (1d, 2s) shell. The lower mass nuclei in this shell have received considerable theoretical attention. The pioneering investigation by Elliott and Flowers (1955) considered the nuclei ^{18,19}O and ^{18,19}F to consist of an inert ^{16}O core and two or three valence particles in the $1d_{5/2,3/2}$ and $2s_{1/2}$ subshells. More recently, Arima *et al.* (1968), using a more severely truncated shell model space with only the $1d_{5/2}$ and $2s_{1/2}$ subshells and an inert ^{16}O core, successfully treated ^{18,19,20}O, ^{18,19,20}F, and ^{20}Ne. Since the ^{16}O nucleus even in its ground state is known to have considerable particle–hole configuration admixtures [e.g., see Brown and Green (1966)], McGrory and Wildenthal (1973) employed a "large shell model calculation" to study the $A = 18$, 19, and 20 nuclei using a ^{12}C inert core and $A-12$ nucleons in $1p_{1/2}$, $1d_{5/2}$, and $2s_{1/2}$ shell model configurations. By including the $1p_{1/2}$ subshell, most excitations of the ^{16}O core assumed by the previous investigators to be inert are now automatically included. All the troublesome odd-parity ^{19}F states (such as the $\frac{1}{2}^-$ first excited state at 0.11 MeV)[1] are also now successfully accounted for. Although very much more complicated, the procedures were similar in all the above calculations to our discussion of intermediate coupling for ^{6}Li in Chapter V. Comparisons of the model

[1] In Chapter V this state was described as a ^{15}N + α-cluster model state equivalent to the shell model description $(1s)^4(1p)^{11}(1s, 2d)^4$. That contention is now seen to be completely justified.

characteristics in all the above cases (although implying different effective Hamiltonians) with experimental information relating to level schemes, electromagnetic moments, and transition rates, etc., are quite satisfactory.

The surprising result, however, is that all of the above microscopic spherical shell model calculations in a quite natural way account for levels that appear to be strikingly rotational in character. The energy levels of the nucleus ^{20}Ne, for example, can be fitted with six or seven fairly well-developed rotational bands accounting for most of the 20 or so levels below 15 MeV in excitation. The ground-state rotational band has also been discussed from a cluster model point of view [e.g., Chang (1969)]. For a recent reference, particularly on possible high-spin rotational states in ^{20}Ne, see Panagiotou *et al.* (1972) and the references cited therein. In all the above cases a simple, deformed-potential, strong-coupling model (particularly if band mixing is considered) can account equally well for the nuclear levels. In these fits the phenomenologically required parameters are all physically reasonable.

Perhaps even more surprising is the suggestion by Clegg (1961) that a similar situation exists in the 1p-shell nuclei, $4 < A \leqslant 16$. More recently El-Batanoni and Kresnin (1966) have successfully parameterized the odd-A nuclei in this shell with physically acceptable rotational parameters. The case of the isobaric

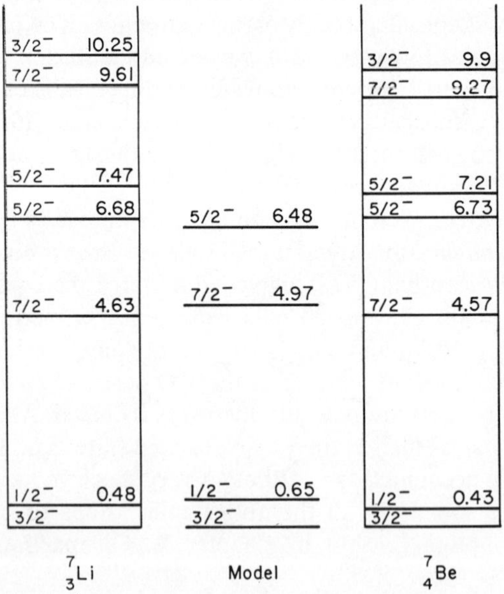

FIG. 6.29. The experimental energy levels of ^7Li and ^7Be, indicating possible collective rotational assignments using the model discussed in the text. All the levels shown have isospin $T = \frac{1}{2}$.

analog nuclei ^{7}Li and ^{7}Be is illustrated in Fig. 6.29. These $A = 7$ nuclei are assumed to have $\beta_0 \approx 0.4$[1] and to have the odd nucleon in a $\frac{1}{2}-[110]$ Nilsson orbital. An appropriate decoupling parameter would be $a \approx -1.40$, inverting the energy ordering of the $K = \frac{1}{2}$ band to yield a ground state of $J^\pi = \frac{3}{2}^-$ followed by the sequence $J^\pi = \frac{1}{2}^-, \frac{7}{2}^-$, and $\frac{5}{2}^-$, just as experimental observation requires. With the value of $\hbar^2/2\mathscr{I}_0 \approx 0.54$ MeV used in (6-124), a satisfactory energy match is obtained (refer to the figure).

The nuclei $A = 6$, 7, 8, 9, and 10 appear to have prolate deformations for ground-state bands with moments of inertia leading to $0.50 \lesssim \hbar^2/2\mathscr{I}_0 \lesssim 0.90$ MeV. The nuclei $A = 12$, 13, and 14 appear to have oblate deformations for ground-state bands with $0.75 \lesssim \hbar^2/2\mathscr{I}_0 \lesssim 1.00$ MeV. The nucleus ^{16}O appears to be spherical in the ground state. The nuclei with $A = 15$ are nearly spherical, being slightly prolate or oblate for various intrinsic rotational bands.

▶**Exercise 6-16** (a) Using the empirical energies shown in Fig. 6.24 for the nucleus ^{25}Al, determine the values of $\hbar^2/2\mathscr{I}_0$ for the various rotational bands through the use of (6-124) and Table 6-2 when necessary.

(b) Is the trend of $\hbar^2/2\mathscr{I}_0$ with band-head energy found in part (a) physically reasonable?

(c) What average deformation β_0 results from a consideration of the $K = \frac{1}{2}$ bands?

(d) Using the average β_0 found in part (b), estimate $\hbar\omega_0$ from Fig. 6.19 by using the observed energy difference between the $\frac{1}{2}-[330]$ orbital and that of the average of the $\frac{5}{2}+[202]$ and $\frac{1}{2}+[211]$ orbitals.

(e) Verify that the stated values in the text, $\beta_0 \approx 0.40$, $a \approx -1.40$, and $\hbar^2/2\mathscr{I}_0 \approx 0.54$ MeV, result in the model energy level scheme for the $A = 7$ nuclides illustrated in Fig. 6.29.

(f) Compare the above calculated values of $\hbar^2/2\mathscr{I}_0$ with the predictions from the irrotational liquid drop model using (6-86) with $R = 1.38A^{1/3}$ fm. Comment on your results.

(g) Calculate $\hbar\omega_0$ using (6-141) or (4-33a), and (4-33b). Compare these values with the result obtained in (d). ◀

In view of the above discussions, the question again can be raised as to which of several models is most appropriate. The models we have been discussing are of course interrelated, and one or another may prove simply a more convenient first-order starting point. Their equivalence becomes apparent in the higher order approximations. The large shell model calculations permitting configuration mixing with many particle–hole states

[1] The prolate core shape assumed here leads to a positive core contribution to the ground-state static quadrupole moment. The loose-particle contribution is negative, however, and leads to the rather small observed total value of -0.05 b.

produced by nucleon–nucleon residual interactions such as pairing begin to incorporate collective features into the first-order extreme single-particle picture. On the other hand, rotational band mixing involving different intrinsic particle states begins to include individual-particle effects into the simpler, pure rotational model. Finally, the imposition of physically required symmetries and invariance properties tends to blur out any sharp semiclassical interpretation of the different model solutions. Thus ^{20}Ne, ^{24}Mg, ^{28}Si, etc. can be treated from the viewpoint of a spherical shell model, a cluster model, an α-particle model, or a deformed rotational model. Reconciliation of the various models on a microscopic basis can be sought in a self-consistent Hartree–Fock calculation, obeying required symmetries. Banerjee *et al.* (1969) have shown that Hartree–Fock calculations under time-reversal invariance, reflection symmetry through the x, y plane, and rotational invariance under a rotation by π about the z axis lead, on general grounds, to predictable nuclear shapes. For example, they predict ^{20}Ne to have a prolate axially symmetric shape, ^{28}Si and ^{36}Ar to have oblate axially symmetric shapes, and ^{24}Mg and ^{32}S to have triaxially asymmetric shapes. These shapes are also roughly those to be expected from the close packing or energy-favored packing of α-particle clusters. Refer also to Abgrall *et al.* (1972), Kurath (1972), and Reid *et al.* (1974). The discussion offered by Hecht *et al.* (1972) on a model calculation showing the transition from shell model to collective behavior in the $(g_{7/2}, d_{5/2,3/2}, \text{ and } s_{1/2})^k$ configuration is particularly instructive. Additionally, a review paper aptly entitled "The many facets of nuclear structure" (Bohr and Mottelson, 1973) offers profitable reading for the intercomparison of a number of nuclear models.

Finally, it should be mentioned that a complete detailed reconciliation of the Nilsson rotational model and the intermediate coupling scheme of Chapter V is possible for the simple 1p-shell and some of the (1d, 2s)-shell nuclei. Notable among many papers on the subject are Elliot (1958), Kurath and Pičman (1959), and Banerjee *et al.* (1963). Essentially, it is shown that either a nucleus can be considered as a group of nucleons in a spherical potential well with correlations in their motions produced by the residual interactions or, equally well, a nucleus can be considered a group of noninteracting nucleons in a quadrupole-deformed potential well with the correlated motions thus introduced. The overlap of wave functions derived by the two procedures is strikingly good. Specifically, one can generate the spherical shell-model-like states from the rotational model states by computing the projection integral over the Eulerian angles $\mathscr{R}$, thus

$$\psi^J_{M(K)}(x) = N(J, K, \beta_0) \int d\mathscr{R}\, D^J_{KM}(\mathscr{R})\, \chi_K(\mathscr{R}^{-1}x). \qquad (6\text{-}157)$$

The coordinate x refers to the laboratory frame of reference. The body-centered coordinates in the particle wave function χ_K are written symbolically as $\mathscr{R}^{-1}x$ to indicate the rotation of the coordinates through the Euler angles. The overlap of (6-157) then can be obtained with the intermediate-coupling-derived shell model states. A strong connection is found between the strength of the a/K parameter of intermediate coupling and the deformation β_0 of the rotational model. This is not surprising when we realize that small a/K and large β_0 (with asymptotic rotational states $\Omega = K, \pi[nn_3\Lambda]$) both correspond to the LS coupling limit, while large a/K and small β_0 essentially correspond to the jj-coupled limit.

References

Abgrall, Y., Morand, B., and Caurier, E. (1972). *Nuclear Phys.* **A192**, 372.

Alaga, G. (1969). *Proc. Nuclear Structure Nuclear Reactions* Course XL. Academic Press, New York.

Albrecht, K. (1973). *Nuclear Phys.* **A207**, 225.

Andersen, B. L. (1971). *Nuclear Phys.* **A162**, 208.

Andersen, B. L., Dickmann, F., and Dietrich, K. (1970). *Nuclear Phys.* **A159**, 337.

Arima, A., Cohen, S., Lawson, R. D., and MacFarlane, M. H. (1968). *Nuclear Phys.* **A108**, 94.

Baranger, M. (1960). *Phys. Rev.* **120**, 957.

Banerjee, M. K., Levinson, C. A., and Meshkov, S. (1963). *Phys. Rev.* **130**, 1036, 1064.

Banerjee, M. K., Levinson, C. A., and Stephenson, G. J. (1969). *Phys. Rev.* **178**, 1709.

Belyaev, S. T. (1968). "Collective Excitations in Nuclei." Gordon and Breach, New York.

Bohr, A., and Mottelson, B. R. (1973). *Ann. Rev. Nuclear Sci.* **23**, 363.

Bohr, A., Mottelson, B. R., and Pines, D. (1958). *Phys. Rev.* **110**, 936.

Brack, M., Damgaard, J., Jensen, A. S., Pauli, H. C., Strutinsky, V. M., and Wong, C. Y. (1972). *Rev. Modern Phys.* **44**, 320.

Brown, G. E., and Green, A. M. (1966). *Nuclear Phys.* **75**, 401.

Bunker, M. E., and Reich, C. W. (1971). *Rev. Modern Phys.* **43**, **348**.

Chang, F. C. (1969). *Phys. Rev.* **178**, 1725.

Clegg, A. B. (1961). *Phil. Mag.* **6**, 1207.

Dasso, C., and Klein, A. (1974). *Nuclear Phys.* **A222**, 445.

Davidson, J. P. (1968). "Collective Models of the Nucleus." Academic Press, New York.

Davydov, A. S., and Chaban, A. A. (1960). *Nuclear Phys.* **20**, 499.

Davydov, A. S., and Filippov, G. F. (1958). *Nuclear Phys.* **8**, 237.

Davydov, A. S., and Rostovsky, V. S. (1959). *Nuclear Phys.* **12**, 58.

deBoer, J., and Eichler, J. (1968). *Advan. Nuclear Phys.* **1**, 1.

DeMille, G. R., Kavanagh, T. M., Moore, R. B., Weaver, R. S., and White, W. (1959) *Can. J. Phys.* **37**, 1036.

Eisenberg, J. M., and Greiner, W. (1970). "Nuclear Models." North-Holland Publ., Amsterdam.

El-Batanoni, F., and Kresnin, A. A. (1966). *Nuclear Phys.* **89**, 577.

Elliott, J. P. (1958). *Proc. Roy. Soc. London* **A245**, 128, 562.

Elliott, J. P., and Flowers, B. H. (1955). *Proc. Roy. Soc. London* **A229**, 536.

Erb, K. A., Holden, J. E., Lee, I. Y., Saladin, J. X., and Saylor, T. K. (1972). *Phys. Rev. Lett.* **29**, 1010.

Fabian, W., Horlacher, G. E., and Albrecht, K. (1972). *Nuclear Phys.* **A190**, 533.

Faessler, A., and Greiner, W. (1962). *Z. Phys.* **168**, 425; **170**, 105.

Frankel, S., and Metropolis, N. (1947). *Phys. Rev.* **72**, 914.

Gneuss, G., and Greiner, W., (1971). *Nuclear Phys.* **A171**, 449.

Götz, U., Pauli, H. C., Alder, K., and Junker, K. (1972). *Nuclear Phys.* **A192**, 1.

Goodman, A. L., and Goswami, A. (1974). *Phys. Rev.* **C9**, 1948.

Gorfinkiel, J. I., Mariscotti, M. A. J., and Pomar, C. (1974). *Phys. Rev.* **C9**, 1243.

Grodzins, L. (1972). *Ann. Rev. Nuclear Sci.* **22**, 291.

Grosse, E., Stephens, F. S., and Diamond, R. M. (1973). *Phys. Rev. Lett.* **31**, 840.

Grosse, E., Stephens, F. S., and Diamond, R. M. (1974). *Phys. Rev. Lett.* **32**, 74.

Gupta, R. K. (1973). *Nuclear Phys.* **C7**, 2476.

Gustafson, C., Lamm, I. L., Nilsson, B., and Nilsson, S. G. (1968). *Ark. Fys.* **36**, 613.

Hamilton, J. H., *et al.* (1974). *Phys. Rev. Lett.* **32**, 239.

Hecht, K. T., McGrory, J. B., and Draayer, J. P. (1972). *Nuclear Phys.* **A197**, 369.

Hendrie, D. L., Glendenning, N. K., Harvey, B. G., Jarvis, O. N., Duhm, H. H., Saudinos, J., and Mahoney, J. (1968). *Phys. Lett.* **26B**, 127.

Inglis, D. R. (1973). *Nuclear Phys.* **A215**, 189.

Johnson, A., and Szymański, Z. (1973). *Phys. Rep.* **7C**, 181.

Johnson, A., Ryde, H., and Hjorth, S. A. (1972) *Nuclear Phys.* **A179**, 753.

Jones, H. D., Onishi, N., Hess, T., and Sheline, R. K. (1971). *Phys. Rev.* **C3**, 529.

Khoo, T. L., Bernthal, F. M., Boyno, J. S., and Warner, R. A. (1973). *Phys. Rev. Lett.* **31**, 1146.

Kisslinger, L. S., and Sorensen, R. A. (1960). *K. Danske Vid. Selsk. Mat. Fys. Medd.* **32**; No. 9.

Krumlinde, J., ana Szymański, Z., (1974). *Nuclear Phys.* **A221**, 93.

Kurath, D. (1972). *Phys. Rev.* **C5**, 768.

Kurath, D., and Pičman, L. (1959). *Nuclear Phys.* **10**, 313.

Lane, A. M. (1964). "Nuclear Theory." Benjamin, New York.

Leander, G. (1974). *Nuclear Phys.* **A219**, 245.

Lemmer, R. H. (1960). *Phys. Rev.* **117**, 1551.

Lemmer, R. H., and Green, A. E. S. (1960). *Phys. Rev.* **119**, 1043.

McGrory, J. B., and Wildenthal, B. H. (1973). *Phys. Rev.* **C7**, 974.

Mantri, A. N., and Sood, P. C. (1973). *Phys. Rev.* **C7**, 1294.

Mantri, A. N., and Sood, P. C. (1974). *Phys. Rev.* **C9**, 2076.

Mariscotti, M. A., Scharff-Goldhaber, G., and Buck, B. (1969). *Phys. Rev.* **178**, 1864.

Marshalek, E. R. (1972). *Phys. Lett.* **38B**, 367.

Merzbacher, E. (1970). "Quantum Mechanics." Wiley, New York.

Möller, P., Nilsson, S. G., and Sheline, R. K. (1972). *Phys. Lett.* **40B**, 329.

Mottelson, B. R., and Valatin, J. G. (1960). *Phys. Rev. Lett.* **5**, 511.

Nakagawa, K., Obinata, T., and Sasaki, K. (1972). *Nuclear Phys.* **A191**, 535.

Nilsson, S. G. (1955). *K. Danske Vid. Selsk. Mat. Fys. Medd.* **29**, No. 16.

Nilsson, S. G., Tsang, C. F., Sobiczewski, A., Szymański, Z., Wycech, S., Gustafson, C. Lamm, I. L., Möller, P., and Nilsson, B. (1969). *Nuclear Phys.* **A131**, 1.

Ogle, W., Wahlborn, S., Piepenbring, R., and Fredriksson, S. (1971). *Rev. Modern Phys.* **43**, 424.

Panagiotou, A. D., Gove, H. E., and Harar, S. (1972). *Nuclear Phys.* **C5**, 1995.

Pauli, H. C., (1973). *Phys. Rep.* **7C**, 35.

Rebel, H., Schweimer, G. W., Specht, J., Schatz, G., Löhken, R., Habs, D., Hauser, G., and Klewe-Nebenius, H. (1971). *Phys. Rev. Lett.* **26**, 1190.

Rebel, H., Hauser, G., Schweimer, G. W., Nowicki, G., Wiesner, W., and Hartmann, D. (1974). *Nuclear Phys.* **A218**, 13.
Reid, N. E., Davison, N. E., and Svenne, J. P. (1974). *Phys. Rev.* **C9**, 1882.
Rose, M. E. (1957). "Elementary Theory of Angular Momentum." Wiley, New York.
Ross, C. K., and Nogami, Y. (1973). *Nuclear Phys.* **A211**, 145.
Rowe, D. J. (1970). "Nuclear Collective Motion." Methuen, London.
Saethre, Ø., Hjorth, S. A., Johnson, A., Jägare, S., Ryde, H., and Szymański, Z. (1973). *Nuclear Phys.* **A207**, 486.
Sakai, M. (1972). "Nuclear Data Tables," Vol. 10, p. 511. Academic Press, New York.
Saylor, T. K., Saladin, J. X., Lee, I. Y., and Erb, K. A. (1972). *Phys. Lett.* **42B**, 51.
Scharff-Goldhaber, G., and Goldhaber, A. S. (1970). *Phys. Rev. Lett.* **24**, 1349.
Schwalm, D., Bamberger, A., Bizzeti, P. G., Povh, B., Engebertink, G. A. P., Olness, J. W., and Warburton, E. K. (1972). *Nuclear Phys.* **A192**, 449.
Sheline, R. K. (1960). *Rev. Modern Phys.* **32**, 1.
Sheline, R. K. (1972). *Nuclear Phys.* **A195**, 321.
Siegbahn, K. (1965). "Alpha, Beta, and Gamma-Ray Spectroscopy," Vol. I and Vol. II. North-Holland Publ., Amsterdam.
Sioshansi, P., Garber, D. A., King, W. C., Scharenberg, R. P., Steffen, R. M., and Wheeler, R. M. (1972). *Phys. Lett.* **39B**, 343.
Sørensen, B. (1973). *Nuclear Phys.* **A211**, 565.
Stephens, F. S., Lark, N. L., and Diamond, R. M. (1965). *Nuclear Phys.* **63**, 82.
Stephens, F. S., Kleinheinz, P., Sheline, R. K., and Simon, R. S. (1974). *Nuclear Phys.* **A222**, 235
Swiatecki, W. J. (1956). *Phys. Rev.* **104**, 993.
Temmer, G. M., and Heydenburg, N. P. (1956). *Phys. Rev.* **104**, 967.
True, W. W., Ma, C. W., and Pinkston, W. T. (1971). *Phys. Rev.* **C3**, 2421.
Zamick, L. (1973). *Phys. Lett.* **45B**, 313.

Chapter VII

ELECTROMAGNETIC INTERACTIONS WITH NUCLEI

A. INTRODUCTION

In this chapter we discuss the nature of the interaction of nuclear systems with electromagnetic fields. It is possible to formulate the electromagnetic interaction with great accuracy, and such interactions therefore form a very important tool for the determination of nuclear properties. While the literature is rich in ingenious experimental techniques designated to facilitate such studies, we shall limit our discussion to include only the salient features of the more basic experimental observations. Thus many important topics, such as polarization, angular correlations, Coulomb excitation, and the Mössbauer effect, are left to other numerous references in the literature.

Fortunately for our purposes, much of the classical formalism pertaining to Maxwell's equations can be readily transformed into a proper quantum mechanical description without the necessity of introducing a de novo formal quantization of the electromagnetic field, or considering the full implications of the theory of relativistic quantum electrodynamics. Indeed, we shall treat only the nucleon motions in the nonrelativistic approximation and consider quantities such as charge and magnetic moment as known empirical constants.

For completeness and convenience, we first describe the "free" electromagnetic field and then proceed to discuss the interaction of such fields with nuclear charges and currents. The description of the electromagnetic field intensities $\mathscr{E}$ and $\mathscr{H}$ as vector fields implies a quantum mechanical behavior conveniently discussed in terms of *vector spherical harmonics*. We thus begin with certain necessary mathematical preliminaries.

B. VECTOR SPHERICAL HARMONICS

Suppose that a vector field solution $\mathbf{V}(\mathbf{r})$ exists for some particular rotationally invariant physical problem. Rotational invariance implies that the field components V_x, V_y, V_z transform under arbitrary rotations of the coordinate frame in the same manner as the components x, y, z of the coordinate vector $\mathbf{r}$ itself. Consider first a description in an original Cartesian frame O and subsequently in some rotated frame O'. The very same point in space has the components x, y, z in the original frame and x', y', z' in the rotated frame. Similarly the field components at this point in space are $V_x(x,y,z)$, $V_y(x,y,z)$, $V_z(x,y,z)$ in the original frame and $V_x'(x',y',z')$, $V_y'(x',y',z')$, $V_z'(x',y',z')$ in the rotated frame.

An alternate point of view is possible in which the coordinate frame is kept fixed, but in which the vector field is bodily rotated through an angle that formally carries $\mathbf{r}'$ back into $\mathbf{r}$. This new rotated field is found also to be an equivalent solution to the original problem. As an example, consider a simple rotation through an angle θ about the z axis. The unitary transformation matrix U that carries $\mathbf{r}'$ into $\mathbf{r}$ is

$$U = \begin{pmatrix} \cos\theta & \sin\theta & 0 \\ -\sin\theta & \cos\theta & 0 \\ 0 & 0 & 1 \end{pmatrix}, \tag{7-1}$$

viz., $\mathbf{r} = U\mathbf{r}'$ or[1]

$$\begin{pmatrix} x \\ y \\ z \end{pmatrix} = U \begin{pmatrix} x' \\ y' \\ z' \end{pmatrix}.$$

Similarly, $\mathbf{V}(\mathbf{r}) = U\mathbf{V}'(\mathbf{r}')$. From this last expression we also have $\mathbf{V}'(\mathbf{r}') = U^{-1}\mathbf{V}(\mathbf{r})$ and since $U^{-1} = U^{\dagger}$, we obtain further $\mathbf{V}'(\mathbf{r}') = U^{\dagger}\mathbf{V}(U\mathbf{r}')$. Since this equation applies to any point, we can drop the prime on $\mathbf{r}'$ and finally

[1] We will find it convenient to represent the usual vector by a three-element column matrix.

obtain

$$\mathbf{V}'(\mathbf{r}) = U^{\dagger}\mathbf{V}(U\mathbf{r}), \tag{7-2}$$

or, explicitly,

$$\begin{aligned} V_x'(x,y,z) &= (\cos\theta)V_x(x\cos\theta + y\sin\theta,\ y\cos\theta - x\sin\theta,\ z) \\ &\quad - (\sin\theta)V_y(x\cos\theta + y\sin\theta,\ y\cos\theta - x\sin\theta,\ z) \\ V_y'(x,y,z) &= (\sin\theta)V_x(x\cos\theta + y\sin\theta,\ y\cos\theta - x\sin\theta,\ z) \\ &\quad + (\cos\theta)V_y(x\cos\theta + y\sin\theta,\ y\cos\theta - x\sin\theta,\ z) \\ V_z'(x,y,z) &= V_z(x\cos\theta + y\sin\theta,\ y\cos\theta - x\sin\theta,\ z). \end{aligned} \tag{7-3}$$

For θ equal to an infinitesimal rotation $\theta = \varepsilon$, we have from (7-3), to linear order in ε, and with V and V' evaluated at x, y, z,[1]

$$\begin{pmatrix} V_x' \\ V_y' \\ V_z' \end{pmatrix} = \left[1 + \varepsilon\left(y\frac{\partial}{\partial x} - x\frac{\partial}{\partial y}\right) + \varepsilon\begin{pmatrix} 0 & -1 & 0 \\ 1 & 0 & 0 \\ 0 & 0 & 0 \end{pmatrix}\right]\begin{pmatrix} V_x \\ V_y \\ V_z \end{pmatrix}. \tag{7-4}$$

The fundamental definition of the angular momentum operator $\tilde{\mathbf{J}}$ as the generator of infinitesimal rotations is embodied in the rotation operator $\tilde{R}$ through the expression

$$\tilde{R} = 1 - (i/\hbar)\,\varepsilon\hat{\mathbf{n}}\cdot\tilde{\mathbf{J}}, \tag{7-5}$$

where $\hat{\mathbf{n}}$ defines the direction of the rotation axis. In the present instance

[1] It is to be understood that terms not explicitly written as matrices carry the implied presence of the 3×3 identity or unit matrix I_3. Thus the first two terms inside the bracket are understood to be

$$\begin{aligned} 1 + \varepsilon\left(y\frac{\partial}{\partial x} - x\frac{\partial}{\partial y}\right) &\equiv \left\{1 + \varepsilon\left(y\frac{\partial}{\partial x} - x\frac{\partial}{\partial y}\right)\right\} I_3 \\ &\equiv \begin{bmatrix} \left\{1 + \varepsilon\left(y\frac{\partial}{\partial x} - x\frac{\partial}{\partial y}\right)\right\} & 0 & 0 \\ 0 & \left\{1 + \varepsilon\left(y\frac{\partial}{\partial x} - x\frac{\partial}{\partial y}\right)\right\} & 0 \\ 0 & 0 & \left\{1 + \varepsilon\left(y\frac{\partial}{\partial x} - x\frac{\partial}{\partial y}\right)\right\} \end{bmatrix}. \end{aligned}$$

$\hat{\mathbf{n}} \cdot \mathbf{J} = J_z$ and $\mathbf{V}'(\mathbf{r}) = R\mathbf{V}(\mathbf{r})$. Thus we have

$$J_z = \frac{\hbar}{i}\left(x\frac{\partial}{\partial y} - y\frac{\partial}{\partial x}\right) + \hbar\begin{pmatrix} 0 & -i & 0 \\ i & 0 & 0 \\ 0 & 0 & 0 \end{pmatrix}. \tag{7-6}$$

The first term in (7-6) is immediately recognized to be the operator $\tilde{L}_z$ for orbital angular momentum. It is natural to associate the second term with the intrinsic spin of the vector field and write

$$\tilde{S}_z = \hbar\begin{pmatrix} 0 & -i & 0 \\ i & 0 & 0 \\ 0 & 0 & 0 \end{pmatrix}. \tag{7-7a}$$

When the above procedure is repeated for infinitesimal rotations about the x and y axes as well, the result is $\mathbf{J} = \mathbf{L} + \mathbf{S}$ with

$$S_x = \hbar\begin{pmatrix} 0 & 0 & 0 \\ 0 & 0 & -i \\ 0 & i & 0 \end{pmatrix} \tag{7-7b}$$

and

$$S_y = \hbar\begin{pmatrix} 0 & 0 & i \\ 0 & 0 & 0 \\ -i & 0 & 0 \end{pmatrix}. \tag{7-7c}$$

In the present form none of the spin component matrices is diagonal. This situation is in many respects quite similar to the isovector problem discussed in Chapter I in connection with (1-33)–(1-36). Review of that situation at this time may prove profitable, although for the sake of continuity and completeness we include some repetitions. A simple rotation will carry these matrices into the more familiar form with S_z diagonal. The eigenvalue problem for S_z, namely $S_z\mathbf{V} = \mu\hbar\mathbf{V}$, is readily solved using (7-7a), viz.,

$$S_z\begin{pmatrix} V_x \\ V_y \\ V_z \end{pmatrix} = \hbar\begin{pmatrix} -iV_y \\ +iV_x \\ 0 \end{pmatrix} = \mu\hbar\begin{pmatrix} V_x \\ V_y \\ V_z \end{pmatrix}. \tag{7-8}$$

For (7-8) to have nontrivial solutions the secular determinant must vanish or

$$\begin{vmatrix} \mu & +i & 0 \\ -i & \mu & 0 \\ 0 & 0 & \mu \end{vmatrix} = 0, \tag{7-9}$$

whence $\mu(\mu^2 - 1) = 0$ and $\mu = 0, \pm 1$. The corresponding normalized eigenvectors are[1]

$$\mu = +1: \qquad \chi_1 = -\frac{1}{\sqrt{2}}\begin{pmatrix} 1 \\ i \\ 0 \end{pmatrix} = -\frac{1}{\sqrt{2}}\begin{pmatrix} 1 \\ 0 \\ 0 \end{pmatrix} - \frac{i}{\sqrt{2}}\begin{pmatrix} 0 \\ 1 \\ 0 \end{pmatrix}$$

$$= -\frac{1}{\sqrt{2}}(\hat{e}_x + i\hat{e}_y), \tag{7-10a}$$

$$\mu = 0: \qquad \chi_0 = \begin{pmatrix} 0 \\ 0 \\ 1 \end{pmatrix} = \hat{e}_z, \tag{7-10b}$$

$$\mu = -1: \qquad \chi_{-1} = \frac{1}{\sqrt{2}}\begin{pmatrix} 1 \\ -i \\ 0 \end{pmatrix} = \frac{1}{\sqrt{2}}(\hat{e}_x - i\hat{e}_y). \tag{7-10c}$$

Finally, the rotated spin matrices become

$$S_x' = (\hbar/\sqrt{2})\begin{pmatrix} 0 & 1 & 0 \\ 1 & 0 & 1 \\ 0 & 1 & 0 \end{pmatrix}, \tag{7-11a}$$

$$S_y' = (\hbar/\sqrt{2})\begin{pmatrix} 0 & -i & 0 \\ i & 0 & -i \\ 0 & i & 0 \end{pmatrix}, \tag{7-11b}$$

[1] To avoid possible confusion in the notation, we write the usual unit vectors **i**, **j**, and **k** in the form

$$\mathbf{i} \equiv \hat{e}_x = \begin{pmatrix} 1 \\ 0 \\ 0 \end{pmatrix}, \qquad \mathbf{j} \equiv \hat{e}_y = \begin{pmatrix} 0 \\ 1 \\ 0 \end{pmatrix}, \qquad \text{and} \qquad \mathbf{k} \equiv \hat{e}_z = \begin{pmatrix} 0 \\ 0 \\ 1 \end{pmatrix}.$$

The unnormalized solution for $\mu = +1$ requires simply

$$V_x = -iV_y \qquad \text{and} \qquad V_z = 0, \qquad \text{or} \qquad \chi_1 = V_x e^{i\alpha}\begin{pmatrix} 1 \\ i \\ 0 \end{pmatrix}$$

with α an arbitrary phase. The customary normalization to unit length imposes the condition $|V_x|^2 + |V_y|^2 + |V_z|^2 = 1$. The possible arbitrary phase factors in (7-10) are adjusted to reduce the intrinsic spin eigenvectors to the normal spherical basis vectors; thus, for example, in the case under discussion we take $\alpha = \pi$. Finally suitable solutions for the space vector $\mathbf{V(r)}$ for the case $L = 0$ would be

$$\mathbf{V(r)} = V_{-1}(\mathbf{r})\chi_{-1} + V_0(\mathbf{r})\chi_0 + V_{+1}(\mathbf{r})\chi_{+1}.$$

and

$$S_z' = \hbar\begin{pmatrix} 1 & 0 & 0 \\ 0 & 0 & 0 \\ 0 & 0 & -1 \end{pmatrix}. \qquad (7\text{-}11c)$$

▶**Exercise 7-1** (a) Verify (7-8)–(7-10).

(b) The rotated spin matrices are obtained through the unitary transformation $\mathbf{S}' = U\mathbf{S}U^{-1}$. In the representation with S_z' diagonal, the unit vectors are simply

$$\chi'_{+1} = \begin{pmatrix} 1 \\ 0 \\ 0 \end{pmatrix}, \quad \chi_0' = \begin{pmatrix} 0 \\ 1 \\ 0 \end{pmatrix}, \quad \text{and} \quad \chi'_{-1} = \begin{pmatrix} 0 \\ 0 \\ 1 \end{pmatrix}.$$

Thus the matrix elements of U are $U_{k,l} = \langle \chi_k | \chi_l' \rangle$. From the above and (7-10), show that

$$U = (1/\sqrt{2})\begin{pmatrix} -1 & i & 0 \\ 0 & 0 & \sqrt{2} \\ 1 & i & 0 \end{pmatrix}$$

and that (7-11) follows.

(c) Show that for either the (7-7) or (7-11) representation

$$\mathbf{S}^2 = S_x^{\,2} + S_y^{\,2} + S_z^{\,2} = 2\hbar^2 I_3$$

and hence $S(S+1) = 2$.

(d) Show that in either representation $[S_k, S_l] = i\hbar S_m$ with $k, l, m \equiv x, y, z$ and their cyclic permutations.

(e) We define the operator *curl* $= \tilde{\nabla}\times$ as

$$\tilde{\nabla}\times = \begin{pmatrix} 0 & -\partial/\partial z & \partial/\partial y \\ \partial/\partial z & 0 & -\partial/\partial x \\ -\partial/\partial y & \partial/\partial x & 0 \end{pmatrix}.$$

Show that $\tilde{\nabla}\times$ commutes with $\mathbf{J}$ (i.e., all three components) and $\mathbf{J}^2$. Does $\tilde{\nabla}\times$ commute with $\mathbf{L}$ alone? ◀

From the above it is evident that all vector fields possess an intrinsic spin of $1\hbar$ in addition to any rotational angular momentum associated with the operator $\mathbf{L}$. Thus the eigenvectors $\chi_{0,\pm 1}$ simply correspond to z components of the spin $0, \pm 1$. It is instructive to note that the structure of (7-10), for the selected phase convention, closely resembles the Cartesian representation of the spherical harmonics

$$Y_1^{\,0} = \left(\frac{3}{4\pi}\right)^{1/2} \cos\theta = \left(\frac{3}{4\pi}\right)^{1/2} \frac{z}{r}$$

and

$$Y_1^{\pm 1} = \mp\left(\frac{3}{8\pi}\right)^{1/2} e^{\pm i\varphi} \sin\theta = \mp\left(\frac{3}{8\pi}\right)^{1/2} \frac{x \pm iy}{r},$$

as is entirely appropriate. In the following we shall continue to use (7-10) as the spin basis states, leaving the various angular momentum operators as defined in the matrices (7-6) and (7-7).

The procedure we have followed above is analogous to the standard treatment that shows that two-component spinor fields have intrinsic spin angular momentum $\hbar/2$ [e.g., Merzbacher (1970), Chapter 12)]. Again in an entirely analogous manner to the spin-$\frac{1}{2}$ case it can be shown that all operators on vector fields can be represented by 3×3 matrices.

By identifying the three space vectors χ with the intrinsic spin vectors of a spin-one particle (or field), we can introduce the *vector spherical harmonics* (in analogy to the treatment of the spin-$\frac{1}{2}$ case in Section II, H that resulted in the introduction of the tensors $\mathcal{Y}^M_{J,S,L}$), viz.,

$$\mathfrak{Y}^M_{J,S=1,L}(\theta,\varphi) = \sum_{m'=-L}^{+L} \sum_{m=-1}^{+1} (L, S=1, m', m|JM)\, Y_L^{m'}(\theta,\varphi)\chi_m \tag{7-12a}$$

or since $m+m' = M$,

$$\mathfrak{Y}^M_{J,S=1,L}(\theta,\varphi) = \sum_{m=-1}^{+1} (L, S=1, M-m, m|JM)\, Y_L^{M-m}(\theta,\varphi)\chi_m. \tag{7-12b}$$

There are three such functions for a given J except when $J=0$, two of the same parity with $L = J \pm 1$ and one with opposite parity $L = J$. It readily follows that[1]

$$\begin{aligned} \mathbf{J}^2\mathfrak{Y}^M_{J,1,L} &= J(J+1)\hbar^2\mathfrak{Y}^M_{J,1,L}, & J_z\mathfrak{Y}^M_{J,1,L} &= M\hbar\mathfrak{Y}^M_{J,1,L} \\ \mathbf{L}^2\mathfrak{Y}^M_{J,1,L} &= L(L+1)\hbar^2\mathfrak{Y}^M_{J,1,L}, & \mathbf{S}^2\mathfrak{Y}^M_{J,1,L} &= 2\hbar^2\mathfrak{Y}^M_{J,1,L} \end{aligned} \tag{7-13a}$$
$$P^{(0)}\mathfrak{Y}^M_{J,1,L} = (-1)^L\mathfrak{Y}^M_{J,1,L}$$

[1] The parity operator $P^{(0)}$ inverts only the coordinate variables appearing in the functional dependence of the various vector field components (i.e., $x \to -x$, $y \to -y$, $z \to -z$) and thus only determines whether the field is an odd or even function of position. It is not intended to apply to any *intrinsic parity* the field might possess. Parity operators such as $P^{(0)}$ may be referred to as *spatial parity* or *orbital parity* operators. This distinction will be further clarified subsequently.

Also,

$$\int \mathfrak{Y}_{J,1,L}^{M*} \cdot \mathfrak{Y}_{J',1,L'}^{M'} \, d\Omega = \delta_{JJ'} \, \delta_{LL'} \, \delta_{MM'}. \tag{7-13b}$$

In view of (7-13b), we see that the vector spherical harmonics form an orthonormal set on the unit sphere. Any arbitrary vector field can thus be expanded as follows:

$$\mathbf{V}(\mathbf{r}) = \sum_{J=0}^{\infty} \sum_{M=-J}^{+J} \mathbf{V}(J, M; \mathbf{r})$$

with

$$\mathbf{V}(J, M; \mathbf{r}) = r^{-1}\{f(J, M; r)\, \mathfrak{Y}_{J,1,J}^{M} + g(J, M; r)\, \mathfrak{Y}_{J,1,J+1}^{M} + h(J, M; r)\, \mathfrak{Y}_{J,1,J-1}^{M}\}, \tag{7-14}$$

where

$$r^{-1} f(J, M; r) = \int \mathfrak{Y}_{J,1,J}^{M*} \cdot \mathbf{V}(\mathbf{r}) \, d\Omega$$

and there are two similar expressions for g and h. The functions f, g, and h are functions of $r = |\mathbf{r}|$ alone. Each such $\mathbf{V}(J, M; \mathbf{r})$ is referred to as a pure *multipole field* characterized by J, M, and spatial parity. The expansion of $\mathbf{V}(\mathbf{r})$ in (7-14) thus is referred to as a *multipole expansion*.

It is particularly useful in what follows to examine explicitly the vector spherical harmonic for $L = J$ and to introduce a special notation for this case,

$$\mathfrak{X}_J^{\,M} = \mathfrak{Y}_{J,1,J}^{M} = (J, 1, M-1, | JM)\, Y_J^{M-1} \chi_1 + (J, 1, M, 0 | JM)\, Y_J^{\,M} \chi_0 + (J, 1, M+1, -1 | JM)\, Y_J^{M+1} \chi_{-1}$$

or

$$\mathfrak{X}_J^{\,M}(\theta, \varphi) = \{1/\hbar [J(J+1)]^{1/2}\} \, \tilde{\mathbf{L}} Y_J^{\,M}(\theta, \varphi). \tag{7-15}$$

We note that $\mathfrak{X}_0^{\,0}$ is identically zero since $\tilde{\mathbf{L}} Y_0^{\,0} = 0$.

▶**Exercise 7-2** (a) Using Appendix C, show that

$$\mathfrak{X}_J^{\,M} = [2J(J+1)]^{-1/2} \{ -[(J+M)(J-M+1)]^{1/2} Y_J^{M-1} \chi_1 + \sqrt{2}\, M Y_J^{\,M} \chi_0 + [(J-M)(J+M+1)]^{1/2} Y_J^{M+1} \chi_{-1} \}.$$

(b) Using $\mathbf{L} = L_x \hat{e}_x + L_y \hat{e}_y + L_z \hat{e}_z$, show that (7-15) follows. (*Hint:* Write $\mathbf{L}$ in terms of the raising and lowering operators $L_+ = L_x + iL_y$ and $L_- = L_x - iL_y$ and use the spherical basis vectors.) ◀

C. THE ELECTROMAGNETIC FIELD IN FREE SPACE

The Maxwell equations in free space using mixed Gaussian units are

$$c\nabla \times \mathscr{H} = \partial\mathscr{E}/\partial t, \qquad c\nabla \times \mathscr{E} = -\partial\mathscr{H}/\partial t, \qquad \nabla \cdot \mathscr{H} = 0, \qquad \nabla \cdot \mathscr{E} = 0. \tag{7-16}$$

The most general solution to these equations are the Fourier integrals for the (real) vector fields,

$$\begin{aligned} \mathscr{E} &= \int [\mathscr{E}(\mathbf{r}, \omega)\, e^{-i\omega t} + \mathscr{E}^*(\mathbf{r}, \omega)\, e^{i\omega t}]\, d\omega \\ \mathscr{H} &= \int [\mathscr{H}(\mathbf{r}, \omega)\, e^{-i\omega t} + \mathscr{H}^*(\mathbf{r}, \omega)\, e^{i\omega t}]\, d\omega. \end{aligned} \tag{7-17}$$

It is sufficient for our purposes to examine the periodic solutions

$$\mathscr{E}(\mathbf{r}, \omega; t) = \mathscr{E}(\mathbf{r})\, e^{-i\omega t} + \mathscr{E}^*(\mathbf{r})\, e^{i\omega t}, \qquad \mathscr{H}(\mathbf{r}, \omega; t) = \mathscr{H}(\mathbf{r})\, e^{-i\omega t} + \mathscr{H}^*(\mathbf{r})\, e^{i\omega t}. \tag{7-18}$$

The vector fields $\mathscr{E}(\mathbf{r})$ and $\mathscr{H}(\mathbf{r})$ appearing on the right-hand side of (7-18) evidently can be expanded in vector spherical harmonics using (7-14). These fields of course need not be real quantities.

▶**Exercise 7-3** (a) Show that the fields $\mathscr{E}(\mathbf{r})$ and $\mathscr{H}(\mathbf{r})$ of (7-18) obey the equations

$$c\nabla \times \mathscr{H} = -i\omega\mathscr{E} \qquad \text{and} \qquad c\nabla \times \mathscr{E} = +i\omega\mathscr{H}.$$

(b) We can introduce the vector potential $\mathbf{A}$ in the solenoidal or Coulomb gauge $\nabla \cdot \mathbf{A} = 0$, to give $\mathscr{H} = \nabla \times \mathbf{A}$ and $\mathscr{E} = -(1/c)\, \partial\mathbf{A}/\partial t$. Show that for (7-18), $\mathbf{A}(\mathbf{r}) = -(i/k)\, \mathscr{E}(\mathbf{r})$, where $k = \omega/c$.

(c) Show that the above reduce to the wave equations

$$(\nabla \times \nabla \times - k^2)\, \mathscr{H} = 0, \qquad (\nabla \times \nabla \times - k^2)\, \mathscr{E} = 0, \qquad (\nabla \times \nabla \times - k^2)\, \mathbf{A} = 0$$

or since $\nabla \times \nabla \times \mathbf{V} = \nabla(\nabla \cdot \mathbf{V}) - \nabla^2 \mathbf{V}$,

$$(\nabla^2 + k^2)\, \mathscr{H} = 0, \qquad (\nabla^2 + k^2)\, \mathscr{E} = 0, \qquad \text{and} \qquad (\nabla^2 + k^2)\, \mathbf{A} = 0,$$

where

$$\nabla^2 \mathbf{V} = \begin{pmatrix} \nabla^2 & 0 & 0 \\ 0 & \nabla^2 & 0 \\ 0 & 0 & \nabla^2 \end{pmatrix} \begin{pmatrix} V_x \\ V_y \\ V_z \end{pmatrix} = \begin{pmatrix} \nabla^2 V_x \\ \nabla^2 V_y \\ \nabla^2 V_z \end{pmatrix}.$$ ◀

Since the operators $\nabla \times \nabla \times$, $\mathbf{J}^2$, J_z, and $P^{(0)}$ all commute, it is possible to construct vector field solutions that are simultaneously eigenvectors of $(\nabla^2 + k^2)$ (i.e., satisfy the wave equation), $\mathbf{J}^2$, and J_z, and possess a well-defined parity. These are, of course, the multipole fields of (7-14).

It is useful at this point to introduce the standard multipole classification of electromagnetic fields. First the pair of quantum numbers (J, M) is rewritten as (λ, μ).[1] The multipole order is referred to as a 2^{λ}-pole; *dipole* when $\lambda = 1$, *quadrupole* when $\lambda = 2$, *octupole* when $\lambda = 3$, etc. The multipole is further designated as an *electric multipole* (or E mode or simply *electric mode*) when the vector potential has the spatial parity $(-1)^{\lambda+1}$ and as a *magnetic multipole* (or M mode or simply *magnetic mode*) when the vector potential has a spatial parity $(-1)^{\lambda}$. Clearly, the electric field has the same orbital parity as the vector potential and the magnetic field involving $\nabla \times \mathbf{A}$ has opposite orbital parity.

Let us first consider the electric multipole field of order (λ, μ). In this event the magnetic field has the orbital parity $(-1)^{\lambda}$. Since the spatial parity of $\mathfrak{Y}^{M}_{J,1,L}$ is $(-1)^{L}$, clearly $\lambda = J = L$ for the magnetic field and $g = h = 0$ in (7-14). Thus for the E mode, the magnetic field of multipole order $(\lambda = J = L, \mu = M)$ can be written as[2]

$$\mathscr{H}_{\mathrm{E}}(\mathbf{r}) = r^{-1} f(\lambda, \mu; r)\, \mathfrak{X}_{\lambda}{}^{\mu}(\theta, \varphi) \qquad \text{(E mode)} \tag{7-19a}$$

and the electric field as

$$\mathscr{E}_{\mathrm{E}}(\mathbf{r}) = \frac{i}{k} \nabla \times \mathscr{H}(\mathbf{r}) = \frac{i}{k} \nabla \times [r^{-1} f(\lambda, \mu; r)\, \mathfrak{X}_{\lambda}{}^{\mu}(\theta, \varphi)] \qquad \text{(E mode).} \tag{7-19b}$$

Since $\nabla \times$, $\mathbf{J}^2$, and J_z commute and (7-19a) is an eigenvector of $\mathbf{J}^2$ and J_z, both the electric and magnetic fields have $\mathbf{J}^2 = \lambda(\lambda+1)\hbar^2$ and $J_z = \mu\hbar$. From the change in parity implied in (7-19b), it is clear that the expansion (7-14) for the electric field involves the vector spherical harmonics $\mathfrak{Y}^{M'}_{J',1,L'}$ with $L' = \lambda \pm 1$ and hence $f' = 0$ with $g' \neq 0$ and $h' \neq 0$. Carrying out the reduction of (7-19b) would, of course, determine g' and h' once $f(\lambda, \mu; r)$ is known.

We now turn to the question of the parity relationship of the final to the initial state of the radiating nuclear system when a single photon embodied in the coupled electric and magnetic fields under discussion has been emitted. We distinguish between the symmetry properties of the components of a vector field as functions of $\mathbf{r}$ under various coordinate transformations and

[1] This follows standard convention and avoids confusion when the nuclear total angular momentum must be referred to.

It was anticipation of the eventual introduction of the field–particle interaction that prompted the use of sans serif symbols for the angular momentum field operators such as **L**, **J**, and **S**. We reserve the usual symbols for operators acting on the nucleon coordinates and spin.

[2] The subscript E has been attached to $\mathscr{H}$ and $\mathscr{E}$ in (7-19a) and (7-19b) to designate the reference to the electric mode.

the symmetry properties of transforming the vector components themselves. As (7-2) and (7-3) make clear, the "total" behavior is the result of both operations. When rotational invariance was under consideration, the result was the appearance of an intrinsic spin for the vector field. For determining the total parity of a field, we must also inquire about the *intrinsic parity* of the field components themselves [i.e., the behavior of $A_{-\xi}(x, y, z)$, $\xi = x, y, z$, contrasted to $A_{\xi}(x, y, z)$] as well as the orbital parity expressed in (7-13). As per the discussion of parity in Chapter I, we must associate an odd *intrinsic* parity with the polar vectors $\mathbf{A}$ and $\mathscr{E}$ and an even *intrinsic* parity with the axial vector $\mathscr{H}$. The *total parity* π associated with the electric field in an electric 2^λ-pole radiative transition is thus determined by the product of the intrinsic parity of the electric field and the orbital (or spatial) parity of the electric field, and hence $\pi = (-1)(-1)^{\lambda+1} = (-1)^\lambda$. In this respect the analogy with (1-22) and the attendant discussion for particles is both striking and apt.

When the total parity is determined from the magnetic field, we also get $\pi = (-1)^\lambda$, and, of course, the same result follows from considering the vector potential as well. This same result also will be obtained in the form of a selection rule in a later section where we consider the radiative transition probability in some detail. The essential point will be the nature of the interaction Hamiltonian that contains the operator $\mathbf{j} \cdot \mathbf{A}$ for electric transitions, where $\mathbf{j}$ is the nuclear current operator. The current operator $\mathbf{j}$ is a polar vector and hence gives $\mathbf{j} \cdot \mathbf{A}$ a parity opposite that of $\mathbf{A}$, again leading to the above result. Finally, then, an electric 2^λ-pole radiation field carries a total parity $\pi = (-1)^\lambda$.

The fact that in the E mode the magnetic field can be written in terms of $\mathfrak{X}_J{}^M$ gives a further characteristic of such modes. It is readily established and intuitively evident that the operator $\mathbf{r} \cdot \mathbf{L} = 0$. In view of (7-15), therefore, the magnetic vector is seen to be everywhere perpendicular to the radial vector $\mathbf{r}$. Such modes are thus appropriately enough also referred to as *transverse magnetic* or TM modes.

▶**Exercise 7-4** Show by writing

$$\hat{\mathbf{r}} = (\sin\theta \cos\varphi)\,\hat{e}_x + (\sin\theta \sin\varphi)\,\hat{e}_y + (\cos\theta)\,\hat{e}_z$$

and $\mathbf{L} = L_x \hat{e}_x + L_y \hat{e}_y + L_z \hat{e}_z$ with L_x, L_y, L_z expressed in terms of θ, φ, and their derivatives that

$$\mathbf{r} \cdot \mathbf{L} = 0.$$ ◀

Substituting (7-19a) into the wave equation readily yields the differential equation the function $f(\lambda, \mu; r)$ must satisfy, namely

$$\left[\frac{d^2}{dr^2} - \frac{\lambda(\lambda+1)}{r^2} + k^2\right] f(r) = 0. \tag{7-20}$$

The solutions to (7-20) can be obtained in terms of the spherical Bessel and Neumann functions (and linear combinations of them), depending on the boundary conditions. When the electromagnetic field is created by a radiating source at the origin, an appropriate solution is one which at large distances has the asymptotic form of an outgoing radial wave. This function is the spherical Hankel function of the first kind [refer, for example, to Merzbacher (1970, Chapter 10, Section 2)], which in the asymptotic region or "wave zone" is

$$r^{-1}f(r) \rightarrow (1/kr)\exp\{i[kr-\tfrac{1}{2}\pi(\lambda+1)]\}, \qquad kr \gg \lambda. \tag{7-21}$$

▶**Exercise 7-5** (a) Substituting (7-19a) into the wave equation and using the identity $\mathbf{L}^2 = -\hbar^2 r^2 \nabla^2 + \hbar^2 (\partial/\partial r)(r^2\, \partial/\partial r)$, show that (7-20) follows.

(b) Show that in the asymptotic region where (7-21) prevails, the three vectors: the electric field $\mathscr{E}(\mathbf{r}) = (i/k)\, \nabla \times \mathscr{H}$, the magnetic field $\mathscr{H}$, and $\mathbf{k} = k\hat{r}$ are approximately mutually perpendicular. Show also that $|\mathscr{E}(\mathbf{r})| \approx |\mathscr{H}(\mathbf{r})|$.

Care must be exercised in the physical content of this approximation. Thus while the energy flux determined by the Poynting vector is essentially properly described by such complete transversality, the angular momentum carried by the field absolutely requires that $\mathbf{r} \times (\mathscr{E} \times \mathscr{H})$ be considered non-zero even in the asymptotic region.

In the region $kr \lesssim \lambda$ where the so-called *induction field* cannot be neglected compared to the *radiation field* (7-21), of course only the magnetic field is transverse. ◀

The case of the magnetic multipole field follows a similar treatment, with, however, the important difference that in this case the parity considerations give

$$\mathscr{E}_{\mathrm{M}}(\mathbf{r}) = r^{-1} f(\lambda, \mu; r)\, \mathfrak{X}_\lambda{}^\mu(\theta, \varphi) \qquad \text{(M mode)} \tag{7-22a}$$

and

$$\mathscr{H}_{\mathrm{M}}(\mathbf{r}) = -\frac{i}{k}\nabla \times \mathscr{E}(\mathbf{r}) = -\frac{i}{k}\nabla \times [r^{-1} f(\lambda, \mu; r)\, \mathfrak{X}_\lambda{}^\mu(\theta, \varphi)] \qquad \text{(M mode).} \tag{7-22b}$$

It is probably useful to summarize the somewhat complicated parity characteristics of the 2^λ-pole electric and magnetic multipole fields. The results are shown in Table 7-1.

Since both the E-mode and M-mode fields are solutions to the Maxwell equations, the most general solution must be the linear sum (multipole expansion),

$$\mathscr{E}(\mathbf{r}) = \sum_{\lambda=1}^{\infty} \sum_{\mu=-\lambda}^{+\lambda} [a_{\mathrm{E}}(\lambda,\mu)\mathscr{E}_{\mathrm{E}}(\lambda,\mu;\mathbf{r}) + a_{\mathrm{M}}(\lambda,\mu)\mathscr{E}_{\mathrm{M}}(\lambda,\mu;\mathbf{r})]$$
$$\mathscr{H}(\mathbf{r}) = \sum_{\lambda=1}^{\infty} \sum_{\mu=-\lambda}^{+\lambda} [a_{\mathrm{E}}(\lambda,\mu)\mathscr{H}_{\mathrm{E}}(\lambda,\mu;\mathbf{r}) + a_{\mathrm{M}}(\lambda,\mu)\mathscr{H}_{\mathrm{M}}(\lambda,\mu;\mathbf{r})]. \tag{7-23}$$

TABLE 7-1

PARITY CHARACTERISTICS OF ELECTRIC AND MAGNETIC MULTIPOLE FIELDS

Multipole character	Orbital parity of **A** and $\mathscr{E}$	Orbital parity of $\mathscr{H}$	Total parity π
E mode, 2^λ	$(-1)^{\lambda+1}$	$(-1)^\lambda$	$(-1)^\lambda$
M mode, 2^λ	$(-1)^\lambda$	$(-1)^{\lambda+1}$	$(-1)^{\lambda+1}$

The expansion coefficients $a_{\mathrm{E}}(\lambda,\mu)$ and $a_{\mathrm{M}}(\lambda,\mu)$ are determined by the behavior of the nucleon motions within the nucleus. More specifically, when the radiation field corresponds to an emitted photon, these coefficients are expressed equivalently in terms of transition probabilities connecting the initial and final nuclear states.

The field energy written in terms of the electric and magnetic fields of (7-18) is

$$H_{\mathrm{rad}}(t) = (1/8\pi)\int [\mathscr{E}(\mathbf{r},\omega;t)\cdot\mathscr{E}(\mathbf{r},\omega;t) + \mathscr{H}(\mathbf{r},\omega;t)\cdot\mathscr{H}(\mathbf{r},\omega;t)]\, d^3r. \tag{7-24}$$

The time average of the field energy in terms of $\mathscr{E}(\mathbf{r})$ and $\mathscr{H}(\mathbf{r})$ can be written

$$H_{\mathrm{rad}} = (1/4\pi)\int [\mathscr{E}^*(\mathbf{r})\cdot\mathscr{E}(\mathbf{r}) + \mathscr{H}(\mathbf{r})\cdot\mathscr{H}^*(\mathbf{r})]\, d^3r. \tag{7-25}$$

The Poynting vector (energy flux) of the field is

$$\mathbf{S}_{\mathrm{rad}}(t) = (c/4\pi)\,\mathscr{E}(\mathbf{r},\omega;t)\times\mathscr{H}(\mathbf{r},\omega;t), \tag{7-26}$$

the time average of which is

$$\mathbf{S}_{\mathrm{rad}} = (c/4\pi)[\mathscr{E}^*(\mathbf{r})\times\mathscr{H}(\mathbf{r}) + \mathscr{E}(\mathbf{r})\times\mathscr{H}^*(\mathbf{r})]. \tag{7-27}$$

▶**Exercise 7-6** Consider

$$\mathbf{U}(\mathbf{r},\omega;t) = \mathbf{U}(\mathbf{r})\,e^{-i\omega t} + \mathbf{U}^*(\mathbf{r})\,e^{i\omega t}, \qquad \mathbf{V}(\mathbf{r},\omega;t) = \mathbf{V}(\mathbf{r})\,e^{-i\omega t} + \mathbf{V}^*(\mathbf{r})\,e^{i\omega t}.$$

(a) Show that the time average of $\mathbf{U}(\mathbf{r},\omega;t)\cdot\mathbf{V}(\mathbf{r},\omega;t)$ is

$$\langle\mathbf{U}(\mathbf{r},\omega;t)\cdot\mathbf{V}(\mathbf{r},\omega;t)\rangle = \mathbf{U}^*(\mathbf{r})\cdot\mathbf{V}(\mathbf{r}) + \mathbf{U}(\mathbf{r})\cdot\mathbf{V}^*(\mathbf{r}).$$

(b) Show that the time average of $\mathbf{U}(\mathbf{r},\omega;t)\times\mathbf{V}(\mathbf{r},\omega;t)$ is

$$\langle\mathbf{U}(\mathbf{r},\omega;t)\times\mathbf{V}(\mathbf{r},\omega;t)\rangle = \mathbf{U}^*(\mathbf{r})\times\mathbf{V}(\mathbf{r}) + \mathbf{U}(\mathbf{r})\times\mathbf{V}^*(\mathbf{r}).$$ ◀

Although we do not propose to carry out the quantization of the electromagnetic field, we would like to remark that beginning with (7-24), a Hamiltonian for the field can be constructed of the form

$$H_\lambda = 2\pi c^2 P_\lambda{}^2 + (\omega_\lambda{}^2/8\pi c^2)Q_\lambda{}^2. \tag{7-28a}$$

The quantities P_λ and Q_λ are two Hermitian operators related by

$$P_\lambda = (1/4\pi c^2)\dot{Q}_\lambda \tag{7-28b}$$

and thus H_λ can be formally connected with the Hamiltonian for a linear harmonic oscillator of frequency ω_λ with a "coordinate" operator Q_λ and a "momentum" operator P_λ. The field energy eigenvalues are thus

$$E_\lambda = (n_\lambda + \tfrac{1}{2})\hbar\omega_\lambda. \tag{7-28c}$$

The quantity n_λ is interpreted as the occupation number of the field and the individual excitons, or photons, have the energy

$$E_{\text{photon}} = \hbar\omega_\lambda. \tag{7-28d}$$

The interpretation (or actually the "removal") of the zero-point energy $\frac{1}{2}\hbar\omega_\lambda$ per degree of freedom is a difficulty completely disposed of in modern *renormalization* theories.

For a very comprehensive treatment of the above and the electromagnetic field in general emphasizing nuclear applications, the reader is referred to Eisenberg and Greiner (1970, Vol. 2).

Since the linear momentum density is just $\mathbf{S}_{\text{rad}}/c^2$, the time average total angular momentum in the field is

$$\mathbf{G}_{\text{rad}} = (1/4\pi c)\int \mathbf{r}\times[\mathscr{E}^*(\mathbf{r})\times\mathscr{H}(\mathbf{r}) + \mathscr{E}(\mathbf{r})\times\mathscr{H}^*(\mathbf{r})]\,d^3r. \tag{7-29}$$

It is interesting to note that for a pure multipole field of order (λ,μ), the correspondence principle holds exactly since (7-29) can be shown to give

$$G_{z,\text{rad}} = (\mu/\omega)\,H_{\text{rad}} = (\mu/\omega)\,\hbar\omega = \mu\hbar \tag{7-30a}$$

and similarly

$$\mathbf{G}_{\text{rad}}\cdot\mathbf{G}_{\text{rad}} = \lambda(\lambda+1)\hbar^2. \tag{7-30b}$$

The energy radiated into the solid angle $d\Omega$ can be deduced in terms of the Poynting vector (7-27) through the expression

$$U(\theta, \varphi)\, d\Omega = r^2 (\mathbf{S}_{\text{rad}} \cdot \hat{\mathbf{r}})\, d\Omega. \tag{7-31a}$$

Again drawing on the correspondence principle, (7-31a) can be interpreted to give the transition probability per second for emitting a single photon in the direction (θ, φ), viz.,

$$w(\theta, \varphi)\, d\Omega = [U(\theta, \varphi)/\hbar\omega]\, d\Omega. \tag{7-31b}$$

We wish to evaluate (7-31) for a pure multipole field (λ, μ). The calculation is most readily and appropriately performed in the asymptotic region corresponding to the laboratory distances at which radiation detectors are located. By coupling the results of Exercise 7-5 with (7-27) and noting the form of the E-mode solutions (7-19) and the M-mode solutions (7-22), we readily conclude that $U(\theta, \varphi)$ does not depend on whether the multipole (λ, μ) is electric or magnetic in nature. Using (7-21) and carrying out the necessary steps, we get

$$U(\lambda, \mu{:}\, \theta, \varphi) = (c/2\pi k^2)\, Z_{\lambda\mu}(\theta, \varphi)\, |a_\sigma(\lambda, \mu)|^2 \tag{7-32a}$$

with

$$Z_{\lambda\mu}(\theta, \varphi) = \mathfrak{X}_\lambda^{\mu *} \cdot \mathfrak{X}_\lambda^{\mu} = \frac{1}{2}\left[1 - \frac{\mu(\mu+1)}{\lambda(\lambda+1)}\right] |Y_\lambda^{\mu+1}|^2 + \frac{1}{2}\left[1 - \frac{\mu(\mu-1)}{\lambda(\lambda+1)}\right] |Y_\lambda^{\mu-1}|^2$$
$$+ \frac{\mu^2}{\lambda(\lambda+1)} |Y_\lambda^{\mu}|^2, \tag{7-32b}$$

and where $a_\sigma(\lambda, \mu)$ stands for the amplitude of either the electric mode ($\sigma = E$) or the magnetic mode ($\sigma = M$) in the multipole expansion (7-23).

▶**Exercise 7-7** Derive (7-32). ◀

Some typical angular distributions in terms of $Z_{\lambda\mu}$ are

$$\text{dipole} \begin{cases} (\lambda, \mu) = (1, 0); & Z_{1,0}(\theta) = (3/8\pi) \sin^2\theta \\ (\lambda, \mu) = (1, \pm 1); & Z_{1,\pm 1}(\theta) = (3/16\pi)(1 + \cos^2\theta) \end{cases}$$

$$\text{quadrupole} \begin{cases} (\lambda, \mu) = (2, 0); & Z_{2,0}(\theta) = (15/8\pi) \sin^2\theta \cos^2\theta \\ (\lambda, \mu) = (2, \pm 1); & Z_{2,\pm 1}(\theta) = (5/16\pi)(1 - 3\cos^2\theta + 4\cos^4\theta) \\ (\lambda, \mu) = (2, \pm 2); & Z_{2,\pm 2}(\theta) = (5/16\pi)(1 - \cos^4\theta). \end{cases} \tag{7-33}$$

These angular distributions are illustrated in Fig. 7.1.

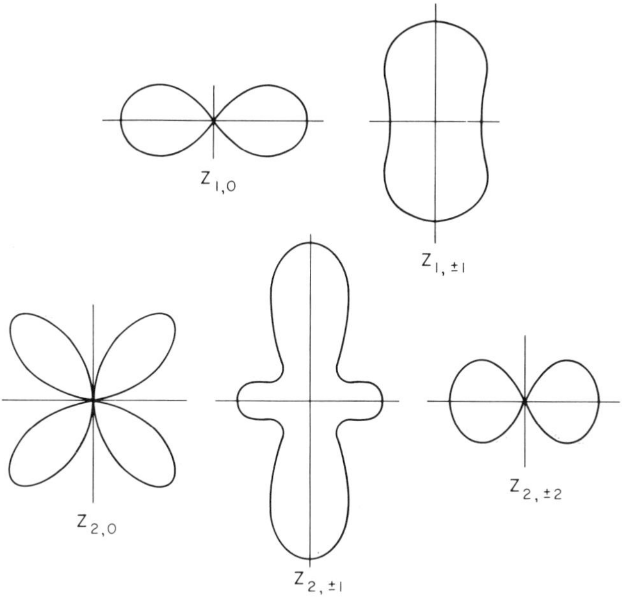

FIG. 7.1. The angular distribution of radiant flux, or equivalently the photon emission probability, for some typical multipole fields.

It is evident from the above that a measurement of the angular distribution of emitted photons from aligned nuclei would determine only the multipole order of the field, but not its parity (i.e., whether it is electric or magnetic in character). For such a determination the *polarization* of the photon also would have to be observed.

▶**Exercise 7-8** The polarization of a field is described in terms of the vector potential **A** (and hence $\mathscr{E}$). The polarization can be linear, circular (both right- and left-handed), or, as a general category, elliptical. Using the results of Exercise 7-2 and (7-10), determine the state of polarization for the electric and magnetic dipole fields in the asymptotic region (where complete transversality can be assumed).

For example, it will be readily found that for $\lambda = 1$, $\mu = 0$, M mode, the vector potential will have the angular dependence of

$$\mathfrak{X}_1{}^0 = i(3/8\pi)^{1/2}(\sin\theta)\begin{pmatrix} -\sin\varphi \\ \cos\varphi \\ 0 \end{pmatrix}.$$

Thus this field is linearly polarized with the polarization vector perpendicular to the plane containing the polar axis and the propagation direction. ◀

We now return to the spin properties of the electromagnetic field. For our purposes it is convenient to examine the case of a simple plane wave propagating in the direction of the polar axis or z axis, $\hat{\mathbf{k}} = \hat{e}_z$, in free space and to ignore any possible static field components. The zero divergence condition on the field vectors results in requiring $\mathbf{A}\cdot\hat{\mathbf{k}} = 0$, $\mathscr{E}\cdot\hat{\mathbf{k}} = 0$, and $\mathscr{H}\cdot\hat{\mathbf{k}} = 0$. Thus only two linearly independent states of polarization are possible for the field. These can be taken to be linearly polarized states customarily in the directions $\hat{e}_x$ and $\hat{e}_y$, or equivalently as right- and left-handed circularly polarized states (RHC or LHC). Using this latter choice, we write

$$\mathbf{A}_{\pm 1}(x, y, z) = \chi_{\pm 1} e^{ikz} = \mp(1/\sqrt{2})(\hat{e}_x \pm i\hat{e}_y)e^{ikz}, \tag{7-34}$$

where the upper algebraic signs refer to RHC and the lower to LHC. We can immediately verify, using (7-7a), that

$$S_z \mathbf{A}_{\pm 1} = \pm\hbar \mathbf{A}_{\pm 1}. \tag{7-35}$$

Although the plane wave states (7-34) are not eigenvectors of $\mathbf{J}^2$, $\mathbf{L}^2$, or parity π, they are eigenvectors of S_z, $\mathbf{S}^2$, and J_z (this latter since $L_z \mathbf{A}_{\pm 1} = 0$).[1] In view of (7-35), only the *longitudinal spin* states having components $\pm\hbar$ along the direction of propagation are possible eigenvectors of spin, $+\hbar$ for RHC and $-\hbar$ for LHC. These spin states are referred to as *helicity states* of the photon and the eigenvalues $\pm\hbar$ correspond to observable physical quantities. Spin components perpendicular to the direction of propagation do not exist.

All electromagnetic field solutions to the general arbitrary problem consist of spin states that are coherent sums of the *two* helicity states defined by the propagation vector $\hat{\mathbf{k}}$ and the spin operator $\hat{\mathbf{k}}\cdot\mathbf{S}$ with

$$\hat{\mathbf{k}}\cdot\mathbf{S} = (\sin\theta_k \cos\varphi_k)S_x + (\sin\theta_k \sin\varphi_k)S_y + (\cos\theta_k)S_z, \tag{7-36}$$

where $(\theta_k, \varphi_k) \equiv \hat{\mathbf{k}}$ and $S_{x,y,z}$ are the spin operators (7-7). The multipole fields we have been discussing are generally complex sums of helicity spin states. A simple case, for example, is for the radiation emitted in the direction of maximum intensity (along the polar axis) for the $\lambda = 1$, $\mu = +1$, M-mode multipole. Such radiation is of pure RHC polarization and hence corresponds to a pure helicity state $+\hbar$. Emission in directions other than $\theta = 0, \pi$ have elliptic polarization and mixed helicity.

[1] These and other relevant relationships can be explicitly demonstrated by using the well-known multipole expansion

$$\mathbf{A}_{\pm 1} = \chi_{\pm 1} e^{ikz} = \sum_{l}^{\infty} i^l [4\pi(2l+1)]^{1/2} j_l(kr) Y_l^0(\theta)\chi_{\pm 1},$$

and

$$Y_l^0(\theta)\chi_{\pm 1} = \sum_{J=l-1}^{l+1} (l, 1, 0, \pm 1 \mid J, \pm 1)\, \mathfrak{Y}_{J,1,l}^{\pm 1}(\theta, \varphi).$$

It might appear that the strange behavior of having only two possible spin components for the photon even though it has spin $S = 1$ is a result of requirements of the Maxwell equations. This behavior can, in fact, be traced to the relativistic requirements for *all* "particles" of any arbitrary spin with zero rest mass. For particles with *nonzero* rest mass, it is always possible to specify an *arbitrary* spin orientation in the *rest frame*. The usual Pauli–Schrödinger treatment of the electron indeed sets the electron spin to any of the states permitted in the rest frame of the electron, i.e., the nonrelativistic limit. This is not possible for zero-rest-mass particles that travel at the velocity c. For such particles only specifications with respect to the direction of propagation are possible, leading always to only two helicity states $\pm S\hbar$, for an arbitrary spin S, which remain relativistically invariant. While helicity states can also be defined for particles with nonzero rest mass, these are not relativistically invariant and depend on the relative velocity of the observer (e.g., it is always possible for the observer to travel faster than the particle and hence have it appear to be traveling backward and thus have opposite helicity).

Neutrinos, for example, are massless, spin $S = \frac{1}{2}$ particles that, like the photon, can only appear in one of two helicity states, having eigenvalues in this case $\pm \hbar/2$. Unlike photons, which are their own antiparticles, the neutrino has only the helicity $-\hbar/2$ (i.e., left-handed), and it is the antineutrino that has the other helicity $+\hbar/2$ (i.e., right-handed). The consequences of this behavior on the part of the neutrino have profound effects in the theory of β-decay as we shall see in Chapter VIII.

While techniques for detecting the state of polarization of visible light are common knowledge, perhaps a few words are in order for the less familiar measurements with nuclear γ rays. The degree of circular polarization for such energetic photons (generally $0.1 \lesssim \hbar\omega \lesssim 10$ MeV) can be measured either by methods that rely on the interaction of the photons with polarized matter or by transferring the helicity of the photons to secondary particles the polarization of which is more directly measurable.

A good example of the first type of arrangement consists in utilizing the Compton scattering from polarized electrons in magnetized ferromagnetic materials. The Compton scattering cross section depends on the state of polarization of the photon. Thus relative photon intensity measurements contrasting the scattering from two different orientations of the electron spins (i.e., two different directions of magnetization) can be used to yield the degree of circular polarization of the photons. Such preferential scattering can be utilized in either a transmission mode or a reflection mode. In the transmission mode the scattering process is used to *remove* photons from a beam of photons impinging on a detector. In the reflection mode only the *scattered* photons are detected. Since the fraction of spin-aligned electrons in ferromagnetic materials magnetized to saturation only amounts to $\sim 8\%$, polarimeters (as such detection devices are called) generally do not exceed a few percent in efficiency.

An example of a technique utilizing the transfer of helicity from the photon to a secondary particle uses the photoelectric effect. While the inner electrons in atomic shells cannot be initially polarized since these shells are filled, the photoejected electrons do become longitudinally polarized if the incident photon is circularly polarized. For example, when $\hbar\omega \gtrsim 0.3$ MeV, the average longitudinal polarization of photoelectrons ejected by completely circularly polarized photons exceeds 50%. The detection of the longitudinal polarization of the electrons, however, requires a further scattering measurement making use of either Mott or Møller scattering. Since the converter and scattering foils must be relatively thin, the polarimeter efficiency is rather low.

An excellent reference discussing the techniques for measurements of the circular polarization of γ rays is by Schopper (1958).

D. NUCLEAR γ-RAY TRANSITIONS

A complete and self-consistent solution to the problem of the interaction of charged particles and the electromagnetic field (generally consisting of external fields as well as "self-fields" generated by the particles under discussion) is extremely difficult. This is true in both the classical and the quantum mechanical treatment of the problem. Fortunately the field–particle interaction Hamiltonian in many practical applications in nuclear physics involves only weak contributions from nuclear currents and magnetic moments, which makes perturbation methods suitable. For example, the current–field interaction involves the fine structure constant $\alpha = e^2/\hbar c \approx 1/137$ in first order, α^2 in second order, and proportionately higher powers in subsequent orders, thus leading to rapid convergence for perturbation methods. Refined perturbation calculations are capable of great precision and have been successfully applied to a very large number of important problems. We limit our requirements to only the simple task of obtaining a general insight into the question of radiative transition rates.

1. Semiclassical Treatment

An early approach to the problem of nuclear radiative decay rates was based on the straightforward extension of the use of the correspondence principle in the semiclassical form that we have employed in the previous sections. We give only a brief description of this technique since we shall rely on a somewhat more formal approach that is more in keeping with the concepts of elementary quantum mechanics. The results obtained in lowest order are identical for both methods and indeed the differences in the two points of view are perhaps more apparent than real.

The semiclassical calculation begins by modifying the Maxwell equations (7-16) to include the presence of nuclear charges, currents, and intrinsic magnetic moments that are assumed sinusoidal functions of time. The static Coulomb and magnetic interactions acting between the nucleons are presumed to have been included in the equations of motion describing the nuclear state. The time-dependent charges, currents, and magnetic moments are taken to be the sources of the radiating electromagnetic fields. The result is the inhomogeneous wave equations, which in terms of the fields $\mathscr{E}(\mathbf{r})$ and $\mathscr{H}(\mathbf{r})$ of (7-18) are[1]

$$\nabla\times\nabla\times\mathscr{H}(r) - k^2\mathscr{H}(r) = (4\pi/c)[\nabla\times\mathbf{j}(\mathbf{r}) + ck^2\mathscr{M}(\mathbf{r})] \qquad (7\text{-}37a)$$

and

$$\nabla\times\nabla\times\mathscr{E}(\mathbf{r}) - k^2\mathscr{E}(\mathbf{r}) = (4\pi ik/c)[\mathbf{j}(\mathbf{r}) + c\nabla\times\mathscr{M}(\mathbf{r})] \qquad (7\text{-}37b)$$

with the usual added subsidiary condition on the conservation of charge

$$\nabla\cdot\mathbf{j}(\mathbf{r}) = i\omega\rho(\mathbf{r}). \qquad (7\text{-}37c)$$

As might be expected, the solutions to (7-37) can be written again in the form (7-19), (7-22), and (7-23). For example, in the case of the E mode, substituting (7-19a) into (7-37a) and taking the scalar product of the resulting equation with $\mathfrak{X}_\lambda^{\mu*}(\theta,\varphi)$ and then integrating over θ and φ (over Ω but not r) gives

$$\left[\frac{d^2}{dr^2} - \frac{\lambda(\lambda+1)}{r^2} + k^2\right] f_{\mathrm{E}}(r) = -\frac{4\pi r}{c}\int [\mathfrak{X}_\lambda^{\mu*}(\theta,\varphi)]\cdot[\nabla\times\mathbf{j}(\mathbf{r}) + ck^2\mathscr{M}(\mathbf{r})]\, d\Omega, \qquad (7\text{-}38)$$

which should be contrasted to (7-20). The solution to (7-38) is most readily obtained by use of the Green's function technique. We wish to use the Green's function $G_\lambda(r,r')$ that satisfies the differential equation

$$\left[\frac{d^2}{dr^2} - \frac{\lambda(\lambda+1)}{r^2} + k^2\right] G_\lambda(r,r') = -\frac{1}{r}\delta(r-r'). \qquad (7\text{-}39)$$

[1] The actual charge density $\rho(\mathbf{r},\omega;t)$, current density $\mathbf{j}(\mathbf{r},\omega;t)$, and the magnetization $\mathscr{M}(\mathbf{r},\omega;t)$ are of course real and written in the manner of (7-18), viz.,

$$\rho(\mathbf{r},\omega;t) = \rho(\mathbf{r})e^{-i\omega t} + \rho^*(\mathbf{r})e^{i\omega t}$$
$$\mathbf{j}(\mathbf{r},\omega;t) = \mathbf{j}(\mathbf{r})e^{-i\omega t} + \mathbf{j}^*(\mathbf{r})e^{i\omega t}$$
$$\mathscr{M}(\mathbf{r},\omega;t) = \mathscr{M}(\mathbf{r})e^{-i\omega t} + \mathscr{M}^*(\mathbf{r})e^{i\omega t}.$$

Equations (7-37) and subsequent equations refer to $\rho(\mathbf{r})$, $\mathbf{j}(\mathbf{r})$, and $\mathscr{M}(\mathbf{r})$ appearing on the right-hand side of these defining equations. These quantities may be complex.

Of the infinite number of solutions that satisfy (7-39), we select $r^{-1}G_\lambda(r,r')$ to be finite at the origin and to correspond to an outgoing spherical wave in the asymptotic region. The form satisfying these conditions is

$$G_\lambda(r,r') = ikrj_\lambda(kr_<)h_\lambda^{(1)}(kr_>), \tag{7-40}$$

where $r_<$ denotes the smaller of r and r', and $r_>$ denotes the larger of the two. The functions j_λ and $h_\lambda^{(1)}$ are the spherical Bessel function and the Hankel function of the first kind, respectively. We thus have for the solution to (7-38) satisfying our physical boundary conditions

$$f_E(\lambda,\mu;r) = \int_0^\infty G_\lambda(r,r')F_E(\lambda,\mu;r')r'\,dr' \tag{7-41a}$$

$$= ikr\int_0^\infty j_\lambda(kr_<)h_\lambda^{(1)}(kr_>)F_E(\lambda,\mu;r')r'\,dr', \tag{7-41b}$$

where we have written $-F_E(\lambda,\mu;r')$ for the right-hand side of (7-38) with r' replacing r. Finally, in the asymptotic region outside the source (i.e., outside the nuclear volume) we identify $r_<$ with r' and $r_>$ with r to obtain

$$f_E(\lambda,\mu;r) \to \{\exp[i(kr-\tfrac{1}{2}l\pi)]\}\int_0^\infty j_\lambda(kr')F_E(\lambda,\mu;r')r'\,dr'. \tag{7-42}$$

Thus comparison with (7-19a) and (7-23) gives

$$a_E(\lambda,\mu) = (4\pi k/c)\int j_\lambda(kr)\mathfrak{X}_\lambda^{\mu *}(\theta,\varphi)\cdot[\nabla\times\mathbf{j}(\mathbf{r})+ck^2\mathcal{M}(\mathbf{r})]\,d^3r, \tag{7-43}$$

where we have dropped the prime on r' in the integral and have used (7-38) defining $F_E(\lambda,\mu;r)$.

Generally, (7-43) is evaluated in the long-wavelength approximation $kR \ll 1$, where R is the nuclear radius which defines the region of nonvanishing contributions to the integral. If we use a value of $R = 1.20A^{1/3}$ fm, the inequality $kR \ll 1$ implies

$$\hbar\omega \ll \hbar c/R = 165A^{-1/3}\ \text{MeV}. \tag{7-44}$$

Thus the approximation should be valid for $\hbar\omega \lesssim 10$ MeV except perhaps for the heaviest nuclei. In addition, the magnetic moment contribution to (7-43) compared to that of the current contribution is of order $\hbar\omega/m_N c^2$. Since the nucleon rest mass $m_N c^2 \approx 940$ MeV, the magnetic moment contribution to (7-43) can be neglected.

When the defining equation for $\mathfrak{X}_\lambda^{\mu}$, (7-15), is introduced, and (7-43) is evaluated to lowest order in kr, the result is[1]

$$a_E(\lambda,\mu) = -\frac{4\pi}{(2\lambda+1)!!}\left(\frac{\lambda+1}{\lambda}\right)^{1/2}k^{\lambda+2}Q_\lambda^{\mu}, \tag{7-45a}$$

[1] The quantity $(2\lambda+1)!! = (2\lambda+1)!/2^\lambda(\lambda!) = 1\times3\times5\times\cdots\times(2\lambda+1)$.

with

$$Q_\lambda^\mu = \int r^\lambda Y_\lambda^{\mu*}(\theta, \varphi)\, \rho(\mathbf{r})\, d^3r. \tag{7-45b}$$

The quantity Q_λ^μ is referred to as the electric multipole moment of order (λ, μ).

Equations (7-45) give us the field expansion coefficients in terms of the nuclear charge distribution. It now remains to derive an expression relating the radiative decay rate to the expansion coefficients in order to solve the problem completely. To obtain the total radiative transition rate, we use (7-31) and (7-32). Noting that the definition of $\mathfrak{X}_\lambda^\mu(\theta, \varphi)$ gives

$$\int Z_{\lambda\mu}(\theta, \varphi)\, d\Omega = \int \mathfrak{X}_\lambda^{\mu*}(\theta, \varphi) \cdot \mathfrak{X}_\lambda^\mu(\theta, \varphi)\, d\Omega = 1,$$

we obtain the total radiative transition rate for the E mode of multipole order (λ, μ)

$$T_\mathrm{E}(\lambda, \mu) = \int w(\theta, \varphi)\, d\Omega = |a_\mathrm{E}(\lambda, \mu)|^2 / 2\pi k^3 \hbar \tag{7-46a}$$

or

$$T_\mathrm{E}(\lambda, \mu) = \frac{8\pi(\lambda+1)}{\lambda[(2\lambda+1)!!]^2} \frac{k^{2\lambda+1}}{\hbar} |Q_\lambda^\mu|^2. \tag{7-46b}$$

The final step in this semiclassical approach is to "translate" Q_λ^μ into its quantum mechanical equivalent

$$Q_\lambda^\mu = e \sum_{\zeta=1}^{Z} \int \psi_\mathrm{f}^*(\mathbf{r}_1, \mathbf{r}_2, \ldots, \mathbf{r}_A)\, r_\zeta^\lambda Y_\lambda^{\mu*}(\theta_\zeta, \varphi_\zeta)\, \psi_\mathrm{i}(\mathbf{r}_1, \mathbf{r}_2, \ldots, \mathbf{r}_A)\, d^3r_1\, d^3r_2 \cdots d^3r_A, \tag{7-46c}$$

where we have summed over all the protons in the nucleus and ψ_f and ψ_i refer to the final and initial states, respectively (the symbolic representation of the nucleon coordinates $\mathbf{r}_1, \mathbf{r}_2, \ldots, \mathbf{r}_A$ is intended to include the spin and isospin variables as well as the spatial coordinates).

The case for M-mode transition follows an analogous treatment, leading to a source-dependent amplitude

$$a_\mathrm{M}(\lambda, \mu) = (4\pi ik/c) \int j_\lambda(kr)\, \mathfrak{X}_\lambda^{\mu*}(\theta, \varphi) \cdot [\mathbf{j}(\mathbf{r}) + c\,\boldsymbol{\nabla} \times \mathscr{M}(\mathbf{r})]\, d^3r. \tag{7-47}$$

Unlike the case for the E-mode amplitude, where only the first term of (7-43) is found to be important, both terms of (7-47) contribute with the same order of importance to the M-mode amplitude. Obtaining the total radiative transition rate $T_\mathrm{M}(\lambda, \mu)$ for the M mode from (7-47) in the long-wavelength approximation follows a similar course to the E-mode case.

2. Perturbation Treatment

Although the results (7-46) and those obtained from (7-47) are quite correct, we now turn to a perhaps more satisfying approach that is based on elementary quantum mechanics to arrive at the same results before proceeding further with the general discussion. We imagine the nucleus in question and the radiation field to be confined to the interior of a large spherical box of radius R_0 with a perfectly conducting wall. The system Hamiltonian consists of three parts,

$$H = H_n + H_r + H_{nr}, \tag{7-48}$$

where H_n is the Hamiltonian for the nucleus, H_r is the Hamiltonian for the radiation field, and H_{nr} is the interaction Hamiltonian of the nucleus and the field.

In a more limited context, the Hamiltonian for a system of particles in the presence of a field specified by a vector potential **A** is written

$$H_n + H_{nr} = \sum_{\alpha} \frac{1}{2m_\alpha}\left(\mathbf{p}_\alpha - \frac{q_\alpha}{c}\mathbf{A}\right)^2 - \sum_{\alpha} \frac{e\hbar}{2m_\alpha c}\boldsymbol{\mu}_\alpha \cdot (\nabla \times \mathbf{A}) + \frac{1}{2}\sum_{\substack{\alpha,\alpha' \\ \alpha \neq \alpha'}} v_{\alpha\alpha'}. \tag{7-49}$$

In our case the indices α (and α') refer to the nucleons of the nucleus, individually having mass, charge, and magnetic moment (in nuclear magnetons), m_α, q_α, and $\boldsymbol{\mu}_\alpha$.[1] The first term in (7-49) can be expanded to give

$$\frac{1}{2m_N}\left(\mathbf{p}_\alpha - \frac{q_\alpha}{c}\mathbf{A}\right)^2 = \frac{\mathbf{p}^2}{2m_N} + \frac{q_\alpha}{2m_N c}(\mathbf{p}_\alpha \cdot \mathbf{A} + \mathbf{A}\cdot\mathbf{p}_\alpha) + \frac{q_\alpha^2}{2m_N c^2}\mathbf{A}^2. \tag{7-50}$$

Since we shall be using a weak field perturbation calculation, and the order of the quadratic term in the vector potential is $e/\hbar c = 1/137$ times the linear term which itself is small, we neglect the quadratic term. Further, we continue to use the Coulomb gauge $\nabla \cdot \mathbf{A} = 0$, and hence **p** and **A** commute, so that $\mathbf{p}_\alpha \cdot \mathbf{A} + \mathbf{A}\cdot\mathbf{p}_\alpha = 2\mathbf{A}\cdot\mathbf{p}_\alpha$. We thus have for (7-49)

$$H_n + H_{nr} = \sum_{\alpha} \frac{\mathbf{p}^2}{2m_N} + \frac{1}{2}\sum_{\substack{\alpha,\alpha' \\ \alpha \neq \alpha'}} v_{\alpha\alpha'} - \sum_{\alpha}\left[\frac{q_\alpha}{m_N c}\mathbf{A}\cdot\mathbf{p}_\alpha + \frac{e\hbar}{2m_N c}\boldsymbol{\mu}_\alpha \cdot (\nabla \times \mathbf{A})\right]. \tag{7-51}$$

[1] The electromagnetic fine structure of the nucleons is generally not considered; they are simply assumed to be pointlike particles with certain intrinsic properties. Thus, while coordinate variables are not explicitly exhibited in (7-49), the assumed point or zero-range field–particle interaction requires computing the field in each case at the nucleon coordinate. We can take the mass of the neutron and proton the same and simply write $m_\alpha = m_N$, and clearly $q_\alpha = +e$ for protons and zero for neutrons, and $\boldsymbol{\mu}_\alpha = +2.793\boldsymbol{\sigma}_\alpha$ for protons and $\boldsymbol{\mu}_\alpha = -1.913\boldsymbol{\sigma}_\alpha$ for neutrons.

We identify the first two terms with H_n and the last term with H_{nr} and write

$$H_n = \sum_{\alpha} \frac{\mathbf{p}_\alpha^{\;2}}{2m_N} + \frac{1}{2} \sum_{\substack{\alpha, \alpha' \\ \alpha \neq \alpha'}} v_{\alpha\alpha'} \tag{7-52a}$$

and

$$H_{nr} = -\sum_{\alpha} \left[\frac{q_\alpha}{m_N c} \mathbf{A} \cdot \mathbf{p}_\alpha + \frac{e\hbar}{2m_N c} \boldsymbol{\mu}_\alpha \cdot (\nabla \times \mathbf{A}) \right]. \tag{7-52b}$$

It is assumed that the unperturbed nuclear eigenstates of H_n are known. The potential energy term $v_{\alpha\alpha'}$ is presumed to include the static electromagnetic interactions of the nucleons with each other through Coulomb forces and magnetic moment interactions. The vector potential $\mathbf{A}$ is taken to be the transverse radiation field we have been discussing and thus excludes any static or longitudinal components.

It should be noted that (7-52b) is the quantum mechanical operator associated with the classical interaction

$$\Delta E(\text{classical}) = -[(1/c)\, \mathbf{A}(\mathbf{r}) \cdot \mathbf{j}(\mathbf{r}) + \mathcal{M}(\mathbf{r}) \cdot \mathcal{H}(\mathbf{r})].$$

Here again the Coulomb component does not appear, and as before $\mathbf{j}(\mathbf{r})$ is the current density and $\mathcal{M}(\mathbf{r})$ is the magnetization or magnetic dipole moment per unit volume.

The radiation field Hamiltonian H_r is written symbolically

$$H_r = (1/8\pi) \sum_{\lambda} \int (\mathcal{E}_\lambda^{\;2} + \mathcal{H}_\lambda^{\;2})\, d^3 r. \tag{7-53}$$

The unperturbed eigenstates of H_r are taken to be a superposition of just those multipole fields we have discussed previously, albeit with new boundary conditions. In general the solution to (7-53) would be the simultaneous presence of many photons n_λ of each frequency ω_λ (the occupation numbers of a more precisely formulated field theory, the details of which need not concern us in the following).[1] The interaction H_{nr} will lead to either an absorption or an emission of a single photon in lowest order, changing one of the occupation numbers to $n_{\lambda,f} = n_{\lambda,i} \pm 1$. We shall assume that our spherical cavity contains only the one nucleus and only a weak field (i.e., essentially empty except for the photon to be emitted or absorbed). We ignore the phenomenon of *induced emission* and hence do not discuss possible nuclear laser action and similar cooperative effects. We focus our attention on simple *spontaneous emission*. For a more comprehensive treatment of the problem including all effects, the reader is referred to other texts [e.g., Merzbacher (1970)].

[1] In all essential aspects the formalism here is identical with that discussed in connection with the nuclear vibrational phonons of Chapter VI.

We shall use a simple time-dependent perturbation calculation for the spontaneous decay rate of an excited state of a nucleus. The nucleus is initially a quasi-stationary excited state (i.e., eigenstate of H_n) ψ_i, and the field is specified by the occupation numbers $n_{\lambda,i}$. The perturbing or interaction Hamiltonian H_{nr} induces the transition to a lower (possibly still excited) nuclear state with the emission of a photon. The field in the cavity is taken to be so weak that photon absorption leading to a nuclear state at an excitation still higher than the initial state is negligible. The explicit time dependence of H_{nr}, appearing through $\mathbf{A}(\mathbf{r}, t)$, is taken as $e^{i\omega t}$ appropriate for the emission process.[1]

The transition probability (decay rate for spontaneous emission) is then given by the *Golden Rule*,

$$T_{if} = (2\pi/\hbar)\,|\langle f|H_{nr}|i\rangle|^2 \rho(E_f)\,\delta(E_f - E_i + \hbar\omega), \tag{7-54}$$

where $\rho(E_f)$ refers to the density of final states (per unit energy). The appearance of the delta function ensures the conservation of energy in the transition and thus only the occupation number n_λ corresponding to a photon energy $\hbar\omega$ is affected (i.e., increased by one). If this point is kept in mind, the states $|f\rangle$ and $|i\rangle$ in (7-54) can be taken to refer explicitly to only the nuclear eigenstates of H_n.[2] The radiation field still enters through $\rho(E_f)$ and the normalization of $\mathbf{A}$.

[1] It is well known that the application of a perturbing Hamiltonian $H_1 \exp(\pm i\omega t')$ having matrix elements $\langle k|H_1|s\rangle$ and acting during a time interval $0 < t' < t$ leads to a final state amplitude of

$$c_f(\omega, t) = -\langle f|H_1|i\rangle \frac{\exp[(i/\hbar)(E_f - E_i \pm \hbar\omega)t] - 1}{E_f - E_i \pm \hbar\omega}.$$

The frequency dependence of c_f is thus seen to be given by essentially a delta function peaked at the energy $E_f - E_i \pm \hbar\omega = 0$. The upper and lower signs correspond to the upper and lower signs in the perturbing Hamiltonian. Clearly, $e^{+i\omega t}$ giving $E_f - E_i + \hbar\omega = 0$ corresponds to the emission process.

In the quantum theory of the radiation field, the component of

$$\mathbf{A}_\lambda(\mathbf{r}, t) = q_{0\lambda}\mathbf{A}_\lambda(\mathbf{r})\,e^{-i\omega t} + q_{0\lambda}^*\mathbf{A}_\lambda^*(\mathbf{r})e^{i\omega t}$$

with the time dependence $e^{i\omega t}$ is associated with a *creation operator* for a photon of the λ mode, while the term with the $e^{-i\omega t}$ factor is associated with a corresponding *annihilation operator*. The analogy with the preceding is immediate. It should be noted that in such a formulation no conceptual difficulty arises for spontaneous emission even in the weak field limit of a cavity completely empty of radiation in the initial state. The creation operator contained in the interaction Hamiltonian operating on the "vacuum" state (i.e., no radiation field at all) gives rise to a final state with one photon. The creation operator behaves like a raising operator for the occupation number. Again recall the corresponding discussion in Chapter VI.

[2] More precisely $|f\rangle$ and $|i\rangle$ are the nuclear and radiation field product states, $|f\rangle = |f_n\rangle|f_r\rangle$ and $|i\rangle = |i_n\rangle|i_r\rangle$. The interaction Hamiltonian properly contains the creation operator $a_\lambda^\dagger$, which, operating on the initial radiation vacuum state, gives the radiation final state, viz., $a_\lambda^\dagger|i_r\rangle = a_\lambda^\dagger|n_\lambda = 0\rangle = |n_\lambda = 1\rangle$, with the result that $\langle f_r|a_\lambda^\dagger|i_r\rangle = \langle n_\lambda = 1|a_\lambda^\dagger|n_\lambda = 0\rangle = 1$ if $|n_\lambda = 1\rangle$ is normalized to a one-photon state.

The normalization of the vector potential **A** is taken to give a single photon of energy $\hbar\omega$ of multipole order permitted by the selection rules of the nuclear matrix elements. This photon state is for the *unperturbed* Hamiltonian H_r. Thus the full Hermitian forms must be used for the field quantities given by (7-18) and

$$\mathbf{A}(\mathbf{r},\omega;t) = \mathbf{A}(\mathbf{r})e^{-i\omega t} + \mathbf{A}^*(\mathbf{r})e^{i\omega t}, \tag{7-55}$$

even though in H_{nr} only the second term of (7-55) will be used.

Again singling out the emission process involving the E mode, we write for the field quantities (7-18) and (7-55)

$$\mathscr{H}_E(\lambda,\mu;\mathbf{r}) = N(\lambda,\mu)j_\lambda(kr)\mathfrak{X}_\lambda{}^\mu(\theta,\varphi), \tag{7-56a}$$

$$\mathscr{E}_E(\lambda,\mu;\mathbf{r}) = (i/k)\,\nabla\times\mathscr{H}_E(\lambda,\mu;\mathbf{r}), \tag{7-56b}$$

and

$$\mathbf{A}_E(\lambda,\mu;\mathbf{r}) = (1/k^2)\,\nabla\times\mathscr{H}_E(\lambda,\mu;\mathbf{r}). \tag{7-56c}$$

Here we have used $r^{-1}f(\lambda,\mu;r) = j_\lambda(kr)$ to give a solution to the unperturbed Hamiltonian H_r that is a standing wave finite at the origin. With a large cavity radius $kR_0 \gg 1$, the electromagnetic field assumes complete transversality at the cavity wall. Since $\hat{\mathbf{r}}\cdot\mathscr{E}(\mathbf{r}) = 0$ and since the tangential component of $\mathscr{E}(r = R_0)$ must vanish at the perfectly conducting wall, all outgoing spherical waves are perfectly reflected, giving rise to ingoing waves of equal amplitude. This superposition constitutes the assumed standing wave.

The normalization constant $N(\lambda,\mu)$ is determined from (7-25) by using the asymptotic form of the fields for $kr \gg 1$, if R_0 is taken large enough. Noting that for $kr \gg 1$, $\mathscr{E}^*\cdot\mathscr{E} \approx \mathscr{H}^*\cdot\mathscr{H}$, we have

$$\hbar\omega = [N^2(\lambda,\mu)/2\pi]\int_0^{R_0} j_\lambda{}^2(kr)r^2\,dr\int_\Omega \mathfrak{X}_\lambda^{\mu *}\cdot\mathfrak{X}_\lambda{}^\mu\,d\Omega. \tag{7-57}$$

The radial integral is

$$\int_0^{R_0} j_\lambda{}^2(kr)r^2\,dr \approx k^{-2}\int_0^{R_0}\cos^2[kr - \tfrac{1}{2}\pi(\lambda+1)]\,dr = R_0/2k^2, \tag{7-58}$$

since $j_\lambda(kR_0) \approx (1/kR_0)\cos[kR_0 - \frac{1}{2}\pi(\lambda+1)] = 0$. The angular integral is just unity; hence

$$N(\lambda,\mu) = (4\pi\hbar k^3 c/R_0)^{1/2}. \tag{7-59}$$

Finally, we have

$$\mathscr{H}_E(\lambda,\mu;\mathbf{r}) = (4\pi\hbar k^3 c/R_0)^{1/2} j_\lambda(kr)\mathfrak{X}_\lambda{}^\mu(\theta,\varphi), \tag{7-60a}$$

$$\mathscr{E}_E(\lambda,\mu;\mathbf{r}) = (i/k)\,\nabla\times\mathscr{H}_E(\lambda,\mu;\mathbf{r}), \tag{7-60b}$$

and

$$\mathbf{A}_E(\lambda,\mu;\mathbf{r}) = (1/k^2)\,\nabla\times\mathscr{H}_E(\lambda,\mu;\mathbf{r}). \tag{7-60c}$$

Before applying (7-52b) it will be useful to reduce the expression (7-60c) for $\mathbf{A}$ to a more manageable form. We use the definition (7-15) for $\mathfrak{X}_\lambda{}^\mu(\theta,\varphi)$, and since $\mathbf{L}$ does not operate on the variable r, the result is

$$\mathbf{A}_\mathrm{E}(\lambda,\mu;\mathbf{r}) = \frac{1}{\hbar k}\left(\frac{4\pi\hbar kc}{\lambda(\lambda+1)\,R_0}\right)^{1/2} \nabla\times\mathbf{L}u_\lambda{}^\mu(r,\theta,\varphi), \tag{7-61a}$$

with

$$u_\lambda{}^\mu = j_\lambda(kr)\,Y_\lambda{}^\mu(\theta,\varphi). \tag{7-61b}$$

We note the vector operation identity

$$\nabla\times\mathbf{L}/\hbar = i[\nabla(1+r\,\partial/\partial r) - \mathbf{r}\,\nabla^2], \tag{7-62}$$

and the recurrence relationship for the spherical Bessel functions

$$(1+r\,\partial/\partial r)j_\lambda(kr) = (\lambda+1)j_\lambda(kr) - krj_{\lambda+1}(kr). \tag{7-63}$$

Applying (7-62) and using (7-63), (7-61) becomes

$$\mathbf{A}_\mathrm{E} = \frac{i}{k}\left(\frac{4\pi\hbar kc}{\lambda(\lambda+1)\,R_0}\right)^{1/2} \{\nabla[(\lambda+1)\,u_\lambda{}^\mu - krj_{\lambda+1}(kr)\,Y_\lambda{}^\mu] + \mathbf{r}k^2u_\lambda{}^\mu\}, \tag{7-64}$$

where we have also used $\nabla^2 u_\lambda{}^\mu = -k^2u_\lambda{}^\mu$.

The nonvanishing contributions to (7-54) occur only within the nuclear volume, and we again make use of the long-wavelength approximation discussed in connection with (7-44). The asymptotic form for the Bessel function with $kr \ll \lambda$ is

$$j_\lambda(kr) = (kr)^\lambda/[(2\lambda+1)!!]. \tag{7-65}$$

In this approximation only the first term in the curly bracket of (7-64) is important, and hence

$$\mathbf{A}_\mathrm{E} = \frac{i}{k}\left(\frac{4\pi\hbar kc}{R_0}\,\frac{(\lambda+1)}{\lambda}\right)^{1/2} \frac{1}{(2\lambda+1)!!}\,\nabla v_\lambda{}^\mu \tag{7-66a}$$

with

$$v_\lambda{}^\mu = (kr)^\lambda\,Y_\lambda{}^\mu. \tag{7-66b}$$

We also need to evaluate the density of final states $\rho(E_\mathrm{f})$. In this connection we consider only the factor introduced by the radiation field and assume the transition relates to specific nuclear substates of $|\mathrm{f}\rangle$ and $|\mathrm{i}\rangle$. If this should not be the case, the usual averaging over initial nuclear states and summing over nuclear final states would introduce a further multiplicative factor resulting from computing $(2J_\mathrm{i}+1)^{-1}\sum_{M_\mathrm{i}}\sum_{M_\mathrm{f}}$. The assumption of perfectly conducting cavity walls gives

$$j_\lambda(kR_0) = 0 \tag{7-67a}$$

or

$$kR_0 - \tfrac{1}{2}\pi\lambda = n\pi, \qquad \text{with} \quad n \text{ an integer.} \tag{7-67b}$$

Now

$$\rho(E_{\mathrm{f}}) = dn/dE_{\mathrm{f}} = (1/\hbar c)\, dn/dk \qquad (\text{since} \quad dE_{\mathrm{f}} = \hbar c\, dk). \tag{7-68a}$$

Therefore from (7-67b) the result is

$$\rho(E_{\mathrm{f}}) = R_0/\pi\hbar c. \tag{7-68b}$$

It is convenient at this point to digress briefly to introduce a useful general theorem. The charge density and current density operators for pointlike charges are

$$\tilde{\rho}(\mathbf{r}) = \sum_{\alpha=1}^{A} q_\alpha\, \delta(\mathbf{r}-\mathbf{r}_\alpha) \tag{7-69a}$$

and

$$\tilde{\mathbf{j}}(\mathbf{r}) = \tfrac{1}{2} \sum_{\alpha=1}^{A} q_\alpha [(\mathbf{p}_\alpha/m_\alpha)\, \delta(\mathbf{r}-\mathbf{r}_\alpha) + \delta(\mathbf{r}-\mathbf{r}_\alpha)(\mathbf{p}_\alpha/m_\alpha)], \tag{7-69b}$$

where the current operator is written in Hermitian form. If the pointlike charge assumption is lifted, $\rho(\mathbf{r})$ and $\mathbf{j}(\mathbf{r})$ can be suitably modified; however, they must always satisfy the continuity equation for charge. Consider any two eigenstates, such as $|a\rangle$ and $|b\rangle$, of the same Hamiltonian. The standard quantum mechanical definition of the time-dependent matrix element of the current is

$$\langle b|\mathbf{j}(\mathbf{r})|a\rangle_{\mathrm{t}} = \langle b|\mathbf{j}(\mathbf{r})|a\rangle \exp[i(E_b - E_a)t/\hbar].$$

The continuity of charge then requires

$$\nabla \cdot \langle b|\mathbf{j}(\mathbf{r})|a\rangle_{\mathrm{t}} = -(\partial/\partial t)\langle b|\rho(\mathbf{r})|a\rangle_{\mathrm{t}},$$

which immediately yields

$$\nabla \cdot \langle b|\mathbf{j}(\mathbf{r})|a\rangle = i[(E_a - E_b)/\hbar]\langle b|\rho(\mathbf{r})|a\rangle. \tag{7-70}$$

If for the electric transitions under discussion we again use only the first term of the interaction Hamiltonian (7-52b), we can rewrite this term in the form

$$H_{\mathrm{nr}} = -(1/c)\,\mathbf{A}(\mathbf{r}) \cdot \mathbf{j}(\mathbf{r}). \tag{7-71}$$

The Golden Rule then becomes

$$T_{\mathrm{E,if}} = (2\pi/\hbar c^2) \left| \int \langle \mathrm{f}|\mathbf{j}(\mathbf{r})|\mathrm{i}\rangle \cdot \mathbf{A}(\mathbf{r})\, d^3 r \right|^2 \rho(E_{\mathrm{f}}). \tag{7-72}$$

It should be noted that the matrix element of $\mathbf{j}(\mathbf{r})$ involves integration on $d^3r_1\, d^3r_2 \cdots d^3r_A$ and the integration on d^3r explicitly exhibited in (7-72) refers to the field coordinate. Of course, in the delta-function limit (7-69b) the current–field interaction becomes zero range and the field coordinate is evaluated at each charge center in running through the sum in (7-69b).

The fact that $\mathbf{A}_E(\mathbf{r})$ in (7-66) is approximately the gradient of a scalar now leads to the theorem referred to as Siegert's theorem. We write, with $\nabla \equiv \nabla_r$,

$$\mathbf{A}(\mathbf{r}) = \nabla\Phi \tag{7-73a}$$

where

$$\Phi = \frac{i}{k}\left(\frac{4\pi\hbar kc}{R_0}\frac{(\lambda+1)}{\lambda}\right)^{1/2}\frac{1}{(2\lambda+1)!!}v_\lambda{}^\mu. \tag{7-73b}$$

The transition matrix element of (7-72) is

$$\int \langle \mathrm{f}|\mathbf{j}(\mathbf{r})|\mathrm{i}\rangle \cdot \nabla\Phi\, d^3r = \int [\psi_\mathrm{f}{}^*\mathbf{j}(\mathbf{r})\psi_\mathrm{i}] \cdot \nabla\Phi\, d^3r_1\, d^3r_2 \cdots d^3r_A\, d^3r. \tag{7-74}$$

Using the general identity $\nabla\cdot(\Phi\mathbf{v}) = \mathbf{v}\cdot\nabla\Phi + \Phi\nabla\cdot\mathbf{v}$, we have

$$\int [\psi_\mathrm{f}{}^*\mathbf{j}(\mathbf{r})\psi_\mathrm{i}]\cdot\nabla\Phi\, d\tau = \int \nabla\cdot[\Phi\psi_\mathrm{f}{}^*\mathbf{j}(\mathbf{r})\psi_\mathrm{i}]\, d\tau - \int \{\nabla\cdot[\psi_\mathrm{f}{}^*\mathbf{j}(\mathbf{r})\psi_\mathrm{i}]\}\,\Phi\, d\tau. \tag{7-75}$$

Using Green's theorem, the first term on the right-hand side of (7-75) can be converted into a surface integral which vanishes if the surface is selected comfortably outside the nuclear volume but still satisfying the long-wavelength approximation. Then using (7-70) with $E_\mathrm{i} - E_\mathrm{f} = \hbar\omega$ and (7-73), we obtain for (7-72)

$$T_{\mathrm{E,if}} \equiv T_\mathrm{E}(\lambda,\mu) = \frac{8\pi(\lambda+1)}{\lambda[(2\lambda+1)!!]^2}\frac{k^{2\lambda+1}}{\hbar}|Q_\lambda{}^\mu|^2, \tag{7-76a}$$

with

$$Q_\lambda{}^\mu = \int \langle \mathrm{f}|\rho(\mathbf{r})|\mathrm{i}\rangle\, r^l Y_\lambda^{\mu *}(\theta,\varphi)\, d^3r. \tag{7-76b}$$

The possibility of writing $Q_\lambda{}^\mu$ in terms of the charge operator instead of the current operator is Siegert's theorem.

A few observations concerning Siegert's theorem are perhaps in order at this point. Clearly, when the charge operator is set equal to the delta-function form (7-69a), integration over the field coordinates d^3r just gives (7-46) of the semiclassical treatment, if, as there, we ignore center-of-mass effects and sum the index α of (7-69a) only over protons. We shall return to this point later.

From the earlier chapters we realize that nucleons are not simple intrinsic pointlike entities. The strongly coupled meson fields produce a charge distribution possessing fine structure within dimensions of the order of 0.8 fm. In addition, the existence of charge exchange forces between nucleons implies the presence of a current associated with the exchange in charge between the interacting nucleons that is not included in $\mathbf{j}(\mathbf{r})$ of (7-69b) even in the delta-function limit of pointlike nucleons. While exchange current contributions can be formulated to give total currents that obey the continuity equation, such efforts are complicated by the need to give a model for the exchange current distribution such as, for example, a current filament along a straight line connecting the interacting nucleons. It is of great convenience, therefore, that in the presence of such meson effects Siegert's theorem can be employed to involve only the nuclear charge distribution expressed in terms of the nucleon degrees of freedom, at least in the long-wavelength approximation. Magnetic moments do not obey a conservation law and the large anomalous moments clearly indicate that strong meson effects are present. These points introduce an essential complexity into magnetic moment effects that is not present for the charge-current density operator. For further discussion on the above points, the reader is referred to the text by Sachs (1953), and for more recent commentary, to Gari and Huffman (1973) and the references contained therein.

The M-mode transition rates are computed in a manner quite similar to that for the E mode. We simply cite the results for both modes in a compact form for the case of the pointlike nucleon assumption,

$$T_\sigma(\lambda, \mu) = \frac{8\pi(\lambda+1)}{\lambda[(2\lambda+1)!!]^2} \frac{k^{2\lambda+1}}{\hbar} |\langle \mathrm{f}|\tilde{O}_\lambda{}^\mu|\mathrm{i}\rangle|^2 \tag{7-77a}$$

where for $\sigma \equiv$ E mode

$$\tilde{O}_\lambda{}^\mu(\mathrm{E}) = \tilde{Q}_\lambda{}^\mu = \sum_{\alpha=1}^{A} \left[q_\alpha r_\alpha{}^\lambda Y_\lambda^{\mu *}(\theta_\alpha, \varphi_\alpha) - \frac{i\mu_0 g_\alpha{}^s k}{2(\lambda+1)} (\boldsymbol{\sigma}_\alpha \times \mathbf{r}_\alpha) \cdot \nabla_\alpha (r^\lambda Y_\lambda{}^\mu)_\alpha{}^* \right], \tag{7-77b}$$

and for $\sigma \equiv$ M mode

$$\tilde{O}_\lambda{}^\mu(\mathrm{M}) = \tilde{M}_\lambda{}^\mu = \mu_0 \sum_{\alpha=1}^{A} \left(\frac{2}{\lambda+1} g_\alpha{}^l \boldsymbol{l}_\alpha + \frac{g_\alpha{}^s}{2} \boldsymbol{\sigma}_\alpha \right) \cdot \nabla_\alpha (r^\lambda Y_\lambda{}^\mu)_\alpha{}^*, \tag{7-77c}$$

where $\mu_0 = e\hbar/2m_\mathrm{N}c$, the g's are the g-factors of Section II,G, and $\boldsymbol{l}_\alpha = (1/\hbar)(\mathbf{r}_\alpha \times \mathbf{p}_\alpha)$ is the individual-nucleon orbital angular momentum operator in units of $\hbar$. In the above we have included the small magnetic moment term in the electric matrix element for the sake of completeness. We should again remark that Siegert's theorem does not apply to the magnetic mode transitions.

However Eqs. (7-77a)–(7-77c) can be somewhat further simplified by noting the identity

$$\nabla_r[r^\lambda Y_\lambda{}^\mu(\theta,\varphi)] = [\lambda(2\lambda+1)]^{1/2} r^{\lambda-1} \mathfrak{Y}^\mu_{\lambda,1,\lambda-1}(\theta,\varphi). \tag{7-78}$$

When the transitions are between unspecified magnetic substates M_i and M_f of the initial and final nuclear states, we introduce the *reduced transition probability* $B(\sigma,\lambda)$, defined as

$$B(\sigma,\lambda) = (2J_i+1)^{-1} \sum_{M_i, M_f} |\langle f|\, O_\lambda{}^\mu(\sigma)\, |i\rangle|^2, \tag{7-79a}$$

in terms of which the transition rate is

$$T_\sigma(\lambda) = \frac{8\pi(\lambda+1)}{\lambda[(2\lambda+1)!!]^2} \frac{k^{2\lambda+1}}{\hbar} B(\sigma,\lambda). \tag{7-79b}$$

We note from (7-77)–(7-79) that successive multipole terms of a given mode type (electric or magnetic) are progressively smaller by the factor $(kR)^2$. Thus, when several values of λ are permitted for a particular mode type, the lowest multipole will be dominant by generally several orders of magnitude [refer to (7-44)]. This effect is readily traced to the progressively smaller amplitude of the higher multipole radiation fields near the origin $r \leqslant R$ where the nuclear currents exist. The reduced interaction that results progressively reduces the transition probability.

E. MATRIX ELEMENTS AND SELECTION RULES

We now turn our attention to the center-of-mass effect that arises when nuclear state wave functions are expressed in laboratory coordinates $\mathbf{r}_\alpha$ instead of center-of-mass (c.m.) coordinates $\mathbf{r}_\alpha'$. In Chapter V we encountered some of the spurious effects introduced by this procedure. In the present instance the effect is to include the Thomson scattering of the photon from the nucleus as a whole. The interaction of the photon with the true internal degrees of freedom of the nucleus should properly be described by the transition $\psi_i(\mathbf{r}_1', \mathbf{r}_2', \ldots, \mathbf{r}_A') \to \psi_f(\mathbf{r}_1', \mathbf{r}_2', \ldots, \mathbf{r}_A')$.

The situation for electric dipole transitions is quite simple and we consider this case as an example. Using only the first term of (7-77b), we obtain the electric dipole operator as

$$\sum_\alpha^A q_\alpha r_\alpha Y_1^{\mu *} = (3/4\pi)^{1/2} D_\mu, \tag{7-80a}$$

where D_μ is the μ component (i.e., $\mu = 0, \pm 1$) of the dipole operator

$$\mathbf{D} = \sum_\alpha^A q_\alpha \mathbf{r}_\alpha. \tag{7-80b}$$

Properly, the required dipole operator should be in terms of the c.m. coordinates $\mathbf{r}_\alpha' = \mathbf{r}_\alpha - \mathbf{R}$, where $\mathbf{R} = (1/A)\sum_\alpha^A \mathbf{r}_\alpha$. In this event we obtain

$$\mathbf{D}' = \sum_\alpha^A q_\alpha \mathbf{r}_\alpha' = \sum_\alpha^A q_\alpha(\mathbf{r}_\alpha - \mathbf{R}) = e\sum_\alpha^Z \mathbf{r}_\alpha - \frac{eZ}{A}\sum_\alpha^A \mathbf{r}_\alpha$$

$$= e\left(1 - \frac{Z}{A}\right)\sum_\alpha^Z \mathbf{r}_\alpha - \frac{eZ}{A}\sum_\alpha^N \mathbf{r}_\alpha \tag{7-81}$$

or

$$\mathbf{D}' = \frac{eN}{A}\sum_\alpha^Z \mathbf{r}_\alpha - \frac{eZ}{A}\sum_\alpha^N \mathbf{r}_\alpha,$$

where as usual Z and N refer to the number of protons and neutrons respectively, and $Z+N = A$. The neutron contribution with an *effective charge* $-eZ/A$ results from a recoil effect of the remaining (charged) nucleons when a neutron undergoes a dipole transition. The proton effective charge is correspondingly *reduced* to $+e[1-(Z/A)]$. Note that for a nucleus consisting of only protons no electric dipole transition would be possible! We also note that (7-80) and (7-81) are related through $\mathbf{D} = \mathbf{D}' + eZ\mathbf{R}$, and indeed the second term $eZ\mathbf{R}$ can be identified with Thomson scattering from the nucleus as a whole.

The case for the general 2^λ-pole electric transition is more complicated but results in effective charges

$$q_Z = +e\left[\left(\frac{A-1}{A}\right)^\lambda + (-1)^\lambda \frac{Z-1}{A^\lambda}\right] \tag{7-82a}$$

and

$$q_N = -eZ(-1/A)^\lambda. \tag{7-82b}$$

Thus, to reasonable accuracy, we write

$$\begin{aligned} q_Z(\lambda=1) &= +e[1-(Z/A)] \approx +\tfrac{1}{2}e \qquad & q_Z(\lambda \geq 2) &\approx +e \\ q_N(\lambda=1) &= -eZ/A \approx -\tfrac{1}{2}e \qquad & q_N(\lambda \geq 2) &\approx 0. \end{aligned} \tag{7-83}$$

Evidently, when these values are used in (7-77) and (7-79), we can use wave functions in terms of simple laboratory coordinates $\mathbf{r}_\alpha$. A recent attempt to deduce empirical values for q_Z and q_N in the ^{208}Pb region by Hamamoto gives $q_Z \approx (0.13\text{–}0.35)e$ and $q_N \approx -(0.04 \text{ to } 0.14)e$ [see Hamamato (1973)]. These values are to be compared to (7-82), (7-83), and theoretical estimates $q_Z = +0.18e$ and $q_N = -0.12e$ based on the nuclear polarizability. A value of $q_N = -0.11e$ is suggested for neutrons in ^{29}Si [see Johnstone and Castel (1973)]. The concept of effective charges is not found useful for magnetic transitions, although empirical effective magnetic-moment operators for M1 transitions are discussed in the literature [e.g., Petrovich (1973)].

For given states $|\mathrm{i}\rangle$ and $|\mathrm{f}\rangle$ characterized by $(J_\mathrm{i}, M_\mathrm{i})$ and $(J_\mathrm{f}, M_\mathrm{f})$, nonvanishing transition rates determined from (7-77) or (7-79) occur only for those multipole operators that obey certain selection rules. The selection rules on angular momentum and parity can be deduced from (7-77) when (7-78) is kept in mind. From the dominant term in (7-77b) for electric transitions, the operator $r^\lambda Y_\lambda^{\mu *}$ only gives nonvanishing matrix elements if

$$|J_\mathrm{f} - J_\mathrm{i}| \leqslant \lambda \leqslant J_\mathrm{f} + J_\mathrm{i}, \qquad M_\mathrm{i} + \mu = M_\mathrm{f}, \qquad \pi_\mathrm{i}\pi_\mathrm{f} = (-1)^\lambda \qquad \text{(E mode).} \tag{7-84}$$

The situation for magnetic transitions is somewhat more complicated since we are dealing essentially with the operator $\mathbf{V} \cdot \mathfrak{Y}^\mu_{\lambda, 1, \lambda - 1}$, where $\mathbf{V}$ is an axial vector. It can be shown that this inner product operator is a tensor of degree λ just as is the essential operator for the electric mode of decay, however, with opposite parity; thus

$$|J_\mathrm{f} - J_\mathrm{i}| \leqslant \lambda \leqslant J_\mathrm{f} + J_\mathrm{i}, \qquad M_\mathrm{i} + \mu = M_\mathrm{f}, \qquad \pi_\mathrm{i}\pi_\mathrm{f} = (-1)^{\lambda+1} \qquad \text{(M mode).} \tag{7-85}$$

These results are not surprising when we note that the conservation of J and J_z requires the radiation field to carry the appropriate differences in $(J_\mathrm{i}, M_\mathrm{i})$ and $(J_\mathrm{f}, M_\mathrm{f})$ [i.e., obey the triangle and sum conditions of (7-84) and (7-85)]. The difference in the parity rule of (7-84) and that of (7-85) essentially follows from the fact that $\boldsymbol{l}$ and $\boldsymbol{\sigma}$ appearing in (7-77c) are axial vectors. Although we have neglected the second term in (7-77b), note that $(\boldsymbol{\sigma} \times \mathbf{r})$ is a polar vector and thus also gives the expected parity rule of (7-84).

When $\lambda = |J_\mathrm{f} - J_\mathrm{i}|$ is permitted by the parity condition of (7-84), thus allowing electric radiation of this multipolarity, one speaks of the transition as being *parity favored*. If this lowest value of λ, $\lambda = |J_\mathrm{f} - J_\mathrm{i}|$, satisfies instead the parity condition (7-85), leading to magnetic radiation of this multipolarity, the transition is described as *parity unfavored*. This terminology arises from the fact that transition rates for (Eλ) compared to (Mλ) decay modes (for transitions differing in parity change but otherwise identical) can be roughly estimated to be

$$T_\mathrm{E}(\lambda)/T_\mathrm{M}(\lambda) \approx \tfrac{1}{10}(m_\mathrm{N} cR/\hbar)^2 = 3.3A^{2/3}. \tag{7-86}$$

Thus $T_\mathrm{E}(\lambda)$ is seen to be several orders of magnitude greater than $T_\mathrm{M}(\lambda)$ under comparable circumstances.

When γ transitions are parity favored, they are almost always pure electric transitions of multipolarity $\lambda = |J_\mathrm{f} - J_\mathrm{i}|$. However, when the transitions are parity unfavored, mixtures of magnetic radiation of multipolarity $\lambda = |J_\mathrm{f} - J_\mathrm{i}|$

and electric radiation of multipolarity $\lambda+1$ are commonly encountered. This follows from the fact that $T_E(\lambda+1)$ can be comparable to $T_M(\lambda)$ when the $(kR)^2$ factor relating successive multipole orders for electric radiation is considered relative to (7-86). The generally smaller $T_E(\lambda+1)$ is most competitive with $T_M(\lambda)$ for A large and $\hbar\omega$ large, viz.,

$$T_E(\lambda+1)/T_M(\lambda) \approx 1.2 \times 10^{-4}(\omega\hbar/1\ \text{MeV})^2 A^{4/3}.$$

A common example is the mixing of (M1) and (E2) radiations. In this case a further enhancement of the (E2) contribution can be expected in distorted nuclei having large quadrupole deformations. For such deformed nuclei the electric quadrupole rate can in fact easily exceed the magnetic dipole rate. The angular distribution of such mixed radiation fields contains contributions from terms Z_λ and $Z_{\lambda+1}$ of the type (7-32) *and* interference terms between these as well.

Some relatively obvious points concerning (7-84) and (7-85) in special cases should be mentioned. When either J_f or J_i is zero (but not both) λ is uniquely equal to the nonzero J value. When both J_i and J_f are $\frac{1}{2}$ only $\lambda=1$ or dipole radiation is possible. When both J_i and J_f are zero, no single-photon radiation is possible at all since no monopole radiation fields exist. Such $0 \to 0$ transitions are rigorously forbidden and are not a consequence of the long-wave approximation.[1]

Selection rules pertaining to the nuclear isospin can also be formulated. As suggested in Exercise 1-1, we can write the electromagnetic interaction Hamiltonian in terms of the neutron and proton isospin projection operators to obtain

$$H_{nr} = H_{nr}^{s} + H_{nr}^{v}, \tag{7-87a}$$

where

$$H_{nr}^{s} = -\frac{1}{2}\sum_{\alpha}\left[\frac{e}{m_N c}\mathbf{A}\cdot\mathbf{p}_\alpha + \frac{e\hbar}{2m_N c}(\boldsymbol{\mu}_p+\boldsymbol{\mu}_n)\cdot(\nabla\times\mathbf{A})\right] \tag{7-87b}$$

is an isoscalar and

$$H_{nr}^{v} = -\frac{1}{2}\sum_{\alpha}\tau_{3\alpha}\left[\frac{e}{m_N c}\mathbf{A}\cdot\mathbf{p}_\alpha + \frac{e\hbar}{2m_N c}(\boldsymbol{\mu}_p-\boldsymbol{\mu}_n)\cdot(\nabla\times\mathbf{A})\right] \tag{7-87c}$$

is third component of an isovector. The electromagnetic transition rate involves the matrix elements of H_{nr}^{s} and H_{nr}^{v} between the states $|i\rangle = |a, T_i\rangle$

[1] Other modes of electromagnetic decay for the $J_i = J_f = 0$ case are possible, however, such as internal conversion or internal pair production in addition to a very weak two-photon decay mode.

and $|\mathrm{f}\rangle = |\ell, T_\mathrm{f}\rangle$, where a and ℓ refer to the characterizing quantum numbers other than those of isospin. The general selection rules are then clearly the triangle and "sum" conditions

$$\Delta T = T_\mathrm{f} - T_\mathrm{i} = 0, \pm 1 \tag{7-88a}$$

$$\Delta T_3 = T_{3\mathrm{f}} - T_{3\mathrm{i}} = 0. \tag{7-88b}$$

We should note that when $T_\mathrm{f} = T_\mathrm{i} = 0$ the transition proceeds entirely through the isoscalar operator H^s_nr. Beyond (7-88) not much can be said in general except for the important case of dipole transitions.

A very strong selection rule exists for the electric dipole transitions in the long-wavelength approximation. The dipole operator (7-81), when written in terms of the isospin projection operators, becomes

$$\mathbf{D}' = \sum_\alpha^A \left\{ \frac{eN}{A}\left(\frac{1+\tau_3}{2}\right)_\alpha - \frac{eZ}{A}\left(\frac{1-\tau_3}{2}\right)_\alpha \right\} \mathbf{r}_\alpha = \frac{e(N-Z)}{2A}\sum_\alpha^A \mathbf{r}_\alpha + \frac{\mathrm{e}}{2}\sum_\alpha^A (\mathbf{r}\tau_3)_\alpha. \tag{7-89}$$

Thus for self-conjugate nuclei (i.e., $N = Z$, and $T_{3\mathrm{f}} = T_{3\mathrm{i}} = 0$) the isoscalar portion of $\mathbf{D}'$ identically vanishes, leaving only the isovector portion, for which the transition between the states $T_\mathrm{i} = T_\mathrm{f} = 0$ is clearly not allowed by the usual triangle condition. An additional, not immediately evident, restriction on the isovector coupling produced by the condition $T_{3\mathrm{f}} = T_{3\mathrm{i}} = 0$ is to require $\Delta T = \pm 1$ in general; thus all $T_\mathrm{i} = T_\mathrm{f}$ transitions are in fact forbidden. The resulting strong selection rule summarized by the requirement $\Delta T = \pm 1$ is particulary important in the light elements where numerous applications occur. The main violation of this rule occurs because of the failure of the states involved to be pure isospin eigenstates.[1] Hence observing (E1) transitions in self-conjugate nuclei serves as a sensitive test for isospin mixing.

The isospin selection rules for (M1) transitions are also interesting. We have from (7-78) for $\lambda = 1$

$$\boldsymbol{\nabla}_\mathrm{r}[rY_1{}^\mu] = \sqrt{3}\,\mathfrak{Y}^\mu_{1,1,0} = (3/4\pi)^{1/2}\chi_\mu. \tag{7-90}$$

Thus the (M1) operator of (7-77c) becomes

$$M_1{}^\mu = \mu_0(3/4\pi)^{1/2}\sum_\alpha^A (g_\alpha{}^l \mathbf{l}_\alpha + \tfrac{1}{2} g_\alpha{}^s \boldsymbol{\sigma}_\alpha)_\mu, \qquad \mu = 0, \pm 1. \tag{7-91}$$

When (7-91) is written with the isospin projection operators, and we use

[1] Failure of the selection rule to hold due to the higher terms in kR omitted in the long-wavelength approximation is generally a negligible effect, as are the effects of the neutron–proton mass difference, the inclusion of the second term in (7-77b), and meson effects not included in Siegert's theorem equivalence.

$g_\alpha^s/2 = \mu_\alpha$, we find

$$M_1{}^\mu = e\mu_0(3/4\pi)^{1/2}\sum_\alpha^A\left\{\left[\left(\frac{1+\tau_3}{2}\right)\boldsymbol{l}\right]_\alpha + \mu_p\left[\left(\frac{1+\tau_3}{2}\right)\boldsymbol{\sigma}\right]_\alpha + \mu_n\left[\left(\frac{1-\tau_3}{2}\right)\boldsymbol{\sigma}\right]_\alpha\right\}_\mu. \tag{7-92}$$

It is convenient to introduce the total angular momentum operator

$$J_\mu = \sum_\alpha^A (\boldsymbol{l}_\alpha + \tfrac{1}{2}\boldsymbol{\sigma}_\alpha)_\mu. \tag{7-93}$$

Since the initial and final states are eigenfunctions of $\mathbf{J}^2$, the angular integration of this operator gives

$$\langle \ell | J_\mu | a\rangle = \delta_{J_i J_f}[J_i(J_i+1)]^{1/2}(J_i, 1, M_i, \mu | J_f M_f)\langle \ell | a\rangle = 0. \tag{7-94}$$

This last result follows from the orthogonality of the two nuclear states. Using (7-93) and (7-94) in (7-92), we get

$$M_1{}^\mu = \tfrac{1}{2}e\mu_0(3/4\pi)^{1/2}\left\{(\mu_p+\mu_n-\tfrac{1}{2})\sum_\alpha^A(\boldsymbol{\sigma}_\alpha)_\mu + \sum_\alpha^A[\boldsymbol{l}_\alpha + (\mu_p-\mu_n)\boldsymbol{\sigma}_\alpha]_\mu\tau_{3\alpha}\right\}. \tag{7-95}$$

With $\mu_p = +2.793$ and $\mu_n = -1.913$, we find the factors $\mu_p+\mu_n-\frac{1}{2} = 0.380$ and $\mu_p-\mu_n = 4.706$. Thus, even when $\boldsymbol{l}_\alpha$ does not contribute significantly, the isoscalar transition rate is unfavored compared to the isovector rate by the ratio

$$(0.380/4.706)^2 \approx 6.5 \times 10^{-3}.$$

We thus have a de facto selection rule operating against $T_i = T_f$, (M1) transitions in self-conjugate nuclei. On the other hand, the $\Delta T = \pm 1$ transitions can proceed without hindrance through the second term of (7-95) whether the nucleus is self-conjugate or not. These rules are commonly called the Morpurgo rules after the author to first call attention to the effects [see Morpurgo (1958)].

►**Exercise 7-9** (a) Verify the results (7-90)–(7-95).

(b) The application of the Wigner-Eckart theorem to the matrix elements of the operator $O(\mathbf{r})\tau_3$, where $O(\mathbf{r})$ affects only the space coordinates, is

$$\langle \ell, T_f, T_{3f} | \sum_\alpha^A [O(\mathbf{r})\tau_3]_\alpha | a, T_i, T_{3i}\rangle = (T_i, 1, T_{3i}, 0 | T_f, T_{3f})\langle \ell, T_f || \sum_\alpha^A [O(\mathbf{r})\boldsymbol{\tau}]_\alpha || a, T_i\rangle.$$

Show that for $T_{3i} = T_{3f} = 0$, we require $\Delta T = \pm 1$ for a nonvanishing result. This verifies the broader rules quoted for (E1) and (M1) transitions in self-conjugate nuclei. ◀

1. Extreme Single-Particle Model

It is difficult to carry the general expressions for the transition rates further than (7-77) and (7-79) without a specific model for the nuclear states. A simple model which may be used as a standard of comparison is the extreme single-particle model. The assumption is made that only the valence nucleon (a proton) makes the nuclear transition that generates the radiation field of the emitted photon. Thus the sum on the index α *is not* performed.[1] The nuclear wave function is then given by the single-proton state with $J_i = j_i$, $J_f = j_f$, $M_i = m_i$, and $M_f = m_f$ (we write simply m for m_j in addition to M for M_J). These wave functions are of the form (4-63) in Chapter IV, viz.,

$$\psi_{nljm}(\mathbf{r}, \boldsymbol{\sigma}) = [u_{n,l}(r)/r]\,\mathscr{Y}^m_{j,1/2,l}(\theta, \varphi)$$

or

$$\psi_{nljm}(\mathbf{r}, \boldsymbol{\sigma}) = [u_{n,l}(r)/r] \sum_{m_s=-1/2}^{+1/2} (l, \tfrac{1}{2}, m-m_s, m_s | jm)\, Y_l^{m-m_s}(\theta, \varphi)\, \chi_{1/2}^{m_s}(\sigma), \tag{7-96}$$

where we have chosen the sum on m_s.

We again consider the (Eλ) transitions first. The reduced transition probability (7-79a) is conveniently written as the product of radial and angular integrals, viz.,

$$B(\mathrm{E}\lambda) = (e^2/4\pi) \left| \int_0^\infty u^*_{n',l'}(r)\, r^\lambda u_{n,l}(r)\, dr \right|^2 S_{j',l',\lambda,j,l}, \tag{7-97a}$$

with

$$S_{j',l',\lambda,j,l} = 4\pi(2j+1)^{-1} \sum_{m',m} \Bigg| \sum_{m_s',m_s} (l', \tfrac{1}{2}, m'-m_s', m_s | j'm')(l, \tfrac{1}{2}, m-m_s, m_s | jm)$$
$$\times\, \chi_{1/2}^{\dagger m_s'} \chi_{1/2}^{m_s} \int Y_{l'}^{(m'-m_s')*}\, Y_\lambda^{\mu*}\, Y_l^{(m-m_s)}\, d\Omega \Bigg|^2, \tag{7-97b}$$

where the primed quantities refer to the final state and the unprimed to the initial state. The form (7-97b) leads to some general selection rules for the

[1] The single-particle states of the nonparticipating nucleons (other than to change the effective charge on the proton due to the c.m. correction) are the same in both the initial and final states and if properly normalized, simply give a multiplicative factor to the matrix element of unity.

(Eλ) transitions:

$$
\begin{aligned}
m_s' &= m_s && \text{(i.e., no proton spin–flip)} \\
m' &= m + \mu && \\
|j-\lambda| \leqslant j' &\leqslant j+\lambda, \quad \lambda \neq 0 && \text{(i.e., triangle condition on } j', \lambda, j) \\
|l-\lambda| \leqslant l' &\leqslant l+\lambda, \quad \lambda \neq 0 && \text{(i.e., triangle condition on } l', \lambda, l) \\
l' + \lambda + l &= \text{even integer} && \text{(i.e., parity condition).}
\end{aligned}
\tag{7-98}
$$

The parity condition requires $\pi_f \pi_i = (-1)^\lambda$, as it should (see Table 7-1).

The angular *statistical factor*, as $S_{j',l',\lambda,j,l}$ is sometimes called, can be evaluated to give

$$S_{j',l',\lambda,j,l} \equiv S_{j',\lambda,j} = (2\lambda+1)(j,\lambda,\tfrac{1}{2},0|j',\tfrac{1}{2})^2, \tag{7-99}$$

where $(j,\lambda,\frac{1}{2},0|j',\frac{1}{2})$ is the simple Clebsch–Gordan coefficient of Appendix A, and we thus see that S is in fact independent of l and l' except for the necessity to satisfy the parity requirement.

A very useful special case of the above is for $(j',l') = (\frac{1}{2},0)$ and $(j,l) = (l+\frac{1}{2},l)$, which corresponds to the unique multipole $\lambda = l$, or (El) radiation. The statistical factor is then identically equal to unity. The radial integral can be crudely evaluated on the assumption of a constant-density model for both $u_{n,l}(r)$ and $u_{n',l'}(r)$, viz.,

$$
\begin{aligned}
u_{n,l}(r) = u_{n',l'}(r) &= (4\pi\rho)^{1/2} r, \qquad 0 < r < R \\
&= 0, \qquad r > R.
\end{aligned}
\tag{7-100}
$$

The functions (7-100) are clearly normalized to unity when $\rho = (\frac{4}{3}\pi R^3)^{-1}$, since

$$\int_0^R u^2\,dr = 4\pi\rho \int_0^R r^2\,dr = 1.$$

Using (7-100), we find that the radial integral of (7-97a) becomes

$$\int_0^\infty u_{n',l'}(r)\, r^\lambda u_{n,l}(r)\,dr = 4\pi\rho \int_0^R r^{\lambda+2}\,dr = [3/(\lambda+3)]\,R^\lambda. \tag{7-101}$$

Combining (7-79b), (7-97), and (7-101), and with $S = 1$, we arrive at the well-known Weisskopf estimate,

$$[T_E(\lambda)]_W = \frac{2(\lambda+1)}{\lambda[(2\lambda+1)!!]^2} \left(\frac{3}{\lambda+3}\right)^2 \frac{e^2}{\hbar c} \left(\frac{\omega R}{c}\right)^{2\lambda} \omega, \tag{7-102}$$

where we have, following custom, kept the full proton charge e instead of using the effective charges (7-83). Since (7-102) is intended as a convenient "unit" in terms of which experimental results can be quoted, the same expression is used for neutron transitions as well, although physical justification for $\lambda \geqslant 2$ is clearly dubious in view of (7-83).

Weisskopf gives an estimate for the magnetic transitions by setting the quantity that derives from (7-77c),

$$\tfrac{1}{4}\lambda^2\{g^s - [2/(\lambda+1)]g^l\}^2 \approx 10, \tag{7-103}$$

to arrive at the approximate value [see Blatt and Weisskopf (1952)]

$$[T_{\rm M}(\lambda)]_{\rm W} = \frac{20(\lambda+1)}{\lambda[(2\lambda+1)!!]^2}\left(\frac{3}{\lambda+3}\right)^2 \frac{e^2}{\hbar c}\left(\frac{\hbar}{m_{\rm N} cR}\right)^2 \left(\frac{\omega R}{c}\right)^{2\lambda} \omega. \tag{7-104}$$

The values (7-102) and (7-104) are referred to as the Weisskopf single-particle values for $T_{\rm E}(\lambda)$ and $T_{\rm M}(\lambda)$ transitions.

The so-called Moszkowski values result when a somewhat less crude use of (7-103) is applied. The two "units" are related by

$$[T_{\rm M}(\lambda)]_{\rm M} = \frac{\lambda^2}{10}\left(\frac{\lambda+3}{\lambda+2}\right)^2 \left(\mu_{\rm p} - \frac{1}{\lambda+1}\right)^2 [T_{\rm M}(\lambda)]_{\rm W}.$$

Thus the Moszkowski rates are 0.93, 3.8, 9.1, 16, and 22 times the Weisskopf estimates for $\lambda = 1$, 2, 3, 4, and 5, respectively. Although the Moszkowski magnetic rates are possibly more realistic, the Weisskopf values are more commonly used. Again in either case no distinction is usually made for neutron versus proton transitions, although when using Moszkowski units the factor $[\mu_{\rm p} - (\lambda+1)^{-1}]^2$ should be replaced by $[\mu_{\rm n}]^2$.

The selection rules for the single-particle (Mλ) transitions are again determined by the appropriate statistical factor and give

$$\begin{aligned}
&m_s' - m_s = 0, \pm 1 && \text{(i.e., proton spin–flip possible)}\\
&m' = m + \mu && \\
&|j-\lambda| \leqslant j' \leqslant j + \lambda, \quad \lambda \neq 0 && \text{(i.e., triangle condition on } j', \lambda, j)\\
&|l-\lambda+1| \leqslant l' \leqslant l + \lambda - 1, \quad \lambda \neq 0 && \text{(i.e., triangle condition on } l', \lambda-1, l)\\
&l' + \lambda + l = \text{odd integer} && \text{(i.e., parity condition).}
\end{aligned} \tag{7-105}$$

The Weisskopf values (7-102) and (7-104) are conveniently given as radiation widths Γ_γ in eV,[1] when $E_\gamma = \hbar\omega$ is in MeV and $R = 1.20\, A^{1/3}$ fm, as

[1] The transition rates $T_\sigma(\lambda)$, the mean life τ, and the width Γ_γ are related as $\Gamma = \hbar/\tau = \hbar T$. When $\Gamma = 1$ eV, $\tau = 6.58 \times 10^{-16}$ sec.

follows:

$$\begin{aligned}
\Gamma_\gamma(\mathrm{E1}) &= 6.8 \times 10^{-2} A^{2/3} E_\gamma^{\,3} & \Gamma_\gamma(\mathrm{M1}) &= 2.1 \times 10^{-2} E_\gamma^{\,3}\\
\Gamma_\gamma(\mathrm{E2}) &= 4.9 \times 10^{-8} A^{4/3} E_\gamma^{\,5} & \Gamma_\gamma(\mathrm{M2}) &= 1.5 \times 10^{-8} A^{2/3} E_\gamma^{\,5}\\
\Gamma_\gamma(\mathrm{E3}) &= 2.3 \times 10^{-14} A^{2} E_\gamma^{\,7} & \Gamma_\gamma(\mathrm{M3}) &= 6.8 \times 10^{-15} A^{4/3} E_\gamma^{\,7}\\
\Gamma_\gamma(\mathrm{E4}) &= 6.8 \times 10^{-21} A^{8/3} E_\gamma^{\,9} & \Gamma_\gamma(\mathrm{M4}) &= 2.1 \times 10^{-21} A^{2} E_\gamma^{\,9}\\
\Gamma_\gamma(\mathrm{E5}) &= 1.6 \times 10^{-27} A^{10/3} E_\gamma^{11} & \Gamma_\gamma(\mathrm{M5}) &= 4.9 \times 10^{-28} A^{8/3} E_\gamma^{11}.
\end{aligned} \tag{7-106}$$

It must be remembered that the experimental width is the sum of the widths for all modes of decay, first summed on all possible γ-ray emitting modes, and also summed on all internal conversion deexcitation modes. In the internal conversion process, the nucleus is deexcited by transferring its energy of excitation to the ejection of one of the atomic electrons. Such electrons appear with a kinetic energy equal to the competing photon energy $\hbar\omega$ less the electronic binding energy. When only the most prominent γ-emission mode and the competing internal conversion process are considered, it is convenient to express the observed total transition rate for deexcitation in the form

$$T_{\text{tot}} = (1+\alpha)\, T_\sigma(\lambda), \tag{7-107}$$

where α is referred to as the internal conversion coefficient. We simply cite here for purposes of a rough estimate the approximate value of α for the ejection of the K-shell electrons by electromagnetic radiation of mode Eλ,

$$\alpha_{\mathrm{K}}(\mathrm{E}\lambda) \approx Z^3 \left(\frac{e^2}{\hbar c}\right)^4 \frac{\lambda}{\lambda+1} \left(\frac{2mc^2}{\hbar\omega}\right)^{\lambda+\frac{5}{2}}, \tag{7-108a}$$

and by the electromagnetic mode Mλ,

$$\alpha_{\mathrm{K}}(\mathrm{M}\lambda) \approx Z^3 \left(\frac{e^2}{\hbar c}\right)^4 \left(\frac{2mc^2}{\hbar\omega}\right)^{\lambda+\frac{3}{2}}, \tag{7-108b}$$

where mc^2 is the electron rest mass energy 0.511 MeV. As can be verified using (7-108), when Z is large, high multipolarities can readily give $\alpha_{\mathrm{K}}(\lambda) > 1$ even when $\hbar\omega$ is comparable to mc^2.[1] Thus, particularly for nuclei with $Z \geqslant 30$, observed decay rates must be reduced to $T_\sigma(\lambda)$ using (7-107) before comparisons with the Weisskopf single-particle estimates (7-106) can be made.

Since (7-108) and the more accurate versions for the internal conversion coefficients are very largely nuclear model independent, the experimental observation of the internal conversion process can give a very accurate means

[1] The expressions (7-108) are valid only for $B_{\mathrm{K}} < \hbar\omega < mc^2$, where B_{K} is the K-shell atomic binding energy.

of determining the multipolarity of the nuclear transition. We do not discuss the internal conversion process further, but refer the reader to many excellent works on the subject [e.g., Siegbahn (1965)].

2. Systematics of Empirical Evidence

Experimentally observed radiative widths, corrected according to (7-107), are usually expressed in "units" of the relevant estimate (7-106), and the quotient $\Gamma_\gamma(\text{obs})/\Gamma_\gamma(\text{W})$ is generally labeled $|\mathrm{M}|^2$. Figure 7.2 shows histograms of $|M|^2$ for (E1), (M1), and (E2) transitions in nuclei with $A \leqslant 40$ [after Skorka *et al.* (1966)].

We remark on the various multipole order histograms of Fig. 7.2 separately. We note first that the (E1) $|M|^2$ values are considerably smaller than unity. Not having used the effective charges (7-82) and (7-83) in the definition of the Weisskopf unit, we would expect that $|M|^2$ is reduced to a value of $\frac{1}{4}$ at the outset. When nuclear models are assumed that have strong correlations in the proton and neutron motions (such as the liquid drop model) a further reduction is expected for electric dipole transitions. The specificity of this effect for electric dipoles arises from the simple connection between the electric dipole operator $\mathbf{D}$ and the position of the center of mass $\mathbf{R}$ exhibited in (7-80) and (7-81). If in such high correlation models components of nuclear matter move together in some sense, the effect is as if several nucleons were collectively involved having some effective charge. A nucleus consisting only of such block entities all with the same effective charge would have zero dipole moment just as for the case of a nucleus consisting of only protons, a point that we have already noted. For example, the only way to get a dipole excitation of a highly correlated α-particle-type nucleus (such as ^{16}O, ^{40}Ca) is by breaking up an α-particle subunit. The excitation energies for the giant dipole resonances in such nuclei are therefore rather similar and approximately correspond to the breakup energy of an α-particle. The average value of $|M|^2$ for the isospin-allowed transitions depends to some extent on the mass range considered and values ranging from 2.6×10^{-3} to 5.5×10^{-2} are reported in the literature. The smallness of these observed $|M|^2$ values thus supports the existence of rather strong correlations in nuclear matter.

The isospin-forbidden (E1) transitions (the shaded portion of the histogram) are a factor of 10 to 100 times slower on the average than the allowed transitions. The fact that they are as *large* as they are can be attributed to the isospin impurity of the involved nuclear states. The mixing of isospin states is generally assumed to occur from the operation of Coulomb forces. For example, consider the isospin mixing of two pure nuclear states $|T=0\rangle$ and $|T=1\rangle$ by the Coulomb potential V_c giving the mixed states

$$|\mathrm{I}\rangle = u|T=0\rangle + v|T=1\rangle \qquad \text{and} \qquad |\mathrm{II}\rangle = v|T=0\rangle - u|T=1\rangle.$$

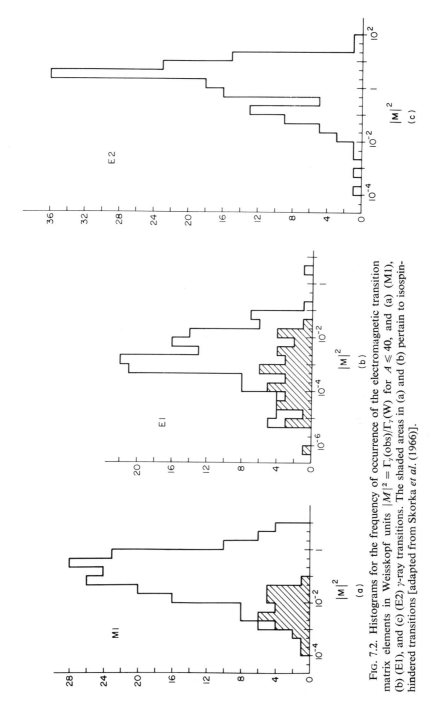

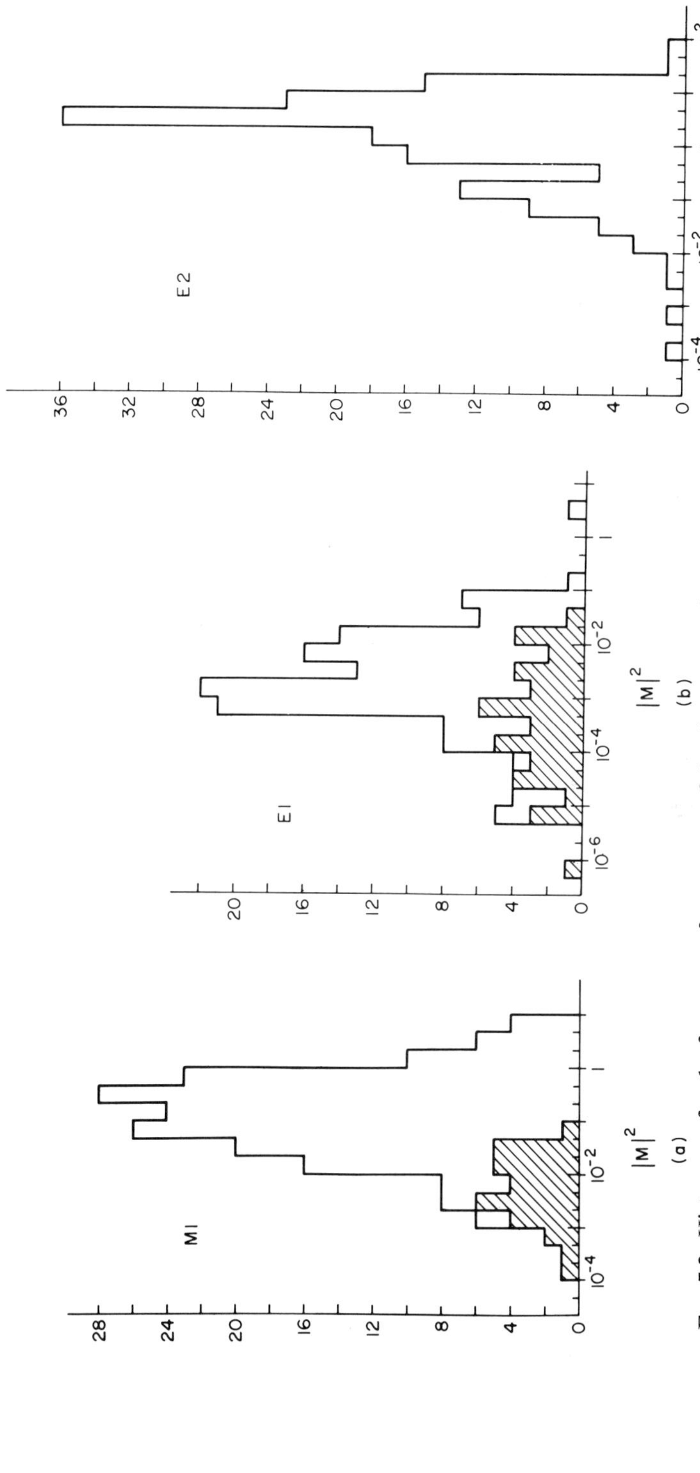

FIG. 7.2. Histograms for the frequency of occurrence of the electromagnetic transition matrix elements in Weisskopf units $|M|^2 = \Gamma_\gamma(\text{obs})/\Gamma_\gamma(\text{W})$ for $A \leqslant 40$, and (a) (M1), (b) (E1), and (c) (E2) γ-ray transitions. The shaded areas in (a) and (b) pertain to isospin-hindered transitions [adapted from Skorka *et al.* (1966)].

From Appendix D, Eq. (D-12), we have

$$v/u \approx \langle T=0|\, V_c \,|T=1\rangle/(E_0 - E_1) \tag{7-109}$$

where we have associated the state $|\mathrm{I}\rangle$ primarily with $|T=0\rangle$, and $|\mathrm{II}\rangle$ primarily with $|T=1\rangle$, i.e., $v/u < 1$; further, $E_{\mathrm{I}} \approx E_0$ and $E_{\mathrm{II}} \approx E_1$, and we have assumed all quantities real. Thus, if the (E1) transition in question is between the state $|\mathrm{I}\rangle$ and the ground state in a self-conjugate nucleus, the transition would occur only by virtue of the $v|T=1\rangle$ admixture in $|\mathrm{I}\rangle$, and would give a value of $|M|_{\mathrm{F}}^2 = |v/u|^2 \, |M|_{\mathrm{A}}^2$, where $|M|^2_{\mathrm{F,A}}$ refers to "forbidden" or "allowed," and $|M|_{\mathrm{A}}^2$ would be determined by a hypothetical transition involving $|\mathrm{i}\rangle \equiv |T=1\rangle$ for the initial state. When $|\mathrm{I}\rangle$ and $|\mathrm{II}\rangle$ are particle unbound, $\langle T=0|\, V_c \,|T=1\rangle$ commonly ranges from 30 to 500 keV. When $|\mathrm{I}\rangle$ and $|\mathrm{II}\rangle$ are bound, the Coulomb matrix element is much smaller, being perhaps only of order 1 keV. Since $E_0 - E_1$ can be of order 1 MeV, we expect "hindrance" factors comparable to those actually observed.

The (M1) $|M|^2$ values are seen to be generally less than unity. The average value of $|M|^2$ for isospin-favored transition reported in the literature ranges from 0.08 to 0.15 and again appears to show a slight A-dependent effect. The collective effects discussed for the (E1) transitions are expected to either not apply or be much weaker. The isospin-unfavored transitions (shaded portion of the histogram) again show hindrance factors of 10 to 100 as did the (E1) rates, in agreement with expectations from the Morpurgo rule.

We should now note that in the limit $\omega \to 0$ and $|\mathrm{i}\rangle \to |\mathrm{f}\rangle \equiv |\text{ground state}\rangle$, the electromagnetic multipole transition matrix elements become equivalent to the static moment expectation values. One might hope, therefore, that insofar as the behavior of bulk nuclear matter is concerned, as opposed to the operation of microscopic selection rules for specific levels, the static moments might give an indication of the $|M|^2$ values to be expected for at least low-energy transitions.

The parity condition would cause the limiting (E1) transition matrix element to vanish. This, of course, agrees with the conclusion drawn in Chapter II for static electric dipole moments. In each event, static moment or transition rate, the behavior is due to the simple polar vector character of the operator **D**.

The limiting (M1) transition matrix element would correspond to the static magnetic dipole moment. These in general are, within an order of magnitude, equal to the Schmidt values discussed in Chapter II. This again suggests that collective effects should produce only moderate changes in (M1) transition rates. Since the observed magnetic moments are all *within* the Schmidt limits (see Fig. 2.13), the effective $|M|$ values would be expected to be *reduced*, in agreement with observation. The selection rule on l and l' contained in (7-105) would require $l' = l$ for magnetic dipole transitions, for which $\lambda = 1$. Thus, while a $\mathrm{d}_{3/2}$ to $\mathrm{s}_{1/2}$ single-particle transition would be forbidden, such

"l-forbidden" $\Delta l = 2$ transitions in fact occur with rather normal matrix elements. It is probable that this effect can be traced to the presence of velocity-dependent terms in the Hamiltonian. The simple spin–orbit interaction

$$V_{ls}(r)\,\boldsymbol{l}\cdot\boldsymbol{\sigma} = V_{ls}(r)(\mathbf{r}\times\mathbf{p}/\hbar)\cdot\boldsymbol{\sigma}$$

is, of course, one such velocity-dependent term. The gauge invariance condition requires replacing $\mathbf{p}$ by $\mathbf{p}-(e/c)\mathbf{A}$ in the presence of a field. While we have considered this for the kinetic energy, the above spin–orbit or other velocity-dependent potential also gives a field coupling term to $H_{\rm nr}$, the interaction Hamiltonian, and this we have not included. Such additional coupling terms quite naturally allow such "l-forbidden" transitions to occur. Whether one considers such effects modifying simple single-particle estimates as collective or not is a matter of choice.

The lowest nonvanishing static electric moment is the quadrupole moment. Here collective effects lead in some instances to quadrupole moments many times greater than single-particle estimates (refer to Fig. 2.11 and the general discussion in Chapter VI). The corresponding collective effects for (E2) transition rates would also be expected to lead to enhanced values of $|M|^2$. Reference to Fig. 7.2(c) indeed shows a prominent component of the histogram events with $1 < |M|^2 < 100$ for $A \leqslant 40$. When (E2) transitions in heavy nuclei are considered, this component of the histogram is greatly increased and $|M|^2$ is extended for the group to $1 < |M|^2 < 10^3$. These large values of $|M|^2$ in fact are among the most striking manifestations of collective effects.

We can readily evaluate $B(\text{E2})$ for transitions in spherical vibrational nuclei. For example, $B(\text{E2})$ for the transition from the first 2^+ (one-phonon state) to the 0^+ ground state in even–even nuclei is

$$B(\text{E2})(2^+\to 0^+) = \left(\frac{3ZeR^2}{4\pi}\right)^2 \frac{\hbar}{2B_2\omega_2} = \left(\frac{3ZeR^2}{4\pi}\right)^2 \frac{a_{20}^2}{5}. \tag{7-110}$$

For rotational nuclei the radiative transition within the same rotational band with quantum number K yields a $B(\text{E2})$ value[1]

$$B(\text{E2}) = (5/16\pi)\,e^2 Q_0{}^2 |(J_{\rm i}, 2, K, 0\,|\,J_{\rm f} K)|^2,$$

and

$$B(\text{E2})\,(2^+\to 0^+) = (5/16\pi)\,e^2 Q_0{}^2. \tag{7-111}$$

These predicted values are found to agree rather well with the observed γ-ray transition rates. A rather simple empirical relationship has been found to connect the quantity Q_0 appearing in (7-111) with moment of inertia $\mathscr{I}_0$

[1] Q_0 is the intrinsic quadrupole moment, written Q' in Chapter II, in units of cm^2.

derived from level energies associated with the rotational band in question. Various analytic expressions are discussed by Mantri and Sood (1972).

▶**Exercise 7-10** Derive (7-110), noting from (6-38) that

$$\tilde{Q}_2{}^\mu = \frac{3Ze}{4\pi} R^2 \left(\frac{\hbar}{2B_2\omega_2}\right)^{1/2} [b_{2\mu} + (-1)^\mu b^\dagger_{2,-\mu}]$$

and from (7-79a) that

$$B(\mathrm{E2}) = \tfrac{1}{5} \sum_{m,\mu} |\langle 1,2,m|\tilde{Q}_2{}^\mu|0,0,0\rangle|^2.$$

Also note that $|1,2,m\rangle = b^\dagger_{2m}|0,0,0\rangle$ or $\langle 1,2,m| = \langle 0,0,0|b_{2m}$, and use (6-41). ◀

For a recent review article on electromagnetic transitions in nuclei emphasizing the present concepts underlying the explanations of the observed characteristics, see Yoshida and Zamick (1972) and the references cited therein.

F. RADIATIVE n–p CAPTURE

In the preceding sections we have considered the radiative decay of excited nuclear states. A very natural extension of these considerations would lead us to examine radiative transitions from states in the continuum (i.e., particle unbound states). A particularly simple yet instructive example is the case of the radiative n–p capture process at low energies. The preponderant n–p interaction at low energies is the elastic scattering process discussed in Chapter I. The capture cross section we now discuss is some two to three orders of magnitude smaller in the *thermal* region (i.e., at a neutron energy of ≈0.025 eV, corresponding to a kinetic energy of thermal motion at room temperature). A typical experimental arrangement might utilize a hydrogenous target and incident neutrons coming from the thermal column of a nuclear reactor.

The initial state of the unbound n–p system can be considered to good accuracy to be in an S-wave (i.e., $L = 0$) scattering state. The coupling of the nucleon spins would give either a singlet or triplet system-spin state. Thus an unpolarized initial state would be an incoherent mixture of a $^3\mathrm{S}_1$ and a $^1\mathrm{S}_0$ state. As with the discussion pertaining to (1-128), we would apply a statistical weight of $\frac{3}{4}$ to the triplet state and $\frac{1}{4}$ to the singlet state. The final state would of course be the $^3\mathrm{S}_1 + {}^3\mathrm{D}_1$ deuteron ground state discussed in Chapter II.

The capture cross section is obtained by evaluating the transition matrix elements $\langle \mathrm{f}|\,O_\lambda{}^\mu(\mathrm{E\ or\ M})\,|\mathrm{i}\rangle$. From the consideration of the selection rules (7-98) and (7-105), we readily conclude that only (M1) and (E2) electromagnetic transitions are possible. The electric quadrupole transition induced by the leading term in the (E2) operator (7-77b) would give radiative contributions only from the $^3S_1$ scattering state decaying to the $^3D_1$ portion of the deuteron ground state. Although this transition rate is nonvanishing, it yields a cross section only $\approx 2\times 10^{-8}$ as large as that for the (M1) rate. We therefore confine our attention solely to the (M1) capture process.

In view of (7-91) and (2-80), we can write the (M1) operator as

$$\tilde{M}_1{}^{\mu} = \tfrac{1}{2}\mu_0(3/4\pi)^{1/2}[(\mu_n+\mu_p)(\boldsymbol{\sigma}_n+\boldsymbol{\sigma}_p)+(\mu_n-\mu_p)(\boldsymbol{\sigma}_n-\boldsymbol{\sigma}_p)+\mathbf{L}]_{\mu}. \tag{7-112}$$

The triangle condition on $\boldsymbol{l}$ contained in (7-105) forbids the S → D magnetic dipole transitions, leaving only the S → S transitions. The $^3S_1 \to {}^3S_1$ decay rate also vanishes, as can be seen from (7-112). The orbital angular momentum portion of the transition matrix clearly vanishes since **L** operating on an S state gives zero. The spin operators leave the orbital or space part of the wave function unaltered, and since the states $^3S_1(E_f>0)$ and $^3S_1(E_i=-E_B)$ are two orthogonal solutions of the same Hamiltonian, this portion of the transition matrix also vanishes. In considering the $^1S_0 \to {}^3S_1$ transition, the orbital angular momentum contribution again vanishes; however, the spin operators do not necessarily give zero contributions since the presence of spin dependence in the n–p force requires the triplet spin and singlet spin Hamiltonians to be different, and hence the above orthogonality argument fails. We are finally left with calculating $\langle {}^3S_1(f)|M_1{}^{\mu}(\boldsymbol{\sigma})|{}^1S_0(i)\rangle$, where $M_1{}^{\mu}(\boldsymbol{\sigma})$ is the spin portion of (7-112).

The first spin term in (7-112) is just proportional to the total spin operator, since $S=\frac{1}{2}(\boldsymbol{\sigma}_n+\boldsymbol{\sigma}_p)$. The total spin operator **S** operating on a singlet spin state gives zero. The spin–flip operator $(\boldsymbol{\sigma}_n-\boldsymbol{\sigma}_p)$ is so named since operating on a spin singlet state yields a spin triplet state and vice versa (see Exercise 2-15). The transition matrix elements of the spin–flip operator do not vanish between the $^3S_1(f)$ and $^1S_0(i)$ states. We need only consider the one component contribution, say for $\mu=0$, since the angular distribution of the emitted radiation is uniform, and in the incoherent sum on μ we simply get a multiplicative factor of three. The isotropy results from the fact that we are dealing with an S state.

We take for the initial n–p state the $^1S_0$ scattering state with unit incident flux as

$$\psi_i = \{C[\exp(i\delta_{0s})]/\kappa r\}\sin(\kappa r+\delta_{0s})\,\chi_0{}^m, \tag{7-113}$$

where δ_{0s} is the singlet phase shift and we have written $\hbar\kappa$ for the n–p momentum in the c.m. system (to avoid confusion with $k=\omega/c$ for the photon). The expression (7-113) follows directly from (1-101) for $r>b$ with $B=C[\exp(i\delta_{0s})]/\kappa$. The normalization to unit incident flux requires $|C|^2=1/v=m_N/2\hbar\kappa$. For the final state we use the approximate deuteron

$^3S_1$ wave function for $r > b$, namely,[1]

$$\psi_f = \left(\frac{\alpha}{2\pi}\right)^{1/2} \frac{e^{-\alpha r}}{r} \chi_1^{m'} \tag{7-114}$$

[see (2-71)]. The transition probability T, which with the unit flux normalization of ψ_i is also simply the cross section, is[2]

$$\sigma_c = T = \tfrac{1}{4} \sum_{\mu=0,\pm 1} T_M(1,\mu) = \tfrac{3}{4} T_M(1,0), \tag{7-115a}$$

and

$$T_M(1,0) = (16\pi/9)(k^3/\hbar)\,|\langle f|\,M_1{}^0(\boldsymbol{\sigma})'\,|i\rangle|^2 \tag{7-115b}$$

with

$$M_1{}^0(\boldsymbol{\sigma})' = \tfrac{1}{2}\mu_0 (3/4\pi)^{1/2} (\mu_n - \mu_p)(\boldsymbol{\sigma}_n - \boldsymbol{\sigma}_p)_z. \tag{7-115c}$$

Direct calculation gives

$$\sigma_c = \pi \left(\frac{e^2}{\hbar c}\right)\left(\frac{\hbar}{m_N c}\right)^2 (\mu_n - \mu_p)^2 (1 - \alpha a_{0s})^2 \left(\frac{E_B}{m_N c^2}\right)\left(\frac{2E_B}{E_n}\right)^{1/2}, \tag{7-116}$$

where E_B is the binding energy of the deuteron, E_n is the *laboratory* energy of the incident neutron, and a_{0s} is the singlet ($L = 0$) scattering length.

Attention should be called to the fact that σ_c of (7-116) is a striking example of a "$1/v$" or $E_n^{-1/2}$ cross section not uncommonly encountered in neutron-induced reactions. It can be interpreted as a dynamic weighting factor favoring those n–p encounters where the interacting pair dwell in close proximity the longest time.

Another interesting point is to note that the radial integration in a proposed $^3S_1 \rightarrow {}^3S_1$ transition would have contained the factor $(1 - \alpha a_{0t})^2$ in place of $(1 - \alpha a_{0s})^2$. Using $\alpha = 0.232\ \text{fm}^{-1}$ and $a_{0s} = -23.71$ fm gives $(1 - \alpha a_{0s})^2 = 42.3$, while on the other hand the value $a_{0t} = +5.42$ fm gives $(1 - \alpha a_{0t})^2 = 0.066$. Of course the near vanishing of the latter quantity is simply the result of the orthogonality of two eigenstates (and the long-wavelength approximation in the electromagnetic operator) previously cited.

[1] To simplify the present calculations, we use the exponential portion of (2-71) for all r even though it is strictly valid only for $r > b$, but we normalize this $u(r)$ to give $4\pi \int_0^\infty u(r)\,dr = 1$, which has the effect of dropping the $e^{\alpha b/2}$ term in (2-71).

[2] Since we perform the sum on μ (hence m'; note $m = 0$, thus $m' = \mu$) the density of final states is unity. The factor $\tfrac{1}{4}$ is obtained by averaging over initial states, i.e., the statistical weight of the singlet state alone. Note that T will involve $|C|^2 = 1/v$ through the normalization of ψ_i to one neutron/cm² sec. With this normalization, T (and hence σ_c) has the dimensions cm². In the more common situations, normalizing ψ_i and ψ_f to unity over all space, T has the dimensions of sec^{-1}. Had ψ_i been normalized with $|C|^2 = 1$, we would have to write $T = v\sigma_c$.

▶**Exercise 7-11** (a) Verify in detail the assertions made concerning the possible (M1) and (E2) electromagnetic transitions and the contributing portions of relevant matrix elements of (7-112). In this latter connection you should review the results of Exercise 2-15.

(b) The variable r in (7-113) and (7-114) is the n–p separation, $\mathbf{r} = \mathbf{r}_n - \mathbf{r}_p$; discuss the role of the coordinate of the center of mass, $\mathbf{R} = \frac{1}{2}(\mathbf{r}_n + \mathbf{r}_p)$.

(c) Derive (7-116) from (7-115). In this derivation you will have to evaluate the integral

$$\mathrm{I} = \int_0^\infty e^{-\alpha r} \sin(\kappa r + \delta_{0s})\, dr = (\kappa \cos\delta_{0s} + \alpha \sin\delta_{0s})/(\alpha^2 + \kappa^2).$$

Note that

$$\lim_{\kappa\to 0} \mathrm{I} \to \kappa(1 - \alpha a_{0s})/\alpha^2. \tag{7-117}$$

Also note that neutron laboratory energy is $E_n = 2\hbar^2\kappa^2/m_N$.

(d) Discuss the above calculation for the $^3S_1 \to {}^3S_1$ transition, noting (1-135). ◀

The calculated value obtained for (7-116) at $E_n = 0.02526$ eV (i.e., neutron velocity 2200 m/sec), using $\mu_p - \mu_n = 4.706$, $E_B = 2.225$ MeV, $\alpha = 0.232\ \mathrm{fm}^{-1}$, $a_{0s} = -23.71$ fm, is

$$\sigma_c = 298\ \mathrm{mb}. \tag{7-118}$$

When more exact theoretical deuteron wave functions are used, the predicted value of σ_c becomes (302.5 ± 4) mb; however, the experimental value is some 10% higher, namely $\sigma_c(\mathrm{obs}) = (334.2 \pm 0.5)$ mb. Considerable theoretical attention has been given to resolving this discrepancy [e.g., see Adler *et al.* (1970), Malik (1972), and Colacci *et al.* (1973)]. The conclusion we arrived at leading to only the $^1S_0 \to {}^3S_1$ transition as the sole (M1) capture source has been questioned in the literature. In particular there may be about a 9% capture via the $^3S_1 \to {}^3S_1$ transition (see Breit and Rustgi, 1970). Meson (pion) exchange currents increase the calculated cross section, but only by 2–3%. Heavy meson currents can also induce $^1S_0 \to {}^3D_1$ transitions and the $\Delta(\frac{3}{2}, \frac{3}{2})$ intermediate state (see Chapter I) also gives contributions. All the meson current contributions add coherently and result in a 9–10% total effect.

Recall also that the observed circular polarization of the capture γ rays, cited in Chapter I, indicates the presence of parity violation in the process.

REFERENCES

Adler, R. J., Chertok, B. T., and Miller, H. C. (1970). *Phys. Rev.* **C2**, 69.
Blatt, J. M., and Weisskopf, V. F. (1952). "Theoretical Nuclear Physics." Wiley, New York.
Breit, G., and Rustgi, M. L. (1971). *Nuclear Phys.* **A161**, 337.
Colacci, M., Mosconi, B., and Ricci, P. (1973). *Phys. Lett.* **B45**, 224.
Eisenberg, J. M., and Greiner, W. (1970). "Nuclear Theory." North-Holland Publ., Amsterdam.
Gari, M., and Huffman, A. H. (1973). *Nuclear Phys.* **C7**, 531.
Hamamoto, I. (1973). *Nuclear Phys.* **A205**, 225.
Johnstone, I. P., and Castel, B. (1973). *Nuclear Phys.* **A213**, 341.
Malik, S. S. (1972). *Phys. Rev.* **C5**, 1807.
Mantri, A. N., and Sood, P. C. (1972). *Phys. Rev.* **C5**, 1422.
Merzbacher, E. (1970). "Quantum Mechanics." Wiley, New York.
Morpurgo, G. (1958). *Phys. Rev.* **110**, 721.
Petrovich, F. (1973). *Nuclear Phys.* **A203**, 65.
Sachs, R. G. (1953). "Nuclear Theory." Addison-Wesley, Reading, Massachusetts.
Schopper, H. (1958). *Nuclear Instrum.* **3**, 158.
Siegbahn, K. (1965). "Alpha, Beta, and Gamma-Ray Spectroscopy," Vols. I and II. North-Holland Publ., Amsterdam.
Skorka, S. J., Hertel, J., and Retz-Schmidt, T. W. (1966). *Nuclear Data* **2**, 347.
Yoshida, S., and Zamick, L. (1972). *Ann. Rev. Nuclear Sci.* **22**, 121.

Chapter VIII

BETA-DECAY

A. INTRODUCTION

We shall not attempt to chronicle here the role of β-decay in the long and interesting history of the study of radioactivity. Suffice it to say that by 1920 Chadwick, among others, had established that, in addition to monoenergetic conversion electrons, β-decay also involved the presence of a broad spectrum of energetic electrons of nuclear origin. These continuous β-spectra displayed a characteristic endpoint energy that corresponded exactly to level energy differences (usually between ground states) in the parent and daughter nuclides. However, electron energies below the endpoint seemed to involve "missing energy," thus violating energy conservation. Heavy-walled calorimeter measurements that should have responded to the absorbed energy from all forms of the known radiations also confirmed this failure of energy balance. It had also been established that β-decay left the total number of nucleons in the daughter nuclide the same as in the parent. Since the electron has an intrinsic mechanical spin $\hbar/2$ and orbital angular momenta are integral multiples of $\hbar$, as are the changes in intrinsic mechanical spin of parent–daughter nuclides, difficulty is encountered in conserving total angular momentum as well.

Pauli suggested in 1930 that the conservation of both energy and angular momentum could be retained if the β-decay involved the emission of a second particle in addition to the electron, called the neutrino.[1] Thus the β-decay of radium E (^{210}Bi, with half-life $t_{1/2} = 5.01$ days) is written

$$^{210}\text{Bi} \rightarrow {}^{210}\text{Po} + \text{e}^- + \bar{\nu} + 1.160 \text{ MeV}, \tag{8-1}$$

where we have noted that β-decay involving the emission of an electron is accompanied by an antineutrino $\bar{\nu}$, a point to be discussed later, and have also explicitly exhibited the "Q value" of 1.160 MeV. The Q value is the difference in neutral *atomic* rest mass energies of ^{210}Bi and ^{210}Po (each in its nuclear ground state).[2] The observed endpoint energy of the β-spectrum is $E_{\beta^-}(\text{max}) = 1.160$ MeV, a value consistent with the zero rest mass energy postulated for the neutrino (or the antineutrino). When the β-decay electron appears with a kinetic energy less than 1.160 MeV, the antineutrino carries away the remaining kinetic energy.

Following up the suggestion of Pauli, Fermi in 1934 developed a theory of β-decay involving the neutrino that has been remarkably successful. This theory in its modern form endows the neutrino with certain properties in addition to its zero rest mass and zero charge already mentioned. The neutrino interacts with other elementary particles only through *weak* interactions. The antineutrino absorption cross section in hydrogen, for example, is approximately 10^{-43} cm^2. The resulting long absorption mean free path in ordinary matter thus explains the failure of the calorimeter measurements to detect the kinetic energy of the neutrino. The neutrino is postulated to be a fermion with intrinsic mechanical spin $\hbar/2$. It has an antiparticle, the antineutrino, both of which obey the Dirac equation.

In Appendix C we introduce the helicity operator $h = \boldsymbol{\sigma}\cdot\mathbf{p}/p$. Since the helicity and the Hamiltonian commute, i.e., $[H, h] = 0$, the helicity is a constant of the motion for relativistic fermions. However, the helicity for

[1] The word *neutrino* is of Italian origin, meaning "little neutral one." The neutrino is assumed to have zero charge.

[2] The neutral Ra E atom ($Z = 83$) with its full complement of orbital atomic electrons decays to ^{210}Po ($Z = 84$), which is then left once ionized with only 83 orbital electrons present. When the negative β-particle, here the electron emitted from the ^{210}Bi nucleus, is also considered, both sides of Eq. (8-1) contain the proper number of electrons corresponding to the respective neutral atoms. The rearrangement energy of the orbital atomic electrons and the eventual pickup of an electron from the surrounding matter to neutralize the ^{210}Po ion are *not* included in the Q-value calculation. Thus, even though the energetic β-particle escapes the entire atomic system of the parent–daughter atom, neutral atomic masses can be used in the Q-value calculation. Hence, in general for β-decay involving electron emission, $Q = [M(Z, A) - M(Z+1, A)]c^2$. For typical papers discussing atomic effects accompanying β-decay, see Isozumi and Shimizu (1971), der Mateosian and Thieberger (1971), and Porter *et al.* (1971).

particles with finite rest mass does not result in a Lorentz-invariant description since, for example, a frame of reference with $\mathbf{p} \to -\mathbf{p}$ can be readily conceived. However, for zero-rest-mass entities, definite helicity does result in an invariant description, as we have seen in the case of the photon discussed in Chapter VII. The helicity of the neutrino is taken as -1 and that of the antineutrino as $+1$. Thus one can speak of the "handedness" of neutrinos; the neutrino is "left-handed" with its spin orientation always such as to be antiparallel to its momentum, while the antineutrino is "right-handed" with its spin orientation always such as to be parallel to its momentum.[1] The consequence of this property of the neutrino will be discussed later in connection with parity conservation in β-decay. In β-decay, as we shall see, two of the basic processes consist of nucleons within a nucleus undergoing the reactions

$$\mathrm{n} \to \mathrm{p} + \mathrm{e}^- + \bar{\nu} \tag{8-2a}$$

$$\mathrm{p} \to \mathrm{n} + \mathrm{e}^+ + \nu\,. \tag{8-2b}$$

The first equation involves the emission of an antineutrino, while the second involves the emission of a neutrino.

It took over 25 years after its postulation to prove conclusively the existence of the neutrino (antineutrino) by observing the reaction

$$\mathrm{p} + \bar{\nu} \to \mathrm{n} + \mathrm{e}^+. \tag{8-3}$$

Reines and co-workers (1960), in a series of experiments with hydrogenous targets, used antineutrinos from a high-flux reactor in which β-decay processes from the neutron-rich fission products generate copious antineutrinos. They observed simultaneously (i.e., in a coincidence measurement) the annihilation radiation from the positron and the capture of the moderated neutron, both primary particles on the right-hand side of (8-3). The reaction (8-3) is evidently related to (8-2b). Attempts to detect the nucleon reaction[2]

$$\mathrm{n} + \bar{\nu} \not\to \mathrm{p} + \mathrm{e}^- \tag{8-4a}$$

in the nuclide embodiment of this reaction,

$$^{37}\mathrm{Cl} + \bar{\nu} \not\to {}^{37}\mathrm{Ar} + \mathrm{e}^-, \tag{8-4b}$$

also using antineutrinos from a reactor, *failed* [see Davis (1955, 1956)]. An upper limit to the cross section for (8-4a) can be set to be less than that for (8-3) by a factor of over 100. The reactions (8-4) *are not* the analogs of (8-2a), as

[1] We, of course, are referring only to the component of spin along the direction of **p** being $\pm\hbar/2$.

[2] A slash through the arrow in a reaction equation is used to indicate that the reaction *does not* occur.

the proper analog of (8-2a) is

$$n + \nu \rightarrow p + e^- . \tag{8-5}$$

If the neutrino and antineutrino were identical, (8-4a) and (8-5) would be equivalent. Since (8-4a) is not observed, we conclude that neutrino and antineutrino are not identical. It is noteworthy to recall that photons and neutral mesons (π^0, ρ^0, η, etc.) *are identical* with their antiparticles.

More precisely, the preceding reactions should be written with a subscript e on the symbol for the neutrino and antineutrino. Thus (8-2) should be written

$$n \rightarrow p + e^- + \bar{\nu}_e \tag{8-6a}$$

$$p \rightarrow n + e^+ + \nu_e . \tag{8-6b}$$

This is because neutrinos produced in reactions involving the muon seem to be different from those involved in β-decay. These muon-associated neutrinos are labeled ν_μ and typical reactions involving them are

$$\pi^+ \rightarrow \mu^+ + \nu_\mu \tag{8-7a}$$

$$\pi^- \rightarrow \mu^- + \bar{\nu}_\mu \tag{8-7b}$$

$$K^+ \rightarrow \mu^+ + \nu_\mu \tag{8-7c}$$

$$K^- \rightarrow \mu^- + \bar{\nu}_\mu . \tag{8-7d}$$

We have already designated neutrinos (both ν_e and ν_μ), electrons, and muons (and each of the antiparticles) as *leptons* in Chapter I. *Leptonic numbers* λ can be assigned to elementary particles, which leads to the conservation of lepton number in reactions, as follows:

$$\text{lepton number:} \qquad \lambda = \begin{cases} +1 & \text{for} \quad e^-, \mu^-, \nu_e, \text{and } \nu_\mu \\ -1 & \text{for} \quad e^+, \mu^+, \bar{\nu}_e, \text{and } \bar{\nu}_\mu \\ 0 & \text{all other elementary particles.} \end{cases} \tag{8-8}$$

We see that reactions (8-6) and (8-7) as well as some other commonly observed reactions, such as

$$e^+ + e^- \rightarrow 2\gamma \tag{8-9a}$$

$$\mu^+ \rightarrow e^+ + \nu_e + \bar{\nu}_\mu \tag{8-9b}$$

$$\mu^- \rightarrow e^- + \nu_\mu + \bar{\nu}_e , \tag{8-9c}$$

all conserve lepton number by having the algebraic sum of λ the same on the left- and right-hand sides of the equations.

It is known that the competing reactions to (8-9b) and (8-9c), namely,

$$\mu^+ \not\rightarrow e^+ + \gamma \tag{8-9b$'$}$$

$$\mu^- \not\rightarrow e^- + \gamma , \tag{8-9c$'$}$$

occur, if at all, with a branching ratio less than 10^{-7}. If only lepton number conservation were required, there would be no need to distinguish ν_e and ν_μ (or $\bar{\nu}_e$ and $\bar{\nu}_\mu$) and the reactions (8-9b′) and (8-9c′) need not be strongly inhibited. The direct demonstration that ν_e and ν_μ are not identical came through the 1962 experiments by a Columbia–Brookhaven group [see Danby *et al.* (1962)]. They found that a neutrino beam produced by decaying pions and kaons [Eqs. (8-7)] could produce only muons (and not electrons) when they impinged on an aluminum spark chamber. Thus reactions

$$\nu_\mu + \mathrm{n} \rightarrow \mathrm{p} + \mu^- \tag{8-10a}$$

$$\bar{\nu}_\mu + \mathrm{p} \rightarrow \mathrm{n} + \mu^+ \tag{8-10b}$$

in nuclei *did* occur, while the reactions

$$\nu_\mu + \mathrm{n} \nrightarrow \mathrm{p} + \mathrm{e}^- \tag{8-10a′}$$

$$\bar{\nu}_\mu + \mathrm{p} \nrightarrow \mathrm{n} + \mathrm{e}^+ \tag{8-10b′}$$

did not occur. If the two types of neutrinos were identical, the reactions (8-10a′) and (8-10b′) would simply be variants of the β-decay reactions (8-6a) and (8-6b). Thus we conclude that muon-associated neutrinos are different from electron-associated neutrinos. In a somewhat ad hoc way, we might introduce a *muonic number* equal to $+1$ for μ^- and ν_μ, -1 for μ^+ and $\bar{\nu}_\mu$, and zero for all other particles; whence a conservation law would result allowing (8-10a) and (8-10b), but forbidding (8-10a′) and (8-10b′). The reactions (8-9b′) and (8-9c′) would then also be forbidden.

In view of the above discussion, it might be worthwhile to remark on whether the electronlike (and positronlike) particles observed in β-decay, more properly labeled β^- and β^+ in Eqs. (8-6a) and (8-6b), are, in fact, *identical* with orbital atomic electrons (and their antiparticles). While early experiments quickly established that β-particle charge and mass were the same as for the ordinary electron, it remained for the experiments of Goldhaber and Scharff-Goldhaber (1948) to give the definitive answer. They essentially attempted to perform the analogous experiments to the muonic-atom X-ray experiments discussed in Chapter II, but with stopped β^--particles. If β^--particles were distinguishable from ordinary electrons, their capture on stopping would have resulted in cascading X-ray emission as for muonic atoms. However, if they were identical with ordinary electrons, the operation of the Pauli principle would prevent X-ray transitions since the appropriate final-state orbitals would already be occupied by electrons. No X rays were, in fact, observed, hence offering proof that β^--particles and ordinary atomic electrons are identical particles. A further indication of this equivalence is offered by the observation that β^+-particles annihilate with ordinary electrons via reaction (8-9a).

The free neutron is unstable to β-decay and, in fact, decays via the transition (8-6a) with a mean life of (15.6 ± 0.2) min with a Q value of (782.4 ± 0.1) keV. Needless to say, the nucleon transition (8-6b) can take place only when the decaying proton is one of the nucleons of a suitable nuclide. An example of β-decay with the emission of a positron is

$$^{13}\mathrm{N} \rightarrow {}^{13}\mathrm{C} + \mathrm{e}^{+} + \nu_{\mathrm{e}} \tag{8-11}$$

with a Q value of 1.199 MeV[1] and a half-life of 10.0 min. This β-decay, incidentally, played a key role in the discovery of artificially induced radioactivity by the Curie-Joliots.

In addition to β-decay resulting in the emission of nuclear electrons $\beta^{\pm}$, yet another type of β-decay called *electron capture* is possible. In electron capture an atomic electron, in an orbital allowing a finite probability for the electron being within the nuclear interior, is absorbed by a nuclear proton, leading to a transition yielding a neutron and neutrino in the final state. As a nucleon transition, we write in an analogous manner to (8-6)

$$\mathrm{p} + \mathrm{e}^{-} \rightarrow \mathrm{n} + \nu_{\mathrm{e}}. \tag{8-12}$$

An actual example of electron capture, mostly involving one of the K-electrons and hence called K-capture, is

$$^{7}\mathrm{Be} + \mathrm{e}^{-} \rightarrow {}^{7}\mathrm{Li} + \nu_{\mathrm{e}} \tag{8-13}$$

with a half-life of 53 days and a Q value (nuclear ground state to ground state, which is the dominant 89.7% branch) of 861.6 keV.[2]

[1] If neutral atomic masses were used in determining the Q value in (8-11), it will be noted that the right-hand side then would have *two* extra electronic masses; thus for positron decay $Q = [M(Z, A) - M(Z-1, A)]\,c^2 - 2m_e c^2$.

[2] While (8-13) explicitly exhibits on the left-hand side of the equation the orbital atomic electron of ^{7}Be which will be absorbed within the nucleus, the total number of electrons is still just the four of the neutral beryllium atom. It is customary to define the Q value for electron capture as $Q = [M(Z, A) - M(Z-1, A)]\,c^2$. Although usual limits of accuracy do not require taking atomic rearrangement energies into consideration for $\beta^{\pm}$-decay, the case of electron capture actually removes an orbital electron for which the full binding energy must be accounted for, not just the change in binding energy for adjacent atoms Z and $Z \pm 1$. For example, ^{80}Br (which largely decays by β^{-}-decay) decays to ^{80}Se by a 2.6% β^{+}-decay branch, as well as by a 5.7% electron capture branch. The K-binding energy for bromine is 13.47 keV, while the difference in the K-binding energies for bromine and selenium is only 820 eV. The atomic rearrangement energy involved in this β^{+}-decay is seen to be an order of magnitude smaller than the energy correction required for the electron capture process (mostly K-capture in the present case). For energy measurements accurate to a few keV, the one correction could be ignored while the other (i.e., for K-capture) could not. Thus for K-capture in general the kinetic energy of the neutrino and the recoiling atom must be written

$$E_{\nu} + E_{\mathrm{R}} = Q - B_{\mathrm{K}}, \tag{8-14}$$

where B_{K} is the K atomic binding energy. It can also happen that $0 < Q < B_{\mathrm{K}}$ as for ^{193}Pt and ^{205}Pb. In these cases only L-capture occurs.

Several points concerning electron capture are worth noting. When the Q value is large compared to the K-shell ($1s_{1/2}$) electronic binding energy, the L_I-shell (classically $2s_{1/2}$) electron capture probability is approximately $\frac{1}{8}$ of the K-shell electron capture probability. Since for ^7Be ($Z=4$), these L-electrons are the *valence* electrons, one might expect to slightly influence the β-decay rate by chemical binding effects when the beryllium is in different chemical compounds. Indeed effects as large as 0.07% were observed in the lifetimes for different chemical states. While $\beta^{\pm}$-decays such as (8-1), (8-2a), and (8-11) involve continuous electron (and hence neutrino) energy spectra, K-capture such as (8-13) gives only a monoenergetic neutrino (and hence a monoenergetic recoil atom). Conservation of energy and momentum dictates that the recoiling lithium atom in (8-13) for the ground-state transition should be 57 eV. Such a prediction has been experimentally confirmed. The ratio of K-capture to β^+-decay probabilities when both are energetically possible is a rather complicated function of the nuclear charge Z and the Q value, particularly when certain interference terms are taken into account. But for large Q values this ratio is[1]

$$w_K/w_{\beta^+} \approx 60\pi(\alpha Z m_e c^2/Q_K)^3. \tag{8-15}$$

Thus it will be seen that positron emission is generally the dominant mode of decay in light elements (when energetically possible), whereas K-capture is more likely in heavy elements. Examples of nuclei in which all three processes, β^-- and β^+-decay as well as K-capture, can occur are ^{64}Cu, ^{74}As, and ^{80}Br, just to cite a few.

Following K-capture, the atomic deexcitation in heavy elements is generally through the emission of X rays, usually the K_α X ray, while in light elements, the deexcitation is generally through the emission of Auger electrons.

Finally, in this general survey of β-decay, a word or two is in order for the possibility of *double β-decay*. A hypothetical example of double β-decay involving the emission of two electrons on a nucleon transition basis is

$$2n \to 2p + 2e^- + 2\bar{\nu}_e. \tag{8-16}$$

An embodiment of (8-13) involving specific nuclides would be

$$^{130}Te \to {}^{130}Xe + 2e^- + 2\bar{\nu}_e. \tag{8-17}$$

The reaction (8-16) or (8-17) is a second-order process not to be confused with a simple two-step process. Indeed in (8-17) this would be impossible since $^{130}Te \to {}^{130}I + e^- + \bar{\nu}_e$ followed by $^{130}I \to {}^{130}Xe + e^- + \bar{\nu}_e$ is not allowed because $M(^{130}I) > M(^{130}Te)$ leads to a negative Q value. Carefully investigated nuclides (among over 50 energetically possible candidates) are ^{48}Ca,

[1] The symbol α represents the fine structure constant, $\alpha = e^2/\hbar c$.

^{124}Sn, ^{150}Nd, and ^{238}U, all with negative results to date. Lower limits on the experimental lifetime for these cases have been set to 10^{18}–10^{19} yr [e.g., see Smith (1971)]. Theoretical estimates for the process (8-16) predict lifetimes 10^{20}–10^{22} yr. Indirect evidence, from mass spectroscopic analysis of geological samples, does exist for double β-decay with lifetimes in the calculated range. Incidentally, if there were no difference between ν_e and $\bar{\nu}_e$, one might expect a reaction such as (8-16) to occur with no neutrinos or antineutrinos being present on the right-hand side of the equation. Such a process, of course, would violate the law of lepton conservation. It is precisely the lepton number which is different for ν_e and $\bar{\nu}_e$. The predicted lifetimes when ν_e and $\bar{\nu}_e$ are identical range from 10^{15} to 10^{16} yr. Since the experimental lower limits are considerably larger, this can be taken as additional proof that the neutrino and antineutrino are not identical.

In addition to double β^--decay, the somewhat longer lived double β^+-decay and double K-capture are also possible.

We have presented only a brief outline of the principal features of the phenomena of β-decay. More detailed information is contained in the excellent works of Wu and Moszkowski (1966), Konopinski (1966), and Schopper (1966). In the interesting special area of macroscopic effects influencing β-decay rates, see the review article by Emery (1972).

B. THEORETICAL FORMULATION (WEAK INTERACTIONS)

In this section we will restrict our considerations to a simplified presentation of the weak interaction theory of electronic β-decay. When specification is necessary, we shall use β^--decay embodied in (8-2a) as the example. There are excellent works that consider β-decay within the framework of elementary particle physics and the general theory of weak interactions. While this broader context represents one of the more exciting developments in modern physics, it is prudent to restrict our attention to the specifically nuclear aspects. Examples of broader treatments are Marshak *et al.* (1969), Okun (1965), Gasiorowicz (1966, Part IV), and DeBenedetti (1964, Section 8). Two somewhat dated but very instructive works are those of Fermi (1951) and Feynman (1962).

1. General Considerations

Fermi in his original formulation of the theory of β-decay was guided by the form of the electromagnetic interaction of a charged current j_μ and a field specified by the vector potential A_μ. In the classical formulation of this interaction using the general four-vectors

$$A_\mu = (\mathbf{A}, i\phi), \qquad j_\mu = (\mathbf{j}, ic\rho), \tag{8-18}$$

we have the familiar form of the interaction Hamiltonian

$$H_{\text{int}} = -c^{-1}\sum_{\mu=1}^{4}\int j_\mu(\mathbf{r},t)\,A_\mu(\mathbf{r},t)\,d^3r$$
$$= -c^{-1}\int \mathbf{j}(\mathbf{r},t)\cdot\mathbf{A}(\mathbf{r},t)\,d^3r + \int \phi(\mathbf{r},t)\,\rho(\mathbf{r},t)\,d^3r. \qquad (8\text{-}19)$$

The translation of these equations into the appropriate quantum mechanical equivalents was discussed in Chapter VII.[1] In particular, if we consider a nucleon interacting with a field, we would write

$$H_{\text{int}} = \tfrac{1}{2}e(1+\tau_3)(\bar{\psi}_{\text{f}}\gamma_\mu\psi_{\text{i}})A_\mu. \qquad (8\text{-}20)$$

In (8-20) we have noted from Appendix C [Eq. (C-68)] that the probability current four-vector for a fermion is

$$j_\mu = ic\bar{\psi}\gamma_\mu\psi = ic\psi^\dagger\gamma_4\gamma_\mu\psi = c\begin{pmatrix}\psi^\dagger\alpha_\mu\psi \\ i\psi^\dagger\psi\end{pmatrix} \quad \begin{matrix}\mu = 1,2,3 \\ \mu = 4.\end{matrix} \qquad (8\text{-}21)$$

Fermi supposed that in β-decay the field potential A_μ in (8-20) should be replaced by a lepton field of four-vector form. The interaction Hamiltonian, being a product of two four-vectors, would be a scalar as required. Before writing down the lepton field in its present elegant form, a slight digression is necessary.

In the β^--decay of the neutron (8-2a), the initial state consists of the neutron while the final state is seen to consist of three positive-energy particles, the proton, the electron, and the antineutrino. An alternate view would have the initial state consist of a neutron and a filled sea of negative-energy neutrinos. The interaction of the neutron and one such neutrino results in the annihilation of the neutron and neutrino and the creation of a proton and an electron. The "hole" left behind in the negative-energy sea of neutrinos is the antineutrino. The Feynman diagrams showing this equivalence, recalling that antiparticles propagate backward in time, are given in Fig. 8.1a,b. In this section, largely devoted to β-decay, we shall drop the label e on the neutrino (and antineutrino) for convenience, except where contrasting reactions involving muons also appear and thus confusion might result.

With such equivalences in mind, a particle (antiparticle) on one side of a reaction can be replaced by its antiparticle (particle) on the other side; thus the same interaction Hamiltonian could describe

$$\begin{matrix} \text{n} \rightarrow \text{p} + \text{e}^- + \bar{\nu}, & \text{n} + \text{e}^+ \rightarrow \text{p} + \bar{\nu} \\ \text{n} + \nu \rightarrow \text{p} + \text{e}^-, & \bar{\text{p}} \rightarrow \bar{\text{n}} + \text{e}^- + \bar{\nu}. \end{matrix} \qquad (8\text{-}22)$$

[1] The difference in sign for the scalar and vector contributions to (8-19) was also considered in connection with (1-31) in Chapter I.

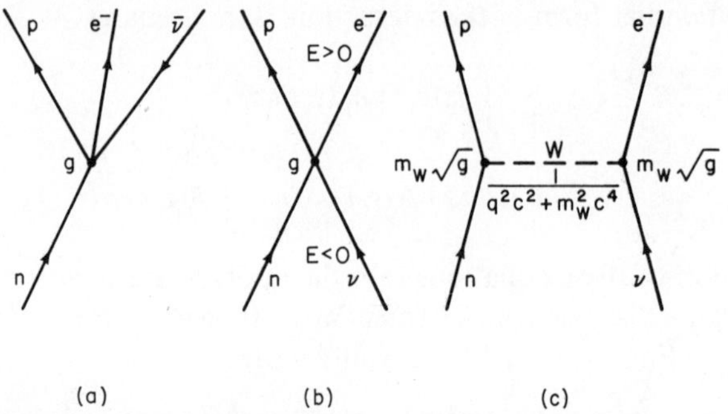

FIG. 8.1. (a) Diagrammatic representation of reaction $n \rightarrow p + e^- + \bar{\nu}$ with zero-range interaction and coupling constant g. (b) Equivalent representation to (a), involving the annihilation of a neutron and negative-energy neutrino and the creation of a proton and an electron. (c) The weak interaction coupling through the intermediate boson with a range $\hbar/m_W c$.

The Hermitian conjugate of this interaction Hamiltonian would describe essentially the inverse of (8-22), namely

$$\begin{aligned} &p \rightarrow n + e^+ + \nu, \qquad && p + e^- \rightarrow n + \nu \\ &p + \bar{\nu} \rightarrow n + e^+, && \bar{n} \rightarrow \bar{p} + e^+ + \nu. \end{aligned} \tag{8-23}$$

It was for these reasons that (8-3) was described as "related" to (8-2b) in discussing the Reines experiment in Section A. The symmetric form with two fermions on each side of the equation permits us to write simply

$$n + \nu \rightleftarrows p + e^- \qquad \text{and} \qquad n + e^+ \rightleftarrows p + \bar{\nu}. \tag{8-24}$$

Returning to the discussion of (8-20) and (8-21), we would then write the vector–vector interaction in time independent form as

$$H_{\text{int}} = C_V \sum_{\mu=1}^{4} \{\bar{\psi}_p(\mathbf{r}), \gamma_\mu t_+ \psi_n(\mathbf{r})\} \{\bar{\phi}_{e^-}(\mathbf{r}), \gamma_\mu t_+{}^L \phi_\nu(\mathbf{r})\}. \tag{8-25}$$

Here t_+ is the raising operator (1-7) converting the isospin state of the neutron to the proton, $t_+{}^L$ an equivalent operator changing the state of a neutrino to an electron, and C_V a *weak coupling constant* giving the strength of the interaction.

Our interaction Hamiltonian (8-25) is seen to be *zero range*, i.e., *all* coordinates (including those of the leptons) are the nucleon coordinate **r**. This assumption of a zero-range interaction corresponds to Fig. 8.1a, b, and has the virtue of simplicity and an exact analogy to the electromagnetic case. There is reason to believe that the true interaction may, in fact, involve an intermediate virtual-particle-like quantum, the *weakon*.[1] This mechanism is illustrated in Fig. 8.1c. The vertex interaction involving $\sqrt{g}$ is referred to as "semiweak." The existence of the weakon or intermediate vector boson[2] would require us to rewrite (8-24) as[3]

$$\mathrm{n} + \bar{\mathrm{p}} \rightleftarrows \mathrm{W}^- \rightleftarrows \mathrm{e}^- + \bar{\nu}_\mathrm{e}, \qquad \bar{\mathrm{n}} + \mathrm{p} \rightleftarrows \mathrm{W}^+ \rightleftarrows \mathrm{e}^+ + \nu_\mathrm{e}. \tag{8-26}$$

As Eq. (8-26) indicates, the weakon is a boson; it is assumed to have spin $1\hbar$ and to have baryon and lepton numbers equal to zero. It is assumed to have a mass greater than the kaon to explain the absence of the decay $\mathrm{K}^+ \not\leftrightarrow \mathrm{W}^+ + \gamma$. Various negative and indirect experimental results [e.g., Burns *et al.* (1965), Marshak *et al.* (1969, Chapter 7), and a more recent result (Barish *et al.*, 1973)] suggest $4 \lesssim m_\mathrm{W} c^2 \lesssim 75$ GeV. One theoretical estimate places the mass at 37.3 GeV [see Lee (1971)]. The lifetime is assumed to be shorter than 10^{-17} sec. To date no clear experimental evidence has been found establishing the existence of the weakon. However, its possible existence and the resulting reactions

$$\mathrm{W}^+ \rightleftarrows \mu^+ + \nu_\mu, \qquad \mathrm{W}^- \rightleftarrows \mu^- + \bar{\nu}_\mu \tag{8-27}$$

lend a simple interpretation to the idea of a universal Fermi interaction. It is experimentally observed that nucleon decay (8-6a)–(8-6b), muon decay (8-9b)–(8-9c), and μ^--capture,

$$\mu^- + \mathrm{p} \to \mathrm{n} + \nu_\mu, \tag{8-28}$$

involve the *same* coupling constants to 2–3% equality. [The reaction (8-28) is the counterpart of ordinary electron capture with, however, muonic atoms.] If the weakon is the common intermediate agency, a simple graphical representation of the situation is possible by means of the "Puppi triangle" (Puppi, 1962) shown in Fig. 8.2.[4]

This diagram links, with the same universal coupling constant, reactions that take a particle–antiparticle pair from one corner of the triangle into another particle–antiparticle pair at either of the other two vertices. In this manner it pictorially summarizes reactions such as (8-6a)–(8-6b), (8-9b)–(8-9c),

[1] A pointlike interaction of the neutrino (antineutrino) in a reaction such as (8-3) would lead to unphysically large cross sections at very high energies. This difficulty is ameliorated if the interaction is taken to be of nonzero range, i.e., through the weakon.

[2] This particle has also been occasionally referred to as the W-boson, the *schizon*, or the IVB-particle.

[3] At least the two charge states of the weakon are required. If neutral leptonic currents should be found to exist in nature, then a neutral weakon W^0 would also have to be introduced. The unified quark model of Pati and Salam (1974) referred to in Chapter I would include such a weakon.

[4] Strangeness-changing leptonic processes also occur.

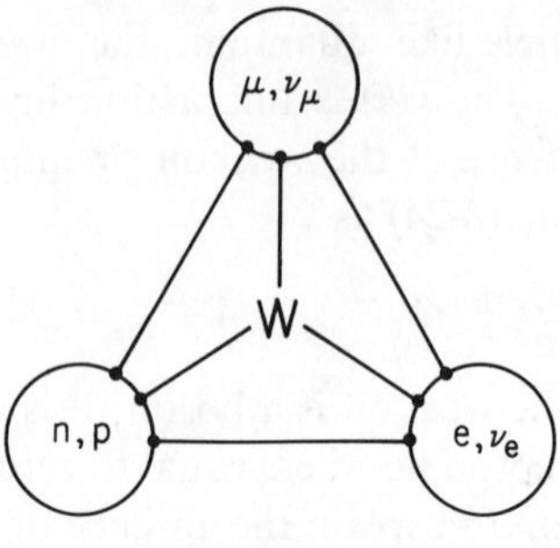

FIG. 8.2. The Puppi triangle and the role of the intermediate vector boson, the weakon W, representing diagrammatically the universal Fermi interaction.

and (8-28) when these are rewritten in terms of reacting pairs having baryon, lepton, and muonic numbers equal to zero.

In ordinary β-decay the momentum transfer (q in Fig. 8.1c is actually the four-momentum transfer) measured in MeV/c is small compared to the term $m_W c^2$. Thus, even if the virtual process involving the weakon did occur, it can be largely ignored without great error. This is equivalent to assuming zero range for the interaction (8-25). The same is less true for processes involving the muons, with the generally much larger momentum transfers encountered.

Particularly in view of the possibility of a universal Fermi interaction, another aspect of the interaction Hamiltonian (8-25) requires attention. Leptons, as far as is known, behave as if they were pointlike particles with no significant anomalous magnetic moments associated with complex internal structure.[1] The nucleons, of course, have considerable structure as we saw in Chapter I. It is then perhaps surprising that reactions (8-26) and (8-28) and the reactions represented in Figs. 8.1c and 8.2 are found experimentally to involve the laboratory or "clothed" nucleons with a universal coupling rather than the bare nucleons. It might have been supposed that this universal character pertained to bare fermions. This unexpected behavior is resolved by the introduction of the *conserved weak vector current* hypothesis. In this connection, it is well to recall that it is *not* the electromagnetic charge current of the bare nucleon that is conserved, but rather that of the *sum* of the nucleon and the virtual or real pion currents. Thus in the virtual process $p \rightleftarrows n + \pi^+$ it is essential to consider the charged pion current as well as that of the bare nucleons in constructing a true divergenceless (or conserved) electric current vector. The interaction of the physical nucleon with an electromagnetic field properly requires the use of this conserved electric current including the pion's participation. In the conserved vector current theory, the pions are

[1] Current theoretical speculation would ascribe an extended structure to the leptons of order 10^{-16} cm.

imagined to play a similar role in the interaction of the nucleon with the weak field. This, of course, implies a coupling of the pion to the leptonic field. Indeed, the pion decay mode characteristics of (8-7a),

$$\pi^+ \to \mu^+ + \nu_\mu,$$

as well as those of the modes

$$\pi^+ \to e^+ + \nu_e \quad \text{and} \quad \pi^+ \to \pi^0 + e^+ + \nu_e, \tag{8-29}$$

are all in accord with the conserved vector current theory.

We now return to the discussion of the basic nature of the interaction Hamiltonian responsible for β-decay. The interaction matrix element corresponding to (8-25) can be written more generally as

$$H_{\text{int}} = \langle \psi_f \phi_f | \tilde{H}_{\beta,V} | \psi_i \phi_i \rangle, \tag{8-30a}$$

with

$$\tilde{H}_{\beta,V} = C_V\{\tilde{N}_V t_+\} \odot \{\tilde{L}_V t_+{}^L\} + \text{h.c.} \tag{8-30b}$$

In this generalized form the interaction operator (8-30b) consists of a portion inducing the β^--decay of (8-25), and, through its Hermitian conjugate, h.c. $= (\tilde{N}_V{}^\dagger t_-) \odot (\tilde{L}_V{}^\dagger t_-{}^L)$, a portion giving rise to β^+-decay and electron capture. The symbol $\odot$ represents an appropriate generalized scalar product of the nucleon and lepton operators, which in the present case is the scalar product of two vector operators. Thus[1]

$$\{\tilde{N}_V\} \odot \{\tilde{L}_V\} = \sum_{\mu=1}^{4} \{\gamma_4 \gamma_\mu\}_N \{\gamma_4 \gamma_\mu\}_L. \tag{8-30c}$$

The direct product space $|\psi\phi\rangle = |\psi\rangle\,|\phi\rangle$ implied in (8-30a) is introduced as in connection with (1-11) in Chapter I, and it is understood that direct product operators such as in (8-30b) have the nucleon operator N operate only on the nucleon field $|\psi\rangle$ and the lepton operator L operate only on the lepton field $|\phi\rangle$. The fermion state functions $|\psi\rangle$ and $|\phi\rangle$ are of course, in principle, expressed in terms of the four-component Dirac spinors, although the nonrelativistic reduction for the nucleon is often used. We have also noted that $\bar{\psi} = \psi^\dagger \gamma_4$ (see Appendix C).

In (8-30a), $|\psi_i\rangle$, the initial nucleon state, can represent either a proton or a neutron. If $|\psi_i\rangle = |\psi_n\rangle$, a neutron state, then $t_+|\psi_n\rangle = |\psi_p\rangle$, in principle allowing β^--decay. However, $t_-|\psi_n\rangle = 0$, indicating essentially that charge conservation prevents β^+-decay of a neutron. In applying (8-30) to a complex nucleus, a sum over all nucleons must be performed whence, if energetically

[1] The operators $t_\pm$, operating as they do in a space different from the relativistic coordinate space, can be readily moved outside the brackets of (8-30b).

possible, the neutrons originally present can contribute to β^--decay and *simultaneously* [through the h.c. of (8-30b) involving t_-] protons originally present can contribute to β^+-decay or electron capture. Orthogonality of the neutron and proton isospin states automatically ensures that proper initial and final charge states for a particular nucleon are involved by using $t_\pm$.

While originally we have taken guidance for the form of the weak interaction expressed in (8-25) from the electromagnetic case (8-19), a wide variety of other bilinear forms coupling the initial and final lepton fields is possible. The most general form,

$$F_L(\mathbf{r}) = \sum_{\eta=1}^{4} \sum_{\zeta=1}^{4} \{\phi_{f\eta}^\dagger, \tilde{O}_{\eta\zeta}\phi_{i\zeta}\}, \tag{8-31}$$

involves an arbitrary linear operator $\tilde{O}_{\eta\zeta}$ which may be expanded in terms of 16 independent (four by four) constant matrices. [Derivative or velocity-dependent coupling is also possible—recall the discussion in connection with (1-46) in Chapter I—but at present there is no experimental evidence for the need to introduce such terms.] These matrices can be grouped into precisely five quantities that, in the form (8-31), obey proper Lorentz transformations. The vector operator

$$\tilde{L}_V = \gamma_4\gamma_\mu, \qquad \mu = 1, 2, 3, 4 \tag{8-32}$$

is one such operator. Its use in (8-31), of course, results in four "components" of $F_{L,\mu}$ that transform like those of a (polar) four-vector.

In Appendix C, it is shown that $\bar{\psi}\psi = \psi^\dagger\gamma_4\psi$ transforms like a true scalar under Lorentz transformations[1]; hence another of these operators is

$$\tilde{L}_S = \gamma_4. \tag{8-33}$$

In our discussion of the pseudoscalar meson theory in Chapter I, we saw that using

$$\tilde{L}_P = \gamma_4\gamma_5 \tag{8-34}$$

leads to a quantity analogous to F_L which behaved like a pseudoscalar. L_P is also one of the five basic Lorentz-invariant β-decay operators. There are also quantities that transform like an axial vector and a second rank antisymmetric tensor. Table 8-1 summarizes the characteristics of all these basic β-decay operators.

Exactly similar considerations apply to the possible nucleon operators N_K. Equations (8-30) can be still further generalized to yield a sum over scalar

[1] The quantity $\psi^\dagger\psi$ is, in fact, the "scalar" or fourth component of the current four-vector; see (8-21).

TABLE 8-1

POSSIBLE β-INTERACTION OPERATORS

Type, K	Number of independent matrices	Form of operator	Nonrelativistic form
S	1	γ_4	1
V	4	$\gamma_4\gamma_\mu$	$1, \mu = 4$ $0, \mu = 1, 2, 3$
T	6	$\gamma_4\gamma_\mu\gamma_\nu, \mu \neq \nu$	$\boldsymbol{\sigma}, \mu, \nu = 1, 2, 3$ $0, \mu$ or $\nu = 4$
A	4	$\gamma_4\gamma_\mu\gamma_5$	$\boldsymbol{\sigma}, \mu = 1, 2, 3$ $0, \mu = 4$
P	1	$\gamma_4\gamma_5$	~ 0

direct products of the operators L_K and N_K, with $K = \text{S, V, T, A, P}$, viz.,

$$H_{\text{int}} = \langle \psi_f \phi_f | \sum_K \tilde{H}_{\beta, K} | \psi_i \phi_i \rangle \tag{8-35a}$$

$$\tilde{H}_{\beta, K} = C_K \{\tilde{N}_K t_+\} \odot \{\tilde{L}_K t_+{}^{\text{L}}\} + \text{h.c.} \tag{8-35b}$$

One can now proceed by using the most general forms (8-35) to predict observables that, when compared to experimental results, reveal the appropriate values of the coupling constants C_K.[1] Since by this date experimental evidence has completely eliminated many of the numerous possibilities, we follow a more direct route, referring the purist to the previously cited references.

We assume that it is known that weak interactions are parity violating; hence we admit direct product nucleon and lepton operators from Table 8-1 that produce both true scalars as in (8-35) and pseudoscalars as well. Thus if $N = N_S = \gamma_4$, we could have $L = C_S L_S + C_S' L_P = \gamma_4(C_S + C_S'\gamma_5)$ with the result that $N \odot L = \{\gamma_4\}_N\{\gamma_4(C_S + C_S'\gamma_5)\}_L$. Also, if $N = N_P = \gamma_4\gamma_5$, we could have $L = C_P L_P + C_P' L_S = \gamma_4\gamma_5(C_P + C_P'\gamma_5)$ since $\gamma_5{}^2 = 1$, and then $N \odot L = \{\gamma_4\gamma_5\}_N\{\gamma_4\gamma_5(C_P + C_P'\gamma_5)\}_L$. Again if the nucleon operator is a

[1] There are, in fact, ten such constants, not five as would appear from the discussion up to this point. Basically this arises from the spinor phase ambiguity in reflection operations, leading to either even- or odd-parity terms in the Hamiltonian. The experimental observation of parity violation in β-decay requires using a mixture of true scalar and pseudoscalar such elements in the Hamiltonian. We do not pursue these points further since the alternate treatment followed hereafter is closely equivalent.

polar vector $N = N_V = \sum_{\mu=1}^{4} \gamma_4 \gamma_\mu$, a scalar product could be formed with

$$L = C_V L_V + C_V' L_A = \sum_{\mu=1}^{4} \gamma_4 \gamma_\mu (C_V + C_V' \gamma_5),$$

and we would have

$$N \odot L = \sum_{\mu=1}^{4} \{\gamma_4 \gamma_\mu\}_N \{\gamma_4 \gamma_\mu (C_V + C_V' \gamma_5)\}_L.$$

Clearly when N is an axial vector operator, we can have the term

$$N \odot L = \sum_{\mu=1}^{4} \{\gamma_4 \gamma_\mu \gamma_5\}_N \{\gamma_4 \gamma_\mu \gamma_5 (C_A + C_A' \gamma_5)\}_L.$$

The last and fifth possibility requires noting that while only one type of tensor operator exists, we can write

$$N \odot L = \sum_{\mu \neq \nu = 1}^{4} \{\gamma_4 \gamma_\mu \gamma_\nu\}_N \{\gamma_4 \gamma_\mu \gamma_\nu (C_T + C_T' \gamma_5)\}_L,$$

in which the quantity multiplied by C_T behaves as a true scalar operator while the quantity with C_T' behaves as a pseudoscalar operator. We thus arrive at the still more general form

$$H_{\text{int}} = \langle \psi_f \phi_f | \left[\sum_K \{\tilde{N}_K t_+\} \odot \{\tilde{L}_K (C_K + C_K' \gamma_5) t_+{}^L\} + \text{h.c.} \right] | \psi_i \phi_i \rangle. \tag{8-36}$$

We note that (8-36) requires for its complete specification the values of *ten* coupling constants. The terms with unprimed constants are true scalars and conserve parity; those with primed constants are pseudoscalars and change parity.[1]

2. The V–A Theory

It is convenient at this point to introduce the *two-component neutrino hypothesis* and definite helicity states. Using Appendix C and partitioning the Dirac four-component spinor into two Pauli spinors u_l and u_s, we have for a fermion of mass m the coupled equations

$$c\boldsymbol{\sigma}^P \cdot \mathbf{p} u_l - mc^2 u_s = E u_s \tag{8-37a}$$

$$c\boldsymbol{\sigma}^P \cdot \mathbf{p} u_s + mc^2 u_l = E u_l. \tag{8-37b}$$

For the neutrino we postulate $m = 0$, whence from (8-37),

$$u_s = \pm u_l. \tag{8-38}$$

[1] The assumption is made here that the intrinsic parities of the electron and neutrino are the same. If the relative parities were opposite, the roles of C_K and C_K' would be reversed.

Thus, since the small component and large component differ at most only in sign, the Dirac spinor is in effect only a two-component wave function.

We now suppose that the free neutrino state is also a simultaneous eigenvector of the Dirac helicity operator of Appendix C, $h = \boldsymbol{\sigma}^{\mathrm{D}} \cdot \hat{\mathbf{p}}$, with $|\mathbf{p}|\hat{\mathbf{p}} = p\hat{\mathbf{p}} = \mathbf{p}$. The eigenvalue problem becomes, using (8-37) with $m = 0$,

$$(\boldsymbol{\sigma}^{\mathrm{D}} \cdot \hat{\mathbf{p}})\phi = \left(\begin{array}{c|c} \boldsymbol{\sigma}^{\mathrm{P}} \cdot \hat{\mathbf{p}} & 0 \\ \hline 0 & \boldsymbol{\sigma}^{\mathrm{P}} \cdot \hat{\mathbf{p}} \end{array}\right)\begin{pmatrix} u_l \\ \hline u_s \end{pmatrix} = \frac{E}{cp}\begin{pmatrix} u_s \\ \hline u_l \end{pmatrix} = h\begin{pmatrix} u_l \\ \hline u_s \end{pmatrix}. \tag{8-39}$$

When $E > 0$, $E/cp = +1$; hence $h = +1$ corresponds to $u_s = +u_l$ and $h = -1$ corresponds to $u_s = -u_l$. Of these two possible helicity states, we now have definite experimental evidence to assign the state $h = -1$ for $E > 0$ to the neutrino, and the state $h = +1$ for $E > 0$ to the antineutrino. Thus we say neutrinos with $E > 0$ have negative (and *only* negative) helicity or are left-handed leptons, described by spinors with $u_s = -u_l$. Conversely, antineutrinos with $E > 0$ have positive helicity or are right-handed leptons, whence

$$|\nu\rangle_{E>0} \equiv \phi^{\mathrm{L}} = \begin{pmatrix} u \\ \hline -u \end{pmatrix}, \qquad |\bar{\nu}\rangle_{E>0} \equiv \phi^{\mathrm{R}} = \begin{pmatrix} u \\ \hline u \end{pmatrix}. \tag{8-40}$$

When $E < 0$, $E/cp = -1$; hence $h = +1$ corresponds to $u_s = -u_l$ and $h = -1$ corresponds to $u_s = +u_l$.

As an example, for massless fermions propagating along the positive z axis [see Appendix C, (C-26) and (C-28), $\mathbf{p} = p\hat{z}$] we identify four Dirac spinors

$$\begin{aligned}
E > 0: \quad & \phi_+{}^{\alpha} = 2^{-1/2}\begin{pmatrix} 1 \\ 0 \\ 1 \\ 0 \end{pmatrix}, \quad h = +1; \\
& \phi_+{}^{\beta} = 2^{-1/2}\begin{pmatrix} 0 \\ 1 \\ 0 \\ -1 \end{pmatrix}, \quad h = -1 \\
E < 0: \quad & \phi_-{}^{\beta} = 2^{-1/2}\begin{pmatrix} 0 \\ 1 \\ 0 \\ 1 \end{pmatrix}, \quad h = -1; \\
& \phi_-{}^{\alpha} = 2^{-1/2}\begin{pmatrix} -1 \\ 0 \\ 1 \\ 0 \end{pmatrix}, \quad h = +1.
\end{aligned} \tag{8-41}$$

The real neutrino ($E > 0$) is identified with $\phi_+{}^\beta$, $h = -1$, and the real antineutrino ($E > 0$) with $\phi_+{}^\alpha$, $h = +1$. The spinor for the antiparticle to the neutrino is the charge conjugate spinor

$$[\phi_+{}^\beta]_C = 2^{-1/2}\begin{pmatrix}1\\0\\1\\0\end{pmatrix}$$

(see Appendix C), which, while equal to $\phi_-{}^\alpha$ with $E < 0$ and $\mathbf{p} \to -\mathbf{p}$, is also just $\phi_+{}^\alpha$ or the antineutrino spinor for $E > 0$. Clearly when $m \neq 0$, this simplification does not obtain, in effect requiring both right- and left-handed particles for $E > 0$ and in addition both right- and left-handed antiparticles for $E > 0$ as well.

We now introduce the useful projection operators $\tilde{\Lambda}_+$ and $\tilde{\Lambda}_-$ defined by[1]

$$\tilde{\Lambda}_\pm = (1/\sqrt{2})(1 \mp \gamma_5). \tag{8-42}$$

▶**Exercise 8-1** (a) Show that for the states (8-40)

$$\Lambda_+\phi^R = \sqrt{2}\,\phi^R, \qquad \Lambda_+\phi^L = 0, \qquad \Lambda_-\phi^R = 0, \qquad \Lambda_-\phi^L = \sqrt{2}\,\phi^L. \tag{8-43}$$

(b) Since $\gamma_5{}^2 = 1$, show that

$$\Lambda_\pm^2 = \sqrt{2}\,\Lambda_\pm \qquad \text{and} \qquad [\Lambda_\pm, \Lambda_\mp] = 0. \tag{8-44}$$

(c) Using the γ-matrices from Appendix C, show that

$$L_K\Lambda_- = \Lambda_+ L_K, \quad K = \mathrm{S, T, P}; \qquad L_K\Lambda_- = \Lambda_- L_K, \quad K = \mathrm{V, A}. \tag{8-45}$$

Hence also show that

$$\begin{aligned}\sqrt{2}L_K\Lambda_- &= \Lambda_+ L_K \Lambda_-, \quad K = \mathrm{S, T, P}\\ \sqrt{2}L_K\Lambda_- &= \Lambda_- L_K \Lambda_-, \quad K = \mathrm{V, A}.\end{aligned} \tag{8-46}$$ ◀

Returning to (8-36), we note that the factor $C_K + C_K'\gamma_5$ in the leptonic operator can be written in terms of $\Lambda_\pm$ by defining two new coupling constants C_{KL} and C_{KR}, as

$$C_K + C_K'\gamma_5 = C_{KR}\Lambda_+ + C_{KL}\Lambda_-, \tag{8-47}$$

with

$$\sqrt{2}\,C_{KR} = C_K - C_K', \qquad \sqrt{2}\,C_{KL} = C_K + C_K'. \tag{8-48}$$

[1] It is convenient to use the factor $1/\sqrt{2}$ rather than the more natural $\frac{1}{2}$ to allow for ready correspondence of our coupling constants with those of early parity-conserving theories where *both* helicity states for the neutrino were admitted.

The entire leptonic operator then becomes

$$L_K(C_K+C_K'\gamma_5)t_+{}^{\mathrm{L}} = C_{K\mathrm{R}}L_K\Lambda_+t_+{}^{\mathrm{L}} + C_{K\mathrm{L}}L_K\Lambda_-t_+{}^{\mathrm{L}}. \tag{8-49}$$

Consider β^--decay with the lepton operator (8-49) inserted in (8-36). The term in the lepton operator with Λ_- would, in view of (8-43), contribute a nonzero term to the transition probability by "annihilating" a left-handed neutrino, which in effect corresponds to the "emission" of a right-handed antineutrino, while Λ_+ would "annihilate" a right-handed neutrino and correspond to the "emission" of a left-handed antineutrino. Since we now know that only left-handed neutrinos and right-handed antineutrinos are involved, the term in the transition probability connected with Λ_+ is not present; conveniently, then, $C_{K\mathrm{R}} = 0$.[1]

It has been conclusively established by experimental results that the electrons emitted in β^--decay have left-handed helicity more often than right-handed helicity. Indeed, at elevated relativistic electron energies the state of the emitted electron asymptotically approaches pure left-handed helicity. In view of (8-46), and considering the first operator on the right-hand side of this equation in each instance operating on the final state, only terms involving $\Lambda_-L_K\Lambda_-$ can survive. Thus the β-decay interaction can be considered to be of vector and axial vector character only.

▶**Exercise 8-2** (a) Using (C-32) of Appendix C to define $u^{\mathrm{R,L}}$, show that for nonzero-mass leptons (such as the electron), eigenstates of helicity $h = \pm 1$ and $E > 0$ are

$$h = +1: \qquad \phi^{\mathrm{R}} = \begin{pmatrix} u^{\mathrm{R}} \\ \hdashline [cp/(E+mc^2)]u^{\mathrm{R}} \end{pmatrix} \tag{8-50a}$$

$$h = -1: \qquad \phi^{\mathrm{L}} = \begin{pmatrix} u^{\mathrm{L}} \\ \hdashline -[cp/(E+mc^2)]u^{\mathrm{L}} \end{pmatrix}. \tag{8-50b}$$

(b) In contrast to the massless neutrino states (8-40) for which the results (8-43) obtain, show that for the nonzero-mass states (8-50),

$$\Lambda_-\phi^{\mathrm{R}} = \frac{1}{\sqrt{2}}\left(1 - \frac{cp}{E+mc^2}\right)\begin{pmatrix} u^{\mathrm{R}} \\ \hdashline -u^{\mathrm{R}} \end{pmatrix} \tag{8-51a}$$

and

$$\Lambda_-\phi^{\mathrm{L}} = \frac{1}{\sqrt{2}}\left(1 + \frac{cp}{E+mc^2}\right)\begin{pmatrix} u^{\mathrm{L}} \\ \hdashline -u^{\mathrm{L}} \end{pmatrix}. \tag{8-51b}$$

[1] In the early days of β-decay theory, allowance was made for the possibility of both right-handed and left-handed neutrinos. Experimental evidence then *compelled* $C_{K\mathrm{R}} = 0$.

(c) Repeat part (b) for $\Lambda_+\phi^{\mathrm{R,L}}$. If the experimental observation requires that the emitted electrons for $v/c \to 1$ be purely left-handed in helicity, demonstrate that only the operator $\Lambda_- L_K \Lambda_-$ can survive in the lepton β-decay operator as stated in the text.

(d) If we define the polarization to be

$$P = \frac{(+1)|\Lambda_-\phi^{\mathrm{R}}|^2 + (-1)|\Lambda_-\phi^{\mathrm{L}}|^2}{|\Lambda_-\phi^{\mathrm{R}}|^2 + |\Lambda_-\phi^{\mathrm{L}}|^2},$$

show that

$$P = -cp/E = -v/c. \quad ◀ \tag{8-52}$$

We note that $C_{K\mathrm{R}} = 0$, imposed by the two-component neutrino theory, implies from (8-48) that $C_K = C_K'$. The presence of both C_K and C_K' in (8-36) results in parity violation in the β-decay process. The condition $C_K = C_K'$ is referred to as *maximum parity violation.* If we had both right-handed and left-handed neutrinos (and antineutrinos) in nature and if they were emitted (absorbed) with equal a priori probability resulting in $C_{K\mathrm{R}} = C_{K\mathrm{L}}$ in (8-49), then from (8-48) we see that C_K' would be zero, which in turn leads to (8-36) consisting of only operators that are true scalars and hence conserve parity. Thus the existence of only one helicity type of neutrino ensures parity violation in the β-decay processes.

As a further convenience, we introduce the more universally used coupling constant designation $g_K = C_{K\mathrm{L}} = \sqrt{2}\, C_K$. Taking $g_{\mathrm{S,T,P}} = 0$, we arrive at[1]

$$H_{\mathrm{int}} = 2^{-1/2}\langle \psi_{\mathrm{f}}\phi_{\mathrm{f}}|\delta(\mathbf{r}_{\mathrm{N}}-\mathbf{r}_{\mathrm{L}}) \sum_{\mu=1}^{4} [\{\gamma_4\gamma_\mu(g_{\mathrm{V}}+g_{\mathrm{A}}\gamma_5)\}_{\mathrm{N}}\{\Lambda_-\gamma_4\gamma_\mu\Lambda_-\}_{\mathrm{L}}\, t_+ t_+{}^{\mathrm{L}}] + \mathrm{h.c.}\,|\psi_{\mathrm{i}}\phi_{\mathrm{i}}\rangle. \tag{8-53}$$

▶**Exercise 8-3** (a) Verify (8-53).

(b) Show that

$$2^{-1/2}\sum_{\mu=1}^{4}\{\gamma_4\gamma_\mu(g_{\mathrm{V}}+g_{\mathrm{A}}\gamma_5)\}_{\mathrm{N}}\{\Lambda_-\gamma_4\gamma_\mu\Lambda_-\}_{\mathrm{L}} = g_{\mathrm{V}}[\{I\}_{\mathrm{N}}\{\Lambda_-\}_{\mathrm{L}} - \{\boldsymbol{\alpha}\}_{\mathrm{N}}\cdot\{\boldsymbol{\alpha}\Lambda_-\}_{\mathrm{L}}] + g_{\mathrm{A}}[\{\gamma_5\}_{\mathrm{N}}\{\Lambda_-\}_{\mathrm{L}} - \{\boldsymbol{\sigma}^{\mathrm{D}}\}_{\mathrm{N}}\cdot\{\boldsymbol{\sigma}^{\mathrm{D}}\Lambda_-\}_{\mathrm{L}}], \tag{8-54}$$

where I is the 4×4 identity or unit matrix. ◀

For an important class of β-decay transitions, the lepton wave functions in lowest order can be considered constant over the region of the nuclear interior. In these cases simple selection rules for the nuclear states can be directly inferred from (8-54). Note that in the form (8-54) the nucleon operators I, $\boldsymbol{\alpha}$, γ_5, and $\boldsymbol{\sigma}^{\mathrm{D}}$ appear. The matrices $\boldsymbol{\alpha}$ and γ_5 have only off-diagonal elements (see Appendix C) and therefore connect large and small components of the

[1] The relative signs of g_{V} and g_{A} depend on the forms used for L_{V} and L_{A} in Table 8-1.

nucleon initial and final states. For the nonrelativistic *nucleon* case, the contribution to the transition matrix therefore involves the factor $\{v/c\}_N$ (the approximate ratio of the small to large component). On the other hand both I and $\boldsymbol{\sigma}^D$ have diagonal elements and couple the large components of the nucleon wave functions in the initial and final states. Another important distinction to be made between these two pairs of operators is the fact that both $\{I\}_N$ and $\{\boldsymbol{\sigma}^D\}_N$ couple only initial and final nucleon states of the same parity (i.e., $\{I\}_N$ is a true scalar operator and $\{\boldsymbol{\sigma}^D\}_N$ is an axial vector operator; see Section I, B, 2), while $\{\boldsymbol{\alpha}\}_N$ and $\{\gamma_5\}_N$ couple only initial and final states of opposite parity (i.e., $\{\boldsymbol{\alpha}\}_N$ is a polar vector operator and $P\gamma_5 P^{-1} = -\gamma_5$). Thus, to the order of approximation under discussion here, nonzero contributions to (8-53) occur from either one pair of operators or the other. The nuclear parity-changing transitions involving $\{\boldsymbol{\alpha}\}_N$ and $\{\gamma_5\}_N$ are of order v^2/c^2 ($\sim 10^{-2}$) compared to the nuclear non-parity-changing transitions involving $\{I\}_N$ and $\{\boldsymbol{\sigma}\}_N$, since transition rates are proportional to $|H_{\text{int}}|^2$.

It is also convenient in evaluating the contribution to the transition matrix by the term[1] $\{\boldsymbol{\sigma}^D\}_N \cdot \{\boldsymbol{\sigma}^D\}_L$ to use spherical tensor components (see Exercise 1-9) and write

$$\{\boldsymbol{\sigma}\}_N \cdot \{\boldsymbol{\sigma}\}_L = \sum_{\mu = -1, 0, +1} (-1)^{\mu} \{\sigma_{\mu}\}_N \{\sigma_{-\mu}\}_L, \tag{8-55}$$

with $\sigma_{\pm 1} = \mp(1/\sqrt{2})(\sigma_x \pm i\sigma_y)$ and $\sigma_0 = \sigma_z$.[2] The operators $\sigma_{\pm 1}$ then allow coupling initial and final states with M_J differing by one unit, or in general $(M_J)_f = (M_J)_i + \mu$.

Thus, keeping only the terms in $\{I\}_N$ and $\{\boldsymbol{\sigma}^D\}_N$ in (8-54) and (8-53), we obtain the so-called *allowed β-decay transitions* that have the following nuclear selection rules:

$$\text{vector interaction:} \quad \Delta J = 0, \quad \Delta M_J = 0, \quad \Delta\pi = \text{no} \tag{8-56a}$$

$$\text{axial vector interaction:} \quad \Delta J = 0, \pm 1 \text{ (except no } 0 \to 0), \quad \Delta M_J = 0, \pm 1, \quad \Delta\pi = \text{no}. \tag{8-56b}$$

For historical reasons vector interactions are also referred to as *Fermi interactions*, and the axial vector interactions are also referred to as *Gamow-Teller interactions.*

[1] While for the lepton states the full 4×4 matrix representation of $\boldsymbol{\sigma}^D$ must be used, for the nonrelativistic nucleon we have

$$\lim_{u_s \to 0} \left\{ (u_{lf}^* \,\vdots\, u_{sf}^*) \begin{pmatrix} \boldsymbol{\sigma}^P & 0 \\ 0 & \boldsymbol{\sigma}^P \end{pmatrix} \begin{pmatrix} u_{li} \\ u_{si} \end{pmatrix} \right\} \to u_{lf}^* \boldsymbol{\sigma}^P u_{li},$$

where the u's are appropriate Pauli spinors.

[2] As in Chapter I, we distinguish between raising and lowering operators $\sigma_{\pm} = \sigma_x \pm i\sigma_y$ and the spherical components $\sigma_{\pm 1}$.

In the procedure outlined above, (8-53) is generally applied using model-inspired nuclear wave functions, and meson-exchange effects and other elementary particle aspects of the nucleons are ignored. Formulations paralleling (8-53) are possible that do consider the full elementary particle aspects of the problem. Essentially equivalent results are obtained by both methods for allowed and certain forbidden transitions. For an illuminating, albeit advanced, discussion of these points, refer to Armstrong and Kim (1972).

The approximation that we have been using in the above, that lepton fields are constant over the nuclear interior, corresponds to the emitted electron and antineutrino pair being treated as an S-wave. Thus the orbital angular momentum involved in the lepton current is $l = 0$. This allows a straightforward interpretation of rules (8-56) for such allowed transitions. The parity selection rule immediately follows, since $\pi_i \pi_f = (-1)^l = +1$, and hence $\Delta\pi =$ no. The electron and antineutrino spins can be added to give either $s = 0$ (the singlet spin state) or $s = 1$ (the triplet spin state). The first state, $s = 0$, is formed in vector coupling, while the second state, $s = 1$, is formed in axial vector coupling. Hence it follows that for the Fermi interaction, $\Delta J \leqslant j = l + s = 0$, while for the Gamow–Teller interaction, $\Delta J \leqslant j = l + s = 1$ except no $0 \to 0$ (the triangle condition for $J_f = J_i = 0$ and $j = 1$ cannot be satisfied in this case).

▶**Exercise 8-4** In Appendix C it is convenient to refer to electron Dirac spinors as spin states "α and β" even though these states *are not eigenspinors* of $\boldsymbol{\sigma}_z^{\mathrm{D}}$. The convenience arises from the fact that in the low-energy, nonrelativistic limit they are equivalent to the Pauli states α and β. It is also instructive to examine the lepton current in β-decay in the nonrelativistic limit. We then specify the Pauli spinors for the electron and neutrino as

$$\phi_e = \begin{pmatrix} e_1 \\ e_2 \end{pmatrix} = e_1 \begin{pmatrix} 1 \\ 0 \end{pmatrix} + e_2 \begin{pmatrix} 0 \\ 1 \end{pmatrix} = e_1 \alpha + e_2 \beta$$

and

$$\phi_\nu = \begin{pmatrix} \nu_1 \\ \nu_2 \end{pmatrix} = \nu_1 \begin{pmatrix} 1 \\ 0 \end{pmatrix} + \nu_2 \begin{pmatrix} 0 \\ 1 \end{pmatrix} = \nu_1 \alpha + \nu_2 \beta.$$

For β^--decay, the leading term in (8-54) for the Fermi interaction (vector interaction) yields the lepton field $\phi_e^{\dagger}\phi_\nu$. Noting (8-55), the Gamow–Teller (axial vector interaction) term gives the current components $\phi_e^{\dagger}(-1)^\mu \sigma_\mu \phi_\nu$. In the present context we can replace the operator Λ_- with I_2, the Pauli unit matrix. By use of the Pauli operator

$$B = \begin{pmatrix} 0 & 1 \\ -1 & 0 \end{pmatrix}$$

we can also write these fields in terms of the *two emitted particles*, namely the electron and antineutrino. These are $\phi_e^\dagger\phi_{\bar{\nu}}$ and $\phi_e^\dagger(-1)^\mu\sigma_\mu\phi_{\bar{\nu}}$, respectively, with $\phi_{\bar{\nu}} = B\phi_\nu^*$ [see Blatt and Weisskopf (1952, p. 697ff.)].

TABLE 8-2

LEPTON FIELDS

L_K	Neutrino initial state	Antineutrino final state
1	$e_1^*\nu_1 + e_2^*\nu_2$	$e_1^*\nu_2^* - e_2^*\nu_1^*$
$-\sigma_{+1}$	$\sqrt{2}\,e_1^*\nu_2$	$\sqrt{2}\,e_1^*\nu_1^*$
$+\sigma_0$	$e_1^*\nu_1 - e_2^*\nu_2$	$e_1^*\nu_2^* + e_2^*\nu_1^*$
$-\sigma_{-1}$	$\sqrt{2}\,e_2^*\nu_1$	$\sqrt{2}\,e_2^*\nu_2^*$

Show that Table 8-2 holds for the lepton fields. It is evident that for the two emitted particles the axial vector components imply a spin triplet ($s = 1$) state, while for the vector coupling we have a singlet spin ($s = 0$) state [see Chapter I, Eq. (1-12)]. ◀

3. Leptonic Matrix Elements

While the most general evaluation of (8-53) is rather laborious, the case in which both the initial and final nuclear states are $J^\pi = 0^+$ is both simple and instructive. Again we shall consider the case of β^--decay. Without any loss in generality we can take the antineutrino to be emitted along the z axis and the electron to be emitted in the xz plane with a polar angle θ (see Fig. 8.3).

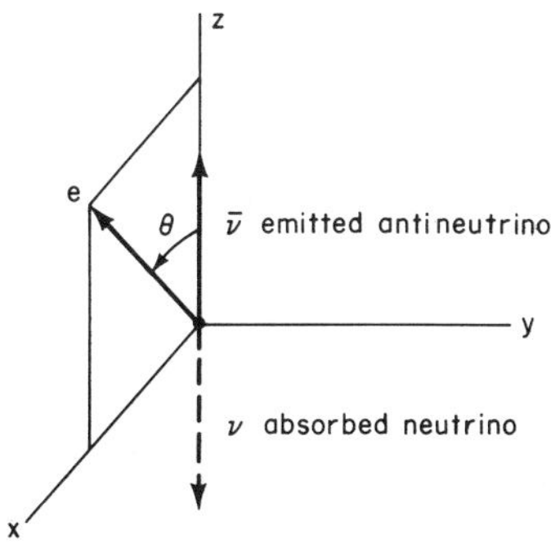

FIG. 8.3. The momentum relationships in β^--decay. The emitted antineutrino can be viewed as an absorbed neutrino with opposite momentum.

The appropriate lepton field for the *initial* state is given by the charge conjugate state of an antineutrino with $E_{\bar{\nu}} > 0$, $h = +1$, and momentum $\mathbf{q}$. The appropriate time-independent wave function is [see (8-41) and Appendix C]

$$\phi_i = \left\{ 2^{-1/2} \begin{pmatrix} 1 \\ 0 \\ 1 \\ 0 \end{pmatrix} \exp[(i/\hbar)(\mathbf{q}\cdot\mathbf{r}_L)] \right\}_C = 2^{-1/2} \begin{pmatrix} 0 \\ -1 \\ 0 \\ 1 \end{pmatrix} \exp[-(i/\hbar)(\mathbf{q}\cdot\mathbf{r}_L)]. \tag{8-57}$$

This wave function is seen to be that of a massless lepton with $E < 0$, $h = -1$, and momentum $-\mathbf{q}$ [see (8-41)], or that of a suitable neutrino. The final-state lepton wave function is that for the electron with either of two possible spin orientations. Since in evaluating H_{int} we must perform the incoherent sum over spin states, and no polarization of a $J = 0$ nucleus is possible, we can express the two required spin states either as the "α and β" states of (C-26) or the $h = \pm 1$ helicity states of (C-33) (see Appendix C). Using the α and β states, we have

$$\phi_f^{\alpha} = [(E+mc^2)^{1/2}/(2E)^{1/2}] \begin{pmatrix} 1 \\ 0 \\ cp(\cos\theta)/(E+mc^2) \\ cp(\sin 0)/(E+mc^2) \end{pmatrix} \exp[(i/\hbar)(\mathbf{p}\cdot\mathbf{r}_L)] \tag{8-58a}$$

$$\phi_f^{\beta} = [(E+mc^2)^{1/2}/(2E)^{1/2}] \begin{pmatrix} 0 \\ 1 \\ cp(\sin\theta)/(E+mc^2) \\ -cp(\cos\theta)/(E+mc^2) \end{pmatrix} \exp[(i/\hbar)(\mathbf{p}\cdot\mathbf{r}_L)]. \tag{8-58b}$$

Here, as in (8-57), $\mathbf{r}_L$ is the lepton coordinate, E refers to the *electron* total energy and m to its rest mass, and $\mathbf{q}$ and $\mathbf{p}$ refer to the antineutrino and electron momenta, respectively. The angle θ is shown in Fig. 8.3. At first we use the approximation of a plane wave for the electronic wave function; at a later point we include the distorting effect of the Coulomb force on the electron by the nuclear charge. We should also note that the role of the leptonic raising operator t_+^{L} in (8-53) is in this instance automatically taken into account by having taken the initial state to be that of a neutrino and the final state that of an electron.

Thus we have, using (8-53) and the large term in (8-54), if the available nuclear energy permits only β^--decay,

$$H^\alpha = g_V\left(\frac{E+mc^2}{2E}\right)^{1/2} H_N\left(\frac{cp\sin\theta}{E+mc^2}\right)$$
$$H^\beta = -g_V\left(\frac{E+mc^2}{2E}\right)^{1/2} H_N\left(1+\frac{cp\cos\theta}{E+mc^2}\right), \tag{8-59}$$

with

$$H_N = \langle\psi_f(\mathbf{r}_1,\mathbf{r}_2,\ldots,\mathbf{r}_A)|\sum_{\xi=1}^{A}\{\exp[-(i/\hbar)(\mathbf{p}+\mathbf{q})\cdot\mathbf{r}_\xi]\}\,t_+^{\xi}\,|\psi_i(\mathbf{r}_1,\mathbf{r}_2,\ldots,\mathbf{r}_A\rangle. \tag{8-60}$$

In (8-60) we finally exhibit explicitly the fact that we must sum over *all the nucleons* in the nucleus since any one of them might have participated in the elementary β-decay process (8-2). The operator t_+^{ξ} ensures the fact that *only neutrons* in the initial nucleus contribute to β^--decay. The exponential factor can be expanded, viz.,

$$\exp[-(i/\hbar)(\mathbf{p}+\mathbf{q})\cdot\mathbf{r}_\xi] = 1-(i/\hbar)(\mathbf{p}+\mathbf{q})\cdot\mathbf{r}_\xi+\cdots. \tag{8-61}$$

Hence we can write

$$H_N = \int I - (i/\hbar)(\mathbf{p}+\mathbf{q})\cdot\int\mathbf{r}+\cdots, \tag{8-62}$$

with

$$\int I = \langle\psi_f(\mathbf{r}_1,\mathbf{r}_2,\ldots,\mathbf{r}_A)|\sum_{\xi=1}^{A} t_+^{\xi}|\psi_i(\mathbf{r}_1,\mathbf{r}_2,\ldots,\mathbf{r}_A)\rangle, \tag{8-63a}$$

and

$$\int \mathbf{r} = \langle\psi_f(\mathbf{r}_1,\mathbf{r}_2,\ldots,\mathbf{r}_A)|\sum_{\xi=1}^{A} \mathbf{r}_\xi t_+^{\xi}|\psi_i(\mathbf{r}_1,\mathbf{r}_2,\ldots,\mathbf{r}_A)\rangle. \tag{8-63b}$$

Generally for Q values of a few MeV, the second term in (8-61) gives $|(p+q)R/\hbar|\approx 10^{-2}$, and consequently H_N can be represented by the quantity $\int I$ of (8-63a). The selection rules (8-56) apply for this case. The quantity $\int\mathbf{r}$ of (8-63b) involves $Y_1^m(\theta_\xi,\varphi_\xi)$ and corresponds to the electron and anti-neutrino pair being emitted in a P-wave. When this term is to be considered, only states ψ_f and ψ_i of opposite parity give nonzero integrals, since $\Delta\pi=(-1)^l=-1$ for this case.

The transition probability in lowest order is determined by

$$|H_{if}|^2 = |H^\alpha|^2+|H^\beta|^2 = |g_V|^2\left|\int I\right|^2[1+(v/c)\cos\theta]. \tag{8-64}$$

▶**Exercise 8-5** (a) Verify Eqs. (8-57)–(8-64).

(b) Show that (8-64) results if the spin sum is taken over electron states of helicity $h = \pm 1$. We note that (8-64) predicts that an electron antineutrino correlation exists favoring small angles between the pair, even though the emission of either electron or antineutrino, if detected alone, would be isotropic. A similar, although much more complicated, derivation for the more general case of S, V, T, A, and P interactions simultaneously acting (as permitted by selection rules) shows that the electron antineutrino or positron neutrino correlation can also be written in the form

$$1 + a(v/c)\cos\theta. \tag{8-65}$$

As we have seen in (8-64), the *anisotropy coefficient* a is $+1$ for a pure V interaction. When possible interference effects between the several interactions are included, $-1 < a < +1$.

Although the coincidence measurement between the electron and antineutrino required to establish the angular correlation (8-65) is not feasible, the electron recoil-atom correlation from which (8-65) can be deduced has been experimentally determined. Experiments with the noble gas sources ^{6}He, ^{19}Ne, ^{23}Ne, and ^{35}Ar, and the β-decay of the neutron clearly indicate a pure (V, A) interaction. See Wu and Moszkowski (1966, Sections 3–4) or Schopper (1966, Chapter 5). ◀

▶**Exercise 8-6** The classic experiment that demonstrated parity nonconservation in nuclear β-decay involved the electron emission of the ground state of spin-aligned ^{60}Co$(J^\pi = 5^+)$ decaying to the 2.506-MeV excited state (99$^+$% branch) of ^{60}Ni$(J^\pi = 4^+)$. Such $J \to J-1$, $\Delta\pi = 0$ transitions are pure Gamow–Teller or axial vector transitions. A simple derivation for this case of β-decay for extreme relativistic electrons (i.e., $E^2 \approx c^2p^2$) is readily given. Consider the case for the aligned initial nuclear state $|J_i, M_i = J_i\rangle$ and the geometry shown in Fig. 8.4.

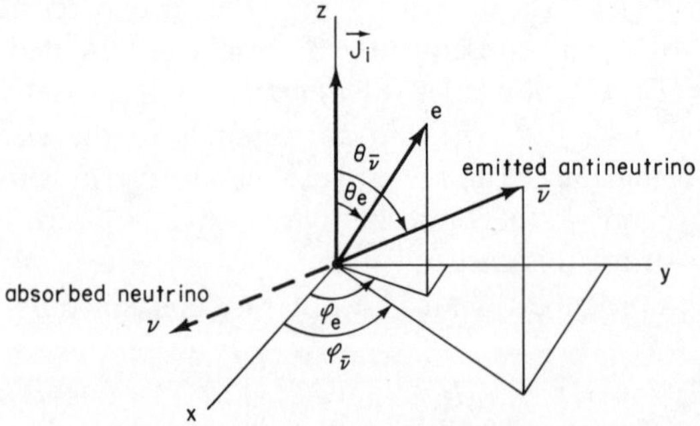

FIG. 8.4. The momentum relationships in β^--decay for an oriented nucleus. The emitted antineutrino can be viewed as an absorbed neutrino with opposite momentum.

(a) Using the leading axial vector interaction term in (8-54), show that (8-53) for β^--decay can be written

$$H_{\text{int}} = 2^{-1/2} g_{\text{A}} \left\{ \int \sigma_{-1} \right\}_{\text{N}} \langle \phi_{\text{f}} | \Lambda_- \sigma_{+1}^{\text{D}} \Lambda_- | \phi_{\text{i}} \rangle. \tag{8-66}$$

(b) Show that in view of (8-66), the sum over the final electron helicity states involves only (note we are taking $E^2 \approx p^2c^2$)

$$\phi_{\text{f}} = 2^{-1/2} \begin{pmatrix} -[\exp(-i\varphi_{\text{e}})] \sin \frac{1}{2}\theta_{\text{e}} \\ \cos \frac{1}{2}\theta_{\text{e}} \\ [\exp(-i\varphi_{\text{e}})] \sin \frac{1}{2}\theta_{\text{e}} \\ -\cos \frac{1}{2}\theta_{\text{e}} \end{pmatrix} \exp[(i/\hbar)(\mathbf{p} \cdot \mathbf{r}_{\text{L}})]. \tag{8-67}$$

(c) Show that the proper initial lepton state is given by the charge conjugate state of an antineutrino with $E_{\bar{\nu}} > 0$, $h = +1$, and momentum $\mathbf{q}$, namely,

$$\phi_{\text{i}} = 2^{-1/2} \begin{pmatrix} [\exp(-i\varphi_{\bar{\nu}})] \sin \frac{1}{2}\theta_{\bar{\nu}} \\ -\cos \frac{1}{2}\theta_{\bar{\nu}} \\ [\exp(-i\varphi_{\bar{\nu}})] \sin \frac{1}{2}\theta_{\bar{\nu}} \\ \cos \frac{1}{2}\theta_{\bar{\nu}} \end{pmatrix} \exp[-(i/\hbar)(\mathbf{q} \cdot \mathbf{r}_{\text{L}})]. \tag{8-68}$$

(d) Evaluate the square of the absolute value of (8-66) and show that the result is

$$|H_{\text{if}}|^2 = |g_{\text{A}}|^2 \left| \int \sigma_{-1} \right|^2 (1 - \cos\theta_{\text{e}})(1 + \cos\theta_{\bar{\nu}}), \tag{8-69}$$

with

$$\int \sigma_{-1} = \left\langle \psi_{\text{f}}(\mathbf{r}_1, \mathbf{r}_2, \ldots, \mathbf{r}_A) \middle| \sum_{\xi=1}^{A} \sigma_{-1}^{\xi} t_+^{\xi} \middle| \psi_{\text{i}}(\mathbf{r}_1, \mathbf{r}_2, \ldots, \mathbf{r}_A) \right\rangle, \tag{8-70}$$

if in lowest order we again take $\exp[-(i/\hbar)(\mathbf{p}+\mathbf{q}) \cdot \mathbf{r}_{\text{L}}] \approx 1$.

The generalized matrix element $|\int \boldsymbol{\sigma}|^2$ is defined as

$$\left| \int \boldsymbol{\sigma} \right|^2 = \sum_{(M_J)_{\text{f}}} \sum_{\mu} \sum_{\xi} |\langle \psi_{\text{f}} | \sigma_{\mu}{}^{\xi} t_{\pm}{}^{\xi} | \psi_{\text{i}} \rangle|^2. \tag{8-71}$$

For a specific $(M_J)_{\text{i}}$ only the particular terms $(M_J)_{\text{f}} = (M_J)_{\text{i}} + \mu$ appear in the sum. Further, it can be shown that (8-71) is independent of $(M_J)_{\text{i}}$.

In the present case we have $(M_J)_{\text{i}} = J_{\text{i}}$ coupled only to $(M_J)_{\text{f}} = (M_J)_{\text{i}} - 1$, since $J_{\text{f}} = J_{\text{i}} - 1$; thus

$$\left| \int \boldsymbol{\sigma} \right|^2 = \left| \int \sigma_{-1} \right|^2. \tag{8-72}$$

When the nuclear orientation is not complete, so that states other than $(M_J)_i = J_i$ are partially populated, the expression (8-69) becomes

$$|H_{if}|^2 = |g_A|^2 \left| \int \sigma \right|^2 [1 - \cos\theta_e \cos\theta_{\bar{\nu}} - (\langle J_z \rangle_i / J_i)(\cos\theta_e - \cos\theta_{\bar{\nu}})]. \tag{8-73}$$

If only the angular distribution of the electrons is observed, we must average over the antineutrino direction of emission. Noting that $\langle \cos\theta_{\bar{\nu}} \rangle = 0$, we have, for the angular distribution of the electrons in the extreme relativistic case,

$$I(\theta_e) = 1 - (\langle J_z \rangle_i / J_i) \cos\theta_e. \tag{8-74}$$

Finally, for the case that $v/c < 1$, a parallel (although more complicated) derivation to (8-74) can be shown to give

$$I(\theta_e) = 1 - (\langle J_z \rangle_i / J_i)(v/c) \cos\theta_e. \quad ◀ \tag{8-75}$$

We note that the result (8-74) or (8-75) in Exercise 8-6 leads to an anisotropic angular distribution for the emitted electron relative to the nuclear spin orientation, and in effect corresponds to a nonzero value for $\langle \mathbf{J}_i \cdot \mathbf{p} \rangle$. Since this latter quantity is the expectation value of a pseudoscalar, conservation of parity would require it to be zero, yielding an isotropic distribution instead. Thus the verification of (8-75) would, in fact, confirm parity nonconservation in β-decay processes. It is illuminating to perform the more involved calculation paralleling Exercise 8-6, but keeping the constants C_A and C_A' unspecified in (8-36) and in subsequent derivations. The result then becomes

$$I(\theta_e) = 1 + A(\langle J_z \rangle_i / J_i)(v/c) \cos\theta_e \tag{8-76a}$$

with

$$A = -2[\mathrm{Re}(C_A{}^* C_A')]/(|C_A|^2 + |C_A'|^2). \tag{8-76b}$$

Clearly, if the non-parity-conserving factor C_A' in the interaction Hamiltonian were zero, A would vanish and $I(\theta_e)$ would indeed result in isotropy.

In a transition with $\Delta J = 1$ and $\Delta M_J = 1$, the electron and antineutrino pair is in a parallel spin state. Since the antineutrino helicity is $h = +1$, while for the fast electron the helicity is preferentially $h = -1$, the angular distribution of the two leptons must be shifted by 180°. Thus with the electron distribution being backward peaked as per (8-74) or (8-75), the antineutrino distribution should be forward peaked. Indeed, we note that, averaging over the electron direction, with $\langle \cos\theta_e \rangle = 0$, (8-73) generates an antineutrino distribution similar to (8-74), however, with a positive sign for the cosine term.

The classic experiment that first indicated $A = -1$ (i.e., $C_A = C_A'$) was performed by Wu *et al.* (1957) using oriented ^{60}Co β^--radioactive nuclei. The paramagnetic cobalt atoms were aligned in a magnetic field at low temperature ($T \approx 0.01°K$), and the atomic magnetic field at the nucleus in turn aligned the nuclei. The low-temperature nuclear alignment was determined to be $\langle J_z \rangle_i / J_i = 0.65$. Pulse height selection on the electron scintillation detector set $v/c \approx 0.6$. The observed anisotropy of 0.25 when electron backscattering corrections were applied yielded $A = -1$ in (8-76a).

C. THE β-SPECTRUM AND DECAY RATES

The matrix elements $|H_{if}|^2$ can be converted into radioactive decay probabilities by use of the well-known "Golden Rule,"

$$N(E)\, dE = (2\pi/\hbar)|H_{if}|^2 \rho(E), \tag{8-77}$$

with $\rho(E)$ equal to the density of final states available to the system. In the following discussion it is a great convenience to use the approximation of an infinitely heavy nucleus. Such a nucleus, while contributing whatever recoil momentum is required by the leptons, carries away essentially no energy.

The number of states available to an electron (again considering β^--decay) in the momentum interval p to $p+dp$, per unit volume, is[1]

$$dN_e = 4\pi p^2\, dp/h^3, \tag{8-78a}$$

and similarly the number of states available to an antineutrino in the momentum interval q to $q+dq$, per unit volume, is[1]

$$dN_{\bar{\nu}} = 4\pi q^2\, dq/h^3. \tag{8-78b}$$

The density of final states is

$$\rho(E) = \frac{dN}{dE_0} = \frac{dN_e\, dN_{\bar{\nu}}}{dE_0} = \frac{16\pi^2}{h^6} p^2 q^2 \frac{dp\, dq}{dE_0} \tag{8-79}$$

with $E_0 = E_e + E_{\bar{\nu}} + E_R$. In our approximation, we take $E_R = 0$ and for simplicity drop the subscript on E_e; hence we write $E_0 = E + E_{\bar{\nu}}$ for the sum of the total lepton energies including their rest masses.

Since the careful examination of experimental electron spectra constitutes an important measure of the antineutrino (neutrino) mass, we shall distinguish by m_e and $m_{\bar{\nu}}$ the electron and antineutrino masses. For the antineutrino then,

$$q^2c^2 + m_{\bar{\nu}}^2 c^4 = E_{\bar{\nu}}^2 = (E - E_0)^2,$$

or

$$qc^2\, dq = (E_0 - E)\, dE_0$$

[1] The sum over spin states is included in $|H_{if}|^2$.

and

$$q^2 \frac{dq}{dE_0} = \frac{E_0 - E}{c^2} q = \frac{E_0 - E}{c^3} [(E_0 - E)^2 - m_{\bar{\nu}}^2 c^4]^{1/2}. \tag{8-80}$$

Hence

$$\rho(E) = (16\pi^2/c^3 h^6) p^2 (E_0 - E) [(E_0 - E)^2 - m_{\bar{\nu}}^2 c^4]^{1/2} \, dp. \tag{8-81}$$

It is useful to write $\rho(E)$ in terms of electron energy alone; hence we use $pc^2 \, dp = E \, dE$ and arrive at

$$N(\varepsilon) \, d\varepsilon = (m_e{}^5 c^4 / 2\pi^3 \hbar^7) |H_{if}|^2 \varepsilon (\varepsilon_0 - \varepsilon)(\varepsilon^2 - 1)^{1/2} [(\varepsilon_0 - \varepsilon)^2 - \mu^2]^{1/2} \, d\varepsilon, \tag{8-82}$$

where $\varepsilon = E/m_e c^2$, $\varepsilon_0 = E_0/m_e c^2$, and $\mu = m_{\bar{\nu}}/m_e$.

A suitable spectrum for attempting to determine μ is the triton decay spectrum. Figure 8.5 shows $N(\varepsilon)$ for various values of μ, holding the *electron*

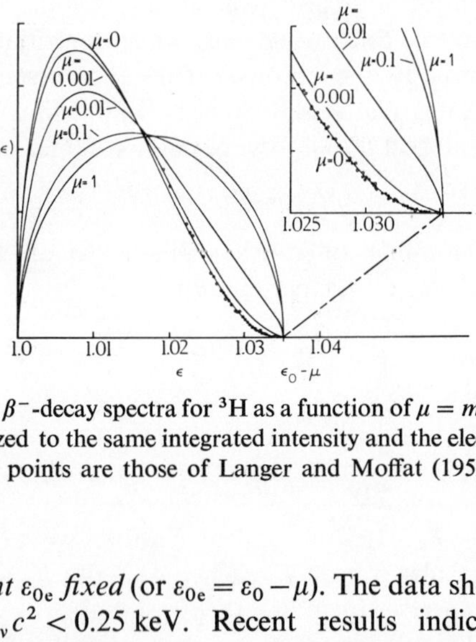

FIG. 8.5. Theoretical β^--decay spectra for ^3H as a function of $\mu = m_{\bar{\nu}}/m_e$. In each instance all spectra are normalized to the same integrated intensity and the electron endpoint energy is held fixed. The data points are those of Langer and Moffat (1952) [figure taken from Konopinski (1966)].

total energy endpoint ε_{0e} *fixed* (or $\varepsilon_{0e} = \varepsilon_0 - \mu$). The data shown in Fig. 8.5 are consistent with $m_\nu c^2 < 0.25$ keV. Recent results indicate $m_\nu c^2 < 60$ eV (Particle Data Group, 1973). For an interesting cosmological argument placing an upper limit of 8 eV on the neutrino rest mass, see Cowsik and McClelland (1972). For some other interesting speculations for nonzero-mass neutrinos, see Bahcall *et al.* (1972) and Pakvasa and Tennakone (1972) and the references cited therein. In all that follows we take $\mu = 0$ and $\varepsilon_{0e} = \varepsilon_0$, whence (dropping the subscript e on m_e)

$$N(\varepsilon) = (m^5 c^4 / 2\pi^3 \hbar^7) |H_{if}|^2 \varepsilon (\varepsilon_0 - \varepsilon)^2 (\varepsilon^2 - 1)^{1/2}. \tag{8-83}$$

We can further write for the maximum β-kinetic energy $E_\beta(\max) = Q = E_0 - mc^2$, and $\varepsilon_0 = (Q/mc^2)+1$.

An important effect so far ignored is the action of the Coulomb force on the electron (or positron). This effect is particularly important at low electron (positron) energies and for high-Z nuclei. For most purposes a nonrelativistic treatment is accurate enough and results in a multiplicative factor $F(Z,\varepsilon)$ appearing in (8-83), viz.,

$$N(\varepsilon) = (m^5c^4/2\pi^3\hbar^7)\,|H_{\text{if}}|^2 F(Z,\varepsilon)\,\varepsilon(\varepsilon_0-\varepsilon)^2(\varepsilon^2-1)^{1/2} \tag{8-84a}$$

with[1]

$$F(Z,\varepsilon) \approx x/(1-e^{-x}), \tag{8-84b}$$

and

$$x = \mp 2\pi Zc/137v \qquad \text{for} \quad \beta^{\pm}\text{-decay}. \tag{8-84c}$$

As cited in the introduction to this chapter, ^{64}Cu is a nucleus that undergoes both electron and positron decay and therefore affords a good illustration

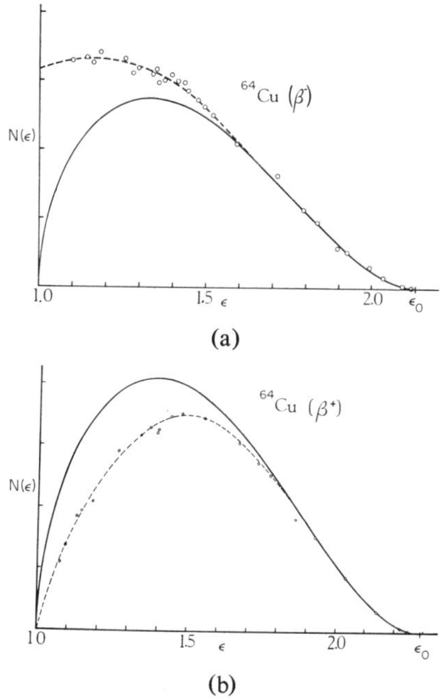

FIG. 8.6. The $\beta^{\pm}$-spectra of ^{64}Cu. The solid curves in each case are (8-83) fitted to the high-energy data points. The dashed curves include the factor $F(Z,\varepsilon)$ of (8-84) to correct for Coulomb effects [taken from Konopinski (1966)].

[1] This estimate also ignores the screening effect of the atomic electrons, which introduce important corrections at the very low-energy end of the β-spectrum.

of the effects of $F(Z, \varepsilon)$. Figure 8.6 clearly shows the *decelerating effect* of the nuclear Coulomb field on electrons which results in *more low-energy* electrons appearing in the spectrum than for $Z = 0$. The positrons are, of course, *accelerated* and hence have *fewer low-energy* positrons.

Since for allowed transitions, $|H_{if}|^2$ is energy independent, (8-84a) is conveniently rewritten as

$$\{N(\varepsilon)/F(Z,\varepsilon)\,\varepsilon(\varepsilon^2-1)^{1/2}\}^{1/2} = \text{const} \times (\varepsilon_0 - \varepsilon). \tag{8-85}$$

A plot of the data $N(\varepsilon)$, according to the left-hand side of (8-85), is referred to as a *Kurie plot* and is seen to result in a straight line as a function ε with an ε intercept of ε_0. Figure 8.7 shows the data of Fig. 8.6 replotted according to (8-85). The linear behavior is manifest.

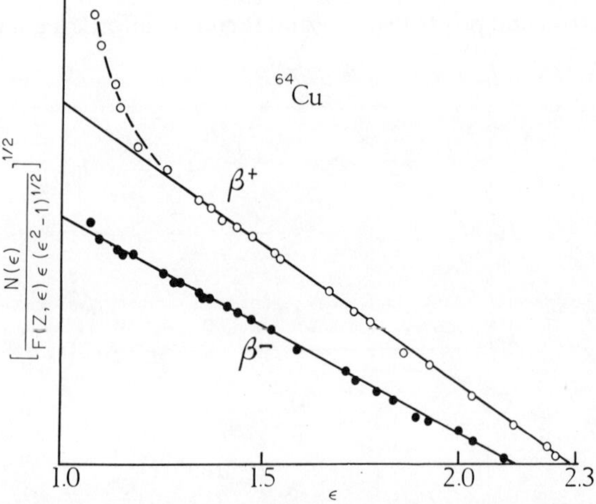

FIG. 8.7. The Kurie plots for ^{64}Cu [taken from Konopinski (1966)].

In β-transitions of the type $J_i \to J_f = J_i$ (but $J_i \neq 0$) and $\Delta\pi = $ no, both the Fermi and Gamow–Teller couplings can contribute to $|H_{if}|^2$. While in certain important polarization and correlation experiments interference terms involving $g_V g_A$ appear, when only the electron spectrum is observed we have for $|H_{if}|^2$ of (8-83) in its most general form the incoherent sum

$$|H_{if}|^2 = |g_V|^2 \left|\int I\right|^2 + |g_A|^2 \left|\int \boldsymbol{\sigma}\right|^2. \tag{8-86}$$

Custom has it to write g_V and g_A in terms of a single universal coupling constant g, viz.,

$$g_V = gC_F, \qquad g_A = gC_{GT}, \tag{8-87}$$

or

$$|H_{\text{if}}|^2 = g^2\left\{|C_{\text{F}}|^2\left|\int I\right|^2 + |C_{\text{GT}}|^2\left|\int \boldsymbol{\sigma}\right|^2\right\}. \tag{8-88}$$

The total transition probability per second, or decay rate, requires integrating over the energy spectrum; thus

$$w = (m^5c^4/2\pi^3\hbar^7)g^2\left\{|C_{\text{F}}|^2\left|\int I\right|^2 + |C_{\text{GT}}|^2\left|\int \boldsymbol{\sigma}\right|^2\right\}f(Z,\varepsilon_0) \tag{8-89a}$$

with [1]

$$f(Z,\varepsilon_0) = \int_1^{\varepsilon_0} F(Z,\varepsilon)(\varepsilon_0-\varepsilon)^2(\varepsilon^2-1)^{1/2}\varepsilon\, d\varepsilon. \tag{8-89b}$$

In terms of the half-life $t_{1/2}$,[2] we define a product, the so-called *ft value*, which removes the kinematic effects from the decay rate and is more reflective of the relevant nuclear matrix elements; thus

$$ft = f(Z,\varepsilon_0)t_{1/2} = f(Z,\varepsilon_0)(\ln 2)/w \tag{8-90a}$$

or

$$\frac{1}{ft} = \frac{1}{(ft)_{\text{F}}} + \frac{1}{(ft)_{\text{GT}}} = \frac{1}{\tau_0}\left\{|C_{\text{F}}|^2\left|\int I\right|^2 + |C_{\text{GT}}|^2\left|\int \boldsymbol{\sigma}\right|^2\right\} \tag{8-90b}$$

with

$$1/\tau_0 = g^2m^5c^4/2\pi^3\hbar^7 \ln 2. \tag{8-90c}$$

Although the preceding discussion was entirely for the case of *allowed* β-transitions, it is customary to define an *ft* value for *any* β-transition, whether allowed or not, once the endpoint energy ε_0 or Q value and the half-life are known, by simply using (8-89b) and (8-90a). This procedure is followed even though forbidden spectral shapes can vary markedly from (8-85). Figure 8.8 shows some calculated but "uncorrected" Kurie plots for unique first, second, and third forbidden β-transitions. In these cases the matrix elements are energy dependent and thus produce the deviations from linearity for Kurie plots based on (8-85). For example, the expression (8-61) to second order generated the term $(i/\hbar)(\mathbf{p}+\mathbf{q})\cdot\int \mathbf{r}$. Transitions of the type $\Delta J = 2$, $\Delta\pi =$ yes

[1] For large values of ε_0 this integral is approximately $\varepsilon_0{}^5/30$. This approximation is the basis for the early historical observation of the so-called "fifth-power law of β-decay" exhibited in "Sargent diagrams" plotting $\log w$ vs. $\log \varepsilon_0$, i.e., the variation $w \sim \varepsilon_0{}^5$ for transitions having similar nuclear matrix elements.

[2] The half-life $t_{1/2}$, the mean life $\bar{t}$, and the decay rate w are simply related as

$$\bar{t} = t_{1/2}/\ln 2 = 1/w.$$

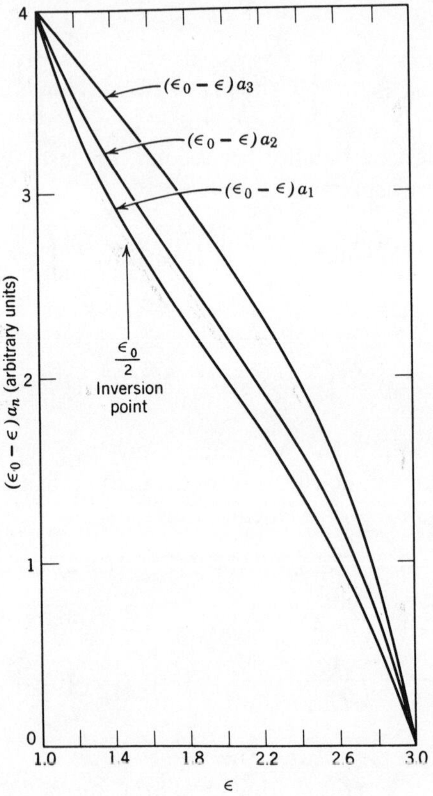

FIG. 8.8. Uncorrected Kurie plots for unique first, second, and third forbidden β-transitions. The curves are calculated for $\varepsilon_0 = 3.0$ [taken from Wu and Moszkowski (1966)].

proceed predominantly by virtue of this interaction term. In this case $|H_{\text{if}}|^2$ gives an additional energy dependence proportional to $|(\mathbf{p}+\mathbf{q})|^2_{\text{av}} = p^2+q^2$ [when we average over the unobserved antineutrino direction, giving $(\mathbf{p}\cdot\mathbf{q})_{\text{av}} = 0$]. In dimensionless units this introduces the shape factor $a_1 = (\varepsilon^2-1)+(\varepsilon_0-\varepsilon)^2$ on the right-hand side of (8-85) and generates the shape labeled $(\varepsilon_0-\varepsilon)a_1$ in Fig. 8.8. When such energy variations are corrected for, linearity for the Kurie plot is regained.

1. Systematics of Empirical Evidence

A typical distribution of ft values calculated by the above procedure of using uncorrected shape values is shown in Fig. 8.9. Among the more rapid or allowed decays $\Delta J = 0, \pm 1$ and $\Delta\pi = \text{no}$, we distinguish a particularly fast group, the so-called "superallowed" transitions $3 < \log ft < 4$, the average allowed group $4 < \log ft \lesssim 6$, and the longer-lived tail of the allowed group designated as "unfavored" with $\log ft \gtrsim 6$. These decays all take place by virtue of the first-order terms in the β-decay matrix elements, and the differences reflect the nature of the nuclear wave functions.

In the next section we shall devote some attention to the behavior of the nuclear matrix elements, but at this point it is useful to single out some accurately determined particular transitions. Table 8-3 lists some selected values for $0^+ \to 0^+$ transitions from the work of Towner and Hardy (1973), Hardy *et al.* (1974), and Clark *et al.* (1973). These, of course, are pure Fermi decays.

Present precision and theoretical interest require that many corrections (generally of order $\lesssim 1\%$) be induced in (8-90) before any fundamental or universal characteristics of β-decay can be examined. The quantity $\bar{f}t$ appearing in Table 8-3 is obtained by altering the definition of f in (8-89b) to include in the integral an energy-dependent shape correction term (deviating from unity by a few percent at most) that includes relativistic effects, a somewhat model-sensitive second forbidden correction depending on $\int \mathbf{r}$, and the

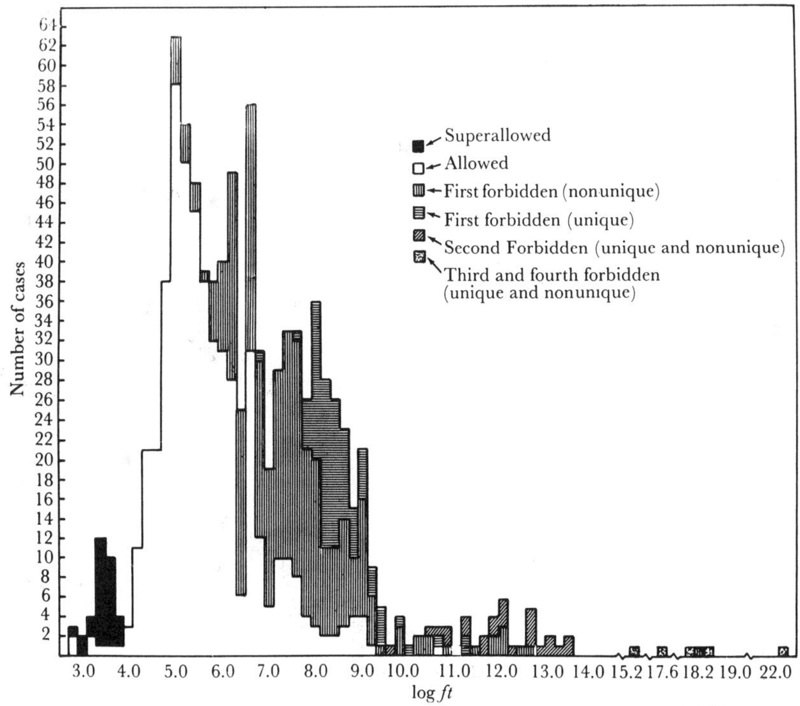

FIG. 8.9. The distribution of ft values as compiled by Gleit *et al.* (1963).

TABLE 8-3

SELECTED ft VALUES FOR $0^+ \to 0^+$, $\Delta\pi = $ no, SUPERALLOWED β-TRANSITIONS[a]

Decaying nucleus	$t_{1/2}$[b] (sec)	E_β(max) (keV)	ft (sec)	$\mathscr{F}t = \bar{f}t(1+\delta_R)(1-\delta_c)$ (sec)
^{14}O	71,085 ± 30	1809.9 ± 0.4	3046 ± 3	3083 ± 3
^{26}Al	6352 ± 5	3211.0 ± 0.5	3040 ± 3	3079 ± 3, 3040 ± 3[c]
^{34}Cl	1532 ± 2	4468.2 ± 1.4	3048 ± 8	3092 ± 8, 3085 ± 9[c]
^{34}Ar	894.7 ± 3.4[c]	5037.2 ± 3.4[c]	3027 ± 17[c]	3078 ± 17[c]
^{42}Sc	684.4 ± 0.9	5409.1 ± 2.3	3074 ± 7	3116 ± 7
^{46}V	426.1 ± 0.7	6032.3 ± 2.2	3077 ± 7	3123 ± 7
^{50}Mn	285.9 ± 0.5	6609.1 ± 2.6	3066 ± 8	3112 ± 8
^{54}Co	193.6 ± 0.6	7227.8 ± 3.8	3064 ± 12	3108 ± 12

[a] Taken from Tower and Hardy (1973) and Clark *et* al. (1973) except where noted otherwise.
[b] Corrected for electron-capture decay branch.
[c] Taken from Hardy *et al.* (1974).

variation of f over the nuclear size. The expression (8-90b) then takes the form

$$\bar{f}t(1+\delta_R) = \tau_0/(|C_F'|^2 |\textstyle\int I|^2 + |C_{GT}'|^2 |\textstyle\int \boldsymbol{\sigma}|^2), \tag{8-90d}$$

where δ_R is the so-called *outer* radiative correction term and $|C_F'|^2$ and $|C_{GT}'|^2$ are *effective* coupling constants. The outer term is an energy-dependent and interaction-independent radiative correction (i.e., independent of the details of the strong interaction and independent of whether the weak interactions are mediated by an intermediate boson), and is given by a universal function that only depends on ε_0. The effective coupling constants can be written

$$|C_F'|^2 = |C_F|^2(1+\Delta_F) \qquad \text{and} \qquad |C_{GT}'|^2 = |C_{GT}|^2(1+\Delta_{GT}).$$

The quantities Δ_F and Δ_{GT} are the so-called *inner* radiative correction terms, which do depend on details of the interactions (e.g., depend on the mass of the intermediate boson, or an equivalent nonlocality cutoff), but are independent of the nuclear wave functions. For the pure Fermi decays the matrix element, assuming that strict analog states are involved, will be shown later to be $|\int I|^2 = 2$. However, this matrix element also requires a correction factor; thus we write

$$\left|\int I\right|^2 = 2(1-\delta_c).$$

The factor δ_c includes a number of effects: charge-dependent effects altering the precise isospin equivalence of the analog nuclear states, "rearrangement" effects of those nucleons that do not participate in the β-decay, etc. The quantity $\mathscr{F}t$ appearing in Table 8-3 is defined as $\bar{f}t(1+\delta_R)(1-\delta_c)$.

While there exist some differences in the literature in treating the calculation of $\bar{f}$ and δ_R, the quantities $\bar{f}t(1+\delta_R)$ and $\mathcal{F}t$ invariably include all the relevant effects. For a more detailed discussion of the above corrections see Sirlin (1967), Blin-Stoyle and Freeman (1970), Int. Conf. (1972, p. 191 ff.), Wilkinson and Marrs (1972), and Towner and Hardy (1973).

Figure 8.10 shows all the corrected values of $\mathcal{F}t$ evaluated by Towner and Hardy and additionally includes a number of less accurately known values than those we selected to give in Table 8-3.

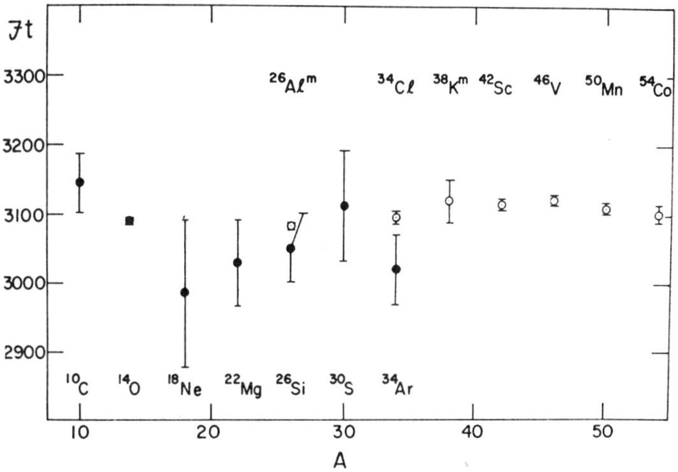

FIG. 8.10. Superallowed $0^+ \rightarrow 0^+$ β-decays. The corrected experimental $\mathcal{F}t$ values for various nuclides [taken from Towner and Hardy (1973)].

While we might quote an average effective experimental $\bar{f}t$ value of (3053 ± 5) sec from Table 8-3, theoretical significance, in view of the previous discussion, should be centered on an average $\mathcal{F}t$ value (particularly for the lighter nuclei where the radiative correction can be made with greater assurance). The above authors suggest a "best" value[1]

$$\mathcal{F}t = (3082.1 \pm 2.1) \text{ sec.} \tag{8-91a}$$

From this value we can deduce, using (8-90), that

$$g_V{}' = gC_F{}' = (1.4129 \pm 0.0005) \times 10^{-49} \text{ erg cm}^3, \tag{8-91b}$$

where the Δ_F correction has *not* yet been applied (hence the primes).

[1] Recent (^{3}He, t) Q-value determinations yield new values of $E_\beta(\max)$ for most of the entries in Table 8-3 [see Hardy *et al.* (1974a)]. The recommended average value of $\mathcal{F}t$ by these authors is nonetheless (3082 ± 3) sec, in good agreement with (8-91a).

Table 8-4 lists some selected values for superallowed β-transitions.

TABLE 8-4

SELECTED ft VALUES FOR SUPERALLOWED $J_i^\pi = J_f^\pi$, β-TRANSITIONS

Decaying nucleus	$t_{1/2}$ (sec)	E_β(max) (keV)	$ft(1+\delta_R)$ (sec)	Reference
n	648 ± 10	782.44 ± 0.07	1114 ± 16	Blin-Stoyle and Freeman (1970), Chemtob and Rho (1971)[a]
^3H	$(3.8695 \pm 0.0013)10^8$	18.7 ± 0.1	1154 ± 10	Chemtob and Rho (1971)[a]
^{15}O	122.24 ± 0.16	1731.8 ± 0.7	4392 ± 8	Ajzenberg-Selove (1970), Wilkinson (1973)
^{17}F	64.50 ± 0.25	1738.0 ± 6	2294 ± 10	Alburger and Wilkinson (1972)
^{39}Ca	0.8604 ± 0.003	5500.0 ± 5	4283 ± 25	Alburger and Wilkinson (1973)
^{41}Sc	0.5963 ± 0.0017	5472.9 ± 2.0	2849 ± 9	Alburger and Wilkinson (1973)

[a] With $\delta_{R,n} = +0.015$ and $\delta_{R,T} = +0.018$.

We note that (8-90d) can be written in the form

$$\frac{2\pi^3\hbar^7 \ln 2}{(g_V')^2 m^5 c^4} = \Lambda' = (\bar{f}t)(1+\delta_R)\left(\left|\int I\right|^2 + \frac{|C_{GT}'|^2}{|C_F'|^2}\left|\int \boldsymbol{\sigma}\right|^2\right). \tag{8-92}$$

In this expression we use the primed quantities to indicate that only outer radiative corrections have been made; thus the coupling constants, as in (8-90d), still have the inner radiative corrections Δ_F and Δ_{GT} to be applied. Also, the corrections corresponding to δ_c must be applied to the matrix elements $\int I$ and $\int \boldsymbol{\sigma}$ before using them in (8-92).

The superallowed transitions of Table 8-4 involve initial and final nuclear states that are believed to be single-particle or single-hole mirror states. For these transitions the matrix elements $|\int I|^2$ and $|\int \boldsymbol{\sigma}|^2$ (with δ_c-type corrections ignored) can be readily deduced (see below). Treating the quantity $\lambda' = C_{GT}'/C_F'$ as a variable in (8-92), we can plot Λ' as a function of $|\lambda'|^2$ and expect a common intersection of the resulting curves. Figure 8.11 shows such a plot in which the average value of $\Lambda' = 2 \times 3082$ for the $0^+ \to 0^+$ transitions is also included. The failure of these curves to intersect at a single point suggests that significant departures from a simple particle or hole state prevail for ^{15}O, ^{17}F, ^{39}Ca, and ^{41}Sc. The average of the $0^+ \to 0^+$ and n (i.e., neutron

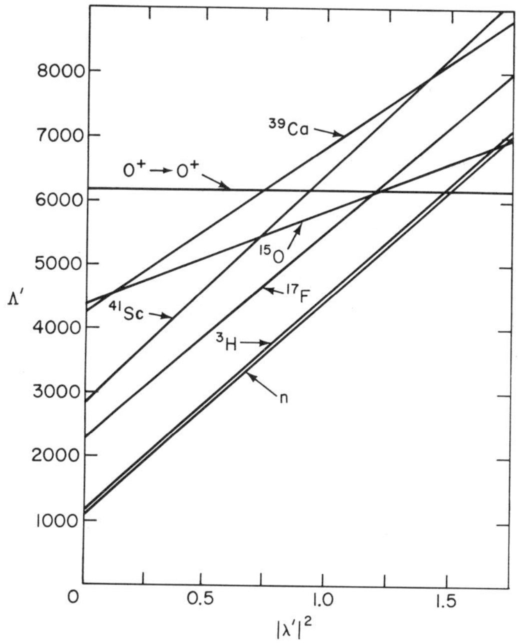

FIG. 8.11. Λ' vs. $|\lambda'|^2$ diagram for selected β-transitions. The common intersection of the curves labeled n and $0^+ \rightarrow 0^+$ determines $|\lambda'|^2 = |C'_{GT}|^2/|C_F'|^2 = 1.51$ and $\Lambda' = 6164$ sec (see text).

decay) intersection gives a value[1]

$$|\lambda'| = 1.23 \pm 0.01, \tag{8-93a}$$

and we also deduce

$$\Lambda' = 6164 \pm 5 \text{ sec.} \tag{8-93b}$$

Electron–neutrino correlations determine the phase of λ. The most recent value (Particle Data Group, 1973) is

$$\lambda = |\lambda| e^{i\eta}$$

with[2]

$$\eta = 1.1° \pm 1.3°. \tag{8-94}$$

[1] In view of the fact that g_A (and hence λ) can require different renormalization corrections in different nuclei, it is prudent to define a reference value for λ based on the neutron decay parameters. Recent literature values quote $|\lambda'| = 1.248 \pm 0.010$ (Particle Data Group, 1973), 1.243 ± 0.011 (Chemtob and Rho, 1971), and 1.226 ± 0.011 (Blin-Stoyle and Freeman, 1970).

[2] Some authors define the tensor and axial vector interactions in Table 8-1 as $O_T = i\gamma_4 \gamma_\mu \gamma_\nu$, $\mu \neq \nu$, and $O_A = i\gamma_4 \gamma_\mu \gamma_5$. This has the effect of changing the sign of the g_A term in (8-53) and in subsequent expressions. Authors using this convention would quote $\eta = 181.1° \pm 1.3°$.

We should note that time-reversal invariance requires $\eta = 0 + n\pi$, $n =$ integer, whence, to within experimental error, (8-94) is in agreement with time-reversal invariance in β-decay processes. The additional minus sign in (8-54) associated with the $\boldsymbol{\sigma}_N \cdot \boldsymbol{\sigma}_L$ term results in describing the situation $\eta = 0°$ as giving a basic V$-$A ("V minus A") β-interaction.

The understanding of the difference of 23% between the values of $|g_V|$ and $|g_A|$ requires a renormalization calculation to account for the strong coupling between nucleons and the virtual pion fields present for "dressed" nucleons. Such calculations assume that without mesonic effects the vector and axial vector coupling constants would, in fact, be equal. The conserved vector current (CVC) theory, as we have discussed earlier, requires that the bare and dressed nucleon vector coupling constants be identical. This is not, however, the case for the axial vector coupling constant. Indeed, calculations indicate that $g_A \approx 1.2 g_V$, in good agreement with the experimental value [see Adler (1965), Weisberger (1965), and Bincer (1966)].

A consequence of the CVC theory is to require the decay of the charged pion to have, in addition to the dominant branch $\pi^+ \to \mu^+ + \nu_\mu$, the weak branch [branching ratio $(1.07 \pm 0.02) \times 10^{-8}$] $\pi^+ \to \pi^0 + e^+ + \nu_e$. This weak branch decay is a $0^+ \to 0^+$ decay, very analogous to the $^{14}O \to {}^{14}N$ β-decay, and results from a pure vector interaction. Again using $|\int I|^2 = 2$, a recent determination gives $\Lambda = (6380 \pm 640)$ sec (Depommier *et al.* 1968), which agrees within experimental error with the nuclear value in (8-93b).

A further requirement of the CVC theory is the equality of the muon decay coupling constant and the nucleon β-decay coupling constant. Since muons do not have strong interactions, we would speculate that $g_V{}^\mu = g_A{}^\mu = g_\mu = g_V$. The $\mu^- \to e^- + \nu_\mu + \bar{\nu}_e$ decay probability is readily calculated in a manner analogous to the β^--decay of the neutron[1] and yields

$$w_\mu = (g_\mu{}^2 c^4 / 192\pi^3 \hbar^7) m_\mu{}^5. \tag{8-95}$$

When the values for the experimental lifetime

$$t_{1/2}(\mu) = (1.5231 \pm 0.0002) \times 10^{-6} \text{ sec}$$

and mass

$$m_\mu c^2 = (105.6595 \pm 0.0003) \text{ MeV}$$

are inserted in (8-95) and small radiative corrections are applied, Duclos *et al.* (1973) give the result

$$g_\mu = (1.43577 \pm 0.00012) \times 10^{-49} \text{ erg cm}^3. \tag{8-96}$$

[1] The analogy becomes striking when the muon decay is written $\mu^- + \nu_e \to \nu_\mu + e^-$ and the neutron decay is written $n + \nu_e \to p + e^-$. Both involve Fermi and Gamow–Teller interactions.

This value is within 1.6% of g_V' in (8-91b). Assuming this discrepancy to be real (the validity of the various applied corrections is not above question), Cabibbo (1963) has put forward an explanation based on the breakdown of $SU^{(3)}$ symmetry that would indeed give $g_V \approx 0.98 g_\mu$. According to this theory, $g_V{}^2 = g_\mu{}^2 \cos^2\theta_V$, where θ_V is referred to as the Cabibbo angle. Combining this expression with $(g_V')^2 = g_V{}^2(1+\Delta_F)$ permits us to determine Δ_F from (8-91b), (8-96), and the independently determined Cabibbo angle $\theta_V^{(KG)} = (0.240 \pm 0.019)$ rad (Klein–Gordon equation) or $\theta_V^{(K)} = (0.197 \pm 0.015)$ rad (Kemmer equation) [see Towner and Hardy (1973)]. The results are $\Delta_F^{(KG)} = (2.6 \pm 1.0)\%$ and $\Delta_F^{(K)} = (0.7 \pm 1.0)\%$. These values are inconsistent with any local weak interaction theory and require a mediating boson. The mass of the W-boson or weakon corresponding to these values can be estimated to be

$$7^{(K)} < m_W c^2 < 300^{(KG)}\ \text{GeV}.$$

A Cabibbo angle that would correspond to $m_W c^2 \approx 30$ GeV, i.e., $\Delta_F \approx 1.4\%$, is $\theta_V \approx 0.214$ rad (when the above values of g_V' and g_μ are used).

The interested reader is referred to Wu and Moszkowski (1966, Chapter 7), Schopper (1966, Chapter 9), Konopinski (1966), or Okun (1965) for further discussions of weak interaction topics and more details on the preceding points at an elementary level.

Finally, a word is in order concerning the appellation "weak" interaction operative in β-decay. The decay rate (8-89a) can be rewritten in terms of a new dimensionless coupling constant $G_\beta{}^2/\hbar c$, viz.,

$$w = \frac{mc^2}{\hbar}\frac{G_\beta{}^2}{\hbar c}\frac{1}{2\pi^3}\left\{|C_F|^2\left|\int I\right|^2 + |C_{GT}|^2\left|\int \boldsymbol{\sigma}\right|^2\right\} f(Z, \varepsilon_0) \qquad \text{(8-97)}$$

where we have introduced the elementary factor $mc^2/\hbar$ for inverse time. Thus the relationship between g and G_β is found to be

$$\frac{G_\beta{}^2}{\hbar c} = \frac{g^2}{(mc^2)^2}\left(\frac{mc}{\hbar}\right)^6. \qquad \text{(8-98)}$$

Substituting the value (8-91b) and using the pion mass for m in (8-98), we find $G_\beta{}^2/\hbar c \approx 5 \times 10^{-14}$.[1] This is to be contrasted to $f^2/\hbar c \approx 1$ for strong or hadronic interactions (see Chapter I) and $e^2/\hbar c = 1/137$ for electromagnetic interactions. It is also amusing to note that the dimensionless gravitational coupling constant can be written as $\Upsilon m^2/\hbar c \approx 2 \times 10^{-39}$ when we use the nucleon mass with the value for Newton's gravitational constant $\Upsilon = 6.673 \times 10^{-8}$ dyne cm^2 g^{-2}.

[1] It is customary to follow Marshak and Sudarshan in using the pion mass in (8-98). If the electronic mass is used instead, $G_\beta{}^2/\hbar c \approx 10^{-23}$.

D. NUCLEAR MATRIX ELEMENTS

Allowed Fermi transitions involve the nuclear matrix element $\int I$ defined by (8-63a). To simultaneously allow for both β^--decay and β^+-decay (and electron capture), we use the operator

$$\sum_{\xi=1}^{A} t_{\pm}{}^{\xi} = T_{\pm} = T_1 \pm iT_2. \tag{8-99}$$

For nuclear states that are pure isospin states, $\mathbf{T} = \sum_{\xi=1}^{A} \mathbf{t}_\xi$, and the integral $\int I$ vanishes unless $T_f = T_i$ and $T_{f3} = T_{i3} \pm 1$, in which event[1]

$$\langle T, T_3 \pm 1 | T_{\pm} | T, T_3 \rangle = [T(T+1) - T_3(T_3 \pm 1)]^{1/2}. \tag{8-100}$$

Of course, (8-100) pertains only to the isospin variables of the initial and final state; $\int I$ would also involve the overlap integral in the space coordinates. The β-transitions in Table 8-3 are believed to involve isospin analog states with $T = 1$ and for which only the quantum number T_3 differs for the initial and final state. Therefore, using (8-100), we immediately get $|\int I|^2 = 2$. We have used this result earlier in deducing the value of g_V given in (8-91). It should be noted that while there might be variations in the virtual pion fields associated with the nucleons as one considers the whole range of nuclear species, one consequence of CVC is to keep g_V (and hence ft) a constant. This high degree of constancy is evident in Table 8-3.

It is perhaps instructive to evaluate $\int I$ for the example $^{14}O \to {}^{14}N + e^+ + \nu$, using a microscopic calculation. This pure Fermi β^+-transition is from the ground state of ^{14}O, $J^\pi = 0^+$ and $T = 1$, $T_3 = +1$, to the first excited state of ^{14}N (the corresponding isospin analog state in ^{14}N) with $J^\pi = 0^+$ and $T = 1$, $T_3 = 0$. Treating each state as a two-hole state[2] having the stated isospin characteristics and assuming that the β-transition involves only these holes, we have from (8-63a)

$$\int I = \langle 2^{-1/2}[\nu(1)\pi(2) + \pi(1)\nu(2)] | t_-^{(1)} + t_-^{(2)} | \pi(1)\pi(2) \rangle. \tag{8-101}$$

Then, noting that $t_-^{(1)} | \pi(1)\pi(2)\rangle = |\nu(1)\pi(2)\rangle$, etc., we immediately get $\int I = \sqrt{2}$, in agreement with the macroscopic result using (8-100).

We should also note that admixtures of higher particle (or hole) configurations *would not* alter the results, provided only that the states remain isospin analogs.

[1] Since $T_{\pm}|T, T_3\rangle = [(T \mp T_3)(T \pm T_3 + 1)]^{1/2} |T, T_3 \pm 1\rangle$.

[2] The states involved are rather likely to be pure individual-particle states involving *two holes* in the completed 1p-shell nucleus ^{16}O. The distinction between a two-hole state and a two-particle state need not be made in the present instance since it can be shown formally [see Visscher and Ferrell (1957)] that the hole operator in β-decay differs from that for particles by a trivial sign change.

Transitions of the type $0^+ \to 0^+$ but apparently $\Delta T = 1$ proceed primarily through small isotopic impurities in either the parent or daughter state that, in fact, allow for $\Delta T = 0$ coupling through the impurity. Hindrance factors ranging from 10^4 to 10^8 are observed and offer evidence of impurity amplitudes of 10^{-2}–10^{-4}. An example of such a β-transition is the $0^+ \to 0^+$ ground state to ground state transition in the ^{170}Lu β^+-decay to ^{170}Yb. The isospin of ^{170}Lu is $T_3 = -14$ and largely $T = 14$, that of ^{170}Yb is $T_3 = -15$ and $T = 15$. The ft value 5.7×10^9 sec suggests an amplitude impurity of $\sim 2 \times 10^{-4} |T = 15\rangle$ in the ground state of the parent.

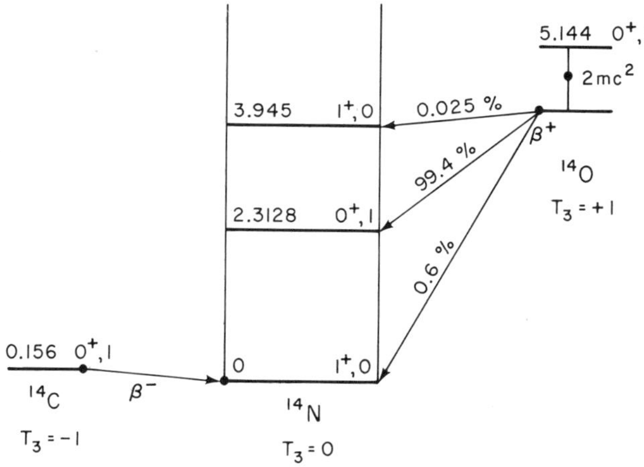

FIG. 8.12. The levels of the $A = 14$ triad involved in β-decay. The levels are identified by energy (in MeV) and the designation J^π, T.

Returning to the $A = 14$ triad, Fig. 8.12 shows the energy level diagram for the relevant states in ^{14}C, ^{14}N, and ^{14}O. The ft values for the ^{14}O β^+ branches are $ft_0 = 2.14 \times 10^7$ sec, $ft_1 = 3100$ sec, and $ft_2 = 1200$ sec involving the ground, first excited, and second excited states of ^{14}N, respectively. The ft value for the ^{14}C β^--decay is 1.12×10^9 sec.[1] A closely related β^--decay is that of ^{6}He, $J^\pi = 0^+$ and $T = 1$, going to the ground state of ^{6}Li with $J^\pi = 1^+$ and $T = 0$. The ft value for this decay is 808 sec. In this latter case we have two particles in the p shell as compared to two p-shell holes for the $A = 14$ cases.

The ^{6}He decay, and the ^{14}O decay to the second excited state of ^{14}N, are clearly pure Gamow–Teller superallowed transitions. The ^{14}O decay to the first excited state of ^{14}N has already been discussed as a typical pure Fermi

[1] The convenient half-life of ^{14}C, $t_{1/2} = 5692 \pm 21$ yr, leads to its use as a powerful archeological tool through "carbon dating" techniques. For a recent review of nuclear applications in general in art and archaeology, see Perlman *et al.* (1972).

transition. The decay of ^{14}C and ^{14}O to the ground state of ^{14}N would be expected to be superallowed pure Gamow–Teller; instead both are seen to have anomalously large *ft* values.

▶**Exercise 8-7** Assume that the ground states of both ^{6}He and ^{6}Li can be approximated as pure *LS*-coupled two-particle states. Defining

$$\phi(l_1 l_2|LM_L) = \sum_{m_1} (l_1 l_2 m_1, M_L - m_1|LM_L) Y_{l_1}^{m_1}(1) Y_{l_2}^{M_L - m_1}(2) \tag{8-102}$$

for the spatial wave function coupling $\mathbf{l}_1 + \mathbf{l}_2 = \mathbf{L}$ and $M_L = m_1 + m_2$, we have for two p-shell particles coupled to an S state ($L = 0$)

$$\phi(11|00) = 3^{-1/2}[Y_1{}^1(1)Y_1^{-1}(2) - Y_1{}^0(1)Y_1{}^0(2) + Y_1^{-1}(1)Y_1{}^1(2)]. \tag{8-103}$$

(a) Verify (8-103). Show that if the ground state of ^{6}He, $J^\pi = 0^+$ and $T = 1$, is a pure $^{13}\mathrm{S}_0$, *LS*-coupled state,[1] then

$$\psi(^{13}\mathrm{S}_0) = \phi(11|00)\chi_0{}^0(1,2)\zeta_1^{-1}(1,2). \tag{8-104}$$

If the ground state of ^{6}Li, $J^\pi = 1^+$ and $T = 0$, is a pure $^{31}\mathrm{S}_1$, *LS*-coupled state, show that

$$\psi(^{31}\mathrm{S}_1)_{0,\pm 1} = \phi(11|00)\chi_1^{0,\pm 1}(1,2)\zeta_0{}^0(1,2). \tag{8-105}$$

(b) Show that

$$\begin{aligned} (\sigma_{\pm 1}^{(1)} t_+^{(1)} + \sigma_{\pm 1}^{(2)} t_+^{(2)})\chi_0{}^0\zeta_1^{-1} &= \sqrt{2}\,\chi_1^{\mp 1}\zeta_0{}^0 \\ (\sigma_0^{(1)} t_+^{(1)} + \sigma_0^{(2)} t_+^{(2)})\chi_0{}^0\zeta_1^{-1} &= \sqrt{2}\,\chi_1{}^0\zeta_0{}^0. \end{aligned} \tag{8-106}$$

(c) In view of (8-104)–(8-106), show that the Gamow–Teller matrix element (8-71) is

$$\left|\int \boldsymbol{\sigma}\right|^2 = \sum_{(M_J)_f} \sum_{\mu} \sum_{\xi=1}^{2} |\langle\psi_f|\sigma_\mu{}^\xi t_+{}^\xi|\psi_i\rangle|^2 = 6. \quad ◀ \tag{8-107}$$

In view of the result of Exercise 8-7 and using (8-92) with $|\lambda'|^2 = 1.51$, we predict $ft = 6164/(6 \times 1.51) = 680$ sec for the β-decay of ^{6}He based on pure *LS* states. This predicted value is 15% lower than the experimental value. As we saw in Chapter V, *LS* coupling was approximately valid only for the lightest p-shell nuclei with an increasing tendency for *jj* coupling to become dominant as *A* increased. Since the allowed nuclear matrix elements involve only spin and isospin operators, it is convenient to expand nuclear wave functions in *LS* states. Restricting ourselves to the p shell, the most general

[1] As in Chapters I and V, we use the notation $^{(2S+1)(2T+1)}L_{J,M_J}$ and $\chi_S^{M_S}(1,2)$ and $\zeta_T^{T_3}(1,2)$ for the coupled mechanical spin and isospin two-particle states. For convenience, the radial dependence of the wave function can be suppressed.

wave functions for ^{6}He and ^{6}Li ground states are[1]

$$\psi(^6\text{He}, 0^+, T=1) = [a_S\psi(^1S_0)+a_P\psi(^3P_0)]\zeta_1^{-1}(1,2) \tag{8-108a}$$

and

$$\psi(^6\text{Li}, 1^+, T=0)_{M_J} = [a_S'\psi(^3S_1)+a_P'\psi(^1P_1)+a_D'\psi(^3D_1)]_{M_J}\zeta_0^{\,0}(1,2). \tag{8-108b}$$

The Gamow–Teller nuclear matrix element in this case is

$$\left|\int \boldsymbol{\sigma}\right|^2 = 6\,|a_S a_S' - 3^{-1/2}a_P a_P'|^2. \tag{8-109}$$

▶**Exercise 8-8** (a) Verify that

$$\psi(^3P_0) = 3^{-1/2}[\phi(11|1,1)\chi_1^{-1} - \phi(11|1,0)\chi_1^{\,0} + \phi(11|1,-1)\chi_1^{\,1}] \tag{8-110a}$$

and

$$\psi(^1P_1)_{M_J} = \phi(11|1, M_L = M_J)\chi_0^{\,0}. \tag{8-110b}$$

(b) Using the orthonormality of the functions $\phi(l_1 l_2|LM_L)$ in the quantum numbers l_1, l_2, L, and M_L, derive (8-109).

[*Hint:* Use the fact that $(M_J)_i = 0 = (M_J)_f - \mu$ to identify nonvanishing contributions to $\int \boldsymbol{\sigma}$.] ◀

Had the wave functions for ^{6}He and ^{6}Li been pure *jj*-coupled states with each particle in the $p_{3/2}$ subshell, we would have (see Table 5-2)

$$\psi\{^6\text{He}; (\tfrac{3}{2})^1(\tfrac{3}{2})^1 J=0, T=1\} = 3^{-1/2}[\sqrt{2}\,\psi(^1S_0) - \psi(^3P_0)]\zeta_1^{-1}(1,2) \tag{8-111a}$$

and

$$\psi\{^6\text{Li}; (\tfrac{3}{2})^1(\tfrac{3}{2})^1 J=1, T=0\}_{M_J} = 54^{-1/2}[2\sqrt{5}\,\psi(^3S_1) + \sqrt{30}\,\psi(^1P_1) - 2\psi(^3D_1)]_{M_J}\zeta_0^{\,0}(1,2). \tag{8-111b}$$

Then using (8-109), we find $|\int \boldsymbol{\sigma}|^2 = \frac{10}{3}$, which leads to a predicted value for *ft* that is 50% too large.

▶**Exercise 8-9** (a) Verify the above statements involving (8-111).

(b) If the intermediate coupling wave function for the ^{6}He ground state is $0.973\,|^{13}S_0\rangle - 0.231\,|^{33}P_0\rangle$ and that for the ground state of ^{6}Li is $0.996\,|^{31}S_1\rangle + 0.072\,|^{11}P_1\rangle + 0.039\,|^{31}D_1\rangle$, show that $|\int \boldsymbol{\sigma}|^2 = 5.75$. [See Chapter V, Table 5-8 and Eq. (5-61).] ◀

The results of Exercise 8-9 predict an *ft* value that is only 12% lower than the experimental value, using $|\lambda'|^2 = 1.51$. Alternately, coupling the result $|\int \boldsymbol{\sigma}|^2 = 5.75$ with $\Lambda = 6164$ sec yields $|\lambda'|^2 = 1.33$ from the observed *ft* value.

[1] The notation is $^{(2S+1)}L_J$, and again the radial dependence is suppressed.

Turning to the superallowed β^+-transition from ^{14}O to the second excited state of ^{14}N, we might try to approximate these states by pure *jj* states coupling two holes in the p shell. Validity for such an approximation might be inferred from Exercise 5-12. The ground state of ^{14}O would then be taken as two $p_{1/2}$ holes, leading to (see Table 5-2)

$$\psi\{^{14}\text{O}; (\tfrac{1}{2})^{-1}(\tfrac{1}{2})^{-1}J=0, T=1\} = 3^{-1/2}[\psi(^1\text{S}_0)+\sqrt{2}\,\psi(^3\text{P}_0)]\zeta_1^{-1}(1,2), \tag{8-112a}$$

while the second excited state in ^{14}N, presumed to be one $p_{3/2}$ hole coupled to a $p_{1/2}$ hole, leads to

$$\psi\{^{14}\text{N}^*; (\tfrac{3}{2})^{-1}(\tfrac{1}{2})^{-1}J=1, T=0\}_{M_J}$$
$$= 27^{-1/2}[-4\psi(^3\text{S}_1)+\sqrt{6}\,\psi(^1\text{P}_1)-\sqrt{5}\,\psi(^3\text{D}_1)]_{M_J}\zeta_0{}^0(1,2). \tag{8-112b}$$

The use of (8-109) gives $|\int\boldsymbol{\sigma}|^2 = \frac{8}{3}$ and hence $ft = 6164/(1.51\times 8/3) = 1530$ sec, in fair agreement with the experimental value of 1200 sec.

The ground state of ^{14}N might be conjectured to be a pure *jj* state consisting of two coupled $p_{1/2}$ holes. In this event

$$\psi\{^{14}\text{N}; (\tfrac{1}{2})^{-1}(\tfrac{1}{2})^{-1}J=1, T=0\}_{M_J}$$
$$= 54^{-1/2}[-\sqrt{2}\,\psi(^3\text{S}_1)+2\sqrt{3}\,\psi(^1\text{P}_1)+2\sqrt{10}\,\psi(^3\text{D}_1)]_{M_J}\zeta_0{}^0(1,2). \tag{8-113}$$

The value of $|\int\boldsymbol{\sigma}|^2$ is $\frac{2}{3}$ for either the ^{14}O or ^{14}C decay to the state (8-113). The *ft* value predicted is 6100 sec, in gross disagreement with the very large experimental values. If instead the ground states for ^{14}C (or ^{14}O) and ^{14}N are taken to be the pure *LS* states $^{13}\text{S}_0$ and $^{31}\text{S}_1$, respectively, we have $|\int\boldsymbol{\sigma}|^2 = 6$, and the predicted *ft* value is even smaller than for *jj* coupling. Figure 8.13 illustrates the relative behavior of $|a_S'|^2$, $|a_P'|^2$, and $|a_D'|^2$ as well as $|\int\boldsymbol{\sigma}|$ for intermediate coupling as a function of the spin–orbit strength in the presence of central interactions, but without any tensor or quadratic spin–orbit interactions (see Chapter V). It is thus clear that intermediate coupling with such simple residual interactions cannot account for the large observed *ft* values.

Visscher and Ferrell (1957) were in fact able to account for the observed *ft* value for the ^{14}C decay by using an additional tensor two-body interaction in intermediate coupling to yield $|a_S a_S' - (a_P a_P'/\sqrt{3})| = (8.3\pm 0.6)\times 10^{-4}$ for (8-109). Such an "accidental cancellation" accounts for the anomalously large *ft* value. The required calculated values of $a_S' = 0.173$, $a_P' = 0.355$, and $a_D' = 0.920$ for ^{14}N and $a_S = 0.764$ and $a_P = 0.646$ for ^{14}C [in wave functions analogous to (8-108)] also give reasonable values for the static moments and low-lying energy levels. We nrte that with these coefficients the relative S- and

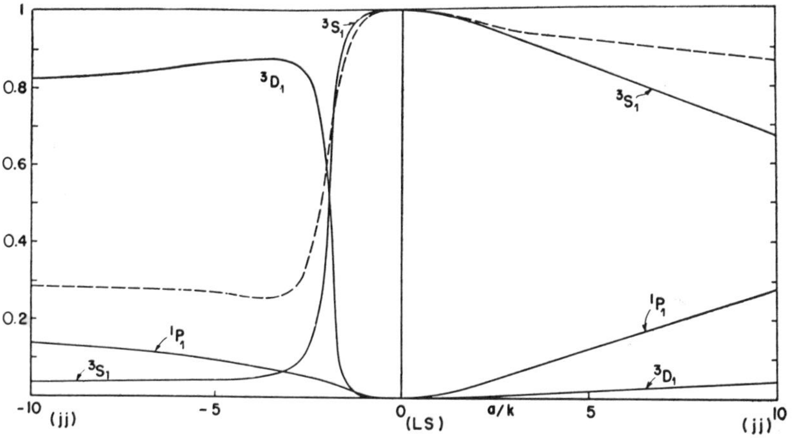

FIG. 8.13. The square of the amplitudes $^3S_1$, $^1P_1$, and $^3D_1$ in the wave function for the ground state of ^6Li (to the right with $a/K > 0$), and for the ground state of ^{14}N (to the left with $a/K < 0$). The parameter a/K measures the relative strength of the ordinary one-body spin–orbit interaction in intermediate coupling (see Chapter V). No tensor or quadratic spin–orbit interactions are included. The dashed curve indicates the relative magnitude of $|\int \boldsymbol{\sigma}|$ (for absolute values, unity along the vertical scale corresponds to $\sqrt{6}$) [taken from Inglis (1953)].

P-state signs are now proper to give cancellation in (8-109), whereas the signs in (8-112a)[1] and (8-113) do not. Also note the large D-state admixture in the ground state of ^{14}N. If the ground state were a *pure* $^3D_1$ state, the β-decay from ^{14}C and ^{14}O would, in fact, be *L-forbidden*. The cancellation spoken of is then required only for the inherently smaller S- and P-state contributions for the realistic case and can be accomplished by an acceptably small tensor force. In Chapter V, the intermediate coupling parameter Set (5-62) yields $a_S' = 0.100$, $a_P' = 0.195$, and $a_D' = 0.976$ for ^{14}N and $a_S = 0.748$ and $a_P = 0.665$ for ^{14}C. These values also result in $|\int \boldsymbol{\sigma}|^2$ essentially vanishing, and perhaps are preferable since they involve the matrix element $\langle ^{31}S_1 | O_{12} | ^{31}D_1 \rangle = -1.30$ MeV rather than -2.58 MeV as estimated by Visscher and Ferrell (1957). The smaller magnitude is more in keeping with the phenomenological parameters obtained from the two-nucleon problem. Thus, while intermediate coupling *without* a tensor or quadratic spin–orbit interaction may be very successful in predicting the values for level energies, predicting those physical properties of nuclei that are sensitive to exact details of the wave function (such as the relative S- and P-state phase in the present instance) requires greater sophistication.

[1] The ^{14}C ground-state wave function is the same as (8-112a) with only the isospin function changed to $\zeta_1{}^1(1,2)$ to correspond to $T_3 = +1$ in this case.

In the preceding examples we have considered various $A = 6$ and $A = 14$ β-decay processes. In each instance, we calculated nuclear matrix elements as if "only the valence particles (or holes) contributed" to the transition. Before considering other β-decay cases we should inquire into the validity of this assertion in view of the fact that both the Fermi and Gamow–Teller operators require a sum over *all A* nucleons. In each instance we employed wave functions restricted to a truncated 1p-shell configuration. In a simple core plus valence particles (or holes) description, we are concerned with the $(1s)^4$ and $(1s)^4(1p)^{12}$ core states. For such closed-shell or saturated configurations the partition diagram of Fig. 3.1 is particularly appropriate and rigorously valid. Specifically, each of the spatial states designated by a partition row is occupied by four nucleons, two protons and two neutrons each with "spin up" and "spin down." The core wave function is uniquely the totally antisymmetrized state of this partition array of individual-particle states. Contributions to the nuclear transition matrix elements involve totally or in part the sum of isospin operators $t_\pm(i)$. This operator when applied to the core nucleons either gives zero directly [e.g., $t_+(i)\pi(i) = 0$] or results in a partition violating the Pauli principle (e.g., leading to either three neutrons or protons within a partition row), which in turn leads to zero overlap with any physically acceptable final state. Thus nonzero contributions to the nuclear matrix element can come only from the valence particles (or holes) alone.

The extension of the concept of an inert or nonparticipating spectator core to one involving saturated partition rows in midshell regions is of dubious usefulness since in these cases nuclear states are not characterizable with definite partitions under the operation of realistic nucleon–nucleon forces. Nonetheless it is instructive to calculate the Fermi and Gamow–Teller matrix elements for β-decay between nuclear pairs in an extreme single-particle treatment involving a model of inert saturated partition rows of four nucleons each plus a single particle (or hole). The entire β-decay interaction is attributed to the single "valence" nucleon (or hole). Such nuclear pairs correspond to the mirror nuclei.[1] In the spirit of the extreme single-particle model, we also couple the core nucleons to a $J^\pi = 0^+, T = 0$ core state. The total nuclear wave function is thus determined by the single-particle (or hole) state.

We assume the single-particle state is generated by a Hamiltonian containing central and spin–orbit potential terms. Thus the appropriate constants of the motion are j, j_z, l, and s. In addition we have $\mathbf{j} = \boldsymbol{l} + \mathbf{s}$; hence $j = l \pm \frac{1}{2}$;

[1] Clearly, ab initio we are dealing with the $|N - Z| = 1$ or $T_3 = \pm\frac{1}{2}$, odd-A nuclei. Recall our earlier definition of mirror nuclei, one member of the pair with $Z_1 = \frac{1}{2}(A+1)$ and $N_1 = \frac{1}{2}(A-1)$ and the other with $Z_2 = \frac{1}{2}(A-1)$ and $N_2 = \frac{1}{2}(A+1)$.

also $m = m_l \pm \frac{1}{2}$.[1] A typical wave function is (see Chapter IV)

$$\psi(n,l,j,m,t_3;\mathbf{r}) = R_{n,l}(r) \sum_{m_l} (l,\tfrac{1}{2},m_l,m-m_l|jm)\, Y_l^{m_l}(\theta,\varphi)\, \chi_{1/2}^{m-m_l}(\sigma)\, \zeta_{1/2}^{t_3}(\tau). \tag{8-114}$$

In view of our assumptions and the form of (8-114), evidently the Fermi matrix element becomes simply

$$\left|\int I\right|^2 = \delta_{nn'}\, \delta_{ll'}\, \delta_{jj'} \tag{8-115}$$

for the mirror transitions.

Contributions to the β-decay from the Gamow–Teller interaction for the cases under discussion clearly involve $\delta_{nn'}$ and $\delta_{ll'}$. Since the operator $\boldsymbol{\sigma}$ can flip spin, however, $\Delta j = 0, \pm 1$ transitions are possible. The value for $|\int \boldsymbol{\sigma}|^2$ is readily found to be

$$\left|\int \boldsymbol{\sigma}\right|^2 = (j+1)/j \qquad j_\mathrm{i} = j_\mathrm{f} = l + \tfrac{1}{2} \tag{8-116a}$$

$$\left|\int \boldsymbol{\sigma}\right|^2 = j/(j+1) \qquad j_\mathrm{i} = j_\mathrm{f} = l - \tfrac{1}{2} \tag{8-116b}$$

$$\left|\int \boldsymbol{\sigma}\right|^2 = (2j_\mathrm{f}+1)/(l+\tfrac{1}{2}), \qquad \begin{cases} j_\mathrm{i} = l+\tfrac{1}{2}, \quad j_\mathrm{f} = l-\tfrac{1}{2} \\ \text{or} \\ j_\mathrm{i} = l-\tfrac{1}{2}, \quad j_\mathrm{f} = l+\tfrac{1}{2}. \end{cases} \tag{8-116c}$$

▶**Exercise 8-10** The results (8-116) can be obtained in a variety of ways [see Konopinski (1966) or Mayer and Jensen (1955)]; however, a straightforward calculation using (8-114) is instructive.

The decay of unpolarized nuclei requires $|\int \boldsymbol{\sigma}|^2$ to be independent of m_i; therefore it is convenient to assume $m_\mathrm{i} = j_\mathrm{i}$. Consider the case $j_\mathrm{i} = l+\frac{1}{2}$, so that the initial state has only one term in the expansion (8-114).

(a) Show that for $j_\mathrm{f} = j_\mathrm{i}$,

$$\begin{aligned}\left|\int \boldsymbol{\sigma}\right|^2 &= \{2|(l,\tfrac{1}{2},l,-\tfrac{1}{2}|l+\tfrac{1}{2},l-\tfrac{1}{2})|^2 + |(l,\tfrac{1}{2},l,\tfrac{1}{2}|l+\tfrac{1}{2},l+\tfrac{1}{2})|^2\}\, \delta_{nn'}\, \delta_{ll'} \\ &= [(j+1)/j]\, \delta_{nn'}\, \delta_{ll'},\end{aligned}$$

with the first term resulting from the transition operator having $\mu = -1$ and the second from $\mu = 0$.

(b) Show that for $j_\mathrm{i} = l+\frac{1}{2}$ and $j_\mathrm{f} = l-\frac{1}{2}$ only the transition operator

[1] We write m for m_j.

having $\mu = -1$ contributes and

$$\left|\int \boldsymbol{\sigma}\right|^2 = 2|(l, \tfrac{1}{2}, l, -\tfrac{1}{2}|l-\tfrac{1}{2}, l-\tfrac{1}{2})|^2 \delta_{nn'} \delta_{ll'}$$

$$= [4l/(2l+1)] \delta_{nn'} \delta_{ll'}.$$ ◀

▶**Exercise 8-11** Show that $|\int \boldsymbol{\sigma}|^2$ for the decay of the parents n, ^3H, ^{15}O, ^{17}F, ^{39}Ca, and ^{41}Sc is 3, 3, $\frac{1}{3}$, $\frac{7}{5}$, $\frac{3}{5}$, and $\frac{9}{7}$, respectively, and that $|\int I|^2 = 1$ for all cases. These nuclei might be presumed to be closed-shell configurations plus a particle or hole. ◀

The curves shown in Fig. 8.11 were generated using the results of Exercise 8-11, and by ignoring the small corrections to $\int I$ and $\int \boldsymbol{\sigma}$, which in any case would not be detectable on the scale of the figure. The failure to have a common intersection at $|\lambda'|^2 = 1.52$ is most probably attributable to the inaccuracy of the above-estimated matrix elements due to the presence of configuration mixing.

It is possible to show that under the assumption of the charge independence of nuclear forces and again for mirror nuclei [see Konopinski (1966) or Mayer and Jensen (1955)] we obtain

$$\left|\int \boldsymbol{\sigma}\right|^2 = [(J+1)/J] \left|\left\langle \sum_{\xi=1}^{A} \sigma_z{}^\xi \tau_3{}^\xi \right\rangle_{M_J=J}\right|^2, \tag{8-117}$$

where the expectation value refers to *either* the parent or daughter state. This equation has greater generality than the extreme single-particle model discussed up to this point, and requires only that parent and daughter states are related by $T_+\psi(T_3 = -\frac{1}{2}) = \psi(T_3 = +\frac{1}{2})$. Here J refers to the total angular momentum of the parent and daughter mirror states.

By using (8-117) in an extreme *jj*-coupling model, we can relate $|\int \boldsymbol{\sigma}|^2$ to the nuclear magnetic moment μ. The magnetic moment operator [see Chapter II, (2-41) and Exercise 2-7] can be written, using neutron and proton projection operators, as[1]

$$\boldsymbol{\mu} = \sum_{\xi=1}^{A} \{\tfrac{1}{2}(1+\tau_3{}^\xi)(\mu_p \boldsymbol{\sigma}^\xi + \mathbf{l}^\xi) + \tfrac{1}{2}(1-\tau_3{}^\xi)\mu_n \boldsymbol{\sigma}^\xi\}. \tag{8-118}$$

We assume that the nucleons can be grouped into saturated shell or subshell core groups giving zero contribution and a valence group in a partially filled subshell with a particular j, either $j = l \pm \frac{1}{2}$. The resulting total angular momentum is written J. Then the magnetic moment becomes

$$\mu = \tfrac{1}{2}[(\mu_p - \mu_n - \tfrac{1}{2}) \pm (l+\tfrac{1}{2})]$$
$$\times \left\langle \sum_{\xi} \sigma_z{}^\xi \tau_3{}^\xi \right\rangle_{M_J=J} + \tfrac{1}{2}J \pm \tfrac{1}{2}(\mu_p + \mu_n - \tfrac{1}{2}) J (l+\tfrac{1}{2})^{-1}, \tag{8-119}$$

[1] All quantities μ are in units of μ_0, the nuclear magneton.

where the signs $\pm$ refer to $j = l \pm \frac{1}{2}$. We can thus relate the matrix element $|\int \boldsymbol{\sigma}|^2$ to the observed value of the magnetic moment, with the result for *mirror nuclei* that

$$\left|\int \boldsymbol{\sigma}\right|^2 = \frac{J+1}{J}\left|\frac{2\mu - J[l+\frac{1}{2}\pm(\mu_p+\mu_n-\frac{1}{2})](l+\frac{1}{2})^{-1}}{\mu_p-\mu_n-\frac{1}{2}\pm(l+\frac{1}{2})}\right|^2 \quad \text{for} \quad j = l \pm \frac{1}{2}, \tag{8-120a}$$

or using $\mu_n = -1.9131$ and $\mu_p = +2.7928$,

$$\left|\int \boldsymbol{\sigma}\right|^2 = \frac{J+1}{J}\left|\frac{2\mu - J(l+\frac{1}{2}\pm 0.380)(l+\frac{1}{2})^{-1}}{4.206\pm(l+\frac{1}{2})}\right|^2 \quad \text{for} \quad j = l \pm \frac{1}{2}. \tag{8-120b}$$

The magnetic moment μ appearing in (8-120) is for *either* the parent or the daughter state.

▶**Exercise 8-12** (a) Show that for the extreme single-particle model with $J_i = J_f = j$ and for the case that μ is given exactly by the Schmidt limit, (8-120) reduces to (8-116). Discover that this is so whether μ refers to the parent or the daughter.

(b) The magnetic moments for ^3H and ^3He are $+2.97885$ and -2.12755, respectively. Show that the average value of $|\int \boldsymbol{\sigma}|^2$, deduced for the parent and daughter using (8-120), is $|\int \boldsymbol{\sigma}|^2 = 3.532$. Contrast the difference for the parent and daughter values with the deviation of the average from the extreme single-particle model value.

(c) Using (b) and Λ' of (8-93b), show that the *ft* value in Table 8-4 yields $|\lambda'|^2 = 1.53$.

(d) Using the magnetic moment of -1.59460 for ^{41}Ca, (8-93b), and Table 8-4, contrast the values of $|\int \boldsymbol{\sigma}|^2$ and $|\lambda'|^2$ obtained by using (8-120) and the extreme single-particle model for the β-decay $^{41}Sc \rightarrow {}^{41}Ca$. ◀

The ground-state structure of ^3H and ^3He has received considerable attention [see the review paper by Delves and Phillips (1969)]. Even so, the exact values of the magnetic moments and the β-decay rate still remain to be accounted for. Important meson exchange current contributions are required, particularly for the isovector moment and the axial vector β-decay transition probability. Indeed, meson exchange current effects of the right magnitude can be estimated to account for the observed values [see Delves and Phillips (1969), Blomqvist (1970), Chemtob and Rho (1971), and Gerstenberger and Nogami (1972)].

An interesting suggestion for the trinucleon problem (as well as the more general nuclear matter problem) is to include inelastic nucleon–nucleon scattering effects in which one nucleon is raised to a resonance or excited state (i.e., N* or Δ). The relevant graphs for these interactions are shown in Fig. 1.18

of Chapter I. Two typical papers on the subject are by Ichimura *et al.* (1972) and Green and Schucan (1972). The $(\frac{3}{2}, \frac{3}{2})$ $\Delta(1236)$ resonance is found to be most important in modifying the trinucleon ground-state properties through both one-Δ (3–6% probability) and two-Δ (~2% probability) contributions. Calculated values for the isoscalar and isovector magnetic moments are

$$\mu^{S} = \tfrac{1}{2}[\mu(^{3}\mathrm{H})+\mu(^{3}\mathrm{He})] = 0.42\text{–}0.43$$

and

$$\mu^{V} = \tfrac{1}{2}[\mu(^{3}\mathrm{H})-\mu(^{3}\mathrm{He})] = 2.40\text{–}2.67;$$

these may be compared to the experimental values $\mu^{S} = 0.426$ and $\mu^{V} = 2.553$. The effect of the presence of the Δ component is to raise the Gamow–Teller matrix element by approximately 2% over what it would be without it.

While the results calculated in Exercise 8-12 indicating an A dependence for $|\lambda'|^2$ based on (8-120) are questionable, particularly in view of the known small but significant higher configuration admixtures in the ground-state wave functions, neither can the constancy of g_A be unambiguously established. Also, whether the meson effects that can be present are used to reevaluate g_A or $|\int \boldsymbol{\sigma}|$ is perhaps a matter of choice. The constancy of g_V, on the other hand, as CVC would require, seems to be clearly established. Wilkinson (1973) has carefully examined the use of combining static magnetic moment data with β-decay ft values in order to determine the matrix element $\langle \sigma_z \tau_3 \rangle$. These values can then be compared to the predictions from theoretical wave functions [see Wilkinson (1973) for details]. A further consideration requires including a proper relativistically defined magnetic moment operator [see Mukhopadhyay and Miller (1973)]. Nonnegligible effects totaling several percent are found, clearly influencing efforts to determine a fundamental g_A/g_V ratio.

Before it is possible to examine the empirical evidence relevant to the above points it is necessary to be able confidently to include the small corrections δ_R and δ_c. In fact a study of these correction factors is of interest in and of itself. In order to sort out these factors from the inner corrections Δ_F and Δ_{GT}, it is found desirable to determine the empirical values for a ratio of ft values in cases where the uncorrected nuclear matrix elements and the effective coupling constants can be sensibly taken as identical. The quantity thus determined is written

$$[(ft)_1/(ft)_2] - 1 = \delta_{c1} - \delta_{c2} = D^{c}_{1,2}. \tag{8-121}$$

Examples of the applicability of the (8-121) might involve a β^--decay and a mirror β^+-decay going to the same daughter state, e.g.,

$$^{8}\mathrm{Li}(2^+, T=1) \xrightarrow{\beta^-} {}^{8}\mathrm{Be}(2.94\ \mathrm{MeV};\, 2^+, T=0)$$

and

$$^{8}\mathrm{B}(2^{+}, T=1) \xrightarrow{\beta^{+}} {}^{8}\mathrm{Be}(2.94 \text{ MeV}; 2^{+}, T=0)$$

[see Wilkinson and Alburger (1971) and Tribble and Garvey (1974)]; or ^{12}B and ^{12}N each decaying to ^{12}C (4.44 MeV level) [see Alburger (1972)]. In the general case of odd-A nuclei the final states for the β^-- and β^+-decays are different but are each other's analogs, e.g.,

$$^{13}\mathrm{O}(\tfrac{3}{2}^{-}, T=\tfrac{3}{2}) \xrightarrow{\beta^{+}} {}^{13}\mathrm{N}(\tfrac{1}{2}^{-}, T=\tfrac{1}{2})$$

and

$$^{13}\mathrm{B}(\tfrac{3}{2}^{-}, T=\tfrac{3}{2}) \xrightarrow{\beta^{-}} {}^{13}\mathrm{C}(\tfrac{1}{2}^{-}, T=\tfrac{1}{2})$$

[see Blin-Stoyle *et al.* (1971)]. Finally, decay chains or cascades can also be compared, such as $^{34}\mathrm{Ar} \to (\beta^{+})^{34}Cl$ and $^{34}Cl \to (\beta^{+})^{34}\mathrm{S}$. From the entries in Table 8-3 we see (using an average value of $\mathscr{F}t = 3089$ sec for ^{34}Cl) that $D_{1,2}^{c}(A=34) = (0.35 \pm 0.6)\%$ which in this case suggests little charge-dependent mixing.

For general references on analyzing the available data, see also *Int. Conf.* (1972, p. 191 ff.), Wilkinson (1972), Wilkinson *et al.* (1972), and Towner (1973). The theoretical analysis of the nonzero values of $D_{1,2}^{c}$ involves allowing for electromagnetic effects, meson exchange effects, and possible second-class current effects. An interesting meson exchange effect is that considered by Lipkin (1971). It is diagrammatically shown in Fig. 8.14, and involves the

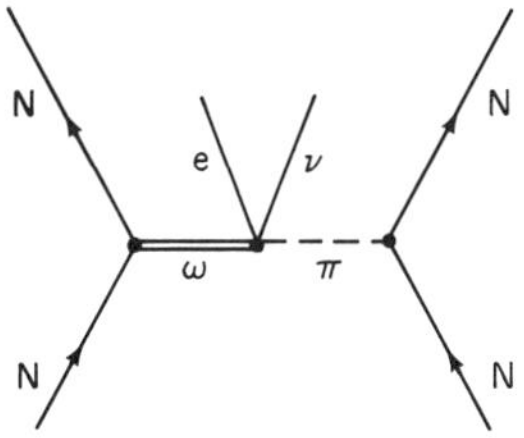

FIG. 8.14. Meson-exchange contribution to β-decay through the ω–π decay.

ω–π decay. Such effects could lead to an energy-independent δ, favored by the data concerning $A = 8$. The second-class currents are vector and axial vector weak currents that have opposite G-parity from those generally considered (as we have throughout). These can appear giving both energy-dependent and energy-independent contributions. Unfortunately, a discussion of these points is beyond the intended scope of our treatment. The interested reader can refer to Holstein and Treiman (1971), Kim and Fulton (1971), Wolfenstein and Henley (1971), and *Int. Conf.* (1972, p. 221 ff.).

References

Adler, S. L. (1965). *Phys. Rev. Lett.* **14**, 105.
Ajzenberg-Selove, F. (1970). *Nuclear Phys.* **A152**, 1.
Alburger, D. E. (1972). *Phys. Rev.* **C6**, 1167.
Alburger, D. E., and Wilkinson, D. H. (1972). *Phys. Rev.* **C6**, 2019.
Alburger, D. E., and Wilkinson, D. H. (1973). *Phys. Rev.* **C8**, 657.
Armstrong, L., and Kim, C. W. (1972). *Phys. Rev.* **C6**, 1924.
Bahcall, J. N., Cabibbo, N., and Yahil, A. (1972). *Phys. Rev. Lett.* **28**, 316.
Barish, B. C., Bartlett, J. F., Bucholz, D., Humphrey, T., Merritt, F. S., Nagashima, Y., Sciulli, F. J., Shields, D., Suter, H., Krafczyk, G., and Maschke, A. (1973). *Phys. Rev. Lett.* **31**, 180.
Bincer, A. M. (1966). *Phys. Rev. Lett.* **16**, 754.
Blatt, J. M., and Weisskopf, V. F. (1952). "Theoretical Nuclear Physics." Wiley, New York.
Blin-Stoyle, R. J., and Freeman, J. M. (1970). *Nuclear Phys.* **A150**, 369.
Blin-Stoyle, R. J., Evans, J. A., and Khan, A. M. (1971). *Phys. Lett.* **36B**, 202.
Blomqvist, J. (1970). *Phys. Lett.* **32B**, 1.
Burns, R., Goulianos, K., Hyman, E., Lederman, L., Lee, W., Mistry, N., Rettberg, J., Schwartz, M., Sunderland, J., and Danby, G. (1965). *Phys. Rev. Lett.* **15**, 42.
Cabibbo, N. (1963). *Phys. Rev. Lett.* **10**, 531.
Chemtob, M., and Rho, M. (1971). *Nuclear Phys.* **A163**, 1.
Clark, G. J., Freeman, J. M., Robinson, D. C., Ryder, J. S., Burcham, W. E., and Squire, G. T. A. (1973). *Nuclear Phys.* **A215**, 429.
Cowsik, R., and McClelland, J. (1972). *Phys. Rev. Lett.* **29**, 669.
Danby, G., Gaillard, J. M., Goulianos, K., Lederman, L. M., Mistry, N., Schwartz, M., and Steinberger, J. (1962). *Phys. Rev. Lett.* **9**, 36.
Davis, R. (1955). *Phys. Rev.* **97**, 766.
Davis, R. (1956). *Bull. Amer. Phys. Soc.* Ser II:1, 219.
DeBenedetti, S. (1964). "Nuclear Interactions." Wiley, New York.
Delves, L. M., and Phillips, A. C. (1969). *Rev. Modern Phys.* **41**, 497.
Depommier, P., Duclos, J., Heitze, J., Kleinknecht, K., Riesberg, H., and Soergel, V. (1968). *Nuclear Phys.* **B4**, 189.
der Mateosian, E., and Thieberger, P. (1971). *Phys. Rev. Lett.* **27**, 1816.
Duclos, J., Magnon, A., and Picard, J. (1973). *Phys. Lett.* **47B**, 491.
Emery, G. T. (1972). *Ann. Rev. Nuclear Sci.* **22**, 165.
Fermi, E. (1934). *Z. Phys.* **88**, 161.
Fermi, E. (1951). "Elementary Particles." Yale Univ. Press, New Haven, Connecticut.
Feynman, R. P. (1962). "The Theory of Fundamental Processes." Benjamin, New York.
Gasiorowicz, S. (1966). "Elementary Particle Physics." Wiley, New York.
Gerstenberger, R. V., and Nogami, Y. (1972). *Phys. Rev. Lett.* **29**, 233.
Gleit, C. E., Tang, C. W., and Coryell, C. D. (1963). Nuclear Data Sheets, Appendix 5.
Goldhaber, M., and Scharff-Goldhaber, G. (1948). *Phys. Rev.* **73**, 1472.
Green, A. M., and Schucan, T. H. (1972). *Nuclear Phys.* **A188**, 289.
Hardy, J. C., Schmeing, H., Benenson, W., Crawley, G. M., Kashy, E., and Nann, H. (1974). *Phys. Rev.* **C9**, 252.
Hardy, J. C., Ball, G. C., Geiger, J. S., Graham, R. L., Macdonald, J. A., and Schmeing, H. (1974a). *Phys. Rev. Lett.* **33**, 320.
Holstein, B. R., and Treiman, S. B. (1971). *Phys. Rev.* **C3**, 1921.
Ichimura, M., Hyuga, H., and Brown, G. E. (1972). *Nuclear Phys.* **A196**, 17.
Inglis, D. R. (1953). *Rev. Modern Phys.* **25**, 390.

Int. Conf. Few Particle Probl. (1972). Univ. of California. North-Holland Publ., Amsterdam.
Isozumi, Y., and Shimizu, S. (1971). *Phys. Rev.* **C4**, 522.
Kim, C. W., and Fulton, T. (1971). *Phys. Rev.* **C4**, 390.
Konopinski, E. J. (1966). "The Theory of Beta Radioactivity." Oxford Univ. Press (Clarendon), London and New York.
Langer, L., and Moffat, D. (1952). *Phys. Rev.* **88**, 689.
Lee, T. D. (1971). *Phys. Rev. Lett.* **26**, 801.
Lipkin, H. J. (1971). *Phys. Rev. Lett.* **27**, 432.
Mayer, M. G., and Jensen, J. H. D. (1955). "Elementary Theory of Nuclear Shell Structure." Wiley, New York.
Marshak, R. E., Riazuddin, Ryan, C. P. (1969). "Theory of Weak Interactions in Particle Physics." Wiley, New York.
Mukhopadhyay, N. C., and Miller, L. D. (1973). *Phys. Lett.* **47B**, 415.
Okun, L. G. (1965). "Weak Interaction of Elementary Particles." Addison-Wesley, Reading, Massachusetts.
Pakvasa, S., and Tennakone, K. (1972). *Phys. Rev. Lett.* **28**, 1415.
Particle Data Group (1973). *Rev. Modern Phys.* **45**, S1.
Pati, J. C., and Salam, A. (1974). *Phys. Rev.* **D10**, 275.
Perlman, I., Asaro, F., and Michel, H. V. (1972). *Ann. Rev. Nuclear Sci.* **22**, 383.
Porter, F. T., Freedman, M. S., and Wagner, F. (1971). *Phys. Rev.* **C3**, 2246.
Puppi, G. (1962). *Proc. Int. Conf. High Energy Phys., CERN* p. 713 ff.
Reines, F., Cowan, C. L., Harrison, F. B., McGuire, A. D., and Kruse, H. W. (1960). *Phys. Rev.* **117**, 159.
Schopper, H. F. (1966). "Weak Interactions and Nuclear Beta Decay." North-Holland Publ., Amsterdam.
Sirlin, A. (1967). *Phys. Rev.* **164**, 1767.
Smith, G. R. (1971). *Phys. Rev.* **C4**, 1344.
Towner, I. S. (1973). *Nuclear Phys.* **A216**, 589.
Towner, I. S., and Hardy, J. C. (1973). *Nuclear Phys.* **A205**, 33.
Tribble, R. E., and Garvey, G. T. (1974). *Phys. Rev. Lett.* **32**, 314.
Visscher, W. M., and Ferrell, R. A. (1957). *Phys. Rev.* **107**, 781.
Weisberger, W. I. (1965). *Phys. Rev. Lett.* **14**, 1047.
Wilkinson, D. H. (1972). *Nuclear Phys.* **A179**, 289.
Wilkinson, D. H. (1973). *Phys. Rev.* **C7**, 930.
Wilkinson, D. H., and Alburger, D. E. (1971). *Phys. Rev. Lett.* **26**, 1127.
Wilkinson, D. H., and Marrs, R. E. (1972). *Nuclear Instrum. Methods* **105**, 505.
Wilkinson, D. C., Goosman, D. R., Alburger, D. E., and Marrs, R. E. (1972). *Phys. Rev.* **C6**, 1664.
Wolfenstein, L., and Henley, E. M. (1971). *Phys. Lett.* **36B**, 28.
Wu, C. S., and Moszkowski, S. A. (1966). "Beta Decay." Wiley, New York.
Wu, C. S., Ambler, E., Hayward, R. W., Hoppes, D. D., and Hudson, R. P. (1957). *Phys. Rev.* **105**, 1413.

Appendix A

COUPLING OF TWO ANGULAR MOMENTA, CLEBSCH–GORDAN COEFFICIENTS

The use of an explicit wave function for a state involving several nucleons that is an eigenstate for the system total angular momentum and its z component is frequently required. Usually such a system state is constructed from the individual single-particle states of the type discussed in Chapters IV and V. The problem encountered here is part of the general problem of compounding (or coupling) two or more of any of several angular momentum-like quantities, e.g., spin, orbital angular momentum, total angular momentum, or isospin, to yield a resultant that also has the characteristics of an appropriately corresponding angular momentum-like quantity.

The simplest nontrivial example is the coupling of two equal spins of $\frac{1}{2}$ each. As is well known, the four linear combinations of the "spin-up" and "spin-down" individual spin states α and β result in four system states, three comprising the triplet-spin state for the system (with $S = 1$ and $M_S = +1, 0, -1$) and one comprising the singlet-spin state for the system (with $S = 0$ and $M_S = 0$). These states are all cited in (1-12).

In this Appendix, we shall consider only the simple case of coupling two angular momenta. The separate momentum states are designated $\phi(j_1, m_1)$ and $\phi(j_2, m_2)$ and the system state $\psi(J, M)$. We require that $\psi(J, M)$ be an eigenstate of both the operators $\tilde{\mathbf{J}}^2$ and $\tilde{J}_z$, and seek the quantum mechanical equivalent to the classical vector addition

$$\mathbf{j}_1 + \mathbf{j}_2 = \mathbf{J}. \qquad \text{(A-1)}$$

The first implication of (A-1) is that

$$|j_1 - j_2| \leqslant J \leqslant j_1 + j_2 \qquad \text{(A-1a)}$$

and carries over to the quantum mechanical case as well; it is referred to as a *triangle condition*.

If the state $\psi(J, M)$ is to be composed of simple linear combinations of product states $\phi(j_1, m_1)\phi(j_2, m_2)$, then rotation of the coordinate frame about the z axis through an angle θ would give

$$\psi'(J, M) = \psi(J, M) e^{iM\theta}, \qquad \text{(A-2)}$$

while for the product states we would have

$$\begin{aligned}[\phi(j_1, m_1)\phi(j_2, m_2)]' &= \phi(j_1, m_1)\exp(im_1\theta)\phi(j_2, m_2)\exp(im_2\theta) \\ &= \phi(j_1, m_1)\phi(j_2, m_2)\exp[i(m_1 + m_2)\theta]. \qquad \text{(A-3)}\end{aligned}$$

Hence we conclude the well-known result that

$$M = m_1 + m_2. \qquad \text{(A-4)}$$

It generally happens that for a particular value of M several combinations of m_1 and m_2 satisfy (A-4). When this occurs $\psi(J, M)$ would be composed of a linear combination of such product states, and while $\psi(J, M)$ would be a simultaneous eigenfunction of the operators $\mathbf{j}_1{}^2$ and $\mathbf{j}_2{}^2$ as well as of $\mathbf{J}^2$ and J_z, it would not be an eigenfunction of either j_{1z} or j_{2z}. The general expression for $\psi(J, M)$ is

$$\psi(J, M) = \sum_{\substack{m_1, m_2 \\ m_1 + m_2 = M}} C(j_1 j_2 m_1 m_2 | JM)\phi(j_1, m_1)\phi(j_2, m_2), \qquad \text{(A-5)}$$

where the coefficients of the linear expansion are referred to as the *Clebsch–Gordan coefficients*.[1] The case when only one product state occurs is for the so-called "stretched" configuration, namely $J = j_1 + j_2$ and $M = \pm J$. In this case $m_1 = +j_1$ and $m_2 = +j_2$ or $m_1 = -j_1$ and $m_2 = -j_2$ and only these possibilities exist for satisfying (A-4).

The phases for the Clebsch–Gordan coefficients are chosen to make all the

[1] We shall drop the C designation and simply write $(j_1 j_2 m_1 m_2 | JM)$ for the Clebsch–Gordan expansion coefficients.

elements real. These coefficients are also constructed to conserve normalization, i.e., the sum of absolute values squared of the Clebsch–Gordan coefficients appearing in (A-5) is unity. Also note that condition (A-4) reduces (A-5) to a simple sum on either m_1 or m_2.

Although we have written (A-5) in terms of symbols suggesting total angular momentum, j_1 and j_2 could also stand for any other angular momentum-like quantities, e.g., each referring to spin, or one referring to spin and the other to orbital angular momentum, and so on. The only constraint in addition to the triangle condition is that j_1+j_2-J should always be an integer.

Expressions for $(j_1 j_2 m_1 m_2|JM)$ in closed analytic form are rather cumbersome for the general case; however, for $j_2 = \frac{1}{2}$ or 1 and $j_1 =$ anything, rather simple relationships exist as shown in Tables A-1 and A-2.

TABLE A-1

CLEBSCH–GORDAN COEFFICIENTS $(j_1 j_2 m_1 m_2|JM)$ FOR $j_2=\frac{1}{2}$

	$m_2=+\frac{1}{2}$; thus $m_1 = M-\frac{1}{2}$	$m_2=-\frac{1}{2}$; thus $m_1 = M+\frac{1}{2}$
$J=j_1+\frac{1}{2}, M$	$[(j_1+M+\frac{1}{2})/(2j_1+1)]^{1/2}$	$[(j_1-M+\frac{1}{2})/(2j_1+1)]^{1/2}$
$J=j_1-\frac{1}{2}, M$	$-[(j_1-M+\frac{1}{2})/(2j_1+1)]^{1/2}$	$[(j_1+M+\frac{1}{2})/(2j_1+1)]^{1/2}$

The Clebsch–Gordan coefficients can also be used to generate the inverse transformation of states, viz.,

$$\phi(j_1 m_1)\phi(j_2 m_2) = \sum_{\substack{J,M \\ M=m_1+m_2 \\ |j_1-j_2|\leq J\leq j_1+j_2}} (j_1 j_2 m_1 m_2|JM)\psi(J,M). \quad \text{(A-6)}$$

It is clear that the state described by (A-6) is *not* an eigenfunction of $\mathbf{J}^2$, although it is of J_z. It will be observed that the same table is used in (A-6) as in (A-5), albeit involving different elements. For (A-5) one employs the elements appearing in an appropriate *row*, while for (A-6) one employs the elements appearing under an appropriate *column*.

For example, with $j_1 = j_2 = \frac{1}{2}$, e.g., two intrinsic spins of $\frac{1}{2}$, we have for $J=1$ and $M=+1$, using the first row of Table A-1,

$$\psi(1,+1) = \chi_1{}^1(1,2) = \phi_1(\tfrac{1}{2},+\tfrac{1}{2})\phi_2(\tfrac{1}{2},+\tfrac{1}{2}) = \alpha(1)\alpha(2) \quad \text{(A-7)}$$

since

$$[(j_1+M+\tfrac{1}{2})/(2j_1+1)]^{1/2} = 1 \quad \text{and} \quad [(j_1-M+\tfrac{1}{2})/(2j_1+1)]^{1/2} = 0.$$

TABLE A-2

CLEBSCH–GORDAN COEFFICIENTS $(j_1 j_2 m_1 m_2|JM)$ FOR $j_2 = 1$

	$m_2 = +1$; thus $m_1 = M-1$	$m_2 = 0$; thus $m_1 = M$	$m_2 = -1$; thus $m_1 = M+1$
$J = j_1+1, M$	$\left[\frac{(j_1+M)(j_1+M+1)}{(2j_1+1)(2j_1+2)}\right]^{1/2}$	$\left[\frac{(j_1-M+1)(j_1+M+1)}{(2j_1+1)(j_1+1)}\right]^{1/2}$	$\left[\frac{(j_1-M)(j_1-M+1)}{(2j_1+1)(2j_1+2)}\right]^{1/2}$
$J = j_1, M$	$-\left[\frac{(j_1+M)(j_1-M+1)}{2j_1(j_1+1)}\right]^{1/2}$	$\frac{M}{[j_1(j_1+1)]^{1/2}}$	$\left[\frac{(j_1-M)(j_1+M+1)}{2j_1(j_1+1)}\right]^{1/2}$
$J = j_1-1, M$	$\left[\frac{(j_1-M)(j_1-M+1)}{2j_1(2j_1+1)}\right]^{1/2}$	$-\left[\frac{(j_1-M)(j_1+M)}{j_1(2j_1+1)}\right]^{1/2}$	$\left[\frac{(j_1+M+1)(j_1+M)}{2j_1(2j_1+1)}\right]^{1/2}$

Also we clearly have

$$\psi(1,0) = \chi_1{}^0(1,2) = 2^{-1/2}[\alpha(1)\beta(2)+\beta(1)\alpha(2)], \tag{A-8}$$

etc. We also note that employing the elements appearing in the second column of Table A-1 for the case $m_1 = +\frac{1}{2}$ and $m_2 = -\frac{1}{2}$, and hence $M = 0$, we have

$$\phi_1(\tfrac{1}{2},\tfrac{1}{2})\,\phi_2(\tfrac{1}{2},-\tfrac{1}{2}) = \alpha(1)\beta(2) = 2^{-1/2}[\chi_1{}^0(1,2)+\chi_0{}^0(1,2)]. \tag{A-9}$$

The results (A-8) and (A-9) are, of course, the familiar results for the coupling of two spin-$\frac{1}{2}$ particles, but they would also equally well apply to the coupling of two $1\mathrm{p}_{1/2}$ single-particle shell model states in *jj* coupling. For example, two $1\mathrm{p}_{1/2}$ nucleons could couple to form a state with $J = 1$ and $M = 0$, viz.,

$$\begin{aligned}\psi[(1\mathrm{p}_{1/2})^1, (1\mathrm{p}_{1/2})^1 &: J=1,\, M=0] \\ &= 2^{-1/2}[\phi_1(1\mathrm{p}_{1/2}, m=+\tfrac{1}{2})\,\phi_2(1\mathrm{p}_{1/2}, m=-\tfrac{1}{2}) \\ &\quad + \phi_1(1\mathrm{p}_{1/2}, m=-\tfrac{1}{2})\,\phi_2(1\mathrm{p}_{1/2}, m=+\tfrac{1}{2})].\end{aligned} \tag{A-10}$$

As the final example consider coupling a spin $\frac{1}{2}$ with the orbital angular momentum $l = 2$ (i.e., a nucleon in a d state) to form the wave function $\psi(\mathrm{d}_{5/2}, m_j = +\frac{1}{2})$. We have, using the first row of Table A-1,

$$\psi(\mathrm{d}_{5/2}, m_j = +\tfrac{1}{2}) = \sqrt{\tfrac{3}{5}}\,Y_2{}^0(\theta,\varphi)\,\alpha + \sqrt{\tfrac{2}{5}}\,Y_2{}^1(\theta,\varphi)\,\beta. \tag{A-11}$$

Table A-3 gives some of the Clebsch–Gordan coefficients of Tables A-1 and A-2 in evaluated form for a few selected cases. For convenience explicit expressions for some of the spherical harmonics are also included.

Useful properties of Clebsch–Gordan coefficients and the underlying theory of angular momentum are available in excellent specialized texts suitable for students [e.g., Rose (1957), or Edmonds (1957)]. We cite here only three useful expressions:

$$(j_1 j_2 m_1 m_2|JM) = (-1)^{j_1+j_2-J}(j_2 j_1 m_2 m_1|JM), \tag{A-12}$$

$$(j_1 j_2 m_1 m_2|JM) = (-1)^{j_1+j_2-J}(j_1 j_2, -m_1, -m_2|J, -M), \tag{A-13}$$

and

$$(j_1 j_2 m_1 m_2|JM) = (-1)^{j_2+m_2}\left(\frac{2J+1}{2j_1+1}\right)^{1/2}(j_2 J, -m_2 M\,|\,j_1 m_1). \tag{A-14}$$

Many additional properties are best exhibited by a change in notation, giving the Clebsch–Gordan coefficients in terms of the *Wigner 3-j symbols*

$$(j_1 j_2 m_1 m_2|JM) = (-1)^{j_1-j_2-M}(2J+1)^{1/2}\begin{pmatrix} j_1 & j_2 & J \\ m_1 & m_2 & M \end{pmatrix}. \tag{A-15}$$

TABLE A-3

SELECT CLESBCH–GORDAN COEFFICIENTS AND SPHERICAL HARMONICS[a]

Notation:

J	J	...
M	M	...
m_1 m_2	Coefficients	
m_1 m_2		
. .		

1/2 × 1/2

m_1	m_2	1 +1
+1/2	+1/2	1

m_1	m_2	1 0	0 0
+1/2	-1/2	1/2	1/2
-1/2	+1/2	1/2	-1/2

m_1	m_2	1 -1
-1/2	-1/2	1

1 × 1/2

m_1	m_2	3/2 +3/2
+1	+1/2	1

m_1	m_2	3/2 +1/2	1/2 +1/2
+1	-1/2	1/3	2/3
0	+1/2	2/3	-1/3

m_1	m_2	3/2 -1/2	1/2 -1/2
0	-1/2	2/3	1/3
-1	+1/2	1/3	-2/3

m_1	m_2	3/2 -3/2
-1	-1/2	1

2 × 1

m_1	m_2	3 +3
+2	+1	1

m_1	m_2	3 +2	2 +2
+2	0	1/3	2/3
+1	+1	2/3	-1/3

m_1	m_2	3 +1	2 +1	1 +1
+2	-1	1/15	1/3	3/5
+1	0	8/15	1/6	-3/10
0	+1	6/15	-1/2	1/10

m_1	m_2	3 0	2 0	1 0
+1	-1	1/5	1/2	3/10
0	0	3/5	0	-2/5
-1	+1	1/5	-1/2	3/10

m_1	m_2	3 -1	2 -1	1 -1
0	-1	6/15	1/2	1/10
-1	0	8/15	-1/6	-3/10
-2	+1	1/15	-1/3	3/5

m_1	m_2	3 -2	2 -2
-1	-1	2/3	1/3
-2	0	1/3	-2/3

m_1	m_2	3 -3
-2	-1	1

1 × 1

m_1	m_2	2 +2
+1	+1	1

m_1	m_2	2 +1	1 +1
+1	0	1/2	1/2
0	+1	1/2	-1/2

m_1	m_2	2 0	1 0	0 0
+1	-1	1/6	1/2	1/3
0	0	2/3	0	-1/3
-1	+1	1/6	-1/2	1/3

m_1	m_2	2 -1	1 -1
0	-1	1/2	1/2
-1	0	1/2	-1/2

m_1	m_2	2 -2
-1	-1	1

2 × 1/2

m_1	m_2	5/2 +5/2
+2	1/2	1

m_1	m_2	5/2 3/2	3/2 +3/2
+2	-1/2	1/5	4/5
+1	+1/2	4/5	-1/5

m_1	m_2	5/2 +1/2	3/2 +1/2
+1	-1/2	2/5	3/5
0	+1/2	3/5	-2/5

m_1	m_2	5/2 -1/2	3/2 -1/2
0	-1/2	3/5	2/5
-1	+1/2	2/5	-3/5

m_1	m_2	5/2 -3/2	3/2 -3/2
-1	-1/2	4/5	1/5
-2	+1/2	1/5	-4/5

m_1	m_2	5/2 -5/2
-2	-1/2	1

3/2 × 1/2

m_1	m_2	2 +2
+3/2	+1/2	1

m_1	m_2	2 +1	1 +1
+3/2	-1/2	1/4	3/4
+1/2	+1/2	3/4	-1/4

m_1	m_2	2 0	1 0
+1/2	-1/2	1/2	1/2
-1/2	+1/2	1/2	-1/2

m_1	m_2	2 -1	1 -1
-1/2	-1/2	3/4	1/4
-3/2	+1/2	1/4	-3/4

m_1	m_2	2 -2
-3/2	-1/2	1

3/2 × 1

m_1	m_2	5/2 +5/2
+3/2	+1	1

m_1	m_2	5/2 +3/2	3/2 +3/2
+3/2	0	2/5	3/5
+1/2	+1	3/5	-2/5

m_1	m_2	5/2 +1/2	3/2 +1/2	1/2 +1/2
+3/2	-1	1/10	2/5	1/2
+1/2	0	3/5	1/15	-1/3
-1/2	+1	3/10	-8/15	1/6

m_1	m_2	5/2 -1/2	3/2 -1/2	1/2 -1/2
+1/2	-1	3/10	8/15	-1/6
-1/2	0	3/5	-1/15	-1/3
-3/2	+1	1/10	-2/5	1/2

m_1	m_2	5/2 -3/2	3/2 -3/2
-1/2	-1	3/5	2/5
-3/2	0	2/5	-3/5

m_1	m_2	5/2 -5/2
-3/2	-1	1

Note: A $\sqrt{}$ is to be understood over every coefficient; e.g., for -8/15 read $-\sqrt{8/15}$.

$$Y_0{}^0 = (1/4\pi)^{1/2}$$

$$Y_1{}^0 = (3/4\pi)^{1/2} \cos\theta;\ Y_1{}^1 = -(3/8\pi)^{1/2} \sin\theta\, e^{i\varphi}$$

$$Y_2{}^0 = (5/4\pi)^{1/2}(\tfrac{3}{2}\cos^2\theta - \tfrac{1}{2});\ Y_2{}^1 = -(15/8\pi)^{1/2}\sin\theta\cos\theta\, e^{i\varphi};\ Y_2{}^2 = \tfrac{1}{4}(15/2\pi)^{1/2}\sin^2\theta\, e^{2i\varphi}$$

$$Y_3{}^0 = (7/4\pi)^{1/2}(\tfrac{5}{2}\cos^3\theta - \tfrac{3}{2}\cos\theta);\ Y_3{}^1 = -\tfrac{1}{4}(21/4\pi)^{1/2}\sin\theta(5\cos^2\theta - 1)e^{i\varphi}$$

$$Y_3{}^2 = \tfrac{1}{4}(105/2\pi)^{1/2}\sin^2\theta\cos\theta\, e^{2i\varphi};\ Y_3{}^3 = -\tfrac{1}{4}(35/4\pi)^{1/2}\sin^3\theta\, e^{3i\varphi}$$

$$(Y_l{}^m)^* = (-1)^m Y_l{}^{-m}$$

[a] From Particle Data Group (1973).

Extensive tabulations of the 3-*j* symbols appear in many works; one of the more useful is Rotenberg *et al.* (1959). For a rather practical closed-form general expression for Clebsch–Gordan coefficients see Sato and Kaguei (1972).

Finally, a simple vector diagram illustrating (A-5) can be constructed as shown in Fig. A.1. This diagram, the basis of the semiclassical model, portrays the system state in terms of the coupled vectors $\mathbf{j}_1$ and $\mathbf{j}_2$ precessing about the resultant vector $\mathbf{J}$, which in turn precesses about the z axis. At all times the magnitude of the total angular momentum, i.e., $\mathbf{J}^2 = (\mathbf{j}_1 + \mathbf{j}_2)^2$, and the value

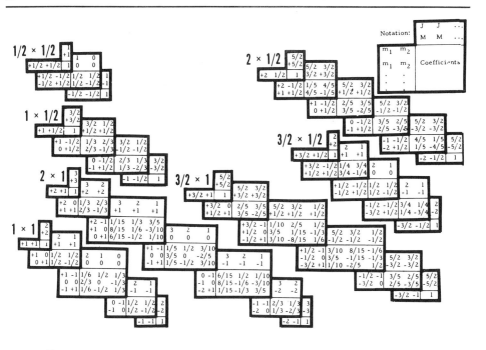

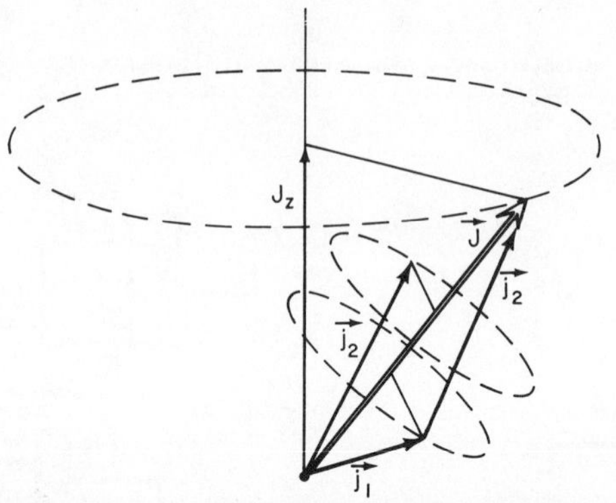

FIG. A.1. The vector model description of the coupled angular momentum $\mathbf{J}=\mathbf{j}_1+\mathbf{j}_2$, illustrating the precessional motion of the relevant quantities.

of J_z are constants of the motion. Clearly the projections of j_1 and j_2 onto the z axis, i.e., j_{1z} and j_{2z}, cannot be constants. This is taken to correspond to the requirement of several terms in the sum in (A-5). Only in the classical correspondence principle limit $J = M$ and $J \gg 1$ would j_{1z} and j_{2z} become constants of the motion. This corresponds to the analog of the "stretched" case discussed in connection with (A-5).

REFERENCES

Edmonds, A. R. (1957). "Angular Momentum in Quantum Mechanics." Princeton Univ. Press, Princeton, New Jersey.

Particle Data Group (1973). *Rev. Mod. Phys.* **45**, S1.

Rose, M. E. (1957). "Elementary Theory of Angular Momentum." Wiley, New York.

Rotenberg, M., Bivins, R., Metropolis, N., and Wooten, J. K. (1959). "The 3-j and 6-j Symbols." Technology Press, MIT, Cambridge, Massachusetts.

Sato, M., and Kaguei, S. (1972). *Phys. Lett.* **42B**, 21.

Appendix B

THE WIGNER–ECKART THEOREM

The necessity often arises of evaluating matrix elements of tensor operators, an important subclass of which includes the vector operators. These calculations are often considerably simplified by using a particularly useful theorem, referred to as the *Wigner–Eckart theorem*. While many derivations of this theorem exist for a generalized tensor operator, a particularly illuminating treatment for the special case of vector operators is given by Bethe and Jackiw (1968). We shall closely follow their presentation.

One basic definition of a vector operator $\tilde{\mathbf{A}}$ is that the expectation values of its components $\langle A_x \rangle$, $\langle A_y \rangle$, and $\langle A_z \rangle$ transform under the rotations of the coordinate frame in the same way as corresponding classical vector components. It is a standard demonstration in texts on quantum mechanics that this definition, coupled with the representation of the general rotation operator $U_{\mathrm{R}} = \exp[-(i/\hbar)\,\hat{\mathbf{n}}\cdot\mathbf{J}\theta]$ to describe a rotation about an axis $\hat{\mathbf{n}}$ through an angle θ, gives the result

$$[A_\alpha, J_\beta] = [J_\alpha, A_\beta] = i\hbar A_\gamma, \tag{B-1a}$$

and

$$[A_\alpha, J_\alpha] = 0 \tag{B-1b}$$

with α, β, and γ standing for x, y, z and their cyclic permutations.[1] We additionally introduce the raising and lowering operators $J_\pm$ and $J_0 = J_z$ as well as the *modified* spherical vector components for **A**, namely $A_\pm$ and A_0, defined as[2]

$$J_\pm = J_x \pm iJ_y, \qquad J_0 = J_z \tag{B-2a}$$

$$A_\pm = A_x \pm iA_y, \qquad A_0 = A_z. \tag{B-2b}$$

It readily follows from (B-1) and (B-2) that

$$[A_\pm, J_0] = \mp \hbar A_\pm \tag{B-3a}$$

$$[A_\pm, J_\pm] = 0 \tag{B-3b}$$

$$[A_\pm, J_\mp] = \pm 2\hbar A_0 \tag{B-3c}$$

$$[A_0, J_0] = 0 \tag{B-3d}$$

$$[A_0, J_\pm] = \pm \hbar A_\pm. \tag{B-3e}$$

We shall consider a state vector $|\lambda JM\rangle$ which is the simultaneous eigenvector of the Hamiltonian H, the angular momentum operators $\mathbf{J}^2$ and J_z, and other relevant commuting operators Λ. The matrix elements of these operators are thus diagonal in this representation and

$$H|\lambda JM\rangle = E|\lambda JM\rangle \tag{B-4a}$$

$$\mathbf{J}^2|\lambda JM\rangle = J(J+1)\hbar^2|\lambda JM\rangle \tag{B-4b}$$

$$J_z|\lambda JM\rangle = M\hbar|\lambda JM\rangle \tag{B-4c}$$

$$\Lambda|\lambda JM\rangle = \lambda|\lambda JM\rangle. \tag{B-4d}$$

We also presume a λ dependence such that the raising and lowering operators $J_\pm$ produce the standard result

$$J_\pm|\lambda JM\rangle = C|\lambda J, M\pm 1\rangle \tag{B-5a}$$

with

$$C = \hbar[(J\mp M)(J\pm M+1)]^{1/2}, \qquad \text{independent of } \lambda. \tag{B-5b}$$

Since

$$\langle J'M'|J_\pm|JM\rangle = C\,\delta_{J'J}\,\delta_{M',M\pm 1}, \tag{B-6}$$

[1] The commutation relationships (B-1) can also be used as an equivalent definition for a vector operator **A**.

[2] In what is to follow, we essentially relate the rotational properties of the vector operator **A** to that of the total angular momentum. It is thus important to treat these two quantities on the same footing. We therefore use the modified spherical components (B-2b) rather than the standard $A_{\pm 1} = \mp(1/\sqrt{2})(A_x \pm iA_y)$ and $A_0 = A_z$, and emphasize this difference by writing $A_\pm$ instead of the usual $A_{\pm 1}$.

we can conveniently write

$$J_\pm|\lambda JM\rangle = \langle J, M\pm 1|J_\pm|JM\rangle|\lambda J, M\pm 1\rangle \tag{B-7a}$$

$$\langle\lambda JM|J_\pm = \langle JM|J_\pm|J, M\mp 1\rangle\langle\lambda J, M\mp 1| \tag{B-7b}$$

$$J_0|\lambda JM\rangle = \langle JM|J_0|JM\rangle|\lambda JM\rangle = M\hbar|\lambda JM\rangle. \tag{B-7c}$$

Now consider taking the matrix elements of $[A_\pm, J_0]$ [we need consider only $J' = J$ from here on, in view of (B-6) and the obvious corresponding result for J_0]; then

$$\langle\lambda' JM'|A_\pm J_0 - J_0 A_\pm|\lambda JM\rangle = (M-M')\hbar\langle\lambda' JM'|A_\pm|\lambda JM\rangle. \tag{B-8}$$

Using (B-3a), we have

$$\mp\hbar\langle\lambda' JM'|A_\pm|\lambda JM\rangle = (M-M')\hbar\langle\lambda' JM'|A_\pm|\lambda JM\rangle, \tag{B-9}$$

and hence $M-M' = \mp 1$ or $M' = M\pm 1$, for otherwise the matrix element vanishes. We therefore conclude that $A_\pm$ has nonvanishing matrix elements precisely for those cases where the matrix elements of $J_\pm$ are nonvanishing. The ratio of such corresponding matrix elements (at this point possibly M dependent) can be written

$$\xi_\pm = \langle\lambda' J, M\pm 1|A_\pm|\lambda JM\rangle/\langle J, M\pm 1|J_\pm|JM\rangle. \tag{B-10}$$

Next we take the matrix elements of $[A_\pm, J_\pm]$. In view of (B-3b), we get

$$\langle\lambda' JM'|A_\pm J_\pm - J_\pm A_\pm|\lambda JM\rangle = 0, \tag{B-11}$$

or using (B-7a) and (B-7b),

$$\langle\lambda' JM'|A_\pm|\lambda J, M\pm 1\rangle\langle J, M\pm 1|J_\pm|JM\rangle$$
$$= \langle\lambda' J, M'\mp 1|A_\pm|\lambda JM\rangle\langle JM'|J_\pm|J, M'\mp 1\rangle. \tag{B-12}$$

In view of (B-9) and (B-10), the matrix elements $\langle\lambda' JM'|A_\pm|\lambda J, M\pm 1\rangle$ and $\langle\lambda' J, M'\mp 1|A_\pm|\lambda JM\rangle$ appearing in (B-12) are nonzero only if the magnetic quantum number of the bra is one unit larger than the magnetic quantum number of the ket for the case A_+, and one unit smaller for the case A_-. Hence, $M' = (M+1)+1$, $M' = (M-1)-1$, $M'-1 = M+1$, and $M'+1 = M-1$, all of which are consistent with $M' = M\pm 2$ for the two cases $A_\pm$. Substituting $M' = M\pm 2$ into (B-12) gives

$$\langle\lambda' J, M\pm 2|A_\pm|\lambda J, M\pm 1\rangle\langle JM\pm 1|J_\pm|JM\rangle$$
$$= \langle\lambda' J, M\pm 1|A_\pm|\lambda JM\rangle\langle J, M\pm 2|J_\pm|J, M\pm 1\rangle,$$

which we can rewrite as

$$\frac{\langle \lambda' J, M\pm 2|A_\pm|\lambda J, M\pm 1\rangle}{\langle J, M\pm 2|J_\pm|J, M\pm 1\rangle} = \frac{\langle \lambda' J, M\pm 1|A_\pm|\lambda JM\rangle}{\langle J, M\pm 1|J_\pm|JM\rangle} = \xi_\pm. \tag{B-13}$$

We thus observe that $\xi_\pm$ is independent of M.

Finally, take matrix elements of $[A_\pm, J_\mp]$ and use (B-3c) to obtain

$$\begin{aligned}\pm 2\hbar\langle \lambda' JM'|A_0|\lambda JM\rangle &= \langle \lambda' JM'|A_\pm|\lambda J, M\mp 1\rangle\langle J, M\mp 1|J_\mp|JM\rangle \\ &\quad - \langle \lambda' J, M'\pm 1|A_\pm|\lambda JM\rangle\langle JM'|J_\mp|J, M'\pm 1\rangle.\end{aligned} \tag{B-14}$$

Again the matrix elements involving $A_\pm$ on the right-hand side of (B-14) require that $M' = M$, otherwise they each vanish. Inserting $M' = M$ in (B-14) and using (B-13), we get

$$\begin{aligned}\pm 2\hbar\langle \lambda' JM|A_0|\lambda JM\rangle &= \xi_\pm\{\langle JM|J_\pm|J, M\mp 1\rangle\langle J, M\mp 1|J_\mp|JM\rangle \\ &\quad - \langle J, M\pm 1|J_\pm|JM\rangle\langle JM|J_\mp|J, M\pm 1\rangle\}.\end{aligned} \tag{B-15}$$

Evaluating the quantity inside the curly bracket using (B-5a) and (B-5b) gives, after a little algebra,

$$\pm 2\hbar\langle \lambda' JM|A_0|\lambda JM\rangle = \pm 2\xi_\pm\hbar^2 M = \pm 2\hbar\xi_\pm\langle JM|J_0|JM\rangle. \tag{B-16}$$

Thus

$$\xi_0 = \langle \lambda' JM|A_0|\lambda JM\rangle/\langle JM|J_0|JM\rangle = \xi_\pm. \tag{B-17}$$

Clearly, ξ does not depend either on M or on whether we are dealing with A_+, A_-, or A_0, although, of course, it may depend on λ, λ', and J. It is convenient to rewrite ξ, referred to as the *reduced matrix element*, in standard notation as

$$\xi = \langle \lambda' J||\mathbf{A}||\lambda J\rangle, \tag{B-18}$$

and to summarize (B-10), (B-13), and (B-17) for the various spherical components of (B-2) by writing

$$\langle \lambda' JM'|\tilde{\mathbf{A}}|\lambda JM\rangle = \langle \lambda' J||\mathbf{A}||\lambda J\rangle\langle JM'|\tilde{\mathbf{J}}|JM\rangle. \tag{B-19}$$

Equation (B-19) is the specialized version of the Wigner–Eckart theorem appropriate for vector operators. Some special cases involving (B-19) are of interest. Since (B-19) applies to any component of $\mathbf{A}$, consider the projection on the vector $\mathbf{J}$.[1] Then

$$\begin{aligned}\langle \lambda' JM'|\mathbf{A}\cdot\mathbf{J}|\lambda JM\rangle &= \langle \lambda' J||\mathbf{A}||\lambda J\rangle\langle JM'|\mathbf{J}^2|JM\rangle \\ &= J(J+1)\hbar^2\delta_{MM'}\langle \lambda' J||\mathbf{A}||\lambda J\rangle.\end{aligned} \tag{B-20}$$

[1] Note that $\mathbf{A}\cdot\mathbf{J} = A_x J_x + A_y J_y + A_z J_z$ and also $[A_\alpha, J_\alpha] = 0$, $\alpha = x, y, z$, hence $\mathbf{A}\cdot\mathbf{J} = \mathbf{J}\cdot\mathbf{A}$. Thus the inner product is defined unambiguously.

Also, we can further write

$$\langle \lambda' JM' | A_z | \lambda JM \rangle = M\hbar\, \delta_{MM'} \langle \lambda' J || \mathbf{A} || \lambda J \rangle. \tag{B-21}$$

Combining (B-20) and (B-21) to eliminate the reduced matrix element, we obtain

$$\langle \lambda' JM | A_z | \lambda JM \rangle = [M/J(J+1)\hbar] \langle \lambda' JM | \mathbf{A} \cdot \mathbf{J} | \lambda JM \rangle. \tag{B-22}$$

Equation (B-22) is also the basis for the semiclassical precessing vector model and represents a form of the Landé formula. It can be interpreted to state that for a system characterized by a definite angular momentum and z component of angular momentum, the time average of the z component of an arbitrary vector observable A_z can be obtained by considering **A** to precess about **J** and to simultaneously have **J** precess about the z axis. The component A_z is then obtained by first projecting **A** onto **J**, obtaining

$$A_J = \frac{\mathbf{A} \cdot \mathbf{J}}{|\mathbf{J}|} = \frac{\mathbf{A} \cdot \mathbf{J}}{\hbar [J(J+1)]^{1/2}},$$

and then projecting A_J onto the z axis through the cosine factor $M/[J(J+1)]^{1/2}$, giving the final result

$$A_z = \frac{M}{J(J+1)\hbar} \mathbf{A} \cdot \mathbf{J}. \tag{B-22a}$$

This situation is illustrated in Fig. B.1.

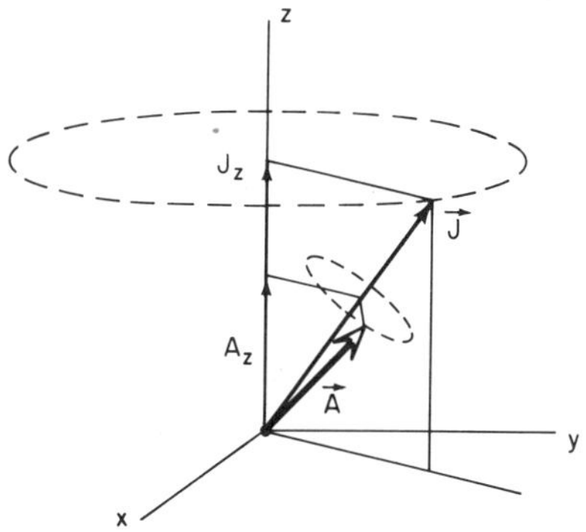

FIG. B.1. The semiclassical precessing vector model.

Finally, we should point out that $\tilde{\mathbf{A}}$ is, of course, any vector operator, such as the magnetic dipole moment operator discussed in Chapter II. In the *jj*-coupling scheme with $\mathbf{J} = \sum_i \mathbf{j}_i$, **A** could be any one of the vector operators $\mathbf{j}_i$. In the *LS* or Russell–Saunders coupling scheme with $\mathbf{L} = \sum_i \boldsymbol{l}_i$ and

$\mathbf{S} = \sum_i \mathbf{s}_i$, we can identify $\mathbf{L}$ with $\mathbf{J}$ and any one of the $\boldsymbol{l}_i$ with $\mathbf{A}$, and similarly for $\mathbf{S}$ and any one of the $\mathbf{s}_i$.

The results (B-19)–(B-22) can be generalized for the case $J' \neq J$ and indeed to the case of more complicated operators than simply vector operators. Excellent thorough discussions for these cases can be found in numerous references [e.g., de-Shalit and Talmi (1963) or Rose (1957)]. For our purposes, it is sufficient to define the relevant larger class of operators, the irreducible spherical tensor operators T_l^m of rank l, as a set of $2l+1$ operators $(m = -l, -l+1, \ldots, l)$ which transform under infinitesimal rotations according to the relationship

$$\delta T_l^m(\mathbf{r}) = T_l^m(\mathbf{r}) - T_l^m(\mathbf{r}') = (i/\hbar)\, \delta\theta(\hat{\mathbf{n}}\cdot\mathbf{J})\, T_l^m(\mathbf{r}). \tag{B-23}$$

Such operators can be shown to obey the commutation rules

$$[J_\pm, T_l^m] = \hbar[(l \mp m)(l \pm m+1)]^{1/2} T_l^{m\pm 1} \tag{B-24a}$$

and

$$[J_0, T_l^m] = \hbar m T_l^m. \tag{B-24b}$$

Identifying $T_1^{\,0} = A_0$, $T_1^{\,1} = A_{+1}$, and $T_1^{-1} = A_{-1}$,[1] we see that (B-24) reduces to the corresponding equation (B-3).

Following a procedure somewhat similar to the foregoing, it develops that a reduced matrix element $\langle J' || T_l || J \rangle$ can be defined such that the matrix elements of the tensor operator T_l^m, $\langle J'M' | T_l^m | JM \rangle$, are simply appropriate Clebsch–Gordan coefficients multiplying this reduced matrix element, viz.,

$$\langle \lambda' J' M' | T_l^m | \lambda J M \rangle = (JlMm | J'M') \langle \lambda' J' || T_l || \lambda J \rangle. \tag{B-25}$$

Here we have also made explicit reference to the additional quantum numbers λ. Equation (B-25) is the general statement of the Wigner–Eckart theorem. One very important consequence of (B-25) is that, unless the triangle condition $\mathbf{J} + \boldsymbol{l} = \mathbf{J}'$ and the condition $M + m = M'$ can be met, the matrix element identically vanishes.

The angular momentum operator components can also be written in the spherical tensor form $T_1^{\pm 1} = J_{\pm 1}$ and $T_1^{\,0} = J_0$ and (B-25) applied. In particular for the $J_0 = J_z$ component,

$$\langle \lambda' J' M' | J_z | \lambda J M \rangle = M\hbar\, \delta_{JJ'}\, \delta_{MM'}\, \delta_{\lambda\lambda'} = (J1M0 | J'M') \langle \lambda' J' || \mathbf{J} || \lambda J \rangle. \tag{B-26}$$

From Table A-2 we have for $J = J'$ and $M = M'$,

$$(J1M0 | JM) = M/[J(J+1)]^{1/2},$$

[1] $\mathbf{A}$ expressed in this form is an irreducible first-rank spherical tensor. We must recall further that $A_{\pm 1} = \mp(1/\sqrt{2})A_\pm$, connecting the standard spherical components with the modified ones of (B-3). For example, $[J_\mp, A_{\pm 1}] = \sqrt{2}\hbar A_0$; hence $[A_\pm, J_\mp] = \pm 2\hbar A_0$, etc.

and hence[1]

$$\langle \lambda' J' || \mathbf{J} || \lambda J \rangle = [J(J+1)]^{1/2} \hbar \, \delta_{\lambda\lambda'} \, \delta_{JJ'}. \tag{B-27}$$

If (B-25) is applied to a vector operator **A** as well as to **J**, we have

$$\langle \lambda' J' M' | A_\mu | \lambda J M \rangle = (\langle \lambda' J' || \mathbf{A} || \lambda J \rangle / \langle \lambda' J' || \mathbf{J} || \lambda J \rangle) \langle \lambda' J' M' | J_\mu | \lambda J M \rangle, \tag{B-28}$$

where $\mu = 0, \pm 1$. For the special case of $J = J'$ and the particular component of **A** given by the scalar product $\mathbf{J} \cdot \mathbf{A}$, we expect from (B-25) that

$$\langle \lambda' J M' | \mathbf{J} \cdot \mathbf{A} | \lambda J M \rangle = C \langle \lambda' J || \mathbf{A} || \lambda J \rangle, \tag{B-29}$$

where the constant C must be independent of the nature of **A**. Hence **C** can be evaluated for $\mathbf{A} = \mathbf{J}$, whence from (B-27) and the matrix element of $\mathbf{J}^2$ we have

$$C = [J(J+1)]^{1/2} \hbar \, \delta_{MM'}.$$

This gives

$$\langle \lambda' J || \mathbf{A} || \lambda J \rangle = \{1/[J(J+1)]^{1/2} \hbar\} \langle \lambda' J M | \mathbf{J} \cdot \mathbf{A} | \lambda J M \rangle.$$

Finally, when this result is substituted into (B-28), we arrive at the important relationship

$$\langle \lambda' J M' | A_\mu | \lambda J M \rangle = [\langle \lambda' J M | \mathbf{J} \cdot \mathbf{A} | \lambda J M \rangle / \hbar^2 J(J+1)] \langle J M' | J_\mu | J M \rangle. \tag{B-30}$$

This result, when specialized for $\mu = 0$, reduces precisely to (B-22).

References

Bethe, H. A., and Jackiw, R. (1968). "Intermediate Quantum Mechanics." Benjamin, New York.
de-Shalit, A., and Talmi, I. (1963). "Nuclear Shell Theory." Academic Press, New York.
Rose, M. E. (1957). "Elementary Theory of Angular Momentum." Wiley, New York.

[1] Not infrequently, the factor $\hbar$ is omitted in equations such as (B-3)–(B-27) and the following. A less trivial difference in convention (e.g., de-Shalit and Talmi, 1963) defines the reduced matrix elements with an addition factor $(2J'+1)^{1/2}$. Referring to such reduced matrix elements by the notation $(J'||T_l||J)$, i.e., enclosed in parentheses, we have $(2J'+1)^{1/2} \langle J' || T_l || J \rangle = (J' || T_l || J)$. The reader must simply be on his guard in these matters.

Appendix C

BRIEF REVIEW OF DIRAC THEORY

The purpose of this appendix is to remind the reader of the salient features of the relativistic theory for fermions, and in view of the various conventions used in the literature to select one consistent set for the purposes of this text. For a suitable first treatment, reference to any of the numerous texts dealing with the subject in a thorough manner is a necessity.

A direct translation of the relativistic energy equation for the field-free particle of rest mass m

$$E^2 = p^2c^2 + m^2c^4 \tag{C-1}$$

into its quantum mechanical equivalent yields the Klein–Gordon equation

$$(\Box^2 - \mu^2)\psi = 0 \tag{C-2a}$$

with

$$\Box^2 = \nabla^2 - (1/c^2)(\partial^2/\partial t^2) \qquad \text{and} \qquad \mu^{-1} = \hbar/mc. \tag{C-2b}$$

The linearization of the "square-root equation" obtained from (C-2a) is accomplished by attempting to construct an operator for the Hamiltonian of the form

$$\tilde{H}_{\mathrm{D}} = c\tilde{\boldsymbol{\alpha}} \cdot \tilde{\mathbf{p}} + \tilde{\beta}mc^2. \tag{C-3}$$

It can be shown that $\tilde{\boldsymbol{\alpha}} = (\tilde{\alpha}_x, \tilde{\alpha}_y, \tilde{\alpha}_z)$ and $\tilde{\beta}$, and hence $\tilde{H}_D$, are at the minimum 4×4 matrices that are essentially unique up to a similarity transformation. The Dirac first-order linear differential equation results in

$$i\hbar(\partial/\partial t)\psi = \tilde{H}_D\psi = (c\tilde{\boldsymbol{\alpha}}\cdot\tilde{\mathbf{p}} + mc^2\tilde{\beta})\psi, \tag{C-4}$$

in which $\tilde{\boldsymbol{\alpha}}$ and $\tilde{\beta}$ are 4×4 constant matrices (i.e., independent of $\mathbf{r}$ and t) and ψ is a four-element column matrix or Dirac spinor.[1] In order for the square of (C-4) to be identical with (C-2a), the commutation relations follow, viz.,

$$\begin{aligned}
\alpha_i^2 &= I, \quad i = x,y,z; \qquad \beta^2 = I \\
\alpha_i\alpha_j + \alpha_j\alpha_i &= 0, \qquad i,j = x,y,z; \quad i \neq j \\
\alpha_i\beta + \beta\alpha_i &= 0, \qquad i = x,y,z.
\end{aligned} \tag{C-5}$$

Since the matrices α_x, α_y, α_z, and β do not commute among themselves, only one at most can appear as a diagonal matrix in any representation. The so-called *standard representation* results when β is taken as diagonal. Then it develops that

$$\alpha_x = \begin{pmatrix} 0 & 0 & 0 & 1 \\ 0 & 0 & 1 & 0 \\ 0 & 1 & 0 & 0 \\ 1 & 0 & 0 & 0 \end{pmatrix} = \begin{pmatrix} 0 & \sigma_x^{\mathrm{P}} \\ \sigma_x^{\mathrm{P}} & 0 \end{pmatrix} \tag{C-6a}$$

$$\alpha_y = \begin{pmatrix} 0 & 0 & 0 & -i \\ 0 & 0 & i & 0 \\ 0 & -i & 0 & 0 \\ i & 0 & 0 & 0 \end{pmatrix} = \begin{pmatrix} 0 & \sigma_y^{\mathrm{P}} \\ \sigma_y^{\mathrm{P}} & 0 \end{pmatrix} \tag{C-6b}$$

$$\alpha_z = \begin{pmatrix} 0 & 0 & 1 & 0 \\ 0 & 0 & 0 & -1 \\ 1 & 0 & 0 & 0 \\ 0 & -1 & 0 & 0 \end{pmatrix} = \begin{pmatrix} 0 & \sigma_z^{\mathrm{P}} \\ \sigma_z^{\mathrm{P}} & 0 \end{pmatrix} \tag{C-6c}$$

[1] Operators such as $\tilde{\mathbf{p}}$ are trivially written

$$\tilde{\mathbf{p}} = \tilde{\mathbf{p}}I_4 = \tilde{\mathbf{p}}\begin{pmatrix} 1 & 0 & 0 & 0 \\ 0 & 1 & 0 & 0 \\ 0 & 0 & 1 & 0 \\ 0 & 0 & 0 & 1 \end{pmatrix} = \begin{pmatrix} \tilde{\mathbf{p}} & 0 & 0 & 0 \\ 0 & \tilde{\mathbf{p}} & 0 & 0 \\ 0 & 0 & \tilde{\mathbf{p}} & 0 \\ 0 & 0 & 0 & \tilde{\mathbf{p}} \end{pmatrix}.$$

The identity matrix $I \equiv 1 \equiv$ unit matrix, when its dimensionality is to be identified, can be written for the 4×4 matrices as

$$I_4 = \begin{pmatrix} 1 & 0 & 0 & 0 \\ 0 & 1 & 0 & 0 \\ 0 & 0 & 1 & 0 \\ 0 & 0 & 0 & 1 \end{pmatrix}.$$

$$\beta = \begin{pmatrix} 1 & 0 & 0 & 0 \\ 0 & 1 & 0 & 0 \\ 0 & 0 & -1 & 0 \\ 0 & 0 & 0 & -1 \end{pmatrix} = \left(\begin{array}{c|c} I_2 & 0 \\ \hline 0 & -I_2 \end{array}\right). \tag{C-6d}$$

The last term in each of these equalities has been written in *partitioned form* where it is understood that the operators such as σ_x^{P} (the x component of the Pauli spin matrix) are 2×2 submatrices, including the zero $\left(\begin{smallmatrix}0 & 0\\ 0 & 0\end{smallmatrix}\right)$ and the 2×2 identity or unity operator $I_2 = \left(\begin{smallmatrix}1 & 0\\ 0 & 1\end{smallmatrix}\right)$. The four-component spinors can also be partitioned so that

$$\psi = \begin{pmatrix} a \\ b \\ c \\ d \end{pmatrix} = \left(\begin{array}{c} u \\ \hline v \end{array}\right) \quad \text{with} \quad u = \begin{pmatrix} a \\ b \end{pmatrix}, \quad v = \begin{pmatrix} c \\ d \end{pmatrix}; \tag{C-7}$$

where u and v are two-element Pauli spinors.[1]

The set of equations (C-6) can be compactly written as

$$\boldsymbol{\alpha} = \left(\begin{array}{c|c} 0 & \boldsymbol{\sigma}^{\mathrm{P}} \\ \hline \boldsymbol{\sigma}^{\mathrm{P}} & 0 \end{array}\right), \qquad \beta = \left(\begin{array}{c|c} 1 & 0 \\ \hline 0 & -1 \end{array}\right). \tag{C-8}$$

We also have the operator[2]

$$\boldsymbol{\alpha}\cdot\mathbf{p} = \sum_{i=x,y,z} \alpha_i p_i = \sum_{i=x,y,z} \left(\begin{array}{c|c} 0 & \sigma_i^{\mathrm{P}} \\ \hline \sigma_i^{\mathrm{P}} & 0 \end{array}\right)\left(\begin{array}{c|c} p_i & 0 \\ \hline 0 & p_i \end{array}\right)$$

$$= \left(\begin{array}{c|c} 0 & \sigma_x^{\mathrm{P}} p_x + \sigma_y^{\mathrm{P}} p_y + \sigma_z^{\mathrm{P}} p_z \\ \hline \sigma_x^{\mathrm{P}} p_x + \sigma_y^{\mathrm{P}} p_y + \sigma_z^{\mathrm{P}} p_z & 0 \end{array}\right)$$

$$= \left(\begin{array}{c|c} 0 & \boldsymbol{\sigma}^{\mathrm{P}}\cdot\mathbf{p} \\ \hline \boldsymbol{\sigma}^{\mathrm{P}}\cdot\mathbf{p} & 0 \end{array}\right), \tag{C-9}$$

[1] We note that β can be used to define the projection operators $\frac{1}{2}(1\pm\beta)$. The operator $\frac{1}{2}(1+\beta)$ projects ψ onto the subspace $\beta = +1$ and gives

$$\tfrac{1}{2}(1+\beta)\psi = \begin{pmatrix} a \\ b \\ 0 \\ 0 \end{pmatrix} = \left(\begin{array}{c} u \\ \hline 0 \end{array}\right),$$

similarly

$$\tfrac{1}{2}(1-\beta)\psi = \left(\begin{array}{c} 0 \\ \hline v \end{array}\right).$$

Thus u and v relate to the eigenspinors of the diagonal operator β with eigenvalues $\beta = \pm 1$.

[2] Matrix multiplication is so defined that the product of two fully developed matrices is exactly equal to the product of their partitioned matrices using the same rules of multiplication and treating the submatrices as the elements of the partitioned matrices.

where, of course,

$$\sigma_x{}^{\mathrm{P}} p_x = \begin{pmatrix} 0 & 1 \\ 1 & 0 \end{pmatrix} p_x \begin{pmatrix} 1 & 0 \\ 0 & 1 \end{pmatrix} = \begin{pmatrix} 0 & 1 \\ 1 & 0 \end{pmatrix} \begin{pmatrix} p_x & 0 \\ 0 & p_x \end{pmatrix}$$

$$= \begin{pmatrix} 0 & p_x \\ p_x & 0 \end{pmatrix}, \qquad \text{etc.}$$

When a particle with charge q is simultaneously acted on by an electromagnetic potential characterized by the four-vector $(\mathbf{A}, iA^0)$ and perhaps by some unspecified additional force derivable from a potential V, the Dirac Hamiltonian becomes

$$\tilde{H}_{\mathrm{D}} = c\tilde{\boldsymbol{\alpha}} \cdot \tilde{\boldsymbol{\pi}} + mc^2\tilde{\beta} + q\tilde{A}^0 + \tilde{V}, \tag{C-10}$$

where we write[1]

$$\tilde{\boldsymbol{\pi}} = \tilde{\mathbf{p}} - (q/c)\tilde{\mathbf{A}}.$$

In partitioned form the corresponding time-independent wave equation $H_{\mathrm{D}}\psi = E\psi$ becomes

$$\left(\begin{array}{c|c} (qA^0 + V + mc^2) & c\boldsymbol{\sigma}^{\mathrm{P}} \cdot \boldsymbol{\pi} \\ \hline c\boldsymbol{\sigma}^{\mathrm{P}} \cdot \boldsymbol{\pi} & (qA^0 + V - mc^2) \end{array}\right) \left(\begin{array}{c} u_l \\ \hline u_s \end{array}\right) = E \left(\begin{array}{c} u_l \\ \hline u_s \end{array}\right). \tag{C-11}$$

This equation is equivalent to the two coupled Pauli equations

$$(qA^0 + V + mc^2)\, u_l + c(\boldsymbol{\sigma}^{\mathrm{P}} \cdot \boldsymbol{\pi})\, u_s = E u_l \tag{C-12a}$$

and

$$c(\boldsymbol{\sigma}^{\mathrm{P}} \cdot \boldsymbol{\pi})\, u_l + (qA^0 + V - mc^2)\, u_s = E u_s. \tag{C-12b}$$

It is important to introduce two additional operators, $\boldsymbol{\sigma}^{\mathrm{D}}$ and ρ. These are defined in partitioned form as the Dirac spin operator

$$\boldsymbol{\sigma}^{\mathrm{D}} = \left(\begin{array}{c|c} \boldsymbol{\sigma}^{\mathrm{P}} & 0 \\ \hline 0 & \boldsymbol{\sigma}^{\mathrm{P}} \end{array}\right), \tag{C-13}$$

and

$$\rho = -i\alpha_x \alpha_y \alpha_z = \left(\begin{array}{c|c} 0 & 1 \\ \hline 1 & 0 \end{array}\right). \tag{C-14}$$

Evidently $\boldsymbol{\sigma}^{\mathrm{D}} = \rho\boldsymbol{\alpha} = \boldsymbol{\alpha}\rho$ and $\boldsymbol{\alpha} = \rho\boldsymbol{\sigma}^{\mathrm{D}} = \boldsymbol{\sigma}^{\mathrm{D}}\rho$; also $\rho^2 = 1$. The spin angular momentum character of $\boldsymbol{\sigma}^{\mathrm{D}}$ is made manifest by the "commutation relations" $\boldsymbol{\sigma}^{\mathrm{D}} \times \boldsymbol{\sigma}^{\mathrm{D}} = 2i\boldsymbol{\sigma}^{\mathrm{D}}$ for the Dirac 4×4 spin matrices in exact analogy to the

[1] In the present sense the charge on an electron is $q = -e$.

relations $\boldsymbol{\sigma}^{\mathrm{P}} \times \boldsymbol{\sigma}^{\mathrm{P}} = 2i\boldsymbol{\sigma}^{\mathrm{P}}$ for the Pauli 2×2 spin matrices. It is readily discovered that while neither the orbital angular momentum operator

$$\mathbf{L} = \mathbf{r} \times \mathbf{p} = \left(\begin{array}{c|c} \mathbf{L} & 0 \\ \hline 0 & \mathbf{L} \end{array}\right) \tag{C-15}$$

nor the spin operator $\boldsymbol{\sigma}^{\mathrm{D}}$ commutes with any general Dirac Hamiltonian H_{D}, the operator $\mathbf{J}$ defined as

$$\mathbf{J} = \mathbf{L} + \tfrac{1}{2}\hbar\boldsymbol{\sigma}^{\mathrm{D}} = \left(\begin{array}{c|c} \mathbf{L}+\frac{1}{2}\hbar\boldsymbol{\sigma}^{\mathrm{P}} & 0 \\ \hline 0 & \mathbf{L}+\frac{1}{2}\hbar\boldsymbol{\sigma}^{\mathrm{P}} \end{array}\right), \tag{C-16}$$

does.[1]

There are important differences in the behavior of angular momentum in the relativistic and nonrelativistic cases. Consider the simple case of the ordinary central potential $\mathbf{A} = 0$, $A^0 = 0$, and $V(\mathbf{r}) = V(|\mathbf{r}|) = V(r)$ in (C-10), so that

$$H_{\mathrm{D}}(\text{central}) = c\boldsymbol{\alpha} \cdot \mathbf{p} + mc^2\beta + V(r). \tag{C-17}$$

Contrary to the nonrelativistic case, we now find

$$[\mathbf{L}, H_{\mathrm{D}}] \neq 0, \qquad [\mathbf{L}^2, H_{\mathrm{D}}] \neq 0, \qquad [\boldsymbol{\sigma}^{\mathrm{D}}, H_{\mathrm{D}}] \neq 0. \tag{C-18}$$

On the other hand,

$$[\mathbf{J}, H_{\mathrm{D}}] = 0, \qquad [(\boldsymbol{\sigma}^{\mathrm{D}})^2, H_{\mathrm{D}}] = 0 \;[\text{since } (\boldsymbol{\sigma}^{\mathrm{D}})^2 = 3], \qquad [\mathbf{J}^2, H_{\mathrm{D}}] = 0. \tag{C-19}$$

Thus, while for the nonrelativistic case of the central potential, $\mathbf{L}$, $\mathbf{L}^2$, $\mathbf{S}$, $\mathbf{S}^2$, $\mathbf{J}$, and $\mathbf{J}^2$ were all constants of the motion, in the relativistic case, only $\mathbf{J}$, $\mathbf{J}^2$, and $\mathbf{S}^2$ are.

It is very useful to examine the simple case of a freely propagating particle in some detail. It would appear natural to attempt to solve the time-dependent wave equation (C-4) in this case by trying to write $\psi(\mathbf{r}, t)$ in the form

$$\psi(\mathbf{r}, t) = \phi \exp[i(\mathbf{k}\cdot\mathbf{r} - \omega t)] \tag{C-20}$$

with $\hbar\mathbf{k} = \mathbf{p}$ and

$$\phi = \begin{pmatrix} u_1 \\ u_2 \\ u_3 \\ u_4 \end{pmatrix} = \left(\begin{array}{c} u_l \\ \hline u_s \end{array}\right).$$

[1] Even for the field-free case $\mathbf{A} = 0$, $A^0 = 0$, $V = 0$, we have, for example, $[L_z, H_{\mathrm{D}}] = \hbar^2 c[\alpha_x(\partial/\partial y) - \alpha_y(\partial/\partial x)]$, which in general is not zero. However, since $\frac{1}{2}\hbar[\sigma_z{}^{\mathrm{D}}, H_{\mathrm{D}}]$ is just the negative of $[L_z, H_{\mathrm{D}}]$, the operator J_z defined by (C-16) does commute with H_{D}. A similar result obtains for the other two components.

If $\psi(\mathbf{r}, t)$ is to be an eigenspinor of the Hamiltonian, it readily follows that $E = \hbar\omega$ and

$$E^2 = p^2c^2 + m^2c^4 \qquad \text{or} \qquad \omega^2/c^2 = \mathrm{k}^2 + \mu^2,$$

with $\mu^{-1} = \hbar/mc$. We then arrive at the famous conclusion that for a particle of mass m propagating with a real momentum $\mathbf{p}$, the total energy E can be either greater or less than zero, viz.,

$$E = \pm(p^2c^2 + m^2c^4)^{1/2} \qquad \text{or} \qquad \omega/c = \pm(\mathrm{k}^2 + \mu^2)^{1/2}. \tag{C-21}$$

We wish to examine the consequence of these two possibilities for the state (C-20). With $\mathbf{A} = 0$, $A^0 = 0$, and $V = 0$, (C-12) becomes

$$(E - mc^2)\, u_l = c(\boldsymbol{\sigma}^{\mathrm{P}} \cdot \mathbf{p})\, u_{\mathrm{s}} \tag{C-22a}$$

$$(E + mc^2)\, u_{\mathrm{s}} = c(\boldsymbol{\sigma}^{\mathrm{P}} \cdot \mathbf{p})\, u_l. \tag{C-22b}$$

We distinguish the two cases $E > 0$ and $E < 0$. For the positive-energy state, the use of (C-22b) is most convenient in relating u_{s} in terms of u_l. This avoids the difficulty of the singularity that would result if (C-22a) were used to write u_l in terms of u_{s} and $E \to mc^2$ (i.e., the nonrelativistic limit). Thus we take

$$u_{\mathrm{s}} = [c(\boldsymbol{\sigma}^{\mathrm{P}} \cdot \mathbf{p})/(E + mc^2)]\, u_l. \tag{C-23}$$

For the low-velocity, nonrelativistic case, the relative order of magnitude of $|u_{\mathrm{s}}|$ and $|u_l|$ can be seen to be

$$|u_{\mathrm{s}}| \approx (cp/2mc^2)\, |u_l| = \tfrac{1}{2}(v/c)\, |u_l|.$$

For $E > 0$, then, we can speak of u_{s} as the "small component" of ϕ, and u_l as the "large component" of ϕ. To be precise,

$$\phi = \begin{pmatrix} u_l \\ \hdashline [c(\boldsymbol{\sigma}^{\mathrm{P}} \cdot \mathbf{p})/(E + mc^2)]\, u_l \end{pmatrix}, \tag{C-24}$$

with u_l an arbitrary Pauli spinor, $u_l = \binom{u_1}{u_2}$. We note that

$$\boldsymbol{\sigma}^{\mathrm{P}} \cdot \mathbf{p} = \begin{pmatrix} 0 & 1 \\ 1 & 0 \end{pmatrix} p_x + \begin{pmatrix} 0 & -i \\ i & 0 \end{pmatrix} p_y + \begin{pmatrix} 1 & 0 \\ 0 & -1 \end{pmatrix} p_z = \begin{pmatrix} p_z & p_- \\ p_+ & -p_z \end{pmatrix}, \tag{C-25}$$

with $p_\pm = p_x \pm i p_y$.

Two obvious choices for u_l are the two orthogonal states $u_l = \alpha$, the Pauli eigenspinor of $\sigma_z = +1$, and $u_l = \beta$, the Pauli spinor for the state $\sigma_z = -1$.

Thus we get, with N a suitable normalization constant,

$$E>0,\ \alpha; \qquad \psi_+^{\alpha}(\mathbf{r},t) = N\exp[i\hbar^{-1}(\mathbf{p}\cdot\mathbf{r}-Et)]\begin{pmatrix} 1 \\ 0 \\ cp_z/(E+mc^2) \\ c(p_x+ip_y)/(E+mc)^2 \end{pmatrix} \tag{C-26a}$$

and

$$E>0,\ \beta; \qquad \psi_+^{\beta}(\mathbf{r},t) = N\exp[i\hbar^{-1}(\mathbf{p}\cdot\mathbf{r}-Et)]\begin{pmatrix} 0 \\ 1 \\ c(p_x-ip_y)/(E+mc^2) \\ -cp_z/(E+mc^2) \end{pmatrix}. \tag{C-26b}$$

When the particle is in the negative-energy state, we use (C-22a) to relate u_l in terms of u_s, viz.,

$$u_l = -[c(\boldsymbol{\sigma}^{\mathrm{P}}\cdot\mathbf{p})/(|E|+mc^2)]u_s. \tag{C-27}$$

For $E<0$ and $E\to -mc^2$, we see that this time $|u_l|\approx\frac{1}{2}(v/c)|u_s|$, and hence u_s is now the larger component. Thus the labels and the convention of "large" and "small" refer only to the case $E>0$. Taking $u_s=\alpha$ or β gives

$$E<0,\ \alpha; \qquad \psi_-^{\alpha}(\mathbf{r},t) = N\exp[i\hbar^{-1}(\mathbf{p}\cdot\mathbf{r}+|E|t)]$$

$$\times\begin{pmatrix} -cp_z/(|E|+mc^2) \\ -c(p_x+ip_y)/(|E|+mc^2) \\ 1 \\ 0 \end{pmatrix} \tag{C-28a}$$

$$E<0,\ \beta; \qquad \psi_-^{\beta}(\mathbf{r},t) = N\exp[i\hbar^{-1}(\mathbf{p}\cdot\mathbf{r}+|E|t)]$$

$$\times\begin{pmatrix} -c(p_x-ip_y)/(|E|+mc^2) \\ cp_z/(|E|+mc^2) \\ 0 \\ 1 \end{pmatrix}. \tag{C-28b}$$

If the Dirac spinors in (C-26) and (C-28) are to be normalized to unity, the normalization constant for all four cases can be written $N^2=(|E|+mc^2)/2|E|$. This normalization is, however, not Lorentz invariant.

We note that in the low-velocity limit $v/c\to 0$, the states (C-26a), (C-26b), (C-28a), and (C-28b) are eigenspinors of the Dirac spin operator σ_z^{D}. For example,

$$\sigma_z{}^D \phi_+{}^\alpha = \begin{pmatrix} 1 & & & \\ & -1 & & \bigcirc \\ & & 1 & \\ \bigcirc & & & -1 \end{pmatrix} \begin{pmatrix} 1 \\ 0 \\ 0 \\ 0 \end{pmatrix} = +1 \begin{pmatrix} 1 \\ 0 \\ 0 \\ 0 \end{pmatrix}.$$

This, however, is not the case when the small component is taken to have its general value rather than its limiting value of zero. In the general case

$$\sigma_z{}^D \phi_+^{\alpha,\beta} = \begin{pmatrix} \sigma_z{}^P & 0 \\ 0 & \sigma_z{}^P \end{pmatrix} \begin{pmatrix} \alpha, \beta \\ [c(\boldsymbol{\sigma}^P \cdot \mathbf{p})/(E+mc^2)]\alpha, \beta \end{pmatrix}$$

$$= \begin{pmatrix} +\alpha, -\beta \\ [c(\boldsymbol{\sigma}^P \cdot \hat{z})(\boldsymbol{\sigma}^P \cdot \mathbf{p})/(E+mc^2)]\alpha, \beta \end{pmatrix}$$

The operators[1] $(\boldsymbol{\sigma}^P \cdot \hat{z})$ and $(\boldsymbol{\sigma}^P \cdot \mathbf{p})$ do not commute in general, i.e., $[\sigma_z{}^P, (\sigma_x{}^P p_x + \sigma_y{}^P p_y + \sigma_z{}^P p_z)] \neq 0$. If they had commuted, these operators on the small component could have been inverted in their order giving just $\pm u_s$, and $\sigma_z{}^D \phi_+^{\alpha,\beta} = \pm \phi_+^{\alpha,\beta}$ would have resulted. Only in the special case $\mathbf{p} = p_z \hat{z}$, i.e., a plane wave traveling in the z direction, is $\phi^{\alpha,\beta}$ an eigenspinor of $\sigma_z{}^D$.[2]

The question then arises concerning the proper invariant spin description for propagation in an arbitrary direction. The preceding situation suggests introducing an operator, the *helicity operator*,

$$h \equiv h(\hat{\mathbf{p}}) = \boldsymbol{\sigma}^D \cdot \hat{\mathbf{p}} = \boldsymbol{\sigma}^D \cdot \mathbf{p}/|\mathbf{p}| = \begin{pmatrix} \boldsymbol{\sigma}^P \cdot \hat{\mathbf{p}} & 0 \\ 0 & \boldsymbol{\sigma}^P \cdot \hat{\mathbf{p}} \end{pmatrix}, \quad \text{(C-29a)}$$

with

$$\boldsymbol{\sigma}^P \cdot \hat{\mathbf{p}} = \begin{pmatrix} p_z/p & p_-/p \\ p_+/p & -p_z/p \end{pmatrix} = \begin{pmatrix} \cos\theta & e^{-i\varphi}\sin\theta \\ e^{i\varphi}\sin\theta & -\cos\theta \end{pmatrix}, \quad \text{(C-29b)}$$

and

$$p_z = p\cos\theta, \quad p_\rho = p\sin\theta, \quad p_\rho \cos\varphi = p_x, \quad p_\rho \sin\varphi = p_y;$$

also, $p \equiv |\mathbf{p}|$.

It is straightforward to show that $[(\boldsymbol{\alpha}\cdot\mathbf{p}), (\boldsymbol{\sigma}^D \cdot \mathbf{p})] = 0$, and that hence $[h, H_D] = 0$. In addition, the useful theorem

$$(\boldsymbol{\sigma}^D \cdot \mathbf{A})(\boldsymbol{\sigma}^D \cdot \mathbf{B}) = \mathbf{A} \cdot \mathbf{B} + i\boldsymbol{\sigma}^D \cdot (\mathbf{A} \times \mathbf{B}), \quad \text{(C-30)}$$

---

[1] As in the main text, the caret symbol over a quantity indicates a unit vector, e.g., $\hat{a} = \mathbf{a}/|\mathbf{a}|$.

[2] The fact that σ_z can be a constant of the motion for the special case of a plane wave propagating in the z direction does not contradict (C-18) since

$$[\sigma_z{}^D, H_D] = 2\hbar c[\alpha_y(\partial/\partial x) - \alpha_x(\partial/\partial y)] = 0$$

for such a wave function, i.e., the wave function is not a function of either x or y.

provided **A** and **B** commute with $\boldsymbol{\sigma}^{\mathrm{D}}$, immediately gives

$$h^2 = 1; \qquad \text{thus} \quad h = \pm 1. \tag{C-31}$$

Thus the eigenvalue of the helicity operator, i.e., the component of the spin in the direction of the linear momentum, can be either ± 1. When the eigenvalue is $+1$, one speaks of a *right-handed* helicity state, and when it is -1, of a *left-handed* state. Figure C.1 illustrates these two situations.

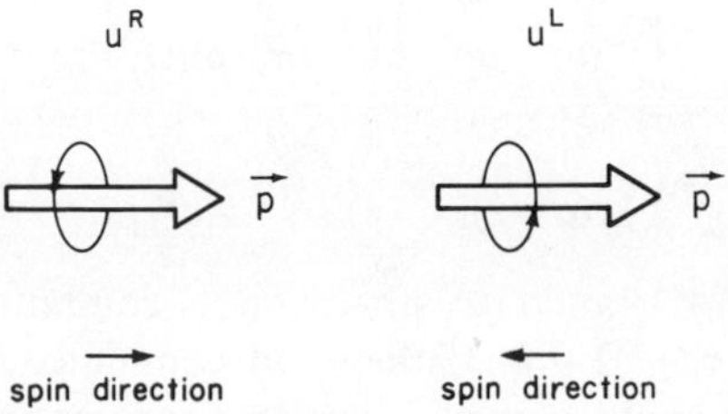

FIG. C.1. Schematic representation of the right-handed and left-handed helicity states for a freely propagating particle.

The construction of the eigenvectors of helicity $h = \pm 1$ can proceed in any one of a number of ways, as, for example, by a rotation of the coordinates to correspond to having the new z axis in the direction of propagation. Perhaps the most straightforward and entirely equivalent method is to employ the obvious 2×2 Pauli projection operators $\frac{1}{2}(1 \pm \boldsymbol{\sigma}^{\mathrm{P}} \cdot \hat{\mathbf{p}})$, which can be used to construct Pauli eigenspinors for $\sigma(\hat{\mathbf{p}}) = \pm 1$ (up to a normalization), viz.,[1]

$$\tfrac{1}{2}(1 + \boldsymbol{\sigma}^{\mathrm{P}} \cdot \hat{\mathbf{p}}) \begin{pmatrix} 1 \\ 0 \end{pmatrix} = \tfrac{1}{2} \begin{pmatrix} 1 + \cos\theta \\ e^{i\varphi} \sin\theta \end{pmatrix} = \cos\tfrac{1}{2}\theta \begin{pmatrix} \cos\frac{1}{2}\theta \\ e^{i\varphi} \sin\frac{1}{2}\theta \end{pmatrix}.$$

We define the normalized Pauli spinors

$$u^{\mathrm{R}}(\hat{\mathbf{p}}) \equiv u^{\mathrm{R}} = \begin{pmatrix} \cos\frac{1}{2}\theta \\ e^{i\varphi} \sin\frac{1}{2}\theta \end{pmatrix} \tag{C-32a}$$

and

$$u^{\mathrm{L}}(\hat{\mathbf{p}}) \equiv u^{\mathrm{L}} = \begin{pmatrix} -e^{-i\varphi} \sin\frac{1}{2}\theta \\ \cos\frac{1}{2}\theta \end{pmatrix}. \tag{C-32b}$$

[1] The multiplicative factor $\cos\frac{1}{2}\theta$ indicates that the α state $\binom{1}{0}$ has positive components in all directions except $\theta = \pi$, i.e., the α state is orthogonal to the β state. The spinor is obviously properly normalized if this factor is omitted. Since spinors are defined only up to an arbitrary phase, the factor $e^{i\varphi}$ can appear in various ways.

These results are well known from elementary Pauli spin theory.[1] Thus for example, for $E > 0$, $h = +1$, we take for (C-24) $u_l = u^{\mathrm{R}}$ and obtain

$$\phi^{\mathrm{R}} = \begin{pmatrix} u^{\mathrm{R}} \\ \hdashline [c(\boldsymbol{\sigma}^{\mathrm{P}}\cdot\mathbf{p})/(E+mc^2)]u^{\mathrm{R}} \end{pmatrix} \quad \text{or} \quad \phi^{\mathrm{R}} = \begin{pmatrix} \cos\frac{1}{2}\theta \\ e^{i\varphi}\sin\frac{1}{2}\theta \\ [cp/(E+mc^2)]\cos\frac{1}{2}\theta \\ [cp/(E+mc^2)]e^{i\varphi}\sin\frac{1}{2}\theta \end{pmatrix}. \tag{C-33}$$

The other state, $h = -1$, for $E > 0$ and the two-helicity states for $E < 0$ can be similarly constructed. The normalization constant in all cases is the same as for (C-26) and (C-28).

Dirac proposed a novel interpretation of the negative-energy states to the wave equation (C-10) which are mathematically required to form a complete set of states. The presence of these states introduces nontrivial considerations. Even if at $t = 0$ we begin with a state $\psi(0)$ which is a superposition of only positive-energy states, $\psi(t)$ in general will have negative-energy states present. For example, if an electron at $t = 0$ in a positive-energy state E_+ is subjected to an oscillating electric field of frequency ω, it will at some later time have a nonzero probability of making a transition to states of energy $E_+ + \hbar\omega$ and $E_+ - \hbar\omega$. Clearly, if $\hbar\omega > E_+ + mc^2$, the latter transition is to a negative energy. Under these circumstances the ground state of the hydrogen atom would not be stable either, but would decay in a transition to a negative-energy state with the emission of a photon.

The historic suggestion put forward was to imagine that what we normally consider the "vacuum" is, in fact, an infinite sea of electrons occupying all the negative-energy states. An electron, if present in a positive-energy state, being a fermion, is then forbidden to make a transition to a negative energy. It was further postulated that physical observables correspond only to *deviations* from this vacuum state. Interactions with one of the electrons in a negative-energy state could induce a transition in which the electron appears in a positive-energy state, while at the same time leaving behind a "hole" or vacancy in the sea of negative-energy states. Dirac interpreted the hole to represent a particle of the same mass as the electron but of opposite charge, namely the positron. The positron is also referred to as the *antiparticle* of the electron. The interaction considered above would then result in the simultaneous appearance of an electron–positron pair; the process is referred to as pair production. Once such a hole exists, an electron having a positive energy could in fact make a transition to this "vacancy" with the emission of a photon. In this process an electron–positron pair annihilate with the emission of a photon.

[1] It is readily verified that for the spinor u^{R}, for example, the expectation values for σ_x, σ_y, and σ_z are $\langle\sigma_z\rangle = \cos\theta$, $\langle\sigma_x\rangle = \sin\theta\cos\varphi$, and $\langle\sigma_y\rangle = \sin\theta\sin\varphi$, as they should be.

Thus an electron (charge $-e$) *missing* from state (C-28a) (i.e., the state with momentum $\mathbf{p}$, $E < 0$, "spin state" α) is equivalent, insofar as physical observables are concerned, to a *real* positron (charge $+e$, or *absence* of negative charge) with momentum $-\mathbf{p}$, $E > 0$ (i.e., *absence* of negative energy), and "spin state" β. If the wave function for a particle is designated by $\psi(\mathbf{r}, t)$, we associate with it a wave function $[\psi(\mathbf{r}, t)]_C$ for the corresponding antiparticle, such that expectation values for various quantities are as shown in Table C-1. The wave function $[\psi(\mathbf{r}, t)]_C$ is referred to as the *charge conjugate* wave function of $\psi(\mathbf{r}, t)$.

TABLE C-1

Particle	$\rightleftharpoons$	Antiparticle
$\psi(\mathbf{r}, t)$	$\rightleftharpoons$	$[\psi(\mathbf{r}, t)]_C$
$\langle \mathbf{p} \rangle = \mathbf{p}$		$\langle \mathbf{p} \rangle = -\mathbf{p}$
$\langle \boldsymbol{\sigma} \rangle = \boldsymbol{\sigma}$		$\langle \boldsymbol{\sigma} \rangle = -\boldsymbol{\sigma}$
$\langle \text{mass} \rangle = m$		$\langle \text{mass} \rangle = m$
$\langle \text{charge} \rangle = q$		$\langle \text{charge} \rangle = -q$
$\langle H(q) \rangle = E$		$\langle H(-q) \rangle = -E$

It is readily verified that corresponding to an electron in state (C-26a), we have

$$[\psi_+{}^\alpha(\mathbf{r}, t)]_C = N \exp[-i\hbar^{-1}(\mathbf{p}\cdot\mathbf{r} - |E|t)] \begin{pmatrix} c(p_x - ip_y)/(|E| + mc^2) \\ -cp_z/(|E| + mc^2) \\ 0 \\ 1 \end{pmatrix}. \tag{C-34}$$

If the same construction, conforming to Table C-1, is attempted for the three other states of (C-26) and (C-28), it is discovered that we can write, up to an overall phase which is arbitrary in any event,[1]

$$[\psi(\mathbf{r}, t)]_C = \pm i\beta\alpha_y [\psi(\mathbf{r}, t)]^* \tag{C-35}$$

and

$$i\beta\alpha_y = \begin{pmatrix} 0 & 0 & 0 & 1 \\ 0 & 0 & -1 & 0 \\ 0 & -1 & 0 & 0 \\ 1 & 0 & 0 & 0 \end{pmatrix}. \tag{C-36}$$

[1] This operation is called *charge conjugation*. Rather straightforward use of (C-10) shows that $[\psi(\mathbf{r}, t)]_C$ satisfies the same Dirac equation as $\psi(\mathbf{r}, t)$ with $q \to -q$. Using the commutation rules (C-5), it follows that $\psi(\mathbf{r}, t) = \pm i\beta\alpha_y [\psi(\mathbf{r}, t)]_C{}^*$. The explicit operator description of charge conjugation depends on the representation used for $\boldsymbol{\alpha}$ and β. The present form is for β diagonal as in (C-6d).

The charge conjugate wave functions representing positrons in the hole theory have to be handled "artfully," and a number of interpretations of these wave functions are possible. For example, the time factor $e^{(i/\hbar)|E|t}$ appearing in (C-34), which was constructed for a negative-energy state, can also be interpreted to correspond to a positive energy, but with the particle propagating backward in time. According to this view a *positron* would be represented by an *electron propagating backward in time.*

Setting aside the precise interpretation of the space–time factors in (C-26) and (C-28) and their charge conjugate wave functions, we can assign all the four Dirac spinors to *real* particles ($E > 0$) corresponding to a momentum $\mathbf{p} = p_x \hat{x} + p_y \hat{y} + p_z \hat{z}$ (up to an arbitrary phase):

$$\phi^{\alpha}_{\text{electron}} = N \begin{pmatrix} 1 \\ 0 \\ cp_z/(|E|+mc^2) \\ c(p_x+ip_y)/(|E|+mc^2) \end{pmatrix} \tag{C-37a}$$

$$\phi^{\beta}_{\text{electron}} = N \begin{pmatrix} 0 \\ 1 \\ c(p_x-ip_y)/(|E|+mc^2) \\ -cp_z/(|E|+mc^2) \end{pmatrix} \tag{C-37b}$$

$$\phi^{\alpha}_{\text{positron}} = +[\phi^{\alpha}_{\text{electron}}]_{\text{C}} = N \begin{pmatrix} c(p_x-ip_y)/(|E|+mc^2) \\ -cp_z/(|E|+mc^2) \\ 0 \\ 1 \end{pmatrix} \tag{C-37c}$$

$$\phi^{\beta}_{\text{positron}} = +[\phi^{\beta}_{\text{electron}}]_{\text{C}} = N \begin{pmatrix} cp_z/(|E|+mc^2) \\ c(p_x+ip_y)/(|E|+mc^2) \\ 1 \\ 0 \end{pmatrix}. \tag{C-37d}$$

It does not serve our purposes to explore further the Dirac hole theory since modern field theory formalism offers a far more satisfactory description, eliminating many of the conceptual difficulties while retaining all the required successes. Those wishing further details of the hole theory of Dirac are referred to Dirac (1947, Section 73) or Messiah (1966, Chapter XX) among numerous other works.

Up to this point the use of the $\boldsymbol{\alpha}$, β, and $\boldsymbol{\sigma}^{\text{D}}$ matrices has permitted a description offering a ready reduction to the low-velocity, nonrelativistic limit. In many applications, particularly those involving relativistic invariance, a second form or covariant form, treating space and time coordinates on a more equal footing, is more suitable. We define the relativistic Minkowski coordinates by a four-vector using the notation[1]

$$x, y, z, t \rightarrow x_\mu = (x_1, x_2, x_3, x_4 = ict) = (\mathbf{r}, ict) \tag{C-38}$$

and note that therefore

$$i\hbar\, \partial/\partial t \equiv -c\hbar\, \partial/\partial x_4.$$

If we multiply the time-dependent Dirac equation (C-4) by $\beta/\hbar c$, we obtain, since $\beta^2 = 1$,

$$[\beta(\partial/\partial x_4) - i\beta\boldsymbol{\alpha}\cdot\nabla + (mc/\hbar)]\,\psi = 0. \tag{C-39}$$

We now introduce the *gamma matrices* γ_k,

$$\boldsymbol{\gamma} \equiv \gamma_k = -i\beta\alpha_k = +i\alpha_k\beta, \quad k = 1, 2, 3 \tag{C-40a}$$

$$\gamma_4 = \beta, \tag{C-40b}$$

and hence write

$$\left(\sum_{\mu=1}^{4} \gamma_\mu(\partial/\partial x_\mu) + (mc/\hbar)\right)\psi = 0, \tag{C-41a}$$

or

$$\left(\sum_{\mu=1}^{4} \gamma_\mu p_\mu - imc\right)\psi = 0. \tag{C-41b}$$

The form (C-41) is the desired covariant form.[2] In (C-41b) we have defined $p_\mu = -i\hbar\,\partial/\partial x_\mu$. It is also very useful to define a quantity $\gamma_5 = \gamma_1\gamma_2\gamma_3\gamma_4 = i\alpha_1\alpha_2\alpha_3$. Noting the definitions of $\boldsymbol{\alpha}$ and β, it is readily established that

$$\boldsymbol{\gamma} = i\begin{pmatrix} 0 & -\boldsymbol{\sigma}^{\mathrm{P}} \\ \boldsymbol{\sigma}^{\mathrm{P}} & 0 \end{pmatrix}$$

$$\gamma_4 = \beta = \begin{pmatrix} 1 & 0 \\ 0 & -1 \end{pmatrix}, \qquad \gamma_5 = -\rho = \begin{pmatrix} 0 & -1 \\ -1 & 0 \end{pmatrix}. \tag{C-42}$$

The Dirac spin operator $\boldsymbol{\sigma}^{\mathrm{D}}$ is defined as before through (C-13). The commutation relationships among the gamma matrices are

$$\gamma_\mu\gamma_\nu + \gamma_\nu\gamma_\mu = 2\delta_{\mu\nu}, \qquad \mu, \nu = 1, 2, 3, 4, 5, \tag{C-43}$$

and we also have

$$\alpha_\mu = -\gamma_5{\sigma_\mu}^{\mathrm{D}}, \qquad {\sigma_\mu}^{\mathrm{D}} = -\gamma_5\alpha_\mu, \qquad \mu = 1, 2, 3. \tag{C-44}$$

[1] A number of different conventions are used in the literature. The reader should beware.

[2] Since all quantities in (C-41) are scalar quantities with respect to arbitrary rotations and reflections in four-space, the results obtained are guaranteed invariant under Lorentz transformations, and the equation is said to be covariant.

In view of (C-43), we readily recover the Klein–Gordon equation from (C-41), i.e.,

$$\left(\sum_{\mu} p_{\mu}^{2} + m^{2}c^{2}\right)\psi = 0,$$

which in time-independent form with the momentum four-vector

$$p_{\mu} = (p_1, p_2, p_3, iE/c) = (\mathbf{p}, iE/c)$$

just gives the equivalent

$$p^2 - E^2/c^2 + m^2c^2 = 0.$$

We now recall a few basic definitions and equalities. The most direct analogy with the ordinary nonrelativistic Schrödinger treatment results when the following interpretations are made. The state vector $|\psi\rangle$ in Dirac representation is a four-component column matrix or spinor

$$\psi(\mathbf{r}) = \begin{pmatrix} u_1(\mathbf{r}) \\ u_2(\mathbf{r}) \\ u_3(\mathbf{r}) \\ u_4(\mathbf{r}) \end{pmatrix}, \tag{C-45a}$$

and the Hermitian conjugate[1] of $\psi(\mathbf{r})$, written $\psi^{\dagger}(\mathbf{r})$, is a four-component row matrix

$$\psi^{\dagger}(\mathbf{r}) = (u_1^{*}(\mathbf{r}), u_2^{*}(\mathbf{r}), u_3^{*}(\mathbf{r}), u_4^{*}(\mathbf{r})). \tag{C-45b}$$

The inner product, or scalar product, of two-state vectors $\langle\phi|\psi\rangle$, with $|\phi\rangle$ similarly represented by a four-component spinor $(v_1(\mathbf{r}), \ldots, v_4(\mathbf{r}))$, is defined as

$$\langle\phi|\psi\rangle = \sum_{\mu=1}^{4} \int v_{\mu}^{*}(\mathbf{r})\, u_{\mu}(\mathbf{r})\, d^3r. \tag{C-46}$$

When $|\phi\rangle = |\psi\rangle$, the inner product (C-46) determines the norm or normalization of $|\psi\rangle$. When (C-46) is identically zero, the two states $|\phi\rangle$ and $|\psi\rangle$ are orthogonal. All operators can be represented by 4×4 matrices, and the

[1] The Hermitian conjugate, sometimes referred to as the *Hermitian adjoint*, of a matrix is defined as the complex conjugate of the transposed matrix, $A^{\dagger} = A_{ij}^{\dagger} = \tilde{A}^{*} = A_{ji}^{*}$. Also, clearly $(AB)^{\dagger} = B^{\dagger}A^{\dagger}$ for two 4×4 matrices. When $A^{\dagger} = A$ for an operator, it is said to be Hermitian. It readily follows, for example, that α_{μ} and β as well as all the gamma matrices are Hermitian, e.g., $\gamma_{\mu}^{\dagger} = (-i\beta\alpha_{\mu})^{\dagger} = i\alpha_{\mu}^{\dagger}\beta^{\dagger} = -i\beta\alpha_{\mu} = \gamma_{\mu}$, $\mu = 1, 2, 3$, etc. The taking of the Hermitian conjugate of the product of matrices can be extended to the product of any pair of conformable matrices (i.e., where the number of columns of the first matrix in the product is equal to the number of rows of the second); hence, we have $(A\psi)^{\dagger} = (\psi^{\dagger}A^{\dagger})$.

average value or expectation value of an operator $\tilde{Q}$ for the state $|\psi\rangle$ is

$$\langle Q\rangle = \langle\psi|\tilde{Q}|\psi\rangle = \sum_{\mu=1}^{4}\sum_{\nu=1}^{4}\int u_\mu^*(\mathbf{r})\, Q_{\mu\nu}\, u_\nu(\mathbf{r})\, d^3r. \tag{C-47}$$

The time dependence of (C-47) is given by

$$i\hbar(d/dt)\langle Q\rangle = \langle[Q,H]\rangle + i\hbar\langle\partial Q/\partial t\rangle, \tag{C-48}$$

the first term being the expectation value of the commutator $[Q,H]$ of Q and the Hamiltonian H, and the second term obtaining in the event Q is an explicit function of time.

Returning to the $\boldsymbol{\alpha},\beta$ convention, we can write (C-4) and its Hermitian conjugate, with $\mathbf{p} = (\hbar/i)\,\boldsymbol{\nabla}$, $\beta^\dagger = \beta$, and $\boldsymbol{\alpha}^\dagger = \boldsymbol{\alpha}$, as

$$i\hbar\,\partial\psi/\partial t = -i\hbar c\boldsymbol{\alpha}\cdot(\boldsymbol{\nabla}\psi) + mc^2(\beta\psi) \tag{C-49a}$$

and

$$-i\hbar\,\partial\psi^\dagger/\partial t = i\hbar c(\boldsymbol{\nabla}\psi^\dagger)\cdot\boldsymbol{\alpha} + mc^2(\psi^\dagger\beta). \tag{C-49b}$$

Multiplying (C-49a) by $\psi^\dagger$ from the left and (C-49b) by ψ from the right, and subtracting the two resulting equations gives

$$(\partial/\partial t)(\psi^\dagger\psi) = -c\boldsymbol{\nabla}\cdot(\psi^\dagger\boldsymbol{\alpha}\psi). \tag{C-50}$$

If we interpret $\psi^\dagger\psi$ as a probability density $\rho(\mathbf{r},t) = \psi^\dagger(\mathbf{r},t)\psi(\mathbf{r},t)$ and $c\psi^\dagger\boldsymbol{\alpha}\psi = \mathbf{j}(\mathbf{r},t)$ as a probability flux or current, Eq. (C-50) simply becomes the continuity equation for probability[1]

$$\partial\rho(\mathbf{r},t)/\partial t = -\boldsymbol{\nabla}\cdot\mathbf{j}(\mathbf{r},t). \tag{C-51a}$$

In the language of the gamma matrices, this can be expressed as the zero divergence of a four-vector

$$\sum_{\mu=1}^{4}(\partial j_\mu/\partial x_\mu) = 0 \tag{C-51b}$$

with $j_\mu = (\mathbf{j}, ic\rho)$ and $\mathbf{j} = ic\psi^\dagger\gamma_4\boldsymbol{\gamma}\psi$, and $\rho = \psi^\dagger\psi$.

The inference from the above, that the operator for velocity $\mathbf{v} = \dot{\mathbf{r}}$ is $\mathbf{v} = c\boldsymbol{\alpha}$, is in fact true. Since the eigenvalues of any component of $\boldsymbol{\alpha}$ are ± 1, it follows that the only possible results of measuring any component of $\dot{\mathbf{r}}$ are $\pm c$. The classical correspondence principle limit $\langle\dot{\mathbf{r}}\rangle$, of course, need not be $\pm c$. If we use (C-48) for the operator $\tilde{Q} = \tilde{\mathbf{r}}$ and apply it to a wave packet of a freely propagating particle, the integrated equation for $\langle\mathbf{r}(t)\rangle$ becomes

$$\langle\mathbf{r}(t)\rangle = \langle\mathbf{r}(0)\rangle + \frac{c^2}{H}\langle\mathbf{p}\rangle t + \frac{\hbar c}{2iH}(e^{2iHt/\hbar}-1)\left\langle\left(\boldsymbol{\alpha}(0) - \frac{c\mathbf{p}}{H}\right)\right\rangle, \tag{C-52}$$

[1] If the particle in question has charge q, then multiplying (C-49) through by q and interpreting $q\psi^\dagger\psi$ as the charge density and $cq\psi^\dagger\boldsymbol{\alpha}\psi$ as the electric current reduces (C-49) and this equation to the condition of charge conservation.

where H is the total energy of the packet and is a constant, and $\mathbf{p}$ is also a constant. The last term connotes a rapid oscillatory motion of amplitude $\hbar/mc$ (i.e., the Compton wavelength) when $H \approx mc^2$ and a frequency determined by $\omega\hbar = 2mc^2$. This high-frequency vibration or *Zitterbewegung* results from the interference of positive- and negative-energy states appearing in the wave packet. If the wave packet is constructed of *only* either pure positive- or pure negative-energy states, this term in (C-52) vanishes. A field-theoretic formulation of the relativistic theory avoids the above conceptual difficulties.

We note the appearance of γ_4 in defining $\mathbf{j}$ in (C-51b) and inquire about its significance. To this end we note that in the four-vector notation of (C-38) proper Lorentz transformations[1] can be written in terms of transformation matrices $a_{\mu\nu}$ as

$$x_\mu' = \sum_{\nu=1}^{4} a_{\mu\nu} x_\nu . \tag{C-53}$$

For example, the transformation matrix

$$a_{\mu\nu} = \begin{pmatrix} \cos\phi & \sin\phi & 0 & 0 \\ -\sin\phi & \cos\phi & 0 & 0 \\ 0 & 0 & 1 & 0 \\ 0 & 0 & 0 & 1 \end{pmatrix} \tag{C-54}$$

corresponds to a rotation about the z axis by an angle ϕ. The matrix

$$a_{\mu\nu} = \begin{pmatrix} \cos\chi & 0 & 0 & \sin\chi \\ 0 & 1 & 0 & 0 \\ 0 & 0 & 1 & 0 \\ -\sin\chi & 0 & 0 & \cos\chi \end{pmatrix}, \tag{C-55a}$$

with $\tan\chi = iv/c$, corresponds to a transformation to a system of coordinates moving with velocity v in the x direction with respect to the original coordinate frame, i.e.,

$$x' = \frac{x - vt}{(1 - v^2/c^2)^{1/2}}, \qquad y' = y, \qquad z' = z, \qquad t' = \frac{t - vx/c^2}{(1 - v^2/c^2)^{1/2}}. \tag{C-55b}$$

It is evident that to first order (i.e., infinitesimal rotations, $\phi \to \varepsilon$, or $iv/c \to \varepsilon$, etc.),

$$x_i' = x_i + \varepsilon x_j, \qquad x_j' = x_j - \varepsilon x_i, \qquad x_k' = x_k, \qquad x_l' = x_l, \tag{C-56}$$

[1] Proper Lorentz transformations include all generalized rotations in Minkowski space, for example, but do not include space reflection of all three space coordinates, the parity transformation, $(\mathbf{r} \to -\mathbf{r})$, or time reversal $(t \to -t)$.

where i and j stand for *any two* generalized Minkowski coordinates x_μ, and k and l for the remaining two; thus in (C-54), $\varepsilon = \phi$, $i = 1$, and $j = 2$, while in (C-55), $\varepsilon = iv/c$, $i = 1$, and $j = 4$. More complicated rotations involving a series of rotations of the type (C-56) are also possible. We know from Pauli spin theory that a rotation of the coordinate system through an angle θ about an axis $\hat{\mathbf{n}}$ corresponds to a unimodular transformation of the spinors

$$\psi' = S\psi \qquad \text{with} \qquad S = \exp(\tfrac{1}{2}i\theta\hat{\mathbf{n}}\cdot\boldsymbol{\sigma}), \tag{C-57}$$

which, for an infinitesimal rotation about the x axis $\theta \to \varepsilon$, is

$$\psi' = (1+\tfrac{1}{2}i\varepsilon\sigma_1)\psi. \tag{C-58}$$

The transformation (C-58) also holds for Dirac spinors. It is readily verified that $i\sigma_x{}^{\mathrm{D}} \equiv i\sigma_1{}^{\mathrm{D}} = \gamma_2\gamma_3$; hence

$$\psi' = (1+\tfrac{1}{2}\varepsilon\gamma_2\gamma_3)\psi.$$

In fact, for *all* simple rotations of the type (C-56), we can write

$$\psi' = (1+\tfrac{1}{2}\varepsilon\gamma_i\gamma_j)\psi, \tag{C-59}$$

with ε real if i and j are a pair from the coordinates 1, 2, 3; and ε pure imaginary if one of i or j is 4, as with (C-54) or (C-55), for example. To show that (C-59) is the proper transformation, we substitute $\psi = (1-\tfrac{1}{2}\varepsilon\gamma_i\gamma_j)\psi'$ into the Dirac wave equation (C-41b) and obtain

$$\sum_{\mu=1}^{4} \gamma_\mu p_\mu(1-\tfrac{1}{2}\varepsilon\gamma_i\gamma_j)\psi' = imc(1-\tfrac{1}{2}\varepsilon\gamma_i\gamma_j)\psi'. \tag{C-60}$$

If we multiply both sides by $(1+\tfrac{1}{2}\varepsilon\gamma_i\gamma_j)$ from the left and make use of (C-43), we arrive at the first-order result,

$$\sum_{\mu=1}^{4} \gamma_\mu p_\mu\psi' + \varepsilon(\gamma_i p_j - \gamma_j p_i)\psi' = imc\psi' \tag{C-61a}$$

or

$$\sum_{\mu=1}^{4} \gamma_\mu p_\mu'\psi' = imc\psi' \tag{C-61b}$$

with

$$p_i' = p_i + \varepsilon p_j, \qquad p_j' = p_j - \varepsilon p_i, \qquad p_k' = p_k, \qquad p_l' = p_l.$$

Clearly these latter conditions simply correspond to the appropriate Lorentz transformations of the four-momentum. We note, of course, that (C-41b) and (C-61b) are both of the same invariant form, i.e., proper Lorentz transformations of the type discussed here, and, in general, leave the Dirac wave equation invariant in form.

Taking the Hermitian conjugate of (C-59), we get

$$\psi'^\dagger = [(1+\tfrac{1}{2}\varepsilon\gamma_i\gamma_j)\psi]^\dagger = \psi^\dagger[1-\tfrac{1}{2}\varepsilon^*\gamma_i\gamma_j]. \tag{C-62}$$

Hence to first order

$$\psi'^\dagger\psi' = \psi^\dagger[1+\tfrac{1}{2}(\varepsilon-\varepsilon^*)\gamma_i\gamma_j]\psi. \tag{C-63}$$

Thus when neither i nor j is 4, and therefore $\varepsilon^* = \varepsilon$, we have $\psi'^\dagger\psi = \psi^\dagger\psi$, and $\psi^\dagger\psi$ is invariant under such transformations. However, when either i or j is 4, $\varepsilon^* = -\varepsilon$; hence $\psi^\dagger\psi$ is *not* invariant under these Lorentz transformations and therefore is not invariant in general.

We now define the Dirac adjoint (usually simply referred to as the adjoint) of ψ by $\bar{\psi}$ as

$$\bar{\psi} = \psi^\dagger\gamma_4. \tag{C-64}$$

Then

$$\begin{aligned}\bar{\psi}' = \psi'^\dagger\gamma_4 &= \psi^\dagger(1-\tfrac{1}{2}\varepsilon^*\gamma_i\gamma_j)\gamma_4 \\ &= \psi^\dagger\gamma_4(1-\tfrac{1}{2}\varepsilon^*\gamma_i\gamma_j), \qquad i\neq 4 \quad \text{and} \quad j\neq 4 \qquad \text{(C-65a)} \\ &= \psi^\dagger\gamma_4(1+\tfrac{1}{2}\varepsilon^*\gamma_i\gamma_j), \qquad i=4 \quad \text{or} \quad j=4. \qquad \text{(C-65b)}\end{aligned}$$

Recalling that when $i\neq 4$ and $j\neq 4$, ε is real, and when either i or j is 4, ε is pure imaginary, we can write (C-65a) and (C-65b) both as a single equation

$$\bar{\psi}' = \psi^\dagger\gamma_4(1-\tfrac{1}{2}\varepsilon\gamma_i\gamma_j) \qquad \text{for any pair } i \text{ and } j$$

or

$$\bar{\psi}' = \bar{\psi}(1-\tfrac{1}{2}\varepsilon\gamma_i\gamma_j). \tag{C-66}$$

Finally, multiplying by ψ' from the right, we obtain

$$\bar{\psi}'\psi' = \bar{\psi}\psi. \tag{C-67}$$

Thus it is the quantity $\bar{\psi}\psi = \psi^\dagger\gamma_4\psi$ which transforms like a scalar under proper Lorentz transformations, not $\psi^\dagger\psi$. Indeed $ic\psi^\dagger\psi$ is simply the fourth component of the current four-vector

$$j_\mu = ic\psi^\dagger\gamma_4\gamma_\mu\psi = c\begin{pmatrix}\psi^\dagger\alpha_\mu\psi \\ i\psi^\dagger\psi\end{pmatrix} \quad \begin{matrix}\mu = 1,2,3, \\ \mu = 4.\end{matrix} \tag{C-68}$$

Further discussion of such topics as space inversion, time reversal, or charge conjugation would carry us beyond the scope of this text and the reader is referred to advanced treatises.

References

Dirac, P. A. M. (1947). "Principles of Quantum Mechanics." Oxford Univ. Press (Clarendon), London and New York.

Messiah, A. (1966). "Quantum Mechanics," Vol. II. Wiley, New York.

Appendix D

ITERATIVE DIAGONALIZATION OF MATRICES

We present a brief outline of an elementary iterative procedure for diagonalizing energy matrices and relate the early steps to perturbation theory. For simplicity we consider first the case of a two-dimensional representation and suppose that the eigenvalues A and B and the corresponding eigenfunctions ψ_A and ψ_B of a suitable zeroth-order Hamiltonian H_0 are known. Then, of course, the energy matrix $H_{ij} = \langle i|\tilde{H}_0|j\rangle$ is diagonal in this representation and we have

$$H_{ij,0} = \begin{pmatrix} A & 0 \\ 0 & B \end{pmatrix}, \tag{D-1}$$

with $A = \langle \psi_A|\tilde{H}_0|\psi_A\rangle$ and $B = \langle \psi_B|\tilde{H}_0|\psi_B\rangle$.

We now consider the effect of adding a Hermitian perturbation V to the Hamiltonian to give $H = H_0 + V$. In the following discussion we give a somewhat more complete treatment of this state of affairs than required for merely seeking the new energy eigenvalues since the general development has intrinsic value per se. We suppose that the matrix elements of V in the representation H_0 are known, with

$$\begin{aligned} a &= \langle \psi_A|\tilde{V}|\psi_A\rangle, & b &= \langle \psi_B|\tilde{V}|\psi_B\rangle \\ c &= \langle \psi_A|\tilde{V}|\psi_B\rangle, & c^* &= \langle \psi_B|\tilde{V}|\psi_A\rangle. \end{aligned} \tag{D-2}$$

The eigenvalue problem for the new Hamiltonian is stated as

$$\begin{pmatrix} (A+a) & c \\ c^* & (B+b) \end{pmatrix}\begin{pmatrix} u \\ v \end{pmatrix} = \lambda\begin{pmatrix} u \\ v \end{pmatrix}, \tag{D-3}$$

with the two possible values of λ, λ_1 and λ_2, corresponding to the new energy eigenvalues and $\psi_1 = u_1\psi_A + v_1\psi_B$ and $\psi_2 = u_2\psi_A + v_2\psi_B$ corresponding to the new eigenfunctions. For (D-3) to have solutions the secular determinant must vanish, or

$$\det\begin{vmatrix} (A+a-\lambda) & c \\ c^* & (B+b-\lambda) \end{vmatrix} = 0. \tag{D-4}$$

This yields the quadratic equation in λ

$$(A+a-\lambda)(B+b-\lambda) - |c|^2 = 0. \tag{D-5}$$

The two roots of (D-5) are the required new eigenvalues λ_1 and λ_2. When (D-3) is solved for (u_1, v_1) and (u_2, v_2), usually with the normalization $|u_1|^2 + |v_1|^2 = 1$ and $|u_2|^2 + |v_2|^2 = 1$, the solution to the eigenvalue problem is completed.

In the following let us associate the solution λ_1 with A in the sense that $\lim_{V\to 0}\lambda_1 = A$ and similarly associate λ_2 with B. Exact expressions obtained from (D-3), which are useful particularly if the matrix elements (D-2) are small, are

$$v_1/u_1 = -c^*/[(B+b) - \lambda_1] \tag{D-6a}$$

and

$$u_2/v_2 = -c/[(A+a) - \lambda_2]. \tag{D-6b}$$

Well-known theorems involving the trace and determinant of the energy matrix give

$$\lambda_1 + \lambda_2 = (A+a) + (B+b) \tag{D-7a}$$

and

$$\lambda_1\lambda_2 = (A+a)(B+b) - |c|^2. \tag{D-7b}$$

From (D-7a) and (D-6) it immediately follows that

$$u_2/v_2 = -(v_1/u_1)^*. \tag{D-8}$$

With ψ_A and ψ_B orthogonal, (D-8) ensures the orthogonality of ψ_1 and ψ_2 as well, as can be readily verified by calculating $\langle\psi_1|\psi_2\rangle$.

The expressions (D-7) can be used to obtain an interesting relationship between λ_1, λ_2 and $(A+a), (B+b)$. From (D-7), we have

$$(\lambda_2 - \lambda_1)^2 = [(B+b) - (A+a)]^2 + 4|c|^2.$$

This equation together with (D-7a) implies that if $c \neq 0$,

$$\lambda_1 < [(B+b) \text{ or } (A+a)] \qquad \text{and} \qquad \lambda_2 > [(B+b) \text{ or } (A+a)].$$

Thus coupled states, i.e., nonvanishing off-diagonal elements, can never have $\lambda_1 = \lambda_2$. Figure D.1 illustrates two typical situations for coupled states as functions of the strength parameter ξ of the interaction potential written as $V = \xi V_0$.

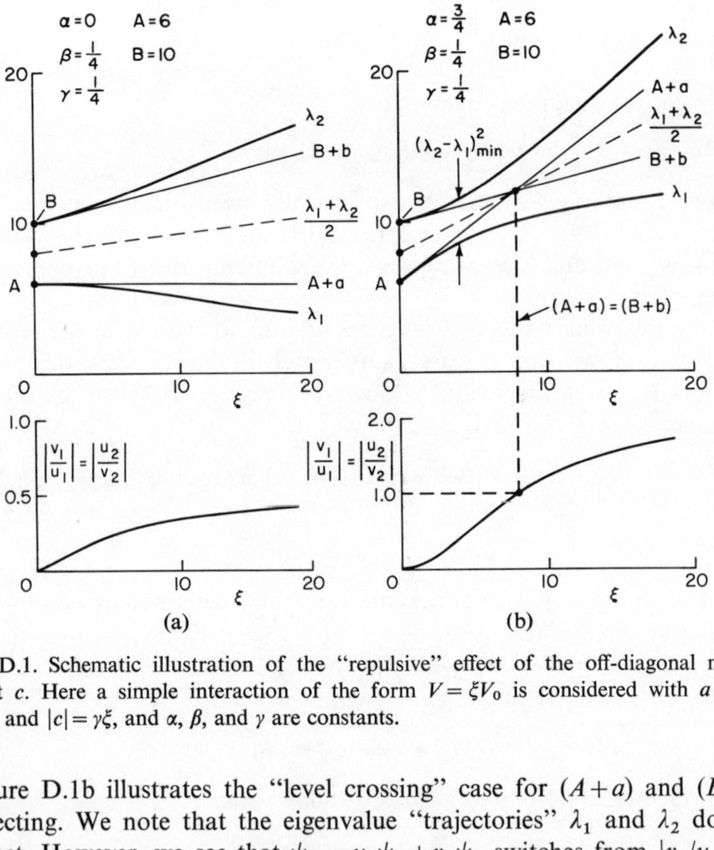

FIG. D.1. Schematic illustration of the "repulsive" effect of the off-diagonal matrix element c. Here a simple interaction of the form $V = \xi V_0$ is considered with $a = \alpha\xi$, $b = \beta\xi$, and $|c| = \gamma\xi$, and α, β, and γ are constants.

Figure D.1b illustrates the "level crossing" case for $(A+a)$ and $(B+b)$ intersecting. We note that the eigenvalue "trajectories" λ_1 and λ_2 do not intersect. However, we see that $\psi_1 = u_1\psi_A + v_1\psi_B$ switches from $|v_1/u_1| < 1$ to $|v_1/u_1| > 1$ at the point where $(A+a) = (B+b)$. The minimum in $(\lambda_2 - \lambda_1)^2$ occurs at a value of

$$\xi = (B-A)(\alpha-\beta)/[(\alpha-\beta)^2 + 4\gamma^2].$$

Obtaining the exact roots of (D-5), of course, offers no difficulty; however, we wish to introduce an iteration procedure that will also be applicable to the

n-dimensional case as well. As a first approximation to (D-5) consider arbitrarily setting $|c|^2 = 0$; thus

$$^1\lambda_1 = A + a, \qquad ^1\lambda_2 = B + b \tag{D-9a}$$

and

$$^1\psi_1 = \psi_A, \qquad ^1\psi_2 = \psi_B. \tag{D-9b}$$

We immediately recognize this to be simple first-order perturbation theory.

Defining the difference between the *true eigenvalue* and the corresponding diagonal element by Δ_i, we have from (D-5)

$$\Delta_1 = A + a - \lambda_1 = |c|^2/(B+b-\lambda_1) \tag{D-10a}$$

$$\Delta_2 = B + b - \lambda_2 = |c|^2/(A+a-\lambda_2). \tag{D-10b}$$

The above first-order approximation contained in (D-9) is equivalent to setting $^1\Delta_i = 0$. (We shall use the superscript q for any quantity $^q\xi$ to indicate the order of a particular approximation.) To a next order of approximation we might substitute for λ_1 in the denominator of (D-10a) its approximate value $^1\lambda_1$ from (D-9a) and obtain

$$^2\Delta_1 = A + a - {}^2\lambda_1 = |c|^2/[(B+b) - (A+a)], \tag{D-11a}$$

and similarly

$$^2\Delta_2 = B + b - {}^2\lambda_2 = |c|^2/[(A+a) - (B+b)]. \tag{D-11b}$$

The solution for $^2\lambda_1$ from (D-11a) is seen to be a variant of the usual second-order perturbation theory for which the energy denominator is generally simply taken to be $B-A$. The expression (D-11) is not only generally more accurate but also avoids the not uncommon difficulty encountered when H_0 is degenerate and $A = B$. When the matrix elements (D-2) are small, the present approximation gives, using (D-6a) and (D-8),

$$\frac{u_2}{v_2} = -\left(\frac{v_1}{u_1}\right)^* = \frac{c}{(B+b) - (A+a)}. \tag{D-12}$$

Again when the denominator is approximated with simply $B-A$, we arrive at the usual second-order perturbation calculation result.

By continued iteration we have for the *exact* solutions,

$$\lambda_1 = A + a - \frac{|c|^2}{(B+b) - \left[(A+a) - \dfrac{|c|^2}{(B+b) - [(A+a) - \cdots]}\right]}. \tag{D-13a}$$

and

$$\lambda_2 = B + b - \cfrac{|c|^2}{(A+a) - \left[(B+b) - \cfrac{|c|^2}{(A+a) - [(B+b) - \cdots]} \right]}. \tag{D-13b}$$

Once an adequately accurate value for λ_1 has been obtained, the solution for the eigenfunction of λ_1 with $|u_1|^2 + |v_1|^2 = 1$ is then readily found to be $\psi_1 = u_1\psi_A + v_1\psi_B$, with

$$u_1 = e^{i\eta}\left[1 + \frac{|c|^2}{(B+b) - \lambda_1}\right]^{-1/2}, \tag{D-14a}$$

and

$$v_1 = -\frac{c^*}{(B+b) - \lambda_1} u_1, \tag{D-14b}$$

where η is an arbitrary phase. The eigenfunction ψ_2 is similarly obtained.

The general n-dimensional problem results in an n-dimensional secular determinant (i.e., λ appears to the power λ^n) with all the resulting complexity this entails. The procedure suggested by the above is however readily applicable to the n-dimensional representation as well, involving at each step only a *linearized* version of the secular determinant.

We introduce the notation ϕ_i for the eigenfunctions of H_0, i.e., $\tilde{H}_0\phi_i = H_{ii}^0\phi_i$, and write the matrix elements as $H_{ij} = \langle\phi_i|\,\tilde{H}_0 + \tilde{V}|\phi_j\rangle$. Clearly, we can write H_{ij} $(i \neq j) = \langle\phi_i|\,\tilde{V}|\phi_j\rangle$ and $H_{ii}^V = \langle\phi_i|\,\tilde{V}|\phi_i\rangle$ with $H_{ii} = H_{ii}^0 + H_{ii}^V$. To first order, all off-diagonal elements H_{ij} $(i \neq j)$ are set equal to zero and the first-order ${}^1\lambda_i$ are determined from the diagonal elements alone, viz., ${}^1\lambda_1 = H_{11} = H_{11}^0 + H_{11}^V$, ${}^1\lambda_2 = H_{22} = H_{22}^0 + H_{22}^V$, etc. To second order, the approximate value ${}^1\lambda_1$ is used for λ in all diagonal elements of the secular determinant, except the one associated with H_{11}, where it is written ${}^2\lambda_1$, and the proper off-diagonal elements are reintroduced. The resulting linearized secular determinant is readily solved for the single unknown quantity ${}^2\lambda_1$. In a similar way ${}^2\lambda_2$ is found by using ${}^1\lambda_2$ for λ in all diagonal positions except the one associated with H_{22}, where it is written ${}^2\lambda_2$, etc. Continuing this process results in obtaining the second-order set of eigenvalues ${}^2\lambda_i$. The procedure of obtaining the new set ${}^{q+1}\lambda_i$ from the set ${}^q\lambda_i$ can then be repeated as often as necessary [in essence paralleling (D-13) viewed as a continued fraction], to obtain as close an approximation to the true set λ_i as required. When the off-diagonal elements are small, this iterative procedure converges quite rapidly.

As an illustration, the obtaining of ${}^{q+1}\lambda_2$ when the set ${}^q\lambda_i$ (in particular, ${}^q\lambda_2$) has already been found, is achieved by solving the following determinant for the single unknown quantity ${}^{q+1}\lambda_2$, viz.,

$$\det \begin{vmatrix} (H_{11} - {}^{q}\lambda_2) & H_{12} & H_{13} & \cdots & H_{1n} \\ H_{21} & (H_{22} - {}^{q+1}\lambda_2) & H_{23} & \cdots & H_{2n} \\ H_{31} & H_{32} & (H_{33} - {}^{q}\lambda_2) & & \vdots \\ \vdots & \vdots & & \ddots & \\ H_{n1} & H_{n2} & \cdots & & (H_{nn} - {}^{q}\lambda_2) \end{vmatrix} = 0. \tag{D-15}$$

The not uncommon case of degeneracy in H_0 again offers no problem. In addition, generalizations of (D-7) to the n-dimensional case can be used either as checks on the set of output values λ_i or to overcome some accidental convergence difficulties that might arise in computing one or more roots (e.g., when $H_{kk}^0 + H_{kk}^V \approx H_{ll}^0 + H_{ll}^V$, or $A + a \approx B + b$ in the example shown in Fig. D.1b for $\xi \approx 8$, and the true eigenvalues λ_k and λ_l are near a "level crossing" region). The generalization of (D-7) is

$$\sum_{i=1}^{n} \lambda_i = \mathrm{Tr}(H_{ij}) \tag{D-16a}$$

and

$$\prod_{i=1}^{n} \lambda_i = \det(H_{ij}). \tag{D-16b}$$

An additional $n-2$ other relationships, somewhat more complicated, exist between the roots that also can be used profitably on occasion.

Extending the notation to the n-dimensional case introduced above, we would write $\psi_i = \sum_{j=1}^{n} C_{ij} \phi_j$. If after finding the set ${}^{q}\lambda_i$ (i.e., stopping with the qth iteration) we wish to determine the approximate eigenfunctions ψ_i, we need to evaluate the expansion coefficients C_{ij}. These are readily obtained from the matrix equation

$$\begin{bmatrix} (H_{11} - {}^{q}\lambda_i) & H_{12} & H_{13} & \cdots & H_{1n} \\ H_{21} & (H_{22} - {}^{q}\lambda_i) & H_{23} & & \vdots \\ H_{31} & H_{32} & (H_{33} - {}^{q}\lambda_i) & & \vdots \\ \vdots & \vdots & & \ddots & \\ H_{n1} & H_{n2} & \cdots & & (H_{nn} - {}^{q}\lambda_i) \end{bmatrix} \begin{bmatrix} C_{i1} \\ C_{i2} \\ C_{i3} \\ \vdots \\ C_{in} \end{bmatrix} = 0. \tag{D-17a}$$

The matrix relationship (D-17a) can also be written in compact form giving the n equations

$$\sum_{j=1}^{n} (H_{kj} - {}^{q}\lambda_i \, \delta_{kj}) \, C_{ij} = 0, \qquad k = 1, 2, \ldots, n. \tag{D-17b}$$

Of the $n-1$ linearly independent relationships involved in (D-17), computational errors resulting from the approximate nature of the solutions ${}^q\lambda_i$ are generally minimized by determining all C_{ij}, $i \neq j$, in terms of C_{ii}. This is accomplished by ignoring the equation containing H_{ii} and considering the remaining $n-1$ linear equations as *inhomogeneous* by virtue of treating C_{ii} as known. The standard application of Cramer's rule or direct elimination can then be used to determine the C_{ij}. Finally, the normalization condition can be used to determine C_{ii}, viz.,

$$C_{ii} = e^{i\eta}\left[1 - \sum_{i \neq j}^{n} |C_{ij}|^2\right]^{1/2}, \tag{D-18}$$

where η is an arbitrary phase constant.

The above procedure is particularly convenient when in fact some (or all) of the eigenvalues of a problem are known from experiment and the question raised concerns the nature and strength of possible "perturbation potentials" that may be added to certain model Hamiltonians H_0, as in Chapter V.

The reader interested in other eigenvalue methods and their convergence properties can consult a wide range of standard works, from the basic treatment afforded by Hildebrand (1952) to the more advanced techniques discussed in Morse and Feshbach, (1953, Part II, Chapter 9). A very useful reference on determinants is by Muir and Metzler (1960). Interesting variants of the useful Brillouin–Wigner perturbation method are available in the literature [see Goldhammer and Feenberg (1956) and Lippmann (1956)]. For a particularly effective third-order perturbation technique, see Nissimov and Elliott (1972).

Finally, we should mention various tridiagonalization techniques, which are particularly useful for calculations involving bases of large dimensionality. These techniques introduce a vector basis obtained by an iterative procedure that reduces the usual $n \times n$ matrix to one containing only nonvanishing tridiagonal elements (i.e., the principal diagonal and one adjacent diagonal on either side; such determinants are also called *continuants*). A good general review of such techniques is given by Whitehead (1972) [see also Muir and Metzler (1960)]. The required computer time for "large" shell model calculations is thereby substantially reduced. The original "Glasgow program" could handle 10,000 basis states. For a typical such calculation see Whitehead and Watt (1972). For possible applications of such techniques to other problems see Nissimov (1973).

References

Goldhammer, P., and Feenberg, E. (1956). *Phys. Rev.* **101**, 1233.

Hildebrand, F. B. (1952). "Methods of Applied Mathematics." Prentice-Hall, Englewood Cliffs, New Jersey.

Lippmann, B. A. (1956). *Phys. Rev.* **103**, 1149.
Morse, P. M., and Feshbach, H. (1953). "Methods of Theoretical Physics." McGraw-Hill, New York.
Muir, T., and Metzler, W. H. (1960). "A Treatise on the Theory of Determinants." Dover, New York.
Nissimov, H. (1973). *Phys. Lett.* **46B**, 1.
Nissimov, H., and Elliott, J. P. (1972). *Nuclear Phys.* **A198**, 1.
Whitehead, R. R. (1972). *Nuclear Phys.* **A182**, 290.
Whitehead, R. R., and Watt, A. (1972). *Phys. Lett.* **41B**, 7.

INDEX

References to individual nuclei are compiled at the end of the index.

D

E

F

G

H

N

T

U

V

W

Nuclei cited as examples

A 5
B 6
C 7
D 8
E 9
F 0
G 1
H 2
I 3
J 4